Practical Aspects of CO_2 Flooding

Perry M. Jarrell
BP

Charles E. Fox
Kinder Morgan

Michael H. Stein
BP

Steven L. Webb
Occidental Petroleum

Editor:
Russell T. Johns
U. of Texas at Austin

Technical Writer:
Linda A. Day

SPE Monograph Series, Volume 22

Henry L. Doherty Memorial Fund of AIME
Society of Petroleum Engineers
Richardson, TX USA

Disclaimer

This book was prepared by members of the Society of Petroleum Engineers and their well-qualified colleagues from material published in the recognized technical literature and from their own individual experience and expertise. While the material presented is believed to be based on sound technical knowledge, neither the Society of Petroleum Engineers nor any of the authors or editors herein provide a warranty either expressed or implied in its application. Correspondingly, the discussion of materials, methods, or techniques that may be covered by letters patents implies no freedom to use such materials, methods, or techniques without permission through appropriate licensing. Nothing described within this book should be construed to lessen the need to apply sound engineering judgment nor to carefully apply accepted engineering practices in the design, implementation, or application of the techniques described herein.

Printed in the United States of America.

ISBN 978-1-55563-096-6

14 15 16 17 18 19 20 / 12 11 10 9 8 7 6 5 4

Society of Petroleum Engineers
222 Palisades Creek Drive
Richardson, TX 75080-2040 USA

http://store.spe.org/
books@spe.org
1.972.952.9393

Preface

This monograph is the work of many people and organizations in addition to the authors, and we gratefully acknowledge their assistance. The material in this monograph was reviewed by SPE Monograph Review Committee and by 25 individual members of the SPE who spent hundreds of hours reading, checking, and critically commenting on all aspects of the material and its presentation. The monograph is a much better volume than it would have been without their help. We especially acknowledge the contributions and support of Jeff Alvarez, Steve Benson, Tim Bradley, Anil Chopra, Laquitta Danford, Steve Hendrickson, Mark Kittridge, Rena Koinis, Steve Melzer, Richard Rogers, Rod Sidle, Jeff Simmons, Pamela Spencer, Herb Wacker, Gary Warden, and Scott Wehner.

Dr. Russell Johns was critical in enabling the completion phase of this monograph. Appointed as the second editor by SPE, he organized new reviewers, re-energized the authors, and steered the project to completion within two years.

We are also most grateful for the assistance of Linda Day, whose technical writing contributions added significantly to the quality of this book.

In addition, we extend our gratitude to the following companies for their support: Amoco Production Co.; BP; Altura Energy Ltd.; Oxy Permian; Shell Oil Co. Inc.; and Kinder Morgan CO_2 Co. L.P.

Perry M. Jarrell
Charles E. Fox
Michael H. Stein
Steven L. Webb

Dedication

We dedicate this book to the people without whose inspiration, help, and support it could never have been completed:

To my father, the late Malcolm Jarrell, who inspired me by carrying out his life's work in the petroleum industry with integrity and delight; to my children, my pride and joy, Mac, Travis, Shea, and William; and especially to the love of my life, Eva, whose patience and encouragement kept the fire burning.—Perry M. Jarrell

To my parents, Gerald and Elizabeth, who encouraged me in all that I have done. To my wife, Sharon, and my children, Amy, David, Joe, and Jenny, who spent many evenings without me as I completed draft after draft after draft after draft after draft…—Charles E. Fox

To my father, the late Jack Stein, who would have loved to have seen this finished product; to my mother, Norma Stein, who will be able to see this monograph completed; and to my wife, Chae, who has willingly sacrificed while encouraging me to spend the time needed to complete this monograph. —Michael H. Stein

To my parents, who taught me the value of hard work; to my sons, Alex and Kevin, who sacrificed their time with me to allow me to finish this book; and especially to my wife, Kelly, who has been a source of encouragement and support as I completed this monograph.—Steven L. Webb

SPE Monograph Series

The Monograph Series of the Society of Petroleum Engineers was established in 1965 by action of the SPE Board of Directors. The Series is intended to provide authoritative, up-to-date treatment of the fundamental principles and state of the art in selected fields of technology. The Series is directed by the Society's Monograph Committee. A committee member designated as Monograph Editor provides technical evaluation with the aid of the Review Committee. Below is a listing of those who have been most closely involved with the preparation of this monograph.

Monograph Committee (2002)

Additional Media Files

Additional media files for this monograph are available online at http://www.spe.org/web/pubs/monograph22/

- Readme file.
- Kinder Morgan CO_2 Scoping Model—The Kinder Morgan model originally was developed by Shell CO_2 Company, Ltd. Shell CO_2 Company, Ltd. was purchased by Kinder Morgan Energy Partners, who revised and renamed the model. It allows the user to perform preliminary economic evaluations of a CO_2 flood, using a few input parameters. (See Chap. 3.)
- Texaco/DOE CO_2 Prophet Scoping Model—The Texaco/DOE model was developed by Texaco as part of a U.S. DOE grant. It allows the user to perform streamtube modeling of simple injection patterns. It requires more input parameters than the Kinder Morgan CO_2 Flood Scoping Model and does not perform an economic evaluation. (See Chap. 3.) The version provided in this edition was recompiled to work with the Microsoft Windows operating system.
- Texaco/DOE CO_2 Prophet Manual—The Texaco/DOE manual is given in three versions: MS Word, WordPerfect, and MS Excel.
- Table F.1. CO_2 properties at various temperatures, °F—The data shown in Table F.1 are given in MS Word format.

Table of Contents

Chapter 1—Introduction **1**
- 1.1 Purpose, Scope, and Organization 1
- 1.2 CO_2 Flood Design Variations 3
- 1.3 Major Influences on CO_2 Flood Success 4
- 1.4 Current and Planned CO_2 Projects 5
- 1.5 CO_2 Sources 6
- 1.6 Industry Projected CO_2 Potential 9

Chapter 2—Review of CO_2 Process Mechanisms **13**
- 2.1 How CO_2 Becomes Miscible with Oil 13
- 2.2 Small-Scale Reservoir Mixing Mechanisms 18
- 2.3 An Overview of Relative Permeability 22
- 2.4 Field Examples of Relative Permeability Effects on Fluid Mobility 24
- 2.5 Modeling Solvent Relative Permeability Under Miscible Conditions 28
- 2.6 Sweep Aspects of a CO_2 Flood 32

Chapter 3—The Technical and Economic Screening Process: Factors To Be Assessed Before Starting a Detailed Study **39**
- 3.1 Reservoir Considerations—Can Incremental Oil Be Recovered? 39
- 3.2 Cursory Prediction of CO_2 Flood Performance—How Much Oil Can Be Recovered? 47
- 3.3 Cursory Investment and Operating Cost Estimates—What Will It Cost? 51
- 3.4 Scoping Economics—Will the Incremental Oil Justify the Costs? 64

Chapter 4—Reservoir Engineering Design Aspects of a CO_2 Flood **69**
- 4.1 Data Requirements 69
- 4.2 History Matching To Develop the Reservoir Description 71
- 4.3 Predicting CO_2 Flood Performance 79
- 4.4 Optimizing the CO_2 Flood Design 87

Chapter 5—Surface Facilities Design **93**
- 5.1 Introduction 93
- 5.2 Specification Methods for Material Selection 93
- 5.3 Production 96
- 5.4 Injection 106
- 5.5 Gas Processing 109

Chapter 6—Well Design **115**
- 6.1 Producing Well: Wellhead to Bottomhole 115
- 6.2 Injection Well: Wellhead to Bottomhole 119

Chapter 7—Implementation: How Can the Design Become Reality? **123**
- 7.1 Planning 123
- 7.2 Preinjection Data Gathering 128
- 7.3 Permits, Contracts, and Agreements 129
- 7.4 Monitoring a CO_2 Flood 132

Chapter 8—Operations: How Can the CO_2 Flood Process Be Managed Effectively? **138**
- 8.1 Reservoir Management 138
- 8.2 Well Management 145
- 8.3 Facility Management 150

Chapter 9—CO_2 Flood Environmental, Health, and Safety Planning **156**
- 9.1 The CO_2 EHS Plan 156
- 9.2 The Dangers of CO_2 156

Appendix A—Worldwide CO_2 Floods **159**
- Active Floods 159
- Inactive Floods 159

Appendix B—Single-Well Cyclic CO_2 Injection Treatment 174
Procedure 174
Mechanisms 174
Applications and Design Considerations 175
Results 176

Appendix C—Case Histories 178

Appendix D—Adjustments to Water Relative Permeability Curves to Account for CO_2 Solubility 183
Extrapolation of the Data for the Original Water Relative Permeability Curve 183
Adjustments for Water Hysteresis 183

Appendix E—Sample Self-Certification Document for Obtaining EOR Tax Credits 186

Appendix F—CO_2 Properties 188

Glossary 273

Author Index 276

Subject Index 278

Chapter 1
Introduction

1.1 Purpose, Scope, and Organization

1.1.1 Purpose. By definition, a monograph is a systematic exposition in writing on a small area of learning, including a methodical discussion of facts and principles involved and conclusions reached. In this monograph on practical CO_2 flooding, we systematically and methodically discuss current and practical CO_2 flooding technologies and industry experiences. This operation-oriented book delineates the application of CO_2 flooding in the field, outlining the entire project development sequence from conception and justification through field implementation and operation. It is intended to serve as a logical guide to the practicing engineer focusing on the "how to" and "why" of miscible and immiscible CO_2 flooding. The emphasis is on *CO_2 flooding*, not other applications such as steam soak or stimulation, carbonated waterflooding, enriched-gas flooding or steam/CO_2 flooding, though we present minor discussions of these.

Our goal is not to re-present published material found in the CO_2 flooding sections of Stalkup's *Miscible Displacement* SPE monograph,[1] or Klins' *Carbon Dioxide Flooding: Basic Mechanisms and Project Design*;[2] rather, we aim to provide an operational complement to these and other CO_2 reservoir engineering-oriented books, just as the book by Rose *et al.*, *The Design Engineering Aspects of Waterflooding*,[3] complements Craig's *The Reservoir Engineering Aspects of Waterflooding*.[4] Of course, this text unavoidably reiterates some fundamental CO_2 flood reservoir engineering concepts because the reader must understand recovery processes to fully appreciate the operational challenges of a CO_2 project.

Monographs have been published on two broad enhanced oil recovery (EOR) process mechanisms—thermal and miscible displacement—but this is the first to concentrate on 1 of 13 specific EOR mechanisms within the broad categories of thermal, gas (miscible and immiscible), and chemical, shown in **Fig. 1.1.**

Focus on CO_2 flooding is appropriate because CO_2 is the second-most applied EOR process in the world, behind steam flooding **(Fig. 1.2).** Carbon dioxide flooding also has significant potential for future applications.

Carbon dioxide floods in North America are the only EOR projects besides hydrocarbon gas floods that have consistently and significantly increased annual EOR production since the 1986 crash in oil prices. According to the 1998 *Oil & Gas J.* CO_2 survey, more than one-half of the current U.S. gas EOR production and one-fifth of all U.S. EOR production come from CO_2 flood projects.[5]

Currently, CO_2 projects are active in at least four other countries worldwide **(Fig. 1.3).** Many more new and old fields throughout the world harbor potential for CO_2 floods that could recover significant incremental oil.

A simple consensus definition for enhanced recovery is elusive, though many have been cited and most differences among them are minor. In this monograph, enhanced recovery means any method of economically recovering oil incremental to that produced by primary or conventional improved-recovery methods. This is not a perfect definition, but it eliminates the common misuse of "enhanced recovery" as a synonym for "tertiary recovery."

This monograph will give readers the information and procedures necessary to bring a project into effective operation. We have coupled international, industrywide experiences with proven and emerging technologies in a logical sequence that easily can be used in new CO_2 flood projects.

1.1.2 Scope and Objectives. We have three basic objectives in presenting this material: to provide an account of the principles and experiences of practical CO_2 flooding in a logical sequence as a comprehensive handbook for the practicing engineer; to assemble extensive reference material for all published CO_2 flood case histories; and to offer a wide spectrum of information, such that the book will appeal to an industry audience worldwide.

The book targets those who are involved in planning, designing, implementing, and operating a CO_2 flood—i.e., practical CO_2 flood engineers—rather than those in research or academia.

We define practical engineering in this book as industry-accepted theory and engineering applied to designing and operating a CO_2 project, from injector bottomhole to producer bottomhole, including all surface facilities. Practical engineers include plant, mechanical, production, reservoir and geophysical engineers, because practical CO_2 flood engineering involves all these disciplines at one stage or another of a project's development. In this sense, the teamwork required to complete a project defines the practical engineering process or system.

This book deals with some reservoir theory and processes, and with the mechanisms involved in the interactions between CO_2 and the reservoir oil, but only to the extent needed for an engineer to competently design and implement a CO_2 flood. Additional detailed information on CO_2 reservoir engineering concepts is admirably documented in the previously mentioned books by Stalkup[1] and Klins,[2] as well as in other books.[6–10]

To better understand CO_2 flooding practices, the reader should have a fundamental understanding of waterflooding operations and management because CO_2 is similar in several operational aspects.[4,11,12] There are particular similarities in the areas of surface fluid measurement, storage and transportation, and downhole injection and production control.

The scope of this text does not allow for a discussion about choosing the best EOR process for a particular field (i.e., whether

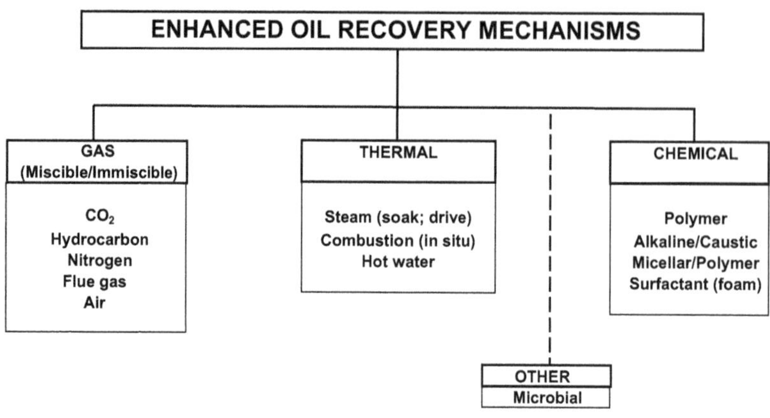

Fig. 1.1—General and specific EOR mechanisms (after Ref. 5).

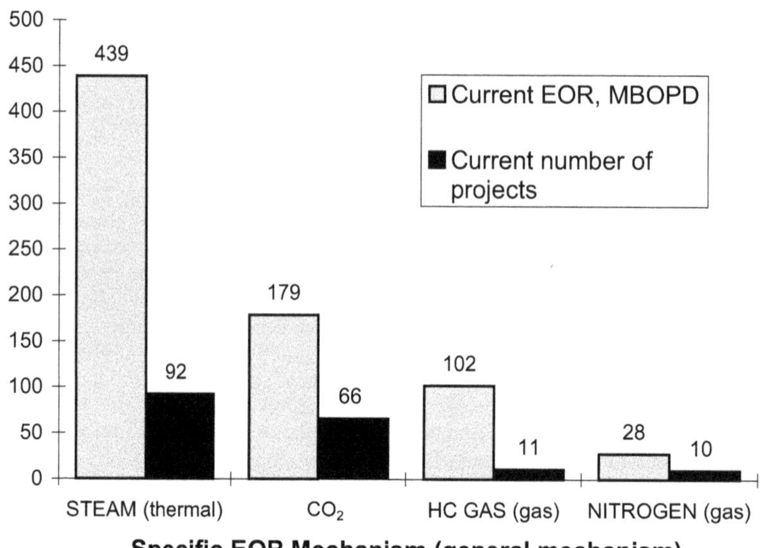

Fig. 1.2—The top four EOR mechanisms in the U.S. (1998 data from Ref. 5, Tables 1 and 2). These mechanisms represent 98% of all EOR production and 90% of all projects.

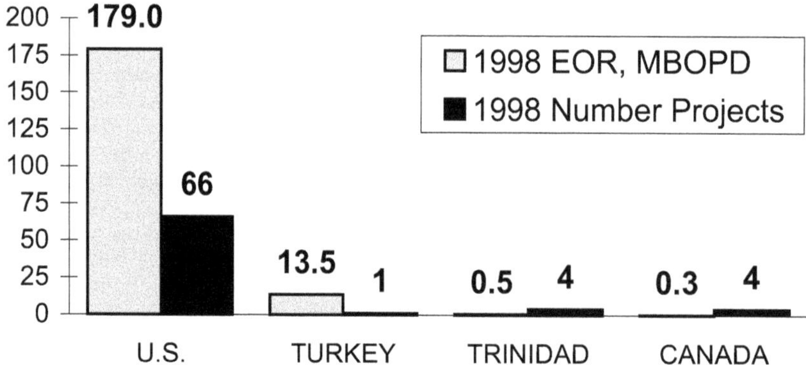

Fig. 1.3—Worldwide active CO_2 floods by country (after Ref. 5, Tables 1, C, E, H).

chemical or steam flooding is more appropriate than CO_2 flooding), although it does address briefly some considerations related to whether CO_2 flooding should be instituted after primary or secondary recovery. We presume that the reader already has some knowledge about which EOR process is appropriate for a target reservoir. Although the book presents CO_2 flooding screening processes, it does not address alternative recovery processes.

Most of the comments throughout the monograph target *miscible* CO_2 floods, though some immiscible flood practices are discussed. Miscible flooding has considerably more documented technology and experience than immiscible, and it offers significantly higher potential for oil recovery. Chap. 2 presents the differences between miscible and immiscible CO_2 flooding.

1.1.3 Organization. We begin this monograph with a review of the fundamental concepts of CO_2 process mechanisms and continue through project operation and management. Progressing sequentially through each chapter will give the engineer the critical aspects of installing and managing an operable flood. Content includes:

Introduction—Chaps. 1 and 2. Chaps. 1 and 2 provide basic information to answer the question, "How does CO_2 flooding work?" Chap. 1 provides general information on the scope and purpose of the book, as well as background information about CO_2 flood types, CO_2 sources, current floods, and projected CO_2 flood potential. Chap. 2 explores the theories behind what happens in the reservoir between the injection and the production well, with a focus on channeling, mobility control, minimum miscibility pressure (MMP), and CO_2/water relative permeability.

Screening—Chap. 3. Chap. 3 deals exclusively with the technical and economic screening process, addressing the basic question, "Which primary factors should be assessed before starting a detailed study?" The first step in planning any reservoir depletion is deciding on the primary factors and then assessing the effect that variances in each of these factors could have on technical and economic success. This chapter explores broad assessment techniques to help optimize the time and money spent in initial feasibility studies. In it we review the following.

- Analysis of reservoir rock and fluid properties to determine if oil can be recovered.
- Cursory estimates of injection and production rates and recovery using "desk-top" calculations or correlations.
- Cursory estimates of investment and operating costs.
- Overall project planning considerations.
- Scoping economics that indicate whether the incremental oil recovered will justify the costs.

Design—Chaps. 4, 5, and 6. In Chaps. 4, 5, and 6 we answer the question, "What is the optimum design and development scenario?" Such a scenario includes project planning and design processes and focuses on building conceptual development options that lead to an optimum design plan.

Chap. 4 focuses on the reservoir studies that underlie a successful CO_2 flood, including reservoir geometry, modeling, prediction, and CO_2 slug sizes. Chap. 5 deals with critical issues related to facilities and fluid process systems, including the CO_2 source and supply system, CO_2 reinjection system, production and injection facilities, and wellhead to wellbore equipment. Other topics pertinent to final design considerations are discussed, as well, such as various startup options to achieve an optimum economic balance for the reservoir and the facilities. Chap. 6 covers production and injection well design considerations.

Implementation, Operation, and Safety—Chaps. 7, 8, and 9. Chaps. 7, 8, and 9 look at the question, "How can the design become reality?"

The key is timing, and successful timing requires defining and coordinating the roles of all involved groups, including contractors, regulatory bodies, company support groups (legal, environmental, data systems), and field and plant operations personnel.

Other important aspects of implementation include installation of surveillance and monitoring systems to make data available to operations management; preinjection data gathering (i.e., pressure falloff tests, step-rate test, key-well liquid and gas composition measurements); and installation and testing of appropriate safety systems, because a CO_2 flood is potentially much more dangerous than a waterflood. Topics related to final operations, such as the necessary approvals from stakeholders (the government, other companies, partners, and contractors), also are examined. Throughout these three chapters, actual operations are reviewed in the context of fundamental operational principles and practices.

Many of the concerns about CO_2 flooding causing large increases in operating costs have been allayed as projects have matured and operators have gained more experience. The sometimes trial-and-error advances made in the past have led to insightful development of cost-effective equipment and practices that can help increase the profitability of existing and future projects. Several of the 27 operators of the 75 active worldwide floods have shared with us their experiences and their suggestions for practices and new technologies to enhance operations.

Appendices. Six appendices provide a variety of supporting information. Appendix A offers pertinent data for each of the known active and inactive worldwide CO_2 floods in tabular form; Appendix B discusses the CO_2 single-well cyclic injection (huff 'n' puff) process and results; Appendix C lists citations for CO_2 flood case histories for ease of reference; Appendix D gives adjustments that should be made to the relative permeability curves for water to account for the solubility of CO_2; Appendix E presents a sample Engineer's Statement to U.S. IRS for EOR Tax Credit, taken from the public record; and Appendix F contains data on CO_2 physical and thermodynamic properties for easy reference.

The Glossary found at the end of the book provides all the distinctive terms and abbreviations used throughout this monograph.

This monograph also comes with additional media files containing two CO_2 flood design scoping models and manuals, plus a computer-readable form of Table F.1, CO_2 Properties at Various Temperatures, °F.

1.2 CO_2 Flood Design Variations

Because most current and potential CO_2 projects are or will be operated as patterned miscible floods, it is important to understand the injectant design variations that are common in such floods. Variations seen in CO_2 flood designs typically stem from the desire to economically improve volumetric efficiency relative to the previous reservoir depletion method.

1.2.1 Injection Designs. There are five basic injection process designs comprising various combinations of continuous, alternating, and chase-fluid injection schemes, as shown in **Fig. 1.4.** The term "chase fluid" refers to the fluid injected into the reservoir immediately following, or chasing, a completed CO_2 volume.

Continuous CO_2 Injection. With continuous CO_2 injection, a predetermined CO_2 slug volume is injected continuously with no other interjected fluid or chase fluid. This approach usually is applied in a gravity-drainage reservoir or a nonwaterfloodable reservoir directly following primary depletion. Sometimes a different gas is used to drive the CO_2 through the reservoir. For example, in vertical downward CO_2 displacement projects, a lighter gas may follow CO_2 to maximize gravity segregation and minimize gravity tonguing and channeling.

Continuous CO_2 Chased With Water. The continuous CO_2 chased with water process is the same as the continuous CO_2 injection method, except that chase water follows the continuous CO_2 slug and immiscibly displaces the mobile, miscible CO_2 oil bank. This approach typically is used in reservoirs with low heterogeneity, which means that the reservoir has more homogeneous layer velocities [the velocities are proportional to the permeability/porosity ratio (k/ϕ)]. Homogeneous reservoirs help minimize gas production rates because they retain more of the injected gas. A detrimental gravity tongue may develop if this design is applied to relatively thick layers in homogeneous floods.

Conventional Alternating CO_2 and Water Chased With Water. In the conventional alternating CO_2 and water chased with water method, a predetermined slug of CO_2 is injected in cycles in which equal volumes of gas and water alternate (known as water alternating gas, or WAG) at a constant gas/water ratio, or WAG ratio. After the total CO_2 slug volume has been injected, a chase of continuous water is started. This injection design is most effective in highly stratified heterogeneous reservoirs because it minimizes gas cycling in high-velocity (k/ϕ) layers by reducing the fraction of injected CO_2 entering those layers. This design improves areal and vertical sweep efficiencies.

Tapered Alternating CO_2 and Water (Sometimes Chased With Water). In the tapered WAG process, CO_2 slugs are injected alternately with ever-increasing water cycle lengths in tapered or unequal cycle volumes until the total CO_2 slug volume has been injected. Sometimes a chase of continuous water follows the taper.

The objective of the tapered approach is to decrease the CO_2 utilization factor (defined as the ultimate surface CO_2 injection volume, Mscf, per ultimate stock tank barrel of incremental oil recovered) by periodically adjusting the WAG ratio and/or cycle slug sizes in flood patterns where utilization is excessive.

Several operators are using the tapered application to manage near-term flood profitability because this approach reduces purchase cost for CO_2 and requires that less gas be recycled. Although

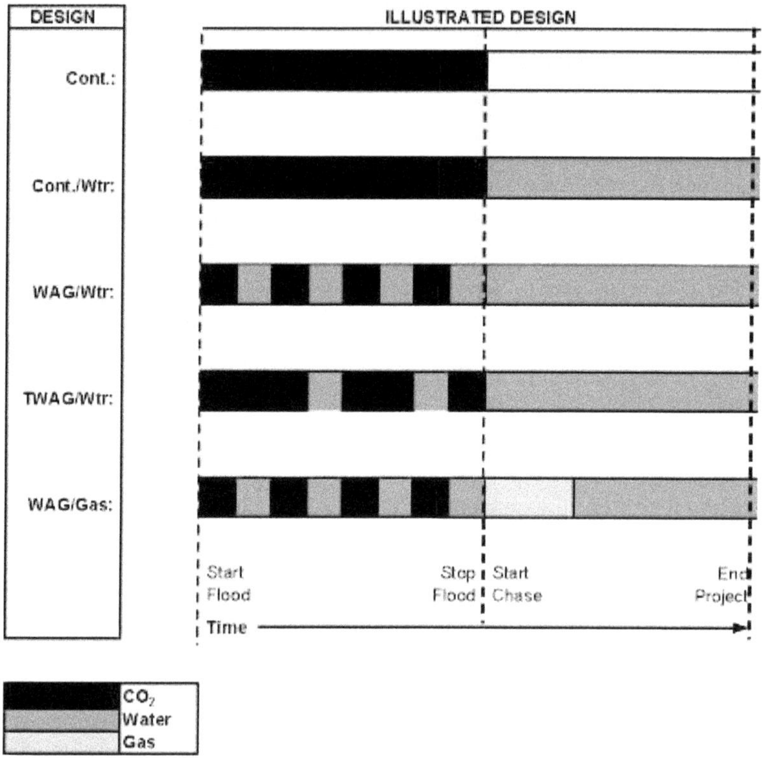

Fig. 1.4—Carbon dioxide flood injectant design variations. "Cont."=continuous CO_2 flood; "Cont./Wtr"=continuous CO_2 chased with water; "WAG/Wtr"=conventional water alternating gas (WAG) CO_2 flood chased with water; "TWAG/Wtr"=tapered alternating CO_2 and water chased with water; and "WAG/Gas"=conventional WAG chased with gas.

this approach may reduce costs, it also may reduce near-term revenues. The near-term oil rate is low because of reduced CO_2 throughput in the faster, mobile, oil-bearing layers, but the oil rate usually increases later in the flood because of the improved areal sweep achieved by decreased total mobility ratios. (In other words, when the CO_2 doesn't sweep through as fast, it does a better job of recovering all the oil.)

Alternating CO_2 and Water Chased With Gas. In the alternating CO_2 and water chased with gas method, CO_2 is injected in cycles alternated with water injection (like a conventional WAG), then chased with a volume of less expensive gas after the total CO_2 slug volume has been delivered. In some cases, water injection may follow the chase gas or be alternated with it. The main purpose of the chase gas is to maintain miscible displacement of the trailing edge of the CO_2 slug while reducing total CO_2 requirements. Sometimes gas is chosen as the chase fluid to avoid using water on water-sensitive lithologies.

1.2.2 Implementation of Flood Designs. Implementation of any flood design varies greatly depending on gas composition, timing relative to waterflooding, and degree of miscibility.

For instance, the injectant may be pure CO_2, or it may contain small percentages of other gases such as hydrogen sulfide (H_2S), hydrocarbon gas, and/or nitrogen. CO_2 floods may be applied preceding or following a waterflood operation, and they may be miscible or immiscible processes. Miscible CO_2 flooding is most effective in recovering medium to high API gravity crudes, whereas immiscible floods work better with lower API gravity crudes and in gravity-drainage reservoirs.

1.3 Major Influences On CO_2 Flood Success

Many different factors influence the success of a CO_2 flood, but most are either technology- or business-related. Technological factors include reservoir conditions that determine whether CO_2 can mobilize significant residual oil in the reservoir, as well as operational factors that determine whether a project can be installed and operated cost effectively. Business factors that affect CO_2 flood success include both economic and political elements.

1.3.1 Technological Factors. A CO_2 flood can succeed or fail for many technological reasons. Carbon dioxide feasibility and operational issues are among the more important technological factors.

The Feasibility of CO_2. In considering CO_2 feasibility, the three most important flood variables to consider are as follows.

- Significant movable oil saturation (which depends on oil properties, remaining oil saturation, reservoir heterogeneity, and reservoir wettability).
- The ability to achieve and maintain thermodynamic MMP in the reservoir (which depends on the average reservoir pressure, fracture parting pressure, injectivity impacts, and oil properties).
- The ability of the CO_2 to contact a large portion of the reservoir, including vertical, areal, and unit displacement (all of which depend on well spacing, mobility ratio, permeability, reservoir heterogeneity and geometry, injection well conformance, areal discontinuity, gas cap, and fracture systems).

Sustaining optimal oil rate and reserve recoveries depends most heavily on maintaining thermodynamic MMP. It also depends on accurately estimating the recoverable oil saturation and on maintaining pattern-by-pattern sweep efficiencies. Lack of attention to any of these properties will lead to lower economic return.

Chap. 3 includes methods for screening reservoirs for CO_2 applicability, including physical testing of the rock and fluids, comparison with actual flood statistics and performance for similar fields, and simplified approximate predictive techniques. Once a reservoir is determined to be a viable CO_2 flood target, the technical operational concerns can be addressed.

Operational Factors. Operational factors relate to controlling the cost of recovering incremental oil. Four significant factors hold the greatest potential for cost control: CO_2 source (which is affected by availability, purity, transmission, and cost); ultimate CO_2 slug requirement (WAG or continuous); CO_2 reinjection facilities (the cost to build or lease a CO_2 removal plant or recompression facility); and production and injection system upgrades (the CO_2 distribution system and gas-gathering system). It is a good idea to start by investigating the possibility of using nearby flood infrastructures. Many existing gas recycle plants, water disposal systems, and CO_2 injection systems are becoming underused as existing CO_2 floods are maturing.

1.3.2 Business Factors. Currently, the greatest uncertainty in CO_2 flooding rests on the business side of operations, where political and economic influences weave a complex tapestry of factors affecting the viability of a CO_2 project. Three important business factors are tax burdens and incentives, government regulations, and oil and gas prices.

Typically, a new CO_2 flood project is burdened by high startup investments, followed by higher operating costs (i.e., cost of CO_2 purchase and recycle), and then a moderately long wait until incremental oil production begins. For a CO_2 project to be economically viable, the incremental oil produced must not only compensate for the additional costs of the project, it must also recover those costs quickly enough to generate an acceptable rate of return.

As more and more operators are reporting project oil rates and reserves that are meeting or exceeding original estimates, industry confidence is growing. Certainly, improvements in several technology areas—such as better reservoir description and prediction methods and improved sweep efficiency techniques—will have a marked effect on future CO_2 flood results. In addition, even moderate improvements in oil prices, government taxes and tax incentives, will offset some of the startup cost burdens enough to allow more operators to risk starting new floods to produce more oil.

1.4 Current and Planned CO_2 Projects

Every two years, the *Oil & Gas J.* publishes an article on worldwide EOR activity, along with pertinent data collected through a survey of operators. Appendix A contains an extensive table of statistics on worldwide activity in CO_2 floods, derived mainly from the last few biennial surveys.[5] Included in this table is a list of all the inactive projects as well. The following discussion cites data from Appendix A, unless otherwise referenced.

Seventy five active CO_2 floods operate in five countries around the world, producing more than 194 MBOPD of incremental enhanced reserves, or about 8.4% of the reported total worldwide EOR production. Field-scale miscible floods are the source of virtually all this production; pilots and/or immiscible CO_2 floods produce only a very small fraction. Of the active CO_2 flood projects, only seven (9%) are pilots.

Projects in the U.S. comprise about 95% of the current worldwide CO_2 EOR production, and about the same fraction of the total number of acres, wells, and total production under flood **(Table 1.1).**

Floods in Canada, Turkey, and Trinidad produce the remaining CO_2 EOR reserves from nine projects, with about half of these operating immiscibly **(Fig. 1.5).**[5] The 15 largest CO_2 floods represent 20% of the total number of active projects and produce 81% of the total EOR incremental BOPD.

Fig. 1.6 shows the 15 largest EOR producing projects sorted by EOR incremental rates, with the Seminole (San Andres), and Wasson (Denver) units exhibiting much higher incremental rates than the others because of their size (number of acres and wells involved), maturity (both were started in 1983 and incremental rates are peaking), and relatively good reservoir quality.

Most CO_2 floods are operated as secondary or tertiary recovery projects—secondary if the flood starts during primary recovery and tertiary if it starts during a conventional improved recovery process (e.g., waterflood).

Although locations of North American CO_2 floods span the continental U.S. and southwestern Canada, they are found mainly along the western Rocky Mountain range into the Permian Basin area **(Fig. 1.7).** Most of the eastern U.S. CO_2 floods are located in

TABLE 1.1—COMPARISON OF U.S. AND WORLD ACTIVE CO_2 FLOOD STATISTICS

	Number of Floods	Number of Acres	Number of Producers	Number of Injectors	Project Total BOPD	Project EOR BOPD
WORLD (including U.S.)	75	316,178	6,738	4,066	312,843	194,259
U.S.	66	300,233	6,504	4,003	297,608	180,024
(% of world)	88%	95%	97%	98%	95%	93%

Projects in the U.S. comprise about 95% of the current worldwide CO_2 EOR production (from Ref. 5, Appendix A.1).

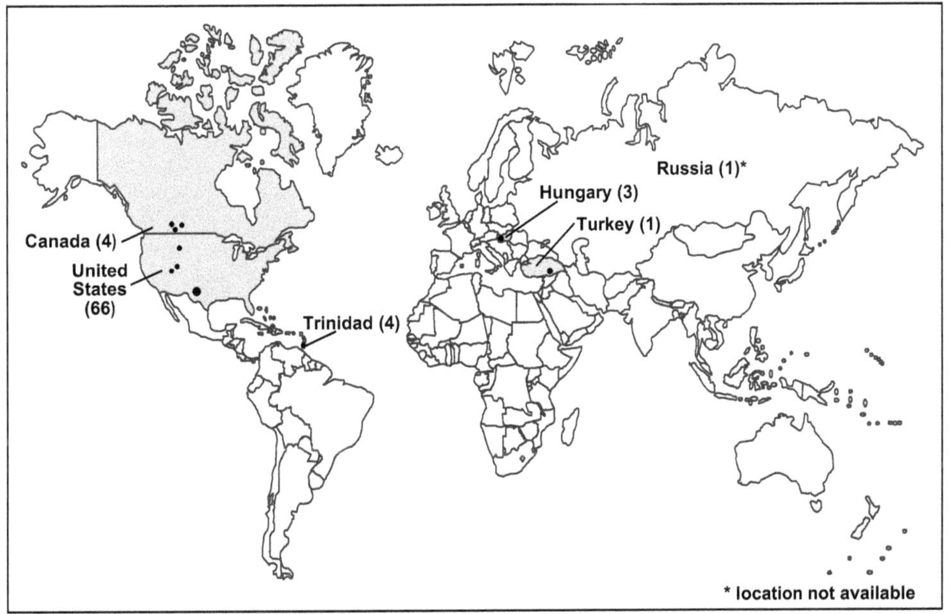

Fig. 1.5—Worldwide locations and number of active CO_2 floods (1998, Ref. 5).

southeastern Texas, southern Louisiana, and southern Mississippi. There are 70 specific North American flood projects in 26 areas, and all are operated miscibly.

As Fig. 1.7 shows, 12 of the 26 CO_2 flood areas in North America are currently inactive and represent potential CO_2 flooding opportunities. Some of these opportunities were announced as "planned EOR projects" in operators' responses to the *Oil & Gas J.* survey.[5]

Of the eight U.S. and one Canadian planned floods expected to start up between 2000 and 2002, about half are in existing or past CO_2 flood areas. As with many potential projects, recent government programs such as the U.S. EOR tax credit are revitalizing previously uneconomic (but technically viable) floods.

As mentioned earlier, CO_2 floods in the U.S. are the only EOR projects (outside of U.S. hydrocarbon gas floods) that have consistently and significantly increased annual EOR production since the 1986 drop in oil prices. More than one-half of the gas and one-fifth of the oil produced by current U.S. EOR projects come from CO_2 flood projects.[5] **Fig. 1.8** illustrates the increasing amount of EOR production from these projects. Note that the 1992 figures include the Seminole (San Andres) unit rates for the first time—this project's production was not reported in previous years. Even without this extra 31 MBOPD, EOR production increased by 16% from 1990 to 1992.

In contrast to miscible floods, the number of active U.S. immiscible floods declined from 28 in 1986 to zero in 1998. This is the largest percentage change for any of the EOR processes during the same time period and is a testimony to the influence of oil prices on marginally profitable EOR projects.

The most active CO_2 flooding area is the U.S. Permian Basin, located in west Texas and eastern New Mexico. Here, 50 projects produce an incremental 145 MBOPD, more than 80% of the current North American enhanced oil produced from CO_2 floods. Most of the Permian Basin projects produce from shallow carbonate reservoirs that average about 5,200 ft deep, 107°F, and 13% porosity. The average crude is 36°API gravity and about 1.4 cp viscosity.

An extensive CO_2 pipeline and CO_2 reinjection infrastructure system exists throughout the Permian Basin, making it attractive for expanding or starting up new projects. High-pressure pipelines supply CO_2 from the natural source fields of Bravo Dome, McElmo Dome, and Sheep Mountain hundreds of miles to the north, and also from process plants that remove CO_2 from the hydrocarbon gas fields of the Delaware-Val Verde Basin hundreds of miles to the south. Several CO_2 removal plants operate throughout the Permian Basin, servicing the recycle needs of most of the CO_2 floods there.

1.5 CO_2 Sources

Access to a nearby, quality CO_2 source is critical to project cost management and success. How close the source should be depends primarily on the designed rates and volumes of CO_2. Generally, higher CO_2 injection-rate projects can tolerate sources farther away because pipeline construction costs do not increase proportionately with line capacity. (Comparison of similar pressure-drop systems shows that the required line diameter does not increase proportionately with flow capacity.) In this basic economy of scale, the installed cost of CO_2 delivery decreases as flood size increases.

The required CO_2 purity will depend primarily on whether the flood is planned as a miscible or immiscible process. Contaminants such as methane and nitrogen can significantly increase the minimum pressure required to achieve miscibility with crude, whereas H_2S gas and lighter hydrocarbons exclusive of methane (C_1) can have the opposite effect. An immiscible process does not depend so heavily on high quality CO_2; in some gravity drainage reservoirs, a CO_2 flood may be successful even when the percentage of CO_2 is rather low. It is important to address issues of water content and noxious gases in CO_2 acquired from industrial sources because of safety issues and the potential for injection line corrosion.

1.5.1 Naturally Occurring CO_2. Carbon dioxide is available throughout much of the world, both from naturally occurring underground reservoirs and from man-made industrial operations. Some CO_2 sources are large enough to serve several large commercial CO_2 floods, while others (e.g., CO_2 recycling plants or small industrial sources) produce only enough for the small requirements of pilot floods. (Pilot floods are generally used to test new areas, although areas adjacent to already developed and successful CO_2 flood projects do not require pilots and await only adequate CO_2 volumes and/or prices in order to start injection.)

Many gas reservoirs around the world produce CO_2 in various percentages of the total gas stream. Most contain relatively low

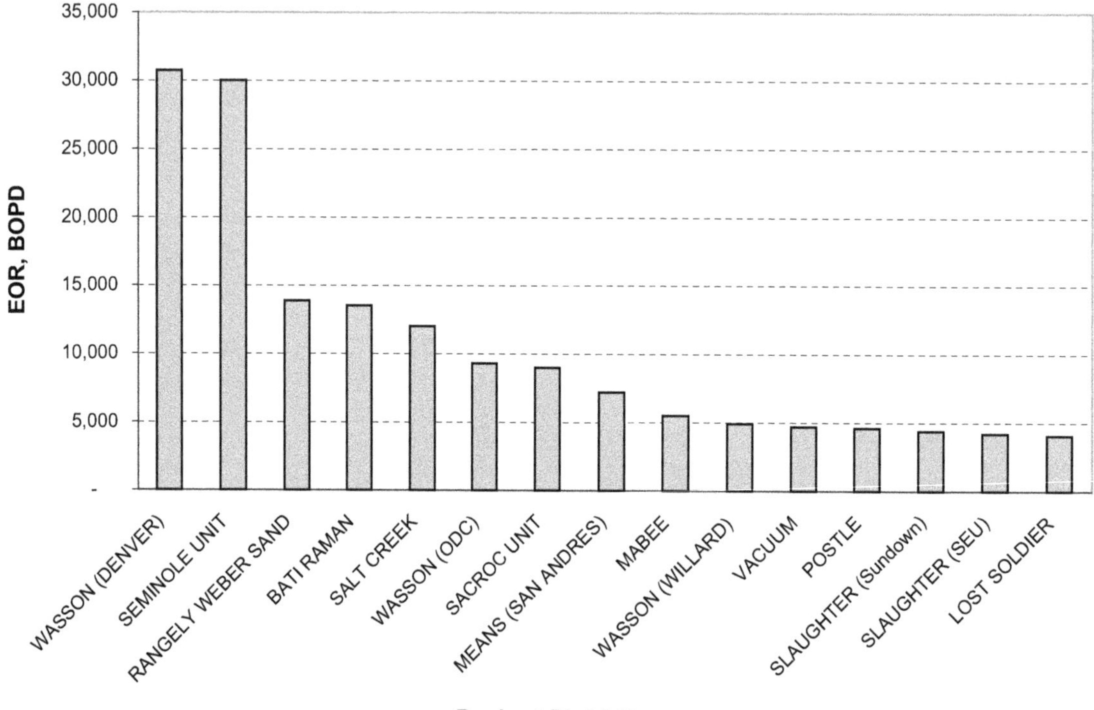

Fig. 1.6—Current worldwide CO_2 projects ranked by EOR incremental production rates (after Ref. 5). The top 15 projects represent 81% of CO_2 EOR production and 20% of the number of projects (from data in Appendix A).

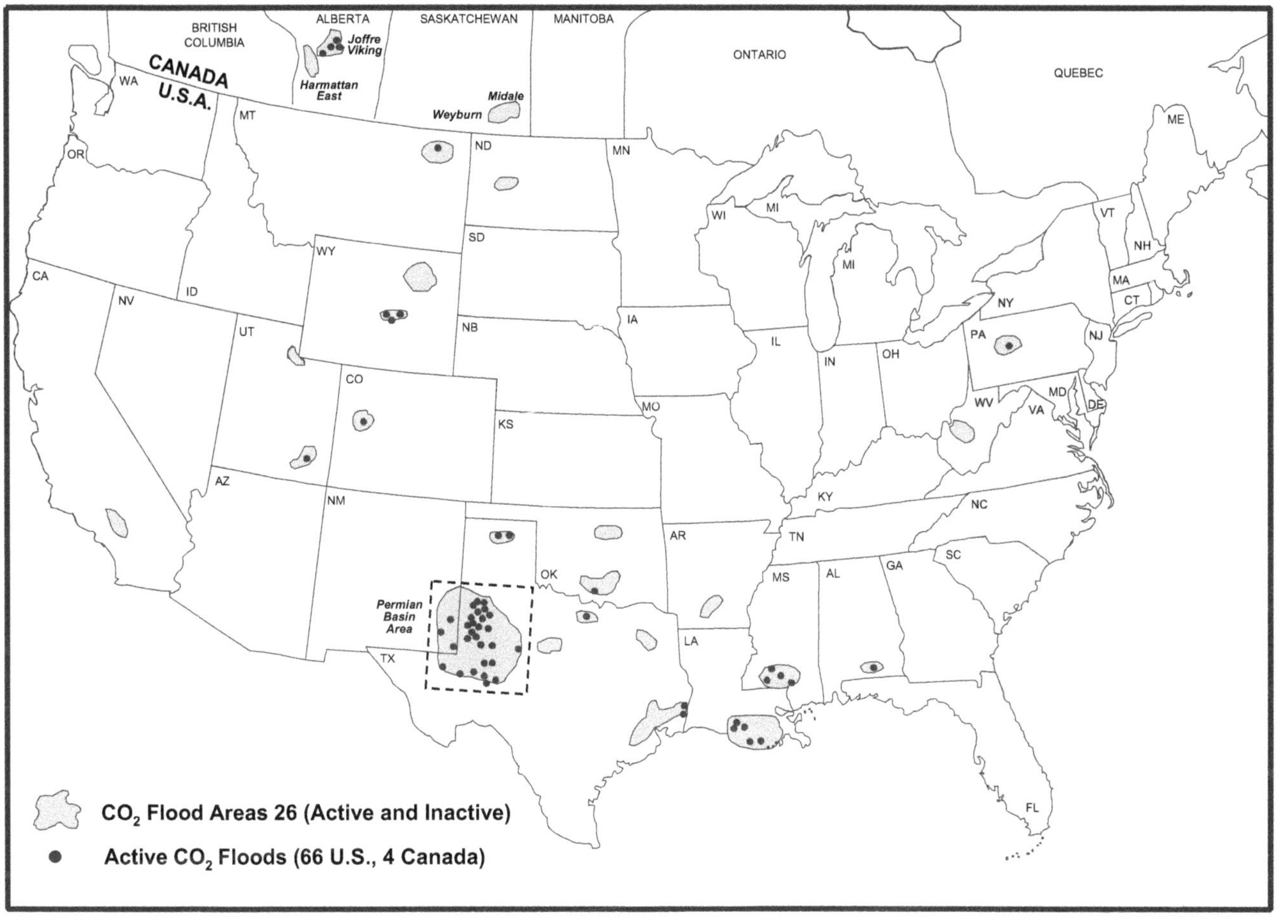

Fig. 1.7—Locations of all North American active and inactive CO_2 floods (1998). There are currently 66 active CO_2 floods in the U.S. and four active floods in Canada.

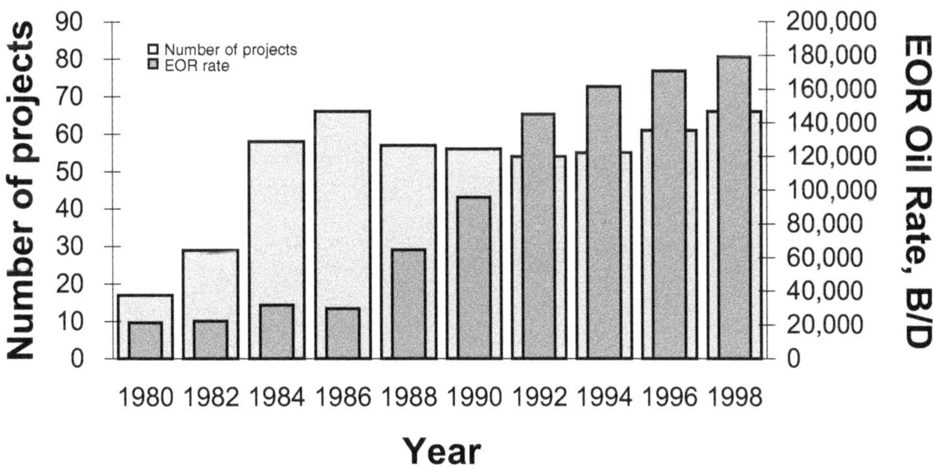

Fig. 1.8—U.S. CO_2 flood activity and EOR rate trends (after Ref 5). The 1992 figures mark the first time that the rates from the Seminole (San Andres) unit were included.

concentrations of CO_2 (less than 25 mol%), and would require processing costs like those required for CO_2 from industrial sources to raise the CO_2 concentration high enough (around 90 to 98%) to become practical for a miscible flood.

To date, the known reservoirs with the largest concentrations of relatively high-purity CO_2 are in the western U.S., eastern Europe, and Indonesia, and in various fields in the mountainous regions of Central and South America. In the former U.S.S.R., United Kingdom, North Sea, Canada, western Europe, Middle East, and Africa, no abundant natural CO_2 resources are known to exist close to oil fields.

Throughout the U.S., more than 50 areas have been identified with reserves of 50 mol% CO_2 or more.[13] **Fig. 1.9** shows several of the more notable CO_2 reservoirs in the U.S., along with their associated pipelines.

The largest of the active natural CO_2 reserve areas, which are listed in **Table 1.2,** are LaBarge-Big Piney in Wyoming (~20 Tcf), Bravo Dome in New Mexico (~12 Tcf), and McElmo Dome in Colorado (~12 Tcf).[14] Several basins in Utah contain from 3 to 6 Tcf of high quality CO_2 but have not been developed to serve active floods.

The latest large CO_2 source field discovery, St. John's, was made in eastern Arizona in 1994. Company press releases report reserves from 6 to 7 Tcf, although field delineation work is still underway. No CO_2 from this field currently is produced for pipeline shipment to any EOR floods, though a pipeline to the Bakersfield, California area and west Texas is under study.

The estimated total CO_2 reserves in the U.S. range from 51 to 70 Tcf, with existing pipeline capacities up to 3.4 Bcf/D.[14] Current pipeline throughput averages about 1.4 Bcf/D. The highest percentages of excess capacity are in the pipeline systems of LaBarge, Wyoming; Jackson Dome, Mississippi; and Sheep Mountain, Colorado. The Bravo Dome and McElmo Dome CO_2 pipelines currently operate at higher capacities, averaging around 65 to 70%. With the installation of additional pumps, an incremental 0.9 Bcf/D of CO_2 could be transmitted along the existing CO_2 pipelines, which would bring the total potential deliverability of CO_2 to 4.3 Bcf/D.

1.5.2 Carbon Sequestration. A newly emerging interest of the petroleum industry is the study of carbon sequestration, which is providing new insights into the use of CO_2 created when fossil fuel is burned. Carbon sequestration is defined by the U.S. Department of Energy (U.S. DOE) as the capture and secure storage of carbon that would otherwise be emitted to or remain in the atmosphere. According to a 1999 carbon sequestration paper by the U.S. DOE,[15] worldwide carbon emissions totaled 7.4 billion tonnes in 1997, or about 360 Bscf/D. More than 80% of the 7.4 billion tonnes comprised CO_2 emissions from fossil-fuel combustion. Although much of this volume came from small (or remote) single sources, it represents a very large potential resource of CO_2 for use in enhanced oil recovery.

The U.S. DOE study states that geologic formations, such as EOR reservoirs, are likely to be the first large-scale ventures for concentrated CO_2 sequestration. Other CO_2 sequestration locations under study are the oceans and certain terrestrial ecosystems.

The study reports that carbon emissions in the U.S. were about 1.8 billion tonnes in 1995. Given the U.S. goal to reduce carbon emissions by a "significant fraction" of 1.0 billion tonnes per year by the year 2025, many in the petroleum industry anticipate that this industry will be called upon to work together to develop novel or advanced geologic sequestration technologies. One of the first such partnerships, formed in 2000 and called the *CO_2 Capture Project,* is being coordinated by BP and includes Chevron, Norsk Hydro, Shell, Statoil, Suncor, and Texaco. This U.S. $20 million project will last three and one-half years and will include research and development of advanced CO_2 separation and geologic storage technology. We estimate that progress in meeting the U.S. goal could provide about 90 MMcf/D of CO_2 available for potential EOR use by 2004.

The above estimate assumes: 1) that the "significant fraction" is 20%, which would yield about 10 Bscf/D of CO_2 in 2025; 2) that progress in achieving 10 Bscf/D sequestration of CO_2 by 2025 would take place in an exponential fashion; and 3) that oil reservoirs will make up about 50% of the locations for sequestration.

There are a few active geological CO_2 sequestration projects. One is in the North Sea and uses an amine absorption process to remove CO_2 from subsurface-produced natural gas. About 50 MMcf/D of CO_2 is compressed and piped to an adjacent platform for injection into the Utsira Formation 3,250 ft below the seabed.[16]

Planned CO_2 sequestration projects include an EOR flood in PanCanadian Petroleum's Canadian Weyburn Field, where about 90 MMcf/D of CO_2 will be injected (350 Bcf in total).[17] The CO_2 will be shipped by pipeline from a Synfuels (coal degasification) plant 200 miles to the south. Another planned project is in BP's Alaskan Schrader Bluff Field, where about 17 MMcf/D of CO_2 from the effluent of field oil and gas processing facilities will be injected.[17,18]

1.5.3 Manufactured CO_2. Sources for artificial, or manufactured, CO_2 vary widely and are abundant. These CO_2 sources can be categorized the following four ways.

• Concentrated, high-pressure—from synthetic fuel plants and gasification combined-cycle power plants.

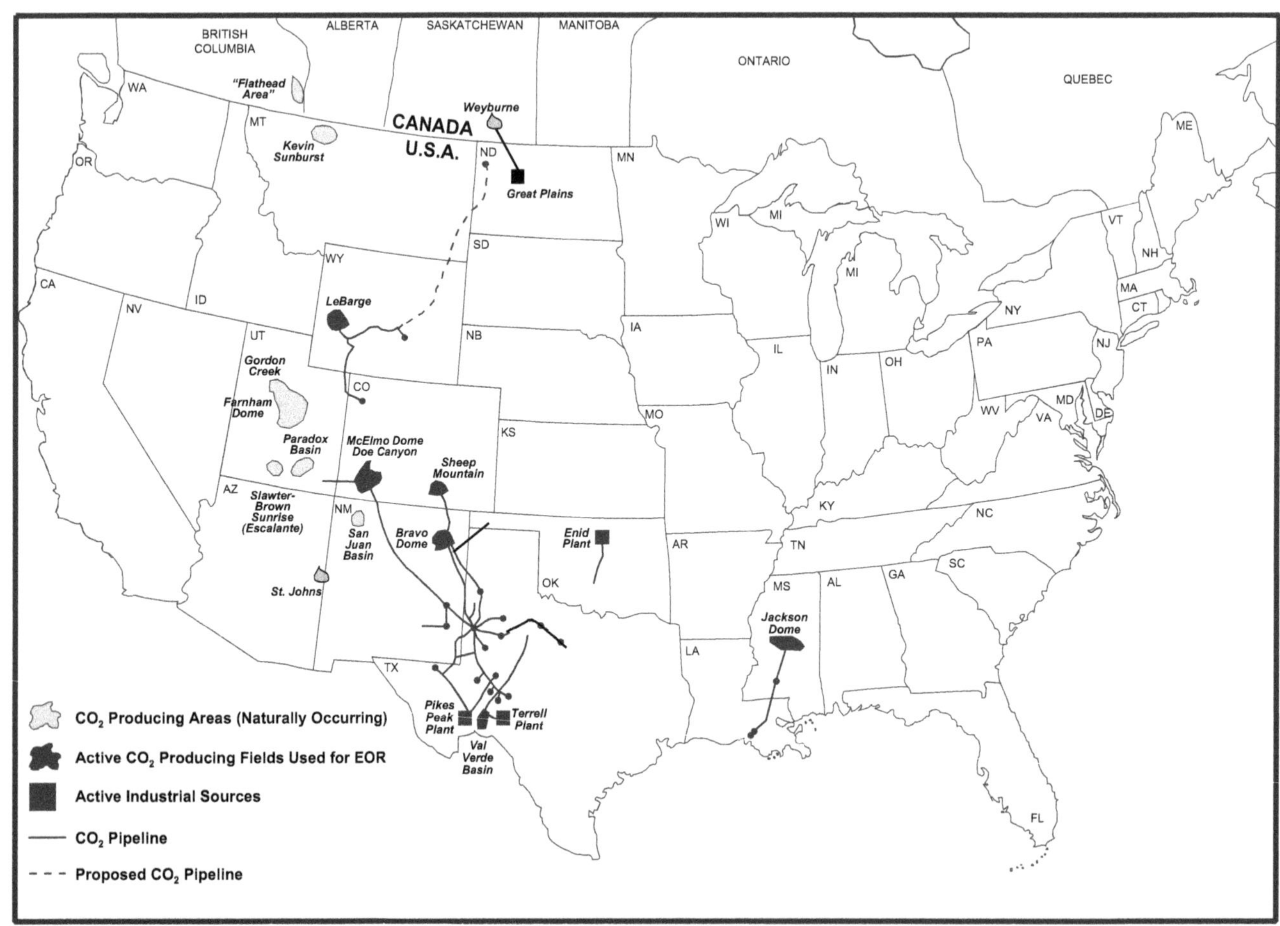

Fig. 1.9—Locations of CO_2 sources and pipelines in the U.S.

TABLE 1.2—NATURAL CO_2 SOURCES IN THE U.S.

Name and Location	Ultimate Recovery (Tcf)	CO_2 %	Pipeline Capacity (MMcf/D)
LaBarge-Big Piney Southwestern Wyoming	20 to 25	63 to 90	250 to 600
Bravo Dome Northeastern New Mexico	8 to 12	99	400 to 700
McElmo Dome Southwestern Colorado	10 to 12	95	650 to 1,000
Sheep Mountain Southeastern Colorado	1 to 2	97	330 to 500
Jackson Dome Central Mississippi	3 to 6	90 to 99	450 to 700
Gordon Creek and Farnham Dome Central Utah	1.5 to 3	98	—
Escalante Anticline and Paradox Basin Southern Utah	1.5 to 3	60 to 98	—
St. Johns Eastern Arizona	6 to 7	70 to 93	—
Totals:	51 to 70		2,080 to 3,500

After Ref. 14 (1990).

• Concentrated, low-pressure—from ammonia plants and ethanol fermentation plants.

• Diluted, high-pressure—from hydrogen plants and refinery fluid catalytic cracker units (FCCU).

• Diluted, low-pressure—from fossil-fuel power plants, cement plants, and anaerobic digestion (biomass and wastes).

The dominant artificial CO_2 source is power plants, mainly because there are so many of them. Approximately 54 Bcf/D of CO_2 is potentially recoverable from existing and planned power plants in the major oil producing regions of the U.S.,[19] although the average byproduct gas content is only 12 to 14% CO_2. The estimated cost to recover and compress flue gas from coal-fired power plants ranges from \$0.75 to \$2.50 per Mcf **(Table 1.3)**, depending on type of feedstock, distance from the flood, and output volume.[14] More competitively priced CO_2 could be produced from gas-fired cogeneration electric power plants, though as yet none of these has been built.

Ammonia plants produce CO_2 (at about 97.5% purity) as a normal byproduct and can deliver competitively priced CO_2 in the range of \$0.50 to \$1.00 per Mscf **(Fig. 1.10)**. Unfortunately, most ammonia plants usually produce less than 14 MMscf/D, and all of them combined produce less than 0.5 Bcf/D of CO_2 in the southern oil producing states of California, Texas, New Mexico, Oklahoma, Mississippi, and Louisiana. Of these states, Louisiana produces the largest share of CO_2.[19] Although most ammonia plants may not be adequate for large field-scale floods, some may be close enough to support a small flood or pilot project.

As Fig. 1.10 shows, other industrial sources such as cement plants and refinery FCCU's produce CO_2 at a much higher cost than that produced at coal-fired power plants and ammonia plants. Because industrial plant CO_2 prices are well above the current price range of naturally occurring CO_2 delivered by pipeline (\$0.80 to \$1.00 per Mscf), it is unlikely that these artificial sources will be major suppliers of CO_2 to full-field floods anytime soon. In light of current and projected CO_2 flood startups and of known CO_2 resources and demand, there appears to be adequate natural supply of CO_2 to serve the industry needs in regions of existing activity. Furthermore, many operators could start new CO_2 floods around the world by developing large, low cost CO_2 sources and/or supply systems, especially in the northern Rockies area of the U.S. and southern Canada, the North Sea, Central and South America, and the former U.S.S.R.

1.6 Industry Projected CO_2 Potential

Current reported worldwide EOR production, including all recovery methods, is about 2.3 MMBOPD. This represents about 3.5% of the total world oil production, and an increase of 4.5% over the rate reported in 1996.[5] Enhanced oil recovery methods are expected to provide an increasingly larger portion of future world oil production as the conventional resource-base rates decline. Some sources predict that worldwide EOR production will rise steadily and reach more than 30 MMBOPD by the year 2020[28] **(Fig. 1.11)**.

Based on current growth trends, CO_2 flooding is expected to be a major contributor to the projected increase in oil production. In order to speculate about future increases in CO_2 flooding activity, we will briefly review historical trends.

A plot of the number of starts per year for CO_2 projects worldwide **(Fig. 1.12)** shows four periods of varying activity. Through 1998, a total of 145 CO_2 floods were started, including active, inactive, miscible, and immiscible floods. The 1970's saw new CO_2 floods at a pace of about one or two per year. From 1980 through 1985, startups soared to about 12 projects per year. From 1986 through 1993, new CO_2 flood starts decreased to a rate of about

TABLE 1.3—SOME RECENT COST ESTIMATES FOR CO_2 RECOVERY FROM POWER PLANT FLUE GASES

Range of Estimated Cost for CO_2 Recovered, $/Mcf	Type of Power Plant	Estimator and Date
$2.14 to $2.58	Modern (pulverized coal) power plant	Alpert, 1988
1.75 to 2.25	Full-size, modern coal-fired power plant	Walter and Ireland, 1988
1.25 to 1.30	Wyoming coal-fired 750 MW, power plant at Glenrock, Wyoming	Rett, 1987
2.13 to 2.47	Retrofitted coal plant for O_x-CO_2 firing, SO_x and NO_x removed, pure CO_2 and N_2 injected for EOR	Park, 1987
0.75 to 1.50	Retrofitted coal plant for O_x-CO_2 firing, SO_x and NO_x injected into reservoir	Wolsky, 1987
1.15	Integrated gasification combined-cycle power plant (IGCC)	Alpert, 1988
1.00 to 1.50	Gas-fired, cogeneration, combined-cycle CO_2 recovery process plants	Miller *et al.*, 1986
<0.00 to 1.11	Site-specific (near oil field) cogeneration, combined-cycle, CO_2 recovery process plants	Miller and Soychak, 1987

After Ref. 14 (1990).

four per year. The latest trend, from 1994 through 1998, shows startups increasing to about five to six per year.[5]

Over the past few decades, more than 70 miscible and immiscible CO_2 floods have been terminated or postponed (rendered "inactive"), which represents about 50% of the total projects. Most of the inactive projects were pilot tests, and about two-thirds were miscible. A few of the historic immiscible projects currently are active, and about 60% of the historic miscible floods are active.[5,29–31]

1.6.1 The 1984 National Petroleum Council (NPC) Study. In 1984, the NPC undertook a study for the U.S. Secretary of Energy that set the standard for U.S. EOR reserve projections. They also created the Tertiary Oil Recovery Information System (TORIS) in Bartlesville, Oklahoma.[8] The EOR study is an update to the original one undertaken in 1976 and has been highly regarded in the industry as the most extensive ever accomplished. It projects a total of 14.5 billion bbl of crude as potential additional EOR reserves. Two SPE papers authored by NPC committee members effectively summarize the total EOR and miscible study results.[32,33]

A subcommittee study on miscible flooding (which included CO_2, hydrocarbon, and nitrogen gases) used a resource base of 600 reservoirs with combined original oil in place (OOIP) of 190 billion bbl, or about 39% of the total U.S. OOIP. These reservoirs were selected based on several accepted screening criteria, such as thermodynamic MMP. The authors of the miscible study recognized CO_2 as the preferred miscible solvent; consequently, a large part of their reserve projections were based on CO_2 flooding.

The NPC study projected a range of 5.5 to 8.5 billion bbl of additional EOR oil from miscible floods. The associated CO_2 requirement to recover this incremental oil was estimated at 30 Tcf, with a CO_2 rate averaging about 4 Bcf/D for several decades. This requirement can be met by current natural CO_2 sources that lie near potential CO_2 flood areas.

1.6.2 Studies Since 1984. Several projections of U.S. CO_2 flooding reserve potential have been published since the 1984 NPC estimate, citing recoveries ranging from 8 billion bbl[34] to 15 billion bbl,[35] and as high as 26 billion bbl.[14] All the above EOR projections depend on future oil prices, implementation of new technology, and economic incentives. If one applies the NPC net CO_2 utilization factor of about 5.5 Mscf/BO to the higher CO_2 EOR estimates, CO_2 requirements could range from 82 to 143 Tscf, with daily rates ranging from 11 to 20 Bcf/D. These rates and volumes are beyond existing known natural CO_2 sources and systems, and would require the discovery of new sources and/or the advancement of competitively priced industrial CO_2 supplies, as well as the upsizing and expansion of existing pipeline systems.

Total U.S. miscible flooding EOR rates historically have matched the NPC $30/bbl case projections very well, as seen in **Fig. 1.13** (based on *Oil & Gas J.* survey data[5]). Because realized oil prices have been lower than $30/bbl, one could conclude that miscible floods (and primarily CO_2 floods) are performing better than expected.

R.S. Park addressed North Sea CO_2 flood needs,[36] estimating that for a typical North Sea oil field, more than 572 MMscf/D of gas injection would be needed to provide an economic EOR project. This economic assessment assumed that about 7.8 Mscf CO_2 (or N_2) per bbl EOR oil would be needed. Because there is no naturally occurring CO_2 source in the U.K., the only alternative is CO_2 from the flue gases of coal-fired power stations, such as the Longannet plant in Edinburgh. Parks concluded that the rate of CO_2 production in this plant would be sufficient and economically feasible, provided that oil prices averaged at least $25/bbl and the plant was retrofitted to burn oxygen instead of air.

The future of CO_2 flooding, as well as any EOR production method, assumes that the existing fields and wells that currently produce will be accessible. A recent study by the U.S. DOE[37] reports that maintaining access to the resource target oil is critical to

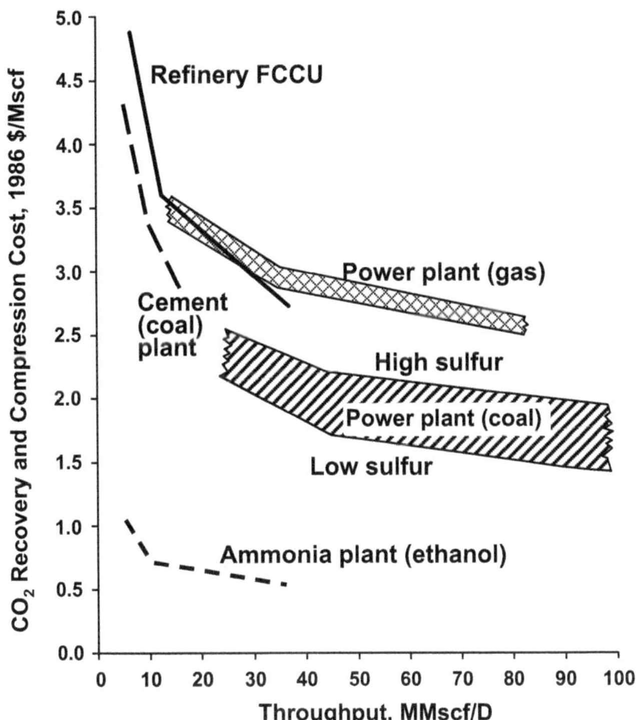

Fig. 1.10—Recovery and compression cost of CO_2 from industrial plant sources (after Ref. 19, 1983.)

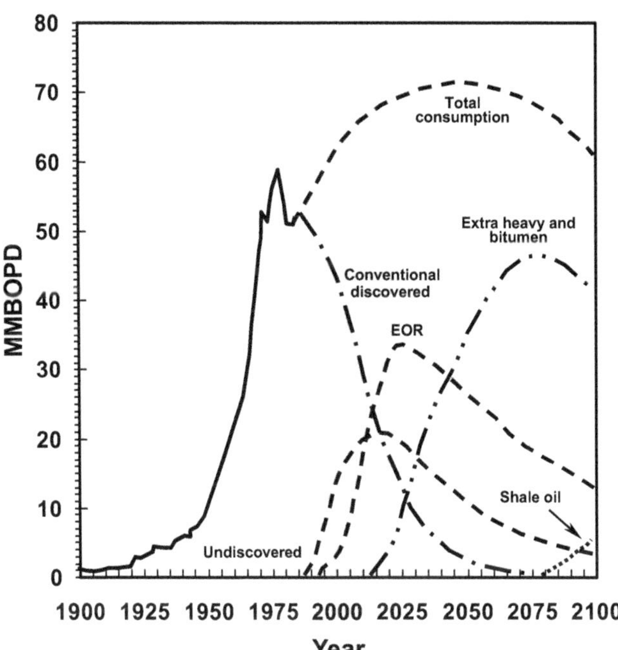

Fig. 1.11—Prediction of world crude oil supply (after Ref. 28, 1990). Based on current trends, CO_2 flooding is predicted to become a major contributor to this increase in production.

the profitability of future EOR projects. As wells reach economic operating limits and are subsequently plugged and abandoned, a significant percentage of the EOR resource base could be lost.

These data support a probability that enhanced recovery project opportunities will not always exist. Unless oil prices, advanced technologies, or other incentives change to allow an attractive profitability, the loss of existing resources will be irreversible.

Nomenclature

k = permeability, L^2, md

ϕ = porosity, dimensionless, fraction

References

1. Stalkup Jr., F.I.: *Miscible Displacement,* Monograph Series, SPE, Richardson, Texas (1983) **8.**
2. Klins, M.A.: *Carbon Dioxide Flooding—Basic Mechanisms And Project Design,* International Human Resources Development Corporation (IHRDC), Boston, Massachusetts (1984).
3. Rose, S.C. *et al.*: *The Design Engineering Aspects of Waterflooding,* Monograph Series, SPE, Richardson, Texas (1989) **11.**
4. Craig, F.F. Jr.: *The Reservoir Engineering Aspects of Waterflooding,* Monograph Series, SPE, Richardson, Texas (1980) **3.**
5. Moritis, G.: "EOR Oil Production Up Slightly," *Oil & Gas J.* (1998) **96,** No. 16, 49.
6. Latil, M. (ed.): *Enhanced Oil Recovery,* Gulf Publishing Co., Paris (1980).
7. "Improved Oil Recovery," Interstate Oil Compact Commission, Oklahoma City, Oklahoma (1983).
8. Bailey, R.E. and Curtis, L.B.: *Enhanced Oil Recovery,* National Petroleum Council, Industry Advisory Committee to the U.S. Secretary of Energy, Washington, DC (1984).
9. Lake, L.W.: *Enhanced Oil Recovery,* Prentice Hall, Englewood Cliffs, New Jersey (1989).
10. Carcoana, A.: *Applied Enhanced Oil Recovery,* Prentice Hall, Englewood Cliffs, New Jersey (1992).
11. Thakur, G.C. and Satter, A.: *Integrated Waterflood Asset Management,* PennWell Publishing, Tulsa (1998).
12. Willhite, G.P.: *Waterflooding,* Textbook Series, SPE, Richardson, Texas (1986).

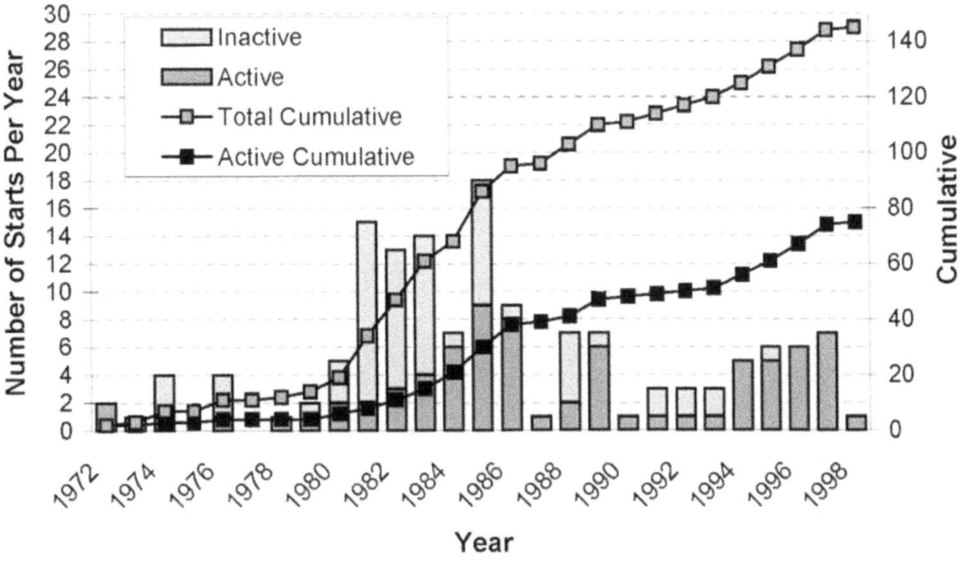

Fig. 1.12—Worldwide CO_2 flood startups (data from Refs. 5, 29–31).

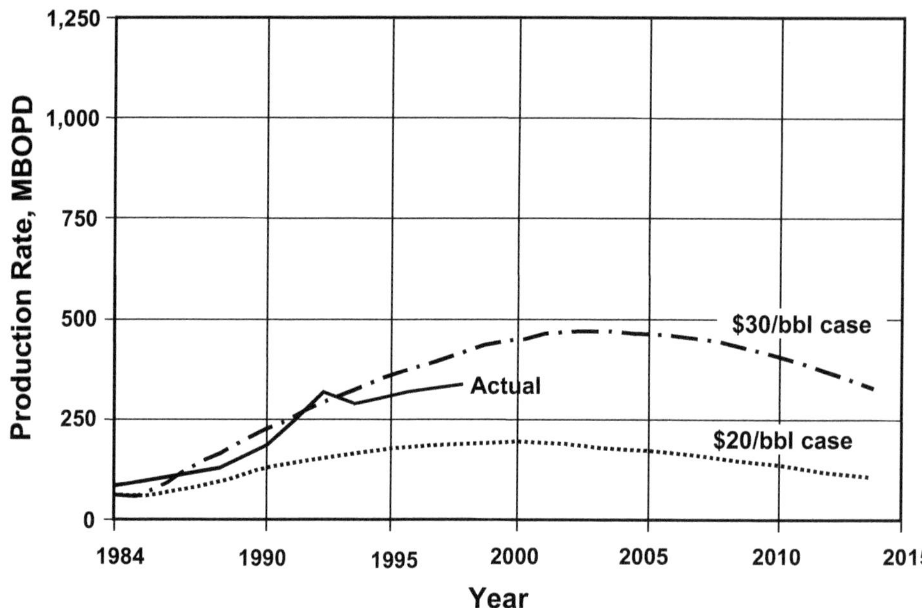

Fig. 1.13—U.S. miscible flood EOR production: NPC forecast vs. actual (after Refs. 5 and 8). The production from U.S. miscible floods has closely tracked the NPC $30/bbl case.

13. Hare, M. *et al.*: "Sources for Delivery of Carbon Dioxide for Enhanced Oil Recovery," Report FE-2512-24 prepared by Pullman Kellogg Co., Houston, for U.S. DOE, Contract EX-76-C-01-2515 (December 1978).
14. Taber, J.J.: "Environmental Improvements And Better Economics In EOR Operations," *In Situ* (1990) **14,** No. 4, 345.
15. *Working Paper On Carbon Sequestration Science and Technology,* Office of Science, Office of Fossil Energy, U.S. DOE, Washington, DC (February 1999).
16. Lien, S.C. *et al.*: "Brage Field, Lessons Learned after Five Years of Production," paper SPE 50641 presented at the 1998 SPE European Petroleum Conference, The Hague, 20–22 October.
17. "CO_2 Capture and Geologic Sequestration: Progress Through Partnership Workshop," BP Amoco, IEA Greenhouse Gas R&D Program, U.S. DOE, and Federal Energy Technology Center, Houston, 28–30 September 1999.
18. McKean, T.A.M. *et al.*: "Conceptual Evaluation Of Using CO_2 Extracted From Flue Gas For Enhanced Oil Recovery, Schrader Bluff Field, North Slope, Alaska," *Proceedings of the 4th International Conference on Greenhouse Gas Control Technologies, 30 August–2 September 1998, Interlaken, Switzerland,* Elsevier Science, Amsterdam (1999), 207–215.
19. Anada, H.R. *et al.*: "Economics of By-Product CO_2 Recovery and Transportation for EOR," *Energy Progress* (1983) **3,** No. 4, 233.
20. Alpert, S.B.: "Options for Dealing with CO_2 Emissions," Electric Power Research Inst., Palo Alto, California (June 1988).
21. Walter, F. B. and Ireland, P.A.: "Coal Fired Power Station CO_2 Recovery," paper presented at the 1988 AIChE Spring National Meeting, New Orleans, 9 March.
22. Rett, J.: "EOR, An Electric Utility's Perspective," *Proceedings of the Third Wyoming Enhanced Oil Recovery Symposium,* R.E. Ewing (ed.), Enhanced Oil Recovery Inst., Laramie, Wyoming (1987) 67–74.
23. Park, R.S.: "Description of a North Sea CO_2 Enhanced Oil Recovery Project," paper presented at the 1987 Argonne National Laboratory Conference on Recovery and Use of Waste CO_2 for Enhanced Oil Recovery, Denver, 19–20 March.
24. Wolsky, A.M.: "A New Method for Recovering CO_2 from Stationary Combustors," paper presented at the 1987 Argonne National Laboratory Conference on Recovery and Use of Waste CO_2 for Enhanced Oil Recovery, Denver, 19–20 March.
25. Miller, D.B. *et al.*: "Economics of Recovery CO_2 from Exhaust Gases," *Chem. Eng. Prog.* (October 1986) 38.
26. Miller, D.B. and Soychak, T.J.: "Additional Wyoming CO_2/EOR Projects May Be Feasible with $1.00–$1.50/MSCF CO_2 Recovered From Flue Gas," *Proceedings of the Second Wyoming EOR Conference: Additional Wyoming CO_2/EOR Projects,* R.E. Ewing (ed.), Enhanced Oil Recovery Inst., Laramie, Wyoming (1986).
27. Miller, D.B. and Soychak, T.J.: "$0.50 to $1.50/MSCF CO_2 from Flue Gas and Cogenerated Low-Cost Power Make More CO_2 EOR Projects Viable," paper presented at the 1987 Gulf Coast Section of SPE EOR Forum, Houston, 24 March.
28. Martin, F.D.: "Enhanced Oil Recovery for Independent Producers," paper SPE 24142 presented at the 1992 SPE/DOE Symposium on Enhanced Oil Recovery, Tulsa, 22–24 April.
29. Moritis, G.: "CO_2 and HC injection lead EOR production increase," *Oil & Gas J.* (1990) **88,** No. 17, 49.
30. Taber, J.J.: "Current Status Of The Use Of CO_2 For Enhanced Oil Recovery," U.S. DOE Argonne National Lab Recovery and Use of Waste CO_2 in EOR Workshop, Denver (1988) 65–99.
31. Cox, B. and Schubert, J.: *1986 EOR Project Sourcebook,* Pasha Publications Inc., Arlington, Virginia (1986), 101–167.
32. Broome, J.H., Bohannon, J.M., and Stewart, W.C.: "The 1984 National Petroleum Council Study on EOR: An Overview," *JPT* (August 1986) 869.
33. Robl, F.W., Emanuel, A.S., and Van Meter, O.E. Jr.: "The 1984 National Petroleum Council Estimate of Potential EOR Miscible Processes," *JPT* (August 1986) 875.
34. Claridge, E.L.: "Review of CO_2 Flooding Projects and Future Prospects," paper presented at the 1984 World Oil and Gas Show and Conference, Dallas, 4–7 June.
35. Holm, L.W.: "Evolution of the Carbon Dioxide Flooding Process," *JPT* (November 1987) 245.
36. Park, R.S.: "Description of a North Sea CO_2 Enhanced Oil Recovery Project," U.S. DOE Argonne National Lab Recovery and Use of Waste CO_2 in EOR Workshop, Denver (1988) 11–20.
37. Brashear, J.P. *et al.*: "Effect of Well Abandonments on EOR Potential," *JPT* (December 1991) 1496.

SI Metric Conversion Factors

acre ×	4.046 856	E + 03	= m^2
acre ×	4.046 856	E − 01	= ha
°API	141.5/(131.5 + °API)		= g/cm^3
bbl ×	1.589 873	E − 01	= m^3
cp ×	1.0*	E − 03	= Pa·s
°F	(°F − 32)/1.8		= °C
ft ×	3.048*	E − 01	= m
ft^3 ×	2.831 685	E − 02	= m^3
tonne ×	1.0*	E + 00	= Mg

*Conversion factor is exact.

Chapter 2
Review of CO_2 Process Mechanisms

In this chapter, we provide practical information on how CO_2 floods recover oil. This will aid readers in understanding subsequent chapters, particularly those that deal with predicting the economic outcome of potential projects. More detailed information on CO_2 process mechanisms can be found in other books on enhanced oil recovery, such as those by Stalkup,[1] Klins,[2] and Lake.[3]

An understanding of the concepts covered in this chapter is a prerequisite for assessing the economic viability of a potential CO_2 flood, determining how CO_2 will be injected, predicting CO_2 flood performance, and providing information for facility design and economic analysis. Topics include the following.

- An overview of miscibility and reservoir mixing mechanisms—Sec. 2.1, *How CO_2 Becomes Miscible With Oil*, and Sec. 2.2, *Small-Scale Reservoir Mixing Mechanisms* focus on laboratory data and explanations about miscibility and the effects of dispersion on oil recovery. These concepts underlie everything qualitative that happens in a CO_2 flood.
- The impact of relative permeability—Sec. 2.3, *An Overview of Relative Permeability*; Sec. 2.4, *How Relative Permeability Affects Fluid Mobility*; and Sec. 2.5, *CO_2 Relative Permeability* discuss the relative permeabilities of oil, water, and CO_2-rich phase in some detail because these three factors greatly affect injection and production rates.
- Sweep aspects of a CO_2 flood—Sec. 2.6 reviews CO_2 sweep control, whether by alternate water/gas injection or by the use of gravity, as in dipping reservoirs.

2.1 How CO_2 Becomes Miscible with Oil

The principles of miscibility are fundamental to understanding how CO_2 floods work. For convenience, the terms used here are defined in the Glossary.

2.1.1 CO_2 Fluid Properties. Carbon dioxide is effective in improving oil recovery for two reasons: density and viscosity. At high pressure, CO_2 forms a phase whose density is close to that of a liquid, even though its viscosity remains quite low **(Figs. 2.1 and 2.2).** Under miscible conditions in west Texas, the specific gravity of this dense CO_2 typically is 0.7 to 0.8 g/cm^3, not much less than for oil and far above that of a gas such as methane, which is about 0.1 g/cm^3. Dense-phase CO_2 has the ability to extract hydrocarbon components from oil more easily than if it were in the gaseous phase (and thus at a lower pressure).[4]

The viscosity of CO_2 under miscible conditions in west Texas (0.05 to 0.08 cp) is significantly lower than that of fresh water (0.7 cp) or oil (1.0 to 3.0 cp). Although the low viscosity of the gas relative to the oil can be detrimental to sweep, CO_2 can improve recovery by reducing the oil viscosity.

2.1.2 Mechanisms for CO_2 Miscibility With Oil. In general, miscibility between fluids can be achieved through two mechanisms: first-contact miscibility and multiple-contact miscibility. When two fluids become completely miscible, they form a single phase; one fluid can completely displace the other fluid, leaving no residual saturation. A minimum pressure is required for two fluids to be miscible.

A clear example of first-contact miscibility is ethanol and water. Regardless of the proportions of the two fluids, they immediately form one phase with no observable interface. Butane and crude oil also are first-contact miscible, and butane might make an ideal solvent for oil were it not for its high cost.

In the multiple-contact miscible process that takes place with CO_2 and crude oils,[4–10] CO_2 and oil are not miscible on first contact, but require many contacts in which components of the oil and CO_2 transfer back and forth until the oil-enriched CO_2 cannot be distinguished from the CO_2-enriched oil **(Fig. 2.3).** Zick[11] calls this process a condensing/vaporizing mechanism. Multiple-contact miscibility between CO_2 and oil starts with dense-phase CO_2 and hydrocarbon liquid. The CO_2 first condenses into the oil, making it lighter and often driving methane out ahead of the "oil bank." The lighter components of the oil then vaporize into the CO_2-rich phase, making it denser, more like the oil, and thus more easily soluble in the oil. Mass transfer continues between the CO_2 and oil until the resulting two mixtures become indistinguishable in terms of fluid properties. At that point, there is no interface between the CO_2 and oil, and one hydrocarbon phase results.

Fig. 2.3 illustrates the condensing/vaporizing mechanism for miscibility. During oil displacement, there is a gradation in composition from pure CO_2 on the left (injection side) to virgin oil on the right (production side). The vaporizing region occurs upstream of the condensing region. Every contact in the process involves a miscible displacement, even though pure CO_2 is not miscible with original oil.

2.1.3 Effect of Pressure on CO_2 Flood Oil Recovery. Miscibility development between CO_2 and oil is a function of both temperature and pressure, but for an isothermal reservoir, the only concern is pressure. As pressure increases, the oil can dissolve more CO_2, and more oil components can be vaporized by the CO_2. At some pressure, when the CO_2 and oil are in intimate contact, they will become miscible. When the contact between CO_2 and oil occurs with little or no reservoir mixing, the pressure at which miscibility happens is defined as the thermodynamic minimum miscibility pressure (thermodynamic MMP). The effects of small-scale reservoir mixing can decrease the displacement efficiency of CO_2 and increase the pressure required for miscibility (see Sec. 2.1.8).

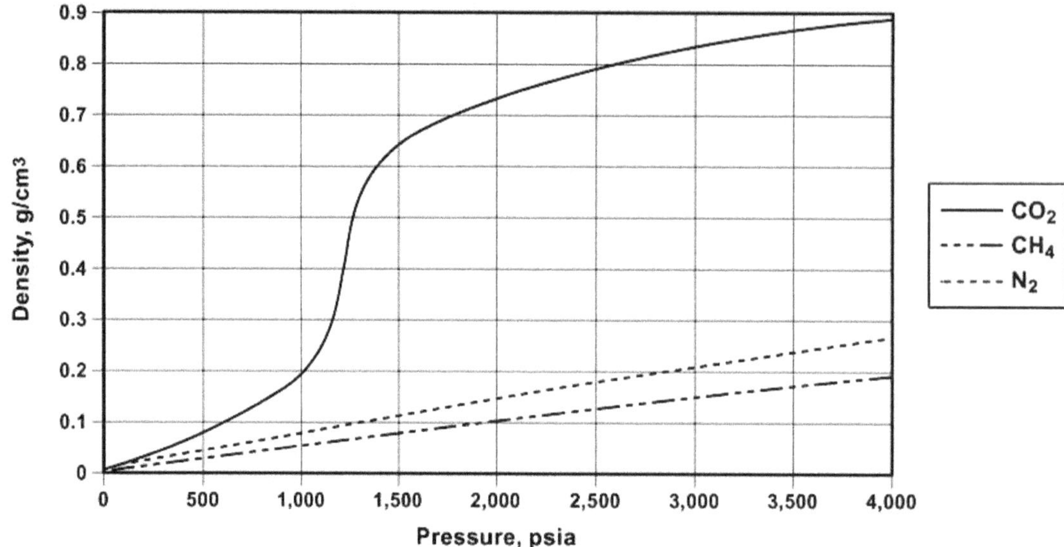

Fig. 2.1—Carbon dioxide, CH_4, and N_2 densities at 105°F. At high pressures, CO_2 has a density close to that of a liquid and much greater than that of either methane or nitrogen. Densities were calculated with an equation of state (EOS).

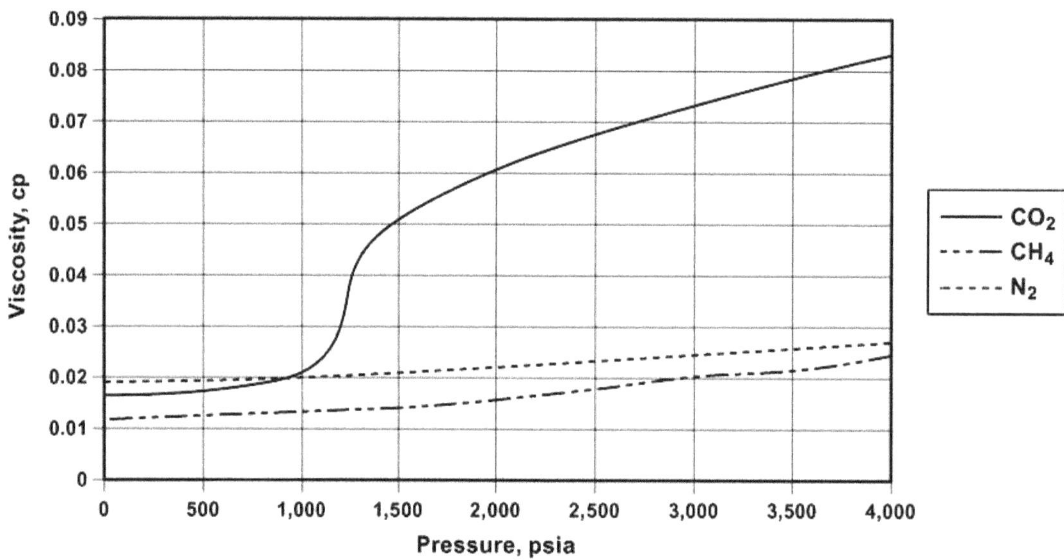

Fig. 2.2—Carbon dioxide, CH_4, and N_2 viscosities at 105°F. At high pressures, the viscosity of CO_2 is also greater than that of methane or nitrogen, although it remains low in comparison to that of liquids. Viscosities were calculated using an EOS.

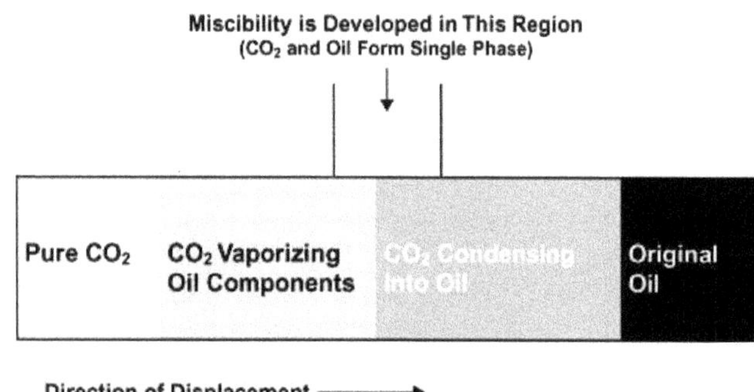

Fig. 2.3—One-dimensional schematic showing how CO_2 becomes miscible with crude oil. At high pressure, CO_2 becomes miscible with the oil through a multiple-contact process in which mass transfers continue between the CO_2 and oil phases until they become a single phase. Zick[11] calls this process a condensing/vaporizing mechanism.

A plot of oil recovery vs. CO_2 pressure (assuming 1D displacement, little reservoir mixing, and constant reservoir pressure) shows oil recovery increasing rapidly with increasing pressure, then flattening out near 100%, when the thermodynamic MMP is reached **(Fig. 2.4).** At pressures above the thermodynamic MMP, there is little increase in oil recovery.

The oil recoveries shown in Fig. 2.4 were obtained by Yellig and Metcalfe[10] from CO_2 displacements of crude oil in slim-tube displacement tests described in the next section. Note that the homogeneous sand used in slim-tube tests limits reservoir mixing that might disturb the multiple-equilibrium-contact process required to develop miscibility. In actual reservoirs, the effect of small-scale reservoir heterogeneities can reduce the ideal slim-tube oil recovery rate shown in Fig. 2.4 (see Sec. 2.1.8).

2.1.4 Measurement of the Thermodynamic MMP by Slim-Tube Tests. The basic laboratory means of determining thermodynamic MMP is the slim-tube test, shown in **Fig. 2.5,** which produces 1D displacements with a very low level of mixing. The slim tube is constructed of stainless steel, typically ¼-in. outside diameter (OD) and 40 ft long. Commonly used packing is 160 to 200 mesh Ottawa sand.

The test begins with a sand pack saturated with oil at a constant temperature. Carbon dioxide is introduced at a given pressure (controlled by a backpressure regulator), and oil displacement is measured as oil is recovered. No water is involved.

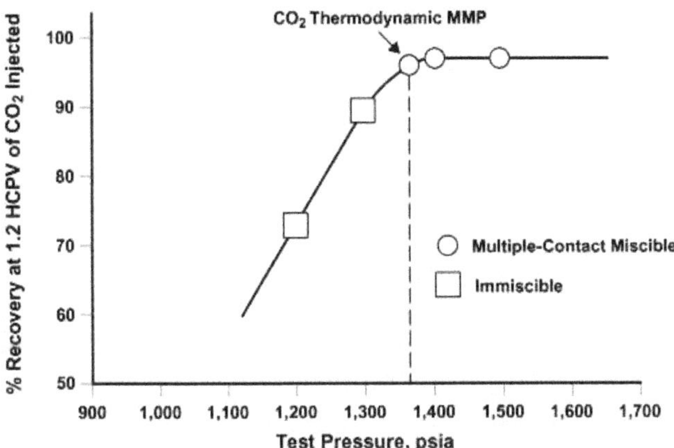

Fig. 2.4—Slim-tube oil recoveries for fixed oil compositions and temperatures (after Ref. 7). Oil recovery increases with pressure until the thermodynamic MMP is reached for a 1D displacement by CO_2 under conditions of limited mixing. Above the thermodynamic MMP of approximately 1,385 psia, CO_2 becomes miscible with the oil and maximum oil recoveries are possible.

A high-pressure sight glass shows the number of phases exiting the slim tube. Below the thermodynamic MMP, the sight glass shows oil with bubbles of CO_2. When the CO_2 has become miscible with the oil, there should be essentially only one phase flowing.

The CO_2 displacements are carried out for a range of pressures, holding the temperature constant at the reservoir temperature. For each pressure, the oil recovery at 1.2 hydrocarbon pore volumes (HCPV's) of CO_2 injected is plotted. An oil recovery factor of at least 90% is often used as a rule of thumb for estimating the thermodynamic MMP.

Uncertainties can arise in identifying the thermodynamic MMP precisely from slim-tube displacement test data. Oil recovery above 90% does not necessarily prove that miscibility has developed because one might occasionally see a few gas bubbles in the sight glass where the fluid exits the slim tube. Such a displacement might be the result of a highly efficient immiscible gas displacement. From a practical point of view, however, achieving an oil recovery of close to 100% is more important than understanding whether or not miscibility truly was developed.

In spite of experimental uncertainties, slim-tube tests provide valuable data that can contribute to the design of a CO_2 flood. Slim-tube data give the upper limit for oil displacement efficiency, which is defined as the oil recovery for 1D flow (ignoring the effects of areal and vertical sweep).

Slim-tube data are also used in tuning an equation-of-state (EOS) model to better predict both the development and loss of miscibility. An EOS model often is used in a reservoir simulator to predict CO_2 flood performance (see Sec. 4.3.2).

Slim-tube tests provide important data, yet there are other factors that affect actual recovery from the reservoir, including both small-scale reservoir mixing effects (discussed in Sec. 2.1.8) and large-scale reservoir heterogeneities, which can greatly impact volumetric sweep and, therefore, recovery.

2.1.5 Prediction of the Thermodynamic MMP for CO_2. So far, we have looked at determining the thermodynamic MMP with slim-tube tests, but slim-tube tests can be expensive. Are there less expensive methods in use to determine the thermodynamic MMP for a particular field?

There are two possible ways to avoid slim-tube tests: mathematical models and thermodynamic MMP correlations. Mathematical models use phase equilibria data and an EOS to estimate the thermodynamic MMP. Significant progress has been made on these models in recent years, and if appropriate data are available they can yield excellent results at low cost.

Useful thermodynamic MMP correlations have been developed by several researchers,[4,6,10,12,13] although these correlations have limitations and should only be used in the absence of slim-tube test data and/or phase equilibrium data that can be input to mathematical models. We start with a discussion of correlations.

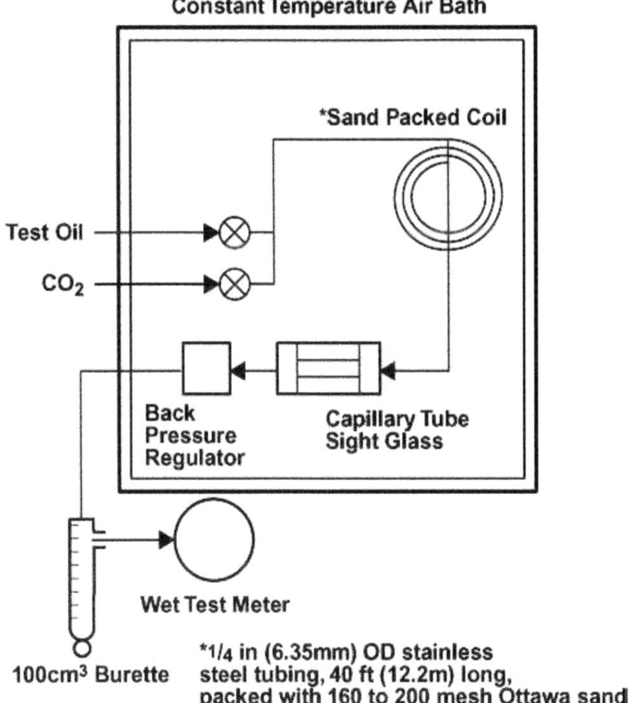

Fig. 2.5—Slim-tube test apparatus (after Ref. 7). The slim-tube test uses CO_2 to displace oil in a sand-packed coil at a range of pressures to determine the thermodynamic MMP.

The Use of Correlations To Determine CO_2 Thermodynamic MMP. Holm and Josendal determined that CO_2 attains dynamic miscibility with crude oil when CO_2 density is high enough to vaporize C_5-through-C_{30} hydrocarbons.[4,6] They found that CO_2 densities at the thermodynamic MMP ranged from 0.40 to 0.65 g/cm^3. They also found that the thermodynamic MMP was related to the average molecular weight of the C_{5+} components of the oil, as well as to the reservoir temperature and pressure. **Fig. 2.6** shows the Holm-Josendal pressure and temperature correlation for oils, taking into account different molecular weights for the C_{5+} components of the oils.[4] The points on these curves represent specific sets of experimental data. Fig. 2.6 also shows extensions developed by Mungan[12] for higher molecular weights. It is clear that heavier oils require substantially higher pressures to become miscible. For example, at 140°F, an oil with a C_{5+} molecular weight of 340 has a thermodynamic MMP above 3,000 psia.

Yellig and Metcalfe[10] developed a similar thermodynamic MMP correlation for CO_2, based on slim-tube tests from a group of light west Texas oils whose average molecular weight distribution for the C_{5+} components is less than 180. Their correlation does not take into account the composition of the oil. Their stated assumption is that if the bubblepoint pressure of the crude oil exceeds their thermodynamic MMP correlation, then the bubblepoint pressure is the thermodynamic MMP.

Fig. 2.7 shows that the Yellig-Metcalfe correlation predicts a lower thermodynamic MMP than does the Holm-Josendal correlation for a C_{5+} molecular weight of 180. Whenever heavy oils are present, the Yellig-Metcalfe correlation may yield inaccurate results, and we recommend the Holm-Josendal correlation with Mungan's extension.[12]

Mathematical Determination of Thermodynamic MMP. In addition to empirically derived correlations, mathematical methods that rely on phase equilibrium can be used to determine thermodynamic MMP. Johns and Orr[14] and Wang and Orr[15] developed a generalized n-component phase equilibrium approach to estimate the thermodynamic MMP for two-phase systems, extending the use of ternary-phase equilibrium diagrams to an arbitrary number of hydrocarbon components. In their work, the oil is characterized by hydrocarbon components, including pseudocomponents for heavier fractions. Their goal is to determine the thermodynamic MMP,

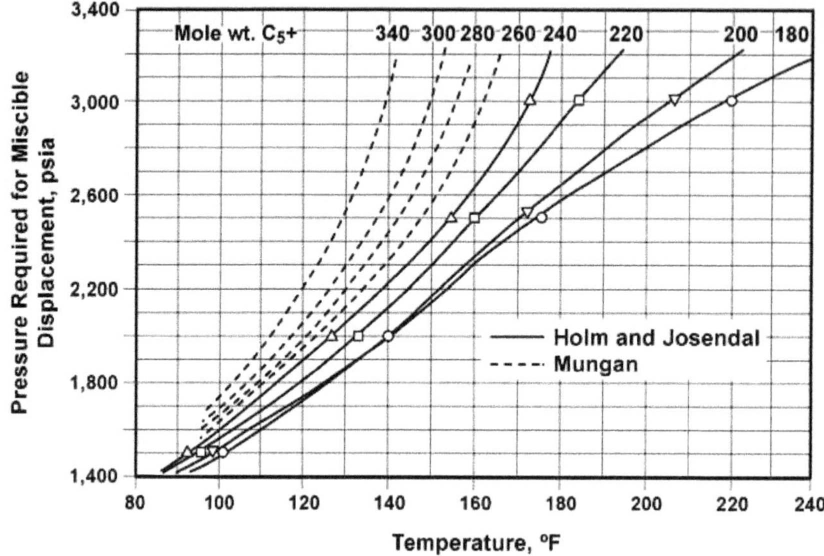

Fig. 2.6—Holm-Josendal CO_2 thermodynamic MMP correlation with Mungan's extension (after Ref. 1). At a given temperature, oils with large molecular weights for their C_{5+} components require higher pressures to become miscible with CO_2.

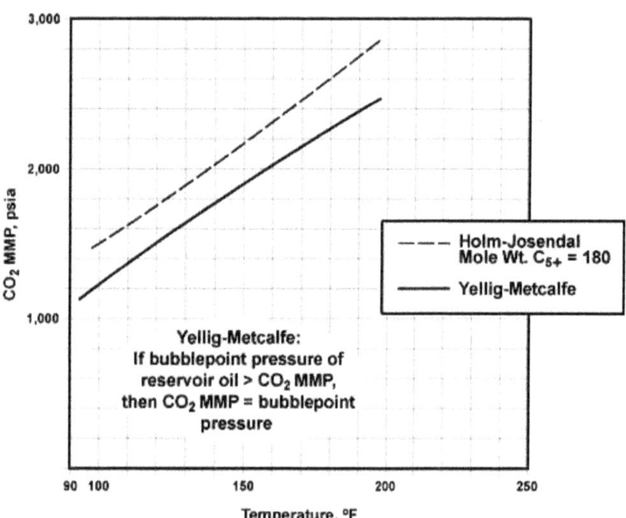

Fig. 2.7—Comparison of CO_2 thermodynamic MMP correlations. The Yellig-Metcalfe correlation predicts lower thermodynamic MMP's than does the Holm-Josendal correlation for a C_{5+} molecular weight of 180. The Holm-Josendal correlation is more realistic.

which is defined under their methods as the pressure required for miscible flow in the absence of dispersion.

The mathematical methods determine the thermodynamic MMP by tracking the 1D compositional path of CO_2 and oil as the two phases exchange components. An EOS is used to calculate the partitioning of the components between the phases that are present. This technique is satisfactory if the EOS has been fine-tuned to reproduce phase behavior near the critical locus, which is a collection of critical points. (A critical point is where oil and vapor are indistinguishable, i.e., the bubblepoint of the liquid is the same as the dewpoint of the vapor.) Johns and Orr[14] and Wang and Orr[15] provide additional detail on the steps used in the mathematical methods.

If sufficient phase equilibrium data between CO_2 and oil do not exist, especially near critical points, the mathematical methods may not satisfactorily predict the thermodynamic MMP. In this case, slim-tube and additional pressure/volume/temperature (PVT) data should be collected to fine-tune the EOS and improve the estimation of thermodynamic MMP. Slim-tube data also are very important when three hydrocarbon phases may exist because the mathematical methods are based on the flow of only two hydrocarbon phases.

2.1.6 Effects of Impurities on Thermodynamic MMP. Impurities in CO_2 affect the thermodynamic MMP because they affect the solubility of CO_2 in oil and the ability of CO_2 to vaporize oil components. For example, Harpole and Hallenbeck[16] reported on laboratory data from the East Vacuum Grayburg San Andres Unit, which showed that when produced gas was mixed with CO_2, the thermodynamic MMP increased from 1,165 psia to 1,465 psia.

Research on the effect of impurities on thermodynamic MMP has been carried out using empirical correlations, although the mathematical techniques of Johns and Orr[14] and Wang and Orr[15] (discussed in Sec. 2.1.5) also can be used, as can slim-tube tests.

Methane, nitrogen, and flue gas usually have higher thermodynamic MMP's with oil than does CO_2,[1] although in reservoirs with very high temperatures, the thermodynamic MMP's for CO_2, methane and nitrogen can be similar. For example, with the Jay/Little Escambia Creek crude, where the reservoir temperature is 285°F, the thermodynamic MMP's for CO_2, methane, and nitrogen all are about 3,600 psia.[17]

Metcalfe tested the effect of impurities on CO_2 flooding,[18] and Sebastian *et al.*[19] developed the following correlation for the effect of impurities on the CO_2 thermodynamic MMP:

$$\frac{p_m}{p_{CO_2}}=1.0-2.13\times10^{-2}(T_{cm}-304.2)+2.51\times10^{-4}(T_{cm}-304.2)^2$$
$$-2.35\times10^{-7}(T_{cm}-304.2)^3, \quad\quad (2.1)$$

where p_m=the mixture MMP, psia; p_{CO_2}=the pure CO_2 MMP, psia; and T_{cm}=the mole-fraction-weighted critical temperature, K.

Sebastian's correlation indicates that impurities with a critical temperature higher than that of CO_2 (which is 304.2 K, or 87.6°F) will reduce the thermodynamic MMP of the mixture, while impurities with a critical temperature lower than that of CO_2 will increase the thermodynamic MMP of the mixture.

2.1.7 Oil Displacement by CO_2 Below the Thermodynamic MMP. Carbon dioxide must be near the thermodynamic MMP to significantly reduce residual oil saturation and increase oil recovery, except with the use of the gravity drainage mechanism discussed in Sec. 2.6.3. Below the thermodynamic MMP, in the region where slim-tube oil recovery is less than 90%, the pressure is not high enough to allow sufficient CO_2 to dissolve into the oil or to vaporize sufficient oil into the CO_2 so that the two phases to become miscible. In this region, CO_2 is not dense enough—it only can vaporize components up to C_6.[4,6] In addition, the CO_2 is not sufficiently soluble in oil to lighten the oil and induce much vaporization.

When the pressure is substantially below the thermodynamic MMP, the primary effect CO_2 has is to swell the oil and reduce its viscosity. Swelling can cause some of the residual oil to become mobile and recoverable. If oil saturation before CO_2-induced swelling is at the residual saturation to water, the swelling caused by CO_2 will increase this oil saturation. Subsequent displacement by water can then reduce oil saturation to the initial level, thus mobilizing additional oil. Oil swelling combined with viscosity reduction allows CO_2 injection to recover heavy oil beyond that which a waterflood can achieve, even though the inability of CO_2 to vaporize much of the hydrocarbon components keeps miscibility from developing.

If the pressure of CO_2 is just below the thermodynamic MMP, CO_2 can reduce residual oil saturation through increased vaporization. With the help of gravity drainage, in which CO_2 or other gases are injected at the top of the reservoir, residual oil saturation can be reduced to near zero in certain cases (see Sec. 2.6.3), even under immiscible conditions.

2.1.8 The Three-Phase Hydrocarbon Region. In addition to the water phase, two or three hydrocarbon phases can exist in the reservoir during the CO_2 flood process. Knowing how many phases are present is important because the number of phases affects relative permeability and injection and production rates (see Secs. 2.3 to 2.5).

At temperatures above 120°F, CO_2 and oil generally form no more than two hydrocarbon phases on initial contact. **Fig. 2.8** is a sample phase diagram for the initial contact between CO_2 and crude oil at a reservoir temperature above 120°F. The solid line represents 100% liquid or 100% vapor. Liquid (L) is to the left of and below the critical point, and vapor (V) is to the right of and above the critical point. Above the solid line (the upper left quadrant), there is only one phase, which is dense gas at higher pressures and liquid at lower pressures. Below and to the right of the solid line are two phases, L and V. The dashed lines represent quality lines (volume percent [vol%] liquid).

Below 120°F, a different type of phase behavior often takes place. **Fig. 2.9** illustrates that a three-phase hydrocarbon region may occur on initial contact between CO_2 and oil. Many fields in west Texas have in-situ temperatures below 120°F and often have three-phase regions in the pressure range of 1,000 to 1,300 psia.

The three phases in the diagram are LL, the "lower" and heavier liquid, which is hydrocarbon-rich and similar to the original oil; UL, the "upper" and typically lighter liquid, which is CO_2-rich; and V, the vapor, which is CO_2-rich. The phases in Fig. 2.9 do not exhibit a critical point on first contact between CO_2 and crude oil (although some CO_2/crude oil systems below 120°F do show a critical point at first contact).

The three-phase region occurs at a high CO_2 mole fraction, generally above 60 mol% CO_2. Above the three-phase region (LL+UL+V) are a CO_2-rich liquid (UL) and an oil-rich liquid (LL). Below the three-phase region is a mixture of oil (LL) and CO_2 vapor (V).

Other complex phase behavior and flow also can occur. For example, Simandoux *et al.*[20] found a fourth phase comprising hydrocarbons in solid form at 185°F.

Henry and Metcalfe[21] have observed reduced mobility in CO_2 corefloods conducted at pressures that favor the formation of a three-phase hydrocarbon region, because the third phase reduces injectivity. The problem of reduced mobility is worsened when mobile water is present (see Sec. 2.4.1).

2.1.9 The Residual Oil Phase in a CO_2 Flood. In 1D homogeneous rock, CO_2 should become miscible with all the oil when the pressure is at or above the thermodynamic MMP. However, small-scale reservoir heterogeneities mix or disperse the CO_2 (see Sec. 2.2) and lower its concentration. Because the CO_2 contacts virgin oil at a reduced concentration, some of the heavier components of the initially contacted oil remain, unvaporized, as residual oil. Subsequent contacts with fresh CO_2 can slowly vaporize these heavier components. **Fig. 2.10** shows that residual oil saturation located behind the miscible region can be reduced further by vaporization.

Because of the additional vaporization, the residual oil saturation after a CO_2 flood is considerably less than the residual oil saturation after waterflooding. The latter is typically 20 to 40% pore volume (PV), while pressure-core data collected after CO_2 field tests have shown residual oil saturation as low as 3 to 5% PV in well-swept areas.[22,23] In some cases, gravity drainage allows CO_2 to displace oil to even lower residual saturations (see Sec. 2.6.3).

Fig. 2.11 shows pressure core data from a San Andres CO_2 field test formation (west Texas carbonate reservoir) and illustrates the

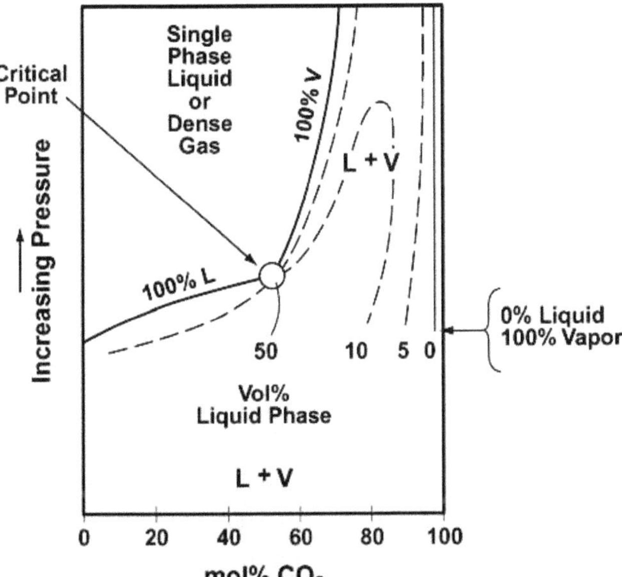

Fig. 2.8—Typical phase behavior for a CO_2/crude oil system above 120°F (after Ref. 1). The pressure-composition diagram shows the phase behavior for different pressures and mole percents of CO_2. The phases here are liquid (L) and vapor (V).

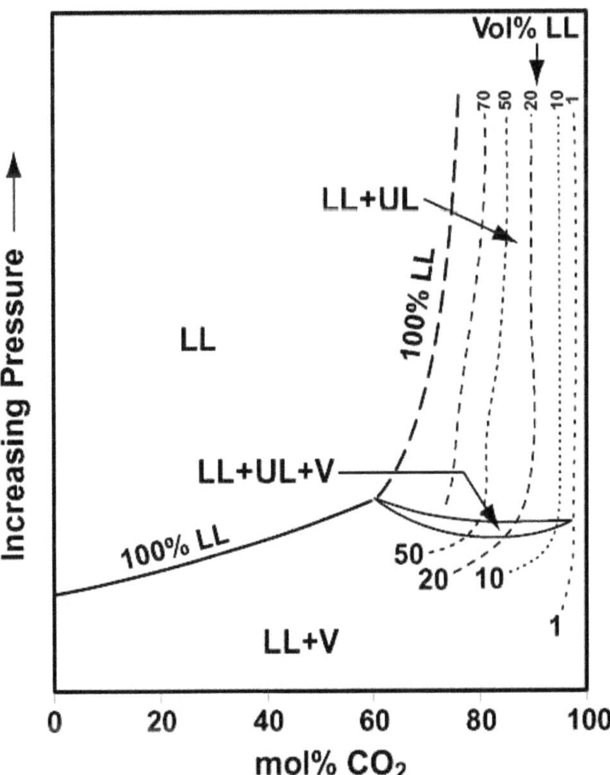

Fig. 2.9—Example of a CO_2/crude oil diagram with three hydrocarbon phases (after Ref. 1). When CO_2 and oil are below 120°F, three hydrocarbon phases may be present. These are the "lower" and heavier liquid (LL), the "upper" and typically lighter liquid (UL), and the CO_2-rich vapor (V) phases.

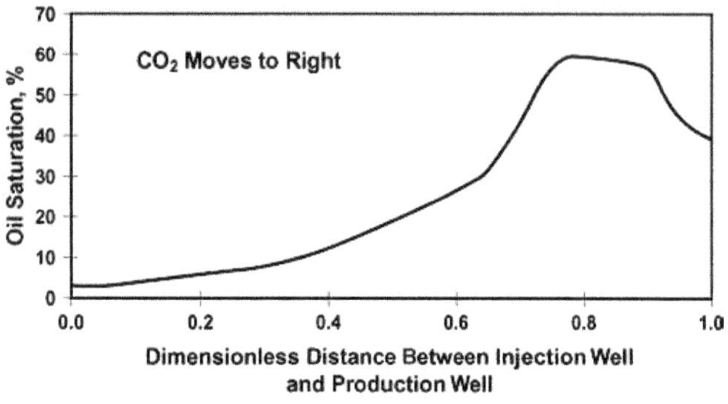

Fig. 2.10—Residual oil saturation in a CO_2 flood with small-scale reservoir mixing effects. Oil that remains because of mixing effects in the reservoir vaporizes slowly, leaving a diminishing tail behind the oil bank mobilized by CO_2.

correlation between the final oil saturation after CO_2 flooding and the concentration of CO_2 collected from the gas in the pressure core. The higher CO_2 mole fractions occurred in the most permeable intervals, which also have the lowest residual oil saturations.

2.1.10 Effect of Water on CO_2/Oil Miscibility. When water is used in a CO_2 flood [a water-alternating-gas (WAG) design], several additional factors must be considered. Although Pollack *et al.*[24] found that water has a small effect on the phase equilibrium between CO_2 and oil, CO_2 dissolution in water may prevent some of the CO_2 from contacting the oil. Dissolved CO_2 also slightly affects the water formation volume factor and viscosity of the water, and can reduce the concentration of CO_2 that contacts the reservoir oil.

The amount of CO_2 that can be dissolved in the reservoir water can be estimated from **Fig. 2.12** (from Refs. 25 and 26) **and Fig. 2.13** (curve fit of data from Refs. 25 through 28). First, use Fig. 2.12 to estimate CO_2 solubility in fresh water as a function of pressure for the appropriate reservoir temperature, interpolating between the different temperatures if necessary. Second, use Fig. 2.13 to correct CO_2 solubility according to the salinity of the water (independent of pressure and temperature). These figures illustrate that CO_2 solubility in water increases with increasing pressure and decreasing salinity.

Chang *et al.*[29] developed correlations for CO_2 solubility in brine, as well as for the effect of dissolved CO_2 on the formation volume factor of water. They also performed reservoir simulations that showed that when water is present, 5 to 10% of the CO_2 is dissolved in the water, which reduces CO_2 incremental oil recovery by 5%.

2.2 Small-Scale Reservoir Mixing Mechanisms

Reservoir mixing occurs on both large and small scales. On a small scale, mixing results from localized variations in permeability and affects the equilibrium contacts between CO_2 and oil. The result is a spreading of components among the various phases, a phenomenon known as convective dispersion.

In practical terms, small-scale mixing reduces the effective CO_2 concentration that contacts virgin oil and slows the development of miscibility. If the magnitude of small-scale reservoir mixing is great, CO_2 may not develop miscibility even though the reservoir pressure is above the thermodynamic MMP. Because CO_2 can vaporize hydrocarbons when pressure exceeds the thermodynamic MMP, it will recover more oil than if the pressure were below the thermodynamic MMP, but not as much oil as if there had been no small-scale mixing. The presence of small-scale dispersive mixing indicates that pressures higher than the thermodynamic MMP may be required for good oil recovery. Whether an overpressurized displacement is needed depends on the level of dispersion and the composition of the reservoir oil and injection gas.

2.2.1 Convective Dispersion. Unlike molecular diffusion, which takes place whether a fluid is flowing or not, convective dispersion only happens during transport. Dispersion results from the tortuous flow path caused by small-scale heterogeneities that are not specifically accounted for in macroscopic models (which have larger length scales than do pore-network models). As illustrated in **Fig. 2.14,** the nonuniform nature of rock pores causes CO_2 to become dispersed and elongates the displacement front.

Fig. 2.15 (after Ref. 30) shows an example of how mixing can affect oil recovery by CO_2 at pressures above the thermodynamic MMP. The lines in Fig. 2.15 represent simulated oil recoveries, and the symbols are actual slim-tube test oil recoveries. The solid lines represent a relatively small amount of dispersion, while the dashed lines represent much larger dispersion. Clearly, the effect of dispersion on oil recovery can be significant.

Two additional conclusions can be drawn from Fig. 2.15. First, higher dispersion translates to lower displacement efficiency and, therefore, to lower recovery. Second, oil recovery above the thermodynamic MMP becomes more sensitive to pressure as dispersion increases, and so the dashed lines show greater separation than the solid lines.

The data in Fig. 2.15 exhibit a slight rate dependency, although the rate dependency is secondary to the effect of dispersion on recovery. The rate effect might be attributed to some slight amount of bypassing caused by gravity override by the CO_2 (see Sec. 2.6.3 for additional discussion on gravity override).

Mathematical Model of Dispersion. Dispersion appears in component mass-balance equations in a term analogous to Fick's law of diffusion. Dispersion is often added to the effective molecular diffusion for porous media. The magnitude of dispersion in the direction of flow is usually much greater than the effective diffusion coefficient.

The dispersion term is actually a tensor, in that the level of dispersion is direction dependent. Dispersion is often assumed to increase linearly with the superficial velocity, even though experi-

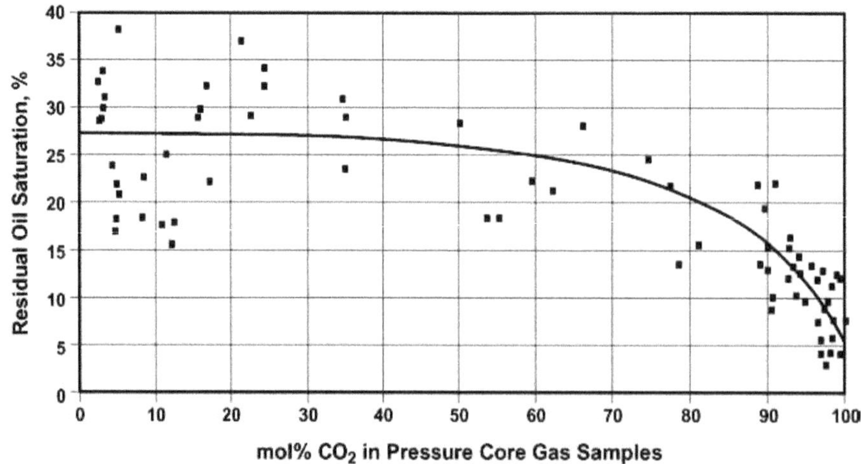

Fig. 2.11—Residual oil saturation as a function of mol%CO_2, from pressurized cores in the Levelland unit 736 CO_2 pilot. As the mol% CO_2 increases, less residual oil remains.

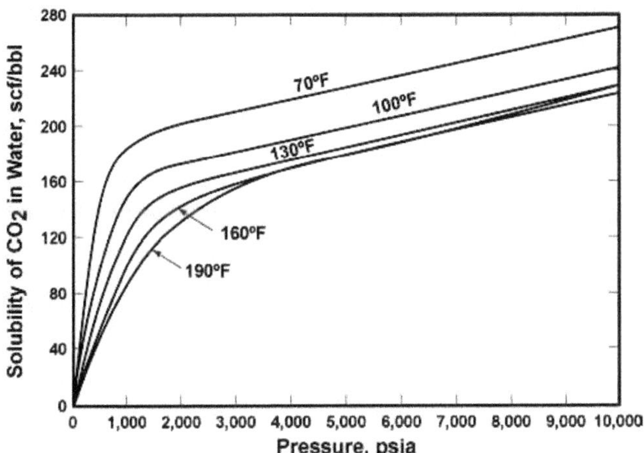

Fig. 2.12—Effect of temperature and pressure on CO_2 solubility in water (after Refs. 25 and 26). Carbon dioxide solubility in water increases with increasing pressure and decreasing temperature. Carbon dioxide dissolved in water is unavailable to contact oil.

mental data by Perkins and Johnston[31] suggest that dispersion increases with velocity to the 1.2 power. The longitudinal component of dispersion (in the primary direction of flow) has the greatest impact on CO_2 flood performance. The transverse component of dispersion is generally about one-tenth the magnitude of the longitudinal dispersion. Although relatively small, transverse dispersion improves the volumetric sweep of the solvent; however, it can also reduce the displacement efficiency.[32]

Longitudinal dispersion is often represented by the following equation for 1D flow.

$$K_l = D_e + \alpha_l v, \qquad (2.2)$$

where K_l=longitudinal dispersion coefficient in ft²/D; D_e=effective molecular diffusion for porous media (which acts equally in all directions); α_l=longitudinal dispersivity in ft; and v=superficial velocity of flow in ft/D. The superficial velocity is the product of porosity and interstitial velocity (the velocity through the pores).

The level of dispersion typically is expressed as a dimensionless Peclet number.[31] The Peclet number is defined as the ratio of convective to dispersive transport.

$$N_{Pe} = L v / K_l \approx L / \alpha_l, \qquad (2.3)$$

where N_{Pe}=Peclet number, and L=length over which the displacement has traveled (slim tube, core, or distance between an injection and production well) in ft.

The approximation in Eq. 2.3 results from neglecting diffusion.[3] Eq. 2.3 shows that the Peclet number is independent of displacement rate.[3] If the longitudinal dispersivity is large compared to the displacement length, then the Peclet number is small; conversely, a small dispersivity typically yields a large Peclet number.

Importance of the Peclet Number on Development of Miscibility. Negahban *et al.*[33] investigated the impact that the Peclet number has on the development of CO_2 miscibility with oil for a west Texas crude. Their research was carried out using a fully compositional reservoir simulator that included convective dispersion (see Sec. 4.3.2 for more information on fully compositional models). Simulation results were matched to previous experimental results by Yellig[34] from Berea sandstone corefloods of different lengths. In each experiment, Negahban *et al.*[33] found that a constant value for effective longitudinal dispersivity sufficed to match Yellig's[34] corefloods: Peclet numbers of 17, 33, and 68 matched core floods of 8 ft, 16 ft, and 32 ft, respectively. The corresponding longitudinal dispersivity used to match all the corefloods was 0.48 ft.

The thermodynamic MMP for CO_2 displacing the subject oil was 1,190 psia, and most of the corefloods were conducted at a pressure of 1,915 psia. Negahban *et al.* noted that miscible residual oil saturation decreased as the core length increased.

Fig. 2.16 compares the final oil saturations predicted by the simulator to Yellig's[34] experimental values after injection of 1.2 HCPV of CO_2. (These oil saturations were swelled about 30% by CO_2.) Smaller Peclet numbers (shorter length cores) produced higher residual oil saturations and reduced oil recovery. Although the corefloods conducted at pressures of 1,900 psia were well above the thermodynamic MMP of 1,190 psia, the simulator indicated that the 8-ft flood (the smallest Peclet number experiment) was not miscible because a single flowing phase did not exist.

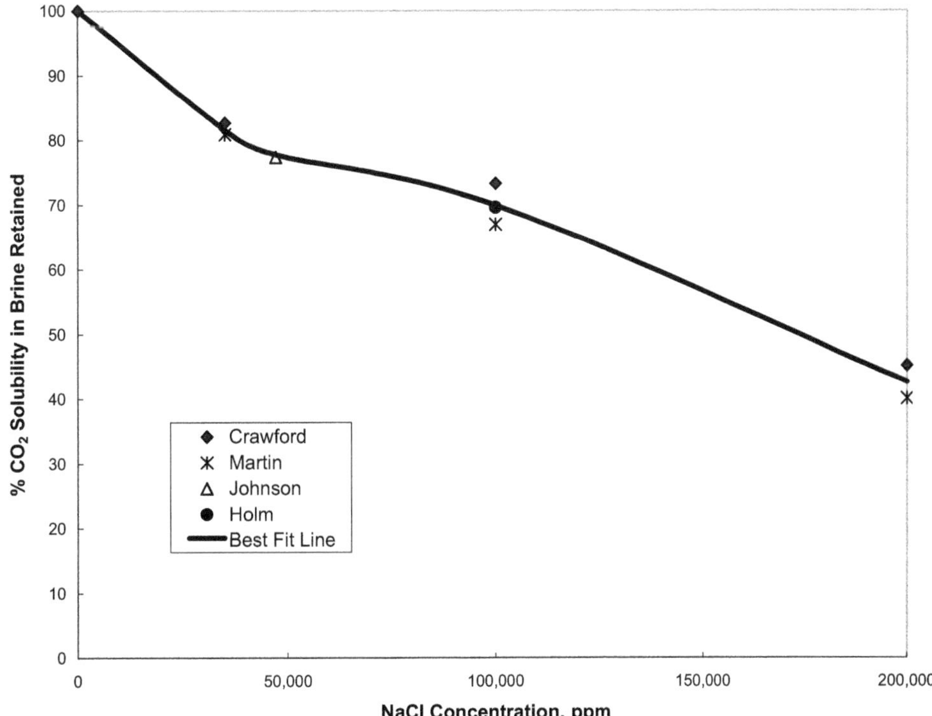

Fig. 2.13—Effect of salinity on CO_2 solubility in water. Carbon dioxide solubility in water depends on salinity (ppm NaCl).

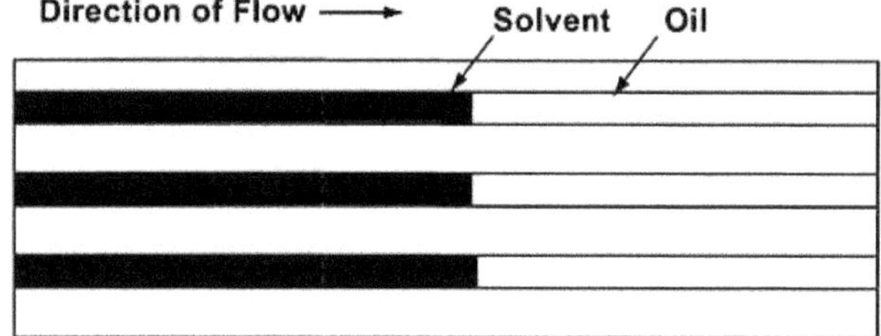

In uniformly sized pores, the solvent travels at the same flow rate through all pores.

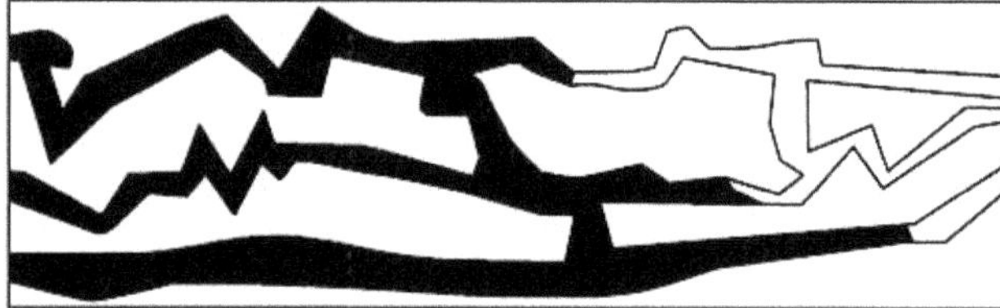

In non-uniformly sized pores, the solvent progresses unevenly.

Fig. 2.14—Uniform and nonuniform pores. Nonuniform pores cause solvent to become dispersed as it flows through a porous medium.

Yellig verified this experimentally by observing the two-phase flow in the sight glass during the 8-ft core displacement.[34]

As shown in Fig. 2.16, the simulator found that for a Peclet number less than 20 (to the left of the dashed line), the CO_2 flood process was not miscible, even though the process was 710 psi above the thermodynamic MMP. The flood became miscible only for core lengths greater than about 10 ft (a Peclet number of 20). Fig. 2.16 shows results for a west Texas oil, but Negahban *et al.*[33] found similar results using oils from other parts of the United States.

Values for Longitudinal Dispersivity. The magnitude of longitudinal dispersivity can greatly impact oil recovery. Unfortunately, values for longitudinal dispersivity vary greatly depending on reservoir heterogeneities and the length scales over which the flood takes place. Slim-tube dispersivities, which are measured for displacements in short cores, typically are about 0.05 ft.[35] Reservoir rock, however, exhibits greater heterogeneities than does slim-tube packing, which means there is a wider range of velocities in the pores of the reservoir and, therefore, increased mixing and dispersivity. In some instances, the dispersivity can increase with the displacement distance, which also is greater in a reservoir than in a slim tube. Thus, slim-tube dispersivities probably are not representative of reservoir dispersivities.

Two types of field tests are used to measure longitudinal dispersivity in the reservoir. These involve mathematical modeling and history matching of tracer flow, either between a pair of injection and production wells, or within a single-well tracer test. (In the single-well tracer test, an oil-soluble chemical is injected into a well, which is shut in for a time before being produced back.)

Pickens and Grisak[36] found that effective longitudinal dispersivity grew with the length traveled through the reservoir. To estimate dispersivities, they used tracer data from fluids that traveled from injection wells to production wells over an entire reservoir, but they did not take into account the effect of producing from commingled reservoir layers (multiple layers produced simultaneously). Inclusion of layer properties in the analysis would have decreased the longitudinal dispersivity required to match the tracer data.

Single-well tracer tests[1,37–39] have provided reliable estimates of longitudinal dispersivity ranging from 0.032 to 5.0 ft. The volume investigated during a single-well tracer test is larger than that of a slim-tube test, but typically is not as large as that of an injection/production pattern. Single-well tracer test volumes can vary greatly and can influence the magnitude of dispersivity.[40]

In another approach to estimating dispersivity, several researchers used geostatistics to examine how permeability varies from point to point within a field. Using both lab and field tracer data, Arya *et al.*[41] found that if the correlation length scale for permeability variations (defined as the average length over which a

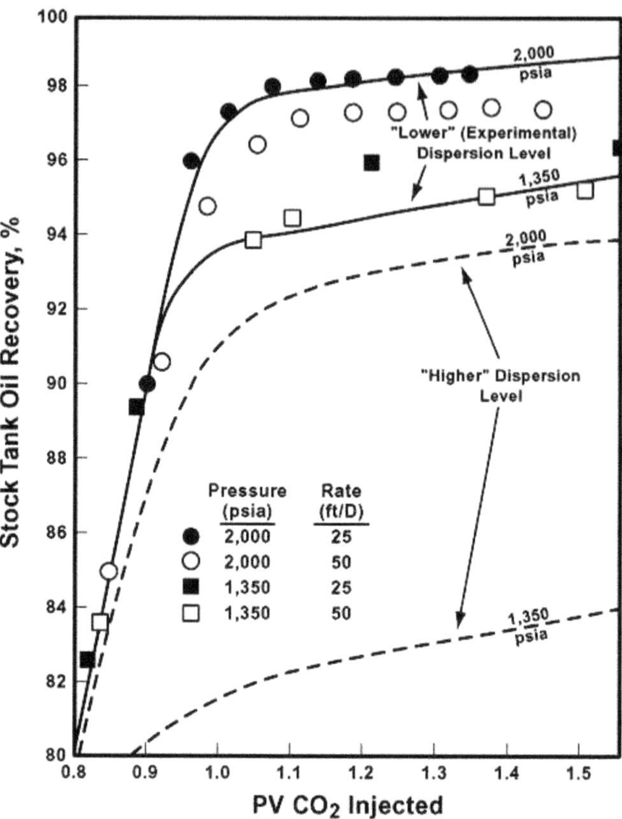

Fig. 2.15—Experimental and calculated oil recoveries: Wasson crude displacement efficiencies at 150°F (after Ref. 30). This graph illustrates how pressure and convective dispersion affect CO_2 flood oil recovery in one dimension. The lines represent simulated oil recoveries at two different pressures. The solid lines show recovery for a small amount of dispersion, and the dashed lines show recovery for large dispersion. The symbols are experimental slim-tube test data. Oil recovery is more sensitive to pressure as dispersion increases. The flood may not be miscible even though the pressure is well above the thermodynamic MMP (lower dashed line).

given permeability correlates with itself) is much smaller than the length of the system (well spacing), the effective dispersivity becomes constant and does not grow with length traveled. In other words, if variations in permeability are random within a layer, dispersivity remains constant. However, if the correlation length scale is not significantly less than the length of the system, dispersivity can grow linearly with length.

Kelkar and Gupta[42] found that random variations in permeability caused dispersivity to increase when the Dykstra-Parsons coefficient exceeded 0.4. Most reservoirs exceed this measure of permeability variation. (The Dykstra-Parsons coefficient is a statistical measure of reservoir heterogeneity defined as the standard deviation from the mean permeability divided by the mean permeability.[43,44] This coefficient approaches unity for extremely heterogeneous rock and zero for extremely homogeneous rock. Typical Dykstra-Parsons coefficient values taken from core data from the Anoton Irish, Slaughter, and Wasson CO_2 projects in west Texas range from 0.48 to 0.81.)

The question then is, "What is the appropriate permeability variation length scale to use for an actual reservoir?" Kittridge *et al.*[45] sampled San Andres outcrops in New Mexico (corresponding to the rock in the Wasson San Andres field) to look for an answer. They found that the correlation length scale was always smaller than the length sampled, and from this they concluded that the correlation length scale for San Andres dolomite-limestone is small—somewhat larger than slim-tube scale but much smaller than typical well spacing. This implies that permeability for this type of reservoir varies randomly within a depositional event and that effective dispersivity should not grow with length.

Why were the results of Kittridge *et al.*[45] so different from those of Pickens and Grisak[36] and Arya *et al.*?[41] The most likely explanation is that these investigators[36,41] used large sampling sizes, as well as zones that included multiple layers. While the statistics derived from their broad collection of data seemed to present a clear picture, they did not accurately reflect the geology within the reservoir.

Stalkup and Crane's[46] results were consistent with those of Kittridge *et al.*,[45] showing that small values of longitudinal dispersivity (0.5 ft) could be used in a fully compositional simulation of an enriched gasflood in Prudhoe Bay (a sandstone reservoir with permeability of several hundred md) to match data from a logging observation well located 525 ft from an enriched gas injection test. Stalkup and Crane[46] based their estimate of effective longitudinal dispersivity on earlier single-well tracer tests.

Counter to the results of both Kittridge *et al.*[45] and Stalkup and Crane,[46] Perez and Chopra[47] found that correlation length was great when they examined horizontal well log data from a crossbedded, laminated sandstone reservoir. One could infer from the work of Perez and Chopra that dispersivities increase with distance traveled.

Additional Factors That Can Increase Longitudinal Dispersivity. Even if longitudinal dispersivity does not grow with length as CO_2 moves through the reservoir, other factors can increase its magnitude. For example, the carbonate reservoirs of west Texas probably have dispersivities greater than 1.0 ft because of dead-end pores, which are common in carbonate rocks.[48–50]

Multiphase flow effects between oil and water also can increase dispersivity. Ramirez *et al.*[51] found that longitudinal dispersivity could be up to 10 times higher when both oil and water are flowing. The single-well tracer tests that provided dispersivity estimates on the order of 1.0 ft were carried out at injection wells where single-phase flow was likely. Away from injection wells and in the middle of the reservoirs, multiphase flow occurs and dispersivity will be greater.

Conclusions. Convective dispersion remains one of the least understood areas of reservoir engineering. The proper magnitude of longitudinal dispersivity likely depends on the level of detail used in characterizing the reservoir. A more detailed characterization of the reservoir dictates that a smaller dispersivity be used to simulate CO_2 flood performance, but it also requires finer gridblocks and more powerful flow simulators. To date, field-scale simulation of CO_2 flood performance has used large but relatively constant dispersivity levels, a numerical artifact of the way that numerical simulators solve equations for flow through porous media. Chap. 4 discusses this practical issue in more detail and addresses specifically the value of using dispersivity in numerical simulation.

2.2.2 Effect of Water Blocking on Small-Scale Reservoir Mixing. In water-wet rocks, a phenomenon takes place in which water shields the globules of residual oil from the CO_2. This "water blocking," as it is called by Tiffin and Yellig,[52] can impact a CO_2 flood in strongly water-wet formations. To understand water blocking, one must understand the important differences between water-wet and oil-wet formations.

Water-Wet vs. Oil-Wet. **Fig. 2.17** illustrates the differences between water-wet and oil-wet mechanisms. Before an oil reservoir forms, the rock contains only water. As oil migrates into the reservoir, the rock either is wetted by the oil over time (weeks in the laboratory), or it remains wetted by the water.

In rock wetted by oil, the oil displaces water in the larger pores and attaches itself to the rock walls. Smaller pores typically remain filled with (connate) water because the capillary pressure is too high for oil to enter. This type of wettability is known both as oil-wet[43] and mixed-wet.[53] Many authors differentiate between oil-wet and mixed-wet formations; however, there is no practical difference between the two. In both cases, the large pores are oil-wet, and this is where most oil recovery takes place. In this book, the term "oil-wet" is used to include both fully oil-wet and mixed-wet formations.

When an oil-wet rock is waterflooded, the water flows easily through the larger pores, pushing oil ahead of it and gradually stripping off oil that remains attached to the rock (a process known as the throughput effect).[53] Thus, water relative permeability is relatively high in oil-wet rock.

In a water-wet formation, the rock remains preferentially attracted to water, and connate water is generally higher. Again oil enters the larger pores upon migration into the reservoir, but this oil does not wet the walls of the pores. After waterflooding, residual oil remains in the form of globules that are too large to squeeze through the small throats of the large pore bodies. In water-wet rock, the water relative permeability tends to be low because the water has to squeeze past trapped oil globules or flow through the small pores, where the water phase is continuous but the pressure drop is greater.

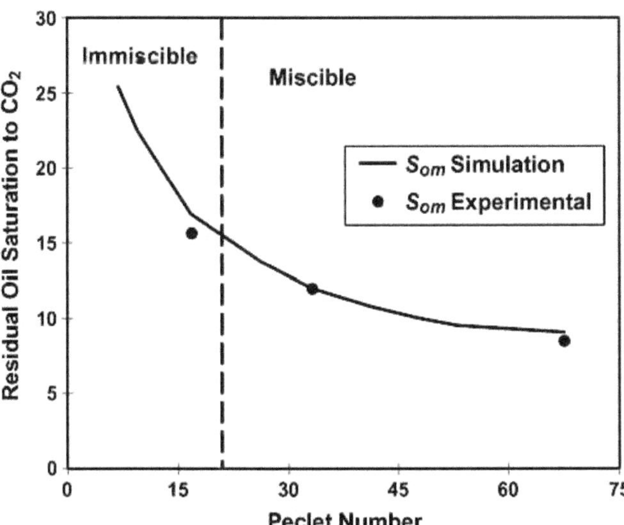

Fig. 2.16—Effect of Peclet number (core length) on residual oil saturation in a 1D CO_2 corefloods above the thermodynamic MMP (after Ref. 33). As the Peclet number (core length) increases, miscible residual oil saturation (S_{om}) decreases. A small Peclet number, which also can result from high dispersivity, can hinder the development of miscibility and decrease oil recovery.

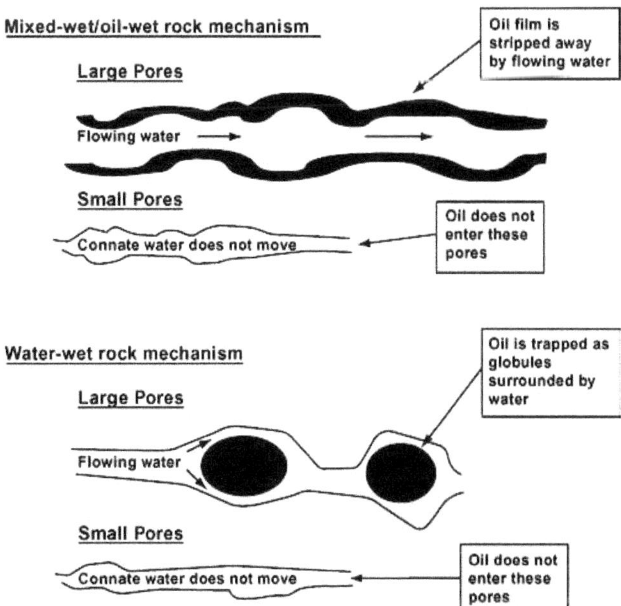

Fig. 2.17—The flow mechanisms for water in mixed-wet/oil-wet and water-wet rock. In mixed-wet/oil-wet rock, some oil remains as a film on the rock and can be gradually stripped away by water, which flows freely through the rock. In water-wet rock, residual oil remains as trapped oil globules surrounded by water. The oil globules can act as choke valves.

Effect of Water Blocking on Oil Recovery. Tiffin and Yellig[52] found that strongly water-wet cores exhibit reduced oil recovery when subjected to simultaneous CO_2/water injection (as shown in **Fig. 2.18**), whereas oil-wet cores do not, and they attribute this reduced recovery to water blocking. When water and CO_2 are injected simultaneously (which is generally performed only in the lab), a fixed water saturation develops. In this situation, the CO_2-rich phase might not have sufficient capillary pressure to enter all the pores. As a result, oil in the smaller pores might not be contacted by CO_2 and could remain trapped, while CO_2 that enters the larger pores may develop miscibility with the oil in these pores. Stern[54] refers to this type of oil bypassing as "capillary induced bypassing."

If the rock is water-wet and CO_2 is injected in WAG mode, the CO_2-rich phase saturation will be limited by the presence of injected water. As a result, the CO_2-rich phase will not have sufficient capillary pressure to enter the pores containing the remaining oil. If CO_2 is injected continuously (rather than WAG), the mobile water can be displaced and additional oil can be contacted and recovered.

Shelton and Schneider[55] found that reducing the injection rate yielded additional oil recovery in water-wet cores when CO_2 and water were injected simultaneously. They attributed the increased oil recovery to the additional time available for the CO_2 to diffuse through the water phase to reach trapped oil.

Water blocking has not been a significant problem in miscible gas/water coreflood experiments that use fresh-state cores (cores obtained with bland water—water with no chemical additives).[56–58] Water blocking has only been a problem in strongly water-wet rocks.[52,54]

CO_2 Flooding in Water-Wet Formations. The conclusion to be drawn here is that it may be wise to use continuous CO_2 flooding in strongly water-wet formations. With continuous injection, CO_2 may overcome capillary pressure effects and miscibly contact all the oil. If water is continually added to the reservoir in short WAG cycles, the CO_2 may never succeed in moving enough water out of the way to allow efficient oil displacement. The Ford Geraldine[59] and Twofreds fields[60] are two examples of successful CO_2 projects in which CO_2 was injected continuously into strongly water-wet rock.

In a comprehensive literature review of CO_2 flooding, Hadlow[61] points out that alternate water/gas injection has not been found to be detrimental to CO_2 flood performance, and he suggests that water blocking is not a significant reservoir problem. However, for those reservoirs that *are* strongly water-wet, continuous CO_2 flooding—rather than WAG—has been used and should probably continue to be used.

2.2.3 Viscous Fingering in CO_2 Floods. For many years, mixing caused by viscous fingering was thought to be an important design consideration for miscible floods. This is not surprising because viscous fingering has been observed in laboratory experiments

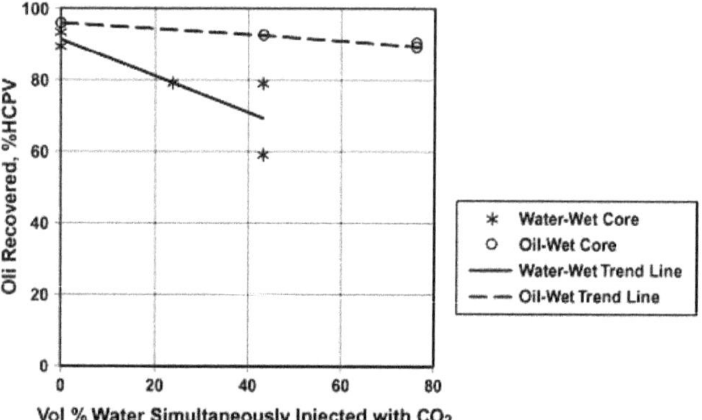

Fig. 2.18—Effect of mobile water on oil recovery for miscible CO_2 floods in water-wet cores and oil-wet cores. Strongly water-wet cores exhibit reduced oil recovery when subjected to simultaneous CO_2/water injection, while oil-wet cores show little effect. Mobile water in the water-wet rock shields the oil from the CO_2.

involving first-contact miscible solvents when the solvent is less viscous than the oil. It is important to note that the laboratory research on viscous fingering was performed using homogeneous porous media (sandpacks with fairly uniform particle size). Reservoirs are never this homogeneous.

In their landmark work, Chang[62] and Chang *et al.*[63] performed high-resolution, fully compositional reservoir simulations of multiple-contact miscible oil displacements by CO_2. They showed that viscous fingering only occurs in the homogeneous porous media used in laboratory experiments and concluded that real reservoir rock has enough variation in small-scale permeability to cause viscous fingers to coalesce. Thus, viscous fingering does not need to be considered in field applications.

Viscous fingering can be confused with channeling of injected CO_2 through high-permeability layers. Channeling is an important consideration, but it is not a viscous fingering phenomenon.

2.3 An Overview of Relative Permeability

A typical design for a horizontal miscible CO_2 flood includes injecting water alternately with CO_2 to remedy areal and vertical sweep problems associated with the higher mobility CO_2-rich phase. Because water is more viscous and less mobile than CO_2, it slows CO_2 flow in high-permeability layers and helps achieve more efficient use (the CO_2 utilization factor is defined as Mscf of CO_2 injected/STBO recovered). Under the conditions that prevail in west Texas projects, water typically is 10 times more viscous than CO_2, and CO_2 use typically runs about 10 to 20 Mscf/STBO.

Engineers who have worked with both water and CO_2 floods are aware that water injectivity (defined as injection rate divided by the difference between bottomhole injection pressure and average reservoir pressure) undergoes striking changes after the first cycle of injected CO_2. These changes are related to the effect that CO_2 has on the relative permeability of water. The effect of three-phase relative permeability in an oil/water/CO_2 system can reduce the mobility of both CO_2 and water.[64–66] *Quantifying water mobility is necessary for meaningful simulation of a CO_2 flood because water mobility affects water injection rates, which in turn affect the rate of tertiary oil production and the economic payout period of the project.* The petroleum literature includes examples in which water injection rates significantly decreased following CO_2 injection, as well as examples in which injection rates did not decrease. *Understanding the "why" behind these conflicting results is important for designing an effective CO_2 flood.*

Just as CO_2 affects water relative permeability, so does water affect CO_2-rich phase relative permeability under miscible conditions. The change in CO_2 relative permeability is important because it affects the CO_2 production rate, which in turn affects the operating and capital costs of the CO_2 recycling facilities.

A quantitative understanding of relative permeability is important because relative permeability curves must be input to reservoir simulators to forecast the performance of a CO_2 flood. A flawed understanding of relative permeability can lead to overly optimistic predictions of oil production rates. Secs. 2.3.1 through 2.3.4 look at these issues in detail.

2.3.1 Oil Relative Permeability. Fig. 2.19 shows laboratory data for water/oil relative permeability from the CO_2 project discussed in Chopra *et al.*[64] This graph contains three relative permeability curves for an oil-wet reservoir: an oil curve, a water curve, and a water hysteresis curve (more on hysteresis shortly). The x-axis spans from 20 to 80% water, while the relative permeabilities of both oil and water are shown on the logarithmic y-axis, which spans from 0.001 to 1.0. A logarithmic axis often is used for relative permeability so that the engineer may examine the shape of the curve at low relative permeability values.

Oil relative permeability in a reservoir is defined as the ratio of the effective oil permeability to a base permeability at a given oil saturation. The base permeability used in this monograph is the oil permeability at connate water saturation. (Air permeability and water permeability are alternative base permeabilities.) Using the monograph definition of base permeability, the relative permeability is 1.0 at the connate water saturation (see Fig. 2.19, where the connate water saturation is about 0.25). As an example, if the oil relative permeability at connate water saturation is 100 md, and the oil relative permeability at a particular water saturation is 0.5, then the effective oil permeability at that saturation is 50 md. Fig. 2.19 shows that the relative permeability of oil declines as water takes up more and more of the pore volume and restricts the flow of oil. At 72% water (28% residual oil saturation to water, S_{or}), the oil relative permeability is zero.

The velocity of oil in a reservoir depends on its effective permeability and is equal to the product of effective permeability, pressure drop and cross-sectional area. As oil relative permeability decreases so does velocity, and eventually the oil ceases to flow.

2.3.2 Water Relative Permeability. The relative permeability of oil in an oil-wet rock is assumed a unique function of water saturation. The relative permeability of water in an oil-wet rock, however, is more complex—it is a function both of water saturation and of the previous history of water in the rock. *Moreover, the exact form that water relative permeability takes in a particular rock can have a major impact on the ability of CO_2 to recover oil.*

At connate water saturation, the water relative permeability is zero. As water is injected, the water relative permeability increases, reaching its maximum at the point where oil ceases to flow, which in Fig. 2.19 is 72% water saturation [1.0 minus the residual oil saturation (S_{or})]. In this case, the maximum water relative permeability (k_{rw}) for a waterflood is 0.5 k_{rw}; the k_{rw} cannot be larger because oil remaining in the rock interferes with the flow of water.

During a CO_2 flood, the residual oil saturation can be reduced to near zero. If enough water is injected behind the CO_2 to dissolve it, the reservoir can approach 100% water saturation. At 100% water saturation, the water relative permeability typically is greater than 1.0, probably because of extra flow capacity from pores containing connate water. This increase in water relative permeability as residual saturation decreases is discussed further in Sec. 2.4.2 and in Appendix D.

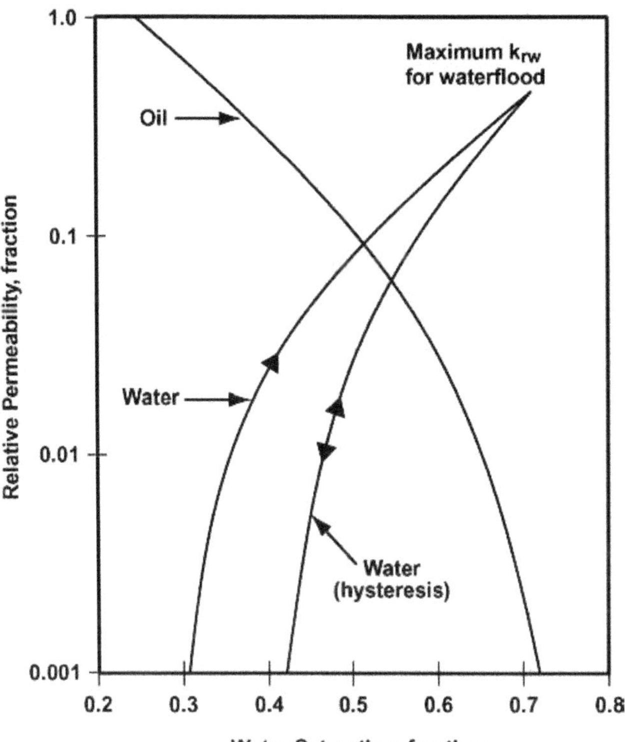

Fig. 2.19—Two-phase relative permeability diagram showing laboratory data for water/oil relative permeability from the CO_2 project discussed in Chopra *et al.*[64] The relative permeability of oil, which is 1.0 at connate water saturation, declines as water is injected, while the water relative permeability (k_{rw}) increases. A water hysteresis relative permeability curve can result when a tertiary oil bank forms during a CO_2 flood.

2.3.3 Water Relative Permeability Hysteresis. When water is injected, its relative permeability follows the water curve in Fig. 2.19. But what happens if oil then is reinjected? This may seem like an arcane question, but the result of reinjecting oil is similar to the increase in oil saturation that occurs when tertiary oil forms an oil bank during a CO_2 flood.

A test of bidirectional water/oil relative permeability is useful for representing this tertiary oil bank development, as well as to clarify water relative permeability and water injectivity. In a bidirectional test, the core is first waterflooded to residual oil saturation conditions using a standard steady-state water/oil relative permeability test, and then oil is injected under steady-state conditions until water ceases to flow. Ideally, the cores would be from a nonwaterflood area of the field using a lease crude with no additives; otherwise, wettability will have to be restored (see Sec. 3.1.3). Coring with water will drive the cores to residual oil saturation, such that the hysteresis effect on relative permeability described below cannot be measured accurately.

For an oil-wet reservoir, we assume that oil can flow back through the same pores that it previously vacated and that the oil relative permeability increases along the same curve that traced its decline (although occasionally some variation from the original curve occurs when oil saturation decreases). The water relative permeability, however, exhibits hysteresis because the curve for increasing water saturation does not necessarily follow the curve of the original water drainage. Hysteresis occurs for an oil-wet rock because some water will be trapped by the same mechanism that entraps oil in a water-wet rock (see Fig. 2.17). Values for water imbibition (increasing oil and decreasing water saturation) relative permeability will be lower than for drainage (decreasing oil and increasing water saturation) water relative permeability, and resulting residual water saturation will be higher than connate water saturation. If additional water cycles occur and water saturation decreases again below ($1.0-S^{or}$), water relative permeability will trace another curve that is parallel and to the left of the original hysteresis curve.[67,68]

The result of water relative permeability hysteresis in an oil-wet reservoir is that water injectivity can be severely reduced. In addition, oil relative permeability cannot return to 1.0 because the increased residual water saturation reduces the maximum oil saturation possible. Consequently, *water hysteresis reduces the peak tertiary oil rate.*

Not all oil-wet reservoirs exhibit the water relative permeability hysteresis shown in Fig. 2.19. Furthermore, if the waterflood takes place in water-wet rock, water will remain on the original relative permeability curve, i.e., little hysteresis will occur.

2.3.4 The Solvent Relative Permeability Concept. When the pressure is sufficiently below the thermodynamic MMP, significant interfacial tension exists between CO_2 and oil, and the flood will be immiscible. In this case, CO_2 will reside in a gas phase that is nonwetting compared to the oil and water, and conventional three-phase relative permeability[69] will apply to the oil, water, and gas relative permeabilities. Conventional three-phase relative permeability models are not discussed in this monograph.

For miscible floods above the thermodynamic MMP, however, the CO_2-rich phase is an effective solvent that has properties similar to the oil. In this case, the solvent is most likely the intermediate wetting phase in oil-wet rock, and solvent relative permeability will be affected by the reduction in pore space occupied by oil and water. If the CO_2-rich phase in oil-wet rock is less wetting than water, then the result is water blocking. There has been no field or laboratory evidence of water blocking in WAG CO_2 projects in oil-wet or mixed-wet rocks; this monograph assumes that the CO_2-rich phase solvent is intermediate wetting for oil-wet and mixed-wet rock.

A first-contact miscible solvent will have properties similar to those of oil and therefore will flow according to the oil relative permeability curve. Solvent injection of CO_2, however, leaves behind a small residual oil saturation. This residual oil interferes with the subsequent flow of solvent and causes the solvent (CO_2-rich phase) relative permeability to be lower than the oil relative permeability. **Fig. 2.20** shows the relationship between the relative permeability curve of the solvent (the CO_2-rich phase that develops miscibility with the oil) and the residual oil saturation. With a residual saturation to CO_2, S_{om}, of 3% (0.03 volume fraction), there is little oil to interfere with flow of the solvent, and the solvent relative permeability is close to that of oil. As S_{om} increases, however, the solvent relative permeability curve moves farther away from the oil relative permeability curve.

Fig. 2.20 illustrates for the example reservoir what happens in practical terms during the life of a CO_2 flood in an oil-wet or mixed-wet reservoir. During the initial waterflood, the relative permeability of water increases along the original water curve to a water saturation of 72%, leaving a residual oil saturation of 28%. During subsequent solvent injection, the solvent relative permeability increases along one of the solvent curves as its saturation increases. (The dashed lines in Fig. 2.20 are the solvent curves for different values of S_{om} reached by the solvent flood.) The residual oil left by the solvent determines which of the solvent curves is used for solvent relative permeability. The oil saturation decreases during solvent injection, and its relative permeability decreases along the oil curve. The water saturation decreases at the same time, and its relative permeability decreases along the water hysteresis curve. When water is injected during the first WAG cycle, the water saturation increases again, and its relative permeability increases along the water hysteresis curve while oil relative permeability continues to decrease along the oil curve. Furthermore, the solvent saturation decreases and its relative permeability decreases along the solvent curve.

Lab tests of CO_2-rich phase relative permeability under miscible conditions have shown that the sum of the final oil saturation to CO_2 plus the residual solvent saturation to water is greater than the residual oil saturation to waterflooding alone.[69–72] Chopra *et al.*[64] theorized similar performance in the presence of residual oil, based on dry-gas/water relative permeability data from Schneider and Owens.[66]

A key concept is that the change in solvent relative permeability indirectly changes the water relative permeability. Both residual solvent and residual oil limit water permeability now in the same way that residual oil saturation (S_{or} in Fig. 2.19) initially limited the maximum water relative permeability. For example, in the case illustrated here, if S_{om}=0.10 (the left-hand dashed line in Fig. 2.20), water saturation is 58% and the volume of trapped solvent is 32% [100%−10% (miscible residual oil)−58% (maximum water saturation)]. Thus, after WAG injection has begun, the maximum water relative permeability is reduced to about 0.1 (the value illustrated by the lower-left dot in Fig. 2.20). This reduced relative permeability, which is sometimes called the "trapped-gas effect," is about five times lower than the preCO_2 value of 0.5. The reduction in the maximum water relative permeability caused by residual solvent and oil is a critical factor in explaining the loss of water injectivity that is so frequently observed in WAG projects, whether they involve miscible CO_2 or miscible enriched-gas projects.[64]

The solvent relative permeability curves illustrated in Fig. 2.20 were calculated using Eq. 2.4. Chopra *et al.*[64] developed the physical concept of the solvent relative permeability model by extending Stone's II three-phase relative permeability model[69] to a system where oil is the most wetting phase, followed by CO_2 solvent, and water. Their model is discussed in detail in Sec. 2.5.

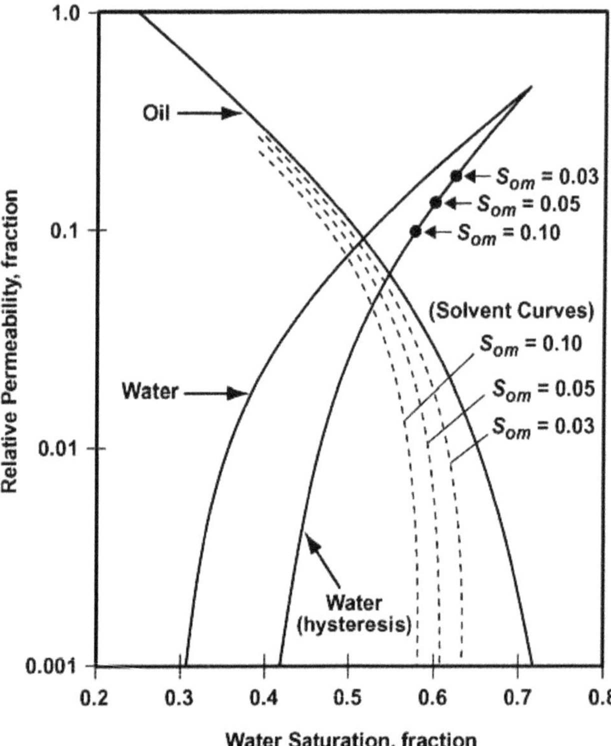

Fig. 2.20—The solvent relative permeability concept (after Ref. 64). The residual oil causes the relative permeability of the solvent (the CO_2-rich phase) to be lower than the relative permeability of the oil. The net effect of the residual oil is that the maximum relative permeability of water (k_{rw}) is reduced. The maximum k_{rw} is on the water hysteresis curve directly above the point where the solvent hysteresis curve intersects the x-axis.

2.4 Field Examples of Relative Permeability Effects on Fluid Mobility

This section presents useful field examples of the relative permeability concepts described in Sec. 2.3 and shows how relative permeability impacts fluid mobility. We examine four topics: the loss of water injectivity typically seen in CO_2 floods; the effect of changing water injection pressures to achieve higher injection rates; the control of inadvertent fracturing caused by higher pressures; and the impact that CO_2 solubility in water has on the injectivity of water during long water cycles such as chase water.

2.4.1 Water and CO_2 Mobility in the Levelland Unit 12-Acre CO_2 WAG Pilot. Henry *et al.*[73] calculated the total fluid mobility during water cycles between a WAG injection well and adjacent reservoir pressure observation wells in the Levelland Unit 12-acre CO_2 pilot. **Fig. 2.21** shows the arrangement of injection, production, and observation wells in that pilot: 601 (injection well), 745

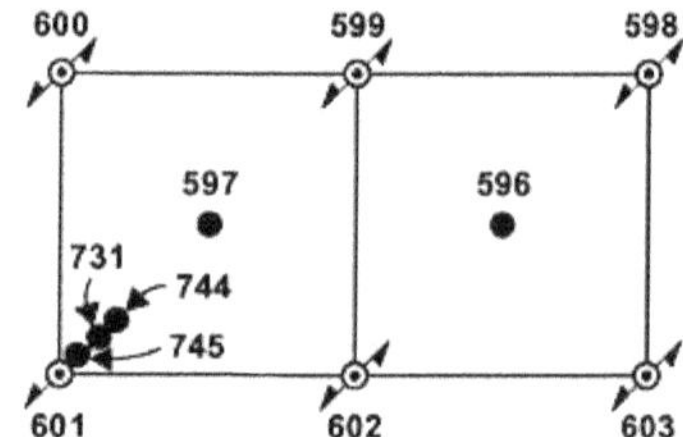

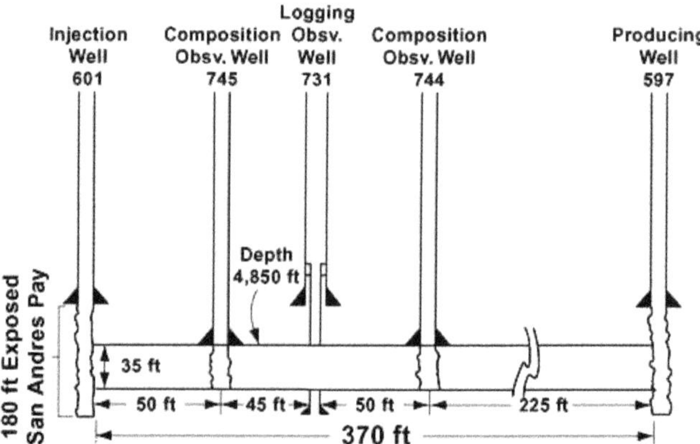

Fig. 2.21—Location and elevation view of observation wells in the Levelland Unit 12-acre CO_2 pilot (after Ref. 73). The top drawing shows the site plan and the injection wells (598–603), production wells (596, 597), and observation wells (745, 731, 744). The bottom drawing shows an elevation view of the completions along a line from the pilot injection Well 601 to the pilot production Well 597.

(observation well for pressure measurements and fluid samples), 731 (logging observation well for saturation measurements), 744 (observation well for pressure measurements and fluid samples), and 597 (production well).

Carbon dioxide injection began in August 1979 at Well 601. The first water cycle commenced in October 1979. During the water-injection phase of the CO_2 flood, the total mobility between the injection well (601) and the nearest observation well (745) was substantially less than it was before CO_2 flooding. Relative mobility (referenced to water mobility before CO_2 injection) increased during the CO_2 cycles and decreased during the water cycles, as shown in **Fig. 2.22.** The large changes in relative mobility close to the injector (shown by the solid line) indicate that not much oil remained between the wells, and that CO_2 and water were the only mobile phases.

The situation was different farther from the injector, as shown by the dashed line representing changes in mobility recorded between Wells 745 and 744. The peaks in relative mobility were delayed because CO_2 took longer to reach these wells. In addition, the mobility curve was not as sensitive to the water/CO_2 cycles. The reduced sensitivity is probably because the three phases present (CO_2-rich phase, oil-rich phase, and water) showed a high degree of mixing, evidenced by fluid sampling data from the observation wells. Moreover, the CO_2 cycles were only 2% HCPV and the second observation well (744) was less likely to be influenced by the CO_2 cycle.

Fig. 2.23, from Henry and Metcalfe,[21] gives the interwell pressure data between observation wells 745 and 744. The reservoir pressures were higher during CO_2 cycles, even though the injection well used identical bottomhole injection pressures (50 psi below formation parting pressure) for both the CO_2 and water cycles. The CO_2 injectivity was greater than the water injectivity and thus boosted reservoir pressure. When water was injected, injectivity declined because the less mobile water flowed more slowly, and the producing well continued to remain pumped-off (low fluid level above the pump intake valve). In effect, the producing well reduced the reservoir pressure. In this case, the reservoir pressure changes of about 300 psi took place between the CO_2 and water cycles. Even though little CO_2 had made it to the more distant pressure-sampling well (744), a strong pressure pulse still reached it.

2.4.2 Water Injectivity and Bottomhole Pressures in CO_2 WAG Projects. The decrease in water relative permeability after CO_2 injection causes water injectivity to decrease as well. (Water injectivity is defined as injection rate divided by the difference between the bottomhole injection pressure and the average reservoir pressure.) When water and CO_2 are injected at the same bottomhole pressure, the water injection rate will decrease unless the water cycle is long enough for the injected water to dissolve enough CO_2 to overcome the decreased injectivity. When the water is injected at a higher bottomhole pressure than the CO_2, to overcome decreased injectivity, the higher bottomhole pressure may induce or extend vertical fractures at the injectors, causing sweep problems. The next five subsections expand on these and other water injectivity issues in CO_2 WAG projects.

Injecting Water and CO_2 at the Same Pressure at the Slaughter Estate Unit. **Fig. 2.24** shows examples of water injection rates for the Slaughter Estate unit CO_2 flood which are lower than CO_2 injection rates (on a reservoir barrel basis).[64,74] The same bottomhole injection pressure was used for the CO_2 and water cycles (25 psi below formation parting pressure). The flood used CO_2 cycles of 1.0% HCPV and water cycles of 0.5% HCPV.

The rate data in Fig. 2.24 show a significant loss of water injectivity of about 40% after the start of CO_2 injection, later stabilizing to a 50% loss in water injectivity. The figure also illustrates that while most of the injectivity loss is observed after the first cycle, as much as 10% HCPV of CO_2 may be required to identify a representative or stabilized loss in water injectivity from field performance (10% HCPV corresponds to 10 cycles of CO_2).

Failure of a CO_2 Flood Pilot in the Grayburg Formation Because of a Large Three-Phase Hydrocarbon Region. The loss in water injectivity can be severe enough that the CO_2 project may fail. A case in point is a double five-spot 12-acre CO_2 WAG pilot in the Grayburg formation that comprised multiple isolated zones of dolomite and sandy dolomite with permeability ranging from 0.3 to 50.0 md. The CO_2 pilot was operated in the late 1980's.

Average reservoir pressure before CO_2 injection ranged from 1,115 to 1,615 psia for the various reservoir zones (based on drill-stem tests in a pilot observation well). The thermodynamic MMP from slim-tube tests was 1,100 psia. At the reservoir temperature of 94°F, a three-phase region existed between 1,000 and 1,400 psia when CO_2 is injected. The injection wells were operated at a constant bottomhole injection pressure of 1,700 psia (50 psi below formation parting pressure) for both the CO_2 and water cycles. After the first cycle, CO_2 was injected at 1,615 psia to promote better water injection performance. The WAG ratio was 0.5 (reservoir barrels (RB) of water)/(RB of gas), with CO_2 cycles of 2% HCPV.

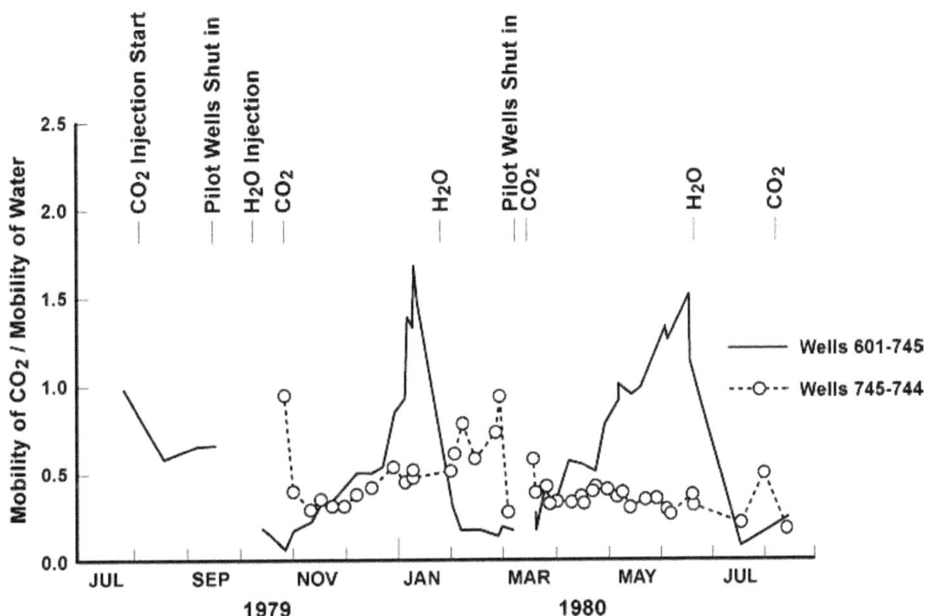

Fig. 2.22—Interwell relative mobilities. The solid line shows changes in the relative mobility between CO_2 and the preCO_2 water injection (expressed as the ratio of the two) between the injector well and the closest observation well. The dashed line shows changes farther away from the injector.

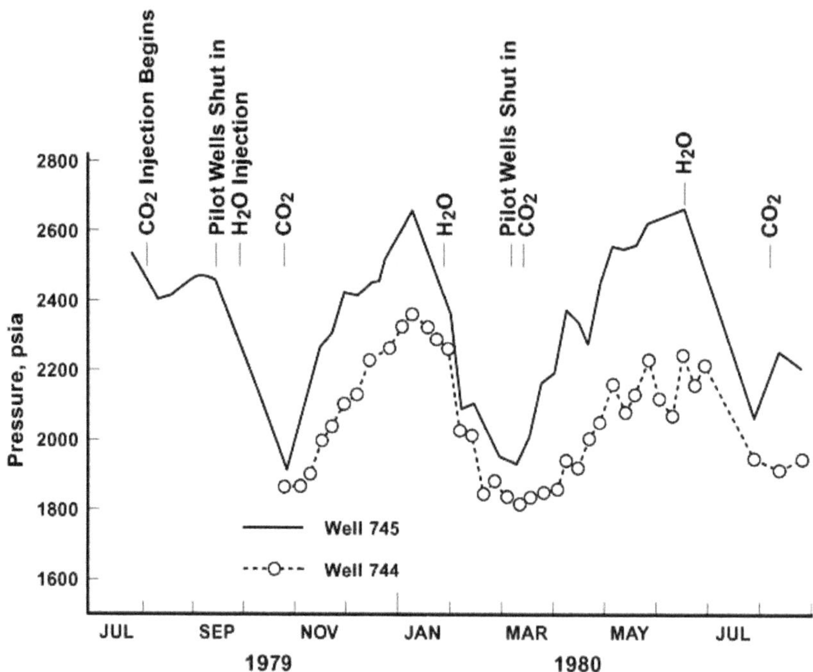

Fig. 2.23—Reservoir pressure measured at the composition observation wells. Unlike relative mobilities, which tend to smooth out farther away from the injection well, reservoir pressure continues to show strong peaks and valleys (from Ref. 21).

During the first water cycle following CO_2 injection, the pilot suffered a 90% loss in water injectivity, and the CO_2 flood was terminated because of its poor performance. Subsequent selective zonal pressure testing at the pilot observation well revealed that the reservoir pressure in the most permeable zone went down 500 psi in the middle of the three-phase region, dropping from 1,615 psia to 1,115 psia. The reservoir pressure in the second highest permeable zone went down 400 psi, dropping from 1,315 psia to 915 psia—below the thermodynamic MMP and the pressure range in which three-phases exist. This suggests that the loss in water injectivity likely occurred because of the development of three hydrocarbon phases.

The loss of injectivity was of greater importance in this case because the three-phase hydrocarbon region occurred close to the injection wells, further reducing mobility. In addition, the loss of injectivity over time reduced the reservoir pressure further below the thermodynamic MMP and decreased the displacement efficiency. Thus, the pilot behaved like an immiscible flood.

Henry and Metcalfe[21] observed reduced mobility in CO_2 corefloods operated in the three-phase hydrocarbon region. The presence of mobile water also can decrease CO_2 mobility by allowing four flowing phases to compete for the pore space.

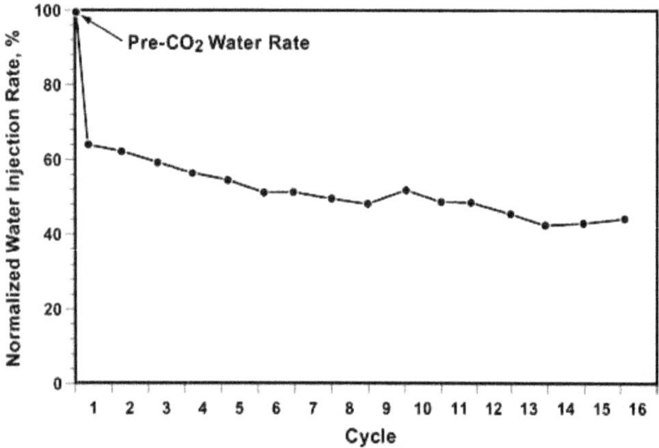

Fig. 2.24—Normalized fieldwide water injection rates for the Slaughter Estate unit CO_2 project. Water injectivity drops dramatically after the first CO_2 cycle (1% HCPV) and continues to decline until about 10% HCPV has been injected.

The poor tertiary oil rate response of this pilot was matched in a reservoir simulator by assuming that the CO_2 was completely immiscible with the oil **(Fig. 2.25).** The solid line in Fig. 2.25 is the simulator prediction of the oil rate for the life of the pilot, while the symbols represent actual oil production rates. Had the pilot not been terminated, the ultimate oil recovery (total slug of 40% HCPV) estimated from the simulator would have been only 2.5% of the original oil in place (OOIP). It is clear that injectivity loss is very important to ultimate recovery.

The injectivity loss was also matched for the immiscible CO_2 displacement by using Stone's II relative permeability correlation[69] for an oil-wet formation. The reservoir description of the pilot was obtained by matching the waterflood performance of all of the pilot and offset wells; this description was not changed to match the CO_2 flood.

Increasing Injection Pressures to Achieve Higher Water Injection Rates. Several WAG injection projects[75,76] have shown that water can be injected at rates as high as the preCO_2 water injection rates if the bottomhole water injection pressure is increased to a level above that of CO_2. **Fig. 2.26** is an example taken from the Wertz Tensleep project.[75] In this project, however, water injection rates were always lower than CO_2 injection rates on an RB basis.

Gorell[77] used a reservoir simulator to demonstrate why water injection rates might not decrease from the preCO_2 levels when CO_2 is injected at a lower bottomhole pressure than the water (while maintaining a reservoir pressure above the thermodynamic MMP). He explained that CO_2 injection pressures could be lower and still allow large CO_2 injection rates because of the increased compressibility of CO_2 (2×10^{-4} psia^{-1} at 2,000 psia and 105°F) compared to that of water (3×10^{-6} psia^{-1} under the same conditions). Carbon dioxide also has a much lower viscosity than does water (Fig. 2.2). A lower bottomhole injection pressure for CO_2 also causes the reservoir pressure to decrease, which causes a larger pressure drop near the well than existed before CO_2 injection. The water injection rate may stay at its preCO_2 level if this larger pressure drop compensates for the decrease in water relative permeability after CO_2 is injected.

Field Examples of Inadvertent Reservoir Fracturing. If injection pressures and rates are not carefully controlled, the bottomhole injection pressure can exceed the formation parting pressure and induce and/or extend a vertical fracture from the wellbore. If a long fracture at an injection well aligns with a production well, areal

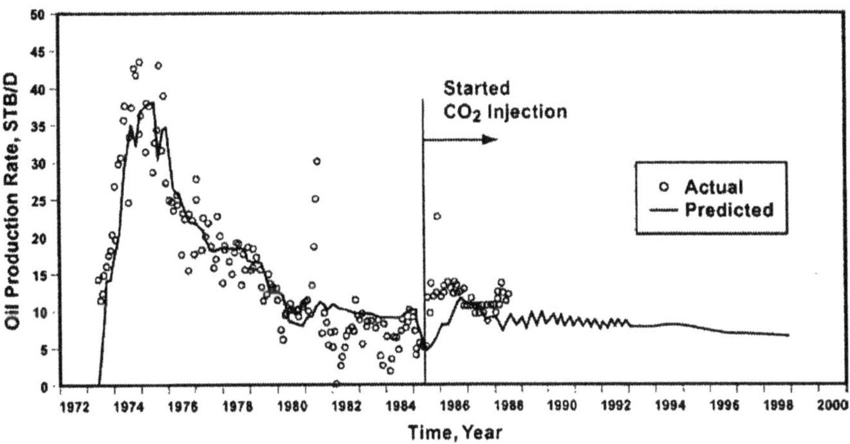

Fig. 2.25—Failure of a CO_2 pilot because of loss of miscibility caused by a large pressure drop associated with a three-phase region. The three-phase hydrocarbon region caused an unusually large loss in water injectivity. Consequently, pressure in much of the pilot dropped below the thermodynamic MMP, and for the most part, the pilot behaved like an immiscible CO_2 flood. Projected ultimate incremental oil recovery was only 2.5% of OOIP.

sweep efficiency can be significantly decreased and the result will be poor oil recovery.

The formation parting pressure depends on both reservoir pressure and mechanical rock properties.[78] If reservoir pressure decreases, fracture parting pressure also decreases. Thus, lowering the injection pressure of CO_2 to achieve higher water injection rates also lowers the reservoir pressure and the formation parting pressure. As a result, inadvertent injection above fracture parting pressure can take place. The authors believe that this phenomenon may have been responsible for the very high water injectivities maintained at the Wertz Tensleep project (Fig. 2.26).

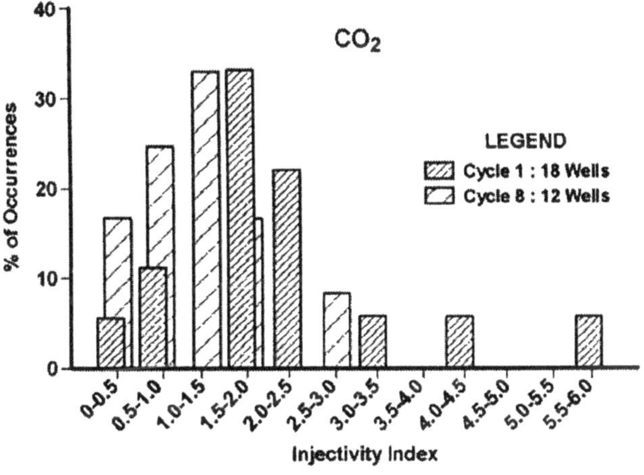

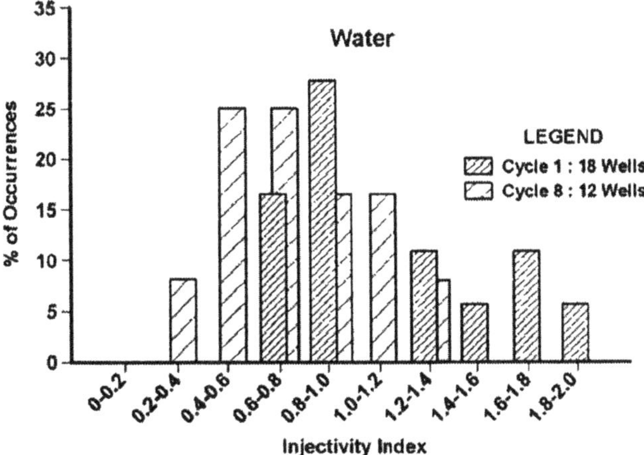

Fig. 2.26—Injectivity index relative to preCO_2 base level. For this injectivity distribution for two WAG cycles at the Wertz Tensleep CO_2 project (after Ref. 75), the histogram shows that CO_2 injectivity was higher than water injectivity. Water injectivity during the CO_2 flood, however, was similar to the preCO_2 water injectivity level because the water cycles used a higher bottomhole injection pressure than did the CO_2 cycles.

Another cause of increased fracture potential is change in the surface injection temperature. If relatively cool water is injected into a higher-temperature formation at high rates, there may be no time for the water to warm up before it reaches the perforated interval. Under these conditions, the injected water could cool the formation in the area of the injection well. Reservoir cooling embrittles the rock, decreases the formation parting pressure, and raises the potential for fracture initiation and growth.[79,80] Such thermal fractures have occurred in the Prudhoe Bay enriched gas WAG project.* Water temperature profiles are useful in determining whether thermal fracturing is an issue.

Fractures that occur during a CO_2 flood can lead to increased water injectivity, but also to reduced oil recovery. In one WAG field test[71], the operator could not inject water at the prior bottomhole pressure. In an attempt to solve the problem, the injection well was shut in and water was injected into the well by a pump truck to displace CO_2 surrounding the well. The pressure from the pump truck injection likely created a fracture by exceeding fracture parting pressure. Roper *et al.*[81,82] matched the performance of this field test with a simulator, using an assumption that a fracture had been created. They showed that the match resulted from two factors: an increase in the effective wellbore radius caused by creating a longer fracture, and a reduction in the reservoir pressure caused by shutting in the well.

Roper *et al.*[81,82] also found that water injection rates following CO_2 will not decrease much when fluid crossflow exists between reservoir layers and all layers are open to injection. Fluid crossflow into another interval allows water to find an alternate route around previously injected CO_2. This phenomenon may also have been a factor in the high water injection rates found at the Wertz Tensleep CO_2 flood.[75]

Effect of CO_2 Solubility on Relative Permeability. The increase in water relative permeability that takes place as water dissolves CO_2 is another factor that increases water injection rates from the low values typically seen after CO_2 injection. Figs. 2.12 and 2.13 show that CO_2 solubility in water can be significant, but as Prieditis *et al.*[71] demonstrated for a west Texas CO_2 project, negligible water is vaporized by CO_2. In the west Texas CO_2 floods, which experienced large losses in water injectivity, operators injected CO_2 and water at the same bottomhole pressures and used very short cycles for CO_2 and water (typically 1 to 2% HCPV for CO_2, and 0.5 to 1.0% HCPV for water). These short cycles make it difficult for the water to dissolve very much CO_2, whereas longer

*Personal Communication with Laurence R. Murray, BP (1999).

water cycles may partially restore the typical water injectivity loss by the end of each water cycle.

Fig. 2.27 illustrates the effect of CO_2 solubility on k_{rw}. The solid lines in this curve are similar to those shown in Fig 2.19, although the linear y-axis better illustrates differences when relative permeability is high. In the illustration, the S_{or} is 30% and the maximum k_{rw} from a water/oil relative permeability test is 0.40. The dotted lines give the k_{rw} after the water has dissolved some of the CO_2-rich phase (solvent) and water saturation has increased. If enough water is injected to dissolve all the solvent and if the S_{or} is zero, the k_{rw} would reach a minimum value of 1.0 at 100% water. To estimate k_{rw} at 100% water saturation, the measured water permeability at 100% water saturation should be divided by the oil permeability at connate water saturation. In the absence of such data, a value of 1.0 should be assumed. West Texas San Andres cores have been found to have water permeability at 100% water saturation that typically is 1.2 times the oil permeability at connate water saturation.

In practical terms, the water saturation in the reservoir after a CO_2 cycle will approach $(1.0-S_{om})$ in the vicinity of the injection well only after injecting a large volume of chase water. The maximum k_{rw} at a water saturation of $(1.0-S_{om})$ should be measured on a core sample in the presence of S_{om} after the water has dissolved CO_2.

Appendix D gives the extrapolated data and explains in detail how to construct the k_{rw} curves in Fig. 2.27. The methodology is useful when simulating the effects of CO_2 solubility in water on CO_2 flood performance. However, when only short cycles and no chase water injection are planned, the effect of CO_2 solubility in water on CO_2 flood performance can be disregarded.

Figs. 2.28 through 2.31 give the results from modified black-oil simulations of a waterflood and CO_2 flood performance with and without the effects of CO_2 solubility in water. The reservoir system in the simulations was one-fourth of a five-spot pattern with seven layers. The cycle sizes used are representative of CO_2 projects that have suffered significant losses in water injectivity. Modified black-oil simulation is discussed in Ref. 64 and Sec. 4.3.1.

These figures show that for short cycles, CO_2 solubility has little effect on well performance, but that during chase water injection, CO_2 solubility does increase water injection and oil production rates. A fully compositional model study by Chang *et al.*,[29] which included CO_2 solubility in water, generated similar results. Large increases in water injection rates also were seen in the Levelland Unit 736 CO_2 pilot during chase water injection. While oil recovery is slightly improved during chase water injection, the increased water rates can increase lifting costs and adversely affect the economics of the final phase of the project.

2.5 Modeling Solvent Relative Permeability under Miscible Conditions

An understanding of CO_2-rich phase relative permeability under miscible conditions, also referred to as the solvent relative permeability, is essential for predicting CO_2 production rates, which itself is prerequisite for designing both the CO_2 processing plant and the surface gathering lines. The CO_2 processing plant usually is the largest part of the total investment in a CO_2 flood. This section presents specific experimental data and a mathematical model used to represent solvent relative permeability in numerical simulators.

2.5.1 Measurements of Solvent Relative Permeability for Mixed-Wet and Oil-Wet Reservoirs. Experiments carried out by three groups—Shyeh-Yung,[70] Prieditis *et al.*,[71] and Wegener and Harpole[72]—have been instrumental in elucidating CO_2-rich phase relative permeability under miscible conditions.

Shyeh-Yung and San Andres Outcrop Cores. Shyeh-Yung[70] measured CO_2-rich phase relative permeability (kr_{CO_2}) in San Andres outcrop cores that were soaked with crude oil until they reached a mixed-wet condition. **Fig. 2.32** shows measurements of kr_{CO_2} (using unsteady-state displacements) at both 1.0 and 6.0 pore volume of CO_2 injected (PVI). At 1.0 PVI, not all the oil bank may have exited the core. Thus, the kr_{CO_2} at 1.0 PVI may be affected by a small amount of mobile oil. The maximum value of CO_2-rich phase endpoint relative permeability is approximately 0.15, which

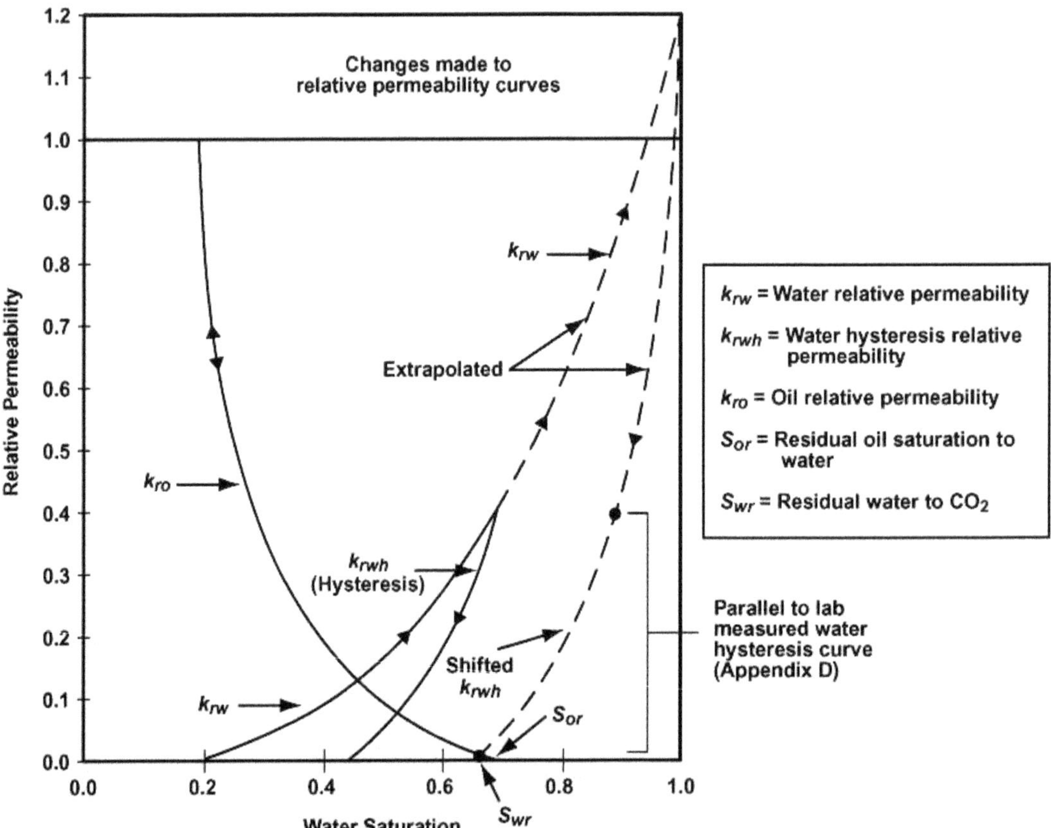

Fig. 2.27—Water relative permeability curves adjusted for CO_2 solubility. The dotted line shows the relative permeability curve for water after it has dissolved CO_2. Appendix D gives the data extrapolated and the methodology used to construct the k_{rw} curves.

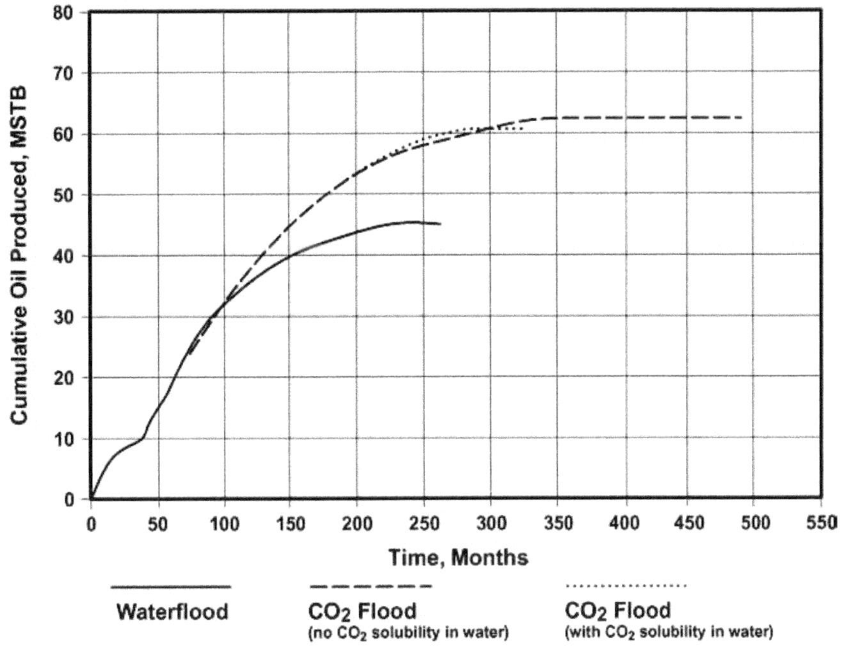

Fig. 2.28—Predicted cumulative oil production with and without CO_2 solubility in water.

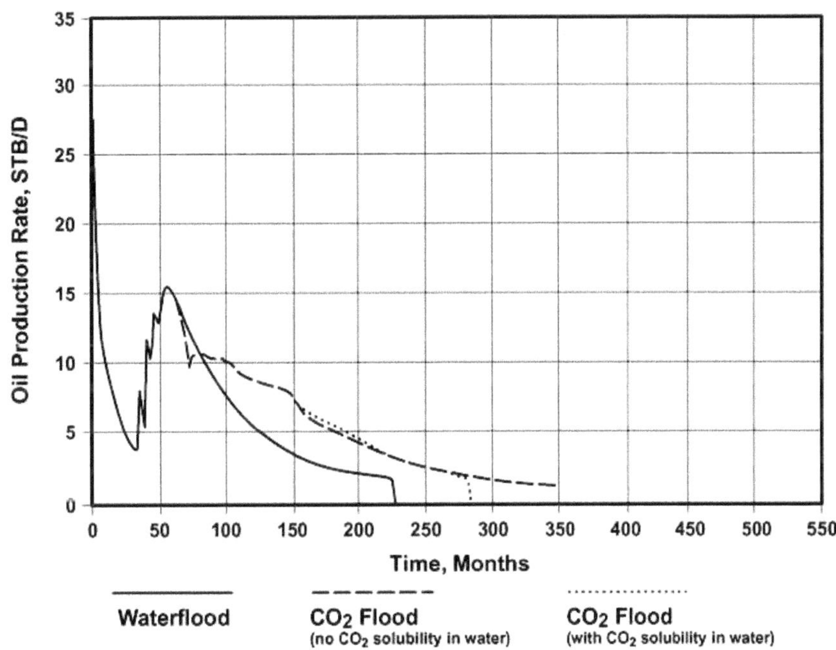

Fig. 2.29—Predicted oil rate with and without CO_2 solubility in water.

is similar to the oil relative permeability after the water saturation has been reduced along the water hysteresis curve (see Fig. 2.19 at about 42% water saturation).

Fig. 2.33 (also from Shyeh-Yung[70]) shows that the highest value for kr_{CO_2} corresponds to the situation where no residual oil is present. As residual oil increases, kr_{CO_2} decreases. For this rock, the maximum kr_{CO_2} in the presence of residual oil was 0.1. Shyeh-Yung and Stadler[83] repeated the San Andres outcrop relative permeability experiments with a hydrocarbon enriched gas and got similar results.

Prieditis and San Andres Reservoir Cores. Prieditis *et al.*[71] also measured CO_2-rich phase and water endpoint relative permeability in San Andres rock, using oil-wet cores from the Mabee field in west Texas. The maximum values found by these experiments were slightly lower than those reported by Shyeh-Yung.[70] With oil permeability at connate water as the base permeability, the maximum kr_{CO_2} ranged from 0.092 to 0.110.

Prieditis found the residual CO_2-rich phase saturation to saturated brine to be about the same as the residual oil saturation to waterflooding. From the high residual CO_2-rich phase saturation in the presence of residual oil saturation to CO_2, a result which Chopra *et al.*[64] found, as well, a reasonable conclusion would be that once CO_2 has been injected, the maximum S_w after a water cycle cannot be as high as it was in the preceding waterflood until the water dissolves the residual CO_2.

Wegener and Harpole, San Andres Formation in South Cowden. Wegener and Harpole[72] also measured CO_2-rich phase endpoint relative permeabilities for an oil-wet reservoir from the San Andres formation of the South Cowden field in west Texas. The measurements showed significant hysteresis in the k_{rw}. Wegener and Harpole found that kr_{CO_2} ranged from 0.022 to 0.083. They also found that the trapped S_{CO_2} was fairly large at 25 to 26%. Because of the hysteresis and large residual CO_2-rich phase saturations, the maximum values of postCO_2 k_{rw} ranged from 0.07 to

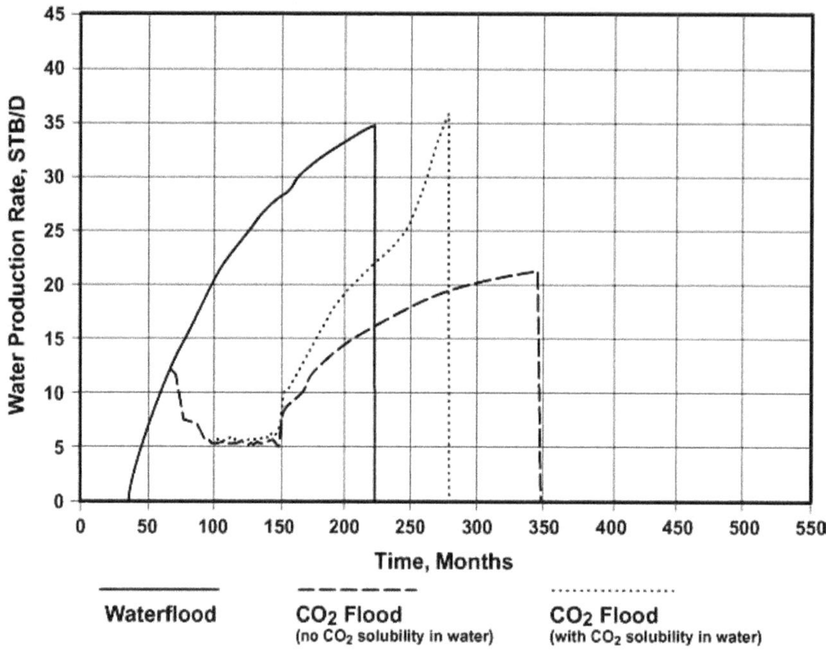

Fig. 2.30—Predicted water production rate with and without CO_2 solubility in water.

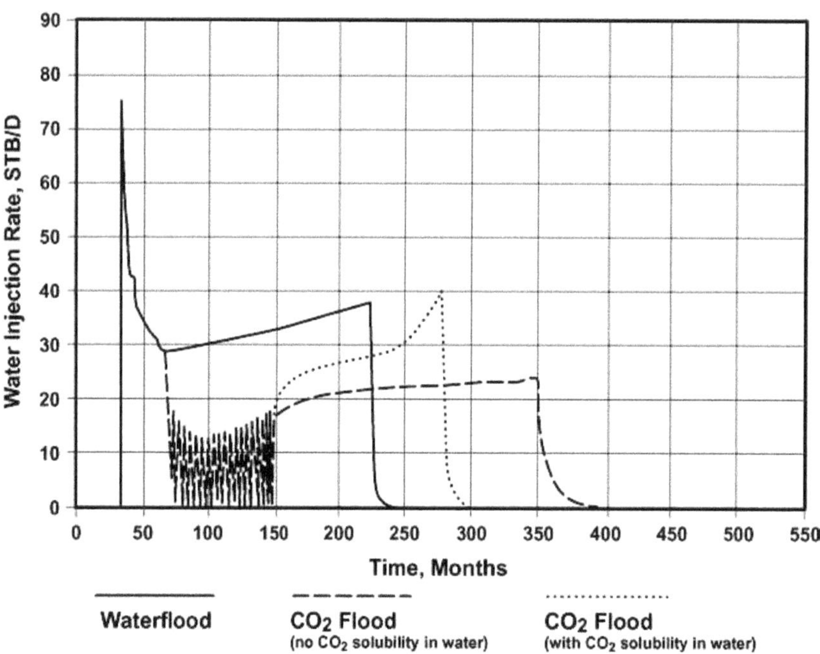

Fig. 2.31—Predicted water injection rate with and without CO_2 solubility in water.

0.15, compared with the preCO_2 values of 0.25 to 0.26. Again, a reasonable conclusion from these results would be that the maximum value of postCO_2 k_{rw} can be significantly lower than preCO_2 values because of the combined effects of water hysteresis and large residual CO_2-rich phase saturations.

2.5.2 Chopra's Mathematical Model for Solvent Relative Permeability. In Sec. 2.3.4, we introduced the concept of solvent (CO_2-rich phase) relative permeability under miscible conditions, which comes from the work of Chopra *et al.*[64] The Chopra model is a modification of Stone's II three-phase relative permeability model.[69]

Chopra *et al.*[64] assume that the rock is preferentially oil-wet. At pressures above the thermodynamic MMP, their model accounts for three phases: the hydrocarbon-rich phase (the residual oil), which is most wetting to the rock; the CO_2-rich phase (the intermediate phase), which is slightly less wetting than oil but more wetting than water; and the water phase, which is the nonwetting phase. As discussed previously, the water relative permeability can increase as water dissolves CO_2 over time.

The solvent relative permeability model of Chopra *et al.*[64] calculates solvent (CO_2-rich phase) relative permeability in a three-phase system, evaluated at S_{om}, as.

$$k_{rCO_2}=[k_{rgo}\,(k_{ro})]+[(k_{rgo}-1)\,(k_{rw})], \quad \text{(2.4)}$$

where k_{rCO_2}=the solvent (CO_2-rich phase) relative permeability; k_{rgo}=the immiscible gas relative permeability with oil at connate water saturation, evaluated at a gas saturation equal to $1.0-S_{om}-S_{wc}$ (after extrapolating the gas relative permeability to a value of 1.0 at a gas saturation equal to $1.0-S_{wc}$); k_{ro}=relative permeability of oil from a water/oil relative permeability test, assumed to be only a function of oil saturation; k_{rw}=relative permeability of water from a water/oil relative permeability test, assumed to be only a

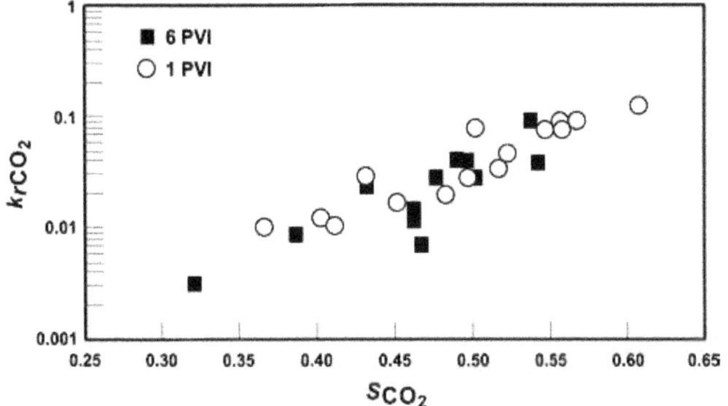

Fig. 2.32—Carbon dioxide-rich phase endpoint relative permeability (k_{rCO_2}) as a function of CO_2-rich phase saturation (S_{CO_2}) in mixed-wet carbonate rock (after Ref. 70). The maximum value of k_{rCO_2} is close to 0.15, about the magnitude of oil relative permeability when k_{rw} has been reduced along the water hysteresis curve (Fig. 2.19). The base permeability is the air permeability, not the oil permeability at connate water saturation.

function of water saturation; S_{om}=residual oil saturation under miscible conditions; and S_w=water saturation.

The solvent relative permeability calculated from Eq. 2.4 has characteristics similar to the CO_2 relative permeability measurements in Refs. 70 through 72. Fig. 2.20 in Sec. 2.3.4 shows examples of calculations of solvent relative permeability using Eq 2.4 for different residual oil saturations to CO_2.

Application of the Chopra Model in a Black-Oil Simulator. The solvent relative permeability model developed by Chopra *et al.*[64] was first used within a modified black-oil model (see Sec. 4.3.1) to predict the performance of CO_2 and enriched-gas WAG pilot projects. The modified black-oil model assumes that above the MMP specified for the model, first-contact miscibility takes place and CO_2 resides in the liquid hydrocarbon phase. Below the specified MMP, CO_2 resides in the gas phase.

In Chopra's solvent relative permeability model, the relative permeability of the liquid hydrocarbon phase is calculated three ways, depending on the concentration of CO_2:

- If the mass fraction of CO_2 in the phase (excluding residual oil) is greater than 90%, the model uses Eq. 2.4 because this phase is the CO_2-rich phase.
- If the CO_2 mass fraction is less than 10%, the model uses oil relative permeability (the limited amount of CO_2 implies that the liquid hydrocarbon phase represents the oil phase).
- For CO_2 mass fractions between 10% and 90%, the model computes relative permeability using linear weighting between CO_2 rich-phase relative permeability (Eq. 2.4) and original oil relative permeability.

When used in a modified black-oil model, Eq. 2.4 has successfully matched every miscible CO_2 pilot and miscible enriched gas pilot to which it has been applied. Furthermore, it has done this without requiring changes to the reservoir description that was determined by matching waterflood performance prior to WAG injection.[64] This is a success record that, to our knowledge, no other model has achieved.

There are, however, two limitations to the Chopra solvent relative permeability model. First, Chopra's model (Eq. 2.4) is correct for displacements *only above the thermodynamic MMP*, where CO_2 acts as a solvent. Below the thermodynamic MMP, CO_2 is generally a vapor phase that is less wetting than water. The CO_2 also will not vaporize the oil significantly. Thus, below the thermodynamic MMP, conventional gas relative permeability should be used, and when three phases exist, published three-phase relative permeability correlations can be used. For example, Stone's II model[69] was used successfully to match the performance of a west Texas oil-wet immiscible CO_2 pilot project (see Sec. 2.4.2).

Second, Chopra's solvent relative permeability model does not address the three-phase hydrocarbon region discussed in Sec. 2.1.8. A model would require four phases to correctly calculate injection and production rates when a three-phase hydrocarbon region existed. To date, there are no reliable laboratory data or models for four-phase flow. Chang[62] tried to account for the three-phase hydrocarbon region by grouping the CO_2-rich vapor and liquid phases together. This approach enabled him to use a three-phase relative permeability correlation when there were actually four phases present (including water). However, Chang's method leads to optimistic rates compared to the results from Henry and Metcalfe[21] (see Sec. 2.1.8).

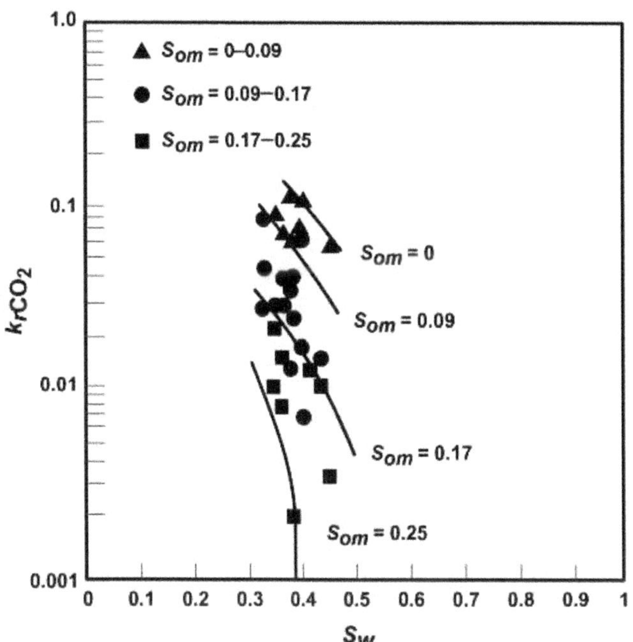

Fig. 2.33—Carbon dioxide-rich phase endpoint relative permeability (k_{rCO_2}) as a function of water saturation (S_w) and final oil saturation (S_{om}) (after Ref. 70). The symbols show k_{rCO_2} measured at different values of S_{om} and S_w; the lines are trend lines. The base permeability is the air permeability, not the oil permeability at connate water saturation.

Verification of Chopra's Model from the Levelland Pilot. The uncertainty in reservoir parameters makes it difficult to verify Chopra's model using field data. Thus, the only legitimate way to validate a CO_2-flood model is to derive and validate the reservoir description by matching simulator results to a waterflood pilot where the physics are better understood; to apply *that validated reservoir description* to a simulator match-up with a CO_2 pilot where the physics are less well understood; and to test different approaches to the physics, *without changing the reservoir description,* until the simulator matches the performance of the actual CO_2 pilot.

Using this approach, Chopra's model was matched to performance data for several pilots[64] and fieldwide[74] projects, all of which were above the thermodynamic MMP. **Figs. 2.34 through 2.36** show the results of this match for the CO_2 flood performance of the Levelland Unit 12-acre CO_2 pilot, in which the residual oil saturation to CO_2 is 15% PV. *The reservoir description of the pilot was derived by history matching waterflood performance[84] and was not changed to match the performance of the CO_2 flood.* In addition, the actual CO_2 and water injection rates were specified in the simulations. The model also matched the actual bottomhole injection pressures.

Chopra's Model With a Fully Compositional Reservoir Simulator. A fully compositional model, as discussed by Hsu *et al.*,[85] is required to address the effects of possibly losing and regaining miscibility. Hsu *et al.* successfully used Chopra's solvent relative permeability model (Eq. 2.4) within a fully compositional reservoir simulator to model CO_2 flood performance at the Wasson Denver Unit CO_2 project in west Texas. Sec. 4.3 offers a more detailed discussion on using a fully compositional reservoir simulator to predict CO_2 flood performance.

2.5.3 Measurements of Solvent Relative Permeability for Water-Wet Reservoirs. For strongly water-wet reservoirs, oil is

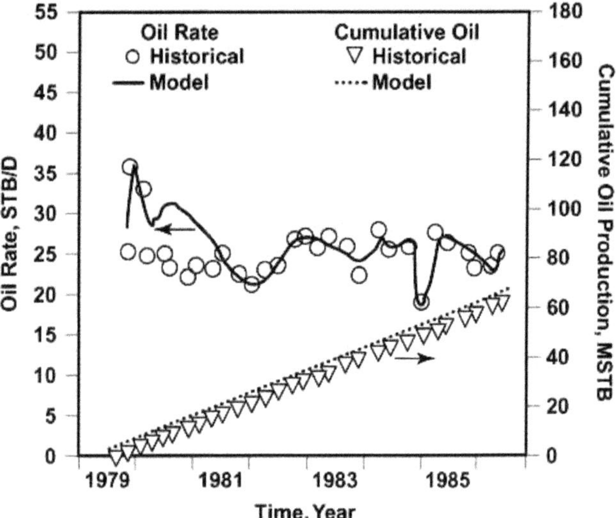

Fig. 2.34—Comparison of actual and predicted oil rates and cumulative oil production using the solvent relative permeability model (after Ref. 64).

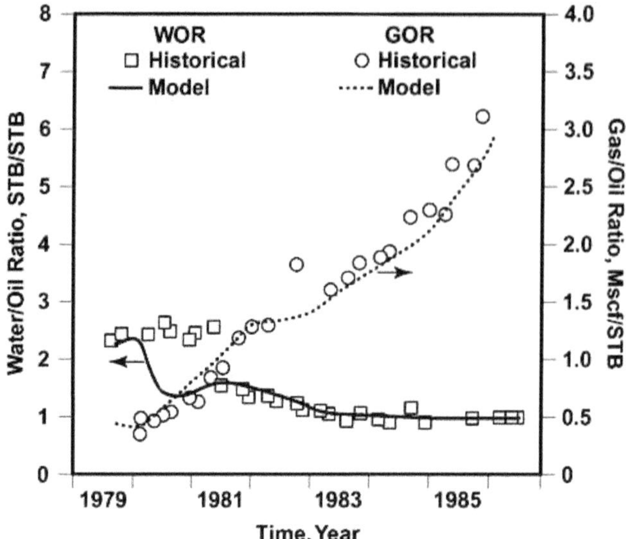

Fig. 2.35—Comparison of actual and predicted water/oil ratios (WOR's) and gas/oil ratios (GOR's) using the solvent relative permeability model (after Ref. 64).

the intermediate wetting phase, and the CO_2-rich phase is the nonwetting phase. Under these conditions, the solvent relative permeability will be reduced by residual oil.

Dria *et al.*[86] measured steady-state relative permeability for solvent, oil, and water for water-wet Baker dolomite with a synthetic oil (decane). They found that the solvent relative permeability was nearly the same as the oil relative permeability at the same fluid saturations. The solvent relative permeability curve, however, was lower than the immiscible gas (nitrogen) relative permeability curve for oil/nitrogen/water experiments. The lower solvent relative permeability indicates that the solvent flows through smaller pores than does the immiscible gas and is therefore more wetting.

Ehrlich *et al.*[87] measured steady-state CO_2/water relative permeability on fresh-state cores from the Little Knife formation in North Dakota. In these experiments, CO_2 and water were injected through a slim tube under miscible conditions. There was no evidence of water blocking because WAG ratios had no effect on residual oil saturations, which ranged from 0.8 to 3.1% (see Sec. 2.2.2). Although such behavior might imply the cores were oil-wet, Ehrlich *et al.*[87] argued that the Little Knife rock changed from water-wet to oil-wet during the CO_2 displacement. (There were, however, indications that the core samples used in their experiments were heterogeneous, which could invalidate the results. The gas/oil relative permeability curves exhibited an immediate decrease in oil relative permeability with the initiation of gas injection, which often indicates the presence of a high-permeability channel. Accurate determination of relative permeability requires that tests be run on samples that are as homogeneous as possible.)

We recommend measuring solvent relative permeability in the presence of mobile water and residual oil. If the rock is water-wet and no measurements are made, then we assume that solvent relative permeability is equal to oil relative permeability.

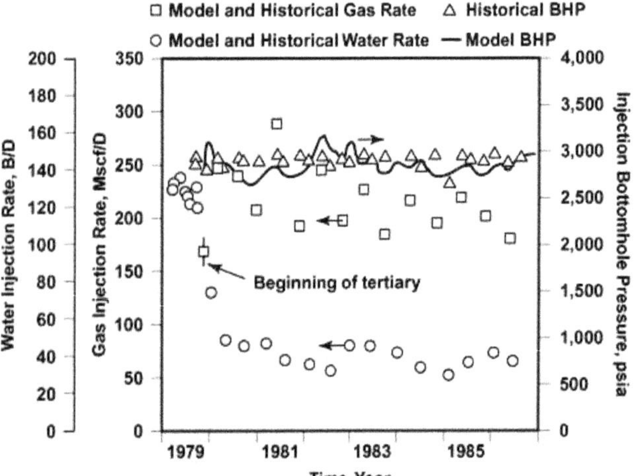

Fig. 2.36—Comparison of actual and predicted water and gas injection rates using the Chopra solvent relative permeability model (after Ref. 64).

2.6 Sweep Aspects of a CO_2 Flood

Large-scale reservoir heterogeneity (large changes in permeability between adjacent geological layers for example) is one of the most important factors determining the success of a CO_2 flood. Heterogeneity affects sweep in a CO_2 flood more than in a waterflood because CO_2 is more mobile than water and can cycle through high-permeability channels more easily.[88] Accurate characterization of reservoir heterogeneity is extremely important for design of large CO_2 floods because the volumetric sweep resulting from reservoir heterogeneity dictates the CO_2 utilization factor (the amount of CO_2 required to recover the desired amount of oil).

2.6.1 Improvement of Sweep by Alternate Water/Gas Injection. Up to this point, water injection in CO_2 floods has been discussed as a means of improving vertical and areal sweep. However, much depends on the way water is used; in some cases, water alone will not provide sufficient control. This section explores various schemes for water injection and other technologies (some more successful than others) that can improve conformance. The use of continuous CO_2 injection in dipping reservoirs also is discussed.

WAG Ratios. Most operators initially inject water and CO_2 at a fixed WAG injection ratio. While water helps improve sweep, a constant WAG ratio does not necessarily provide the optimum sweep or the best economics.[89]

Fig. 2.37 shows the effects of increasing the WAG ratio in the Levelland Unit 1.5-acre CO_2 pilot (736) from 0.22 RB/RB to 0.67 RB/RB. Before the increase, the GOR was about 30 Mscf/STB; after the increase, the GOR decreased to about 10 Mscf/STB, without any decrease in oil rate.

Before increasing the WAG ratio, gas cycled through the high k/ϕ layers with little resistance because the waterflood residual oil from these layers mostly had been produced. Several months before increasing the WAG ratio, gas tracers injected at the pilot injection wells had broken through to the pilot producer within 10 days; by comparison, waterflood tracers injected before the CO_2 flood had required three months to break through to the pilot producer.

In mid-1981, one of the CO_2 cycles accidentally was allowed to exceed its specified injection volume. Although the water cycle was similarly lengthened in an attempt to control breakthrough, gas

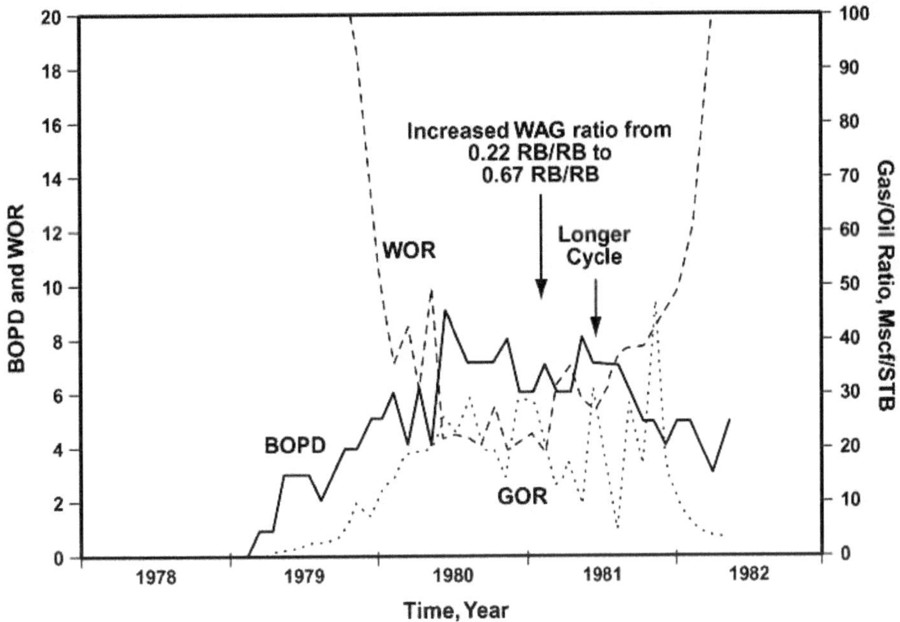

Fig. 2.37—Application of increasing the WAG ratio to improve sweep in the Levelland Unit CO_2 1.5-acre CO_2 pilot (736). In response to an increase in the WAG ratio, the GOR decreased without causing a decrease in the oil production rate.

production started to increase. A further increase in the WAG ratio would have been appropriate to ensure a low gas production rate. Fig. 2.37 shows that in late 1982 the GOR decreased drastically after the pilot went to chase water injection in early 1982.

At other CO_2 floods, the WAG ratio was increased fieldwide to control the gas production rate.[16,74,89–93] **Fig. 2.38** shows the GOR for the Slaughter Estate unit, where plant limitations mandated control of gas production. Larger WAG ratios at these fields also reduced CO_2 purchases and plant operating costs while increasing cash flow, validating the use of water in this way.[16,74,89–93]

The WAG ratio was increased from 0.50 RB/RB to 1.33 RB/RB, which effectively controlled gas production. In some cases, however, the oil rate decreased when the WAG ratio was increased. The gas rate also fell below expected levels. When the WAG ratios were reduced, the oil rates moved back up to their previous levels. These areas did not have high permeability layers where CO_2 could cycle through, so the increased WAG ratios were not needed.

However, increasing water injection is not always effective. Disappointing results from larger WAG ratios occurred in the Joffre Viking CO_2 flood in Canada. Stephenson *et al.*[94] matched the Joffre Viking CO_2 flood performance in a reservoir simulation model by including fluid crossflow within sand bodies. Results showed that the Joffre Viking sand has an average horizontal permeability of 5.0 darcies, and vertical permeability is probably also large. Thus, gravity override probably occurred, and the water failed to flood the same intervals as the CO_2 (see **Fig. 2.39**).

2.6.2 Methods for Improved Conformance When WAG is Not Effective. When WAG has failed to control sweep, other techniques can be used, including surfactant foams, gel polymers, and conventional plugging methods.

Surfactant foams involve injection of a surfactant after a water cycle. The surfactant (dissolved in water) flows preferentially to the highest k/ϕ layers. When CO_2 is injected, it mixes with the surfac-

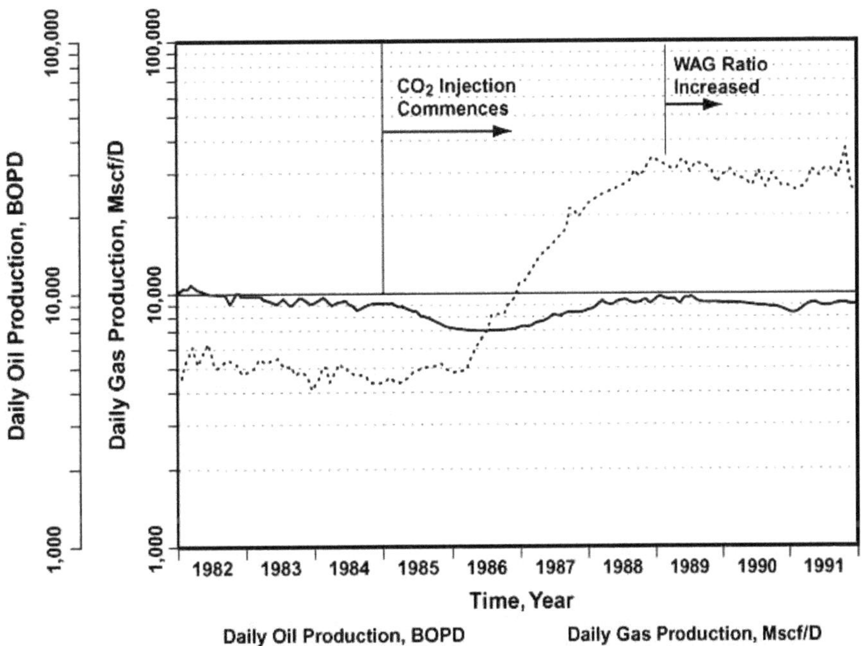

Fig. 2.38—Increasing the WAG ratio to improve sweep in the Slaughter Estate unit CO_2 flood. Increasing the WAG ratio fieldwide reduced CO_2 rates into the plant without decreasing the oil rate.

CO_2 Injection Cycle

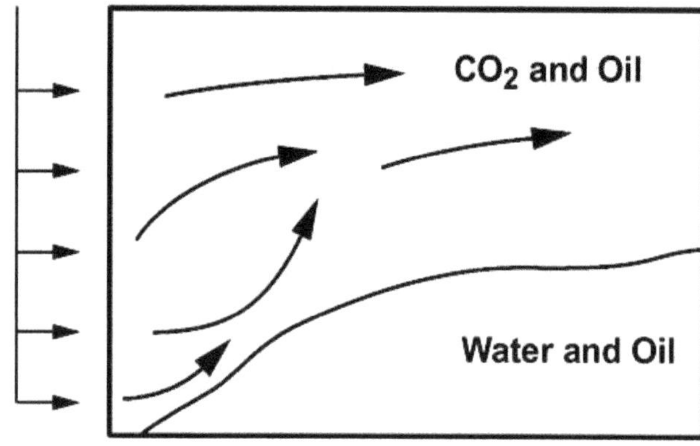

Water Injection Cycle

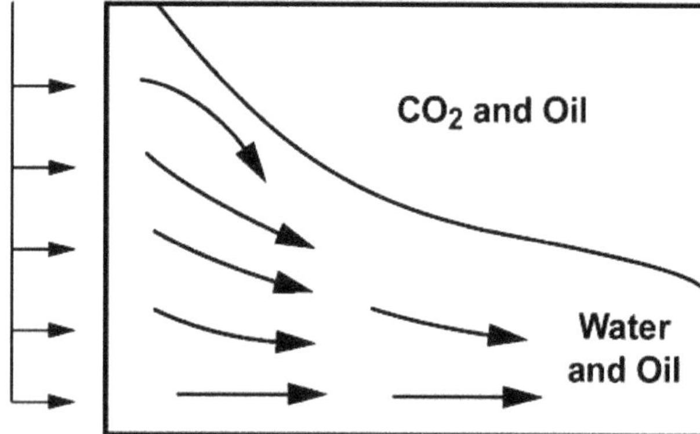

Fig. 2.39—Adverse affect of gravity during WAG injection. In a thick, highly permeable layer during WAG injection, CO_2 may flow to the top of the layer while water gravitates to the bottom, creating a condition that seriously impacts mobility control.

tant in these layers and forms a viscous foam. Theoretically, the foam should move through the reservoir with the water and block the passage of CO_2 by reducing its mobility. However, in field testing, foams have not been found effective in improving CO_2 sweep.[75,94–95]

Phenolic and vinyl polymers have been used at CO_2 WAG injection wells to form an in-situ gel by crosslinking in the presence of chromium ions.[96] The gel must be selectively injected so that it flows to the most permeable zones. After a certain amount of time, the gel can stiffen and block fluid flow through those zones. Gel polymers can remain stable in the presence of CO_2[96] and have been reported to be successful in CO_2 floods.[95,97] Gels are most useful for sealing off thief zones when the well has a good cement job and the gel can be directed toward a given set of perforations within a specific layer of the reservoir. It is important to use an injection profile to make sure the zone can be isolated.

Other inexpensive and simple injection conformance treatments have been used successfully in CO_2 WAG projects. For example, Brokmeyer *et al.*[91] had success with cement squeezes and sand plugbacks in the Lost Soldier Tensleep CO_2 flood in Wyoming. The sand plugbacks were used to stop fluid entry into the lowest pay zone and below the pay. Fullbright *et al.*[98] had similar success in the Rangely Weber Sand Unit CO_2 flood. Unfortunately, these techniques do not perform well when there is significant fluid crossflow between adjacent intervals.

2.6.3 Gravity Effects on CO_2 Flooding. Gravity can aid recovery by improving sweep in dipping reservoirs, often eliminating the need to use water injection for mobility control. Gravity can be an even more important consideration in enriched-gas floods because the density difference between enriched gas and oil is greater than the difference between CO_2 and oil. On the negative side, gravity can counteract the beneficial effects of WAG injection in flat reservoirs.

Effect of Vertical Permeability on Gravity Tongues. When vertical permeability is good and CO_2 is less dense than the oil, CO_2 can override oil in horizontal reservoirs and create "gravity tongues." As CO_2 flows preferentially toward the top portion of a thick, high permeability zone, injected water may flow preferentially toward the lower portion of the zone, as shown in Fig. 2.39. Water normally is injected to control CO_2 mobility and increase the efficiency of the sweep, but in this case, the injection of water may actually reduce vertical sweep.

Good vertical permeability typically occurs when the ratio of average vertical to horizontal permeability is greater than 0.05. The average horizontal permeability for a reservoir should be the arithmetic average of core tests in each layer. The average vertical permeability for the reservoir should be the series average of the core measurements, where the series average is the reciprocal of the sum of the reciprocals for vertical permeability in each layer.[43]

Vertical permeability can be measured in the field using transient pressure tests and repeat formation tests, but minor formation fracturing or cracks in the cement job can render the results of either of these meaningless. For this reason, we recommend using core data to estimate vertical permeability. Care should be taken to identify shale layers in the reservoir because although they have little effect on horizontal permeability, they can substantially reduce the effective vertical permeability. Some reservoirs have naturally low vertical permeability. For example, many (but not all) Permian Basin carbonate reservoirs in west Texas and eastern New Mexico were deposited in a calm, low-energy environment that resulted in layers with low vertical permeability.

Benefits of Gravity in Dipping Reservoirs: The Weeks Island Example. Gravity can be used to help CO_2 flood performance in steeply dipping reservoirs, often eliminating the need for water (WAG) to control mobility. In the Weeks Island "S" Sand Reservoir B, CO_2 was injected continuously in a pilot test, with very good results.[99,100] Reservoir B is made up of highly permeable sandstone (darcy-level permeability) at a 26° dip. Carbon dioxide was injected below the gas/oil contact in one well, while fluid was produced from two wells just above the water/oil contact. Carbon dioxide was injected at rates low enough that gravity could stabilize the CO_2/oil interface.

The injected CO_2 contained 5.0 mol% methane; in addition, the density of the injection stream was adjusted to allow the CO_2 to segregate between the oil in the reservoir and the hydrocarbon gas cap on top. The Weeks Island CO_2 project was operated at a reservoir pressure very close to the thermodynamic MMP.[99,100] A logging observation well showed that the injected CO_2 and oil were segregated, and that gravity helped the CO_2 displace the oil from the top to the bottom.

Ultimately, between 60 and 70% of the oil in place (OIP) at the start of CO_2 injection was recovered in this pilot. The average final oil saturation to CO_2 injection was 1.9% PV in the area completely swept by CO_2. Laboratory corefloods from Perry[99] indicated that gravity drainage would have reduced the residual oil saturation to values considered low even for CO_2 injection under immiscible conditions. **Fig. 2.40** shows that the oil production increased two years after CO_2 injection began in 1978.

Gravity Drainage Can Increase Oil Recovery Even for Immiscible Fluids. Blunt[101] presented data from several investigators that support the feasibility of obtaining low residual oil saturation to immiscible gas under gravity drainage conditions at very low displacement rates. Several investigators have reported residual oil saturations as low as 0.1% PV, although such extremely low oil saturation occurred only in certain water-wet rocks. Blunt explained that the ultralow residual oil saturation is possible if the "spreading coefficient" of the oil is positive and close to zero.

The spreading coefficient is the tendency of the oil to spread between the water coating the rock and the nonwetting gas. The spreading coefficient for water-wet rock is defined as

$$C_{so}=\gamma_{gw}-\gamma_{ow}-\gamma_{go}, \quad \text{(2.5)}$$

where C_{so}=spreading coefficient, dynes/cm^2; γ_{gw}=interfacial tension between the gas and water, dynes/cm^2; γ_{ow}=interfacial tension

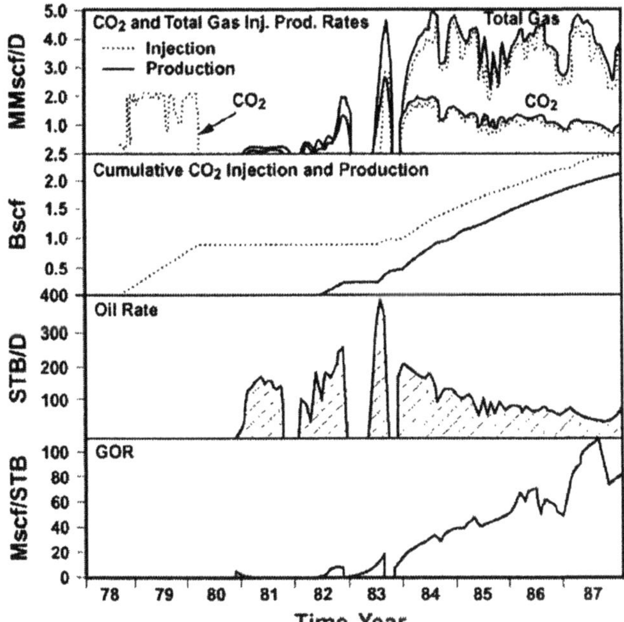

Fig. 2.40—Weeks Island CO_2 pilot performance (after Ref. 99). Gravity effects helped this continuous CO_2 flood achieve good recovery in a dipping reservoir.

between oil and water, dynes/cm^2; and γ_{go}=interfacial tension between the gas and oil, dynes/cm^2.

Corefloods should be performed if the spreading coefficient of the oil in equilibrium with water and gas is positive but small. Gravity drainage experiments, which take about two weeks to complete,[101] show whether good recovery is possible even under immiscible conditions. The reduction in residual oil saturation measured by the experiments (along with the increased oil relative permeability at low oil saturations) should be included in the gas/oil relative permeability model. If good recovery is obtained from the experiments, reservoir simulations should be run to determine what vertical displacement rate should be used to reach economic oil rates.

Fractional flow theory shows that a more steeply dipping reservoir and a low injection rate together provide the highest displacement efficiency for top-down gas injection.[102] However, the improvement in oil recovery must be weighed against the economic performance over the longer production time that may be necessary for this type of recovery.

Mathematical Representation of a Gravity-Stable Displacement Rate. The critical decision in a gravity-assisted flood is which injection rate to use—the optimal rate is inextricably bound to economics. A low rate will yield gravity-stable conditions and give the highest displacement efficiency, but the oil production rate might be too low to be economical. A higher displacement rate will be more likely to bypass oil but may be the only way to achieve economic oil production rates.

Several mathematical expressions can be used to predict the maximum displacement velocity for a first-contact miscible solvent that can displace oil without gravity override. For example, Hill[103] and Deitz[104] developed the following equation for the critical gas velocity, ignoring any mixing between the solvent and the oil:

$$v_c=\frac{0.0000439\,(\rho_o-\rho_s)}{(\mu_o/k_o-\mu_s/k_s)}\sin(\theta), \quad \text{(2.6)}$$

where v_c=critical velocity in ft/D; ρ_o=oil density in lbm/ft^3; ρ_s=solvent density in lbm/ft^3; μ_o=oil phase viscosity in cp; μ_s=solvent phase viscosity in cp; k_o=effective oil permeability in md; k_s=effective solvent permeability in md; and θ=dip angle in degrees.

Dumore[105] offered similar expression for a maximum velocity for a stable gas displacement.

Simulator Approaches. The way to arrive at a realistic and economical oil rate for a gravity-assisted flood is to use a reservoir simulator that incorporates the dip angle. A fully compositional reservoir simulator will give the most accurate characterization of the densities. It can also include the effect that CO_2 mixing with the gases in the gas cap can have on development of miscibility. Because of this, simulation will give the most accurate prediction of gravity override effects.

Simulations of gravity-assisted CO_2 floods must account for gas coning when determining rates, because dip conditions allow gas to be preferentially drawn into the producing well. An accurate determination of vertical permeability is important for assessing how much coning will occur. Because of the difficulty in accurately determining vertical permeability, field experience can be helpful in calibrating simulator predictions. Horizontal production wells also might be useful in minimizing coning problems during gravity-assisted CO_2 floods.

Nomenclature

C_{so} = spreading coefficient, m/t^2, dynes/cm^2
D_e = effective molecular diffusion for porous media, L^2/t, ft^2/sec
k = permeability, L^2, md
K_l = longitudinal dispersion coefficient, L^2/t, ft^2/D
k_o = effective oil permeability, L^2, md
k_{rCO_2} = solvent (CO_2-rich phase) relative permeability, dimensionless
k_{rgo} = immiscible gas relative permeability measured with oil at connate water, dimensionless
k_{ro} = oil relative permeability, dimensionless
k_{rw} = water relative permeability, dimensionless
k_{rwh} = water relative permeability for decreasing water saturation, or water relative permeability hysteresis, dimensionless
k_s = effective solvent permeability, L^2, md
L = length, L, ft
N_{Pe} = Peclet number, dimensionless
p_m = mixture MMP, m/Lt^2, psia
p_{CO_2} = pure CO_2 MMP, m/Lt^2, psia
S_{CO_2} = saturation of CO_2-rich phase, dimensionless, fraction
S_{om} = residual oil saturation to CO_2 under miscible conditions, dimensionless, fraction
S_{or} = residual oil saturation to water, dimensionless, fraction
S_w = water saturation, dimensionless, fraction
S_{wc} = connate water saturation, dimensionless, fraction
S_{wr} = residual water saturation to oil (water saturation at which k_{rwh}=zero), dimensionless
T_{cm} = mole-fraction-weighted critical temperature of a chemical component, T, K
v = superficial velocity of flow, L/t, ft/D
v_c = critical velocity, L/t, ft/D
α_l = longitudinal dispersivity, L, ft
ϕ = porosity, dimensionless, fraction
γ_{go} = gas/oil interfacial tension, m/t^2, dynes/cm^2
γ_{gw} = gas/water interfacial tension, m/t^2, dynes/cm^2
γ_{ow} = oil/water interfacial tension, m/t^2, dynes/cm^2
μ_o = oil phase viscosity, m/Lt, cp
μ_s = solvent phase viscosity, m/Lt, cp
θ = dip angle, dimensionless, degrees
ρ_o = oil density, m/L^3, lbm/ft^3
ρ_s = solvent density, m/L^3, lbm/ft^3

References

1. Stalkup, F.I. Jr.: *Miscible Displacement*, Monograph Series, SPE, Richardson, Texas (1983) 8.
2. Klins, M.A.: *Carbon Dioxide Flooding*, International Human Resources Development Corp., Boston, Massachusetts (1984).
3. Lake, L.W.: *Enhanced Oil Recovery*, Prentice Hall, Englewood Cliffs, New Jersey (1989).
4. Holm, L.W. and Josendal, V.A.: "Effect of Oil Composition on Miscible-Type Displacement by Carbon Dioxide," *SPEJ* (February 1982) 8798.

5. Rathmell, J.J., Stalkup, F.I., and Hassinger, R.C.: "A Laboratory Investigation of Miscible Displacement by Carbon Dioxide," paper SPE 3483 presented at the 1971 SPE Annual Meeting, New Orleans, 3–6 October.
6. Holm, L.W. and Josendal, V.A.: "Mechanisms of Oil Displacement by Carbon Dioxide," *JPT* (December 1974) 1427; *Trans.*, AIME, **257.**
7. Shelton, J.L. and Yarborough, L: "Multiple Phase Behavior in Porous Media During CO_2 or Rich-Gas Flooding," *JPT* (September 1977) 1171.
8. Metcalfe, R.S. and Yarborough, L.: "Effect of Phase Equilibria on the CO_2 Displacement Mechanism," *SPEJ* (August 1979) 242; *Trans.*, AIME, **267.**
9. Orr, F.M. Jr., Yu, A.D., and Lein, C.L.: "Phase Behavior of CO_2 and Crude Oil at Low Temperature Reservoirs," *SPEJ* (August 1981) 480.
10. Yellig, W.F. and Metcalfe, R.S.: "Determination and Prediction of CO_2 Minimum Miscibility Pressures," *JPT* (January 1980) 160.
11. Zick, A.A.: "A Combined Condensing/Vaporizing Mechanism in the Displacement of Oil by Enriched Gases," paper SPE 15493 presented at the 1986 SPE Annual Technical Conference and Exhibition, New Orleans, 5–8 October.
12. Mungan, N.: "Carbon Dioxide-Flooding—Fundamentals," *J. Cdn. Pet. Tech.* (January–March 1981) 87.
13. Cronquist, C.: "Carbon Dioxide Dynamic Miscibility with Light Reservoir Oils," *Proc.*, Fourth Annual U.S. DOE Symposium, Tulsa, 28–30 August, 1978, Vol. 1b-Oil, C5.
14. Johns, R.T. and Orr, F.M. Jr.: "Miscible Gas Displacements of Multicomponent Oils," *SPEJ* (March 1996) 39.
15. Wang, Y. and Orr, F.M. Jr.: "Calculation of Minimum Miscibility Pressure," paper SPE 39683 presented at the 1998 SPE/DOE Improved Oil Recovery Symposium, Tulsa, 19–22 April.
16. Harpole, K.J. and Hallenbeck, L.D.: "East Vacuum Grayburg San Andres Unit CO_2 Flood Ten Year Performance Review: Evolution of a Reservoir Management Strategy and Results of WAG Optimization," paper SPE 36710 presented at the 1996 SPE Annual Technical Conference and Exhibition, Denver, 6–9 October.
17. Christian, L.D. *et al.*: "Planning a Tertiary Oil-Recovery Project for Jay/LEC Fields Unit," *JPT* (August 1981) 1535.
18. Metcalfe, R.S.: "Effects of Impurities on Minimum Miscibility Pressures and Minimum Enrichment Levels for CO_2 and Rich-Gas Displacements," *SPEJ* (April 1982) 219.
19. Sebastian, H.M., Wenger. R.S., and Renner, T.A.: "Correlation of Minimum Miscibility Pressure for Impure CO_2 Streams," *JPT* (November 1985) 2076.
20. Simandoux, P. *et al.*: "State of Art of EOR by CO_2 Flooding," paper presented at the 1983 OAPEC Seminar on EOR, DOHA (November 1983); *Revue Institut Français du Pétrole* (1984) **39,** 403.
21. Henry, R.L. and Metcalfe, R.S.: "Multiple-Phase Generation During Carbon Dioxide Flooding," *SPEJ* (August 1983) 595.
22. Holm, L.W. and O'Brien, L.J.: "Carbon Dioxide Test at the Mead-Strawn Field," *JPT* (April 1971) 431; *Miscible Process II,* Reprint Series, SPE, Richardson, Texas (1985) **18,** 636–647.
23. Bilhartz, H.L. Jr. and Charlson, G.S.: "Coring for In-Situ Saturations in the Willard Unit CO_2 Flood Mini-Test," paper SPE 7050 presented at the 1978 SPE Symposium on Improved Methods of Oil Recovery, Tulsa, 16–19 April.
24. Pollack, N.R., Enick, R.M., and Klara, S.M.: "Effect of CO_2 Solubility in Brine on Compositional Simulation of CO_2 Floods," *SPERE* (May 1992) 253.
25. Crawford, H.R. *et al.*: "Carbon Dioxide—A Multipurpose Additive for Effective Well Stimulation," *JPT* (March 1963) 237.
26. Holm, L.W.: "CO_2 Slug and Carbonated Water Oil Recovery Processes," *Prod. Monthly* (September 1963) 6.
27. Martin, J.W.: "Additional Oil Production Through Flooding with Carbonated Water," *Prod. Monthly* (July 1951) 18.
28. Johnson, W.E., Macfarlane, R.M., and Breston, J.N.: "Changes in Physical Properties of Bradford Crude Oil When Contacted with CO_2 and Carbonated Water," *Prod. Monthly* (November 1952) 16.
29. Chang, Y.B., Coats, B.K., and Nolen, J.S.: "A Compositional Model for CO_2 Floods Including CO_2 Solubility in Water, " paper SPE 35164 presented at the 1996 Permian Basin Oil and Gas Recovery Conference, Midland, Texas, 27–29 March.
30. Gardner, J.W., Orr, F.M., and Patel, P.D.: "The Effect of Phase Behavior on CO_2-Flood Displacement Efficiency," *JPT* (November 1981) 2067.
31. Perkins T.K. and Johnston, O.C.: "A Review of Diffusion and Dispersion in Porous Media," *SPEJ* (March 1963) 70; *Trans.*, AIME, **228.**
32. van der Poel, C: "Effect of Lateral Diffusivity on Miscible Displacement In Horizontal Reservoirs," *SPEJ* (December 1962) 317; *Trans.*, AIME, **225.**
33. Negahban, S., Shiralkar, G.S., and Gupta, S.P.: "Simulation of the Effects of Mixing in Gasdrive Core Tests of Reservoir Fluids," *SPERE* (August 1990) 402.
34. Yellig, W.F.: "Carbon Dioxide Displacement of a West Texas Reservoir Oil," *SPEJ* (December 1982) 805.
35. Negahban, S. and Kremesec, V.J.: "Development and Validation of Equation-of-State Fluid Descriptions for CO_2/Reservoir-Oil Systems," *SPERE* (August 1992) 363.
36. Pickens, J.F. and Grisak, G.E.: "Scale-Dependent Dispersion in a Stratified Granular Aquifer," *Water Resour. Res.* (1981) **17**, No. 4, 1191.
37. Deans, H.A. and Shallenberger, L.K.: "Single-Well Chemical Tracer Method to Measure Connate Water Saturation," paper SPE 4755 presented at the 1974 SPE Improved Oil Recovery Symposium, Tulsa, 22–24 April.
38. Sheely, C.Q.: "Description of Field Tests To Determine Residual Oil Saturation by Single-Well Tracer Method," *JPT* (February 1978) 194.
39. Stalkup, F.: "Predicting the Effect of Continued Gas Enrichment Above the MME on Oil Recovery in Enriched Hydrocarbon Gas Floods," paper SPE 48949 prepared for presentation at the 1998 SPE Annual Technical Conference and Exhibition, New Orleans, 27–30 September.
40. Warren, J.E., and Skiba, F.F.: "Macroscopic Dispersion," *SPEJ* (September 1964) 215; *Trans.*, AIME, **231.**
41. Arya, A. *et al.*: "Dispersion and Reservoir Heterogeneity," *SPERE* (February 1988) 1390.
42. Kelkar, B.G. and Gupta, S.P.: "The Effects of Small Scale Heterogeneities on the Effective Dispersivity of Porous Media," paper SPE 17339 presented at the 1988 SPE/DOE Enhanced Oil Recovery Symposium, Tulsa, 17–20 April.
43. Craig, F.F. Jr.: *The Reservoir Engineering Aspects of Waterflooding,* Monograph Series, SPE, Richardson, Texas (1971) **3.**
44. Dykstra, H. and Parsons, R.L.: "The Prediction of Oil Recovery by Water Flood," *Secondary Recovery of Oil in the United States,* second edition, API, Washington, DC (1950) 160–174.
45. Kittridge, M.G. *et al.*: "Outcrop/Subsurface Comparisons of Heterogeneity in the San Andres Formation," *SPEFE* (September 1990) 233.
46. Stalkup, F.I. and Crane, S.D.: "Reservoir Description Detail Required To Predict Solvent and Water Saturation at an Observation Well," paper SPE 22897 presented at the 1991 SPE Annual Technical Conference and Exhibition, Dallas, 6–9 October.
47. Perez, G. and Chopra, A.K.: "Evaluation of Frontal Models to Describe Reservoir Heterogeneity and Performance," paper SPE 22694 presented at the 1991 SPE Annual Technical Conference and Exhibition, Dallas, 6–9 October
48. Spence, A.P. Jr. and Watkins, R.W.: "The Effect of Microscopic Heterogeneity on Miscible Flood Residual Oil Saturation," paper SPE 9229 presented at the 1980 SPE Annual Technical Conference and Exhibition, Dallas, 21–23 September.
49. Batycky, J.P. *et al.*: "Miscible and Immiscible Displacements Studies in Carbonate Reservoir Cores," *J. Cdn. Pet. Tech.* (January–March 1981) 104.
50. Dai, K.K. and Orr, F.M. Jr.: "Prediction of CO_2 Flood Performance: Interaction of Phase Behavior With Microscopic Pore Structure Heterogeneity," *SPERE* (November 1987) 531.
51. Ramirez, W.F., Shuler, P.J., and Friedman, F.: "Convection, Dispersion, and Adsorption of Surfactants in Porous Media," *SPEJ* (December 1980) 430.
52. Tiffin, D.L. and Yellig, W.F.: "Effects of Mobile Water on Multiple-Contact Miscible Gas Displacements," *SPEJ* (June 1983) 447; *Miscible Process II,* Reprint Series, SPE, Richardson, Texas (1985) **18,** 340–348.
53. Salathiel, R.A.: "Oil Recovery by Surface Film Drainage In Mixed-Wettability Rocks," *JPT* (October 1973) 1216.

54. Stern, D.: "Mechanisms of Miscible Oil Recovery: Effects of Pore-Level Fluid Distribution," paper SPE 22652 presented at the 1991 SPE Annual Technical Conference and Exhibition, Dallas, 6–9 October.
55. Shelton, J.L. and Schneider, F.N.: "The Effects of Water Injection on Miscible Flooding Methods Using Hydrocarbons and Carbon Dioxide," *SPEJ* (June 1975) 217.
56. Spence, A.P. and Ostrander, J.F.: "Comparison of WAG and Continuous Enriched-Gas Injection as Miscible Processes in Sadlerochit Core," paper SPE 11962 presented at the 1983 SPE Annual Technical Conference and Exhibition, San Francisco, 5–8 October.
57. Magruder, J.B., Stiles, L.H., and Yelverton, T.D.: "Review of the Means San Andres Unit CO_2 Tertiary Project," *JPT* (May 1990) 638.
58. Tiffin, D.L., Sebastian, H.M., and Bergman, D.F.: "Displacement Mechanism and Water Shielding Phenomena for a Rich Gas/Crude Oil System," paper SPE 17374 presented at the 1988 SPE/DOE Enhanced Oil Recovery Symposium, Tulsa, 17–20 April.
59. Pittaway, K.R. and Runyan, E.E. "The Ford Geraldine Unit CO_2 Flood: Operating History," *SPEPE* (August 1990) 3330.
60. Thrash, J.C.: "Twofreds Field: a Tertiary Oil Recovery Project," paper SPE 8382 presented at the 1979 SPE Annual Technical Conference and Exhibition, Las Vegas, Nevada, 23–26 September.
61. Hadlow, R.E.: "Update of Industry Experience with CO_2 Injection," paper SPE 24928 presented at the 1992 SPE Annual Technical Conference and Exhibition, Washington, DC, 4–7 October.
62. Chang, Y.: "Development and Application of an Equation of State Compositional Simulator," PhD dissertation, U. of Texas, Austin, Texas (1990).
63. Chang, Y. *et al.*: "Carbon Dioxide Flow Patterns Under Multiphase Flow, Heterogeneous Field Scale Conditions," paper SPE 22654 presented at the 1991 SPE Annual Technical Conference and Exhibition, Dallas, 6–9 October.
64. Chopra, A.K., Stein, M.H., and Dismuke, C.T.: "Prediction of Performance of Miscible-Gas Pilots," *JPT* (December 1990) 1564.
65. Patel, P.D., Christman, P.G., and Gardner, J.W.: "Investigation of Unexpectedly Low Field-Observed Fluid Mobilities During Some CO_2 Tertiary Floods," *SPERE* (November 1987) 507.
66. Schneider F.N. and Owens, W.W.: "Relative Permeability Studies of Gas-Water Flow Following Solvent Injection in Carbonate Rocks," *SPEJ* (February 1976) 23; *Trans.*, AIME, **261.**
67. Land, C.S.: "Calculation of Imbibition Relative Permeability for Two- and Three-Phase Flow From Rock Properties," *SPEJ* (June 1968) 149.
68. Carlson, F.M.: "Simulation of Relative Permeability Hysteresis to the Nonwetting Phase," paper SPE 10157 presented at the 1981 SPE Annual Technical Conference and Exhibition, San Antonio, Texas, 5–7 October.
69. Stone, H.L.: "Estimation of Three-Phase Relative Permeability and Residual Oil Data," *J. Cdn. Pet. Tech.* (October–December 1973) 53.
70. Shyeh-Yung, J-G.J.: "Mechanisms of Miscible Oil Recovery: Effects of Pressure on Miscible and Near-Miscible Displacements of Oil by Carbon Dioxide," paper SPE 22651 presented at the 1991 SPE Annual Technical Conference and Exhibition, Dallas, 6–9 October.
71. Prieditis, J., Wolle, C.R., and Notz, P.K.: "A Laboratory and Field Injectivity Study: CO_2 WAG in the San Andres Formation of West Texas," paper SPE 22653 presented at the 1991 SPE Annual Technical Conference and Exhibition, Dallas, 6–9 October.
72. Wegener, D.C. and Harpole, K.J.: "Determination of Relative Permeability and Trapped Gas Saturation for Predictions of WAG Performance in the South Cowden CO_2 Flood," paper SPE 35429 presented at the 1996 SPE/DOE Symposium on Improved Oil Recovery, Tulsa, 21–24 April.
73. Henry, R.L. *et al.*: "Utilization of Composition Observation Wells in a West Texas CO_2 Pilot Flood," paper SPE 9786 presented at the 1981 SPE/DOE Symposium on Enhanced Oil Recovery, Tulsa, 5–8 April.
74. Stein, M.H. *et al.*: "Slaughter Estate Unit CO_2 Flood: Comparison Between Pilot and Field-Scale Performance," *JPT* (September. 1992) 1026.
75. Kleinstelber, S.W.: "The Wertz Tensleep CO_2 Flood: Design and Initial Performance," *JPT* (May 1990) 630.
76. Hervey, J.R. and Iakovakis, A.C.: "Performance Review of a Miscible CO_2 Tertiary Project: Rangely Weber Sand Unit, Colorado," *SPERE* (May 1991) 163.
77. Gorell, S.B.: "Implications of Water-Alternate-Gas Injection, for Profile Control and Injectivity," paper SPE 20210 presented at the 1990 SPE/DOE Enhanced Oil Recovery Symposium, Tulsa, 22–25 April.
78. Hubbert, M.K. and Willis, D.G.: "Mechanics of Hydraulic Fracturing," *Trans.*, AIME (1957) **210,** 153.
79. Ali, N. *et al.*: "Injection Above-Parting-Pressure Waterflood Pilot, Valhall Field, Norway," paper SPE 22893 presented at the 1991 SPE Annual Technical Conference and Exhibition, Dallas, 6–9 October.
80. Stevens, D.G., Murray, L.R., and Shah, P.C.: "Predicting Multiple Thermal Fractures in Horizontal Injection Wells; Coupling of a Wellbore and a Reservoir Simulator," paper SPE 59354 presented at the 2000 SPE/DOE Improved Oil Recovery Symposium, Tulsa, 3–5 April.
81. Roper, M.K. Jr. *et al.*: "Interpretation of a CO_2 WAG Injectivity Test in the San Andres Formation Using a Compositional Simulator," paper SPE 24163 presented at the 1992 SPE/DOE Symposium on Enhanced Oil Recovery, Tulsa, 22–24 April.
82. Roper, M.K. Jr., Pope, G.A., and Sepehrnoori, K.: "Analysis of Tertiary Injectivity of Carbon Dioxide," paper SPE 23974 presented at the 1992 SPE Permian Basin Oil and Gas Recovery Conference, Midland, Texas, 18–20 March.
83. Shyeh-Yung, J. J. and Stadler, M.P.: "Effect of Injectant Composition and Pressure on Displacement of Oil by Enriched Hydrocarbon Gases," *SPERE* (May 1995) 109.
84. Chopra, A.K., Stein, M.H., and Ader, J.C.: "Development of Reservoir Descriptions To Aid in Design of EOR Projects," *SPERE* (May 1989) 143; *Trans.*, AIME, **261.**
85. Hsu, C-F., Morell, J.I., and Falls, A.H.: "Field-Scale CO_2-Flood Simulations and Their Impact on the Performance of the Wasson Denver Unit," *SPERE* (February 1997) 4.
86. Dria, D.E., Pope, G.A., and Sepehrnoori, K.: "Three-Phase Gas/Oil/Brine Relative Permeabilities Measured Under CO_2 Flooding Conditions," *SPERE* (May 1993) 143.
87. Ehrlich, R., Tracht, J.H., and Kaye, S.E.: "Laboratory and Field Study of the Effect of Mobile Water on CO_2 Flood Residual Oil Saturation," paper SPE 11957 presented at the 1983 SPE Annual Technical Conference and Exhibition, San Francisco, 5–8 October.
88. Chopra, A.K.: "Reservoir Descriptions Via Pulse Testing: A Technology Evaluation," paper SPE 17568 presented at the 1988 SPE International Meeting on Petroleum Engineering, Tianjin, China, 1–4 November.
89. Pariani, G.J. *et al.*: "An Approach To Optimize Economics in a West Texas CO_2 Flood," *JPT* (September 1992) 984.
90. Tanner, C.S. *et al.*: "Production Performance of the Wasson Denver Unit CO_2 Flood," paper SPE 24156 presented at the 1992 SPE/DOE Symposium on Enhanced Oil Recovery, Tulsa, 22–24 April.
91. Brokmeyer, R., Borling, D.C., and Pierson, W.: "Lost Soldier Tensleep CO_2 Tertiary Project, Performance Case History; Balroll, Wyoming," paper SPE 35191 presented at the 1996 SPE Permian Basin Oil and Gas Recovery Conference, Midland, Texas, 27–29 March.
92. Masoner, L.O., Abidi, H.R., and Hild, G.P.: "Diagnosing CO_2 Flood Performance Using Actual Performance Data," paper SPE 35363 presented at the 1996 SPE/DOE Symposium on Improved Oil Recovery, Tulsa, 21–24 April.
93. Sharma, A.K. and Clements, L.E.: "From a Simulator to Field Management: Optimum WAG Application in a West Texas CO_2 Flood—A Case History," paper SPE 36711 presented at the 1996 SPE Annual Conference and Exhibition, Denver, 6–9 October.
94. Stephenson, D.J., Graham, A.G., and Luhning, R.W.: "Mobility Control Experience in the Joffre Viking Miscible CO_2 Flood," *SPERE* (August 1993) 183; *Trans.*, AIME, **295.**
95. Borlin, D.C.: "Injection Conformance Control Case Histories Using Gels at the Wertz Field CO_2 Tertiary Flood in Wyoming," paper SPE 27825 presented at the 1994 SPE/DOE Symposium on Improved Oil Recovery, Tulsa, 17–20 April.
96. Martin, F.D. *et al.*: "Gels for CO_2 Profile Modification," paper SPE 17330 presented at the 1988 SPE/DOE Enhanced Oil Recovery Symposium, Tulsa, 17–20 April.
97. Moffitt, P.D. and Zornes, D.R.: "Postmortem Analysis: Lick Creek Meakin Sand Unit Immiscible CO_2 Waterflood Project," paper SPE 24933 presented at the 1992 SPE Annual Technical Conference and Exhibition, Washington, DC, 4–7 October.
98. Fullbright, G.D. *et al.*: "Evolution of Conformance Improvement Efforts in a Major CO_2 WAG Injection Project," paper SPE 35361 pre-

sented at the 1996 SPE/DOE Symposium on Improved Oil, Tulsa, 21–24 April.

99. Perry, G.E.: "Weeks Island "S" Sand Reservoir B Gravity Stable Miscible CO_2 Displacement, Iberia Parish, Louisiana," paper SPE 10695 presented at the 1982 SPE/DOE Joint Symposium on Enhanced Oil Recovery, Tulsa, 4–7 April.
100. Johnston, J.R.: "Weeks Island Gravity Stable CO_2 Pilot," paper SPE 17351 presented at the 1988 SPE/DOE Enhanced Oil Recovery Symposium, Tulsa, 17–20 April.
101. Blunt, M.J.: "An Empirical Model for Three-Phase Relative Permeability," paper SPE 56474 presented at the 1999 SPE Annual Technical Conference and Exhibition, Houston, 3–6 October.
102. Smith, C.R., Tracy, G.W., and Farrar, R.L.: *Applied Reservoir Engineering,* volume 2, OGCI, Tulsa (1992).
103. Hill. S.: "Genie Chemique," *Chem. Eng. Sci.* (1952) **1,** No. 6, 246.
104. Deitz, D.N.: "A Theoretical Approach to the Problem of Encroaching and Bypassing Edge Water," *Proc.,* Academy Amsterdam B (1953) **56,** 83.
105. Dumore, J.M.: "Stability Considerations in Downward Miscible Displacements," *SPEJ* (December 1964) 356; *Trans.,* AIME, **231.**

SI Metric Conversion Factors

acre	× 4.046 856	E + 03	= m^2
acre	× 4.046 856	E − 01	= ha
°API	141.5/(131.5 + °API)		= g/cm^3
bbl	× 1.589 873	E − 01	= m^3
cp	× 1.0	E − 03	= Pa·s
dyne	× 1.0*	E − 02	= mN
°F	(°F − 32)/1.8		= °C
°F	(°F + 459.67)/1.8		= K
ft	× 3.048*	E − 01	= m
ft^3	× 2.831 685	E − 02	= m^3
in.	× 2.54*	E + 00	= cm
lbm	× 4.535 924	E − 01	= kg
psi	× 6.894 757	E + 00	= kPa
psia	× 6.894 757	E + 00	= kPa
tonne	× 1.0*	E + 00	= Mg

*Conversion factor is exact.

Chapter 3
The Technical and Economic Screening Process: Factors To Be Assessed Before Starting a Detailed Study

The first steps in developing a depletion plan for a reservoir are to identify the primary factors that will have the greatest impact on the CO_2 flood; to carry out a "scoping" assessment (initial appraisal) of those factors' probable impact on the project's technical and economic success; and to use those findings to decide whether to abandon the project, move forward with a detailed investigation, or, if financial risk is not significant, proceed with a CO_2 flood.

One might proceed with a CO_2 flood without further study if it were a small project for which CO_2 processing facilities with excess capacity were available nearby, or if it were a CO_2 pilot test in a case where the technical risk could be sufficiently well defined such that management would be confident in then deciding whether or not to proceed with a full-blown project.

This chapter covers the screening and scoping processes for the above steps. We assume here that a CO_2 source is available.

The scoping process addresses three questions:

1. Can CO_2 recover incremental oil?
2. If so, what rates and volumes of oil can be recovered?
3. What are the estimated investment and operating costs?

The cost question includes expenses related to the integrity of the existing facilities, as well as those related to CO_2 environmental and safety issues that may affect both field operators and nearby towns and cities.

All three of these questions are important. Just because a CO_2 flood may be able to recover a large volume of incremental oil at a high rate does not mean that the flood will be economically successful. Sometimes front-end capital costs are too high to justify flood implementation (although some flood operators now offer creative financing plans to reduce front-end costs and accelerate payoff). If scoping economics indicate that CO_2 flooding might be viable, a more detailed study (described in Chaps. 4 through 7) may be undertaken to refine and optimize the design and economics.

This chapter is divided into four parts:

• Reservoir considerations—Sec. 3.1 describes the reservoir factors that should be considered in determining whether CO_2 flooding could be expected to recover significant incremental oil. Rather than presenting specific screening criteria, this section guides the reader through the thought processes involved in determining whether CO_2 flooding is a viable technical alternative to existing field operations.

• Cursory performance predictions—Sec. 3.2 offers an approach that can be used to generate a high-quality scoping estimate of CO_2 flood performance once a reservoir has been judged a good technical candidate for CO_2.

• Cursory cost estimates—Sec. 3.3 helps determine approximate investment and operating costs for the proposed CO_2 project.

• Scoping economics—Sec. 3.4 combines the results from Secs. 3.2 and 3.3 to generate scoping economics for a CO_2 project. This section also covers fundamental factors that influence scoping economic decisions.

3.1 Reservoir Considerations—Can Incremental Oil Be Recovered?

When CO_2 is injected into a reservoir, there is no guarantee that the CO_2 will recover enough oil to make the project economic. While many CO_2 injection projects have been highly successful, others have not. Sec. 2.4.2 and Fig. 2.25 take a look at an example of a failed project, a CO_2 pilot in which the reservoir pressure fell below the thermodynamic MMP and little incremental oil was recovered.

In considering a CO_2 flood, the first task of the engineer is to investigate each of the following eight reservoir attributes to identify problems that might cause technical and/or economic failure:

• Average reservoir pressure and thermodynamic MMP.
• Well patterns and stage of depletion.
• Residual oil saturation to waterflooding.
• Reservoir wettability.
• Reservoir heterogeneities and interwell continuity.
• Injection well conformance.
• Ability to inject and produce fluids at economical rates.
• Gravity effects.

3.1.1 Average Reservoir Pressure and Thermodynamic MMP. Average reservoir pressure and thermodynamic MMP are two of the more important variables that must be screened before initiating a CO_2 flood. (Sec. 2.1.3 defines the concept of thermodynamic MMP.) The real concern is how much of the reservoir is above the thermodynamic MMP. Unfortunately, this is difficult to know without a detailed reservoir simulation; because of this, in most cases, the average reservoir pressure is used as an estimate of probable success.

The question of thermodynamic MMP is complicated by the presence of small-scale heterogeneities, which can impede the development of miscibility (see Sec. 2.2.1) and raise the reservoir MMP significantly higher than the thermodynamic MMP. Wherever this occurs, the CO_2 process will be less effective, although it still can improve oil recovery. The thermodynamic MMP always should be considered the lower limit for reservoir pressure in a viable CO_2 flood.

An additional factor that can reduce oil recovery is loss of water injectivity in a WAG injection project (see Sec. 2.4.1 and Fig. 2.23), which can decrease the average reservoir pressure during the water injection cycle. To prevent this, we recommend that average reservoir pressure be maintained at least 500 psi above the thermodynamic MMP in WAG injection projects. This 500 psi value is the largest decrease in average reservoir pressure observed so far in a WAG CO_2 flood; however, this is a guideline based on experience in west Texas, and the actual safe pressure difference will vary with reservoir properties and operation.

Reservoir pressure can be maintained safely above thermodynamic MMP by controlling and/or increasing the bottomhole injection pressure while taking care not to exceed the formation parting pressure, or by increasing the ratio of injection to production wells (discussed in Sec. 3.1.2).

If reservoir pressure must be increased to assure that it is well above thermodynamic MMP, the effect on reservoir operation also must be considered. For example, if the resulting injection pressure exceeds surrounding lease injection pressures, injected fluid may leave the intended area.

Remember that CO_2 contaminants such as methane and nitrogen can increase the thermodynamic MMP (Sec. 2.1.6) and decrease oil recovery, while impurities such as C_3 (propane) and heavier hydrocarbons can reduce the thermodynamic MMP. Contaminants can enter a gas cap when continuous CO_2 is injected into the top of a reservoir in a gravity-assisted flood. They also can be introduced when a recycled CO_2 stream includes produced hydrocarbon gases.

Determination of the Average Reservoir Pressure. Reservoir pressure is important because zones that fall much below the thermodynamic MMP will not be successfully CO_2 flooded. Ideally, one would measure average reservoir pressure in different zones by using a repeat formation tester (RFT) on new wells.

When new-well data are unavailable, the best way to determine the average reservoir pressure is to take pressure measurements from observation wells and conduct pressure transient tests on injection or production wells. The advantage of observation wells is that they can be located away from injection or production wells, and hence offer more accurate information about the reservoir. Unfortunately, the disadvantage is the expense of drilling. For this reason, pressure transient tests most often are used.

There are two SPE monographs on pressure transient testing,[1,2] so we provide only a brief overview here. Pressure transient tests conducted on a single well include pressure falloff and pressure buildup tests. A pressure falloff test is run by shutting in an injection well and monitoring the resultant loss of pressure at the wellbore as pressurized injectant dissipates throughout the reservoir (the falloff test is the preferred method if waterflooding has been used in the reservoir). A pressure buildup test is run by shutting in a producing well that is not operated by a pump, and monitoring the resultant increase of pressure at the wellbore.

The data collected in pressure transient tests then are analyzed mathematically to find out pressure and other important information about the well. For example, the rate of pressure decline or increase reveals much about the reservoir and the specific well at which the test is run. Beyond telling about average reservoir pressure, these tests also can provide valuable information about reservoir permeability, "skin" (wellbore damage/plugging or results of wellbore stimulation through acidizing or fracturing), and even fracture length. There are caveats associated with this procedure. Its results can be skewed by the effects of multiphase flow, the shape factor, and the multilayer nature of most reservoirs.

Errors Introduced by Multiphase Flow. The differential equations used in pressure transient analysis are geared strictly to single-phase flow, which, as we have said before, is not realistic if the reservoir is being waterflooded, has water influx, or has gas-cap expansion. In a well where so much water has been injected that the wellbore is surrounded only by water, it might seem that a pressure falloff test would perfectly accurate, but this is not so. Away from the injection well, in the part of the reservoir still occupied by multiple phases, multiphase flow impedes fluid mobility and pressure falloff. In fact, multiphase flow can affect both pressure falloff and pressure buildup tests.

Stein and Shiralkar[3] used a reservoir simulator to evaluate the use of analytical pressure transient methods for determining average reservoir pressure during a mature waterflood. They found that the analytical method yielded estimates that were about 10% higher than those produced by a reservoir simulator. The simulator was the "truth case," revealing the effects of multiphase flow out in the reservoir away from the injection well. This is not to say that the analytical pressure transient methods cannot be used, but only that they may slightly overestimate the average reservoir pressure.

The Effect of the Shape Factor. Shape factors account for the effect of the well pattern, and choosing an incorrect shape factor can cause significant error in the average reservoir pressure calculated by the pressure transient analysis. In their abovementioned paper describing the effects of multiphase flow, Stein and Shiralkar[3] also present shape factor equations for several waterflood patterns, as well as a technique for determining shape factors for irregular patterns. Larsen[4] provides shape factors in various injection patterns.

The Weighted Average. In a reservoir with only one layer (very rare), the average pressure as determined by a pressure transient test is the average reservoir pressure of that layer. However, real-world reservoirs have many layers, each with a different permeability-thickness product. While all layers may start out at the same pressure, the pressure in each layer will be different after primary depletion and secondary recovery. In multilayer reservoirs, pressure buildup and falloff test analysis yields an average reservoir pressure that is weighted according to the formation flow capacity of each layer.[5]

Thus, the data will *not* reveal a thief zone that may ruin the flood. Different layers can be isolated with packers and tested separately—if the well is completed with casing. Such zonal isolation test results will be invalid if the well has a poor cement job or if several layers of the well have been hydraulically stimulated (including acidization), which usually causes vertical fractures.

In summary, pressure transient testing is one of the best ways of determining reservoir pressure—but be aware of its limitations. Running drill stem tests or RFT's on a new well is the best way to determine layer pressures.

Determination of the Thermodynamic MMP. The best source of thermodynamic MMP is the routine laboratory slim-tube test described in Sec. 2.1.4. In the absence of slim-tube test data, thermodynamic MMP correlations can be used (Fig. 2.7). The Yellig and Metcalfe[6] correlation is based entirely on west Texas oils; other correlations (Holm and Josendal,[7,8] Mungan,[9] and Johnson and Pollin[10]) extend Yellig and Metcalfe's work to account for the increase in thermodynamic MMP caused by increased molecular weight of the C_{5+} components. Sebastian *et al.*[11] offer a means (see their Eq. 2.2) for correcting the empirical MMP correlations for impurities in the injected CO_2.

Another way to arrive at the thermodynamic MMP is by using the mathematical EOS techniques described in Sec. 2.1.5. The EOS approach simulates multiple contacts without any reservoir mixing effects (Johns and Orr[12] and Wang and Orr[13]). This approach also can be used to determine how impurities in the CO_2 stream will change the thermodynamic MMP.

The Effects of Viscosity on Sweep. Heavier oils have higher thermodynamic MMP's because it is harder for CO_2 to vaporize oil components to achieve miscibility. Heavier oils (low API gravity) also tend to have higher viscosity, which can have a negative effect on sweep. Even if CO_2 is immiscible with a high-viscosity oil, however, CO_2 still can swell the oil, thereby greatly reducing oil viscosity and helping improve oil recovery.

One way to use CO_2 for a high-viscosity oil reservoir is to apply a single-well cyclic CO_2 injection treatment (see Appendix B). After CO_2 is injected and the gas is allowed to soak into the oil, the well is returned to production. The cyclic injection technique generally is performed two or three times. Note that the level of oil recovery for this treatment will be considerably lower than for a miscible CO_2 flood.

Fig. 3.1 illustrates the relationship between viscosity and miscibility in CO_2 flood projects using data from Moritis in the 1992 *Oil & Gas J.* EOR survey.[14]

Fig. 3.1 suggests some approximate screening criteria for selecting CO_2 projects that could be miscible—oils with gravity above 25°API and viscosity of less than 3 cp. Note, however, that Fig. 3.1 does show several exceptions to the 3-cp oil viscosity limit. Of course, the reservoir pressure must be at least above the thermodynamic MMP for the oil actually to achieve miscibility.

3.1.2 Well Patterns and Stage of Depletion. ***Well Patterns.*** Stein and Shiralkar[3] showed that the ratio of injection wells to production wells is a strong indicator of average reservoir pressure—as the ratio of injectors to producers increases, so does the average reservoir pressure. This assumes that any free gas during the waterflood has been redissolved in the oil during repressurization in preparation for CO_2 flooding.

Field evidence from the Slaughter field in west Texas[3] and the East Vacuum Grayburg San Andres Unit in New Mexico[15] verify this result. Under ideal conditions of single-phase flow and equal effective wellbore radii at all wells (Eq. 4.2), the average reservoir pressure of a five-spot waterflood pattern may be estimated by the following equation. The effective wellbore radius is the actual radius plus the effect of skin, whether skin is a positive quantity (reflecting wellbore damage) or a negative quantity (representing fracturing).

$$\bar{p}_R = p_{bhp} + \tfrac{1}{2}(p_{bhi} - p_{bhp}), \quad \text{(3.1)}$$

where $\bar{p}_R$=the average reservoir pressure of the entire pattern in psia; p_{bhp}=the bottomhole production pressure in psia; and p_{bhi}=the bottomhole injection pressure in psia.

The analogous equation for an inverted nine-spot pattern is

$$\bar{p}_R = p_{bhp} + \tfrac{1}{4}(p_{bhi} - p_{bhp}), \quad \text{(3.2)}$$

The five-spot pattern is preferable because its higher ratio of injection to production wells helps maintain a higher average reservoir pressure.

Whatever type of pattern is used for a CO_2 flood, it is critical that there are no major volumetric sweep problems under current operation. The adage goes: "If it is a good waterflood, it could be a good CO_2 flood, but if it is a bad waterflood, it will be a terrible CO_2 flood." Problems with sweep and low reservoir pressure in a waterflood will worsen during CO_2 flooding, and irregular pattern spacing that is a minor problem in a waterflood may cause more serious problems in distributing CO_2 across the reservoir. If just a few patterns have problems, CO_2 injection should not start in those patterns until the problems are solved.

Stage of Depletion—Primary, Secondary, or Tertiary Flooding. When deciding on the most appropriate time to begin a CO_2 project, several technical variables must be examined: reservoir pressure, dip, wettability, and project economics. For instance, most CO_2 floods are tertiary and use alternate water/CO_2 injection. These projects often take place in relatively flat reservoirs with oil-wet or mixed-wet rock, where water injected during secondary recovery should not hurt the process (see Sec. 2.2.2). As long as the reservoir pressure is maintained above the thermodynamic MMP, the injected CO_2 should help increase oil recovery. One economic advantage of beginning a WAG CO_2 injection project after waterflood is that water injection facilities are already in place.

If the reservoir has a definite dip, it may be advantageous to inject the CO_2 continuously at the top of the reservoir. Once again, the reservoir pressure should exceed the thermodynamic MMP to achieve a satisfactory oil recovery (except with some water-wet reservoirs that have a special gravity-drainage mechanism—see Sec. 2.6.3). Furthermore, care should be taken so that the injected CO_2 is not diluted by methane or nitrogen that might exist in a gas cap (see Sec. 2.1.6). Continuous CO_2 injection into the top of a dipping reservoir that has no initial gas cap could begin immediately after a field's discovery. This avoids dilution by methane and nitrogen that could migrate in and form a secondary gas cap during primary depletion.

Wettability also can affect project timing. If the formation is water-wet, water blocking can occur if water contacts the oil before CO_2 does (see Sec. 2.2.2 for additional information on water blocking). In this case, injecting CO_2 continuously after primary depletion eliminates the water blocking problem, although it could cause poor control of volumetric sweep. Ford Geraldine[16] and Twofreds[17] are examples of successful CO_2 projects where CO_2 was injected continuously in flat, water-wet reservoirs. If continuous CO_2 is used after a waterflood in a water-wet reservoir, the CO_2 can displace all the water surrounding the residual oil, but this process is not quite as efficient as secondary injection.

If WAG injection is used in a water-wet reservoir, CO_2 should be injected continuously at first to stay ahead of the water. WAG injection can be initiated later to improve sweep, although it may also cause some problems; for example, water blocking will cause some oil to be bypassed (see Sec. 2.2.2).

3.1.3 Residual Oil Saturation to Waterflooding. The residual oil saturation to water is an important variable to consider before

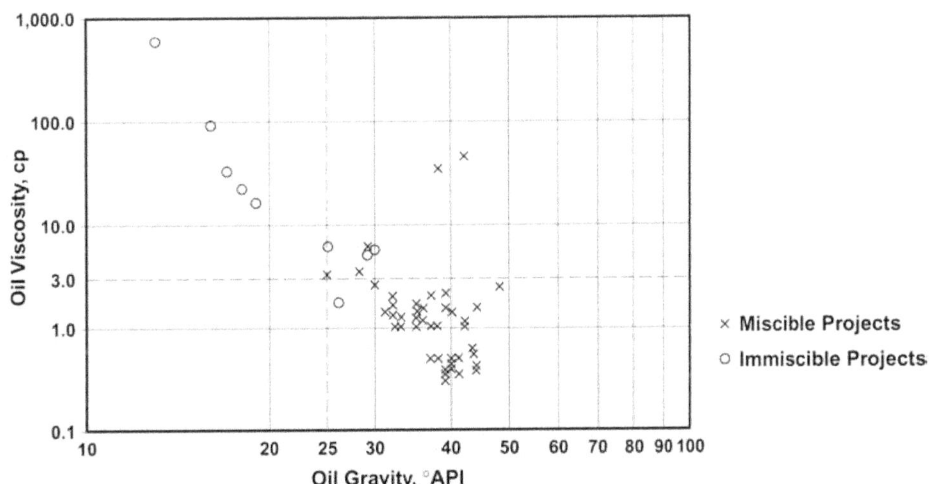

Fig. 3.1—Viscosity and API gravity relative to miscibility for 1992 worldwide CO_2 projects (from Ref. 14). It is clear that miscibility relies heavily on the presence of light oils with relatively low viscosity. In the upper left quadrant (high viscosity, low API gravity), none of the CO_2 floods is miscible.

beginning a CO_2 flood that follows a waterflood or a natural water drive. If the residual oil saturation to water is too low, there may not be enough oil left to yield an economical CO_2 flood. A screening guideline used for CO_2 floods is that the residual oil is equal to 20% of the total PV,[18] although the appropriate value for a particular case will depend on reservoir sweep, project costs and the current price of oil.

Several tests can be used to determine the residual oil saturation to water. In the laboratory, waterflood endpoint tests and water/oil relative permeability tests find out the likely residual oil saturation for individual cores. A disadvantage of these tests is that they do not sample a large portion of the reservoir. In addition, they must be designed such that the capillary pressure in the tests represents reservoir conditions. If the capillary pressure of the test core is too high, the test may show unrealistically low connate water and residual oil saturations. In the field, test methods include single-well tracer tests and pressure and sponge cores.

Waterflood Endpoint Test. In the waterflood endpoint test, the core first is flooded with mineral oil to replace the crude oil and bring the core to connate water saturation, and then is flooded with at least 50 PV water until little or no additional oil is produced. The amount of oil remaining is the residual oil to waterflooding. This test also can be used to determine whether the rock is water- or oil-wet, as discussed below in Sec. 3.1.4.

The waterflood endpoint test requires a sample core at connate water and natural reservoir wetting conditions. Representative wettability can be achieved by coring either with a lease crude or with water, *without* surfactants or other additives that could alter wettability.[19] If wettability already has been altered, cores may be somewhat restored to their natural wettability through four steps:

1. Avoid removal of clays during cleaning of the cores.
2. Saturate with field brine.
3. Flood with crude oil to connate water saturation.
4. Age the core until natural wettability is reached, which may take 300 to 1,000 hours.

The proper aging period is difficult to determine and may require some experimentation. Because of the uncertainties inherent in wettability restoration, it is always better to start with a core that has been obtained using either a lease crude or brine without the use of additives in either fluid.

Special care should be taken with stress-sensitive rock. If permeability is very sensitive to net overburden pressure, then any kind of flow test should be performed under stress conditions that closely resemble the reservoir stress state.

Water/Oil Relative Permeability Tests. Water/oil relative permeability tests are used to determine both residual oil saturation and the relative permeability characteristics of core samples. The core samples should have representative wettability as discussed above.

The test begins with the core at connate water saturation. In a simple unsteady-state displacement, water displaces viscous mineral oil (typically 25 cp), yielding the water/oil relative permeability curve for increasing water saturation. This is the direction of change in water saturation that occurs during waterflooding.

The preferred relative permeability method, however, is the steady-state water/oil relative permeability test, in which simultaneous oil and water injection is repeated until steady-state conditions are reached. The fraction of water injected in each step increases until only water is injected in the final step. Because the steady-state procedure takes place in multiple steps, it yields more datum points for water/oil relative permeability than does the unsteady-state technique.

Water/oil relative permeability tests are recommended over endpoint tests because they provide the relative permeability data required for a detailed design of a CO_2 flood. Sec. 4.1.2 gives more information on these requirements.

Single-Well Tracer Test. The single-well tracer test can be used to determine residual oil saturation averaged over the entire perforated interval tested.[20,21] A slug of oil-soluble chemical tracer is injected into the well; the well is shut in for a time; and the tracer chemical is produced back. A mathematical analysis of the rate at which the tracer is produced back can reveal the residual oil saturation to water, as well as the dispersivity of the formation. The tracer test also can be run with water-soluble tracers to determine connate water.

In both cases, the analytical results must be history-matched with simulations to produce meaningful information. The test is most accurate when conducted over a single reservoir zone of uniform porosity and permeability. If there are multiple layers that differ significantly, reasonably accurate results can be achieved if core data (porosity and permeability) for every foot of the producing zone are incorporated into the mathematical model. Without these detailed data, the results are prone to the same averaging errors encountered in pressure transient testing.

Because the single-well tracer test is complex to perform and analyze, it usually is carried out by a service company or by an oil company research/technology center.

Pressure Core and Sponge Core Data. Pressure coring and sponge coring both are suitable coring techniques for determining the residual oil saturation to water. Conventional coring should not be used because oil escapes as pressure is released during each core's trip to the surface.

With pressure coring, this problem is solved by maintaining the core in a pressurized chamber that later allows gases to be captured and analyzed in the laboratory—along with the oil in the core—on a foot-by-foot basis. With sponge coring, somewhat the same task is accomplished much less expensively by using a special oil-wet sponge that runs the length of the core. The sponge soaks up oil that leaves the core when it is brought up the through the wellbore, but it provides only average values for the escaping oil. Therefore, the sponge core cannot yield a true foot-by-foot record of residual oil saturation. Rather, it provides an average residual oil saturation over each core section. The sponge also does not collect gas because the sponge core loses pressure while rising to the surface.

In neither technique does the residual oil saturation obtained take into account swelling caused by dissolved reservoir gas. To make adjustment for this, the residual oil saturation should be multiplied by the formation oil volume factor.

In both tests, we recommend that new wells be drilled fairly close to existing water injection wells to take advantage of the waterflooding that already has occurred. The drilling fluid—preferably water without additives or wetting agents—also acts as a waterflood ahead of the drill bit. Chemical and radioactive tracers such as tritium can be added to the drilling fluid to show how much of an interval of core has been flushed. This is important because low-permeability intervals are not always flushed by the drilling fluid enough to yield an accurate residual oil saturation reading.

Pressure coring is considered the best method for estimating residual oil saturation over a large interval. However, the cost of drilling a well, coring, and analysis often necessitates choosing a less expensive method such as the single-well tracer test.

3.1.4 Reservoir Wettability. Formation wettability must be known before starting CO_2 flooding because strongly water-wet rock can adversely affect oil recovery during a WAG injection process.[22–25] Water blocking of CO_2 has not been found to be a problem in mixed wet and oil-wet formations[26–28] (see Sec. 2.2.2).

For strongly water-wet rock, the best mode of CO_2 flooding probably is continuous injection (possibly with a tail-end WAG to improve sweep), rather than using WAG for the entire flood. If injected water moves ahead of the CO_2, the water can shield the oil in water-wet rocks, causing CO_2 to bypass the oil and leading to reduced oil recovery. Continuous CO_2 injection projects have been used successfully in strongly water-wet reservoirs (Ford Geraldine, Twofreds, and North Cross Devonian).[16,17,29]

Choosing whether or not to inject water with CO_2 is important in the screening process because with continuous CO_2 the gas production rates will be significantly higher than with alternate CO_2/water injection, thereby raising plant investment and operating costs. Sec. 3.3 discusses facility design shortcuts that can be taken if continuous CO_2 injection is used.

So how does one determine whether or not the formation is so strongly water-wet that continuous CO_2 injection should be used? There are four main methods: the waterflood endpoint test for

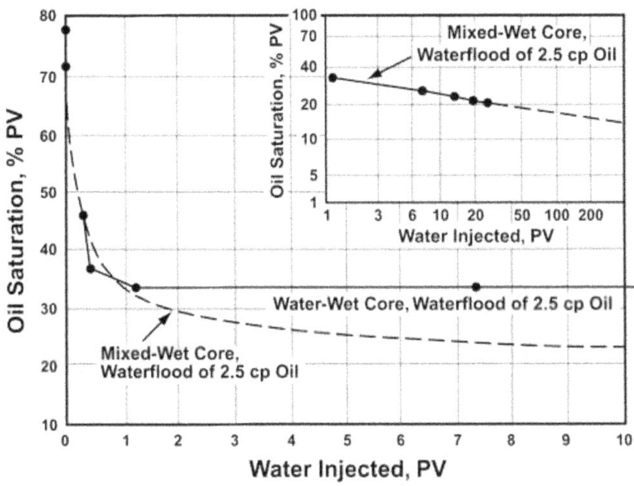

Fig. 3.2—Waterflood behavior for water-wet and oil-wet cores (after Ref. 30). Oil-wet rock shows a continuously declining oil saturation from the throughput effect, while water-wet core does not. The inset shows the results of Salathiel's centrifuge technique.[30]

residual oil saturation, the water/oil relative permeability test, the water fractional flow plot, and the Ammot wettability test.

Waterflood Endpoint Test for Residual Oil Saturation. This is one of the simplest ways of testing wettability, and it is based on the work of Salathiel,[30] who demonstrated that formation wettability can affect the residual oil saturation determined in the laboratory. Salathiel showed that oil-wet or mixed-wet formations (in which the large pores invaded by the oil are oil-wet, while the small pores in which the connate water is not displaced remain water-wet) exhibit a throughput effect on residual oil saturation to waterflooding, while strongly water-wet formations do not.

In mixed-wet rocks (and, by inference, oil-wet rocks), residual oil saturation decreased asymptotically with continued water injection (the throughput effect), as shown by the lower curved line in **Fig. 3.2.** When the same rock was subjected to continued waterflooding by centrifugation (which allowed hundreds of PV's of water to be injected, an amount not practical under reservoir conditions except when very close to an injection well), the residual oil saturation continued to decrease, as shown by the inset in Fig. 3.2. Salathiel's centrifuge technique not only greatly increased the PV's of water injected,[30] it also may have increased the capillary pressure beyond what is representative of reservoir conditions.

The extremely low oil saturations that Salathiel obtained after hundreds of PV water were injected would only occur in the vicinity of an injection well. The key point from Fig. 3.2 is that the existence of a throughput effect indicates mixed-wet behavior.

In contrast, Salathiel found that in strongly water-wet rock, the residual saturation to waterflooding reached a certain value and remained there, exhibiting no throughput effect. If no waterflood throughput effect occurs, the rock is strongly water-wet. With such rock, water blocking could occur were a WAG injection scheme used.

Water/Oil Relative Permeability Test. Wettability also can be determined using relative permeability data from the water/oil relative permeability test. If the water relative permeability exhibits significant hysteresis (see Fig. 2.19), the rock is definitely oil-wet. However, this test's results are not definitive because not all oil-wet rock exhibits a water relative permeability hysteresis effect. In addition, many operators skip hysteresis testing because it doubles the cost of this already expensive test.

In determining wettability, the differentiating result from water/oil relative permeability testing is the maximum water relative permeability (k_{rw}) at residual oil saturation. For a strongly water-wet rock, the maximum will be lower than for mixed-wet or oil-wet rock; in fact, if the maximum k_{rw} is 0.1 or less, the rock is water-wet. **Figs. 3.3 and 3.4,** taken from Ref. 19, illustrate the differences in relative permeability behavior between strongly water-wet and oil-wet rocks.

As discussed in Sec. 2.2.2, water-wet rock has low k_{rw} because the water must flow either through the small pores where the water phase is continuous or around the oil globules trapped in the large pores, where they tend to act as choke valves. Even when water saturation increases, k_{rw} in a water-wet rock increases very little. Note also that in strongly water-wet rock, the connate water saturation often is high, above 30%.

In oil-wet rock, connate water generally is lower, and water in the smaller pores is not well connected. This helps redirect the water to the larger pores, and because the oil is on the walls of these pores (not trapped as globules), the water can go through more easily, such that k_{rw} increases.

In oil-wet rocks with high connate water, the maximum k_{rw} is not as high as it would be if there were less connate water because the fairly continuous water phase in the small pores takes on more

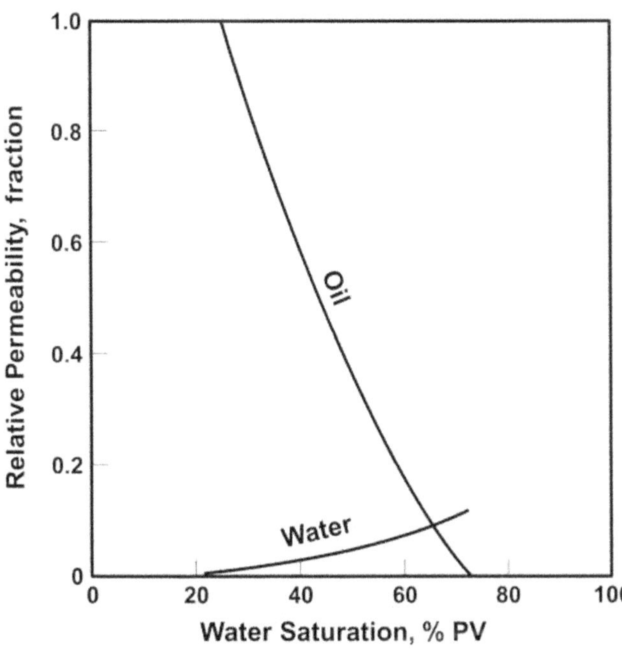

Fig. 3.3—Typical water/oil relative permeability in strongly water-wet rock (taken from Ref. 19). In strongly water-wet rock, relative permeability of the water remains low.

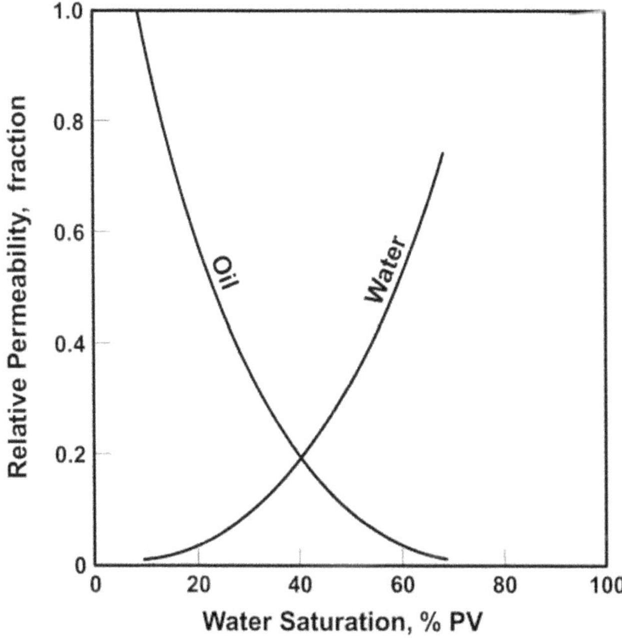

Fig. 3.4—Typical water/oil relative permeability in strongly oil-wet rock (taken from Ref. 19). In strongly oil-wet rock, the relative permeability of the water climbs steeply as water saturation increases.

of the water flow. These smaller pores cause a larger pressure drop that decreases relative permeability.

Water Fractional Flow Plots. Water fractional flow graphs offer another clear indication of a core's wettability and are easy to plot based on relative permeabilities and viscosities calculated using equations presented by Craig.[19]

Figs. 3.5 and 3.6 from Ref. 19 show that strongly water-wet rocks have a J-shaped fractional flow curve, while strongly oil-wet rocks have an S-shaped one. The difference between the two curves—the trailing end of the "S" on the upper right-is the result of the throughput effect, as oil continues to be produced.

The above fractional flow rule of thumb assumes that the oil viscosity is not much higher than the water viscosity. A very high oil viscosity (above 25 cp) can create an "S" shape in the fractional flow curve for a water-wet rock, but reservoirs with oil of this viscosity would not be good candidates for CO_2 flooding anyway. In any event, a J-shaped fractional flow curve is a strong warning against running an alternating CO_2/water injection flood.

Ammot Wettability Test. Wettability also can be determined using an Ammot wettability test, a spontaneous imbibition test[31,32] that standardly is run by service companies. This test requires core at natural wetting conditions. In the test, the core goes through five steps: Flood with mineral oil to replace the crude oil, leaving the core at connate water saturation; flood with at least 50 PV of water until little or no additional oil is produced; immerse in oil to determine how much oil imbibes; flood with oil back to connate water saturation; immerse in water to determine how much water imbibes. The relative volumes of oil and water that imbibe are used to create a wettability index .

If water and oil imbibe into the core in approximately equal measures, this indicates mixed wettability that should pose no water blocking problems. If only water imbibes into the core, the core is strongly water-wet and water blocking should be expected for WAG injection.

3.1.5 Reservoir Heterogeneities. The degree of vertical reservoir heterogeneity can strongly affect CO_2 flood performance in reservoirs where layer effects are more important than gravity effects, i.e., reservoirs where the ratio of vertical to horizontal permeability is low. To repeat the adage stated previously, "If it is a good waterflood, it could be a good CO_2 flood, but if it is a bad waterflood, it will be a terrible CO_2 flood." If several reservoirs with similar reservoir pressures and thermodynamic MMP's have been waterflooded and are candidates for CO_2 flooding, it makes sense to rank them by waterflood sweep efficiency, with highest efficiency reservoirs being the best candidates.

Reservoir Layering. The low viscosity of CO_2 adversely affects sweep. Chopra[33] has used numerical simulation to illustrate how alternate CO_2/water injection is more sensitive to changes in reservoir layer properties (i.e., different permeabilities) than is waterflooding. Continuous CO_2 injection is even more sensitive to reservoir properties because there is no injected water to help control volume sweep by slowing down the movement of CO_2.

Warden* has found that simulated incremental tertiary oil recovery from alternate CO_2/water injection, shown in **Fig. 3.7,** is very sensitive to the Dykstra-Parsons coefficient (defined in Sec. 2.2.1). In Warden's simulations, which used fixed volumes of CO_2 and water injected alternately, larger values of the Dykstra-Parsons coefficient decreased incremental oil recovery. Warden found similar results when using the Lorenz coefficient as a measure of het-

*Personal Communication with G.W. Warden, BP (1989).

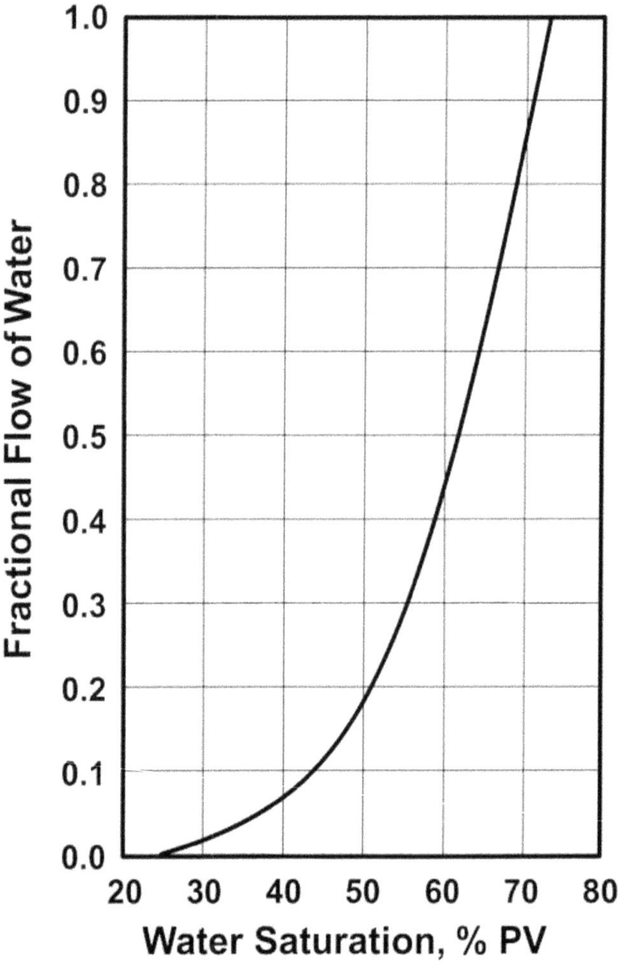

Fig. 3.5—J-shaped fractional flow curve representative of strongly water-wet rock (after Ref. 19). We have used oil viscosity (μ_o) of 1.0 cp and water viscosity (μ_w) of 0.5 cp.

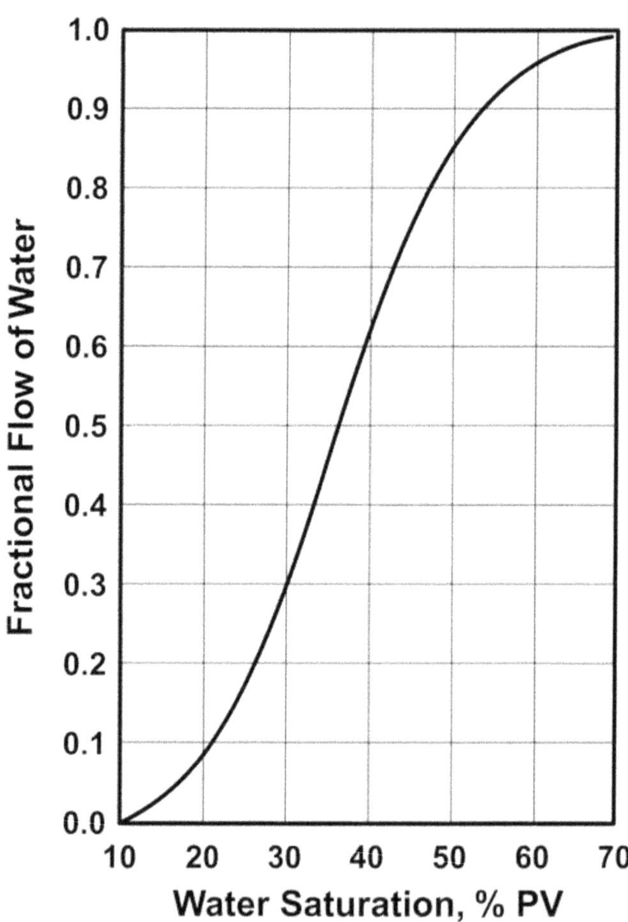

Fig. 3.6—S-shaped fractional flow curve representative of strongly oil-wet rock (after Ref. 19). The top of the "S" is the result of the throughput effect. We have used oil viscosity (μ_o) of 1.0 cp and water viscosity (μ_w) of 0.5 cp.

erogeneity. Craig[19] discusses how to calculate the Lorenz coefficient from a plot of the cumulative kh vs. the cumulative ϕh for all the intervals (all data are plotted in order of decreasing k/ϕ).

Additional Types of Reservoir Heterogeneities to Consider. High-permeability channels, directional permeability trends, faults, and natural fractures all cause sweep problems. For example, the carbonate reservoirs in southern Mexico underwent tremendous folding and resultant faulting that created a large number of fractures with high permeability, mostly parallel to the faults. In an extreme case, the October oil field of Egypt has approximately 100 wells and about the same number of faults. New drilling continues to identify additional faults.

If a reservoir is naturally fractured to the extent that fractures have adversely affected waterflood sweep, a CO_2 flood using the same pattern probably will be unsuccessful, with one exception: If the natural fractures have increased vertical permeability, a gravity-assisted top-down CO_2 flood might be successful (see Sec. 3.1.8). In this case, the CO_2 injection rates would have to be decreased to reduce the effects of coning, but not decreased so much that the oil rate is also decreased and project cash flow is reduced.

Interwell Continuity. If waterflood performance is poor because of poor continuity between injection and production wells, a CO_2 flood also can be expected to perform poorly. Well logs or core data may be used to map zones and determine continuity. If good continuity does not exist, the well spacing may have to be reduced to have a technically successful CO_2 flood.

Well spacing can be reduced by infill drilling, horizontal wells, or long induced vertical fractures. With induced vertical fractures, care must be taken that they not leave the pay zone or cause sweep problems.

While continuity is important, too much continuity in the form of fractures can be detrimental if fractures connect injectors and producers in an existing waterflood. If the direction of the fractures is understood, it may be possible to place injectors and producers to prevent the fractures from connecting them. The cost of well conversions must be included in the CO_2 flood economic evaluation.

3.1.6 Injection Well Conformance. Conformance is a measure of where injected fluids (water or CO_2) are entering the pay zone. Ideally, injected fluids enter the formation only at pay zones and spread out evenly across these zones to avoid early breakthrough. There are three tools used to measure conformance: injection/withdrawal ratios (IWR's), injection profiles, and temperature profiles.

Injection/Withdrawal Ratio. The IWR (injection rate in RB/D divided by the total fluid production rate in RB/D) is a good test of overall injection conformance. If the waterflood IWR is above 1.0 and the reservoir pressure has increased enough to redissolve any free gas saturation, water injection unintentionally may be going into a nonoil-bearing zone or migrating across lease boundaries. If nothing is done to remedy this type of problem, then CO_2 injected

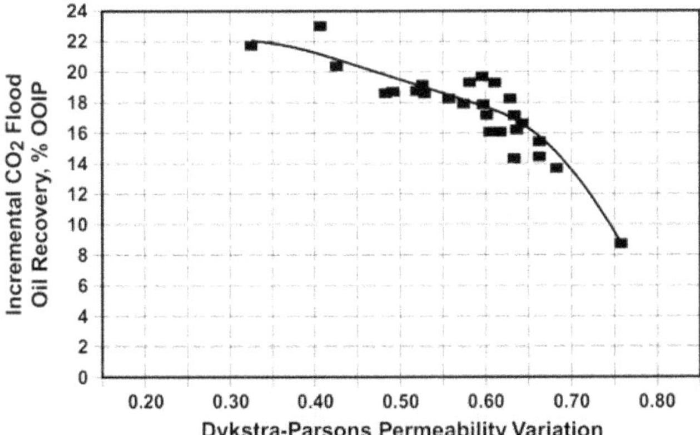

Fig. 3.7—The impact of stratification on incremental oil recovery in a CO_2 flood (after Warden*). Incremental oil recovery decreases as reservoir heterogeneity (indicated by the Dykstra-Parsons coefficient) increases.

into the same wells will not be used effectively, either. Because the purchase and/or recycle costs for CO_2 are about 12 times the cost for water injection, it is important to correct any situation that would waste CO_2 by injecting it out of pay.**

Injection Profiles. An injection profile tool is used to estimate where injected fluid is going in a well. This tool ejects a radioactive tracer from one end and measures it at the other end, typically while the tool is pulled from the bottom of the well to the top.

Two types of measurements are taken. First, the velocity of the fluid in the annular space between the wellbore and the tool is measured. As this tool is moved down the wellbore, a reduction in volumetric flow rate implies that some of the fluid has gone into the formation at intervals above the tool's current location. Second, a material balance is carried out based on the amount of radioactivity detected at the trailing end of the tool. Missing radioactivity also represents fluid that has gone into the formation. A side-by-side plot of the two results should show good agreement about where injection (or undesirable fluid loss) is taking place.

Temperature Profile. To carry out a temperature profile, the well is injected with cool water and shut in for 3 to 24 hours or more, after which a temperature tool pulled from the bottom of the well to the top records the temperatures along its vertical path. In zones where little injection or fluid loss takes place, the temperature of the shut-in fluid will rise to reach ambient reservoir temperatures; in zones where fluid is being injected or lost, the temperature will remain closer to the cool-water injection temperature. Temperature profiles are easy to run, relatively inexpensive, and commonly included with injection profiles.

Examples of Good and Bad Conformance. If injected water moves from the wellbore into zones other than the oil productive zones, the injected CO_2 will also if the same bottomhole injection pressure is used. As a result, the CO_2 utilization factor will be higher than expected.

Fig. 3.8 shows a well where the injected water is confined to the oil production zone. The logs on the left are gamma ray and neutron porosity logs (not part of this discussion), and the next two log traces show velocities and tracers. Three temperature profiles (injection temperature, 3-hour shut-in and 24-hour shut-in) corroborate the velocity and tracer profiles. In addition, the temperature profiles show that the temperature increases dramatically below the well, evidence that no fluid is being lost in that area.

Fig. 3.9 shows a well with very poor conformance. Here the velocity profile of the injection test indicates that 33% of the injected fluid is flowing out the bottom of the well and the tracer profile indicates that 75% of the fluid is leaving from the bottom. (These two different results represent the statistical accuracy of these tests.) The proof is the temperature profile, which actually gets cooler at the bottom of the well.

Sometimes a temperature profile may exhibit only a slight bending if the temperature log cannot reach the bottom of the wellbore, a situation that may occur because of fill at the bottom. It may be necessary to clean out the wellbore to obtain a more certain indication of whether or not injected fluids are going below pay.

Injection below pay is not unusual in a waterflood operated above fracture parting pressure. Kittridge[34] described how monitoring well logging data confirmed injection of CO_2 below pay in the Denver Unit in the Wasson Field in west Texas. The injection below pay apparently is occurring here because the waterflood was operated by injecting above fracture parting pressure. If an oil productive zone is bounded above only by an impermeable zone, any fracture created can propagate below pay.**

Temperature profiles also can be used to detect channeling above pay and behind casing and to find casing leaks in water injection wells. These problems sometimes can be remedied by cement squeezes before CO_2 injection. Hill[35] gives an excellent review of temperature and velocity profile tests.

Temperature and injection profiles were used successfully in the Slaughter Estate Unit CO_2 flood to identify a conformance

*Personal Communication with G.W. Warden, BP (1989).

**Personal Communication with S.C. Wehner, Phillips Petroleum (2000).

problem stemming from injection above formation parting pressure. Problems were suspected at one injection well because water was being injected at a much higher bottomhole pressure than was CO_2. Temperature and injection profiles revealed that 95% of the water was going out the bottom of the well. When the bottomhole injection pressure was reduced, all the water flowed into the pay. Apparently, previous water injection had opened a fracture that closed when pressure was reduced.

3.1.7 Ability to Inject and Produce Fluids at Economical Rates. To yield economical fluid injection and production rates, the reservoir must have sufficient formation flow capacity. If this is not the case, the wells must be stimulated enough to achieve those rates. If a WAG CO_2 flood is being considered and water injection has never been attempted in the reservoir, a pilot water injection test should be performed to reveal potential water injection problems. Possible problems include sensitivity of clays in the formation to injected water (swelling of clays and migration of clay fines to block pores), unusually low formation parting pressure, and a combination of very low permeability and very low water relative permeability.

Note that the presence of water in the reservoir before a waterflood does not preclude water injectivity problems related to clays because clays are very sensitive to salinity, and the salinity of injected water will nearly always differ from that of formation water.

Even in cases in which a waterflood has run successfully, there is no guarantee of a successful WAG CO_2 flood. For example, you may discover water injection problems related to three- or four-phase relative permeability effects (see Secs. 2.5 and 2.6) after the first CO_2 cycle. A pilot WAG test on one injection well using a trucked-in supply of CO_2 can prove invaluable. A minimum injection volume of 10% HCPV of CO_2 generally is required to generate sufficient evidence of future water injection characteristics during CO_2 flooding. If all works as planned, you will have corroborated laboratory results; if the unexpected happens and WAG results do not reach expectations, you may have saved yourself considerable time, energy, and money that otherwise would have been spent in unsuccessful operations. If you discover problems that absolutely prevent water injection, you might want to rethink the idea of doing a CO_2 flood or consider using continuous CO_2 injection rather than WAG.

3.1.8 Gravity Effects. Potential gravity effects on flow and sweep also should be considered in screening a reservoir for CO_2 flooding. Gravity can affect a CO_2 flood in two ways, through significant reservoir dip and by fluid crossflow between reservoir layers.

Gravity Effects in Dipping Reservoirs. In Louisiana Gulf Coast reservoirs with significant dip, continuous CO_2 injection has proved to be a successful oil recovery process.[36,37] In these reservoirs, volumetric sweep of the CO_2 was enhanced by gravity when CO_2 (less dense than oil) was injected at the top of the reservoirs.

Injection rates in gravity-assisted wells depend on reservoir dip and vertical permeability: the higher the dip and the vertical permeability, the greater the stable injection rate. Operating above the stable injection rate causes the CO_2 to begin bypassing oil.

If a gravity-stable rate is not practical, economics may dictate that the rate be exceeded. Even though cumulative oil recovery would be sacrificed to some extent, the flood still may provide a good economic return. Blunt[38] has presented laboratory data from several investigators that demonstrate that at low displacement rates, gravity-assisted CO_2 (even if it is contaminated to the extent that it is no longer miscible) can reduce residual oil saturation in some water-wet sands to less than 1% PV (see Sec. 2.6.3).

Injected CO_2 should be kept away from any gas cap to avoid dilution of the CO_2 with natural gas because methane will raise the thermodynamic MMP and could make the flood process immiscible. High vertical permeability may increase the possibility of CO_2 dilution.

Gravity Effects Caused By Crossflow. One must understand reservoir characteristics thoroughly to accurately assess CO_2 flooding in stratified reservoirs where significant crossflow takes place. When gravity-induced fluid crossflow occurs between layers, special consideration should be given to both the planned WAG ratio and the completion techniques at the wells. For a WAG CO_2 injection project, fluid crossflow can occur between layers. A result may be that water and CO_2 may segregate and not effectively sweep the same layers (see Fig. 2.39), causing each layer to experience drastically different WAG ratios.

Well recompletions and horizontal wells[39] may help improve sweep by capitalizing on gravity effects. Given that CO_2 generally is less dense than either oil or water, an injection well may be perforated selectively at the top of the pay, or a horizontal injection well may be completed near the top of the pay. The producing wells then would be completed closer to the bottom of the pay,

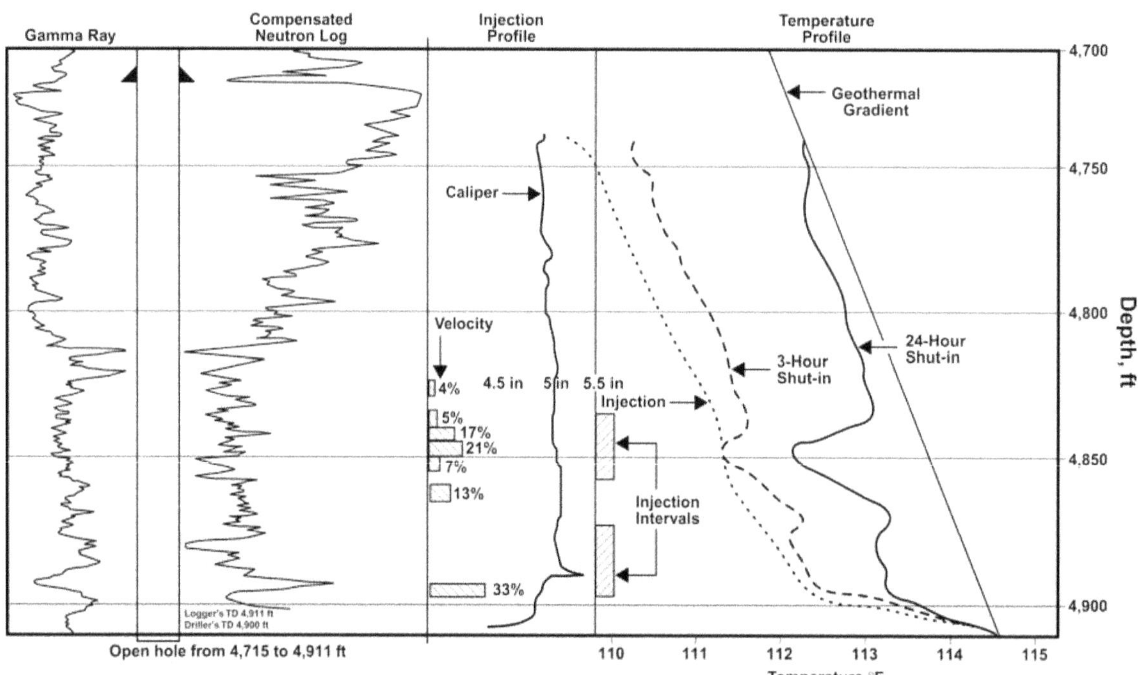

Fig. 3.8—Injection and temperature profiles for a successful water injection well. Note that injection is confined to the oil productive zone, a fact supported by the increasing temperatures at the bottom of the well. Caliper is measured in inches.

such that the CO_2 moves slowly downward to improve sweep. The disadvantage of this technique is that any mobile water present tends to be produced at the bottom of the well along with the oil.

Gravity effects have not been a major factor in the performance of CO_2 WAG injection projects in west Texas reservoirs because of the low vertical permeability there. Higher permeability reservoirs are more likely to have crossflow problems.

3.2 Cursory Prediction of CO_2 Flood Performance—How Much Oil Can Be Recovered?

At this early scoping stage of a CO_2 project design, scaling techniques are a better means of forecasting CO_2 flood performance than are reservoir simulators. To yield meaningful predictions, reservoir simulators require an accurate characterization of both the reservoir and its operation, and characterizing a reservoir accurately can be a time-consuming and somewhat expensive process. If a scoping analysis shows that a CO_2 flood might be economical and if significant financial risk is involved, then using a reservoir simulator to forecast performance would be a useful and economically justified next step. Chap. 4 discusses the use of simulators.

The scoping-quality estimates of CO_2 flood performance discussed in this section can be developed by scaling up the actual or predicted CO_2 flood performance from a similar reservoir. (More will be said in Sec. 3.2.1 about the exact meaning of "similar.")

Hadlow[40] and El-Saleh[41] have presented performance plots of several CO_2 injection projects, which are useful in comparing performance of similar floods during the scoping process. Because published data on CO_2 floods may include projects in early stages of CO_2 flooding, some of the data may require extrapolation before scaling up. If actual CO_2 flood performance from similar fields is not available or if the flood to be scaled up is not mature enough, it may be necessary to use a reservoir simulator to predict performance.

A scoping-quality design for a CO_2 flood may not be accurate enough when the financial risk of CO_2 flooding is great, which usually is the case when capital investments before CO_2 flooding are large. Some items that can increase capital investments are CO_2 recovery plants, sulfur removal plants, compressors for reinjection of produced gas, CO_2 injection facilities, CO_2 pipeline, and modifications to wells and production facilities for handling increased amounts of produced gas resulting from CO_2 injection.

Scoping-quality estimates of CO_2 flooding may suffice when the upfront capital investments are small enough to minimize financial risk. Small CO_2 injection projects (those with no CO_2 processing plant and fewer than 10 patterns) possibly could be designed and implemented from scoping-quality estimates without resorting to reservoir simulation.

3.2.1 Scaling Up CO_2 Flood Performance. The process of scaling up CO_2 flood performance in one reservoir involves taking CO_2 flood performance from a similar reservoir and multiplying the performance by scaling variables. The similar, or analogous, reservoir may be another reservoir already under CO_2 flood, a pilot CO_2 flood in the same reservoir, or even detailed CO_2 flood predictions from another similar reservoir.

Scaling variables are intended to account for approximate differences in reservoir properties. The analogous reservoir must have similar reservoir characteristics and operation for the scaleup process to yield acceptable results at the target reservoir. The results of the scaleup process include target reservoir predictions for incremental oil rate to CO_2 flooding, CO_2 injection rate, CO_2 production rate, water injection rate, and water production rate. If the properties between reservoirs are not similar, the scaleup process can provide misleading results. Chap. 4 discusses more comprehensive methods for predicting CO_2 flood performance without having to scale up from another property.

Scaleup Assumptions. In the absence of representative data, injection rates are sometimes assumed. For instance, the average water-cycle injection rate during a CO_2 flood might be assumed to be as low as one-half the preCO_2 water injection rate, and the CO_2 injection rate might be assumed to be some multiple of the preCO_2 water injection rate. Note that the "one-half" assumption in this example is based on pattern WAG projects (such as the Slaughter field) where CO_2 and water are injected at the same bottomhole pressure and the water cycle lengths are short (about 1% HCPV).

Accurate estimation of the water injection rate is important because the water injection rate affects the timing of the incremental oil response and, hence, the economic payout of the project. Likewise, accurate estimation of both water injection and production rates also is necessary to plan for saltwater disposal needs.

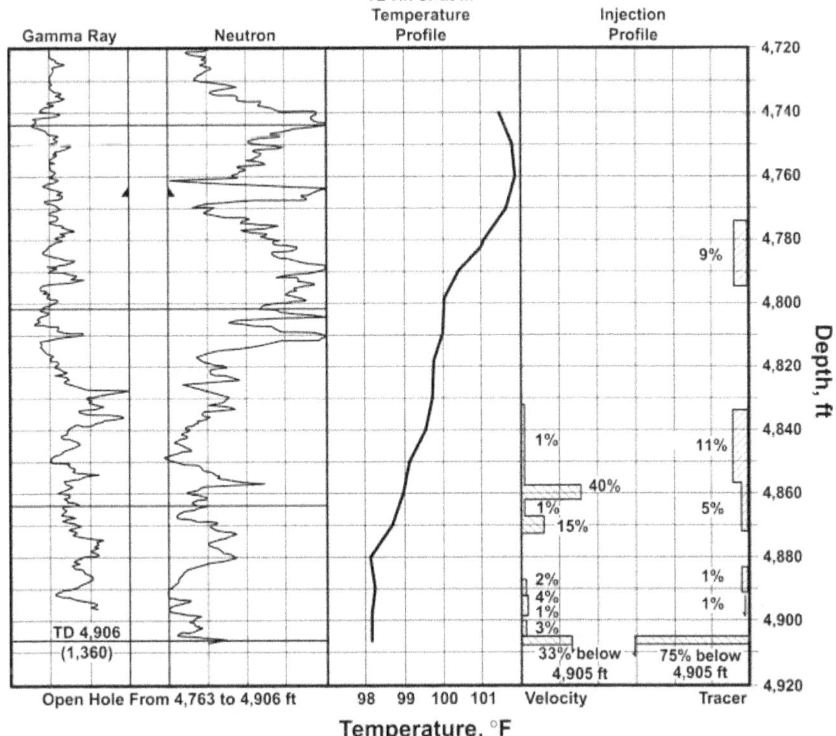

Fig. 3.9—Injection and temperature profiles for a water injection well with poor conformance. Much of the water is leaving the well below the oil productive zone, as shown by the velocity and tracer logs and by a temperature profile that bends to the left.

The Use of Incremental Rates. To simplify the scaleup process, in predicting CO_2 flood performance, the incremental oil production rate to CO_2 injection at the analogous reservoir is used, rather than the total oil recovery rate that is actually occurring. Using the incremental rate removes any ongoing effect of the previous recovery process (e.g., waterflood). Once the incremental oil recovery rate has been scaled up to the target reservoir, this rate then is added to the rate for current operation at the target reservoir (extrapolated at least up to the expected CO_2 flood completion time) to derive the total expected oil rate.

Decline curve analysis[42–44] is the tool most commonly used to extrapolate the oil rate before CO_2 injection. This analysis likely will go beyond the economic limit of the current operation because the CO_2 flood will stretch economic operation beyond the well's current production time.

Reservoir simulation studies in oil-wet and mixed-wet west Texas reservoirs have shown that the rate of incremental tertiary oil recovery to CO_2 flooding is only slightly sensitive to the stage of waterflooding when the CO_2 flood starts, as long as the reservoir pressure is well above the thermodynamic MMP. This conclusion also should apply to a dipping reservoir that has experienced water influx and will be CO_2 flooded from the top down. Again, the reservoir pressure must be above the thermodynamic MMP for this assumption to be valid.

Necessary Similarities in Fluid and Reservoir Properties and Operations. The analogous and target reservoirs must be similar in the following ways for the scaleup to provide reliable rate estimates:

- Reservoir geological depositional environment. Carbonates and sandstones obviously are not similar; neither are sandbar and channel deposits or natural fractures and vugs.
- Depositional order of reservoir layers. For example, if one project has increasing permeability from top to bottom, the other project also should have increasing permeability from top to bottom. Injected fluids will have different vertical sweep characteristics and sweep control mechanisms, depending on depositional order (e.g., do you plug the top or the bottom of the pay zone?), and recovery strategies will differ.
- Relative permeability characteristics—water/oil fractional flow, for example.
- Degree of reservoir stratification—similar values for Dykstra-Parsons coefficient or Lorenz coefficient [both of which are statistical measures of reservoir heterogeneity in which a number approaching unity is extremely heterogeneous and a number approaching zero is extremely homogeneous (see Glossary and Ref. 19)].
- Fluid properties—oil gravities within 1°API.
- Reservoir pressure and temperature—thermodynamic MMP's within 100 psi.
- Types of injection patterns.
- Field operation—similar stages of depletion and similar reservoir and bottomhole pressures.
- Proposed CO_2 flood operation—similar WAG ratio and CO_2 slug size.
- Gravity considerations—for dipping reservoirs and top-down CO_2 injection projects where gravity effects are important, the two projects must have similar dip and a similar ratio of viscous to gravity forces, as well as the previously listed similarities. Gravity effects that cause crossflow within geologic layers are difficult to scale up correctly.

While the above areas should be similar between the two projects, other factors may be different:

- Total formation flow capacity and total porosity-thickness product—may vary because the scaleup process accounts for differences in these two variables. The patterns also can have different areas if the pay is continuous in both, because the scaleup process accounts for different pattern areas when layer continuity exists.
- Bottomhole pressures—may differ between the two projects up to the point that fluid properties would be very different. The greatest effect of different bottomhole pressures is seen when the two projects have substantially different fractions of the reservoir that are below the thermodynamic MMP.
- Connate water saturation—may be different because the scaleup process will handle this. However, widely different connate water saturations may indicate different relative permeability characteristics and violate that similarity requirement.

Accounting for Gravity Effects. In reservoirs where you hope to take advantage of gravity by injecting CO_2 at the top of a dipping reservoir, the analogous and target reservoirs must have similar ratios of viscous forces (forces associated mostly with horizontal movement) to gravity forces. The reason for this is that the scaling equations given below do not account for gravity, such that scaleup calculations will be incorrect if the viscous force/gravity force ratio differs much between the two fields. In such cases, the authors recommend that reservoir simulation studies be used.

An expression for the ratio of viscous to gravity forces has been presented for a linear system, applicable to a top-down gas injection project in an isolated dipping fault block.[18,19] In oilfield units, this ratio is defined as[18,19]

$$R_{v/g} = \frac{2{,}050\, u\, \mu_o\, L}{k\, \Delta\rho\, h_n}, \quad \text{(3.3)}$$

where $R_{v/g}$=the ratio of viscous to gravity forces; u=the Darcy velocity in B/D-ft^2, which is equivalent to the injection rate in RB/D divided by cross-sectional area in ft; μ_o=the oil viscosity in cp; L=the distance in ft between the injection and production wells; k=the horizontal permeability in md (the horizontal direction is defined as parallel to the direction of reservoir dip); $\Delta\rho$=the density difference between oil and CO_2 in g/cm^3; and h_n=the net pay thickness in ft.

Eq. 3.3 does not include vertical permeability. An additional similarity that should be honored in scaling up reservoir performance with gravity effects is that of ratios of vertical to horizontal permeability.

An Overview of the Mathematical Scaleup Process. In scaling, the goal is to use known results for time and incremental oil rates from an analogous project to estimate the time and the incremental oil rates for a target project. The scaling process incorporates the various factors mentioned above that may differ between the two projects. The scaling relationships for CO_2 flooding are extensions of dimensionless equations for waterflooding.[45,46]

Begin the process by selecting an "average" pattern for the analogous and target reservoirs to serve as the basis for the scaleup. The average pattern for both should be chosen from around one injection well because the injection wells control the WAG ratio and CO_2 slug volume.

Next, build into a spreadsheet application a table of times and associated incremental oil production rates for the analogous project. Using these data, plus certain reservoir data, one then can build a scaleup table of times and associated incremental oil production rates for the target pattern.

The Scaling Equation for Time. The first scaling equation determines production timing. In this equation, a table of times for the target reservoir is generated by applying basic reservoir properties for the analogous and target reservoirs to a table of times from the analogous reservoir. Given an analogous project (call it project 1), and a table of times (t_1), each t_1 can be scaled to a corresponding t_2 in the target project (project 2) with the following equation:

$$t_2=(V_{dhc2}/V_{dhc1})\,(kh_{n1}/kh_{n2})\,(\Delta p_1/\Delta p_2)\,(F_{s2}/F_{s1})\,t_1, \quad \text{(3.4)}$$

where subscripts 1 and 2 correspond to projects number 1 and 2 (scaling up from project 1 to project 2); V_{dhc}=the displaceable hydrocarbon pore volume (HCPV) in RB; k=permeability in md; Δp=the pressure drop in psi between injection and production well pairs; and F_s=the mathematical shape factor term for the pattern, which is described as:

$$F_s=0.00708/\ln\,[4.492\,A/(C_{Ao}\,r_{wi}\,r_{wp})], \quad \text{(3.5)}$$

where A=the pattern area in ft^2; r_{wi}=the effective wellbore radii of the injectors in ft; r_{wp}=the effective wellbore radii of the producers in ft; and C_{Ao}=the dimensionless pattern shape factor (e.g., the shape factor, C_{Ao}, for a five-spot pattern equals 30.88). Stein and Shiralkar[3] and Larsen[2] present shape factors for other types of patterns.

In Eq. 3.4, V_{dhc} (displaceable HCPV in RB) is defined as

$$V_{dhc}=(1/5.615)\,[A\ 43{,}560\ (\phi h_n)]\,(1.0-S_{wc}-S_{om}), \quad (3.6)$$

where 5.615 is in ft^3/bbl and 43,560 is in ft^2/acre; A=the pattern area in acres; ϕ=porosity; S_{wc}=the connate water saturation; and Som = the residual oil saturation to CO_2

Note that in Eq. 3.6, the saturations for residual oil and water imply that these components behave as if they are part of the rock; in other words, they take up space that is not available to the flood. While Eq. 3.6 is a valid scaleup approach when there are differences between project areas and porosity-thickness products, the equation is subject to some error when scaling between reservoirs with different connate water saturations.

The Effect of Connate Water Saturation on Scaleup. Field B in **Fig. 3.10** demonstrates that when the difference in connate water saturation is not great, Eq. 3.6 is a valid way to scale simulated multilayer CO_2 flood performance from one connate water saturation to another. The water/oil relative permeability data used in the predictions were rescaled to new values of connate water saturation and residual oil saturation to water. (Simulated CO_2 flood performance was predicted by the method described in Chopra *et al.*[47]

However, Field A in Fig. 3.10 (which was run with the same 10% residual oil saturation to CO_2 as for Field B) shows that when the difference in connate water is larger, the rescaling process is not as accurate. Eq. 3.6 yields results similar to the normalized curves in Fig. 3.10.

The differences in scaling between different connate water saturations might be attributable to the effect of reservoir layering. The lower permeability layers in the simulations in Fig. 3.10 are not much affected by the CO_2 flood, i.e., CO_2 does not break through in the low permeability layers at the producing wells, which means that these layers still are affected by the waterflood run before CO_2 injection. For these layers, Eq. 3.6 really should contain the residual oil saturation to water, S_{or}, rather than the residual oil saturation to CO_2, S_{om}.[45,46]

Accounting for Variations in Effective Wellbore Radius. Shape factors for flood patterns can be used to adjust scaled-up incremental production and injection rates, to compensate for the effect of varying effective wellbore radii. To be rigorous in the use of Eq. 3.6, one should know the effective wellbore radius for each well; however, the effective wellbore radii may not be known, especially in producing wells for which it is difficult to analyze pressure buildup tests in the presence of multiphase flow. In scoping calculations, the logarithm that contains the effective wellbore radii is assumed to have a small effect.

The Scaling Equation for Incremental Oil Rate. The incremental oil rate equation uses a table of incremental oil rates (q_{o1}) from the analogous reservoir to derive the incremental oil rates (q_{o2}) for the target reservoir. The equation is

$$q_{o2}=(kh_2/kh_1)\,(\Delta p_2/\Delta p_1)\,(F_{s1}/F_{s2})\,q_{o1}, \quad (3.7)$$

where subscripts 1 and 2 correspond to projects 1 and 2 (scaling up from project 1 to project 2).

Eq. 3.7 also can be used to scale up any other type of injection or production rates, including water production, water injection, CO_2 injection, CO_2 production, hydrocarbon gas production, and natural gas liquid (NGL) production.

Simplification of the Scaling Equations. In cases where patterns are being scaled up, Eqs. 3.4 and 3.7 can be simplified by replacing the mathematical term ($k_h\ \Delta p/F_s$) with the term $R_{i/w}$, injection/withdrawal ratio. The resultant term is proportional to the water injection rate at watered-out conditions. The use of $R_{i/w}$ rather than the more complex mathematical term ($k_h\ \Delta p/F_s$) also compensates for the fraction of the injection that leaves the pay zone. The scaleup equations then become

$$t_2=(V_{dhc2}/V_{dhc1})\,(q_{wi1}/q_{wi2})\,(R_{i/w2}/R_{i/w1})\,t_1 \quad (3.8)$$

and

$$q_{o2}=(q_{wi2}/q_{wi1})\,(R_{i/w1}/R_{i/w2})\,q_{o1}, \quad (3.9)$$

where subscripts 1 and 2 correspond to projects 1 and 2 (scaling up from project 1 to project 2) and q_{wi}=the water injection rate at watered-out conditions, in BWPD.

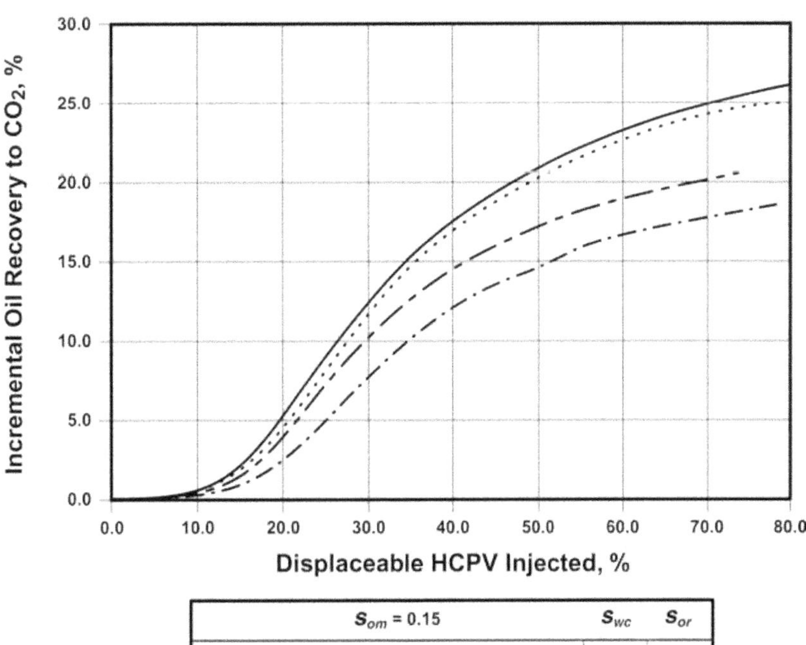

S_{om} = 0.15	S_{wc}	S_{or}
— · — · Field A Rescaled relative permeability	0.215	0.294
— — – Field A Original relative permeability	0.328	0.297
——— Field B Rescaled relative permeability	0.215	0.294
· · · · · · Field B Original relative permeability	0.192	0.286

Fig. 3.10—Dimensionless CO_2 incremental oil recovery. If connate water changes are small, Eq. 3.6 is a valid way to scale simulated multilayer CO_2 flood performance from one connate water saturation to another.

A Hypothetical Example. To illustrate their use, we have applied Eqs. 3.6, 3.8, and 3.9 below in a hypothetical scaling up of an incremental oil rate using the pattern properties in **Table 3.1.**

Using Eq. 3.6, the displaceable HCPV's in RB for patterns 1 and 2 are given, respectively, by the following calculations.

$$V_{dhc1}=(1/5.615)\ (40)\ (43{,}560)\ (10.0)\ (1.00-0.20-0.15)$$
$$=2{,}017{,}026$$

$$V_{dhc2}=(1/5.615)\ (30)\ (43{,}560)\ (12.0)\ (1.00-0.15-0.15)$$
$$=1{,}954{,}963$$

The scaled time t_2, is computed from Eq. 3.8 to be related to the time in **Table 3.2,** t_1, by:

$$t_2=(1{,}954{,}963/2{,}017{,}026)\ (500/750)\ (1.11/1.00)\ t_1=0.711\ t_1$$

The scaled incremental oil rate is computed from Eq. 3.9 to be related to the incremental oil rate of pattern 1 by:

$$q_{o2}=(750/500)(1.00/1.11)\ q_{o1}=1.364\ q_{o1}$$

Table 3.2 and **Fig. 3.11** show the scaled up and original results over time. Fig. 3.11 plots the pattern 1 rate and the scaled up rate. Pattern 2 has an earlier response with a higher peak incremental oil rate, primarily because this pattern has greater formation flow capacity. The pattern 2 CO_2 flood also is completed more quickly because of this greater capacity. To facilitate economic evaluation, the scaled rates in Table 3.2 can be interpolated to even year values.

3.2.2 Scaleup Methods That Use Cumulative Injection and Production. Hsu[48] and Folger[49] scaled up CO_2 flood performance using curves of both cumulative incremental oil recovered (as a percentage of OOIP) and cumulative CO_2 production (as a percentage of original reservoir HCPV), plotted against a type of dimensionless cumulative injection.

For the dimensionless injection, Hsu[48] used the sum of the cumulative CO_2 and water injected (in RB) as a percentage of the original HCPV (also in RB). **Figs. 3.12 and 3.13** show plots using Hsu's scaleup technique.

Unlike that in Hsu, the x-axis used by Folger[49] in her plots (not shown here) includes only the cumulative CO_2 injected in RB (also as a percentage of the original HCPV) and ignores the amount of water injected. Plotting only the injected CO_2 is fine for continuous CO_2, but if the water cycles are large, this type of plot will show an apparent increase in incremental oil recovery being achieved without CO_2 injection. By contrast, a plot that includes both CO_2 and water injected will show that oil is recovered during both the water and CO_2 cycles.

Because CO_2 recovers more oil than does water, plots like those in Figs. 3.12 and 3.13 are sensitive to the WAG ratio. Once again, scaling up requires similar WAG ratios and cycle sizes.

To scale up from plots like those in Figs. 3.12 and 3.13, values for the injection rates of CO_2 and water must be assumed. In most cases they are assumed to be constant; for example, the water injection rate might be assumed to be equal to one-half of the preCO_2 water injection rate, and the CO_2 injection rate on a reservoir barrel basis might be assumed to be equal to the preCO_2 water injection rate. These assumed injection rates are summed over time to arrive at cumulative injection. Cumulative production then is taken from the plots (Fig. 3.12 and 3.13), and production rates are derived from the slopes of the plots.

Each scaleup method—plot and equation—has its advantage. Using plots such as those in Figs. 3.12 and 3.13 requires less reservoir and operational information. The equation approach, on the other hand, is more rigorous, allowing differences in pattern size and formation flow capacity to be included; in addition, the water and CO_2 injection rates can vary over time, rather than being assumed constant.

TABLE 3.1—SAMPLE SCALEUP DATA

	Analogous: Pattern 1	Target: Pattern 2
q_{wi}, water injection rate, BWPD	500	750
A, pattern area, acre	40	30
ϕ_h, porosity-thickness, ft	10.0	12.0
S_{om}, residual oil saturation to CO_2, fraction	0.15	0.15
S_{wc}, connate water saturation, fraction	0.20	0.15
$R_{i/w}$, injection/withdrawal ratio, RB/RB	1.00	1.11

TABLE 3.2—SCALED UP TIMES AND INCREMENTAL OIL RATES FROM EQ. 3.8

Pattern 1 Time (years)	Pattern 1 Incremental Oil Rate (STB/D)	Pattern 2 Time (years)	Pattern 2 Incremental Oil Rate (STB/D)
1.0	0	0.71	0
2.0	0	1.42	0
3.0	10	2.13	14
4.0	30	2.84	41
5.0	60	3.56	82
6.0	90	4.27	123
7.0	100	4.98	136
8.0	90	5.69	123
9.0	80	6.40	109
10.0	70	7.11	95

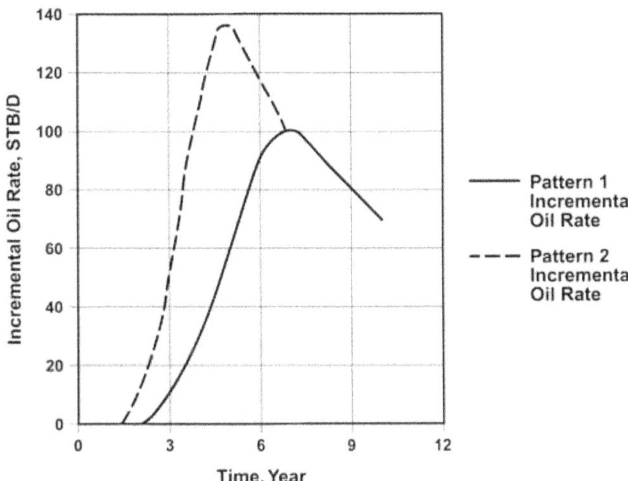

Fig. 3.11—Example of scaling up time and incremental oil rate. Pattern 2 has an earlier response with a higher peak incremental oil rate because of its greater formation flow capacity (see Table 3.2).

3.2.3 CO_2 Purchase Requirements and Reinjection of Produced Gas. At any point in the flood, the estimated CO_2 purchase requirement will be equal to the projected injection rate minus the projected production rate, assuming that produced CO_2 is reused either after separation or in combination with produced gas.

Recycling all the produced gas should have little effect on CO_2 flood performance, as long as enough pure CO_2 is used ahead of the recycled impure CO_2. Reinjected NGL's and H_2S tend to lower the thermodynamic MMP, while reinjected methane and nitrogen tend to raise it (see Sec. 2.1.6).

Because reinjected gas will be augmented with pure CO_2 until the flood is in a mature enough stage that purchased CO_2 is no longer required, there should be a sizeable buffer (~95%) of pure CO_2 ahead of the recycled (impure) CO_2. Because the recycled gas mostly serves to push along the pure CO_2, rather than to contact virgin oil itself, there should be very little net change to the thermodynamic MMP.

Note that using recycled gas to start a CO_2 flood can cause problems. The authors are aware of situations in which the thermodynamic MMP has increased several hundred psi because recycled CO_2 was being used in this way.

Eq. 2.2 can be used to quantify any changes in thermodynamic MMP caused by gas composition.

3.2.4 Estimating the H_2S Production Rate. A greater-than-expected production of H_2S in a CO_2 flood can be caused by two factors: increased vaporization of H_2S at the separators when the CO_2 mole fraction [and the gas/oil ratio (GOR)] increases, and the high solubility of H_2S in water. The vaporization-related increase happens for the same reason that increases in NGL's do; it stems from a rise in the volume of gas that bubbles out of the oil in the separators. The increase related to water solubility is subtler.

Hydrogen sulfide typically is present to some extent during a waterflood, if only because it is created by anaerobic sulfur-reducing bacteria introduced into the injected water. In the separator, H_2S partitions between the oil, the water, and the gas. At high water/oil ratios (WOR's) it prefers gas and water to oil. For example, phase equilibrium calculations for the Slaughter Estate Unit indicated that immediately before CO_2 injection (with a WOR of 11 STB/STB and a GOR of 0.5 Mscf/STB), 45% of the total H_2S was produced in the relatively small amount of gas, 46% in the water, and only 9% in the oil. The ratio of produced gaseous H_2S to oil, R_{H_2S}, was 0.156 Mscf/STB.

Under these conditions, H_2S was not a particular problem because the water was reinjected and the small amount of H_2S produced in the gas was converted easily into elemental sulfur by the sulfur removal unit in the gas processing plant.

After the CO_2 flood was underway, the total amount of gas in the system increased, and some of the H_2S that previously had been dissolved in water now partitions into the gaseous phase. The R_{H_2S} increased by a factor of 2.8. Similar increases in this ratio (factors of at least 2.0) have been observed in other CO_2 floods in west Texas and Wyoming.

For a scoping analysis, one could estimate the peak H_2S production rate to be about two to three times the product of the preCO_2 R_{H_2S} and the peak oil rate expected from the CO_2 flood. Knowing the peak H_2S production rate is important for estimating sulfur removal costs. It also is important from the standpoint of safety and the environment—for example, flaring H_2S creates sulfur dioxide.

If safety problems, legal issues and/or other restrictions can be addressed, it is possible to leave H_2S in the recycled gas. Here it will cause a slight decrease in the thermodynamic MMP and hence a slight increase in oil recovery, as long as the H_2S mole fraction is less than 10% of the gas. Eq. 2.1 in Sec. 2.1.6 can be used to estimate change in the thermodynamic MMP caused by H_2S and other impurities left in the CO_2 stream.

3.3 Cursory Investment and Operating Cost Estimates—What Will It Cost?

The next step in the technical screening process is to estimate the capital investment and incremental operating costs of processing the rates and volumes predicted using the methods outlined in Sec. 3.2. The cost estimating step requires making assumptions about facility and well designs and CO_2 flood operating practices.

There is no simple and easy estimating method because floods operate under so many reservoir and facility conditions. This section considers the main factors in making decisions on broad design issues and operating practices; it also discusses the effect such decisions have on front-end capital investments and incre-

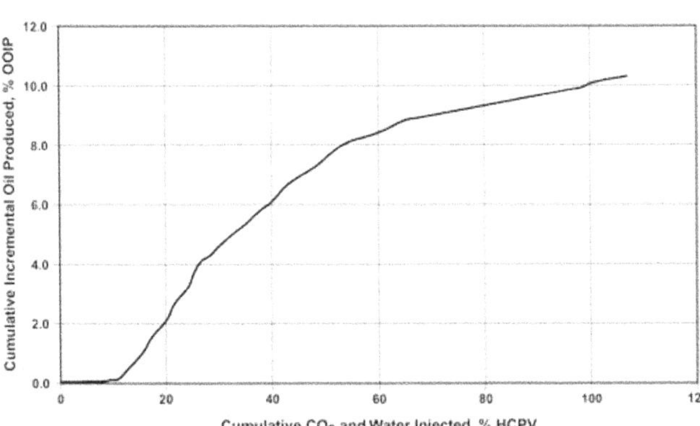

Fig. 3.12—An example of a plot for scaling up dimensionless CO_2 flood performance from one project to another to show incremental oil recovery. The plot itself shows cumulative incremental oil produced vs. cumulative CO_2 and water injected.

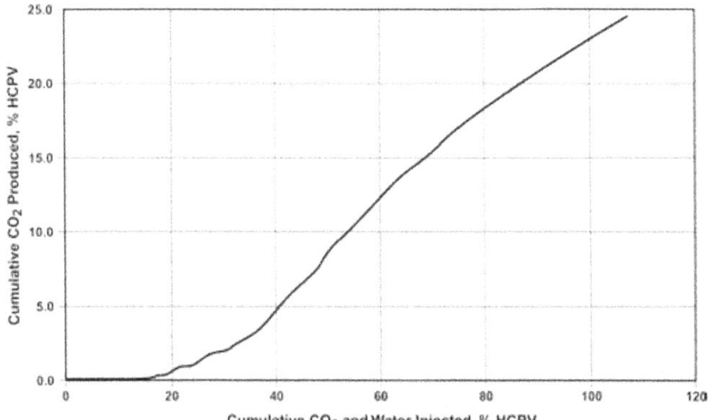

Fig. 3.13—An example scaling up dimensionless CO_2 flood performance from one project to another to show CO_2 produced. The plot itself shows cumulative CO_2 produced vs. cumulative CO_2 and water injected.

mental operating expenses. More specific design considerations are covered in Chaps. 4 through 8.

The order of discussion is:

Sec. 3.3.1—Overall design planning considerations.
Sec. 3.3.2—CO_2 source and transmission.
Sec. 3.3.3—Reinjection facilities.
Sec. 3.3.4—Production and injection facilities.
Sec. 3.3.5—Operating costs.

Table 3.3 outlines some specific expense categories that fall within the middle three areas of investment above.

Table 3.3 does not include methods for estimating drilling and workover costs (such as running liners or conformance plugbacks) because these are not specific to CO_2 flooding. The use of exotic metallurgies is an option for liner design, and costs vary according to the specific corrosion parameters required. The impact that preCO_2-flood well work will have on incremental lease fluid production and injection rates should be included in the scoping study as part of the total investment and operating costs. Drilling and workover costs can account for a significant portion of total CO_2 flood startup investments, especially when a comprehensive infill drilling program is required.

In this monograph, the time value of money is our basis for evaluating near- and long-term costs in terms of relative present value. We recognize that some design choices may have to be made between options that have the same present value but differ in terms of larger upfront (capital costs) or smaller long-term expenditures (operating costs). In these cases, the deciding factors will be company financial and tax positions (e.g., EOR tax credits on CO_2 purchases) and government incentive and disincentive programs (e.g., environmental laws on fluid spills or CO_2 emissions).

For instance, a company installing a corrosion-resistant production system could choose between stainless steel, with its high upfront investment and low operating costs, or carbon steel, with its low upfront investment and higher operating costs from periodic corrosion inhibition treatments. The present value may be the same for either case, but the smaller upfront capital cost case may be advantageous in terms of investment spending efficiency. These business influences on scoping design are not accounted for here, although they should be included in a CO_2 flooding analysis. More about these influences will be covered in Chap. 6.

Rates, Fluids, and Processes. The most important factors in total facility and well design were covered in Sec. 3.2. After working through the reservoir screening process, an engineer should know which of the following apply to the subject field:

- Oil recovery process—miscible or immiscible.
- Flood type—continuous or WAG.
- Reinjection method—separate CO_2, separate CO_2 and H_2S, or reinject full-stream gas.
- General project size—area of flood and number of wells in area.

These parameters dictate the rates, timing, and compositions of fluids moving through the reservoir, in the wellbore, and on the surface. At this stage, the estimates of production and injection rates vs. time should be known or have been estimated using the scaleup method described in Sec. 3.2 or another method, taking into account the differences between the analogous and target fields. Comparing fields with significant differences can seriously affect the reliability of scaleup study results; for example, a large amount of methane in the CO_2 source for the target field (not present in the CO_2 source for the analogous field) could increase the thermodynamic MMP, which might require higher injection pressures or result in lower recoveries than projected by the scaleup.

Facility and Well Designs. Field facility and well designs vary considerably between continuous and WAG processes, and cost planning will depend on which injectant design type is used. **Table 3.4** shows the same investment categories listed in Table 3.3, along with a matrix of likely cost and capacity changes related to use of a WAG or continuous process for a hypothetical secondary or tertiary miscible CO_2 flood.

Many facility and well factors are unaffected by flood injectant design type, including recycle plant processing, drilling and workovers, and automation investment decisions. However, some investment categories can be markedly affected by injectant design, including water handling requirements for a secondary WAG CO_2 flood (startup of water injection always requires new

TABLE 3.3—CURSORY INVESTMENT COMPONENTS, OUTLINED BY DESIGN CATEGORY
CO_2 Source and Transmission
CO_2 recovery process
CO_2 transmission system
Reinjection Plant
Dehydration and compression
CO_2 removal
H_2S removal
Production and Injection Facilities
Production facilities
Central tank battery, production and satellite batteries, fluid gathering, gas gathering (lines and meters)
Water injection facilities
Makeup water, disposal, distribution system
CO_2 injection system
Injection skids (meters and control), flowlines
Production wells
Wellhead, lift equipment
Injection wells
Wellhead, downhole equipment

TABLE 3.4—INVESTMENT CATEGORY COMPARISON FOR DIFFERENT TYPES OF MISCIBLE FLOODS

Investment Categories	Secondary Recovery		Tertiary Recovery	
	WAG	Continuous	WAG	Continuous
Rejection plant	N	N	N	N
Dehydration/compression	ND	ND	ND	ND
CO_2 removal	ND	ND	ND	ND
H_2S removal				
Production facilities				
Central tank battery	U	NC	U	U
Satellite batteries	U	U	U	U
Fluid gathering	U	NC	U	U
Gas gathering (lines, meters)	U	U	U	U
Water injection facilities				
Makeup water	N	X	D	D
Disposal	X	NC	N	N
Distribution system	N	NC	D	NC[1]
CO_2 injection system				
Injection skids (meters, control)	N	N	N	X[1]
Flowlines	N	N	N	X[1]
Production wells				
Wellhead	U/NC	U/NC	U/NC	U/NC
Lift equipment[2]	U (wu/gu)	NC (gu)	D (wd/gu)	D(wd/gu)
Injection wells				
Wellhead	N	N	U/NC	U/NC
Downhole equipment	N	N	U/NC	U/NC

LEGEND
N = new
U = upsize capacity or psia
D = downsize capacity or psia
NC = no change
X = not needed
ND = not dependent on flood type

[1] Assumes existing water injection lines and wellhead can be modified for CO_2 injection
[2] wu = water production up, wd = water production down, gu = gas up

facilities) and CO_2 injection systems for a tertiary continuous CO_2 flood (where the existing water injection system from secondary recovery often can be used for CO_2 injection).

Reservoir Screening. The reservoir screening process described in Sec. 3.1 might reveal problems with one or more of the eight reservoir attributes described there, which would decrease the probability of achieving the oil recovery predicted by the cursory screening process described in Sec. 3.2.

Three of the reservoir screening parameters in Sec. 3.1 cannot be changed: residual oil saturation to waterflooding, reservoir heterogeneity, and reservoir wettability. In addition, two of the scaleup factors in Sec. 3.2 cannot be changed: reservoir geology and relative permeabilities. If any of the target reservoir values fall outside the range of values for the analogous reservoir, the accuracy of oil rate and/or recovery predictions could suffer.

The remaining factors, however, can be changed, although most changes will cause an increase in investment and/or operating costs. For example, reservoirs with extremely poor interwell continuity, irregular (nongeometrical) flood patterns, and/or low injector/producer ratios could require significant increases in investment and operating costs. In these cases, injection well infill, horizontal drilling and/or conversion of producing wells to injection wells would probably be needed to maintain reservoir pressure above the thermodynamic MMP. In reservoirs that require repressurization to raise the average reservoir pressure above thermodynamic MMP, startup costs can be significantly increased.

For each suboptimal factor, the cost/benefit ratio must be calculated to determine whether the extra costs would be justified by projected production gains and/or operating cost reductions.

Examples of Well and Field Costs. In a 1992 DOE publication, Pautz *et al.*[50] reported actual capital investment costs for several CO_2 floods. From this reference and a paper by Magyari,[51] we have selected a representative cross section of lease-scale projects to show the range of costs for past startup CO_2 floods. The projects include miscible and immiscible floods, secondary and tertiary floods, domestic and international floods, and carbonate and sandstone reservoirs. **Fig. 3.14** shows facility capital costs, along with the total cost per well at startup, adjusted to 1986 dollars, for nine projects.

The costs shown in Fig. 3.14 include capital investment for surface and downhole well equipment, but do not include CO_2 purchases or reinjection plant costs (except for the Nagylengyel project, which includes some plant equipment). The wide distribution in startup investment ranged from about $100M to $360M per well.

According to the flood operators, startup costs for implementing many of these floods now probably would be less than the costs actually incurred because so much experience has been gained in predicting and managing project peak rate and pressure performance. Operators today also have a better understanding of requirements for optimum facility and operational performance.

Fig. 3.15 shows a breakdown of the upfront capital investments (based on the categories presented in Table 3.3) for a west Texas tertiary miscible WAG flood. The CO_2/H_2S removal plant represented 62% of the total capital investment. The next largest spending category was production facilities, which accounted for 15% of the total cost. New gas-gathering lines and significant satellite battery upgrades accounted for most of the costs in this category. Virtually all the relatively high costs for the water injection facility (7% of the total) were spent to meet water disposal requirements. Later in the project life, water injection facility costs were positively affected by salvage of some large water injection pumps that were no longer needed. The cost for the CO_2 injection system was only 7% of the total.

The schematic shown in **Fig. 3.16** highlights the major cost categories of Fig. 3.15. Fig. 3.16 will be discussed further in Sec. 3.3.3, which covers cost estimation for production and injection facilities.

3.3.1 Overall Design Planning Considerations. Because of the complexities of CO_2 flooding, special consideration should be given to the flood design planning process to enable CO_2 injection

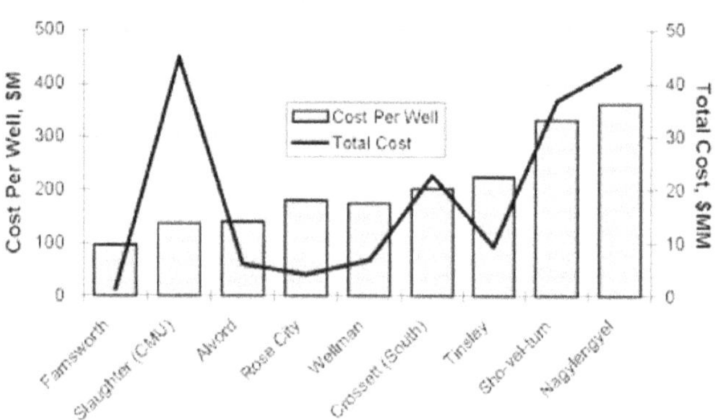

Fig. 3.14—Total startup costs for leasewide CO_2 floods. The bars show the average cost per well for the CO_2 flood in each of these fields, as shown on the left-hand y-axis. The solid line represents the total cost of the CO_2 flood in each project, shown on the right-hand y-axis (after Refs. 50 and 51).

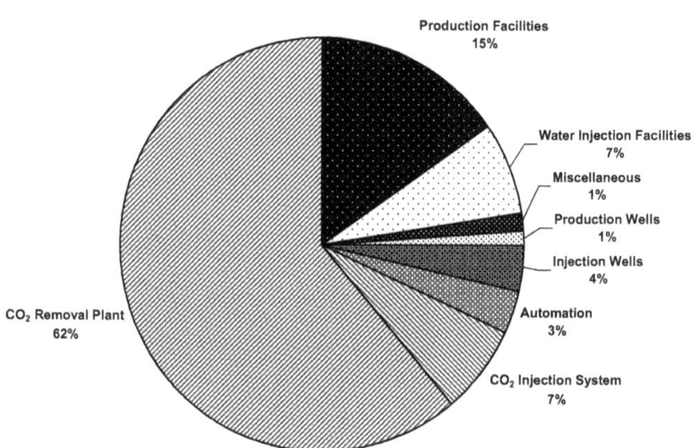

Fig. 3.15—Actual itemized startup investments for a west Texas tertiary miscible WAG CO_2 flood. The CO_2 /H_2S removal plant was the largest cost in this west Texas flood, followed by production facilities (the expense of new gas-gathering lines and battery upgrades) and water treatment. The cost for the CO_2 injection system was only 7% of the total.

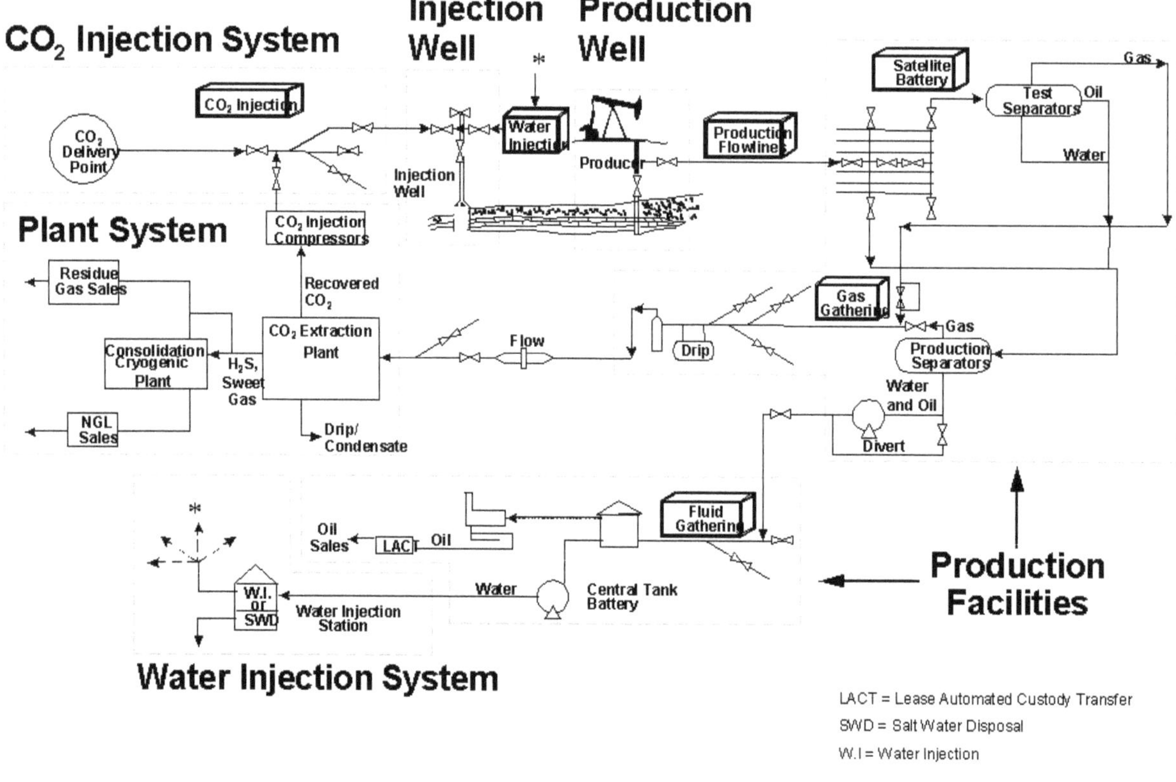

Fig. 3.16—Schematic of the total facility outlay of a west Texas CO_2 flood. This schematic of the west Texas flood represented by Fig. 3.15 shows where the money was spent.

to commence as soon as is practical and to provide more cost efficient, load-leveled production and injection.

The design steps include conducting a reservoir process optimization study; determination of CO_2 separation and/or reinjection plant size and process type (or contract negotiation with a nearby plant that is available for reinjection); deciding on facility design, startup order, and timing; contract negotiation for CO_2 supply and transmission; and securing stakeholder involvement (e.g., partners' technical and financial commitment).

Fig. 3.17 shows the timeline plan (Gantt chart) of a typical design process for a large-scale and small-scale CO_2 flood. The screening and implementation steps are shown, as well, to present the overall conceptual process and illustrate the relatively long time commitment required. A large-scale project—around 200 wells or 5,000 BOPD—can require a total project lead time of 12 to 18 months before CO_2 injection commences.

The small project Gantt chart shows that the lead time can be reduced significantly (to 6 to 12 months) by shortening the reservoir description, modeling, and prediction steps. This assumes that analogous reservoir performance can be scaled up for the final design plan, as is done with a small-scale project or an expansion to an existing field. For both large- and small-scale projects, CO_2 flood injection can begin in half the time if nearby CO_2 processing plants are available or if the full gas stream is reinjected.

In the early 1980's, the cycle time from planning to startup was about twice as long as it is today. Even so, some small-scale floods were screened, designed, and implemented within 15 to 30 months.[52,53] Detailed Gantt charts should be built for each project during the planning stage to better manage individual and team contributions and commitments continuously.

Certain factors can lengthen the project design steps shown in Fig. 3.17. Critical reservoir data required to describe, history-match, and build predictions for the reservoir should be gathered early in the planning process. If these data are incomplete or inadequate, less rigorous reservoir design measures may have to be used, such as those described in Sec. 3.2.1. Alternatives are discussed in Chap. 4. Equipment design specifications and sources also should be identified early, because deficiencies in this area may lengthen the design stage, particularly if a field demonstration pilot is required to test metallurgies, coatings, or inhibitors.

In many ways the design phase of a CO_2 flood is similar to that of a waterflood because waterflooding is a normal constituent of a CO_2 project: One must determine the injection water source and treatment (unless the field is already under waterflood operation), the design of the injection station, the injection water transmission systems, and some well mechanics. These items are not addressed here in much detail because they are documented in other publications.[19,54] We also do not address other nonspecific development planning and design tasks, such as infill drilling.

Because reinjection will account for a large part of project costs, one of the most important planning steps is identifying alternatives for reinjecting the produced CO_2 gas. Even if a nearby reinjection plant is available, it might not be able to receive the gas production from a project, particularly if other area operators are competing for

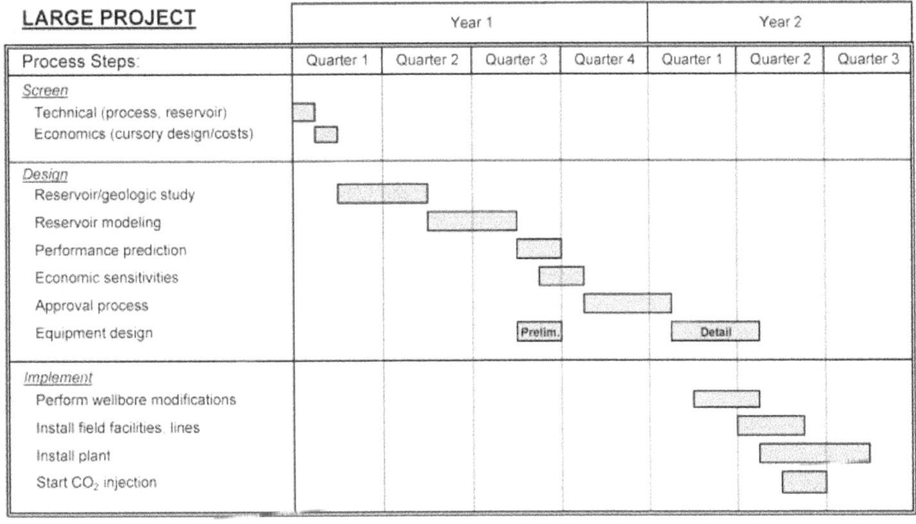

Fig. 3.17—Conceptual CO_2 flood design plan. Small projects move along faster without the detailed reservoir simulations required by larger or riskier projects.

the same plant recycle capacity. This can become a lengthy negotiating phase of the project, and it might become necessary to perform cursory plant design work to use as a base-case position.

An Iterative Process. The most effective way to design and optimize the key components of a CO_2 flood—the reservoir, facility, and economics—is to use an iterative process in which variables are modified by a team of geoscientists, engineers, reservoir modeling experts, economists, gas processing experts, and operations personnel. **Fig. 3.18** illustrates that the planning process is both sequential and a series of iterative loops. Each trip around a loop involves a greater level of detail, from initial assumptions and estimations through continual optimization until the final plan is achieved.

To create a more realistic facilities design plan, all the CO_2 flood business groups should be involved in the process, including members of the legal, administrative, computer systems, land, and plant and field operations areas. Having input from the field and plant operations groups throughout the planning phase can significantly improve the process because these personnel are familiar with operational considerations that could hinder an effective or timely startup.

Collaborative Development of Multiple-Unit Fields. In many cases, the potential for a CO_2 flood covers multiple fields in multiple units and/or lease configurations run by different operators. When this is the case, one of the best ways to improve the economics of a CO_2 project is to find out from the lease operators whether common CO_2 flood facility systems can be shared. When preparing for fieldwide CO_2 flood implementation, good communication such as this among operators can help avoid redundant systems, which make investment and operating costs higher for everyone.

Figs. 3.19a and 3.19b are conceptual illustrations of economies possible through sharing among operators. Fig.3.19a illustrates redundant recycle plants, water production/injection stations, and CO_2 lateral lines for two separate leases in the same field. While this may look unlikely on paper, this situation easily could occur, particularly if the operators of leases A and B start their CO_2 floods in different years.

When adjacent operators do not communicate, several undesirable outcomes are possible. First, the CO_2 lateral line right-of-way costs may be higher than necessary. Second, each project may use a different WAG ratio and timing, causing inefficient water injection and disposal. Third, gas production rates and timing likely will not be coordinated, leading to higher recycling costs. And fourth, cost saving opportunities could be lost, such as the ability to use one project's produced water as another project's injection water.

Fig. 3.19b illustrates the same field under collaborative development, sharing recycle plants, water production/injection stations, and CO_2 lateral lines. The result here is lower investment and operating costs for both operators. If the operators of leases A and B start up at the same time, a number of advantages can be realized. First, the advancement of the flood front will be more uniform, minimizing lease-line oil losses. Second, loads can be leveled across both projects, helping to minimize peak-rate requirements and permit more efficient operations. Third, the water disposal/makeup system can be load-leveled for more efficient use of the total number of water disposal wells; for example, it may be possible to stimulate one disposal well to increase capacity, and to plug and abandon (PxA) other wells. And fourth, CO_2 injection-line costs should be lowered, not only because less pipe is needed but also because it is less expensive (on a \$/Mscf basis) to build one large-diameter pipe than two small-diameter pipes.

Another opportunity to lower costs through communication with nearby operators lies in the size of the CO_2 supply-line lateral. Identifying future CO_2 supply needs for the area could allow planning for a larger lateral than originally expected, which can lower the cost overall because the cost of CO_2 throughput per foot of pipe is lower for larger diameter lines.

Securing a CO_2 Source. It is a good idea to begin early securing a source and price for CO_2 in areas where CO_2 trunkline distribution systems are not well developed. Begin negotiations using the CO_2 injection rates and volumes from the cursory design phase and revise as final design rates become available. One effective way to promote trunkline installation and achieve an equitable and economic CO_2 supply price in an area with great CO_2 flood potential is to secure the interest of all stakeholders—pipelines, government, potential CO_2 flood operators, and CO_2 source owners.

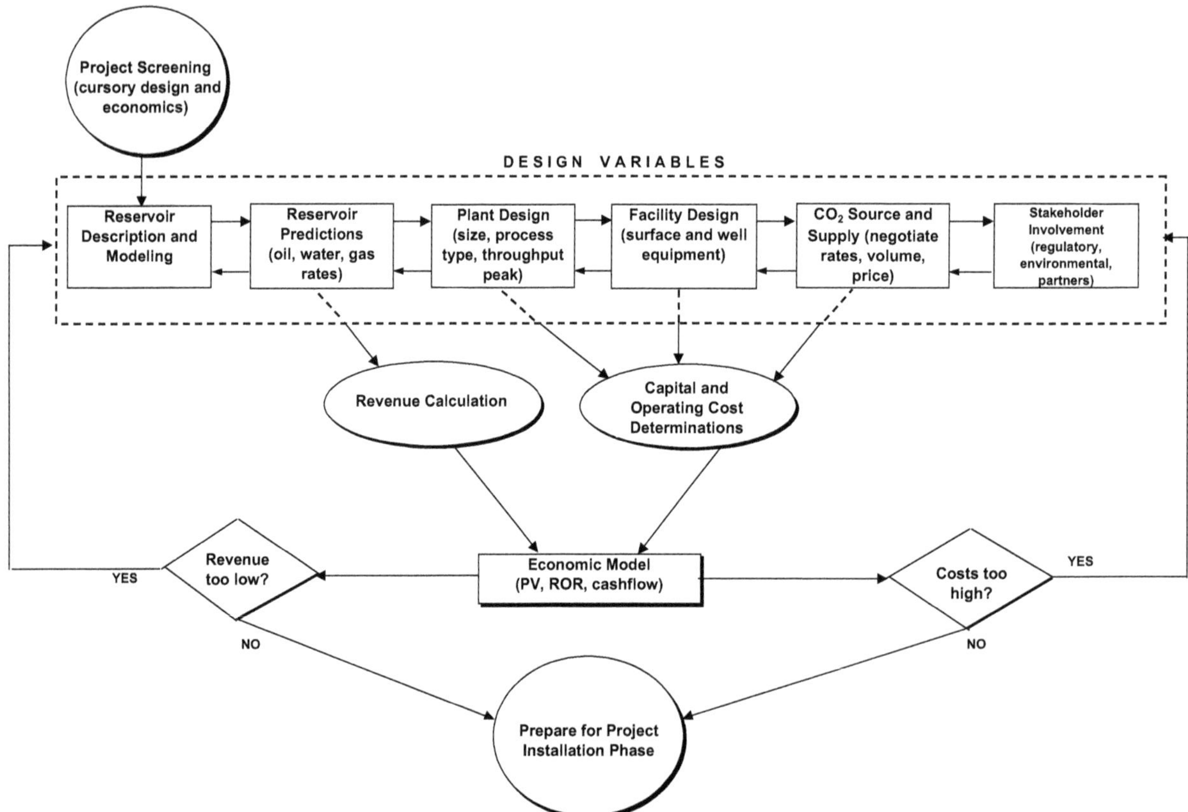

Fig. 3.18—Conceptual design process. This process can be viewed as a series of iterative loops, where each trip around a loop produces more accurate information and a more detailed plan.

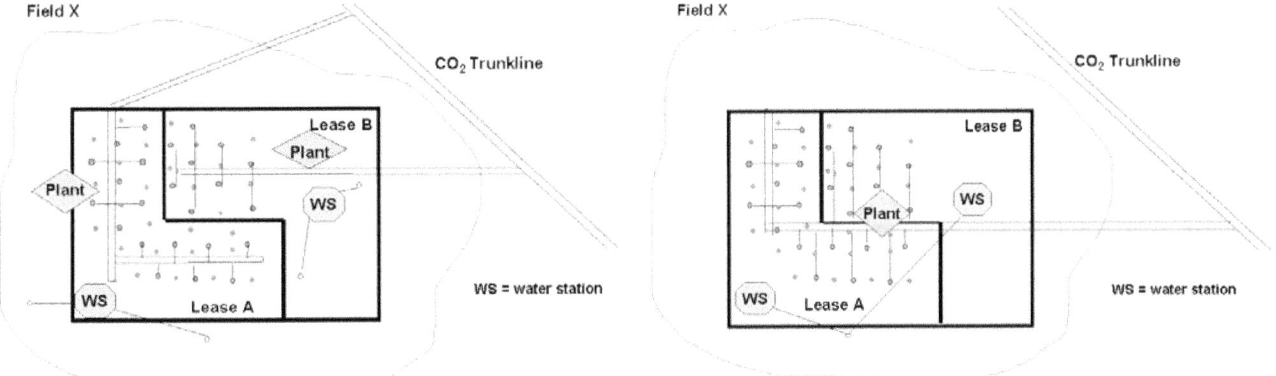

Figs. 3.19a and 3.19b—Independent vs. collaborative lease development. Lack of communication among operators preparing for CO_2 flooding can lead to higher investment and operating costs for everyone. Good communication, on the other hand, can translate into big savings, as shown by Fig. 3.19b.

For large CO_2 projects, developing alternative phase-in or expansion scenarios can have a significant impact on the final design plan and economics.[55] For example, operating a flood in phases may make it possible to build a smaller plant because the project's gas peak production rates will be lower. In addition, oil response from the first phase can help pay upfront installation costs for succeeding phases. Different gas/water injection ratios and CO_2 slug sizes also can have a strong impact in optimizing project economics (as discussed in Sec. 3.4.3).

Stakeholder Involvement. All CO_2 flood stakeholders should be involved as early as is practical in the planning process. In addition to the company personnel and offset operators who are directly involved, you also should include operating partners and the local and federal governments. Early involvement of these partners can shorten approval time and gain you valuable alternative design input. Furthermore, an early and complete understanding of the government applications and forms required for a CO_2 flood can significantly accelerate the project start date.

3.3.2 CO_2 Source and Transmission. The methods and costs of securing a supply of CO_2 are varied. Natural and manufactured sources offer a wide range of delivered CO_2 pressures, temperatures, rates, concentrations, and reliability—all important variables to consider in locating an adequate CO_2 source.

Fig. 3.20 illustrates the cost of CO_2 purchases (CO_2 gas and transport) compared with that of the plant and field equipment for a flood project. In this graph, the first 10 years' costs for several actual and planned Permian Basin WAG CO_2 flood projects are averaged and show the relatively high costs of CO_2 purchases. The CO_2 purchase cost ranged from 55 to 75%, averaging 68% of the total investment for these wells. This underscores the importance of having an accurate estimate of the cost of CO_2 purchases because it is a key component in projecting flood profitability.

A Significant Upfront Cost. To estimate the cost of CO_2, you will need to know the purity required and the planned injection rate and pressure over time, all of which can be estimated from the material presented in Secs. 3.2.1 and 2.1.6. The cost of CO_2 is a major economic component of a flood, and when estimating total project cost, the ultimate volume of purchased CO_2 usually is considered part of the capital investments because most of the CO_2 costs must be borne upfront. (When CO_2 breaks through at production wells, it can be recycled, eliminating the need for continuing large volume purchases.) Like other front-end expenses, CO_2 source and supply costs depend on whether the flood is operated miscibly or immiscibly.

Just as the CO_2 itself is considered a capital investment, so are any costs of preparing it for use in the reservoir, including purifying the gas (a CO_2 removal plant) and increasing its pressure to maintain a miscible flood (compressors). Unlike other capital expense categories, CO_2 source and supply decisions do not depend on the reservoir's stage of depletion (secondary or tertiary flood) or on the CO_2 injectant design type (WAG or continuous). Because of this, the search for a cost effective CO_2 source can begin before final flood design decisions are made.

Naturally Occurring CO_2 Sources and Costs. The most abundant and cost effective source of CO_2 is natural high-concentration reservoirs (see Sec. 1.5 for more on CO_2 sources). For CO_2 floods requiring less than 14 MMscf/D, nearby ammonia plants can be cost-competitive with natural sources. Power plants also could be viable sources of CO_2 for small research and development pilot projects that have no other source nearby, although trucking in CO_2 from other lower-cost sources may be cost-competitive if the source is not very far away.

The supply cost for fieldwide floods can be estimated using two supply systems: as a contractual delivered cost/Mscf from existing transmission systems (as in west Texas, where this cost is market driven) or as a front-end capital investment expense for installing

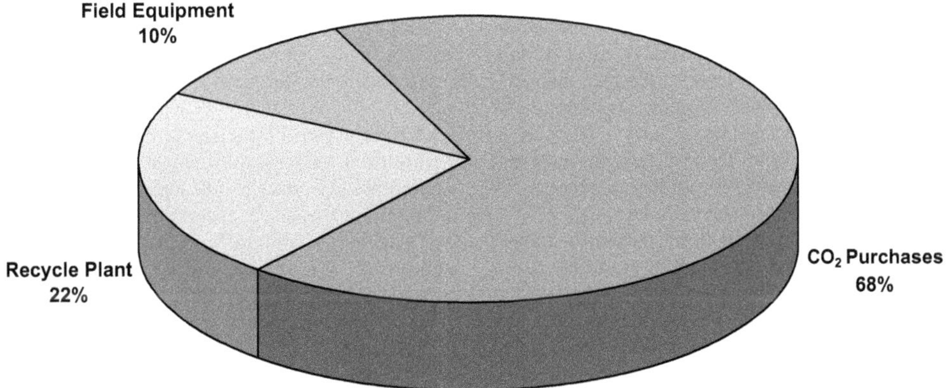

Fig. 3.20—Ten-year cost breakdown for several actual and planned west Texas CO_2 floods. Because CO_2 purchases are the single largest expense in a CO_2 flood project, an accurate estimate of that cost is a key component in projecting flood profitability. On average, the cost for CO_2 for these west Texas projects was 68% of total costs over the first 10 years of each project.

equipment and trunklines to compress and deliver CO_2 (as in Turkey and Hungary). In either system, the CO_2 source could be any of those mentioned in Sec. 1.5. We will discuss both approaches in this section, though most CO_2 needs in the U.S. can be served by tapping into existing high-purity, high-pressure CO_2 pipelines.

Taber[56] published very broad CO_2 cost estimates in 1990. He calculated CO_2 cost per barrel of incremental oil for 1982 and 1990, using pipeline-delivered CO_2 and the best industry estimates of net utilization factors (defined as cumulative CO_2 purchase volume in scf, divided by cumulative tertiary oil recovered in bbl). In 1982, CO_2 cost (injectant cost only) ranged from \$12 to \$30/bbl, and in 1989, it ranged from \$5 to \$15/bbl. The lower cost in 1989 primarily reflects lower oil prices and, to some extent, improving CO_2 use estimates. It is clear that oil prices have a strong effect on the market price of pipeline CO_2

For most leasewide CO_2 floods scheduled for startup in the near term, CO_2 most likely will come through the existing pipeline systems shown in Fig. 1.9. Estimating the cost of delivered CO_2 from these systems is fairly simple—one need only request prices from the pipeline operators. The cost quoted will include the tariff needed to recoup their cost to lay a lateral line from the existing trunkline to the lease line.

One of the first rules of thumb for estimating the cost of delivered CO_2 was published by the National Petroleum Council (NPC)[57] and updated in a 1989 Interstate Oil Compact Commission (IOCC) study.[58] The following NPC equation[57] was developed for Permian Basin projects in west Texas and eastern New Mexico. It reflects the oil-price-driven nature of the CO_2 market:

$$C_{CO_2}=0.50+(0.02\times C_o), \quad (3.10)$$

where C_{CO_2}=the CO_2 price in \$/Mscf and C_o=the oil price in \$/bbl.

For a posted oil price of \$18/bbl, Eq. 3.10 produces an estimate of \$0.86/Mscf. For example, pipeline CO_2 from Val Verde natural gas fields was delivered to the Ford Geraldine Unit for about \$1.00/Mscf (1988 dollars). Recycled CO_2 for the project was quoted at around \$0.45/Mscf.[59]

The IOCC report states that CO_2 prices in the remaining regions of the U.S. can be estimated at double the west Texas price, based on NPC operator surveys.

Year 2000 prices (in \$/Mscf) for high-pressure pipeline CO_2 delivered at the Permian Basin "CO_2 trading center" (Denver City, Texas) are lower and can be better estimated using the following equation:

$$C_{CO_2}=(0.55\times C_s)/16.00, \quad (3.11)$$

where C_s=the west Texas sour posted price in \$/bbl, 0.55=the CO_2 lowest price (floor price), regardless of oil price, in \$/Mscf; and 16.00=the index price in \$/bbl. Using \$18.00 as the posted price of west Texas sour, this method yields a CO_2 cost estimate of \$0.62/Mscf, compared to \$0.86/Mscf given by Eq. 3.10.

A shortcut for estimating CO_2 prices at Denver City (based on Eq. 3.11) is to use 3.5% of the sour oil price.

Manufactured CO_2 Sources and Costs. Understanding the general processes for recovering, compressing, and transmitting natural and manufactured CO_2 helps put the delivered costs into perspective. Fig. 1.10 shows Anada's[60] cost estimates for CO_2 recovery and compression from plant sources. Ammonia and ethanol plants offer the lowest prices for manufactured CO_2, ranging from \$0.44 to \$0.84/Mscf (1986 dollars), depending on plant capacity. The cost is low because no CO_2 removal or recovery is required (CO_2 is a normal byproduct in these plants).

Feinberg[61] used Anada's models to develop data on the cost of recovering and compressing CO_2 from noncommercial natural gas, as is the case with CO_2 from Val Verde/Delaware Basin sources. When CO_2 concentration is greater than 75 mol% in natural gas, recovery costs were assumed to be virtually the same as recovery costs in ammonia plants because no CO_2 separation process would be required. However, for CO_2 concentrations below 75 mol%, recovery and compression cost estimates were considerably higher (about \$2.80/Mscf, assuming potassium carbonate (K_2CO_3) treatment in the recovery step) and the resultant CO_2 concentration was 98 mol%.

According to Feinberg,[61] the 1990 estimated cost to recover and compress CO_2 from power plant flue gas ranged from \$1.50 to \$2.50/Mscf, depending on the type of feedstock and the volume of the output. More recent work puts CO_2 recovery and compression cost from power plant flue gas at \$0.50 to \$2.00/Mscf.[62]

CO_2 Transmission Cost. The cost to transport CO_2 depends not only on the size of the line but also the distance from the source and the terrain over which the line is laid. Small floods and pilots can secure CO_2 by means of truck, rail, or barge, but large projects require pipelines.

High-pressure, supercritical CO_2 pipelines have been shown by several investigators to be the most cost effective means of transporting CO_2.[60,63–65] **Fig. 3.21** presents cost estimates from three of these references. Anada's estimates include updates to Hare's 1978 estimates; both Anada and Hare used a discounted cash flow rate of return (DCFROR) of 15% in their models to recoup the pipeline investment. Lewin and Associates' calculations used pretax economics to reduce pipeline costs to a minimum; they also used a 20% DCFROR. Though direct comparisons cannot be made between the Lewin and Associates model and the others, the scope and trend of costs are similar.

The log-log plot of these data shows that flow in the range of 50 to 500 MMscf/D reasonably can be estimated with a straight-line function. Using Anada's 1982 model, transmission cost in pipelines with these flow rates would range from \$0.11 to \$0.36/Mscf per 100 miles.

In developing their estimates for CO_2 pipeline transmission rates, Lewin and Associates estimated trunkline capital costs using the following equation.

$$C_p=100{,}000+(2{,}008\times q_p)^{0.834}, \quad (3.12)$$

where C_p=the pipeline cost in \$/mile and q_p=the pipeline rate in MMscf/D. This equation covers line diameters of 8 to 18 in. and flow rates of 50 to 500 Mscf/D.

Note that transmission cost will be higher than indicated by the above equation when the line runs through terrain such as swamps, mountains, or rolling hills. In these cases, the cost can be estimated from other pipeline cost equations developed by Lewin and Associates.

The CO_2 supply should be as reliable as possible to minimize delivery interruptions to flood operations. When the Ford Geraldine[66] and North Cross[67] floods incurred such an interruption, it changed their flood operations and caused delays in incremental oil response. The CO_2 source for these floods was the Val Verde/Delaware Basin natural gas plants that provide CO_2 as a secondary product. The interruptions stemmed from weak markets for hydrocarbon gas, the plants' primary product stream.

Pipeline companies typically have borne the cost of pipe installation. To recover their cost plus an acceptable return on their investment, the companies charge tariffs. In 1999, CO_2 transportation cost through one of the main Permian Basin trunklines was reported to be in the range of \$0.10 to \$0.11/Mscf per 100 miles.[68]

At this point in the process, one should know not only the estimated CO_2 injection rates and volumes, but also the annual cost of delivered CO_2. These data now can be incorporated into an economic model for the flood.

3.3.3 Reinjection Plant. The term "reinjection plant" refers to any facility used to reinject either the total produced gas stream or a cleaned-up CO_2 stream into the reservoir. The market for hydrocarbon products and the reservoir MMP will determine whether the total gas stream should be processed for hydrocarbon sales and high-purity CO_2 reinjection, or merely dehydrated and recompressed for reinjection. In either case, a reinjection plant will have to be available and accessible, or one will have to be constructed.

There are a number of pluses and minuses to consider with reinjection:

- Reinjecting CO_2 will reduce purchase cost, but it will require a reinjection facility, which may have to be constructed.
- Reinjecting the total gas stream will eliminate the need to purify the CO_2, which will lower the cost of reinjection facilities,

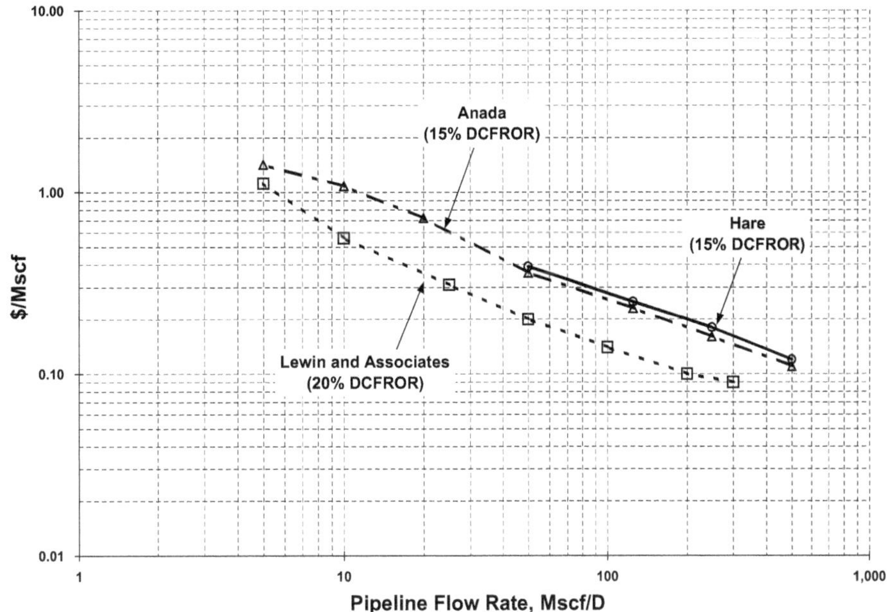

Fig. 3.21—Carbon dioxide pipeline transportation costs per 100 miles. Three different models for estimating CO_2 pipeline transmission cost arrive at similar results. The Anada and Hare models are based on a discounted cash flow rate of return (DCFROR) of 15%, while the Lewin and Associates model uses 20%. Within the range of 50 to 500 MMscf/D, the relationship on this log-log plot is a straight line.

but adding impurities to the reservoir may impair recovery by causing unacceptable increases in the reservoir MMP.

• Purifying the CO_2/hydrocarbon gas stream and selling the hydrocarbon products may be profitable if market prices are favorable. Alternately, it may cause additional problems and expense if there are issues with the amounts of methane, H_2S, and other contaminants allowed by the pipeline.

A decision to build a reinjection plant requires an estimate of capital investment. Major capital expenses include compressors, CO_2 separation equipment, and H_2S removal equipment. If a recycle plant is not needed because a nearby plant has excess available capacity, then one can project incremental operating costs based on the plant's processing fee. Fig. 3.20 shows that in west Texas, averaged CO_2/H_2S removal plant costs represented 22% of total project investments through the first 10 years. In individual projects the costs ranged from 15 to 37%.

Estimating Reinjection Plant Costs. An estimate of the size and type of reinjection plant required and the associated costs requires the following data from the screening steps we covered before, as well as some new information. Previous screening steps include:

• Reservoir conditions—MMP and reservoir pressure (e.g., whether MMP is close to reservoir pressure) and reservoir gas composition (e.g., whether it is high in C_1, H_2S, N_2) determine plant process type.

• Flood design—gas production rates vs. time (peaks) determines plant size; total gas injection composition (e.g., percent CO_2 required) determines plant process type.

Some new information also will be needed:

• Field operations—field gas collection operating pressures determine inlet pressure compression; required maximum CO_2 injection wellhead pressure and distance from plant to lease line determine outlet pressure compression.

• Market conditions—value of current and projected produced hydrocarbon gas and NGL's determines plant process type; associated power costs will be a significant part of the present value of the total plant cost, even though they are not a capital expense *per se*.

• Other factors—potential future CO_2 project implementation timing and associated gas production rates (company and outside-operated); partnering with other operators.

The above information allows the plant size (throughput rate) to be determined and hints at the potential recovery processes that could be used. Now the next scoping design decision can be addressed—the reinjection process type.

Reinjection Plant Process Types and Costs. The simplest and least expensive reinjection plant process is one that only dehydrates and compresses the recovered gas before reinjecting it, without removing the hydrocarbon gases from the stream. Examples of this approach include the projects in Hungary and Turkey and the floods that use the Val Verde gas supply. The next step up in cost is a process that separates CO_2 from the natural gas stream; this choice usually is made when there is a the desire to sell the separated NGL's. The highest cost scenario involves further separation of H_2S and other gases (such as N_2) to produce both purified CO_2 and a saleable hydrocarbon stream.

Several CO_2 removal processes are technically proven, operating on different physical and chemical principals. **Fig. 3.22** shows which process types are economical for different ranges of CO_2 feed content. These general cost comparisons assume operation at similar inlet and outlet pressures and temperatures.

Two of the most widely applied CO_2 separation methods use membranes or the Ryan-Holmes extractive distillation process. Both operate within a wide range of inlet CO_2 content and can be used in tandem with other processes, such as the chemical solvents DEA (Diethanolamine), MEA (Monoethanolamine), MDEA (Methyl Diethanolamine), TEA (Triethanolamine), and DMEA (Dimethylethylamine). When H_2S content is very high, a selective solvent or fractionation system can remove H_2S from the total gas stream. These are the LO-CAT, Claus, SRU, Selectox, and Stretford processes.[69]

Fig. 3.23 shows general investment cost trends developed from actual construction investment costs for Ryan-Holmes and membrane process plants (in combination with tertiary amines systems), both with and without H_2S removal processes. A block schematic of the process for one of these plants is shown in **Fig. 3.24.**

The costs shown in Fig. 3.23 have an accuracy margin of ±20% when used for the outlet pressures shown, and are reported in 1990 dollars. Two inlet pressures are shown, 5 psia and 150 psia, as well as two H_2S removal processes, Claus and LO-CAT. This plot is useful only for scoping economics purposes.

The cost of a plant that does not remove H_2S from the gas stream before reinjection (curves six and seven) is half that of a plant that removes H_2S. Also notice that increasing the inlet plant pressure from 5 psia to 150 psia reduces the investment costs by about 25%, primarily because fewer compressors are required. The higher inlet pressure option also will garner significant long-term savings because of reduced compression costs.

The general trend for the CO_2 and H_2S removal plant costs is about $1.0 million per MMscf of plant capacity. Without H_2S

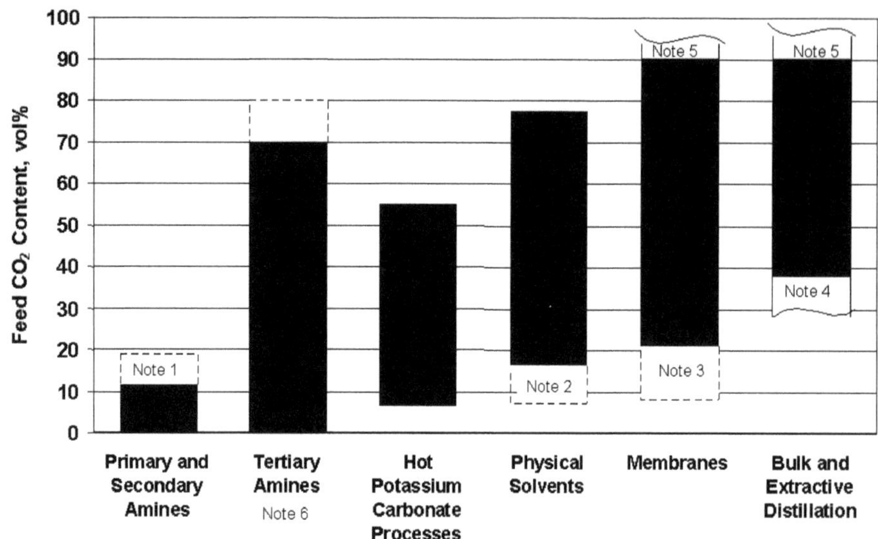

Fig. 3.22—Typical economic range for CO_2 recovery processes (after Ref. 68). There are many proven CO_2 removal processes, each of which is appropriate for a slightly different range of CO_2 in the feed. Notes: (1) chemical solvents may be economical up to approximately 20% acid gas (H_2S+CO_2); (2) physical solvents may be attractive down to 5 to 10% acid gas (50 psia partial pressure); (3) membranes may be attractive down to 8 to 12% acid gas with recycle—strongly dependent on power cost for compression; (4) the lower economic limit for bulk and extractive distillation is not yet well-defined; (5) CO_2 recovery is unlikely to be economical above 90% feed CO_2 content; (6) tertiary amines—TEA, MDEA, DMEA; (7) the range for combination solvent processes (such as sulfinol) is considered to be roughly similar to that of hot potassium carbonate and physical solvent processes.

removal, the capital cost is about $0.5 million per MMscf. Bear in mind, however, that the cost savings achieved by not separating out sour gas must be weighed against the reinjection safety risks of handling this potentially lethal gas, particularly in populated areas.

If there is a nearby process plant with excess capacity, one alternative is to pipe gas to this plant and pay a process fee, typically on an inlet MMscf basis. If this plant does not meet the CO_2 purity or pressure requirements for the CO_2 flood reservoir and operation design, more questions must be answered, such as whether the flood design can accommodate the variance of the plant CO_2 purity or pressure outputs, and whether equipment can be installed downstream of the plant to bring the CO_2 into flood design specifications. In addition, more compression may be required to transport the produced gas to the neighboring plant; even so, the overall cost scenario still may be lower than that for building a new plant.

Operating costs for CO_2 reinjection process plants in Texas in 1993 ranged from $0.22 to $0.30/Mscf for total stream (sweet gas) reinjection. These numbers include both capital amortization and operating costs. Those CO_2 removal plants that operated with H_2S extraction in 1993 cost about $0.40/Mscf per inlet stream.

3.3.4 Production and Injection Facilities. Production and injection facilities discussed in this section include the equipment components outlined under "Production and Injection Facilities" in Table 3.3, including well equipment costs. The diagram in Fig. 3.16 illustrates the process flow in an actual miscible WAG CO_2 flood facility that is representative of a typical low-pressure field gathering system and low-pressure plant inlet. Note that no field compression equipment has been detailed.

The system shown in Fig. 3.16 essentially is a waterflood operation layout, augmented by new systems for CO_2 injection, CO_2 extraction, and water disposal. Systems requiring major upsizing for this CO_2 flood were the gas-gathering lines and the satellite production vessels. As shown in Fig. 3.20, field facility costs—including construction costs for CO_2 removal plants—accounted

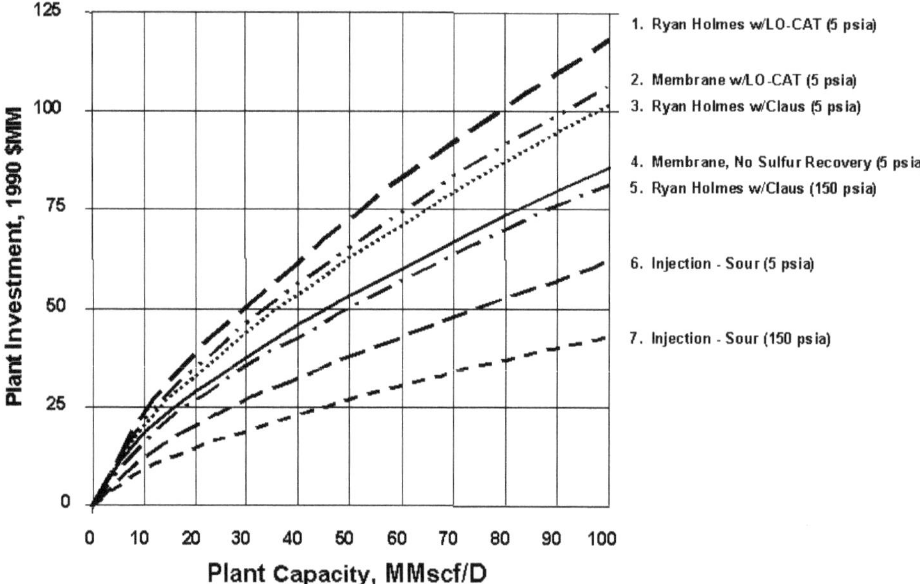

Fig. 3.23—Carbon dioxide separation plant costs (membrane, Ryan-Holmes). The graph shows plant capacity vs. cost for various configurations, with and without H_2S removal.

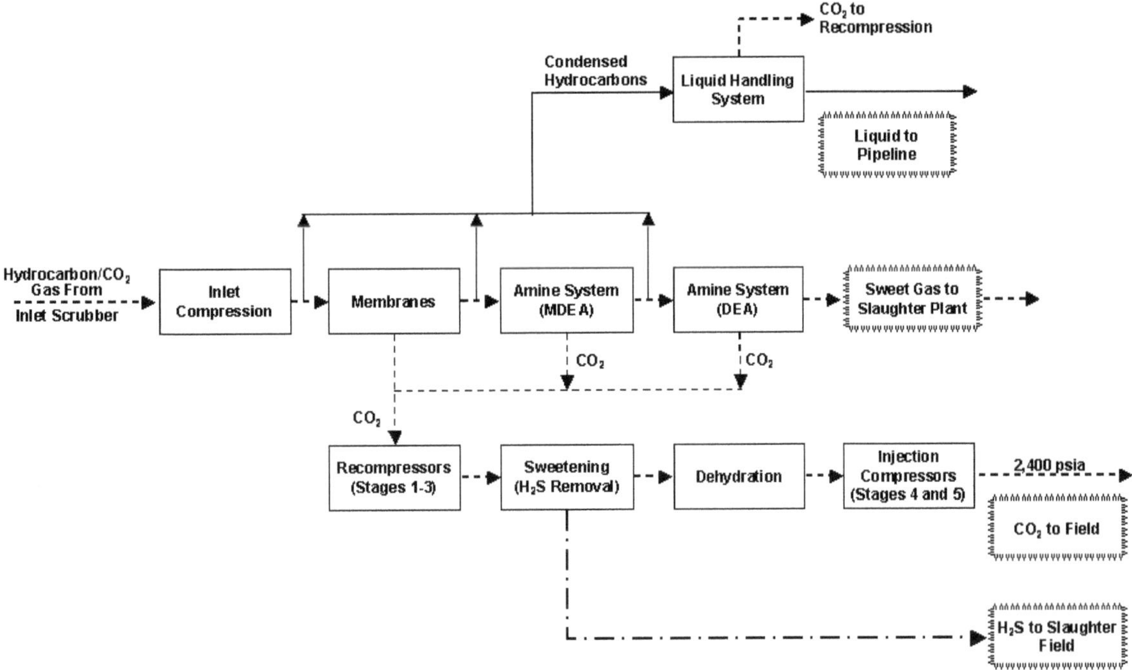

Fig. 3.24—Mallet CO_2 removal plant process flow. This removal plant incorporates membranes, amine systems, and sweetening.

for an average of 10% of total project costs for miscible tertiary five-spot-patterned WAG floods. Individual facility costs ranged from 5 to 15% of the total. More recent studies indicate that facility costs should be closer to 5% of total investment costs for the first 10 years.

Production, water injection, CO_2 injection, and well facilities are the main areas covered in the cursory design and cost estimates in this section. **Fig. 3.25** gives an idea of relative costs for each of these, showing the actual average investment costs for several west Texas CO_2 floods. These costs are shown as 1984 dollars per total well count at the time of CO_2 flood startup.

Investment Costs for Producers and Injectors. Investments indexed on a total well basis generally provide a reasonable estimating tool for each of the main production and injection facility categories. However, if investment costs are indexed by well type (producer or injector), they offer a more accurate result, as shown in **Table 3.5.** This table shows the same Permian Basin flood data presented in Fig. 3.25, but breaks them down into the detailed categories shown in Table 3.3.

Table 3.5 shows that for the Permian Basin floods, the average investment cost per producer was about $42M/well, while injectors averaged about $79M/well, excluding costs for recycle plants and water disposal wells (1984 dollars). The cost of automation systems is shown separately; it is relatively low compared to the total facility costs, and is key to flood control and monitoring, but because some operators will choose not to automate, and because automation is not required for a CO_2 flood, we will not discuss these costs here.

Like the design process for the reinjection plant, the field facilities process requires certain data. In addition to the information gathered for the reinjection plant design, one also should collect data on the oil and water production rate profiles from the scaleup and on the desired maximum production wellhead pressures. Below are the minimum data required for cursory field and well facility design.

• Compositions—liquid properties, gas properties (methane, CO_2 content).

• Rate profiles—oil and water production, hydrocarbon gas production, CO_2 production, water injection, CO_2 injection.

• Pressures—producing wellhead pressures, injection wellhead pressures.

These data may be available only on a total lease basis, but better scoping estimates can be made if data are gathered in smaller areas delineated by production satellites.

The discussions that follow focus on the main field facility design areas shown in Table 3.3.

Production Facilities. Equipment in this category includes the gas-gathering system, fluid-gathering system, production and satellite batteries, and central tank batteries. The incremental costs of upgrading and installing new production equipment will be one

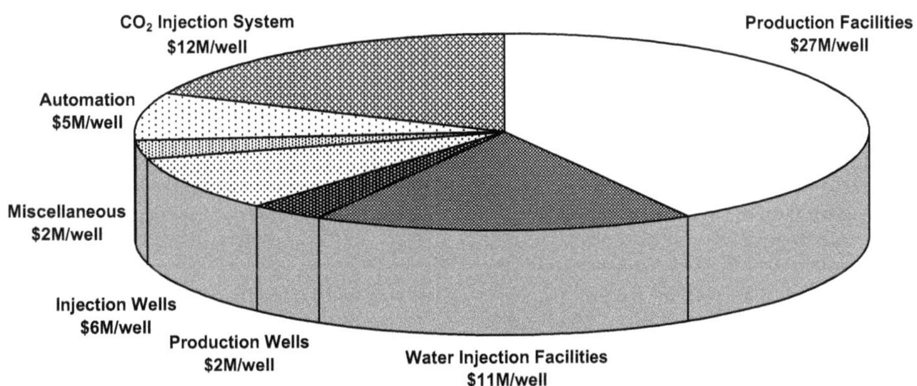

Fig. 3.25—Actual investment costs for several Permian Basin CO_2 floods. This chart gives a good overview of the relative costs of different systems. These costs are in 1984 dollars/total well count at the time of flood startup.

TABLE 3.5—INVESTMENT COSTS, INDEXED BY WELL TYPE

Producer		Injector	
Investment Cost	$M/Well	Investment Cost	$M/Well
Central tank battery	2.0	Wellhead	16.8
Satellite batteries	10.2	Downhole equipment	4.0
Fluid gathering	6.6	Automation	15.7
Gas gathering	18.1	CO_2 injection skids	5.8
Lift	2.9	CO_2 flowlines	34.2
Wellhead	0.0	Miscellaneous	2.0
Miscellaneous	2.0		
Recycle plant	106.4	Recycle plant	106.4
Makeup water	0.0	Makeup water	0.0
Salt water disposal (SWD)	12.8	Salt water disposal (SWD)	12.8
Water distribution	0.0	Water distribution	0.0
Total	160.9	Total 197.6	
Total without plant	54.5	Total without plant	91.3
Total without plant and SWD	41.7	Total without plant and SWD	78.5

All costs are in 1984 dollars.

of the highest facility costs incurred, primarily because large volumes of gas will have to be separated, metered, and transmitted to the reinjection plant.

Most upgrades will relate to gas-gathering lines and satellite batteries. Because of the corrosive nature of CO_2 when water is present, fiberglass or plastic (polyethylene) lining generally is necessary in the pipe. For low-pressure to midpressure floods (plant inlet less than about 150 psia), the pipe itself may be fiberglass or plastic, while higher-pressure gathering lines are usually plastic-lined steel. At midrange pressures, line diameter will be a deciding factor in which type of pipe to use.

At the production batteries, separators usually must be upsized to handle the increased CO_2 GOR mixture, and they may require an internal coating to prevent corrosion.

New gas measurement systems may have to be installed at each satellite location if rates and compositions are to be measured with each well fluid production test. Infrared analyzers are widely used for measurement of CO_2 concentrations in produced gas streams; the alternative is to collect gas samples and send them to a testing laboratory.

Water Injection Facilities. Water injection facilities include disposal systems, makeup systems, and injection distribution systems. Because water system costs are not unique to CO_2 flooding, they will not be covered in detail here; however, we will consider specific changes in water injection needs that should be included in the total design and cost estimate.

Water makeup may be necessary for underpressurized reservoirs that require pressurization above the thermodynamic MMP by using accelerated water injection before CO_2 injection. The additional water can be obtained from newly drilled water-source wells, excess off-lease water production, or natural surface-water sources. These water sources vary widely in cost, but the expense usually is not as high as that of the extra power needed to inject at higher wellhead pressures during repressurization. Water source costs may contribute to higher capital costs if source wells have to be drilled, or they may increase operating costs (usually on a cost/bbl basis) if water is purchased from off-lease sources.

Disposal needs arise from water injectivity reductions in tertiary CO_2 floods during the startup time. Continuous and WAG CO_2 injection into previously waterflooded areas will displace the near equivalent water injection volumes. If these displaced volumes are less than or equal to lease water production, then disposal will have to be arranged either to an existing off-lease system or into disposal wells. For continuous CO_2 floods, estimation of this displaced water rate is obvious, but estimating water injection reductions in WAG floods is not so straightforward. Water injectivity reductions of as much as 50% have been recorded during the first few cycles. One theory for the injectivity reduction is the hysteresis effect of three-phase fluid movement through pore throats that previously were mobilizing two phases. Chap. 2 provided a detailed discussion on the mechanisms of decreased water injectivity.

Once the rate profile for excess produced water is established, the disposal method and cost can be determined. Capital costs can be relatively high if disposal wells must be drilled, especially if they will be deep wells in areas with shallow freshwater aquifers under environmental protection.

For a secondary WAG CO_2 project, the water injection system will be a large investment because there will be no existing water injection system, as there is for a tertiary WAG flood. New water injection pumps, lines, wellheads, and control equipment will have to be installed, and a reliable source of water secured. However, a secondary CO_2 flood has the advantage of potentially higher S_{or} targets, which can mean earlier reserve recovery and lower total capital costs over the project life.

Starting a WAG CO_2 flood in an existing waterflood area usually does not require significant water injection equipment changes, except in cases where long-term water injectivity reductions allow salvage or downsizing of injection pumps.

CO_2 Injection System. One of the higher cost categories in the design of any type CO_2 flood is the CO_2 injection facility, which includes distribution lines and the metering and control skids. Most WAG floods operate with separate injection systems for water and CO_2, which necessitates the installation of a new CO_2 injection system. Relatively low-cost carbon steel pipe can be used for most CO_2 injection lines and valves when the CO_2 source is dry. Injection skids contain the metering and control equipment to maintain the required wellhead rate and pressure, and are usually located 30 to 70 ft from the wellhead to minimize the potential for damage during well servicing.

Shortcuts can be taken in CO_2 injection system design for a tertiary continuous flood because existing high-pressure water injection lines, wellheads, and controls can be used to transmit CO_2 to the wellhead. The chase injection for each area of the lease should be timed to reduce or eliminate substantial rework at the water injection stations. Existing injection systems were used in the Hansford Marmaton[70] and Ford Geraldine fields.[59]

Corrosion prevention costs in the CO_2 injection system are relatively small compared to capital requirements. Only the injection wellhead equipment and the downhole tubulars will need corrosion resistant materials or inhibition when dry CO_2 is transported to the wellhead.

Well Equipment. This section covers production and injection well equipment from the wellhead to the bottomhole.

These costs usually are a small fraction of the total facility investment. Rather than concentrating these costs upfront, one may spread them over the first few years of a project's life by carrying out artificial lift equipment and production wellhead equipment changes on an as-needed basis, as gas breakthrough occurs. Actual rates and pressures will dictate equipment needs at each well.

Lift changes also may be needed because water production (which increases at the start of water injection in a secondary WAG flood) declines after the first cycle of CO_2 injection because of reduced offset well WAG injectivity (see *Water Injection Facilities* in Sec. 3.3.4). Lower production rates can cause a need to downsize lift equipment. Converting to flowing wells may be a cost effective alternative if the reservoir is expected to transmit the consistent gas rates required to maintain steady oil production. This was done in the Seminole San Andres CO_2 flood, as well as in parts of other CO_2 floods. The Central Vacuum Unit in New Mexico and other floods are using plunger lift systems successfully to produce liquids efficiently.

Equipment costs for WAG injection wells are much higher than for normal producing wells because of the higher pressures involved and the increased potential for corrosion. Because of higher-pressure gas injection, upgraded packers often are required. Some operators have installed neutral-set packers with high durometer elastomers to prevent packing elements from swelling because of CO_2 invasion.[71] In most cases, wellhead valves and flanges also must be upgraded for higher pressure ratings and for corrosion resistance.

The Christmas tree assembly has the highest potential for corrosion in a WAG flood because this is where CO_2 and water are most likely to meet. To prevent corrosion from alternating CO_2 and water injection, downhole steel tubulars commonly are plastic-coated with phenolic resin, though some fiberglass line also has been used successfully.

In a 1998 CO_2 symposium,[72] Shell recommended developing scoping startup costs for well preparation for a WAG flood using these relationships:

$$C_{iw}=(5\times D)+35{,}000, \quad (3.13)$$

where C_{iw}=injection well cost in dollars and D=well depth in ft.

$$C_{pw}=40{,}000, \quad (3.14)$$

where C_{pw}=producing well cost in dollars. Of the \$40,000, \$15,000 is for equipment and \$25,000 is for cleanout/repair.

3.3.5 Operating Costs. Scoping study estimates for incremental field and well operating costs generally depend upon the same primary factors as do cursory estimates for the original facility design. Installation of new equipment means that there also will be incremental costs for operation and maintenance. Although CO_2 injectant costs usually are expensed as operating costs, we will not consider them here because estimating them is rather simple once performance profiles are established.

Like facility cost estimates, operating cost projections cannot be "cookbooked"—there is no simple formula that will apply to all (or even most) potential CO_2 floods. A few floods will realize a net reduction in operating costs, while others will see as much as a 40% increase. A study of Permian Basin floods showed that average operating costs of CO_2 flooding increased about 10% over those of prior waterflooding, not including the costs of purchasing CO_2 and handling produced gas.*

The major operating cost categories affected by CO_2 flooding are chemicals, injection and production facilities, well servicing, labor, power, and workovers (remedial type).

Chemical Usage. Chemical usage tends to increase slightly as corrosion-, scale-, and paraffin-inhibition programs are heightened, but in absolute dollars this cost usually is not a large part of the total operating cost. For example, corrosion inhibitor use has not been as high as originally expected in the Slaughter field CO_2 floods; although chemical costs for the CO_2 flood have been about 15 to 20% higher than costs for the waterflood, they still constitute a very small fraction of total field costs, around 4%. Oil is a naturally protective coating, and increased oil production can mitigate chemical needs.

Injection and Production Facilities. Maintenance costs for injection and production facilities may stay flat or even drop if existing equipment is significantly upsized using new material. In general, maintenance costs involve relatively small cost items, such as valves, meters, and small piping at production batteries and injection stations.

Labor Costs. Labor costs should be expected to increase because the high-pressure systems involved in CO_2 floods generally require more attention. Slightly more manpower also will be required to handle operations when wells will be switched frequently between water and CO_2. Some wells switch every six months, some switch every 20 to 30 days, and at the Salt Creek flood in Texas, some wells switch every day. Labor costs also will be affected by extra manpower required should a new reinjection plant be constructed to provide CO_2 to the field, and by the extent to which field automation systems are used.

Well Servicing. Another category that may increase noticeably is well servicing, again because of increased gas production. The biggest contributor to increased costs typically will be failed gas locking pumps and injection packers. Heavier kill fluids also may be required because of an increased gas rate and pressure. For 5,000-ft WAG CO_2 injection wells in the Permian Basin with no downhole mechanical shutoff equipment, incremental servicing costs range from \$8M to \$12M/well per year.

Power Costs. Power costs associated with water injection and artificial lift are likely to be a significant part of total incremental operating costs. If 30 to 50% decreases in water-cycle injectivity are realized, as with some existing CO_2 floods, then water station pump reductions and lift changes from electrical submersible pumps (ESP's) to beam or flowing can significantly lower power costs. Some floods realize as much as 25% in power savings because of reduced water injection pump requirements and conversions to flowing well status.

Workovers. Workovers usually make up a large percentage of field operating cost. Remedial workovers required to handle scale and paraffin problems related to CO_2 flooding are field-specific. Most operators have experienced some increased workover activity but report that workover levels return to near preCO_2 levels after CO_2 production rates peak. As happens with well servicing, workover fluid costs tend to increase because heavier fluids are needed to kill and control the higher gas rate and the high-pressure wells. The extra rig time needed to handle gassier wells safely also increases cost. In gravity-stable floods with high dip reservoirs, several recompletions may be necessary to minimize CO_2 cycling. Before CO_2 injection startup, some workovers may be required to minimize out-of-pay injectivity.

Summary. A recent informal survey of several Permian Basin operators, undertaken by The University of Texas of the Permian Basin's Center for Energy and Economic Diversification (CEED), showed that CO_2 flood operating costs average about 30% higher than waterflood operating costs, including the cost categories enumerated above, as well as the CO_2 recycle cost. **Fig. 3.26** shows the total annual operating cost for an active WAG CO_2 flood, starting a year before CO_2 waterflood conditions. These data also indicate an increase of about 30%.

Fig. 3.27 is a pie chart showing the accumulated costs for each of the specific operating categories discussed in this section (and also including CO_2 recycle cost) for the flood referenced in Fig. 3.26. The relative contributions of all categories except CO_2 recycle costs remained constant through eight years of operation.

Recycle costs accounted for 32% of the cumulative field expense over the eight years, as shown in Fig. 3.27. In general, this fraction will change with time, depending on the control of gas production rates at the project. The main factors driving this change are capacity limitations in the processing or recycle plant and WAG changes implemented to optimize pattern GOR's. The Twofreds CO_2 flood reported that recycle costs were about 48% of total operating costs. Considering only the field operating expenses for Twofreds, repairs and maintenance made up 31% of

*Personal Communication with L. Stiles, Exxon (1995).

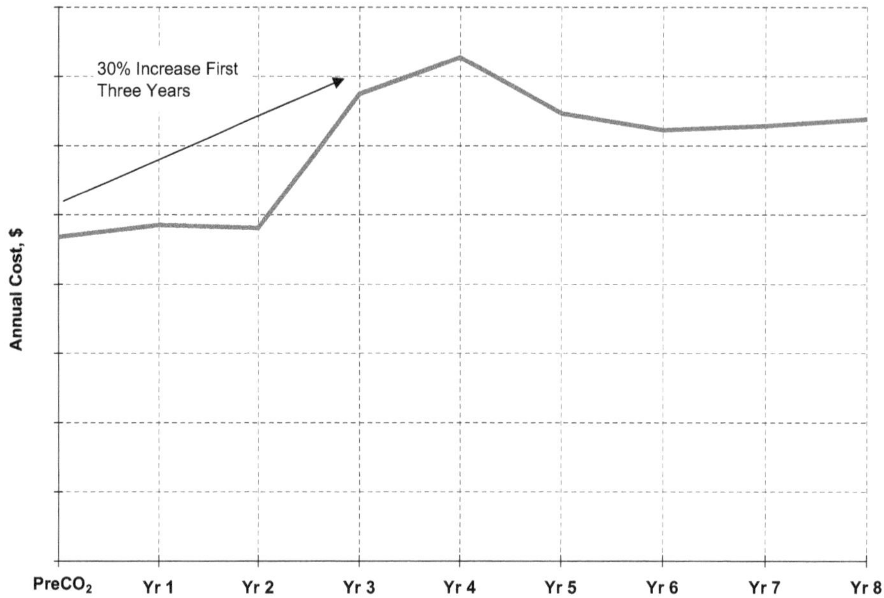

Fig. 3.26—Profile of actual field operating costs for a Permian Basin WAG CO_2 flood, with recycle. This plot shows an increase of about 30% in the early phases.

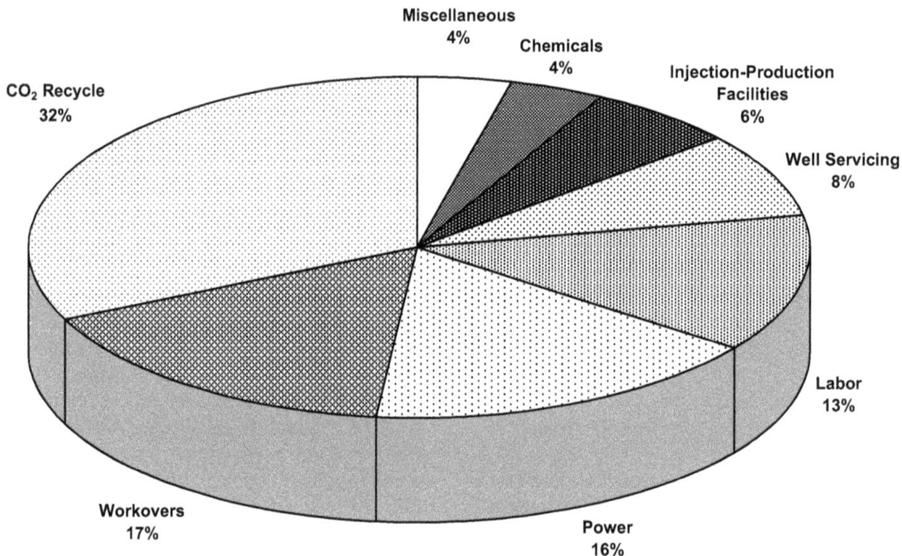

Fig. 3.27—Pie chart showing cumulative field operating costs for the Permian Basin WAG CO_2 flood referenced in Fig. 3.26, with recycle. During the eight years of flood operation, the relative sizes of the individual cost categories remained constant, except for the CO_2 recycle cost

the total; labor, 14%; corrosion chemicals, 13%; and all other expenses, 23%.[17]

3.4 Scoping Economics—Will the Incremental Oil Justify the Costs?

Once the CO_2 flood has been scoped to predict performance, investments, and operating costs, the next step is to determine whether the project might be economical. Detailed CO_2 project evaluation should proceed only after scoping economics calculations indicate that CO_2 flooding might be acceptable under a practical operating scenario.

For a project in which investment costs constitute a high risk to stakeholders, a more detailed CO_2 flood design should always be carried out; however, if the risk is low, scoping economics may be adequate to justify proceeding with implementation. A detailed facility design still will have to be performed before CO_2 injection, as outlined in Chaps. 5 and 6, but the scoping rate predictions should suffice for the design.

Economic calculations usually are based on cumulative incremental cash flow, also known as the net present value of the CO_2 flood based on the company-specific discount rate. The time frame included in the calculations begins at the startup of the flood and ends when the economic limit of the CO_2 flood (the point at which it is no longer profitable) is reached.

The discount rate accounts for the time value of money. A discount rate of 10% means that dollars invested today will grow at a rate of 10% per year; correspondingly, \$100 spent a year from now costs only \$90 in today's money.

3.4.1 Cumulative Incremental Cash Flow. If the CO_2 flood is initiated after a waterflood, the cumulative incremental cash flow is the total cash flow during CO_2 operations minus the cash flow for continued waterflood operations, up to the waterflood's economic limit. **Fig. 3.28** shows an example of cash flow from the scoping analysis of a miscible tertiary WAG CO_2 project. This example includes the construction cost of a CO_2 recycle plant.

Incremental cash flow starts out negative because of the high front-end investments often required. For the CO_2 flood to be economical, the cumulative cash flow eventually will have to become positive at the discount rate chosen by the stakeholders.

Using a higher discount rate will cause higher relative value to be placed on the early cash flow of the project, thereby increasing

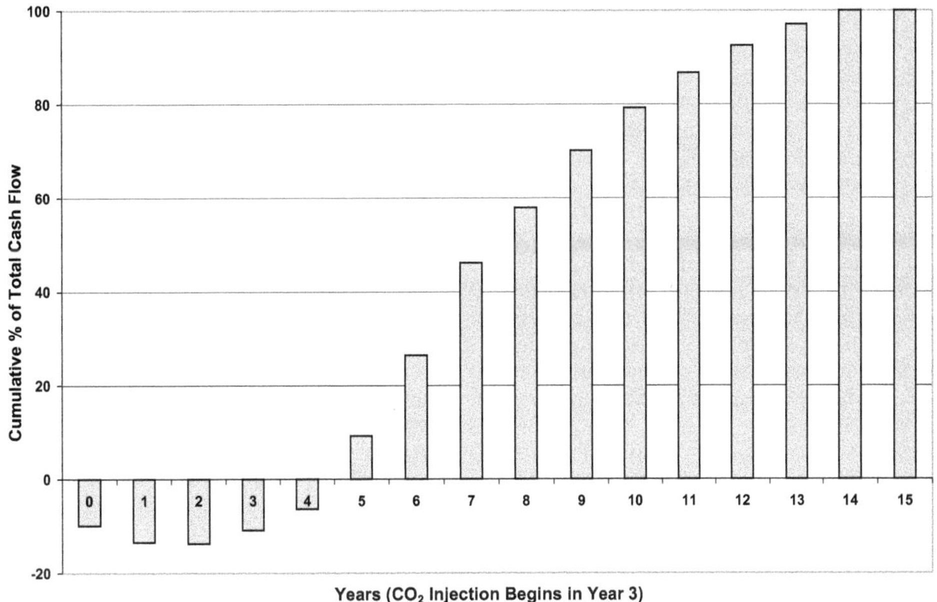

Fig. 3.28—Example of cumulative incremental cash flow in a CO_2 flood. This curve has a typical shape: upfront negative cash flow followed by positive cash flow later. The shape of the curve and the payout period depend heavily on the discount rate used.

the negative impact; in addition, the highly discounted cash flow of the later years can become negligible. At high discount rates, a positive cumulative incremental cash flow can be difficult to achieve.

A low discount rate will help mollify the impact of the early negative cash flow without overly impacting the positive cash flow later. If the discount rate is set too low, however, a project may appear to be profitable when it is not.

Cash flow during CO_2 operation is calculated by the following equation as a function of time (normally on an annual basis):

$$F_c = I - C_w - C_{CO_2} - C_{rCO_2} - C_f - C_g - T, \quad (3.15)$$

where F_c=cash flow; I=revenue (based on oil, NGL's, and gas production rates during CO_2 operation); C_w=well operating costs during CO_2 flood operations; C_{CO_2}=cost for CO_2 purchases; C_{rCO_2}=produced CO_2 recycle cost; C_f=facility investment costs; C_g=produced gas process plant investment costs; and T=taxes (including tax credits). All are in U.S. dollars.

Cash flow for continued operations (i.e., waterflood) as a function of time is usually given by

$$F_{cb} = I - C_w - T, \quad (3.16)$$

where F_{cb}=cash flow base case; I=revenue (based on oil, NGL's, gas production rates during CO_2 operations); C_w=well operating expenses during continued operations; and T=taxes. All are in U.S. dollars.

Note that all components in Eqs. 3.15 and 3.16 should reflect the company's working interest.

Fig. 3.29 plots the costs used to calculate cash flow from the analysis in Fig. 3.28 and includes an annual inflation rate of 4% and no discount rate.

In the operating-cost profile for this project, early annual expenses for CO_2 purchases are about double the total well operating costs, tapering to half the well costs after ten years. At the 10-year mark, produced gas recycling costs become more significant, averaging about the same as CO_2 purchases.

The present value of this incremental cost profile, discounted at 10% for facilities, recycle plant, CO_2 purchases, and CO_2 recycle, are shown in **Fig. 3.30.** Note that the front-end capital costs for the recycle plant are comparable in present value to the long-term CO_2

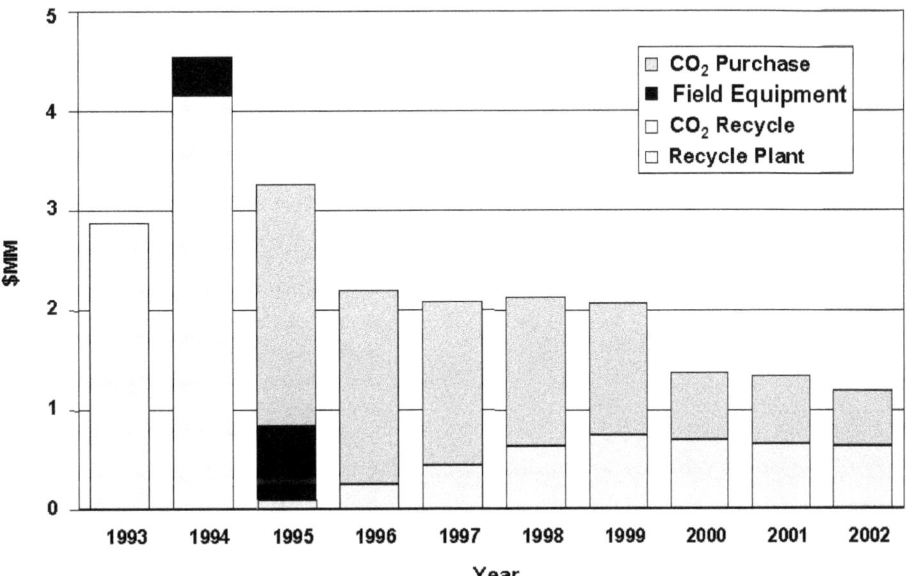

Fig. 3.29—Scoping study 10-year incremental cost projections for a miscible WAG CO_2 flood. Components shown here make up the cumulative costs shown in Fig. 3.28.

purchase costs. Also, the CO_2 recycle costs are about twice the discounted present value of the field equipment.

Field expansion of existing CO_2 projects can have significantly different cash flow profiles. Without the burden of upfront plant costs and with the ability to take advantage of existing CO_2 injection and production infrastructures at minimal cost, the cumulative cash flow could become positive in about half the time shown in the example above.

3.4.2 Ratio of Net Present Value to Investment. Another economic parameter often used to assess financial risk by measuring capital efficiency is the ratio of net present value to investment,[73] which is a discounted return on investment (a return on investment measured in terms of present value). The net present value at a particular discount rate is divided by cumulative investment costs calculated at the same discount rate.

CO_2 flood economics are prone to a low ratio of net present value to investment because of the large front-end investment typically required and the wait for the peak incremental oil rate and its associated revenue, which generally appears several years after the start of CO_2 injection. Yet, Fig. 3.28 shows that a CO_2 flood can act as a "cash cow" once it starts producing, providing a source of income for many years.

Campbell *et al.* pointed out that high discount rates fail to show the value of such cash cows.[74] They suggested using a lower discount rate during the later years of the flood, when the risk is not great, and claimed that this method more realistically represents the value of the flood. In support of this idea, one can note that many U.S. waterfloods now serving as cash cows might never have been undertaken had they been based on calculated discounted cash flow.

3.4.3 Summary. The challenge in obtaining the best financial performance from a CO_2 project is to reduce front-end investments and achieve the earliest possible oil response to CO_2. These same economic sensitivities (concern for the time value of money) should be considered for different gas processing options and different gas/water injection schemes. For example,

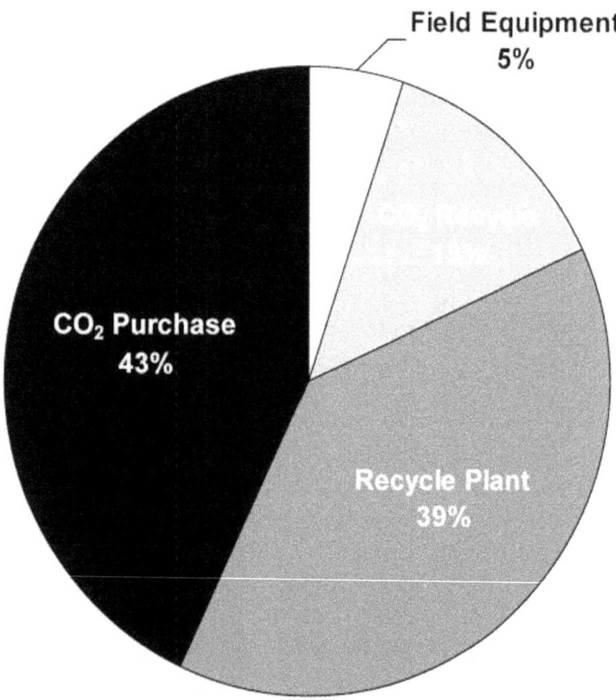

Fig. 3.30—Net present value (10) for the first 10 years of the flood shown in Fig. 3.29. At a discount rate of 10%, the cumulative present value of the components shown in Fig. 3.29 reveals that front-end capital costs for the recycle plant are comparable to long-term CO_2 purchases. In addition, CO_2 recycle costs are about twice the present value of the field equipment.

injecting at a high initial gas/water injection ratio and tapering this injection ratio later to improve sweep has been used successfully to improve financial performance in pattern CO_2 floods.[75] In addition, front-end investments can be reduced significantly if building a recycle plant is unnecessary or can be delayed, and if produced gas can be processed at an existing plant nearby. Reinjecting all the produced gas without separating the CO_2 and hydrocarbon gas also reduces front-end investment. If H_2S is present, the safety risks of high-pressure reinjection of this dangerous gas must be weighed against extraction costs.

Economic sensitivities also should be assessed against the effects of such factors as lower oil recovery, lower peak oil rate, lower oil prices, and higher than expected gas production. The sensitivity of the economic evaluation to reservoir performance is important because the scoping processes discussed in this chapter do not fully account for reservoir heterogeneities and could produce overly optimistic results. Moreover, the instability in oil prices, particularly since 1986, warrants including this factor as part of the economic analysis.

Nomenclature

A = pattern area, L^2, ft^2
C_{Ao} = pattern shape factor, dimensionless
C_{CO_2} = CO_2 price, M/L^3, U.S. \$/bbl
C_f = facility investment costs, M, U.S. \$
C_g = produced gas process plant investment costs, M, U.S. \$
C_{iw} = injection well cost, M, U.S. \$
C_o = oil price, M/L^3, U.S. \$/bbl
C_{pw} = producing well cost, M, U.S. \$
C_{rCO_2} = produced well recycle cost, M, U.S. \$
C_s = west Texas sour posted price, M/L^3, U.S. \$/bbl
C_w = well operating costs, M, U.S. \$
D = depth, L, ft
F_c = cash flow, M, U.S. \$
F_{cb} = cash flow base case, M, U.S. \$
F_s = mathematical shape factor term for a pattern, dimensionless
h_n = net pay thickness, L, ft
I = revenue, M, U.S. \$
k = horizontal permeability, L^2, md
k_{rw} = water relative permeability, dimensionless
L = distance between injector/producer well pairs, L, ft
Δp = pressure difference between injector-producer well pairs, m/Lt^2, psi
$\overline{p}_R$ = average reservoir pressure, m/Lt^2, psia
p_{bhi} = bottomhole injection pressure, m/Lt^2, psia
p_{bhp} = bottomhole production pressure, m/Lt^2, psia
q_o = incremental oil production rate, L^3/t, STB/D
q_{wi} = water injection rate at watered-out conditions, L^3/t, BWPD
r_{wi} = effective wellbore radius of an injection well, L, ft
r_{wp} = effective wellbore radius of an production well, L, ft
R_{H_2S} = ratio of produced gaseous H_2S to oil, dimensionless, Mscf/STB
$R_{i/w}$ = injection/withdrawal ratio, dimensionless
$R_{v/g}$ = ratio of viscous to gravity forces, dimensionless
S_{om} = residual oil saturation to CO_2, dimensionless, fraction
S_{or} = residual oil saturation to water, dimensionless, fraction
S_{wc} = connate water saturation, dimensionless, fraction
t = time, t, yr
T = taxes, M, U.S. \$
u = darcy velocity, $L^3/t\text{-}L^2$, $B/D\text{-}ft^2$
V_{dhc} = displaceable hydrocarbon pore volume, L^3, RB
μ_o = oil viscosity, m/Lt, cp
μ_w = water viscosity, m/Lt, cp
$\Delta\rho$ = density difference between oil and CO_2, m/L^3, g/cm^3
ϕ = porosity, dimensionless, fraction

References

1. Matthews, C.S. and Russell, D.G.: *Pressure Buildup and Flow Tests in Wells,* Monograph Series, SPE, Richardson, Texas (1967) **1.**
2. Earlougher, R.F. Jr.: *Advances in Well Test Analysis,* Monograph Series, SPE, Richardson, Texas (1977) **5.**
3. Stein, M.H. and Shiralkar, G.S.: "Estimation of Average Reservoir Pressure in Chicken Wire and Nine-Spot Patterns Under Mature Waterflood," paper SPE 24399 available from SPE, Richardson, Texas (1992).
4. Larsen, L.: "Wellbore Pressures in Reservoirs With Constant-Pressure or Mixed No-Flow/Constant-Pressure Outer Boundary," *JPT* (September 1984) 1613.
5. Chopra, A.K., Stein, M.H., and Ader, J.C.: "Development of Reservoir Descriptions To Aid in Design of EOR Projects," *SPERE* (May 1989) 143.
6. Yellig, W.F. and Metcalfe, R.S.: "Determination and Prediction of CO_2 Minimum Miscibility Pressures," *JPT* (January 1980) 160.
7. Holm, L.W. and Josendal, V.A.: "Mechanisms of Oil Displacement By Carbon Dioxide," *JPT* (December 1974) 1427.
8. Holm, L.W. and Josendal, V.A.: "Effect of Oil Composition on Miscible-Type Displacement by Carbon Dioxide," *SPEJ* (February 1982) 87.
9. Mungan, N.: "Carbon Dioxide Flooding—Fundamentals," *J. Cdn. Pet. Tech.* (January–March 1981) 87.
10. Johnson, J.P. and Pollin, J.S.: "Measurement and Correlation of CO_2 Miscibility Pressures," paper SPE 9790 presented at the 1981 SPE/DOE Enhanced Oil Recovery Symposium, Tulsa, 5–8 April.
11. Sebastian, H.M., Wenger, R.S., and Renner, T.A.: "Correlation of Minimum Miscibility Pressure for Impure CO_2 Streams," *JPT* (November 1985) 2076.
12. Johns, R.T. and Orr, F.M. Jr.: "Miscible Gas Displacements of Multicomponent Oils," *SPEJ* (March 1996) 39.
13. Wang, Y. and Orr, F.M. Jr.: "Calculation of Minimum Miscibility Pressure," paper SPE 39683 presented at the 1998 SPE/DOE Improved Oil Recovery Symposium, Tulsa, 19–22 April.
14. Moritis, G.: "EOR Increases 24% Worldwide; Claims 10% U.S. Production," *Oil & Gas J.* (1992) **90,** No. 16, 51.
15. Harpole, K.J. and Hallenbeck, L.D. "East Vacuum Grayburg San Andres CO_2 Flood Ten Year Performance Review: Evolution of a Reservoir Management Strategy and Results of WAG Optimization," paper SPE 36710 presented at the 1996 SPE Annual Technical Conference and Exhibition, Denver, 6–9 October.
16. Pittaway, K.R. and Runyan, E.E.: "The Ford Geraldine Unit CO_2 Flood: Operating History," *SPEPE* (August 1990) 333.
17. Kirkpatrick, R.K., Flanders, W.A., and DePauw, R.M.: "Performance of the Twofreds CO_2 Injection Project," paper SPE 14439 presented at the 1985 SPE Annual Technical Conference and Exhibition, Las Vegas, Nevada, 22–25 September.
18. Stalkup, F.I. Jr.: *Miscible Displacement,* Monograph Series, SPE, Richardson, Texas (1983) **8.**
19. Craig, F.F. Jr.: *The Reservoir Engineering Aspects of Waterflooding,* Monograph Series, SPE, Richardson, Texas (1971) **3,** 12–44.
20. Deans, H.A. and Schallenberger, L.K.: "Single-Well Chemical Tracer Method to Measure Connate Water Saturation," paper SPE 4755 presented at the 1974 SPE Improved Oil Recovery Symposium, Tulsa, 22–24 April.
21. Sheely, C.Q.: "Description of Field Tests To Determine Residual Oil Saturation by Single-Well Tracer Method," *JPT* (February 1978) 194.
22. Stalkup, F.I.: "Displacement of Oil by Solvent at High Water Saturation," *SPEJ* (December 1970) 337.
23. Shelton, J.L. and Schneider, F.N.: "The Effects of Water Injection on Miscible Flooding Methods Using Hydrocarbons and Carbon Dioxide," *SPEJ* (June 1975) 217.
24. Tiffin, D.L. and Yellig, W.F.: "Effects of Mobile Water on Multiple-Contact Miscible Gas Displacement," *SPEJ* (June 1983) 447.
25. Lin, E.C. and Huang, E.T.S.: "The Effect of Rock Wettability on Water Blocking During Miscible Displacement," *SPERE* (May 1990) 205.
26. Spence, A.P. and Ostrander, J.F.: "Comparison of WAG and Continuous Enriched-Gas Injection as Miscible Processes in Sadlerochit Core," paper SPE 11962 presented at the 1983 SPE Annual Technical Conference and Exhibition, San Francisco, 5–8 October.
27. Tiffin, D.L., Sebastian, H.M., and Bergman, D.F.: "Displacement Mechanism and Water Shielding Phenomena for a Rich Gas/Crude Oil System," paper SPE 17374 presented at the 1988 SPE/DOE Enhanced Oil Recovery Symposium, Tulsa, 17–20 April.
28. Magruder, J.B., Stiles, L.H., and Yelverton, T.D.: "Review of the Means San Andres Unit CO_2 Tertiary Project," *JPT* (May 1990) 638.
29. Pontious, S.B. and Tham, M.J.: "North Cross (Devonian) Unit CO_2 Flood—Review of Flood Performance and Numerical Simulation Model," *JPT* (December 1978) 1706.
30. Salathiel, R.A.: "Oil Recovery by Surface Film Drainage In Mixed-Wettability Rocks," *JPT* (October 1973) 1216.
31. Bobek, J.E., Mattax, C.C., and Denekas, M.O.: "Reservoir Rock Wettability—Its Significance and Evaluation," *Trans.,* AIME (1958) **213,** 155.
32. Amott, E.: "Observations Relating to the Wettability of Porous Rocks," *Trans.,* AIME (1959) **216,** 156.
33. Chopra, A.K.: "Reservoir Descriptions Via Pulse Testing—A Technology Evaluation," paper SPE 17568 presented at the 1988 SPE International Meeting on Petroleum Engineering, Tianjin, China, 2–4 November.
34. Kittridge, M.G.: "Quantitative CO_2 Flood Monitoring, Denver Unit Wasson (San Andres) Field," paper SPE 24644 presented at the 1992 SPE Annual Technical Conference and Exhibition, Washington, DC, 4–7 October.
35. Hill, A.D.: *Production Logging: Theoretical and Interpretive Elements,* Monograph Series, SPE, Richardson, Texas (1990) **14,** 19–60.
36. Perry, G.E.: "Weeks Island "S" Sand Reservoir B Gravity Stable Miscible CO_2 Displacement, Iberia Parish, Louisiana," paper SPE 10695 presented at the 1982 SPE/DOE Third Joint Symposium on Enhanced Oil Recovery, Tulsa, 4–7 April.
37. Johnston, J.R.: "Weeks Island Gravity Stable CO_2 Pilot," paper SPE 17351 presented at the 1988 SPE/DOE Enhanced Oil Recovery Symposium, Tulsa, 17–20 April.
38. Blunt, M.J.: "An Empirical Model for Three-Phase Relative Permeability," paper SPE 56474 presented at the 1999 SPE Annual Technical Conference and Exhibition, Houston, 3–6 October.
39. Lim, M. T. *et al.*: "Simulation of Carbon Dioxide Flooding Using Horizontal Wells," paper SPE 24929 presented at the 1992 SPE Annual Technical Conference and Exhibition, Washington, DC, 4–7 October.
40. Hadlow, R.E.: "Update of Industry Experience With CO_2 Injection," paper SPE 24928 presented at the 1992 SPE Annual Technical Conference and Exhibition, Washington, DC, 4–7 October.
41. El-Saleh, M.: "Analogy Procedure for the Evaluation of CO_2 Flooding Potential for Reservoirs in the Permian and Delaware Basins," paper SPE 35391 presented at the 1996 SPE/DOE Symposium on Improved Oil Recovery, Tulsa, 21–24 April.
42. Fetkovich, M.J.: "Decline Curve Analysis Using Type Curves," *JPT* (June 1980) 1065.
43. Lijek, S.J.: "Simple Performance Plots Used in Rate-Time Determination and Waterflood Analysis," paper SPE 19847 presented at the 1989 SPE Annual Technical Conference and Exhibition, San Antonio, Texas, 8–11 October.
44. Lo, K.K., Warner, H.R. Jr., and Johnson, J.B.: "A Study of Post-Breakthrough Characteristics of Waterfloods," paper SPE 20064 presented at the 1990 SPE California Regional Meeting, Ventura, California, 4–6 April.
45. Carlson, F.M. and Stein, M.H.: "Automatic Waterflood History Matching Using Dimensionless Performance Curves," paper SPE 24897 presented at the 1992 SPE Annual Technical Conference and Exhibition, Washington, DC, 4–7 October.
46. Stein, M.H. and Carlson, F.M.: "Dimensionless Equations for Waterflood History Matching," paper SPE 25521 available from SPE, Richardson, Texas (1993).
47. Chopra, A.K., Stein, M.H., and Dismuke, C.T.: "Prediction of Performance of Miscible-Gas Pilots," *JPT* (December 1990) 1564.
48. Hsu, C., Koinis, R.L., and Fox, C.E.: "Technology, Experience Speed CO_2 Flood Design," *Oil &Gas J.* (23 October 1995) 51.
49. Folger, L.K. and Guillot, S.N.: "A Case Study of the Development of the Sundown Slaughter Unit CO_2 Flood Hockley County, Texas," paper SPE 35189 presented at the 1996 SPE Permian Basin Oil & Gas Recovery Conference, Midland, Texas, 27–29 March.
50. Pautz, J.F. *et al.*: "Enhanced Oil Recovery Projects Data Base," Report DE92001039 prepared by IIT Research NIPER, Bartlesville, Oklahoma for Contract No. DE-FC22-83FE60149, U.S. DOE, Washington, DC (April 1992).

51. Magyari, D. and Udvardi, G.: "Operation know-how obtained by production units of Nagylengyel CO_2 gas cap recovery," *Oil &Gas J.* (22 July 1991) 99.
52. Moore, J.S.: "Design, Installation, and Early Operation of the Timbalier Bay S-2B(RA)SU Gravity-Stable, Miscible CO_2-Injection Project," *SPEPE* (September 1986) 369.
53. Bears, D.A. *et al.*: "Paradis CO_2 Flood Gathering, Injection, and Production Systems," *JPT* (August 1984) 1312.
54. Rose, S.C., Buckwalter, J.F., and Woodhall, R.J.: *The Design Engineering Aspects of Waterflooding,* Monograph Series, SPE, Richardson, TX (1989) **11.**
55. Wackowski, R.K., *et al.*: "Applying Rigorous Decision Analysis Methodology to Optimization of a Tertiary Recovery Project: Rangely Weber Sand Unit, Colorado," paper SPE 24234 presented at the 1992 SPE Oil and Gas Economics, Finance and Management Conference, London, 28–29 April.
56. Taber, J.J.: "Environmental Improvements And Better Economics In EOR Operations," *In Situ* (1990) **14,** No. 4, 345.
57. Bailey, R.E. and Curtis, L.B.: *Enhanced Oil Recovery,* National Petroleum Council, Industry Advisory Committee to the U.S. Secretary of Energy, Washington, DC (1984).
58. *An Evaluation of the Known Remaining Oil Resource in the State of Texas,* Interstate Oil Compact Commission, Oklahoma City, Oklahoma (1989).
59. Pittaway, K.R. and Rosato, R.J.: "The Ford Geraldine Unit CO_2 Flood—Update 1990," *SPERE* (November 1991) 410.
60. Anada, H.R. *et al.*: "Economics of By-Product CO_2 Recovery and Transportation for EOR," *Energy Progress* (1983) **3,** No. 4, 233.
61. Feinberg, D. and Karpuk, M.: "CO_2 Sources for Microalgae-Based Liquid Fuel Production," Report TP-232-3820 prepared by Solar Energy Research Institute, Golden, Colorado for Contract No. DE-AC02-83CH10093, U.S. DOE, Washington, DC (August 1990).
62. Tontiwachwuthikul, P. *et al.*: "Large Scale Carbon Dioxide Production from Coal-Fired Power Stations for Enhanced Oil Recovery: A New Economic Feasibility Study," *J. Cdn. Pet. Tech.* (July 1998) 48.
63. Hare, M. *et al.*: "Sources for Delivery of Carbon Dioxide for Enhanced Oil Recovery," Report FE-2512-24 prepared by Pullman Kellogg Co., Houston, for Contract No. EX-76-C-01-2515, U.S. DOE, Washington, DC (December 1978).
64. Lewin and Assocs. Inc.: "Economics of Enhanced Oil Recovery," report DOE-/ET/78-C-01-2628, Final Report for U.S. DOE (March 1981).
65. Farris, C.B.: "Unusual Design Factors for Supercritical CO_2 Pipelines," *Energy Progress* (1983) **3,** No. 3, 150.
66. Lee, K.H. and El-Saleh, M.M.: "A Full-Field Numerical Modeling Study for the Ford Geraldine Unit CO_2 Flood," paper SPE 20227 presented at the 1990 SPE/DOE Seventh Symposium on Enhanced Oil Recovery, Tulsa, 22–25 April.
67. Mizenko, G.J.: "North Cross (Devonian) Unit CO_2 Flood: Status Report," paper SPE 24210 presented at the 1992 SPE/DOE Eighth Symposium on Enhanced Oil Recovery, Tulsa, 22–24 April.
68. Wilkinson, C.: "CO_2 Supply, Transportation and Distribution in the Permian Basin," presented at the 1993 SPE Gulf Coast Section Permian Basin—Improved Oil Recovery Study Group Symposium, Houston.
69. Maddox, R.N. and Morgan, D.J.: "Select EOR Processes For CO_2," *Hydrocarbon Processing* (1986) 59.
70. Flanders, W.A., Stanberry, W.A., and Martinez, M.: "CO_2 Injection Increases Hansford Marmaton Production," *JPT* (January 1990) 68.
71. Linn, L.R.: "CO_2 Injection and Production Field Facilities Design Evaluation and Considerations," paper SPE 16830 presented at the 1987 SPE Annual Technical Conference and Exhibition, Dallas, 27–30 September.
72. 1998 Annual Permian Basin CO_2 Conference, Shell Oil Co., Midland, Texas, 8–10 December.
73. Campbell, J.M. Jr., Campbell, J.M. Sr., and Campbell, R.A.: *Analysis and Management of Petroleum Investments: Risk, Taxes, and Time,* CPS, Norman, Oklahoma (1991).
74. Campbell, J.M. Jr., Campbell, J.M. Sr., and Campbell, R.A.: "Economic Shortsightedness: One Cause and Cure," paper SPE 24651 presented at the 1992 SPE Annual Technical Conference and Exhibition, Washington, DC, 4–7 October.
75. Pariani, G.J. *et al.*: "An Approach to Optimize Economics in a West Texas CO_2 Flood," *JPT* (September 1992) 984.

SI Metric Conversion Factors

°API	141.5/(131.5 + °API)		= g/cm^3
acre ×	4.046 856	E + 03	= m^2
acre ×	4.046 856	E − 01	= ha
bbl ×	1.589 873	E − 01	= m^3
cp ×	1.0	E − 03	= Pa·s
°F	(°F − 32)/1.8		= °C
°F	(°F + 459.67)/1.8		= K
ft ×	3.048*	E − 01	= m
ft^2 ×	9.290 304	E − 02	= m^2
ft^3 ×	2.831 685	E − 02	= m^3
in. ×	2.54*	E + 00	= cm
psi ×	6.894 757	E + 00	= kPa
psia ×	6.894 757	E + 00	= kPa
yr ×	1.0*	E + 00	= a

*Conversion factor is exact.

Chapter 4
Reservoir Engineering Design Aspects of a CO_2 Flood

This chapter addresses reservoir engineering aspects of CO_2 projects that involve substantial financial risk, such that the scoping approach presented in Chap. 3 is insufficient.

Moving beyond scoping design calculations requires reservoir simulation, either with a modified black-oil reservoir simulator (whose components are separator oil, separator hydrocarbon gas, CO_2, and water) or a fully compositional reservoir simulator. Once access to an appropriate simulator is gained, there are four steps in accurately predicting CO_2 flood performance:

• Collecting valid input data (Sec. 4.1)—These data include the properties of crude oil and CO_2 at reservoir conditions and the relative permeability characteristics of the phases present.

• History matching (Sec. 4.2)—The history matching process develops and refines the reservoir description to be used to predict CO_2 flood performance. It does this by matching simulator results against actual performance during primary and waterflood production. Reservoir description parameters used in the simulator are systematically modified and fine-tuned until predictions match reality.

• Predicting CO_2 performance (Sec. 4.3)—Once reservoir flow is understood through history matching, its data are input to a simulator to predict CO_2 flood performance. This simulator may or may not be the one used to history match the reservoir. This section covers the pros and cons of modified black-oil simulators vs. fully compositional models.

• Determining the optimum flood design (Sec. 4.4)—In this step, the final design of the flood is determined through economic analysis based on reservoir simulation of various CO_2 injection schemes.

4.1 Data Requirements

This section discusses the additional data required to predict CO_2 flood performance, most of which is obtained in the laboratory. For primary production and waterflood, the data include standard fluid properties (formation volume factor for gas and oil, viscosity for gas and oil, and solution GOR), plus water/oil and gas/oil relative permeability. A CO_2 flood requires additional information about fluid properties and relative permeability relationships.

4.1.1 Fluid Properties. ***Thermodynamic MMP.*** The most important laboratory datum for CO_2 flood evaluation is the thermodynamic MMP. Below the thermodynamic MMP, a CO_2 flood may not be much of an improvement over simple pressure maintenance of a waterflood. The most direct way to measure the thermodynamic MMP is the slim-tube test described in Sec. 2.1.4 and in Yellig and Metcalfe's work on MMP prediction.[1]

If the difference between the reservoir pressure and the expected MMP is at least several hundred psi, one could bypass slim-tube tests and estimate the thermodynamic MMP using correlations and mathematical models that incorporate fine-tuned EOS's, as discussed in Sec. 2.1.5. Calibrating the EOS, however, requires fluid property data near a critical point. These data are not always easy to obtain, and one important source is slim-tube displacements at and above the thermodynamic MMP, which always go through a critical point. In one fashion or another, then, slim-tube tests are warranted for major CO_2 flood projects.

Slim-tube tests are only a starting point. Small-scale reservoir heterogeneities cause convective dispersion of CO_2 in the reservoir, which can hinder the complete development of miscibility between the CO_2 and oil (see Sec. 2.2) and cause the MMP to actually exceed the thermodynamic MMP. To assess the displacement efficiency of a CO_2 flood under actual reservoir mixing conditions, it is important to run 1D fully compositional simulations with convective dispersion. This simulation step will be discussed in Sec. 4.3.2.

Phase Equilibrium, Density, and Viscosity. Reservoir simulation is used to collect data on phase equilibrium, density, and viscosity for CO_2/crude oil mixtures. While some of these data are not needed for a modified black-oil simulator, all are required for a fully compositional simulator. The application of fully compositional simulation can include matching slim-tube test results (both above and below the thermodynamic MMP), simulating 1D CO_2 displacements to assess the effect of dispersion on oil recovery (see Sec. 2.2 about dispersion), and predicting CO_2 flood reservoir performance.

The phase equilibrium data obtained for CO_2 flooding usually are similar to the pressure-composition data shown in Figs. 2.8 and 2.9. These data are collected through a series of constant-composition expansions for various combinations of recombined crude oil hydrocarbon mixtures and CO_2.

The data in Figs 2.8 and 2.9 represent a first-contact miscibility process and may not provide the phase equilibrium information needed to accurately design a CO_2 flood. In particular, because some CO_2/oil systems (like the one in Fig. 2.9) do not exhibit a critical point upon initial contact, the first-contact phase equilibrium data are not always sufficient to calibrate an EOS. It is preferable to obtain phase equilibrium data in a way that better represents the multiple-contact miscibility process, as did Turek *et al.*[2] Multiple-contact experiments can supplement slim-tube data for use in fine-tuning an EOS.

The basic procedure for a multiple-contact experiment is as follows:

• To assess the vaporization behavior of residual oil—Once crude oil and CO_2 come to equilibrium, the vapor phase is removed and the remaining liquid mixture is contacted again with fresh CO_2 or a mixture representative of the expected CO_2 injection stream. This procedure is repeated several times.

• To assess the condensation behavior of CO_2—After crude oil and CO_2 have come to equilibrium, the oil phase is removed and replaced with fresh crude oil, while the CO_2-rich phase remains in the equilibrium cell. This procedure also is repeated several times.

CO_2 Impurities. As we discussed in Sec. 2.1.6, CO_2 purity can affect oil recovery. If recycled gas is added to CO_2 after the CO_2 breaks through (this is the general practice), the resultant level of impurities may not be great enough to affect performance because miscibility with the oil already will have been established; however, if recycled produced gas from a nearby lease, containing CO_2, is used at the start of a new CO_2 flood, it can reduce oil recovery if the thermodynamic MMP is significantly increased, and miscibility might not be developed. If the injection stream to be used is impure enough to affect performance, then a representative blend of CO_2 and impurities should be used for the fluid property and phase equilibrium measurements.

4.1.2 Relative Permeability for CO_2 Flooding. ***Water/Oil Relative Permeability.*** While water/oil relative permeability measurements already may have been made for a preceding waterflood, they may not suffice to predict the performance of the CO_2 flood. This is because the CO_2 process can cause water saturation to decrease (after having increased during a waterflood) and because the reversal in water saturation may be affected by hysteresis of the water relative permeability curve if the reservoir is moderately oil-wet. If there is a water hysteresis effect, bidirectional water/oil relative permeability tests should be run. Such tests normally are conducted under steady-state conditions.

CO_2-Rich Phase Relative Permeability. The CO_2 flood process may exhibit three-phase flow and sometimes even four-phase flow at certain pressures for reservoirs below 120°F (see Sec. 2.1.8). A new phase that often arises after the injected CO_2 develops miscibility with the oil is the CO_2-rich phase.

Historically, reservoir engineers have assumed either that the CO_2-rich phase flows in the same pores as the oil (and thus should be assigned the relative permeability of oil), or that the CO_2 flows like a dry gas (and thus is the least wetting phase). While immiscible CO_2 probably would flow like a dry gas, the flow characteristics of a CO_2-rich phase that has developed miscibility with reservoir oil are not so easily presumed.

Relative permeability for CO_2-rich phase under miscible conditions has been measured for several moderately oil-wet west Texas reservoirs.[3–5] These measurements involved unsteady-state CO_2/water relative permeability in the presence of a residual oil saturation. The residual oil saturation is critical because of the large effect it has on CO_2 relative permeability (see Sec. 2.5).

Unsteady-state measurements are easier to perform in the presence of residual oil than are the often-preferred steady-state measurements, although they have the disadvantage of sometimes yielding only endpoint water relative permeabilities. In between the endpoints (which are residual water saturation and maximum water saturation), water relative permeability can be described with a Corey[6] type of power-law model without affecting the prediction of CO_2 relative permeability significantly. This model works because the endpoint values of relative permeability and saturation are the most important parts of the relative permeability data. This model is given as:

$$k_{rw}=(k_{rw})_{\max}\,\{(S_w-S_{wr})\,/\,[(S_w)_{\max}-S_{wr}]\}^b, \qquad (4.1)$$

where k_{rw}=water relative permeability, dimensionless; $(k_{rw})_{\max}$=maximum water relative permeability attained after water displaces CO_2, dimensionless; S_w=water saturation in volume fraction; $(S_w)_{\max}$=maximum water saturation after a water cycle, volume fraction; S_{wr}=residual water saturation after a CO_2 cycle, volume fraction; and b=an empirical exponent which is always positive (a typical value is 2.0).

Available data indicate that the CO_2-rich phase has lower permeability than does oil; moreover, the presence of a small residual oil saturation appears to interfere with CO_2-rich phase relative permeability. The CO_2-rich phase permeability does not flow along the gas curve either.

Chopra *et al.*[7] theorized that the CO_2-rich phase would have reduced relative permeability. They argued that the CO_2-rich phase became more wetting than water by extracting oil components; as an intermediate wetting phase, it was adversely affected by both the oil and water phases (see Sec. 2.5). Chopra *et al.*[7] also developed a correlation for the CO_2-rich phase relative permeability curve (under miscible conditions) which qualitatively captures the relative permeability characteristics reported by Shyeh-Yung,[3] Prieditis *et al.*,[4] and Wegener and Harpole.[5] Chopra's correlation applies to oil-wet or mixed-wet reservoirs, and has been used in both modified black-oil[7] and fully compositional[8] reservoir simulators.

Occasionally, CO_2 forms a three-phase hydrocarbon region with oil (see Sec. 2.1.8). Adding water yields four phases—in addition to water, there can be a CO_2-rich liquid phase, a CO_2-rich vapor phase, and an oil-rich liquid phase. No relative permeability measurements have been reported for this system. Chang[9] groups the two CO_2-rich phases together as a matter of convenience, and if the three-phase hydrocarbon region does not occupy a large portion of the reservoir (which is the case in many but not all west Texas CO_2 floods), Chang's assumption[9] should have little adverse effect on the simulation results.

We recommend that CO_2 and water relative permeability be measured under conditions in which the CO_2 is miscible with the oil. In absence of these data, Chopra's correlation could be used for oil-wet and mixed-wet reservoirs, and the assumption that the CO_2 would flow along the oil relative permeability curve could be used for water-wet reservoirs. The gas curve could be used as an upper limit for CO_2 relative permeability for water-wet reservoirs.

4.1.3 Representative Core Samples. Ideally, one would conduct all tests using new native-state rock samples cored with a lease crude and without additives that might change wettability.[10] Native-state core is advantageous because wettability characteristics and connate water saturation are preserved, such that relative permeability measurements represent reservoir conditions.

If a core is cut using brine and without using additives that could change wettability, representative relative permeability can be reached, provided that there is no water relative permeability hysteresis effect. **Fig. 4.1** demonstrates how coring with water can flush the core to a residual oil saturation.

Standard laboratory practice is to flush the core with oil to reestablish irreducible water saturation. If there is no water hysteresis curve, the irreducible water saturation will equal connate water saturation. If there is a water relative permeability hysteresis curve, the irreducible water saturation will be higher than connate water, and the effective oil permeability (product of the oil permeability at connate water and the relative permeability) may be considerably less than it would be at connate water. The *apparent* maximum oil relative permeability (Fig. 4.1) could be misinterpreted as the oil permeability at connate water saturation, and the resultant water/oil relative permeability might not accurately represent the relative permeability characteristics of the actual rock.

The problem illustrated in Fig. 4.1 can be avoided if the natural wettability of the core is restored to the proper connate water saturation. Wendel *et al.*[11] discuss the solvent selection process for cleaning the core and recommend that the core be aged with reservoir crude and brine at reservoir pressure and temperature for 1,000 hours. This aging period is based on the time required to achieve equilibrium wetting conditions in contact angle tests.[11]

4.1.4 WAG Injection Tests. As a supplement to measuring laboratory data or using CO_2 rich-phase relative permeability correlations, one can run a field CO_2/water injection test to better understand reservoir flow characteristics. In light of the current unknowns regarding injectivity in CO_2 projects, a single-well injection test, either WAG or continuous CO_2 injection, could be a very wise decision if an inexpensive supply of CO_2 is available.

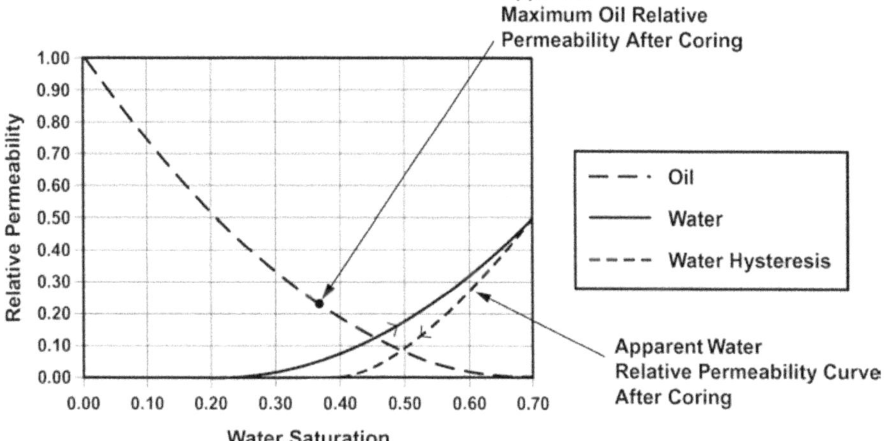

Fig. 4.1—The effect on water/oil relative permeability when water is used to core strongly oil-wet rock. Coring with water flushes the core to residual oil saturation; subsequent flushing with oil may cause water to follow the hysteresis curve for water relative permeability.

Field experience with CO_2 injection projects indicates that one full cycle can show a significant decrease in injectivity, although at least 10% HCPV of CO_2 should be injected to achieve representative injection rates.[12]

4.2 History Matching To Develop the Reservoir Description

For a simulator to accurately predict performance of a CO_2 flood, the reservoir description must realistically represent the geological characteristics of the reservoir. If the reservoir description is realistic, output from the simulator should match all previous reservoir performance. If reservoir simulator predictions of preCO_2 injection performance do not match actual field results, and if the fluid properties and relative permeability data are representative, then the reservoir description must be lacking important features. It is a waste of time to predict CO_2 flood performance with a reservoir description that does not satisfactorily match historical performance.

A reservoir description is developed by collecting and interpreting reservoir properties and then is refined by the history-match process shown in **Fig. 4.2.** Note that while the block containing the "geological, petrophysical, pressure transient, and seismic data analysis" comes after the "grid network" block, the initial analysis of geological, petrophysical, pressure transient, and seismic would be initiated much earlier. The data analysis would be incorporated into the reservoir simulator after the grid network is determined. The grid network may have to be refined further if it cannot incorporate all the important geological features (e.g., faults) of the reservoir.

4.2.1 The Model Study Area. The first step in predicting future CO_2 flood performance is to select a model study area. Simulating an entire field or lease is not always practical. Large fields can require enormous computing time for each simulation run, and the large amount of information to history match is cumbersome. Newer computers are always faster, but history matching an entire field with many wells still can be quite tedious because of the thought involved in changing the reservoir description at every well (including the effects of completions, workovers, and lift revisions).

A more tractable solution is to model a 3D sector of the field. If the sector has a shape that is repeatable across the field, sector predictions can be scaled up to the remainder of the field. The field might be divided into sections based on operations and/or geology. Separate model studies then could be scaled up for each section of the field.[12]

The size of the model study area should be practical from the standpoint of manpower requirements. The sector also should represent the performance, operation, geology, and boundary conditions of the larger area to which it will be scaled up. The number of wells in a sector pattern generally covers multiple patterns. Hsu *et al.*[8] simulated the performance of 29 inverted nine-spot patterns, encompassing 170 wells in the Wasson Denver Unit in west Texas. Another example is a sector model study of the Wasson ODC Unit in the same field, covering nine five-spot patterns, or 36 wells.

A common practice is to avoid the difficulties of accounting for individual wells within a sector by using average patterns to history match the average performance of many wells. However, the reservoir description obtained using average patterns is not as detailed as one developed with a real model study because it includes only vertical variations in reservoir properties by layers, neglecting important areal variations. Our experience is that CO_2 flood predictions based on average patterns overpredict peak oil production rate and cumulative incremental oil recovery. Some operators have attempted to account for the effect of areal heterogeneities by multiplying the incremental oil recovery predicted using average patterns by an empirical correction factor that ranges from 0.80 to 0.85. Again, we believe it is better to start with a real model study area.

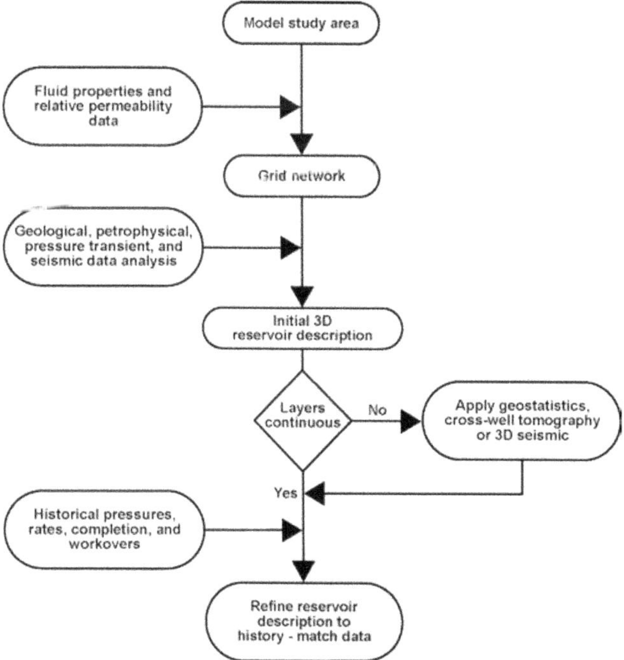

Fig. 4.2—Flowchart of the history-matching process used to refine a reservoir description. In this chart, the block containing "geological, petrophysical, pressure transient, and seismic data analysis" is placed after the "grid network" block. The initial analysis of these four sets of data actually is initiated much earlier, but the data analysis is incorporated into the reservoir simulator only after the grid network is determined.

Boundary Conditions and Flow. When choosing a model study area, it is important to ensure that the fraction of fluid injected from boundary wells is approximately constant with time because an unknown or fluctuating boundary adds uncertainty to the history matching process. This is because a model of an area like the one shown in **Fig. 4.3** assumes that corner wells are quarter wells (one-fourth the rate of the well relates to the model study area), and side wells are half wells (one-half of the rate of the well relates to the model study area).

Monitoring the IWR in a waterflood can be a useful way to determine whether injection is confined to a certain area. A stabilized IWR of 1.0 RB/RB calculated at average reservoir pressure after gas collapse usually is sufficient proof that the boundary condition is stable, i.e., that fluid does not cross the boundary. A stable IWR also may imply that the reservoir pressure is approaching steady-state behavior.

Before gas collapse, the reservoir is not filled with liquid, and an IWR greater than 1.0 RB/RB might imply that a free-gas saturation exists. It might also indicate that fluid is leaving the sector. If the IWR stabilizes to a value other than 1.0 RB/RB, and there is no injection to or production out of pay zones, then the assumed rate allocation factors for boundary wells may be in error because of different reservoir properties inside and outside the model study area. In this case, one should determine whether or not reservoir pressure and injection and production rates have stabilized. Because permeability variations can be localized (adding to the uncertainty about which rate allocation factors are in error), it may be desirable to select a larger and/or different model study area that has a stabilized IWR of 1.0 RB/RB. Two-dimensional streamline modeling also can yield valuable information about the shapes of pattern drainage areas and help in selecting a model study area.

If significant injection or production is known to flow into or out of permeable zones that are not part of the main oil-bearing zones, then these zones may need to be included in the model. Such zones could include capillary transition zones that potentially are important targets for CO_2 flooding, even though they might not be targets for waterflooding.

Ideally, the model study area represented in Fig. 4.3 would have no-flow boundaries on all sides by virtue of symmetric well positions. Fluid may flow parallel to the model study boundary, but should not cross it because any fluid movement across the boundary would violate the no-flow boundary condition common to most reservoir simulators. If there is flow across the boundary of the model study area, the only way to deal with this problem is to expand the model study area until fluid no longer flows across the boundary. It also is useful to check that early well pattern development had well configurations and completion intervals that are consistent within the model study area.

If the reservoir is influenced by water influx, the model study area should include part of the aquifer. A 3D section of the reservoir from the aquifer to the crest of the reservoir might be appropriate for edgewater drive. Bottomwater drive will require that a deeper zone with water be included.

If the reservoir is influenced by a gas cap, the gas cap also should be included. As with edgewater drive, a section of the reservoir may be used. The region should go from the crest of the gas cap to the bottom edge of the reservoir.

Application of Reservoir Cross Sections. Two-dimensional reservoir cross sections occasionally are used in place of 3D reservoir models because cross sections can include many layers and so can better explore reservoir performance in gravity-dominated processes without greatly increasing computer computation time. Cross-sectional models also are sometimes used to evaluate geostatistical reservoir descriptions that may require a large number of gridblocks. The disadvantage of these models is that they neglect areal sweep effects that can be very important in a CO_2 flood. While cross-sectional models can be used to better understand the effects of gravity override in a CO_2 flood, these models cannot be scaled up accurately to 3D because they neglect the areal sweep component.

Hybrid reservoir models have been used in which predicted cross-sectional performance is represented as fractional flow in a 2D areal streamtube model.[13–15] The gravity effects then are incorporated into a pseudofractional flow curve represented as a one-layer streamtube (see Sec. 4.2.4 for discussion of pseudocurves).

4.2.2 Time Period for History Matching. Because initial fluid saturations and pressures strongly affect waterflood performance, the history match should cover the time from initial development of the model study area to current conditions. For example, an interval will have higher water injection rates and lower waterflood oil recovery if the gas saturation is large. The distribution of fluid saturations and pressure usually are better defined at the start of field development because there has been no depletion to alter saturations and pressure from their initial values.

4.2.3 Fluid Properties and Relative Permeability Data for History Matching. History matching past performance in a model study area requires input of representative fluid property and relative permeability data. If there are concerns about which data might be the most representative, we recommend that you prioritize your data as follows:

1. Data collected from the model study area.
2. Data from the same lease, unit, or field.
3. Fluid property data from a nearby reservoir with similar oil gravity (within 1°API).
4. Laboratory-measured relative permeability data from a reservoir with a similar depositional environment and similar oil gravity (similar oil can be important in terms of representative wettability).
5. Correlations[16–23] to predict oil and gas fluid properties from API gravity and gas gravity.

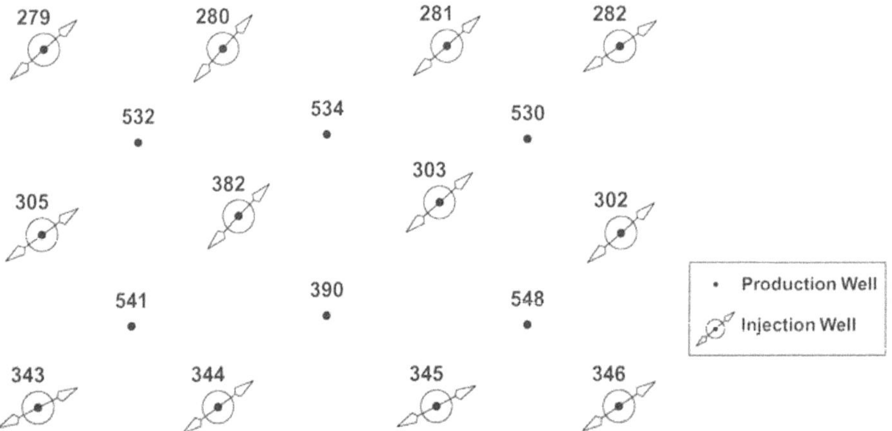

Fig. 4.3—Model study area composed of a group of patterns. Each corner injection well should have one-fourth of its total rate going into the model study area and each side well should have one-half of its rate going into the model study area.

Fluid properties required for history matching include initial bubblepoint pressure, formation volume factor for gas and oil, viscosity for gas and oil, and solution GOR, all as a function of pressure at reservoir temperature. Water/oil and gas/oil relative permeability also are required to match primary oil recovery and waterflood performance.

Relative permeability data commonly are averaged. However, if relative permeability data are available for different geological flow units (depositional events), these data should not be averaged because each geological flow unit could have a different relative permeability.

4.2.4 Grid Sensitivity Study. The grid used in the history match process must be carefully chosen. It must be fine enough to render an accurate determination of reservoir flow equations, but not so fine that it requires excessive computer time. The best way to determine the right grid is to run a grid sensitivity study after the fluid and relative permeability data have been chosen.

Fig. 4.4 shows an example of a one-layer, point-centered grid network with a diagonal grid between injection and production wells. Using properties from the Slaughter Estate Unit, Ader and Stein[24] found that a 6×6 gridblock system **(Fig. 4.5)** provided a fine enough grid to accurately simulate performance of one-quarter of a five-spot waterflood. Increasing the grid fineness beyond a 6×6 gridblock system did not change the results appreciably by engineering accuracy (Fig. 4.5). For parallel-oriented gridblocks in which the injection and production wells line up along a single grid line, Ader and Stein[24] found that a minimum of eight gridblocks between injector-producer well pairs were required to provide accurate numerical solutions **(Fig. 4.6).** The above rules of thumb have been verified for many other waterfloods in west Texas.

Additional gridblock fineness may be necessary to capture important geological features such as discontinuities and faults, or to characterize gas/oil or water/oil contacts. Localized refinement of the grid network can help improve the numerical solutions. Finer grids also are recommended for modeling horizontal wells.

Pseudorelative Permeability Curves and Their Effect on the Grid Network. The use of pseudorelative permeability curves in place of laboratory measured relative permeability curves allows coarser grids to be used. (See Sec. 4.2.6 about constructing grids for the reservoir characterization, Sec 4.2.11 about modifying pseudorelative permeability during history matching, and the SPE monograph on reservoir simulation[25] for more on pseudorelative permeability curves.) The use of pseudorelative permeability curves reduces computing time, allowing additional wells and a larger area to be incorporated into the reservoir model. Scaling up model predictions to the remainder of the field then might not be necessary.

Note, however, that the true representation of fluid flow through porous media relies on solving partial differential equations. Reservoir simulators solve partial differential equations accurately only if the grid network is fine enough for the numerical solution to converge. If too coarse a grid is used (which can be the case with pseudorelative permeability curves), the predicted answers will not converge and the results will be sensitive to the number of gridblocks. In other words, the use of pseudorelative permeability curves substitutes an engineering approximation for correct solution of the partial differential equations for flow through porous media.

Constructing pseudorelative permeability curves for waterflooding incorporates the effects of laboratory measured relative permeability, numerical dispersion, reservoir heterogeneities, and sometimes gravity effects, as well. Pseudorelative permeability curves for CO_2 injection also must capture the phase flow effects of a CO_2 flood. This latter step has been performed only by the use of pseudofractional flow curves.[13–15] In studying enriched gas WAG injection, Jerauld[26] discusses the difficulties in getting coarse-grid simulations with pseudorelative permeability curves to match fine-grid 3D simulations. Given these difficulties, the remainder of this section assumes that pseudorelative permeability curves will not be used.

Reservoir simulators use a grid network to compute either finite differences or finite elements that numerically approximate the partial derivative terms in the reservoir flow equations. In theory, as the gridblock size shrinks to zero, the numerical solutions of the reservoir flow equations become more accurate, and at zero are exact. However, the use of extremely small gridblocks will not be practical if it requires excessive computing memory and time.

4.2.5 Input Data for Reservoir Description. The reservoir description entered for the model study area should integrate the analyses of all available geological, petrophysical, pressure transient test, and seismic data. This large data list is necessary to obtain accurate predictions of CO_2 flood performance because CO_2 flooding is more sensitive than waterflooding to small changes in the reservoir description.[27] In other words, a reservoir description realistic enough to accurately predict waterflood performance may fall short in predicting CO_2 flood performance.

Geological and petrophysical data can provide a 3D distribution of thickness, porosity, and permeability by geological flow units. Permeability can be obtained from core data or from log-core correlations, however, log-core correlations routinely provide air permeability at laboratory conditions. To improve prediction of permeability distribution from logs, neural networks and multivariate analysis have been used.[28]

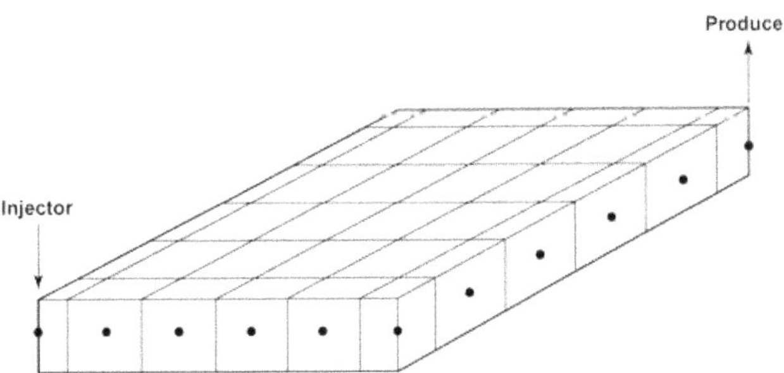

Fig. 4.4—Diagonal grid orientation between injector-producer well pairs.

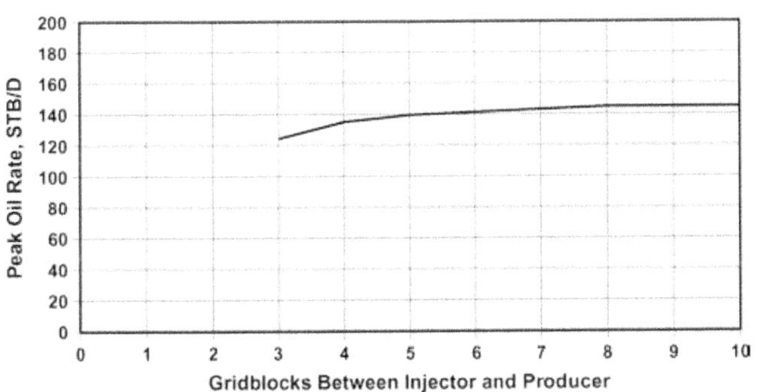

Fig. 4.5—Grid sensitivity for a diagonal grid. A 6×6 grid should provide accurate enough results when producer and injector are on the diagonal.

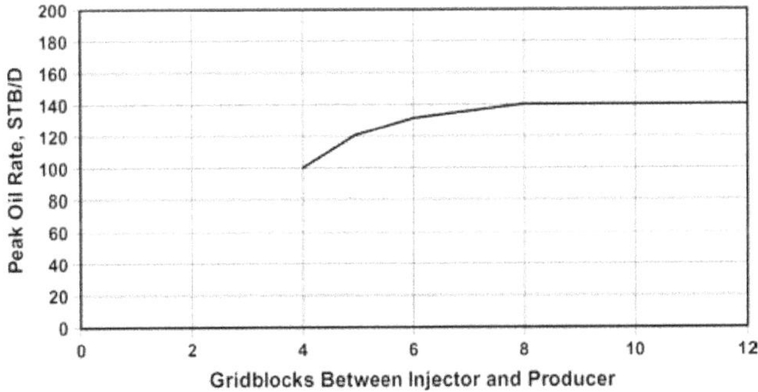

Fig. 4.6—Grid sensitivity for a parallel grid. When injector and producer sit on the same gridline, an 8×8 grid pattern is required.

Pressure transient test data also can be analyzed to relate air permeability to in-situ liquid permeability (oil permeability at connate water saturation). Pressure transient analysis yields a total permeability-thickness product in the vicinity of a well. The air permeability can be summed over every foot, and the total air permeability-thickness product can be scaled to equal the corresponding result from the pressure transient test analysis.

Pressure transient test data also can be used to estimate anisotropic permeability trends through interference testing,[29] vertical permeability from a well completed with partial penetration of the pay[30] or by vertical permeability pulse-tests,[31] distance to faults and faults transmissivity,[32] properties of natural fractures,[33,34] near-wellbore effects (skin), and average reservoir pressure. Pressure transient analysis includes the effect of in-situ stress on permeability. Additionally, an abrupt change in reservoir pressure as revealed by repeat formation test data can indicate a barrier to vertical flow.

There is no single way to tackle reservoir description—the reservoir description for a model study area evolves through a process of refinement. Geological, petrophysical, seismic, and pressure transient test data are important in developing the initial reservoir description, but this description usually requires some refinement before a reservoir simulator prediction of primary and secondary performance will match historical data.[24] The usefulness of the initial reservoir description is limited by the inability to measure the vertical distribution of permeability directly or to accurately predict interwell properties. Although the permeability-thickness product can be determined by layer,[35,36] the accuracy of this procedure is limited by the industry's inability to determine the flow rate into or out of low-permeability layers.[37]

Reservoir Layering. A reservoir is composed of one or more "geological flow units," which are defined on the basis of the depositional environment. In our definition, the geological flow unit is not necessarily homogeneous and may include many layers. In this monograph, we use the term "layer" to mean an interval parallel to the bedding plane and totally included within the geological flow unit. At a particular (x,y) location, a layer has approximately uniform properties (permeability and porosity) over a vertical interval. Layer properties can change areally, and a layer may be discontinuous. When any layer property begins to change significantly in the vertical direction, the layer stops and another layer starts.

The initial reservoir description is mapped into a 3D framework of layers between the wells in the model study area. Such a framework, in the form of a fence diagram,[38] is illustrated in **Fig. 4.7.** These layers will be required as input to a reservoir simulator to describe fluid flow characteristics. Note that not all layers will be continuous between wells—layers can terminate because of a facies change, a fault, or a complete absence of permeability.

Sequence stratigraphy is a useful tool that geologists use to map layers. Three-dimensional seismic data also may reveal information on reservoir continuity. Seismic interpretation can be useful in understanding reservoir geometry, location of faults, and possible high porosity regions commonly identified through seismic inversion. Including geoscientists in the development of the reservoir description is vitally important.

If a layer is very thick, fluids may segregate within it; for example, water injected into a thick layer can partially underrun the oil there. In such cases, the layer should be subdivided into smaller numerical layers so that the fluid segregation can be represented in the reservoir simulator by a series of discrete values for fluid saturations. Thick layers usually are segregated into at least three numerical layers, but more layers are used if warranted by numerical testing.

Porosity variance is not always enough for mapping layers. Permeability can be more useful in mapping because it varies more than porosity. If the reservoir is not subdivided into enough layers to accurately characterize heterogeneities, CO_2 flood predictions are likely to be too high.

Permeability/Porosity Plots to Correlate Layers. Because permeability usually increases with porosity, the ratio of permeability to porosity (k/ϕ) magnifies high permeability streaks. The k/ϕ ratio also appears in the equations for flow through porous media as a direct indicator of how fast fluid flows through a layer.[39] Chopra *et al.*[39] recommend using k/ϕ plots along with gamma ray logs and any other geological markings to map layers within geological flow units.

Initially, geological flow units are mapped across the model study area. Then, intervals of similar k/ϕ within each geological

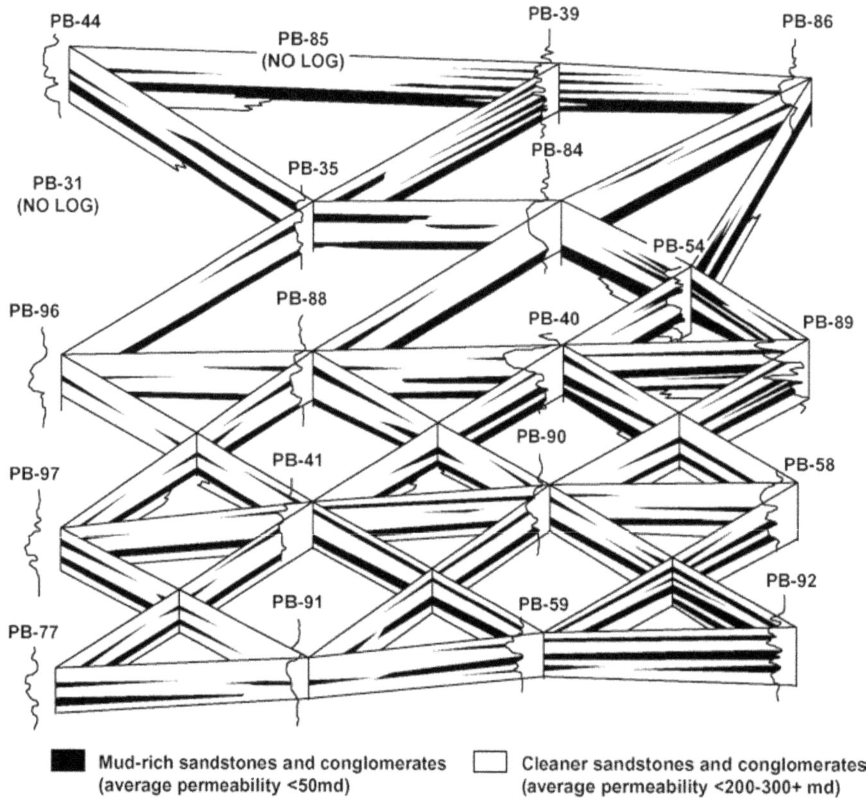

Fig. 4.7—Typical fence diagram (after Ref. 38). Fence diagrams show layer continuity between wells.

flow unit are identified at each well. If layers are continuous between wells, the various k/ϕ intervals should correlate between wells; for example, Layer 8 from 4,840 to 4,848 ft at Well 732 **(Fig. 4.8)** correlates to Layer 8 from 4,842 to 4,848 ft at Well 736 **(Fig. 4.9).** The left side of each of these figures shows the gamma ray log (GRN), and the right side, the k/ϕ ratio.

Aasum *et al.*[40] found that when layers were continuous over the well spacing, the k/ϕ method mapped layers similarly to those mapped by geostatistical methods. Moreover, the use of the two methods yielded similar predictions when input to a reservoir simulator.

Averaging Permeability. As long as layers can be mapped between wells, Dagan[41] found that an arithmetic average was appropriate for horizontal permeability in strongly layered systems. For vertical permeability, we recommend a series average. Depending on the internal distribution of the facies, it might be difficult to map layers between wells, in which case geostatistical techniques[13,40,42–46] might be required to best represent permeability.

4.2.6 The Application of Geostatistics to Reservoir Characterization. This section briefly discusses how geostatistics may be applied to reservoir characterization; however, detailed discussion is beyond the scope of this monograph. The reader should consult cited references for additional information.

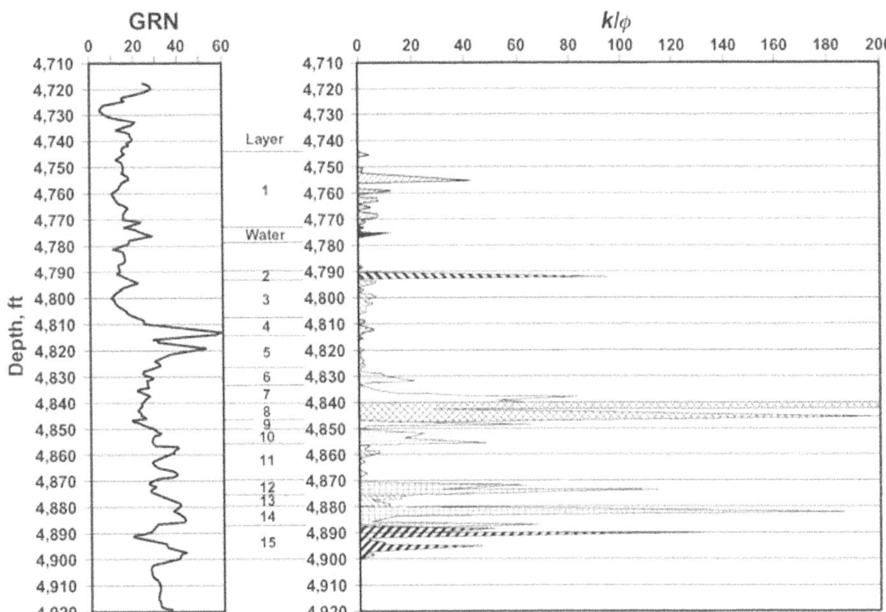

Fig. 4.8—A k/ϕ plot used to map layers for Well 732.

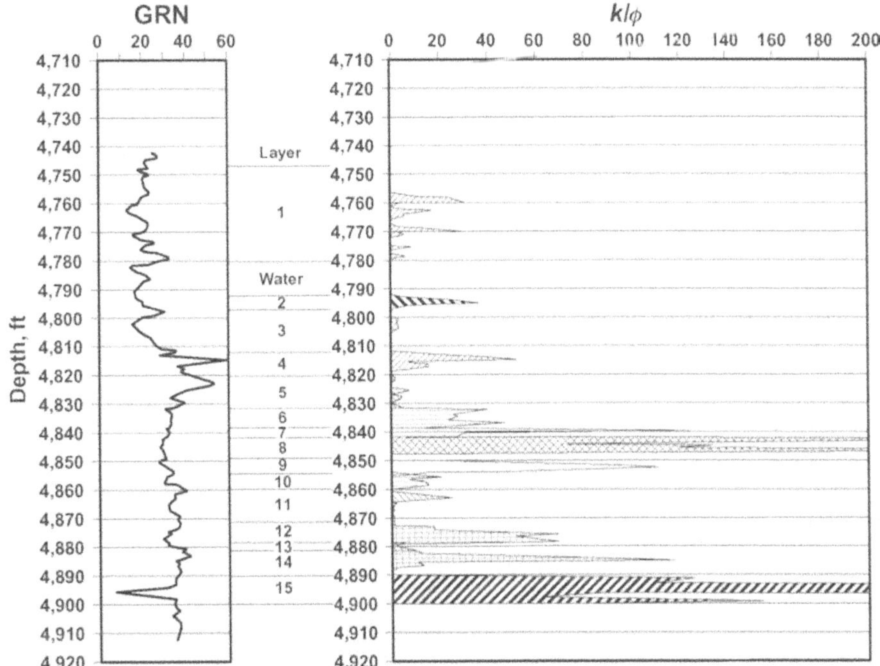

Fig. 4.9—A k/ϕ plot used to map layers for Well 736.

The real advantage of geostatistical techniques over the k/ϕ approach is apparent when the layers are not continuous between wells and when layer properties change more rapidly than the standard gridblock size allows. The standard gridblock size referred to here is consistent with grid sensitivity testing discussed in Sec. 4.2.3, which typically would yield gridblocks with areal dimensions on the order of 100 ft.

Reservoir simulators require that all properties be uniform within a gridblock. Changes in reservoir properties within a gridblock can indicate apparent channels that can adversely affect oil recovery. Geostatistical techniques account for this problem by using a grid that is fine enough to account for the heterogeneities. **Fig. 4.10** illustrates a typical permeability cross section generated with a geostatistical method that uses fractals.[42]

Geostatistics can help to more accurately determine interwell properties[42,43] and to map layers across reservoirs.[13–15,40,42] However, geostatistics should not be viewed as a substitute for geology—it should be used in situations where available geological data are insufficient to describe heterogeneities that can affect performance. Malik and Lake[44] found that applying geostatistics separately to individual geological flow units, rather than to all the geological flow units combined, yielded more realistic layer mapping and predicted reservoir properties more accurately when tested on outcrop data.

There are three steps to applying geostatistical techniques. The first is to use a semivariogram to plot changes in the variance of porosity, permeability, or thickness over distance. **Fig. 4.11** shows a typical semivariogram for thickness, constructed from data.

The semivariogram usually is applied to data on the well spacing scale and indicates how a property such as thickness correlates with itself over distance. In Fig. 4.11, the plot of variance levels off at 3,500 ft, which is termed the correlation length. If the correlation length is very large, the layers are more continuous, while a smaller correlation length implies some loss of continuity. Fractals are used to describe the semivariogram when it does not level out.[13] See Hewett[13] for additional discussion about fractals.

The second step in applying geostatistics is called "kriging," a process used to estimate the spatial variation of a property. Kriging estimates interwell properties by using the data from all wells within an investigation radius equal to the correlation length. It uses a least-squares interpolation technique to develop the best expected values at interwell points.[45] Kriged values vary smoothly between wells and honor known well data. The kriged results still do not contain sufficient information on heterogeneities that are on a smaller scale than the standard gridblock size.

The third geostatistical step is to adjust the expected values at interwell points using conditional simulation[46] (i.e., by adding a stochastic variation or perturbation to reservoir properties at each gridblock). The magnitude of the perturbation is based on the variation found in the semivariogram plot. This step superimposes small-scale heterogeneities on the kriged results. The scale of heterogeneities should be smaller than the standard size simulator gridblock size.

Conditional simulation can be used to yield multiple variations (also called realizations) of small-scale heterogeneities because conditional simulation generates equally probable events based on the determined variance of reservoir properties. Different realizations can be generated by changing the seed value (starting point) for the calculations. In other words, there are myriad ways to perturb the kriged results with small-scale variations.

Fig. 4.12 from Wolcott and Chopra[43] illustrates how conditional simulation can account for reduced oil recovery caused by local-

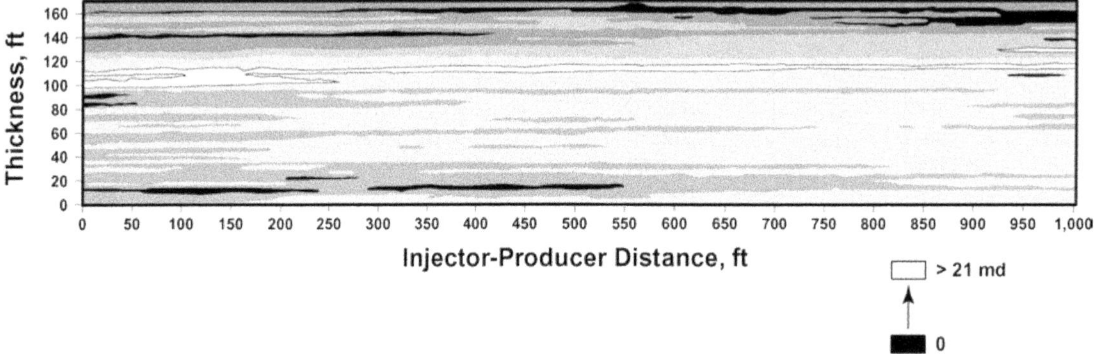

Fig. 4.10—Horizontal permeability cross section generated using geostatistical methods.

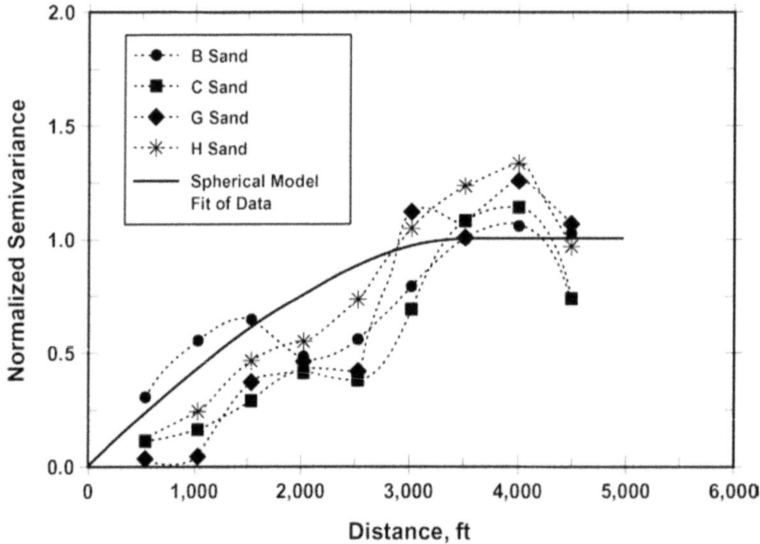

Fig. 4.11—Semivariogram for thickness (after Ref. 43). This plot shows how a layer property (in this case, thickness) correlates with itself over distance.

ized channeling during waterflooding, which is not accounted for by either kriging or standard size gridblocks. The solid line in the graph represents a waterflood simulation for a reservoir description developed by kriging. Each dashed line represents a waterflood simulation for a particular realization in which the kriged reservoir description is perturbed at interwell points by conditional simulation. Nearly all of the realizations reduced the oil recovery from the original unperturbed simulation.

Grid Size in Geostatistical Reservoir Characterization. The gridblock sizes used for conditional simulation can be determined by a grid sensitivity study similar to that described in Sec. 4.2.3. The gridblock size is small enough when it is reduced to the point where negligible change occurs in the predicted performance. Wolcott and Chopra[43] found that 17×17 gridblocks for each quarter of a 40-acre five-spot pattern was sufficient to accurately simulate the effect of areal heterogeneities. Recall that 6×6 gridblocks usually are sufficient for areally homogeneous reservoir properties. The required grid size for conditional simulation depends on the nature of the areal heterogeneities.

Sensitivity studies related to layer thickness also are useful when geostatistics is applied to layer mapping. To conduct sensitivity studies, divide a particular geological flow unit into several layers of arbitrary thickness and simulate performance (possibly in a 2D cross section). Reduce the thickness of each layer (increasing the number of layers), and simulate performance again. The maximum allowable layer thickness (minimum number of layers) is set at the point where successive predictions yield negligible differences.

If geostatistically generated heterogeneities are sufficiently smaller than the standard simulator gridblock size, they can be incorporated into the reservoir simulator by using either a fine-grid conditional simulation or a pseudoflow approach whereby properties such as fractional flow or relative permeability are adjusted for the effects of heterogeneities and input to a coarser grid with fewer layers.

The most computer-intensive approach is to use the grid from conditional simulation. Wolcott and Chopra[43] demonstrated the feasibility of this approach by using 17,160 gridblocks for four 40-acre five-spot patterns. Note that even though larger and larger grids are being used every day, a fine-grid simulation can sometimes be impractical because of the excessive computer memory and computing time involved.

Pseudoflow Characteristics. Geostatistical methods have been applied to hybrid simulation models. Using this approach, waterflooding and CO_2 flooding are simulated using a fine-grid, 2D vertical cross section. The cross section contains all the layers required to reflect small-scale heterogeneities (areal effects are not included because it is 2D). These fine-grid predictions yield a pseudofractional flow curve, which then can be used in a one-layer streamtube model to predict performance.[13–15] The streamtube

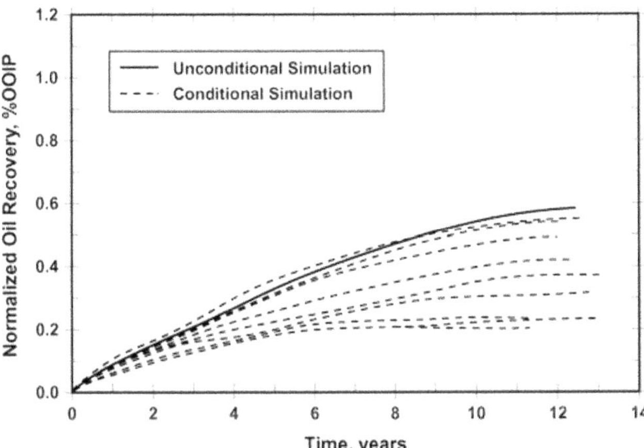

Fig. 4.12—Conditional simulation can account for reduced recovery caused by small-scale heterogeneities. The solid line shows a waterflood simulation developed by kriging, while the dashed lines show realizations generated by perturbing the reservoir description using conditional simulation (from Ref. 43).

fractional flow is used in each of the streamtubes that make up an injection pattern. Martin and Wegner[47] describe how the streamtube model can be updated to account for changes in streamtube shape with time, which is an important consideration because changing mobilities induced by the WAG process can cause streamlines to change between water and gas cycles.

An alternative pseudoflow approach involves the generation of pseudorelative permeability. First, a reservoir simulator is used to predict preCO_2 and CO_2 flood performance using a fine-grid, multilayer reservoir containing, for example, one pattern. The reservoir description may be generated using geostatistical techniques. Then the predictions of a coarse-grid, one-layer reservoir of the same pattern is matched to the fine-grid predictions by modifying relative permeability curves. The coarse one-layer model contains the same geostatistical reservoir description as the fine-grid model, except that the data have been averaged (in a procedure called "upscaling") over the coarser gridblocks. The upscaling process represents the small-scale heterogeneities in a fine-grid model by a pseudofunction in a coarse grid. The relative permeability curves can be transformed into pseudorelative permeability curves that are input to a reservoir model with multiple patterns and a coarse grid.

The pseudorelative permeability exercise can be repeated for each geological flow unit. The coarse-grid model will use several layers, but not as many as the fine-grid model. Additional layers help to more accurately simulate gravity effects.

As reservoir properties (e.g., layer permeability) are refined to match historical performance, the pseudofractional flow or pseudorelative permeability must be recalculated. Pseudofractional flow or pseudorelative permeability is calculated for each realization that is to be simulated with a coarse-grid network.

The pseudofractional flow or pseudorelative permeability curves used in history matching must be generated for two-phase flow of oil and water (waterflooding) and for two-phase flow of oil and gas (volumetric depletion and any dry-gas injection). Additional pseudocurves will have to be generated for CO_2 flooding, and these curves also must be consistent with waterflooding. The work by Jerauld[26] indicates that this is a difficult task and requires further improvements.

As computers become faster, there is less need to use pseudorelative permeability curves as a tool for reducing computing costs. If pseudorelative permeability curves can be avoided, and if the reservoir geology and structure can be more accurately characterized in the description, then more accurate predictions of CO_2 flood performance should be possible.

Conditional Simulation To Estimate Effective Vertical Permeability. Conditional simulation also can be beneficial for estimating effective vertical permeability caused by randomly spaced shales that reduce vertical flow, as demonstrated by Haldorsen and Lake.[48] According to geological data, the shales they studied occupied a small fraction of the reservoir volume. They compared reservoir simulation of randomly distributed shales in a fine grid to a homogeneous system characterized by effective vertical permeability. The effective vertical permeability was determined when the two simulations agreed. They also derived expressions for effective vertical permeability as a function of shale fraction and tortuosity. Richardson *et al.*[49] discuss which geological depositional environments might be expected to have discontinuous shales.

Geostatistical Programs. Hewett,[13] Richardson,[49] and Hohn[50] provide more detail on geostatistics. Commercial computer programs for geostatistical calculations are available. Some of these programs are in the public domain, for example, GSLIB,[51] a PC-based program.

4.2.7 Seismic Applications in Reservoir Characterization.

Seismic imaging also can aid in understanding interwell properties,[52] although current seismic imaging technology yields only approximate values of interwell porosity and permeability. As is the case with logs, with seismic data the value of permeability is an indirect measurement tied to log-core correlations. Seismic imaging can complement geostatistics.

Sometimes 3D seismic attributes can yield useful information about layer continuity and, in limited cases, water/hydrocarbon contacts. Two- and three-dimensional seismic data definitely should be used to characterize faults. Faults that show up in seismic data can be better defined if there are pressure transient test data that indicate no-flow boundaries. Repeating 3D seismic studies over time (known as 4D seismic) can help track the movement of gas and further improve reservoir characterization.

4.2.8 Reality Check for Reservoir Characterization. After detailed reservoir characterization work, one should confirm that the resultant reservoir description is as heterogeneous as the reservoir actually is. Wehner* suggests calculating the Dykstra-Parsons coefficient of the reservoir description to determine whether it is similar to the Dykstra-Parsons coefficient calculated from core data. Of course, the core data must have representative areal and vertical coverage for this step to be meaningful.

4.2.9 Effect of Operations on Input Data. The last step in setting up the input for a reservoir simulator is to account for the effect of operations. All operations, such as completions, lift revisions that affect bottomhole pressure, workovers, and changes in well patterns must be mimicked in the history match. The effects of operations must be separated from uncertainties in the reservoir description (reservoir layer properties).

Pressure transient test results should be used to estimate changes in the skin factor to determine the effective wellbore radius. The effective wellbore radius is related to the actual wellbore radius and the skin factor through the following equation:

$$r_{we}=r_w\, e^{-s}, \quad (4.2)$$

where r_{we}=the effective wellbore radius in ft; r_w=the actual wellbore radius in ft; and s=is the skin factor, dimensionless.

Often, pressure transient tests are not run frequently enough to capture changes in the effective wellbore radius. The skin can increase (thereby decreasing effective wellbore radius) because of formation damage. Similarly, workovers can decrease skin and increase the effective wellbore radius. These effects must be included in the history match.

The effective wellbore radius determined from pressure transient test analysis is slightly affected by the reservoir description. Bennett *et al.*[53] showed that when the effective wellbore radius is determined with a homogeneous assumption, it should be modified by the following equation to correctly account for the heterogeneities related to reservoir layering:

$$(r_{we})_{\text{het}}=(r_{we})_{\text{hom}} / R_{cd} \quad (4.3)$$

and

$$R_{cd}=\Sigma\ \{[(kh)_l/(kh)_t]\ [(\phi h)_l/(\phi h)_t]\}^{0.5}, \quad (4.4)$$

where $(r_{we})_{\text{het}}$=effective wellbore radius in ft, determined with a heterogeneous assumption; $(r_{we})_{\text{hom}}$=effective wellbore radius in ft, determined with a homogeneous assumption; R_{cd}=a mathematical heterogeneity term defined, dimensionless; the subscript l refers to the properties evaluated for each layer; the subscript t refers to the total properties evaluated for all layers; k=horizontal permeability in md; ϕ=porosity, fraction; and h=thickness in ft. Eq. 4.4 is summed over all the layers.

Eqs. 4.2 and 4.3 imply that the proper skin value for a heterogeneous, layered reservoir $[(s)_{\text{het}}]$ is given by:

$$(s)_{\text{het}}=(s)_{\text{hom}}-\ln(R_{cd}), \quad (4.5)$$

where $(s)_{\text{hom}}$=the skin factor for a homogeneous, layered reservoir, dimensionless.

*Personal communication with S.C. Wehner, Phillips Petroleum (2000).

4.2.10 Variables Used in History Matching. The reservoir description is defined during successive iterations of the history-match process. The following data must be history matched in order for the reservoir description to be a valid prediction tool:

- Producing wells: (a) oil rate and cumulative oil recovery; (b) water production rate, water/oil ratio, or water cut; (c) gas production rate or GOR; (d) bottomhole producing pressure; (e) average pressure determined from pressure drawdown or buildup tests; (f) static pressure measurements, and (g) production log flowmeter data.
- Injection wells: (a) water injection rate; (b) gas injection rate (if gas was injected before the CO_2 flood); (c) bottomhole injection pressure; (d) average pressure determined from pressure falloff or injectivity tests; (e) static pressure measurements, and (f) injection profile data.
- Logging observation wells: (a) changes in gas, oil, and water saturations; (b) changes in gas/oil contacts, and (c) changes in water/oil contacts.
- 4D seismic: (a) areal velocity changes and (b) vertical velocity changes.

Matching rates vs. time constitutes a match of volumes vs. time. Combining the match of volumes and average reservoir pressure vs. time satisfies the material balance and confirms the OOIP and aquifer size. Matching both the well rates and the well bottomhole pressures vs. time signifies that inflow performance of the wells has been adequately characterized. A match of the fractions of produced water and gas characterizes reservoir heterogeneities.

Some of the data to be matched will be more accurate than other data, and highest priority should be given to matching the most accurate data. Pressure transient tests sometimes can be affected by multiphase flow, and the average reservoir pressure that is calculated by assuming single-phase flow may have a bias (see Sec. 3.1.1).[54]

4.2.11 Refinement of the Reservoir Description by History Matching. In the process of refining the reservoir description, values for the least well known parameters are adjusted until simulator predictions match historical reservoir performance. The refined reservoir descriptions still should be consistent with the geological depositional environment; otherwise, the history match of past performance might be coincidental and future predictions may deviate from reality.

In the petroleum industry there are two common approaches to history matching. In one approach, waterflood performance is history matched by modifying relative permeability without changing unknown reservoir properties, for example, layer permeabilities. This is equivalent to generating a pseudorelative permeability curve that accounts for unknown geological properties. The problem here is that a separate set of pseudorelative permeability curves is required for CO_2 injection. Unless you systematically generate pseudorelative permeability curves for a particular degree of reservoir heterogeneity, as described in the geostatistical discussion earlier in this chapter, there is no way to generate the correct pseudorelative permeability curves for CO_2 injection.

In the other approach, the least well-known parameters are adjusted first instead of relative permeability. These usually include layer permeability, total permeability-thickness product at wells without analyzable pressure transient test data, interwell properties, and vertical permeability. The more parameters that can be changed to obtain a match, the wider will be the range of uncertainty in the predictions. We recommend that as many parameters as possible be tied to field and laboratory tests.

A lack of relative permeability data may tempt an engineer to change water/oil relative permeability curves to match waterflood performance, but this practice should be avoided for CO_2 floods. Because CO_2 relative permeability often is estimated in a three-phase relative permeability model that uses two-phase relative permeability data, changing the two-phase relative permeability will end up changing the CO_2 relative permeability. In addition, water relative permeability in oil-wet rock may exhibit hysteresis when water saturation decreases during the CO_2 cycle of a WAG flood (see Fig. 2.19). A large difference between the original water relative permeability curve and the hysteresis curve has a strong impact on water injectivity after CO_2 is injected into the reservoir. History matching water relative permeability affects the difference between the initial water relative permeability curve and the water hysteresis curve and thus affects water injectivity. To summarize, modifying relative permeability curves to history match the performance of the reservoir before CO_2 injection can affect CO_2 flood predictions in unexpected ways.

Refining Permeability Distribution. Changes in layer permeability are detected best by observing differences in the timing of secondary oil response and breakthrough of the injectant or water influx from various layers. In a reservoir not dominated by gravity effects, the timings of secondary oil response and water breakthrough in a layer both are strong functions of the layer's k/ϕ. For example, suppose that water breakthrough occurs 12 months after the start of water injection, and the reservoir simulator predicts the earliest breakthrough to occur at eight months. In this case, the k/ϕ of the model layer that broke through first was too high by a factor of 12/8. If the porosity of the subject layer is well known from log analysis, only the permeability should be changed—in this example, multiplied by 8/12 to match the initial water breakthrough. The process of detecting differences in timing of oil response and/or water breakthrough of the various layers should be carried out sequentially in order of decreasing k/ϕ. Plots of predicted layer rates are extremely useful for this purpose. This type of history matching has been performed automatically for layered reservoirs without vertical permeability.[55,56]

When layer permeabilities are adjusted to match timing, permeabilities in layers with lower values of k/ϕ need to be adjusted to conserve the total formation flow capacity, if the total formation flow capacity is known from pressure transient test data. Eq. 4.6 illustrates the conservation of total formation flow capacity:

$$(kh)_t=(kh)_1+(kh)_2+\ldots+(kh)_n, \quad (4.6)$$

where k=permeability in md, h=thickness in ft, and the subscripts 1 through n signify the layer numbers.

Even with the use of geostatistics, the timing of layer responses can be missed. In many cases, the data used to estimate layer permeabilities, even with geostatistics, will be log porosity data used in a correlation that predicts permeability. This measurement of permeability will be indirect, which can cause the magnitude of all the permeabilities in a layer to be in error. Changes in layer permeabilities may be necessary to better match actual timings. Moreover, the value of permeability at the well (honored by geostatistics)—not the interwell permeability—most directly impacts response and breakthrough times because most of the pressure drop between wells is in the near-wellbore regions.

Thus, geostatistically-based reservoir descriptions still may require changes to layer permeabilities to achieve an adequate history match. An added problem when a coarse grid is used is that layer changes require the recalculation of pseudorelative permeabilities.

If a well's cumulative fluid production rates are not closely matched and there are no analyzable pressure transient test data for the well, then the total permeability-thickness product of the well possibly should be changed. Producing wells often have this problem because pressure buildup tests are difficult to analyze correctly when the well produces both oil and water, and the tests cannot be run at all if the well is on pump. A new total permeability-thickness product can be estimated by multiplying the model total permeability-thickness product by the ratio of the actual peak total fluid production rate evaluated at reservoir conditions to the model's predicted rate. After the model total permeability-thickness product is changed, the oil response and water breakthrough times will change for all the layers. The next step is to adjust layer permeabilities.

Using geostatistical realizations is the best method for changing interwell properties while still ending up with a geologically consistent reservoir description.[13–15,42] Geostatistical realizations that do not closely match historical data can be ruled out. The most realistic realization is the one that best fits performance at all the wells. Sometimes a particular realization yields a satisfactory match only at some of the wells. In this case, the interwell components of the realization that helped achieve a match at some of the wells could be incorporated into the kriging step as known data, thereby preserving the favorable aspects of a particular realization. If this is not done, another realization might not contain the favorable aspects.

If interwell properties are blindly changed to achieve a match of past performance, the resultant reservoir description may be inconsistent with surrounding reservoir properties, in which case it would be unlikely that the predicted CO_2 flood performance would match actual performance.

If there is geological and/or petrophysical evidence of vertical permeability, then the vertical permeability should be changed through model sensitivity studies to better match GOR and/or WOR. Very little vertical permeability is required to initiate gravity effects. Vertical barriers, or partial barriers, should be included in the reservoir description when indicated by geological, seismic, core, log, or pressure transient data. The effective vertical permeability of the partial barrier is a parameter that should be history matched.

Operational Effects on History Matching. Sometimes the effects of operations on performance can be difficult to distinguish from reservoir heterogeneities. For instance, the deposition of scale and/or asphaltene in the near-wellbore area can reduce production and affect sweep until the problem is eliminated through a workover. These effects are dynamic not static, as they are with a reservoir description.

The presence of formation damage affects the value assigned to the wellbore radius in the reservoir simulator. Changes in effective wellbore radius in the wells of the model study area must be identified to create an accurate reservoir description and to accurately predict CO_2 flood performance.

The effective wellbore radii should be changed when appropriate to match performance related to formation damage, workovers to repair formation damage, and stimulation workovers. An incorrect effective wellbore radius greatly affects rate and bottomhole pressure and can have some effect on the average reservoir pressure and the timing of oil response and water breakthrough.[13] For a producing well in a waterflood, the natural tendency is for total fluid production rate to increase continuously after water breakthrough, until the rate reaches some maximum value for watered-out conditions.

Fig. 4.13 shows performance for a producing well whose total fluid production rate was adversely affected by scale. In this well, the total fluid production (RB/D) decreased while the waterflood was progressing and the well was pumped-off (constant bottomhole producing pressure with no more than 100 ft of liquid above the pump). The well performance shown here was history matched by continually decreasing the effective wellbore radius until the first workover was performed to remove calcium sulfate scale. Subsequent performance was history matched with an increased effective wellbore radius until scale reformed. The total production rate of the well was specified as a function of time and the effective well radius was adjusted also as a function of time to achieve a constant bottomhole pressure corresponding to pumped-off conditions.

Scale often forms when mineral components that were dissolved by injected fresh water under the high-pressure conditions at the injection well are then precipitated in the low-pressure zone near the producing well. Scale precipitation occurs primarily in water productive zones. In west Texas carbonate reservoirs, workovers to remove calcium sulfate scale have yielded large increases in water production with little or no increase in oil production. In the case shown in **Fig. 4.14,** the effective wellbore radius may be smaller in layers that produce more water, a phenomenon that further complicates the history match process.

4.3 Predicting CO_2 Flood Performance

Once the reservoir description has been refined so that the simulator can match historical performance, then CO_2 flood performance can be predicted. The CO_2 flood predictions should account for operational aspects such as the timing of injection well switches between CO_2 and water and the expected changes in bottomhole injection and production pressures. Bottomhole pump intake pres-

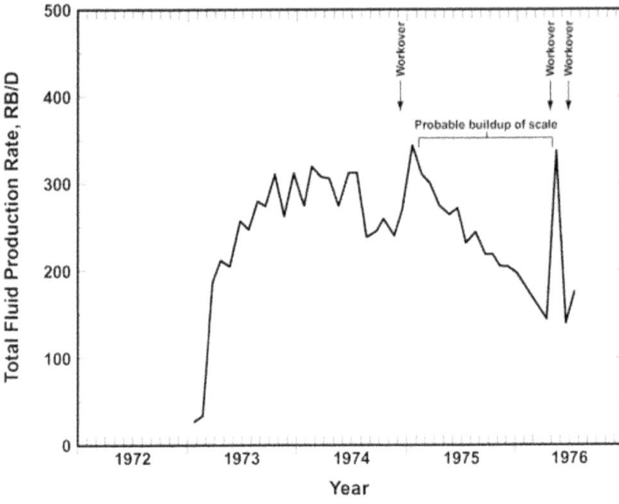

Fig. 4.13—Example of a total fluid production rate adversely affected by scale. The changing effective wellbore radius was input to the simulator to achieve a history match of total fluid production rates at pumped-off conditions.

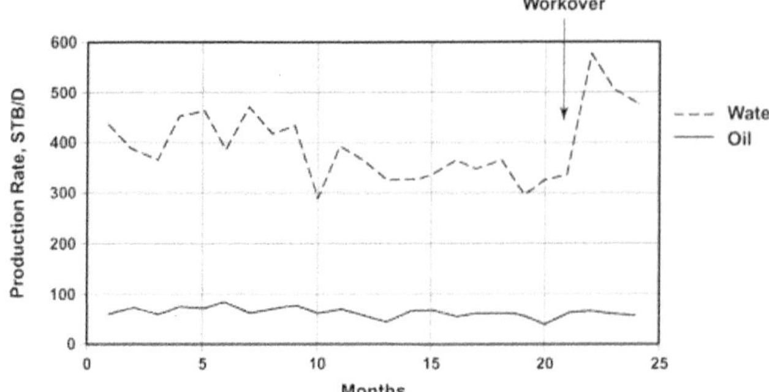

Fig. 4.14—Performance of a production well in which scale primarily affected the watered-out layers. The workover chemicals dissolved calcium sulfate and enabled increased water production.

sures at production wells have been known to increase 300 to 400 psi as CO_2 breaks through, because of a decrease in pump efficiency. Pressures may continue to increase if production wells are converted to natural flow. Past experience with CO_2 flooding in similar reservoirs should be incorporated into well specifications in the reservoir simulator.

The modified black-oil simulator and the fully compositional simulator are the two principal types of reservoir simulators used for predicting CO_2 flood performance. The modified black-oil simulator is an approximate technique to account for oil displacement by CO_2. The fully compositional simulator uses an EOS to predict the development of miscibility between CO_2 and oil. This approach is more rigorous but requires greater computation time.

The choice of simulator depends on the scope of the project and the data available. Both approaches have been used successfully to predict CO_2 flood reservoir performance, although for the fully compositional approach there still are questions about what is the most accurate way to predict both CO_2 relative permeability and physical dispersion.

As faster computers are developed, we expect the use of fully compositional reservoir simulation to increase. The following sections describe how these two simulation approaches predict CO_2 performance.

4.3.1 Modified Black-Oil Reservoir Simulators. Modified black-oil reservoir simulators start with a standard black-oil simulator with three components—stock-tank oil, hydrocarbon gas, and water—and add CO_2 as a fourth component. This type of simulator does not use an EOS to predict the development of miscibility. Instead, it relies on the simple assumption that above the thermodynamic MMP, CO_2 displaces the oil by a first-contact miscibility mechanism. The first-contact miscibility assumption calls for the CO_2 and oil to form one hydrocarbon phase. While CO_2 requires multiple contacts to become miscible with oil (see Sec. 2.1), the first-contact miscibility assumption serves as a mathematical simplification. Because of the physical limitations of the first-contact miscibility assumption, however, the modified black-oil simulator should be viewed as a simplified approach to a complex physical problem.

Below the thermodynamic MMP, the modified black-oil simulator treats the CO_2 as a dry gas. The modified black-oil model uses pressure-dependent CO_2 K-values (equilibrium partition coefficients). At the thermodynamic MMP, the K-value is set at unity. Below the thermodynamic MMP (where CO_2 is treated as a dry gas), the K-values are arbitrarily set to a value greater than unity, and above the thermodynamic MMP, they are arbitrarily set to a value less than unity.[7] The CO_2 K-values below the thermodynamic MMP may be based on actual first-contact CO_2/oil phase equilibrium data.[7]

Residual Oil Saturation to CO_2. The first-contact miscibility assumption leads to zero residual oil saturation and to overprediction of oil recovery. Modified black-oil simulators try to remedy this problem by allowing the engineer to specify a residual oil saturation to CO_2.Typical residual oil saturation to CO_2 levels range from 5 to 15% PV[6,12] and represent residual oil saturation at the injection-well gridblock (the lowest possible oil saturation).

The residual oil left behind in an actual CO_2 flood results from the interaction of phase equilibrium, dispersion, gravity-related mixing, and dead-end pores. Sec. 2.2 contains more information about dispersion and reservoir mixing.

Mixing in Modified Black-Oil Reservoir Simulators. Numerical dispersion in the simulator can smear the oil saturation profile so that residual oil saturation exhibits a CO_2 throughput effect; in other words, oil saturation increases away from the injection well. (The CO_2 throughput effect on residual oil saturation to CO_2 has been observed in pressure core data from CO_2 pilots in the Levelland (see Fig. 2.11) and Slaughter[12] fields in west Texas.)

Note, however, that numerical dispersion in a modified black-oil model does not completely resemble the actual dispersion in the reservoir. Actual dispersion can render an oil displacement process immiscible even when the reservoir pressure is significantly above the thermodynamic MMP (see Sec. 2.2.1). Numerical dispersion in a modified black-oil model will not create this effect.[57]

In short, numerical dispersion used with the first-contact miscibility assumption in modified black-oil simulators lacks a degree of physical meaning. In a modified black-oil model, specifying a higher residual oil saturation to CO_2 to match oil recovery is an approximate method to account for the effect of physical dispersion on displacement efficiency.

Viscous Fingering in Modified Black-Oil Reservoir Simulators. The early modified black-oil reservoir simulators supplied additional smearing of the oil saturation profile through the Todd-Longstaff mixing parameter (ω), which includes the effects of viscous fingering.[58,59] Recall Sec. 2.2.3, however, in which Chang's[9,60] fine-resolution simulation studies proved that viscous fingers only exist in a CO_2 flood when the medium is very homogeneous, e.g., artificial cores of uniform packed beads. In real reservoirs, heterogeneities coalesce viscous fingers as they are formed.

The Prophet model (a user-friendly, PC-based, modified black-oil model recently developed by the U.S. DOE)[61] also includes the Todd-Longstaff mixing parameter.[58] Typically, ω has been set at 0.67 to simulate viscous fingering; the authors recommend setting ω at 1.0 to eliminate the viscous fingering correction.

The Prophet model has other limitations created by its use of streamlines and its calculation of fluid flow only through streamtubes. For example, it cannot account for either the 3D flow associated with fluid crossflow between layers or for the streamline changes in the reservoir caused by the different mobilities and compressibilities of CO_2. Because of its limitations, the Prophet model is recommended only as a scoping tool.

Fluid Properties for Modified Black-Oil Reservoir Simulators. Fluid properties in modified black-oil reservoir simulators are calculated in much the same manner as those in standard black-oil simulators. The components are separator oil, separator gas (gas produced before the CO_2 flood), CO_2, and water. Water properties generally are assumed to be constant, while information about other components is entered into tables that include formation volume factors, viscosities, solution hydrocarbon/gas ratios, and CO_2 K-values, all entered as a function of pressure. Compositional effects are not accounted for and would require the use of a fully compositional model (see Sec. 4.3.2).

Modified black-oil simulators can compensate for the solubility of both natural gas and CO_2 in water. The solubility of CO_2 in water causes about 5% of the CO_2 to be lost to the water during WAG injection. In most cases, the effect on oil recovery is small compared to potential effects from many other process unknowns, although that effect can increase for larger water cycles that would dissolve more CO_2 and thereby increase water injectivity. Carbon dioxide solubility in water becomes an important variable in predicting chase water injection rates because chase water can dissolve a significant portion of CO_2 and can cause significant increases in water relative permeability and water injectivity (see Sec. 2.4.2 and Appendix D). The economic impacts include increased water injection and increased lift costs.

CO_2 and Oil Viscosity. Because the modified black-oil model assumes that both CO_2 and oil reside entirely in the oil phase when the pressure exceeds the thermodynamic MMP, it uses a viscosity mixing rule to calculate the oil-phase viscosity. The following type of equation often is used:

$$\mu_{os}=(x_o\mu_o{}^n+x_s\mu_s{}^n)^{1/n}, \qquad (4.7)$$

where μ_{os}=the viscosity of the oil/solvent (CO_2) mixture in cp; μ_o=the viscosity of the live oil (including dissolved natural gas) in cp; μ_s=the viscosity of the solvent (injected CO_2 stream) in cp; x_o=the fraction of the live oil (containing natural gas in solution), mass or volume fraction; x_s=the fraction of the solvent (CO_2 injected stream) in the oil phase; and n=an empirical viscosity mixing parameter, dimensionless.

Todd and Longstaff[58] recommended using a value of n equal to $-1/4$ (commonly known as the "quarter power mixing rule"), and they used volume fractions in Eq. 4.7 instead of mass fractions.

The quarter power mixing rule causes Eq. 4.7 to slightly underpredict the oil-rich phase viscosity.[2] In reality, above the thermodynamic MMP, when CO_2 mole fraction increases, the viscosity of the oil-rich phase also increases because the oil contains mostly heavier hydrocarbon components (the lighter components having vaporized into the CO_2-rich phase). Eq. 4.7 cannot account for this behavior because the oil is treated as one component.

To compensate for this, Chopra *et al.*[7] recommended adjusting the value of n to match field performance because CO_2 injection rates and oil and gas production rates are most sensitive to the value of n. Typical field-derived values of n have been found to range from 0.75 to 2.25. **Fig. 4.15** shows that using a field-derived value of n overpredicted the oil-rich phase viscosity reported in Turek *et al.*[2]

Chopra *et al.*[7] argued that the overprediction of n was justified because the oil/solvent mixture viscosity in a modified black-oil model represents a pseudoviscosity that takes into account the fact that the actual CO_2 displacement process usually has two hydrocarbon phases (sometimes three) behind the oil bank instead of the one hydrocarbon phase generated by the first-contact miscibility assumption. The fully compositional simulation approach avoids the use of a pseudoviscosity mixing exponent because it predicts one or two CO_2-rich phases (liquid and/or vapor) and an oil-rich liquid phase instead of one oil-solvent phase.

Carbon Dioxide and Oil Density. Modified black-oil reservoir simulators treat the density of the oil/solvent mixture in a manner similar to Eq. 4.7. At conditions of miscibility, the difference in the densities of oil and CO_2 is not as great as the difference in viscosities. Consequently, the error in the predicted oil/solvent mixture density is not as large.

When the reservoir pressure is below the thermodynamic MMP, the CO_2 that is not dissolved in the oil resides in the gas phase, mixing with any hydrocarbon gas not dissolved in the oil. The density and viscosity of the gas phase are calculated with power law equations similar to Eq. 4.7.[7]

Carbon Dioxide Relative Permeability for Modified Black-Oil Reservoir Simulators. The first-contact miscibility assumption implies that CO_2 flows in the same pores as the oil. In others words, the relative permeability of oil applies to both the solvent and the oil. This is the case in most modified black-oil reservoir simulators.[58,59]

However, the implied assumption of equivalent oil and CO_2 relative permeability does not hold true for oil-wet and mixed-wet rock. Applying oil relative permeability to the CO_2-rich phase and using the quarter power viscosity mixing rule ($n=-1/4$ in Eq. 4.7) has led to overprediction of injection and production rates. Even with n in the range of 0.75 to 2.25, the use of oil relative permeability for the CO_2-rich phase still leads to overprediction of water injection rates.[7]

Chopra *et al.*[7] used the concept of solvent relative permeability (see Sec. 2.5.2) to describe the flow of the CO_2-rich phase. The solvent relative permeability concept allows the CO_2-rich phase relative permeability to be less than the oil relative permeability. Because a large volume of CO_2 remains trapped in the reservoir after a CO_2 cycle, the water relative permeability cannot reach the maximum value achieved during the waterflood. Experimental data reported in Shyeh-Yung[3], Prieditis *et al.*,[4] and Wegener and Harpole[5] qualitatively support this concept.

Chopra *et al.*[7] and Hewett[12] found that the solvent relative permeability concept could be used in a modified black-oil reservoir simulator to match actual CO_2 flood performance. In fact, the use of solvent relative permeability has been critical in matching the loss in water injectivity in CO_2 floods of oil-wet and mixed-wet reservoirs in west Texas.

Using solvent relative permeability, water mobilities at injection well gridblocks have been predicted that agree closely with results of pressure falloff tests at the ends of short water cycles.[7] The significance of the short water cycle (less than 3% HCPV injected) is important because the effect of CO_2 solubility in water is still small. Larger water cycles might allow more CO_2 to be dissolved, thereby increasing water injectivity. Solvent relative permeability has a smaller effect on oil rate, cumulative oil production, gas production rate, and gas injection rate than does the viscosity exponent, n.[7]

If there are no laboratory data on CO_2/water relative permeability in the presence of residual oil, we recommend using the sol-

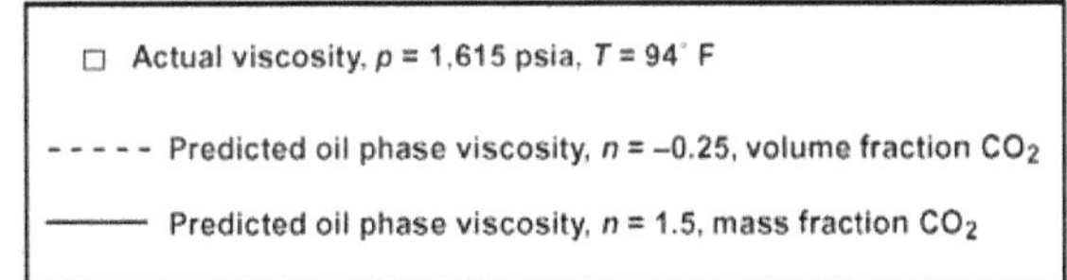

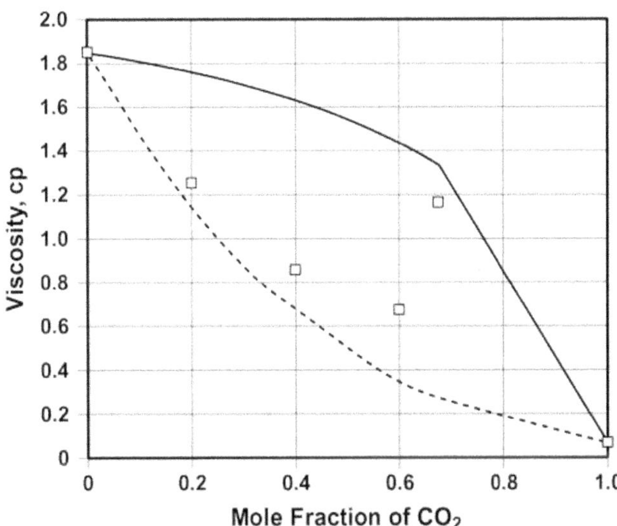

Fig. 4.15—Oil-rich phase viscosity used in a modified black-oil model. The oil-rich phase viscosity required to match field performance with a modified black-oil model generally is greater than the actual oil-rich phase viscosity.

vent relative permeability model of Chopra *et al.*[7] for oil-wet and mixed-wet formations. In a strongly water-wet reservoir, Dria's CO_2 relative permeability experiments with synthetic oil[62] indicate that the assumption of CO_2 relative permeability being equal to the oil relative permeability is most appropriate for simulating a CO_2 flood with a modified black-oil model.

Sec. 2.4 gives a detailed discussion of WAG projects that have not experienced a decrease in water injection rate. Most of these projects used higher bottomhole injection pressures for the water cycle. Note that this practice can be associated with changes in average reservoir pressure and changes in formation parting pressure. Changes in formation parting pressure can cause changes in the skin factor and the effective wellbore radius, which must be included in CO_2 flood predictions to then correctly predict injection rates as a function of bottomhole injection pressure.

Another method that has been used to predict and match observed reductions in water injectivity (as well as reductions for all the other rates in a CO_2 flood) is the introduction of an asphaltene precipitation factor[59] that reduces permeability at each gridblock receiving injected CO_2. Note, however, that asphaltene precipitation has not occurred in all CO_2 floods that experienced reduced injection and production rates.

Three-Phase Relative Permeability Correlations. Above the thermodynamic MMP, the single-contact miscibility assumption yields two-phase flow at most—oil and water. Below the thermodynamic MMP, gas, oil, and water all can exist. Conventional three-phase correlations used in black-oil simulators for waterflood predictions are appropriate.

Accounting for Production of NGL's. Modified black-oil reservoir simulators do not account for production of NGL's, which has a number of important consequences: First, the injection of any gas can increase the GOR dramatically. As gas comes to the surface and bubbles out of solution in the separators, it tends to vaporize light hydrocarbons from the oil, producing additional NGL's in the vapor stream. The loss of these light ends reduces the amount of oil recovered in the separator oil stream and increases the density of the produced oil. Only a fully compositional model (see Sec. 4.3.2) can handle these effects by accounting for phase equilibria between various hydrocarbon components at the separator.

Fig. 4.16 illustrates the effect that an increasing GOR has on the vaporization of light ends in the separator. The plot starts at the beginning of the CO_2 flood, when oil production from the previous waterflood is still declining. As the CO_2 flood progresses, more oil is recovered, followed shortly by an increase in the produced GOR as CO_2 begins breaking through.

The solid line for oil recovery represents the results that might be predicted by a modified black-oil model that does not account for the loss of light ends to the produced gas stream. The dashed line represents what might be predicted with a more rigorous fully compositional model that tracks every component of the oil and gas, handling not only the compositional mass transfer effects in the reservoir but also in the separator. The hatched area between the two lines represents oil that is vaporized in the separators.

The percentage of oil vaporized in the separator during a CO_2 flood can be correlated to the mole fraction of CO_2 in the separator gas, which is directly related to the increase in the GOR. **Figs. 4.17 and 4.18** (for two west Texas fields) illustrate how the percent decrease in oil rate caused by vaporization at the separator increases with mole fraction of CO_2 in the separator gas.

The solid curves in Figs. 4.17 and 4.18 were predicted from a fully-compositional reservoir simulator. These curves should be considered rigorous. The crooked sections of these curves reflect nonuniqueness in correlating only with GOR. The predictions include the effects of the development of miscibility, reservoir heterogeneities, and changing WAG ratio.

The dashed lines in Figs. 4.17 and 4.18 were predicted by phase-equilibrium flash calculations for the separator that were based on the composition of the separator oil and gas during the waterflood and on adding increasing amounts of CO_2. This technique is by no means rigorous because it ignores the effects discussed in the previous paragraph. Yet, the flash calculation technique appears to provide a satisfactory estimate of the oil produced in the vapor stream.

The character of the curves in Figs. 4.17 and 4.18 will vary for different oils, although a 10% reduction in the oil rate at high GOR's appears to be a reasonable estimate based on west Texas experience.

It may not be possible to accurately equate the volume of crude oil lost to the volume of NGL's created because of nonideal mixing conditions. This estimate should be considered approximate.

Grid Size for Modified Black-Oil Simulators. An important consideration is how fine the grid should be to simulate CO_2 flooding. Numerical dispersion is related to gridblock size, and gridblock size acts as pseudodispersion for the first-contact miscibility assumption.

Figs. 4.19 through 4.21 show the results of a one-layer grid sensitivity study for a tertiary CO_2 flood using a modified black-oil reservoir simulator to predict recovery from a quarter of a five-spot waterflood. A nine-point finite difference scheme was used in the simulations to reduce grid orientation effects.[63] The coarser grids had more numerical dispersion in both the longitudinal and transverse directions; the increased longitudinal dispersion reduced the

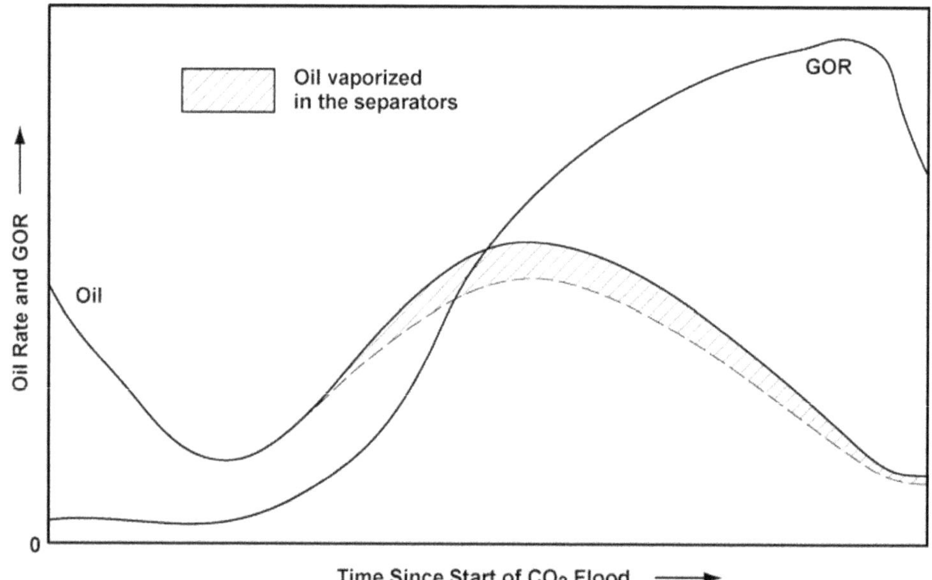

Fig. 4.16—Schematic of oil loss in the separator because of increased GOR. As CO_2 breaks through, the increase in the GOR at the separator causes more light ends to be vaporized from the oil and therefore an unexpected decrease of crude oil in the stock tank. A modified black-oil model (solid line) will not predict this loss, but a fully compositional model (dashed line) will.

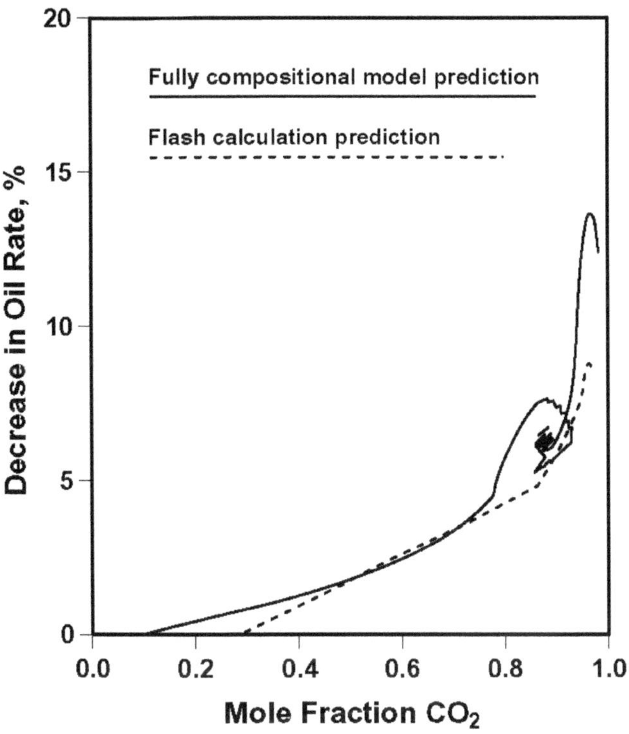

Figs. 4.17—Predicted reduction in oil rate caused by vaporization in the separator. As the mole fraction of CO_2 increases, thereby increasing the GOR, some of the crude oil is vaporized. This plot shows predictions for the Slaughter field.

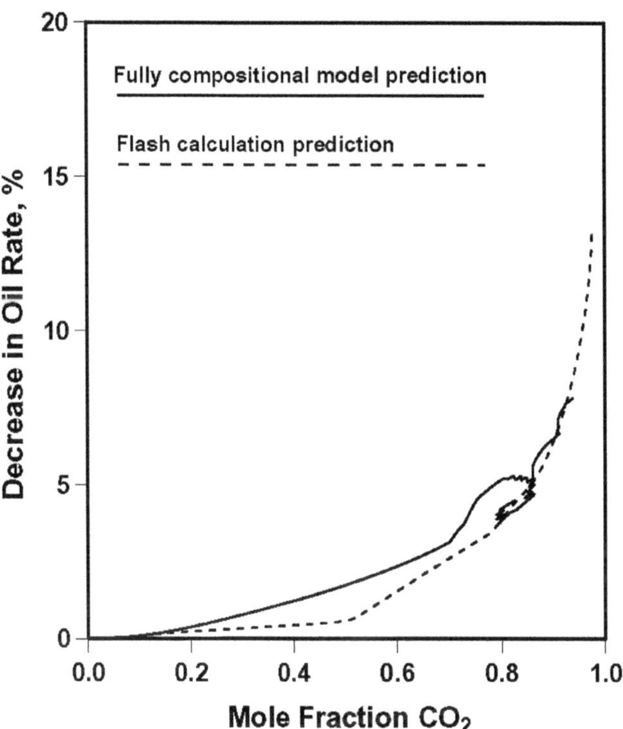

Figs. 4.18—Predicted reduction in oil rate caused by vaporization in the separator. As the mole fraction of CO_2 increases, thereby increasing the GOR, some of the crude oil is vaporized. This plot shows predictions for the Wasson field.

predicted peak oil rate because of smearing, and the increased transverse dispersion improved areal sweep, which in turn led to increased cumulative oil recovery (Fig. 4.21). Increasing the grid fineness from the standard 6×6 gridblock system increased the peak oil rate and GOR (using a constant WAG ratio) and caused the oil rate and GOR to decline more sharply.

The grid sensitivity study also was performed on multilayer grid systems that simulated waterflooding followed by WAG injection. The standard 6×6 grid appears to be satisfactory for multilayer modified black-oil reservoir simulator CO_2 flood predictions—for the problem tested, the differences between the coarse and fine grids were partially canceled out. Apparently, when the oil rate is overpredicted in one layer for a particular grid system, it is underpredicted in another layer. **Fig. 4.22** shows little difference in the peak oil rate for the different grid sizes. The coarsest grid does tend to predict the highest oil rate when the flood is in a tertiary decline.

Predicted cumulative oil production for both waterflooding and CO_2 flooding were similar for the three grids tested **(Fig. 4.23).** Although the finer grid had slightly greater oil recovery for waterflooding, the coarser grid seemed to make up the difference in cumulative oil production during the CO_2 flood **(Fig. 4.24). Fig. 4.25** shows that there is some difference in the peak GOR, with the finest grid predicting the greatest peak. However, the finest grid also had a lower oil rate. The predicted gas production rates were similar, which implies that for the problem tested, the design of the CO_2 removal plant would not be greatly impacted by using the standard 6×6 grid.

The examples show that the modified black-oil simulator exhibits more grid sensitivity in a CO_2 flood than in a waterflood. In this particular case, the errors nearly canceled each other out. However, the fact that grid sensitivity errors are possible implies that the modified black-oil approach is an engineering approximation and not a perfect tool for predicting CO_2 flood performance. The authors recommend conducting a grid sensitivity study to highlight possible errors that could affect economics. Regardless, a grid size that provides satisfactory results for waterflood predictions should not be made coarser for CO_2 flooding.

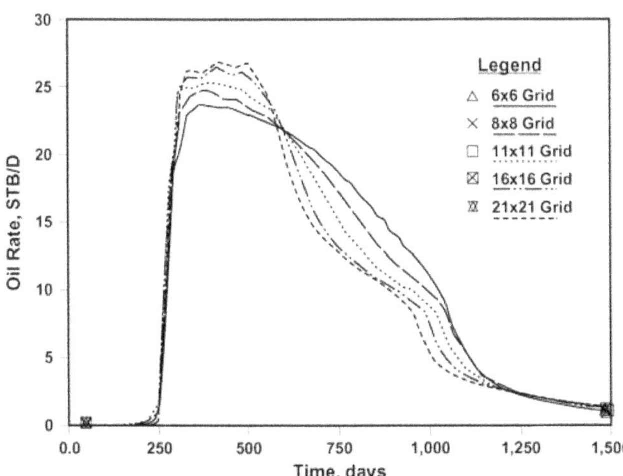

Fig. 4.19—Effect of grid size on predicted oil rate for a single-layer, first-contact miscible CO_2 flood. The finer grid predictions show a higher peak oil rate.

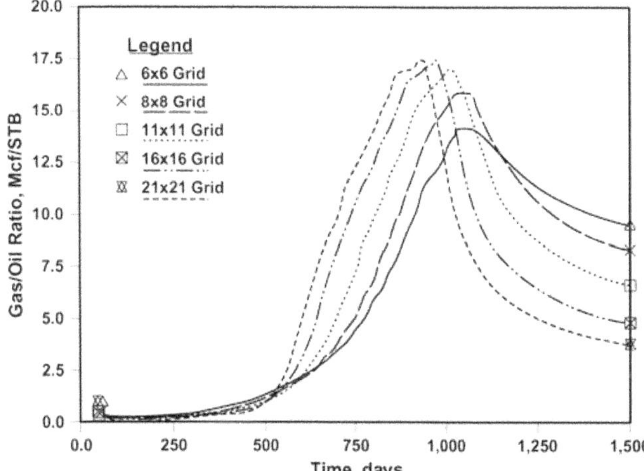

Fig. 4.20—Effect of grid size on predicted GOR for a single-layer, first-contact miscible CO_2 flood. The finer grid predictions show a higher peak GOR.

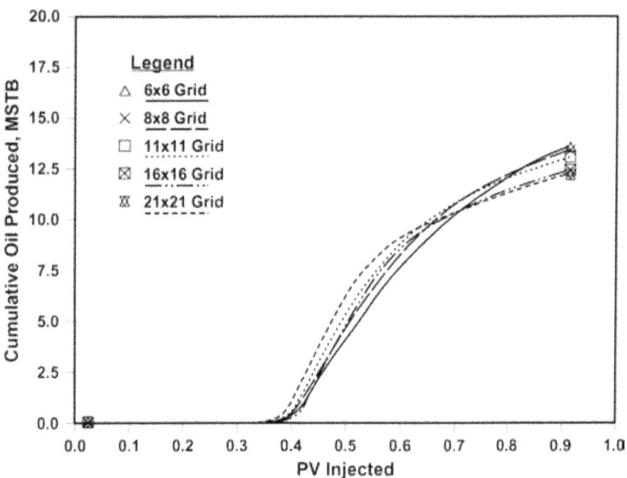

Fig. 4.21—Effect of grid size on predicted cumulative oil recovery for a single-layer, first-contact miscible CO_2 flood. The coarser grid predictions show greater cumulative oil recovery because of improved areal sweep resulting from numerical dispersion.

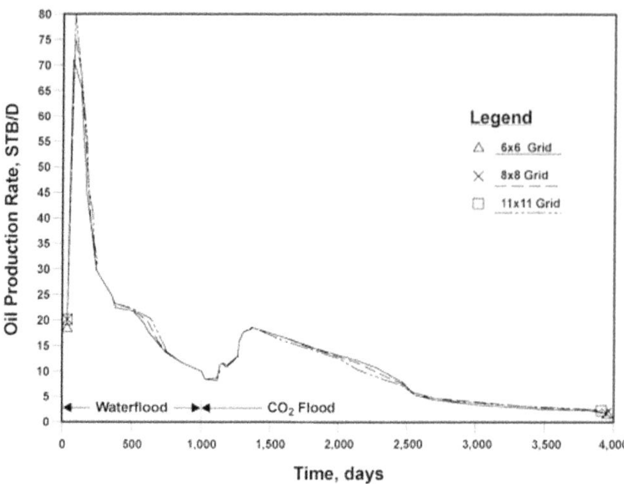

Fig. 4.22—Multilayer grid sensitivity showing the effect of grid size on predicted oil rate for a multilayer, first-contact miscible CO_2 flood. The grid size had very little effect on the multilayer oil rate.

Finer grid systems may be appropriate if geostatistics have been used in developing the reservoir description. The standard 6×6 grid can be used with a geostatistically-derived reservoir description if pseudorelative permeability curves have been determined for both CO_2 flood and waterflood conditions. However, Jerauld[26] has found that for now it is difficult to generate pseudorelative permeability curves that are consistent with fine-scale 3D simulations.

Computation Time. Computation time required for a modified black-oil reservoir simulator can be about 50% greater than the time required for the standard black-oil reservoir simulator used in waterflooding. The increase is caused by the presence of four components (instead of three) and the use of a nine-point finite difference scheme (instead of five-point) to avoid grid orientation problems. The nine-point finite difference scheme is recommended for all CO_2 predictions (both modified black-oil and fully compositional) to avoid grid orientation problems, which are more likely to occur when modeling any kind of gas injection.

4.3.2 Fully Compositional Reservoir Simulators. The development and loss of miscibility caused both by mixing and pressure reductions can be characterized best with a fully compositional reservoir simulator,[8,9,60,64–73] which uses an EOS to calculate phase equilibrium and the mass transfer of components between phases.[74–77] The EOS models used by fully compositional simulators have included those developed by Peng-Robinson,[74] Amoco-Redlich-Kwong,[75] Soave-Redlich-Kwong,[76] and Zudkevitch-Joffe-Redlich-Kwong.[77] All the above EOS models have been proved to be satisfactory for describing CO_2/oil phase behavior.

Fine-Tuning the EOS To Predict the Development of Miscibility. Because of the complexities of crude oils, the EOS commonly used in fully compositional models requires specific fine-tuning of component interaction parameters to calculate realistic phase equilibria.[78] The parameters are fine-tuned to match first-contact phase equilibrium data and any other types of phase equilibrium data, including conventional oil and gas analysis. Fine-tuning is particularly important to match any phase equilibrium data that exhibit a critical point or approach a critical point, to allow the EOS to predict the development of miscibility accurately. For example, in slim-tube tests at or above the thermodynamic MMP, the CO_2 and oil go through a critical point to develop miscibility, and fine-tuning the EOS to match slim-tube test data is crucial if no other source of data near the critical point is available.

Lumping Components. Crude oils have numerous components, which often are grouped into pseudocomponents with similar physical properties to save computation time. Done correctly, this does not sacrifice prediction accuracy. Young[65] recommends dividing the crude oil into 8 to 10 pseudocomponents, while Pederson[79] claims that six is sufficient. Hsu *et al.*[8] used only five pseudocomponents to characterize the oil when they simulated performance of

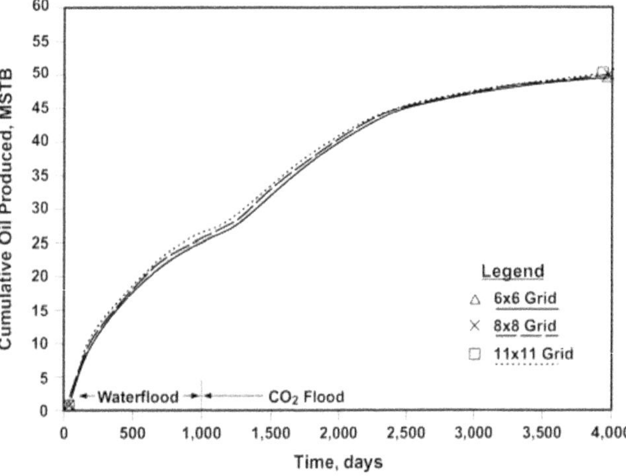

Fig. 4.23—Multilayer grid sensitivity showing the effect of grid size on predicted cumulative oil recovery for a multilayer, first-contact miscible CO_2 flood. The grid size had very little effect on multilayer cumulative oil recovery.

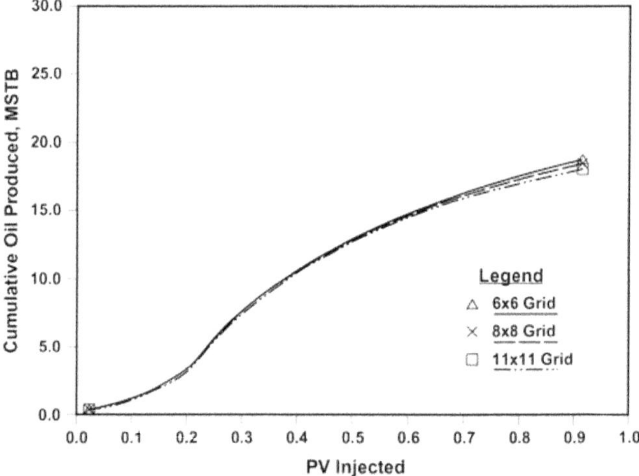

Fig. 4.24—Multilayer grid sensitivity showing the effect of grid size on predicted incremental cumulative oil recovery for a multilayer, first-contact miscible CO_2 flood. The coarser grid predicted slightly greater incremental cumulative oil recovery for a multilayer reservoir because of improved areal sweep resulting from numerical dispersion.

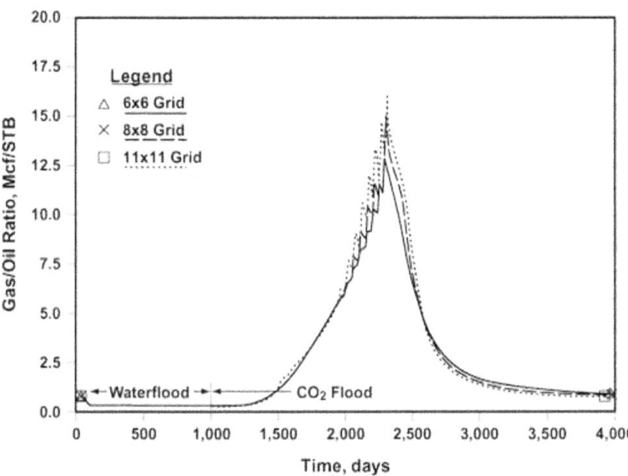

Fig. 4.25—Multilayer grid sensitivity showing the effect of grid size on the predicted GOR for a multilayer, first-contact miscible CO_2 flood. The finer grid predicted a higher peak GOR in a multilayer reservoir.

a 170-well sector of the Wasson Denver Unit in west Texas. Carbon dioxide generally is treated as a separate component.

One common lumping scheme suggested by Chang* includes C_1 and N_2 as one pseudocomponent, C_2–C_4 as a second, C_5 and C_6 as a third, and C_{7+} divided into three , which yields a total of seven hydrocarbon pseudocomponents, counting CO_2.

Complex Phase Behavior in EOS Models. Sec. 2.1.8 discusses how CO_2 can form a three-phase hydrocarbon region with certain oils. This phenomenon has been seen at temperatures of less than 120°F and has been observed in many west Texas oils. The pressure range for the three-phase region varies with oil and temperature (see Fig. 2.9 for a phase diagram example).

Most fully compositional reservoir simulators do not have a three-phase flash routine, which is necessary for calculating the three-phase hydrocarbon region. These models frequently have numerical stability problems when a two-phase flash routine is used to predict phase compositions in a pressure region where three hydrocarbon phases exist.[8,72] Khan *et al.*[72] developed a three-phase flash routine and showed that including the three-phase hydrocarbon region can have an important effect on predicted CO_2 flood performance. Hsu *et al.*[8] found that a two-phase flash routine yielded similar results to the three-phase flash routine for the Wasson field oil, but that the three-phase region occurs only in a small fraction of this reservoir. If the three-phase region is large, the use of a three-phase flash routine is recommended.

The solubility of CO_2 and other hydrocarbon components in water also can be incorporated into the EOS, although inclusion of water solubility for hydrocarbons other than CO_2 usually is not important. Chang *et al.*[73] developed correlations for CO_2 dissolved in brine and for the brine formation volume factor with dissolved CO_2. They concluded that dissolved CO_2 has a negligible effect on water compressibility and viscosity. Chang *et al.*[73] also used the dissolved CO_2 correlations in a fully compositional model to show that during WAG injection, about 5% and 8% of the CO_2 was lost to water in secondary and tertiary applications, respectively. Including CO_2 solubility in water decreased the predicted incremental cumulative oil recovery for CO_2 WAG injection by 5%.[73] Including solubility of CO_2 in water in a fully compositional simulation also will lead to increased water injection during the chase water phase.

Consistent Composition Predictions for Waterflood and CO_2 Flood. To use a fully compositional simulator to predict a CO_2 flood, previous performance must be simulated and proved valid with a fully compositional model. Otherwise, the predicted distribution of oil components at the start of a CO_2 flood will be incorrect. In addition, when CO_2 injection commences, other areas of the reservoir may still be under another form of depletion, so the fully compositional reservoir simulator must be able to accurately predict reservoir performance both before and during CO_2 injection. Because fully compositional simulation can require 10 times the computation time of modified black-oil simulation, the fully compositional simulation of previous performance usually is conducted after the history match of the preCO_2 performance has been completed with a modified black-oil model.

Accurate prediction of oil-phase viscosity is critical to the prediction of primary depletion, waterflood, and CO_2 flood performance.[80–82] Oil-phase viscosities in fully compositional models historically have been calculated with the Lorenz-Bray-Clark (LBC) viscosity correlation,[83] using compositions and densities calculated from the EOS. Note, however, that Lansangan and Smith[80] found that the LBC oil viscosity correlation could do a poor job of predicting oil-phase viscosity when CO_2 is present, and in fact, the LBC correlation can underestimate oil viscosity even when no CO_2 is present. In order for oil viscosity predicted by the LBC correlation to agree with actual oil viscosity, the critical volumes of the various C_{7+} pseudocomponents often are adjusted. Fong *et al.*[80] modified the Orbey-Sandler oil viscosity correlation[81] to predict oil-phase viscosity accurately when CO_2 was present. *We* recommend that the critical volume of the various C_{7+} pseudocomponents always be adjusted to obtain a satisfactory match of the laboratory oil viscosity data, no matter what oil viscosity correlation is used.

CO_2 Relative Permeability. Relative permeability data requirements for a fully compositional reservoir simulator were discussed in Sec. 4.1.2. Two methods have been used to calculate CO_2 relative permeability in a fully compositional model. In the first of these methods, the relative permeability of the CO_2-rich phase is assumed to flow on the gas relative permeability curve. Spivak *et al.*[71] and Chang[9,60] used this approach to predict CO_2 flood performance in a water-wet reservoir, including a gas hysteresis effect that reduced the mobility of the gas phase after water had been injected in the first WAG cycle. However, this approach is not consistent with the water-wet CO_2 relative permeability work by Dria,[62] which indicates that CO_2 relative permeability follows the oil relative permeability curve in a water-wet reservoir.

More recently, Hsu *et al.*[8] used the CO_2-rich phase relative permeability derived by the correlation of Chopra *et al.*[7] for the moderately oil-wet Wasson field. Hsu *et al.*,[8] unlike Chopra *et al.*,[7] did not have to rely on a pseudoviscosity mixing rule such as Eq. 4.7; instead, they used an EOS to predict two hydrocarbon phases so the realistic phase viscosities could be calculated. We recommend this latter approach for oil-wet and mixed-wet reservoirs.

Three- and Four-Phase Relative Permeability. In miscible conditions, fully compositional reservoir simulators require a three-phase relative permeability correlation whenever water exists with the two hydrocarbon phases (e.g., above the thermodynamic MMP when two hydrocarbon phases such as the CO_2-rich and oil-rich liquid phases) exist with water, and below the thermodynamic MMP when there are gas and oil phases in the presence of water.

Carbon dioxide relative permeability measurements made on moderately oil-wet rocks in west Texas[3–5] exhibit behavior consistent with the CO_2 rich-phase relative permeability model that Chopra *et al.*[7] developed by modifying Stone's second three-phase relative permeability model.[84] The west Texas CO_2 relative permeability measurements and the Chopra *et al.*[7] model apply to pressures above the thermodynamic MMP. Sec. 2.5 presents a detailed discussion of CO_2 relative permeability considerations.

Under immiscible conditions, conventional three-phase correlations[84–86] are appropriate. Baker's correlation[86] appears to fit immiscible three-phase relative permeability the best.

Four-phase flow can exist when three hydrocarbon phases are present with water. To date, no experimental four-phase relative permeability measurements exist. Fully compositional simulation studies simply have grouped the CO_2-rich liquid and oil-rich liquids together to form one liquid phase, after which they applied available three-phase relative permeability correlations. This practice tends to underestimate total mobility of the phases and could affect CO_2 flood predictions if the three-phase hydrocarbon region

*Personal communication with Y.B. Chang, ChevronTexaco (2000).

is large. Further research is needed to determine the best and most practical method to predict relative permeability characteristics under both three- and four-phase flow.

Convective Dispersion in Compositional Models. In their work, both Gardner *et al.*[87] and Negahban *et al.*[57] found that increased dispersion could hamper the development of miscibility to the extent that the MMP could increase, residual oil saturation could increase, and oil recovery could decrease. With enough mixing, the CO_2 flood might become an immiscible drive with vaporization of residual oil. Sec. 2.2 provides more detail on how small-scale reservoir heterogeneities disperse CO_2 and reduce its effectiveness in displacing oil by increasing the MMP.

Reservoir mixing through convective dispersion has been included in fully compositional reservoir simulators[9,57,60,88] even though the process of using finite-sized gridblocks in a numerical simulator introduces a numerical error that mimics the dispersion effect. Lantz[89] showed that for first-contact miscible floods, a square gridblock provides a numerical dispersivity that is one-half the size of the gridblock when the time step of the numerical simulation is small. In other words, if dispersivity is not purposely included in the reservoir simulator, numerical errors resulting from a finite grid size will mimic the dispersion effect. The numerical dispersivity is equal in all directions for square gridblocks, while in the reservoir the longitudinal dispersivity usually is much greater than the transverse disparity.

As the gridblock size approaches zero, the numerical dispersivity effect vanishes; however using gridblock sizes close enough to zero to eliminate numerical dispersivity usually is not practical. Consequently, numerical dispersivity is something with which reservoir simulation engineers routinely live.

The following example illustrates numerical vs. physical dispersion. A typical waterflood grid (based on the grid sensitivity study described in Sec. 4.3.1) uses 6×6 gridblocks for a quarter of a five-spot pattern. For a 40-acre five-spot pattern, the gridblock size would be 110 ft, and the numerical dispersivity would be 55 ft. The dispersivity from numerical dispersion will overshadow any physical dispersion specified by an engineer unless the physical dispersivity is larger than 55 ft.

Hsu *et al.*[8] used a numerical dispersivity equal to 75 ft in their Wasson field CO_2 predictions. This value might have been larger than the actual dispersivity, but it still provided a practical reservoir management tool. The coarser grid reduced computing time but also caused the predicted oil recovery to be less efficient. The coarse grids used by Hsu *et al.*[8] imply that longitudinal dispersivity is large and the effectiveness of the CO_2 oil recovery process has been adversely affected by small-scale heterogeneities. Had a very fine grid been used, Hsu most likely would have overpredicted oil recovery.

The fully compositional coarse-grid Wasson CO_2 flood simulations of Hsu *et al.*[8] were based on multiple-contact miscibility between the CO_2 and oil, which meant that the simulator predicted that a residual oil saturation to CO_2 would remain after the CO_2 passed by. The simulator also predicted the throughput effect, in that the residual saturation continually decreased with continued CO_2 throughput until oil saturation in the injection well gridblocks approached zero. Oil saturations away from the injection wells but in gridblocks swept by CO_2 were well above zero but less than the residual saturation to water.

A fully compositional reservoir simulator should be the first choice when studying a WAG CO_2 injection project in which reservoir pressure is near the thermodynamic MMP. As discussed in Sec. 2.4, pressure can oscillate significantly between water and CO_2 cycles during WAG injection, and the reservoir temporarily may go below the thermodynamic MMP. A fully compositional model can be useful in understanding potential loss and redevelopment of miscibility. Dispersion—whether numerical or physical—will affect the development and loss of miscibility and define the amount of oil left behind the injected CO_2.

One-dimensional fully compositional simulations for a range of numerical dispersivities (adjusted by changing the gridblock size) can be useful for addressing the dependence of oil recovery on reservoir pressure.[90,91] The level of longitudinal dispersivity can be adjusted through the gridblock size to better understand the interaction between mixing and miscibility development for a particular CO_2/oil system.

Ramirez *et al.*[92] showed that the presence of mobile water can increase dispersivity by a factor of 10. This implies that continuous CO_2 flooding during secondary recovery could experience less convective dispersion than a WAG CO_2 injection project, which means that a finer (less dispersive) grid would be required to simulate a continuous CO_2 project to avoid pessimistic predictions of flood performance and residual oil saturation. Work by Solano *et al.*[91] indicates that the additional mixing during the WAG process may justify the use a coarser grid for WAG.

Most fully compositional CO_2 simulations automatically incorporate large values for dispersivity because of numerical dispersion. Further work is required to define how large the longitudinal and transverse dispersivities should be, and thus what the appropriate gridblock size should be.

Pseudorelative Permeability in Fine-Grid Fully Compositional Simulation. If geostatistics are used, the resulting fine grid can greatly increase computing time and computer memory requirements. If the fine grid is replaced by a coarser grid with pseudorelative permeability curves, these curves must be consistent with the performance both before CO_2 injection (volumetric depletion and/or waterflood) and during the CO_2 flood. The consistency of the pseudorelative permeability curves should be checked by demonstrating that the course-grid and fine-grid simulation results agree. Jerauld[26] discussed the difficulty in generating valid pseudorelative permeabilities.

Moreover, if the pseudorelative permeability curves are changed later to better match CO_2 flood performance, the changes must be plugged into the history-match run to assure that the match is not affected. If it is, then subsequent future predictions for CO_2 will not be valid. The reservoir description will have to be modified to match both preCO_2 and postCO_2 flood performance.

Other Advantages of Fully Compositional Simulation. Fully compositional models offer distinct advantages in dealing with impure injection streams and accounting for NGL and H_2S production. Impure injection streams result from recycling produced gas into the injected CO_2 stream. If the impurities include methane or nitrogen at a high enough concentration, the MMP can increase, and a fully compositional reservoir simulator is the only way to correctly account for this effect.

Because they use an EOS, fully compositional reservoir simulators also can predict NGL and H_2S production at the separator. This eliminates the postsimulation NGL predictions that must be performed with modified black-oil models (see Sec. 4.3.1). The production of H_2S in a CO_2 flood can be greater than expected because of two factors: vaporization of H_2S at the separators when the CO_2 mole fraction and the GOR increase, and the high solubility of H_2S in water (see Sec. 3.2.4).

Hydrogen sulfide typically is present to some extent during a waterflood, if only because anaerobic sulfur-reducing bacteria introduced in the injected water generate it. In the separator, H_2S partitions between oil, water, and gas; at high WOR's it preferentially resides in the gas phase, but some of the H_2S remains in the oil and water phases. Under these conditions, H_2S does not appear to be a problem if the water is reinjected.

During the CO_2 flood the GOR increases, and H_2S that previously was dissolved in water now preferentially moves to the gaseous phase. The H_2S/oil production ratio has been known to increase by a factor of 2 to 3 in CO_2 floods in west Texas and Wyoming.

Hydrogen sulfide production can be predicted with a fully compositional model by treating the H_2S as a separate component and including the solubility of H_2S in water. Also, an assumption must be made about the initial distribution of H_2S and its origin.

Field Applications of Fully Compositional Simulation. The petroleum literature reports several successful applications of fully compositional reservoir simulators predicting CO_2 flood performance. An example is Roper *et al.*,[93] who matched performance of an alternating CO_2/water injection test in the west Texas Mabee field. Stalkup and Crane[94] matched the saturations at a logging observation well adjacent to an enriched gas (20 mol% CO_2) WAG injection well in the Prudhoe Bay field. Hsu *et al.*[8] used a fully compositional model to match the performance of a 170-well sector of the Wasson Denver Unit that contained 12 layers and a grid network of 83×69 gridblocks per layer. This grid fineness is similar to one that might be used for waterflood predictions. The Wasson model subsequently was scaled up to the rest of the field and is currently used for reservoir management (see Sec. 3.2 for scaleup techniques).

4.3.3 Nonisothermal Effects in Reservoir Simulation. To date, nonisothermal effects have not been addressed by either modified black-oil or fully compositional models. Ali *et al.*[95] found that nonisothermal effects could be important in cases where high rates of cold injection water cooled the reservoir, with reservoir temperature and bed thickness being important parameters that determine the extent of the cooling. If the reservoir temperature changes fluid properties, MMP can also change. Another possibility is that the rock becomes more prone to fracture. Stevens and Murray[96] used a coupled geothermal-reservoir model to predict fracture growth resulting from rock embrittlement during cold water injection.

4.4 Optimizing the CO_2 Flood Design

Once the history match is successful, the simulator is run many more times to determine the optimum CO_2 injection scheme and to size the CO_2 separation plant to maximize the discounted incremental cumulative cash flow (net present value) over the duration of the planned flood. The best financial performance from a CO_2 project will be achieved by reducing front-end investments while achieving the earliest possible oil response to CO_2. Not having to build a recycle plant and being able to process produced gas at a nearby existing plant may increase financial performance of the CO_2 project, although capacity limitations at existing plants can restrict how the project is managed.

Carbon dioxide predictions often are performed to test options related to the following:

- WAG optimization and CO_2 slug size.
- Gravity-dominated floods vs. pattern floods (gravity vs. WAG).
- Pattern modifications.
- Infill drilling coincidental with the start of CO_2 injection.
- Use of horizontal wells.
- Pattern balancing.
- Wellhead/bottomhole injection pressure.
- Production wells—conversion of pumped to flowing and use of bottomhole producing pressures that correspond to different CO_2 plant inlet pressure requirements.
- Use of impure CO_2 injection streams related to CO_2 recycle plant options.
- Phased-in or delayed field development of CO_2 injection

The above list does not explicitly include the question of whether to begin the CO_2 flood in a secondary or tertiary mode. The critical factor for a technically successful secondary CO_2 flood is that the reservoir pressure must be above the thermodynamic MMP at the start of CO_2 injection—or must increase rapidly to above the thermodynamic MMP—to reduce the possibility of an immiscible CO_2 displacement (Hansford Marmaton CO_2 flood).[97] If a secondary CO_2 flood can be technically successful, then the question of secondary vs. tertiary can be answered by studying the economic impact of delaying the startup. (Sec. 3.1.2 provides additional discussion on the timing of CO_2 floods.) In the sections that follow, we cover some of the above topics in greater detail.

4.4.1 WAG Optimization and CO_2 Slug Size. In this section, we discuss WAG optimization and slug size for reservoirs that do not have water-blocking problems, i.e. reservoirs that are oil-wet or mixed-wet (see Sec. 2.2.2 for additional information on water blocking). Also, this section applies only to reservoirs without much dip.

Pariani *et al.*[98] and Hill *et al.*[99] show how net present value can be maximized by paying attention to various WAG injection schemes. Pariani *et al.*[98] found that beginning a pattern CO_2 flood with a low WAG ratio and then increasing the ratio can improve both financial performance and incremental oil recovery. The increasing ratio helps improve sweep, and, perhaps more importantly, helps slow gas production and prevents it from exceeding the capacity of the CO_2 processing plant. **Fig. 4.26** illustrates the economic sensitivity of a flood to different sizes of the CO_2 processing plant (plant capacity is on a pattern basis). The data in this bar graph were generated by using different WAG ratios and increasing the ratios as necessary to maintain the near-constant gas production rates shown on the x-axis. The y-axis is a normalized net present value, and is obtained by dividing each net present value by the net present value of the case plotted to the left of all the other cases.

Once the optimum WAG ratio and plant size were determined, economic sensitivity simulations were performed for CO_2 slug size **(Fig. 4.27)**. Incremental oil recovery estimates continued to increase with larger CO_2 slug sizes, but the normalized present value reaches a maximum and then decreases as oil recovery tails off. Thus the optimum CO_2 slug size must be determined for a specific set of economic conditions. The results of the Pariani *et al.*[98] study were implemented successfully in the Slaughter field in west Texas.

In practice all patterns may start at the same WAG ratio. Later, as CO_2 production increases to the point that volumetric sweep efficiency is being adversely affected, the WAG ratio normally is increased on a pattern-by-pattern basis, starting with the highest GOR patterns. The patterns normally are chosen around one injection well that controls the WAG ratio.

WAG injection schemes derived from reservoir simulator runs should be viewed only as a guide; there always will be some uncertainties in characterizing the reservoir, and a reservoir simulation might not include all patterns because of time and manpower limitations. Production wells always should be monitored

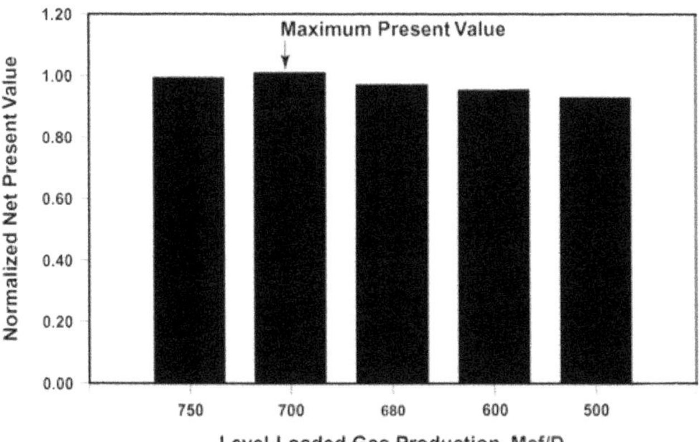

Fig. 4.26—Normalized net present value of a flood for different gas plant processing capacities. In each case, the gas production rate was managed by changing WAG ratios.

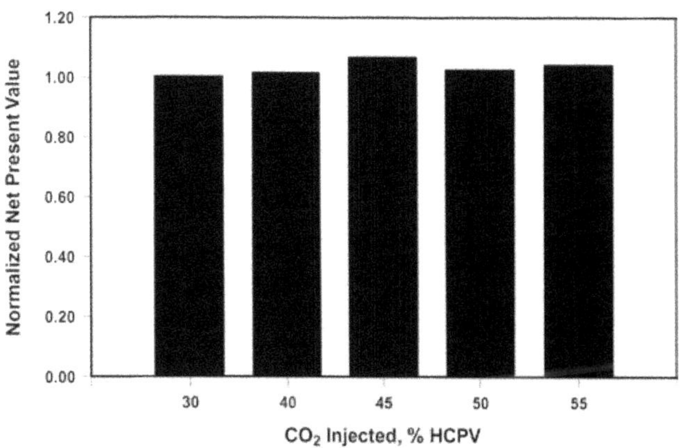

Fig. 4.27—Normalized net present value of a flood for different CO_2 slug sizes. Although oil continues to be recovered as additional CO_2 is injected, there is a point at which it is no longer economical to inject CO_2.

for unexpected CO_2 breakthrough so that the WAG ratio can be changed as necessary. The simulator can provide a first estimate on how to adjust the WAG ratio, but if actual performance deviates from the model prediction, the engineer should change the WAG ratio by trial and error to improve sweep and control the total gas production rate to the CO_2 processing plant. Changing the WAG ratio also can involve changing the size of either the CO_2 or the water cycle.

Controlling the production rate is important because it affects gas plant economics. Maintaining a CO_2 processing plant at a level-loaded gas processing rate improves plant efficiency and reduces operating costs.[98] If the produced gas is sent to a third party for processing, the third party may limit the amount of gas that they will receive. In simple terms, the engineer must balance changes in WAG ratio to control total gas production rate, to reduce gas processing costs and meet facility constraints without unduly impacting oil production rate and revenue.

4.4.2 Gravity-Dominated Floods vs. Pattern Floods. In steeply dipping reservoirs where there are strong gravity effects, CO_2 normally is injected continuously from the top down, as was done in the Weeks Island gravity-stable CO_2 flood.[100] In this type of flood, one of the key goals of the design is to achieve as high an injection rate as possible without losing the stabilizing effect of gravity. While gravity displacement maximizes oil recovery, higher injection rates (and a less stable interface) may improve the economic performance of the flood, even though it is achieved at the expense of long-term oil recovery. Sensitivity to varying the CO_2 injection rate should be analyzed during the design phase to determine the most profitable CO_2 injection rate.

In addition to conducting sensitivity simulations for injection rate, it also might be wise to carry out sensitivity simulations based on pattern flooding for some dipping reservoirs, to determine whether gravity effects will dominate the economic performance. The Weeks Island reservoir has a dip of 27° and permeability of 1.8 darcies, making gravity an almost overwhelming first choice. On the other hand, the Wertz Tensleep CO_2 project—with a reservoir dip of 21°—is a pattern CO_2 flood and was a pattern waterflood.[101] The difference between the two projects is that the Wertz Tensleep reservoir has considerably lower horizontal and vertical permeability than does the Weeks Island reservoir, so that gravity does not have a large influence.[101]

4.4.3 Pattern Modifications. Testing a flood's sensitivity to pattern modifications may be beneficial from both an economic and oil recovery point of view. The optimal pattern alignment depends on the existing patterns and infrastructure.

If the reservoir has water injection sweep problems that can be alleviated by modifying the pattern, then that should be done. However, if a sweep problem cannot be solved for water injection, it is likely to worsen during the CO_2 flood. One novel solution was used in a Canadian CO_2 pilot test in the naturally fractured Midale field, where areal sweep and oil recovery were improved when injection and production wells were aligned perpendicular to the natural fracture trend.[102]

In the Denver Unit of the Wasson field, large losses in water injection rate occurred in the WAG area.[103] The problem was attacked by converting the inverted nine-spot patterns in the WAG area to direct line-drive patterns,[103] which increased the total number of injection wells and injection rates. In addition, the change to line-drive patterns may increase reservoir pressure enough to lead to increased oil recovery[54] because a larger portion of the reservoir will be miscible to CO_2.

In some parts of the Slaughter field CO_2 floods, producing wells in chicken-wire patterns (elongated hexagonal patterns only used in the Slaughter field because of the initial land survey) were modified to increase the ratio of injection wells to producing wells, which increased both reservoir pressure and oil recovery.

4.4.4 Infill Drilling. Drilling additional wells at the start of CO_2 injection also might improve the economics. If reservoir layers lack continuity, infill drilling can improve it and make CO_2 flooding more successful. Simultaneous startup of CO_2 injection and infill drilling successfully increased oil recovery in the Means San Andres Unit,[104] the Dollarhide Unit,[105] and the South Wasson Clearfork Unit.[106]

4.4.5 Horizontal Wells. Horizontal wells may improve the oil production rate and recovery, the sweep, and the financial performance of a CO_2 flood. On the production side, horizontal wells can reduce the effects of gas coning associated with downdip CO_2 injection. On the injection side, horizontal wells can selectively stimulate a low-permeability interval, increasing the CO_2 slug size for that interval.[107]

Horizontal laterals have been drilled in existing vertical CO_2 injection wells. The reduced cost of drilling a horizontal lateral from an existing well—compared to a new horizontal well—can make this an attractive option. In a reservoir simulation study of CO_2 flooding with horizontal wells, Lim *et al.*[108] found that if the vertical/horizontal permeability ratio is at least 0.10, a horizontal injection well can be an excellent means to achieve high injection rates and faster flood response. However, Lim *et al.*[108] also found that the choice of which layer to inject or produce with a horizontal well is very important because the wrong choice can decrease oil recovery rather than increase it.

If horizontal well drilling costs are not too high for a particular field, these wells should be evaluated for CO_2 projects in steeply dipping reservoirs and in reservoirs in which the vertical/horizontal permeability ratio is 0.10 or greater.

4.4.6 Pattern Balancing. Pattern balancing is the art of controlling injection so that each pattern receives an equal volume of CO_2. If CO_2 and water cycles are sized by switching at even time increments, the pattern WAG ratio and cycle sizes will change with time as reservoir mobility changes. Switching CO_2 and water cycles according to fixed injection volumes allows the desired amounts of CO_2 and water to enter the pattern and also controls the WAG ratio.

Regardless of WAG injection scheme, some wells may produce more CO_2 than others do because the formation flow capacity varies from well to well. Balancing CO_2 production requires controlling rates and pressures at the producing wells. In any case, the effects of pattern balancing should be incorporated into the CO_2 predictions and the economics. Note that there is no reason why an equal amount of CO_2 must be injected in each pattern. Some patterns may be more heterogeneous than other patterns and thus might not need the same volume of CO_2 to yield their most economical performance.

In theory, a reservoir simulator can be used to predict pattern allocation factors over time and to make changes to the allocation and WAG injection ratios, managing the reservoir on a daily basis to maximize financial return. However, there always are uncertainties in reservoir predictions, and if the project has a large number of wells, the prediction model might not cover the entire CO_2 project area. Scaling up predictions to other areas (see Sec. 3.2) also has technical limitations.

In summary, the reservoir model can be a useful tool in understanding what key parameters affect performance, and it can provide guidance on how to make pattern allocation and WAG changes. Ultimately, however, actual performance is what will dictate necessary changes in pattern allocation and WAG ratios. These changes most likely will take place in a trial-and-error manner, using reservoir engineering judgment to massage model predictions until they are consistent with actual performance.

4.4.7 Bottomhole Injection Pressure. Some operators reduce the bottomhole injection pressure for a CO_2 project to bring the reservoir pressure closer to the thermodynamic MMP.[104] Part of the reasoning behind this is that the CO_2 formation volume factor (RB/Mscf) increases dramatically above the thermodynamic MMP, and operators want to conserve the CO_2. Reducing reservoir pressure can be an expensive mistake—if there is significant small-scale reservoir mixing, the effective MMP may be several hundred psi above the thermodynamic MMP, and a reduction in pressure can cause the CO_2 flood to fail.

Another factor to consider is that CO_2 injection projects that use small WAG cycles and equal bottomhole injection pressures for both gas and water have experienced large swings in reservoir pressure between water and CO_2 cycles because of phase mobility and compressibility effects (see Sec. 2.4). Such CO_2 projects have experienced poor incremental oil recovery when reservoir pressure was close to thermodynamic MMP at the start of the CO_2 flood.

4.4.8 Production Well Considerations. Several operators have found it economical to convert producing wells on beam pump or electrical submersible pump (ESP) to naturally flowing wells at some time during a CO_2 flood.[103,109] Conversion to flowing reduces lift costs and the need to pull pumps. ESP's, in particular, can exhibit low pump efficiencies during a CO_2 flood because of high gas/liquid ratios.[109]

The downside of converting from pump to flowing is that it also can increase the bottomhole producing pressure and thus decrease the total production rate. While gas processing costs will decrease, so will revenue from produced oil, at least for a while. The most economical time to convert producing wells to flowing can be estimated by incorporating flowing wells into the CO_2 predictions and including potential cost savings in the economics.

Another factor to include when simulating production wells is the increase in pump bottomhole pressure caused by the inefficiency of lifting a gassy fluid when the GOR increases. Bottomhole producing pressures in west Texas CO_2 floods have been known to increase 100 to 200 psi when the GOR is low and 400 to 600 psi when the GOR exceeds 2,000 Scf/BO.

Some operators have used plunger lift successfully in CO_2 floods.* Plunger lift takes advantage of the energy supplied by the produced gas to lift the produced liquids and thereby reduce lift costs. Predictions with plunger lift—or with any other type of lift—should incorporate bottomhole producing pressures that are as realistic as possible.

Finally, it may be useful to study bottomhole producing pressures when considering different CO_2 recycle plant options. In many cases, the inlet pressure to the plant dictates the bottomhole producing pressure. Increasing the inlet pressure can sometimes eliminate a stage of compression and achieve substantial savings in upfront investment and operating costs, but unfortunately, it also can lead to a similar increase in producing bottomhole pressure. Carbon dioxide predictions should be carried out using the higher bottomhole producing pressures to account correctly for the effect that higher producing pressures might have in delaying flood response and project economics. Usually, the type of pressure increase considered at the plant inlet is less than 200 psi and will not affect cumulative oil production very much.

4.4.9 CO_2 Injection Stream Impurities. The purity of the CO_2 injection stream must be included in the predictions because impurities can affect MMP and oil recovery: Methane and nitrogen can increase MMP and decrease oil recovery, while H_2S, propane, and heavier hydrocarbons can decrease MMP and increase oil recovery.

As stated in Sec. 4.3.2, a fully compositional model is the best tool for predicting the effect of impurities on oil recovery. The model must include the timing for injecting impurities—if recycled (impure) gas is injected after CO_2 breakthrough, oil recovery may not be affected if the amount of impurities is small and enough pure CO_2 already has been injected.

Modified black-oil models only allow one MMP, however this is not a problem if either impure CO_2 of a fixed composition is to be injected from the start of the CO_2 flood, or if one is certain that any impure CO_2 to be injected will not contact mobile oil. In the first case, the thermodynamic MMP input to the model should be based on the thermodynamic MMP of the total injection stream, including the effect of impurities. If impurities are introduced after CO_2 breakthrough and gas recycling, the resultant increase in MMP will not be predicted by a modified black-oil simulator. In the second case, if the simulator predictions show that impure CO_2 may contact mobile oil at pressures near the thermodynamic MMP, the predictions may be optimistic. In this case, a fully compositional simulator is recommended.

4.4.10 Phasing and/or Delaying CO_2 Projects. Finally, the development of a CO_2 injection project may be phased in or even delayed. Developing a CO_2 flood in stages can lower front-end investment costs and reduce CO_2 purchases during the early life of the project.[110] In addition, the CO_2 distribution system and produced gas-gathering system may require only minor modifications as CO_2 injection gradually is implemented fieldwide.

By reducing investment costs, a phased approach can allow less attractive areas of a field to be CO_2 flooded economically, and small leases in a field to be deferred until plant capacity is available from nearby, mature CO_2 floods. Delaying a tertiary CO_2 injection project usually has little effect on CO_2 flood performance.

On the downside, delaying a CO_2 project may adversely affect economics because CO_2 injection starts closer to the economic limit of current operations. This reduces the amount of time for which current operating costs are subtracted from total CO_2 operating costs to yield incremental operating costs, thereby increasing the incremental operating costs for the CO_2 project. The incremental net present value for a zero discount rate also will be lower because of the delay.

Sometimes the decision on delaying a CO_2 project depends on forecast oil prices. If the property is small and generates limited cash flow, and if spending is constrained by low oil prices, an operator could be forced into selling a property if the cash flow became negative. In this case, it might be better to CO_2 flood the property instead of deferring.

4.4.11 Summary. Reservoir simulation can be used to generate the economic outlook for various possibilities that exist and to optimize the financial performance of the project. The basic starting input includes the facility and plant investments and operating costs from the scoping step. Contingencies must be built in to account for changes encountered during the detailed design.

In addition to performing economic sensitivity studies of the field and the CO_2 recycling plant, the economic ramifications of options that do not affect CO_2 flood performance should be studied, such as different types of recycle gas facilities. Economic sensitivity studies also should assess the risk associated with such factors as lower oil recovery, lower peak oil rate, lower oil prices, and higher than expected gas production. The details of the economic calculations are discussed in Sec. 3.4.

Nomenclature

k_{rw} = water relative permeability, dimensionless
$(k_{rw})_{max}$ = maximum water relative permeability after water displaces CO_2, dimensionless
b = empirical positive exponent which is always positive
h = thickness, L, ft
k = horizontal permeability, L^2, md
n = empirical viscosity mixing parameter, dimensionless
r_w = actual wellbore radius, L, ft
r_{we} = effective wellbore radius, L, ft
$(r_{we})_{het}$ = effective wellbore radius determined with a heterogeneous assumption, L, ft
$(r_{we})_{hom}$ = effective wellbore radius determined with a homogeneous assumption, L, ft
R_{cd} = mathematical heterogeneity term defined, dimensionless
s = skin factor, dimensionless
$(s)_{het}$ = skin factor for a heterogeneous, layered reservoir, dimensionless
$(s)_{hom}$ = skin factor for a homogeneous, layered reservoir, dimensionless
S_w = water saturation, volume fraction
S_{wr} = residual water saturation after a CO_2 cycle, volume fraction

*Personal communication with S.C. Wehner, Phillips Petroleum (2000).

$(S_w)_{max}$ = maximum water saturation after a water cycle, volume fraction
T = temperature, T, °F
x_o = fraction of the live oil (containing natural gas in solution), mass or volume fraction
x_s = fraction of the solvent (injected CO_2 stream), mass or volume fraction
ϕ = porosity, fraction
μ_o = viscosity of the live oil (including dissolved natural gas), m/Lt, cp
μ_{os} = viscosity of the oil/solvent (CO_2) mixture, m/Lt, cp
μ_s = viscosity of the solvent (injected CO_2 stream), m/Lt, cp
ω = Todd-Longstaff mixing parameter used to represent viscous fingering when set to a positive value less than 1.0, dimensionless

References

1. Yellig, W.F. and Metcalfe, R.S.: "Determination and Prediction of CO_2 Minimum Miscibility Pressures," *JPT* (January 1980) 160.
2. Turek, E.A., Metcalfe, R.S., and Fishback, R.E.: "Phase Behavior of Several CO_2/West Texas-Reservoir-Oil Systems," *SPERE* (May 1988) 505.
3. Shyeh-Yung, J-G.J.: "Mechanisms of Miscible Oil Recovery: Effects of Pressure on Miscible and Near-Miscible Displacements of Oil by Carbon Dioxide," paper SPE 22651 presented at the 1991 SPE Annual Technical Conference and Exhibition, Dallas, October 6–9.
4. Prieditis, J., Wolle, C.R., and Notz, P.K.: "A Laboratory and Field Injectivity Study: CO_2 WAG in the San Andres Formation of West Texas," paper SPE 22653 presented at the 1991 SPE Annual Technical Conference and Exhibition, Dallas, October 6–9.
5. Wegener, D.C. and Harpole, K.J.: "Determination of Relative Permeability and Trapped Gas Saturation for Predictions of WAG Performance in the South Cowden CO_2 Flood," paper SPE 35429 presented at the 1996 SPE/DOE Tenth Symposium on Improved Oil Recovery, Tulsa, 21–24 April.
6. Corey, A.T. *et al.*: "Three-Phase Relative Permeability," *JPT* (November 1956) 63; *Trans.*, AIME (1956) **207,** 349.
7. Chopra, A.K., Stein, M.H., and Dismuke, C.T.: "Prediction of Performance of Miscible-Gas Pilots," *JPT* (December 1990) 1564.
8. Hsu, C-F., Morell, J.I., and Falls, A.H.: "Field-Scale CO_2-Flood Simulations and Their Impact on the Performance of the Wasson Denver Unit," *SPERE* (February 1997) 4.
9. Chang, Y.: "Development and Application of an Equation of State Compositional Simulator," PhD dissertation, U. of Texas, Austin, Texas (1990).
10. Craig, F.F. Jr.: *The Reservoir Engineering Aspects of Waterflooding,* Monograph Series, SPE, Richardson, Texas (1971) **3.**
11. Wendel, D.J., Anderson, W.G., and Myers, J.D.: "Restored-State Core Analysis for the Hutton Reservoir," *SPEFE* (December 1987) 509.
12. Stein, M.H. *et al.*: "Slaughter Estate Unit CO_2 Flood: Comparison Between Pilot and Field-Scale Performance," *JPT* (September 1992) 1026.
13. Hewett, T.A.: "Fractal Distributions of Reservoir Heterogeneity and Their Influence on Fluid Transport," paper SPE 15386 presented at the 1986 SPE Annual Technical Conference and Exhibition, New Orleans, 5–8 October.
14. Emanuel, A.S. *et al.*: "Reservoir Performance Prediction Methods Based on Fractal Geostatistics," *SPERE* (August 1989) 311; *Reservoir Characterization*–2, Reprint Series, SPE, Richardson, Texas (1989), **27,** 38.
15. Tang, R.W., Behrens, R.A., and Emanuel, A.S.: "Reservoir Studies With Geostatistics To Forecast Performance," *SPERE* (May 1991) 253.
16. Standing, M.B. and Katz, D.L.: "Density of Natural Gases," *Trans.*, AIME (1942), **146,** 140.
17. Beal, C.: "The Viscosity of Air, Water, Natural Gas, Crude Oil and Its Associated Gases at Oil Field Temperatures and Pressures," *Trans.*, AIME (1946) **165,** 94.
18. Lasater, J.A.: "Bubble Point Pressure Correlation," *Trans.*, AIME, (1958) **213,** 379.
19. Chew, J-N. and Connally, C.A. Jr.: "A Viscosity Correlation for Gas-Saturated Crude Oils," *Trans.*, AIME (1959) **216,** 23.
20. Lee, A.L.: "The Viscosity of Natural Gases," *JPT* (August 1966) *Trans.*, AIME (1966) 237, 997.
21. Beggs, H.D. and Robinson, J.R.: "Estimating the Viscosity of Oil Systems," *JPT* (September 1975) 1140.
22. Glasø, Ø: "Generalized Pressure-Volume-Temperature Correlations," *JPT* (May 1980) 785.
23. Vazquez, M. and Beggs, H.D.: "Correlations for Fluid Physical Property Prediction," *JPT* (June 1980) 968.
24. Ader, J.C. and Stein, M.H.: "Slaughter Estate Unit Tertiary Miscible Gas Pilot Reservoir Description," *JPT* (May 1984) 837.
25. Mattax, C.C. and Dalton, R.L.: *Reservoir Simulation,* Monograph Series, SPE, Richardson, Texas (1990) **13.**
26. Jerauld, G.R.: "A Case Study in Scaleup of Multi-Contact Miscible Hydrocarbon Gas Injection," paper SPE 39626 presented at the 1998 SPE/DOE Improved Oil Recovery Symposium held in Tulsa, 19–22 April.
27. Chopra, A.K.: "Reservoir Descriptions Via Pulse Testing: A Technology Evaluation," paper SPE 17568 presented at the 1988 SPE International Meeting on Petroleum Engineering, Tianjin, China, 2–4 November.
28. Folger, L.K. and Guillot, S.N.: "A Case Study of the Development of the Sundown Slaughter Unit CO_2 Flood Hockley County, Texas," paper SPE 35189 presented at the 1996 SPE Permian Basin Oil and Gas Recovery Conference, Midland, Texas, 27–29 March.
29. Ramey, H.J. Jr.: "Interference Analysis for Anisotropic Formations—A Case History," *JPT* (October 1975) 1290; *Trans.*, AIME, **259.**
30. Streltsova, T.D.: *Well Testing in Heterogeneous Formations,* Exxon Monograph, John Wiley & Sons, New York City (1988).
31. Hirasaki, G.J.: "Pulse Tests and Other Early Transient Pressure Analysis for In-Situ Estimation of Vertical Permeability," *SPEJ* (February 1974) 75; *Trans.*, AIME, **257.**
32. Yaxley, L.M.: "Effect of a Partially Communicating Fault on Transient Pressure Behavior," *SPEFE* (December 1987) 590.
33. Serra, K., Reynolds, A.C., and Raghavan, R.: "New Pressure Transient Analysis Methods for Naturally Fractured Reservoirs," *JPT* (October 1983) 1902.
34. Stewart, G. and Asharsobbi, F.: "Well Test Interpretation of Naturally Fractured Reservoirs," paper SPE 18173 presented at the 1988 SPE Annual Technical Conference and Exhibition, Houston, 2–5 October.
35. Shah, P.C. *et al.*: "Estimation of the Permeabilities and Skin Factors in Layered Reservoirs With Downhole Rate and Pressure Data," *SPEFE* (September 1988) 555.
36. Shah, P.C. and Spath, J.B.: "Transient Wellbore Pressure and Flow Rates in a Commingled System With Different Layer Pressures," paper SPE 25423 presented at the 1993 SPE Production Operations Symposium, Oklahoma City, Oklahoma, 21–23 March.
37. Bragg, J.R., Roesner, R.E., and Strassner, J.E.: "Measuring Well Injection Profiles of Polymer-Containing Fluids," paper SPE 10690 presented at the 1982 SPE/DOE Joint Symposium on Enhanced Oil Recovery, Tulsa, 4–7 April.
38. Simlote, V.N. *et al.*: "Synergistic Evaluation of a Complex Conglomerate Reservoir for EOR, Barrancas Formation, Argentina," *JPT* (February 1985) 295.
39. Chopra, A.K., Stein, M.H., and Ader, J.C.: "Development of Reservoir Descriptions To Aid in Design of EOR Projects," *SPERE* (May 1989) 143.
40. Aasum,Y., Kelkar, M.G., and Gupta, S.P.: "An Application of Geostatistics and Fractal Geometry for Reservoir Characterization," *SPEFE* (March 1991) 11.
41. Dagan, G.: "Models of Groundwater Flow in Stochastically Homogeneous Porous Formations," *Water Resources Res.* (February 1979) **15,** No. 1, 47.
42. Hewett, T.A. and Behrens, R.A.: "Conditional Simulation of Reservoir Heterogeneity With Fractals," *SPEFE* (September 1990) 217.
43. Wolcott, D.S. and Chopra, A.K.: "Incorporating Reservoir Heterogeneity with Geostatistics to Investigate Waterflood Recoveries," *SPEFE* (March 1993) 26.
44. Malik, M. and Lake, L.W., "Modeling Fluid Flow Through Geologically Realistic Media" Geological Society Special Publication No. 73: *Characterisation of Fluvial and Aeolian Reservoirs,* C.P. North and D.J. Prosser (eds.), U. of Aberdeen, Aberdeen, Scotland, (1993) 367.
45. Journel, A.G. and Huijbregts, C.J.: *Mining Geostatistics,* Academic Press, New York City (1978).

46. Journel, A.G. and Alabert, F.G.: "New Method for Reservoir Mapping," *JPT* (February 1990) 212.
47. Martin, J.C. and Wegner, R.E.: "Numerical Solution of Multiphase, Two-Dimensional Incompressible Flow Using Stream-Tube Relationship," *SPEJ* (October 1979) 313; *Trans.*, AIME, **267.**
48. Haldorsen, H.H. and Lake, L.W.: "A New Approach to Shale Management in Field-Scale Models," *SPEJ* (August 1984) 447.
49. Richardson, J.G. *et al.*: "The Effect of Small, Discontinuous Shales on Oil Recovery," *JPT* (November 1978) 1531.
50. Hohn, M.E.: *Geostatistics and Petroleum Geology,* Van Nostrand Reinhold, New York City (1988).
51. Duetsch, C.V., and Journel, A.G.: *GSLIB: Geostatistical Software Library and User's Manual,* Oxford U. Press, New York City (1992).
52. de Buyl, M., Guidish, T., and Bell, F.: "Reservoir Description From Seismic Lithologic Parameter Estimation," *JPT* (April 1988) 475.
53. Bennett, C.O., Reynolds, A.C., and Raghavan, R.: "Analysis of Finite Conductivity Fractures Intercepting Multilayer Reservoirs," *SPEFE* (June 1986) 259; *Trans.*, AIME, **281.**
54. Stein, M.H. and Shiralkar, G.S.: "Estimation of Average Reservoir Pressure in Chicken Wire and Nine-Spot Patterns Under Mature Waterflooding Conditions," paper SPE 24399 available from SPE, Richardson, Texas (1992).
55. Carlson, F.M. and Stein, M.H.: "Automatic Waterflood History Matching Using Dimensionless Performance Curves," paper SPE 24897 presented at the 1992 SPE Annual Technical Conference and Exhibition, Washington, DC, 4–7 October.
56. Stein, M.H. and Carlson, F.M.: "Dimensionless Equations for Waterflood History Matching," paper SPE 25521 available from SPE, Richardson, Texas (1993).
57. Negahban, S., Shiralkar, G.S., and Gupta, S.P.: "Simulation of the Effects of Mixing in Gasdrive Core Tests of Reservoir Fluids," *SPERE* (August 1990) 402.
58. Todd, M.R. and Longstaff, W.J.: "The Development, Testing, and Application of a Numerical Simulator for Predicting Miscible Flood Performance," *JPT* (July 1972) 74; *Trans.*, AIME, **253.**
59. Chase, C.A. Jr., and Todd, M.R.: "Numerical Simulation of CO_2 Performance," *SPEJ* (December 1984) 597.
60. Chang, Y. *et al.*: "Carbon Dioxide Flow Patterns Under Multiphase Flow, Heterogeneous Field Scale Conditions," paper SPE 22654 presented at the 1991 SPE Annual Technical Conference and Exhibition, Dallas, 6–9 October.
61. Dobitz, J.K. and Prieditis, J.: "A Stream Tube Model for the PC," paper SPE 27750 presented at the 1994 SPE/DOE Ninth Symposium on Improved Oil Recovery, Tulsa, 17–20 April.
62. Dria, D.E., Pope, G.A., and Sepehrnoori, K.: "Three-Phase Gas/Oil/Brine Relative Permeabilities Measured Under CO_2 Flooding Conditions," *SPERE* (May 1993) 143.
63. Shiralkar, G.S. and Stephenson, R.E.: "A General Formulation for Simulating Physical Dispersion and a New Nine-Point Scheme," *SPERE* (February 1991) 115.
64. Coats, K.H.: "An Equation of State Compositional Model," *SPEJ* (October 1980) 363.
65. Young, L.C. and Stephenson, R.E.: "A Generalized Compositional Approach for Reservoir Simulation," *SPEJ* (October 1983) 727.
66. Ács, G., Doleschall, S., and Farkas, É.: "General Purpose Compositional Model," *SPEJ* (August 1985) 543.
67. Watts, J.W.: "A Compositional Formulation of the Pressure and Saturation Equations," *SPERE* (May 1986) 243.
68. Chein, M.C.H., Lee, S.T., and Chen, W.H.: "A New Fully Implicit Compositional Simulator," paper SPE 13385 presented at the 1985 SPE Reservoir Simulation Symposium, Dallas, 10–13 February.
69. Nghiem, L.X. and Li, Y.K.: "Effect of Phase Behavior on CO_2 Displacement Efficiency at Low Temperatures: Model Studies With an Equation of State," *SPERE* (July 1986) 414.
70. Young, L.C.: "Full-Field Compositional Modeling on Vector Processors," *SPERE* (February 1991) 107.
71. Spivak, A., Perryman, T.L., and Norris, R.A.: "A Compositional Simulation Study of the Sacroc Unit CO_2 Project," *Proc.*, Ninth World Pet. Cong., Tokyo (1975) **4,** 187–92.
72. Khan, S.A., Pope, G.A., and Sepehrnoori, K.: "Fluid Characterization of Three-Phase CO_2/Oil Mixtures," paper SPE 24130 presented at the 1992 SPE/DOE Symposium on Enhanced Oil Recovery, Tulsa, 22–24 April.
73. Chang, Y.B., Coats, B.K., and Nolen, J.S.: "A Compositional Model for CO_2 Floods Including CO_2 Solubility in Water," paper SPE 35164 presented at the 1996 SPE Permian Basin Oil and Gas Recovery Conference, Midland, Texas, 27–29 March.
74. Peng, D.Y. and Robinson, D.B.: "A New Two-Constant Equation of State," *I.&E.C Fundamentals* (1976) **15,** No. 1, 59–64.
75. Turek, E.A. *et al.*: "Phase Equilibria in CO_2—Multicomponent Hydrocarbon Systems: Experimental Data and an Improved Prediction Technique," *SPEJ* (June 1984) 308.
76. Soave, G.: "Equilibrium Constants from a Modified Redlich-Kwong Equation of State," *Chem. Eng. Sci.* (1972) **27,** No. 6, 1197.
77. Zudkevitch, D. and Joffe, J.: "Correlation and Prediction of Vapor-Liquid Equilibrium with the Redlich-Kwong Equation of State," *AIChE J.* (1970) **16,** 112.
78. Kremesec, V.J. Jr. and Sebastian, H.M.: "CO_2 Displacements of Reservoir Oils From Long Berea Cores: Laboratory and Simulation Results," *SPERE* (May 1988) 496.
79. Pedersen, K.S., Thomassen, P., and Fredenslund, A.: "Thermodynamics of Petroleum Mixtures Containing Heavy Hydrocarbons: 3 Efficient Flash Calculation Procedures Using the SRK Equation of State," *Ind. Eng. Chem. Process Des. Dev.* (1985) **42,** No. 4, 948.
80. Lansangan, R.M. and Smith, J.L.: "Viscosity, Density, and Composition of CO_2/West Texas Oil Systems," *SPERE* (August 1993) 175.
81. Fong, W.S., Sandler, S.I., and Emanuel, A.S.: "A Simple Predictive Calculation for the Viscosity of Liquid Phase Reservoir Fluids With High Accuracy for CO_2 Mixtures," paper SPE 26645 presented at the 1993 SPE Annual Technical Conference and Exhibition, Houston, 3–6 October.
82. Orbey, H. and Sandler, S.I.: "The Prediction of the Viscosity of Liquid Hydrocarbons and Mixtures as a Function of Temperature and Pressure," *Cdn. J. Chem. Eng.* (June 1993) 437.
83. Lorenz J., Bray, B.G., and Clark, C.R.: "Calculating Viscosities of Reservoir Fluids From Their Compositions," *JPT* (October 1964) 1171; *Trans.*, AIME, **231.**
84. Stone, H.L.: "Estimation of Three-Phase Relative Permeability and Residual Oil Data," *J. Cdn. Pet. Tech.* (October–December 1973) 53–61.
85. Stone, H.L.: "Probability Model for Estimating Three-Phase Relative Permeability," *JPT* (February 1970) 214.
86. Baker, L.E.: "Three-Phase Relative Permeability Correlations," paper SPE 17369 presented at the 1988 SPE/DOE Enhanced Oil Recovery Symposium, Tulsa, 17–20 April.
87. Gardner, J.W., Orr, F.M. Jr., and Patel, P.D.: "The Effect of Phase Behavior on CO_2-Flood Displacement Efficiency," *JPT* (November 1981) 2067.
88. Ewing, R.E., Russell, T.F., and Young, L.C.: "An Anisotropic Coarse-Grid Dispersion Model of Heterogeneity and Viscous Fingering in Five-Spot Miscible Displacement That Matches Experiments and Fine-Grid Simulations," paper SPE 18441 presented at the 1989 SPE Symposium on Reservoir Simulation, Houston, 6–8 February.
89. Lantz, R.B.: "Rigorous Calculation of Miscible Displacement Using Immiscible Reservoir Simulations," *SPEJ* (June 1970) 192.
90. Johns, R.T., Sah, P., and Subramanian, S.: "Effect of Gas Enrichment Above the MME on Oil Recovery in Enriched-Gas Floods," paper SPE 56826 presented at the 1999 SPE Annual Technical Conference and Exhibition, Houston, 3–6 October.
91. Solano, R., Johns, R.T., and Lake, L.W.: "Impact of Reservoir Mixing on Recovery in Enriched-Gas Drives Above the Minimum Miscibility Enrichment," paper SPE 59339 presented at the 2000 SPE/DOE Improved Oil Recovery Symposium, Tulsa, 3–5 April.
92. Ramirez, W.F., Shuler, P.J., and Friedman, F.: "Convection, Dispersion, and Adsorption of Surfactants in Porous Media," *SPEJ* (December 1980) 430.
93. Roper, M.K. Jr. *et al.*: "Interpretation of a CO_2 WAG Injectivity Test in the San Andres Formation Using a Compositional Simulator," paper SPE 24163 presented at the 1992 SPE/DOE Symposium on Enhanced Oil Recovery, Tulsa, 22–24 April.
94. Stalkup, F.I. and Crane, S.D.: "Reservoir Description Detail Required To Predict Solvent and Water Saturations at an Observation Well," paper SPE 22897 presented at the 1991 SPE Annual Conference and Exhibition, Dallas, 6–9 October.

95. Ali, N. *et al.*: "Injection Above-Parting-Pressure Waterflood Pilot, Valhall Field, Norway," paper SPE 22893 presented at the 1991 SPE Annual Technical Conference and Exhibition, Dallas, 6–9 October.
96. Stevens, D.G., Murray, L.R., and Shah, P.C.: "Predicting Multiple Thermal Fractures in Horizontal Injection Wells; Coupling of a Wellbore and a Reservoir Simulator," paper SPE 59354 presented at the 2000 SPE/DOE Improved Oil Recovery Symposium, Tulsa, 3–5 April.
97. Flanders, W.A., Stanberry, W.A., and Martinez, M.: "CO_2 Injection Increases Hansford Marmaton Production," *JPT* (January 1990) 68.
98. Pariani, G.J. *et al.*: "An Approach To Optimize Economics in a West Texas CO_2 Flood," *JPT* (September 1992) 984.
99. Hill, W.J. *et al.*: "CO_2 Operating Plan, South Welch, Dawson County, Texas," paper SPE 27676 presented at the 1994 SPE Permian Basin Oil and Gas Recovery Conference, Midland, Texas, 16–18 March.
100. Perry, G.E.: "Weeks Island "S" Sand Reservoir B Gravity Stable Miscible CO_2 Displacement, Iberia Parish, Louisiana," paper SPE 10695 presented at the 1982 SPE/DOE Joint Symposium on Enhanced Oil Recovery, Tulsa, 4–7 April.
101. Kleinstelber, S.W.: "The Wertz Tensleep CO_2 Flood: Design and Initial Performance," *JPT* (May 1990) 630.
102. Beliveau, D., Payne, D.A., and Mundry, M.: "Waterflood and CO_2 Flood of the Fractured Midale Field," *JPT* (September 1993) 881.
103. Tanner, C.S. *et al.*: "Production Performance of the Wasson Denver Unit CO_2 Flood," paper SPE 24156 presented at the 1992 SPE/DOE Eighth Symposium on Enhanced Oil Recovery, Tulsa, 22–24 April.
104. Magruder, J.B., Stiles, L.H., and Yelverton, T.D.: "Review of the Means San Andres Unit CO_2 Tertiary Project," *JPT* (May 1990) 638.
105. Poole, E.S.: "Evaluation and Implementation of CO_2 Injection at the Dollarhide Devonian Unit," paper SPE 17277 presented at the 1988 SPE Permian Basin Oil and Gas Recovery Conference, Midland, Texas, 10–11 March.
106. Burbank, D.E.: "Early CO_2 Flood Experience at the South Wasson Clearfork Unit," paper SPE 24160 presented at the 1992 SPE/DOE Eighth Symposium on Enhanced Oil Recovery, Tulsa, 22–24 April.
107. Warren, T.M. *et al.*: "Short-Radius Lateral Drilling System," paper SPE 24611 presented at the 1992 SPE Annual Technical Conference and Exhibition, Washington, DC, 4–7 October.
108. Lim, M.T. *et al.*: "Simulation of Carbon Dioxide Flooding Using Horizontal Wells," paper SPE 24929 presented at the 1992 SPE Annual Technical Conference and Exhibition, Washington, DC, 4–7 October.
109. Hervey, J.R. and Iakovakis, A.C.: "Performance Review of a Miscible CO_2 Tertiary Project: Rangely Weber Sand Unit, Colorado," *SPERE* (May 1991) 163.
110. Pittaway, K.R. and Runyan, E.E.: "The Ford Geraldine Unit CO_2 Flood: Operating History," *SPEPE* (August 1990) 333.

SI Metric Conversion Factors

°API	141.5/(131.5 + °API)		= g/cm³
acre ×	4.046 856	E + 03	= m²
acre ×	4.046 856	E − 01	= ha
bbl ×	1.589 873	E − 01	= m³
cp ×	1.0*	E − 03	= Pa·s
°F	(°F − 32)/1.8		= °C
°F	(°F + 459.67)/1.8		= K
ft ×	3.048*	E − 01	= m
ft³ ×	2.831 685	E − 02	= m³
psi ×	6.894 757	E + 00	= kPa
psia ×	6.894 757	E + 00	= kPa

*Conversion factor is exact.

Chapter 5
Surface Facilities Design

This chapter discusses the design of the various surface facilities that make up a CO_2 flood, including material selection and specification. Topics include:

• Introduction—This includes data requirements and initial design steps.

• Specification methods—While field tests are the most accurate means of determining the usefulness of a particular approach, other sources of information also are available.

• Production systems—This section covers design issues for separation, gas gathering, production satellite, liquid gathering, central battery, field compression, and emergency shutdown systems.

• Injection systems—This section covers design issues related to gas repressurization, water injection, and CO_2 distribution systems.

• Gas processing systems-Here we cover basic process selection, H_2S removal issues, and sulfur recovery and disposal systems.

5.1 Introduction

In this chapter we explore surface facility design considerations for CO_2 flooding. We do not intend it to be a comprehensive design manual with a step-by-step method; instead, our purpose is to identify the important (and often unique) issues that must be addressed for a CO_2 flood. There is no single, correct answer, and there are many designs that offer effective solutions.

An important aspect of CO_2 facility design is that many elements must be designed concurrently, involving the simultaneous consideration of many factors, e.g., number and location of separation facilities, amount of compression required, gathering system pressure levels, and type of separation process. Carbon dioxide facility design is a process of evaluating trade-offs and economics. Is it more economically attractive to use a single large facility (with the higher cost of long flowlines but lower cost for pressure) or would it be better to build multiple, smaller, remote facilities for which pressure vessels would cost more but flowlines less? Exhaustive evaluation of alternatives is the only way to arrive at the most cost effective, process-efficient solution.

Several aspects of CO_2 facilities design are similar to those used in design of conventional facilities such as for waterfloods. Key differences include the effects of multiphase flow, the material selection criteria to be used, and the higher pressures that must be handled. **Fig. 5.1** shows a facility overview as a reference for the discussion that follows.

5.1.1 Data Requirements. Designing field facilities for a CO_2 flood requires a large amount of data, which may come from sources such as other design teams (e.g., reservoir, production, and plant teams), operating personnel, design manuals, laboratory testing, company facility files, standard industry practices, and vendor literature. **Table 5.1** shows some of the critical data requirements and sources. This table is not all-inclusive, and each design will have unique requirements.

5.1.2 Initial Design Steps. The first step in designing field facilities for a CO_2 flood is to determine which existing facilities, if any, can be used as-is or with slight modification. The use of existing facilities can reduce capital outlay dramatically, although physical restrictions—such as maximum allowable pressure, capacity limits, and locations available for new facilities—are commonplace. Several questions must be answered to determine whether an existing facility can be used or modified for use in the CO_2 flood:

• Does the facility meet the design requirements for the CO_2 flood? Can it handle the rates, corrosivity, pressures, and temperatures that will be required?

• Does the facility meet safety and reliability requirements for continued operation? Have there been multiple failures of any key components, indicating reliability or safety concerns?

• Is the facility located where it can be used economically?

• Does the location of the facility jeopardize public safety?

• Would there be significant improvements in efficiency, productivity, reliability, and/or safety if the facility were replaced or upgraded?

• Could modifications make otherwise unusable components usable? Would the cost of modification be justified or would replacement be a better alternative?

These questions really are of two types: "Can it be used?" and "Should it be used?" A positive response to the first does not necessarily imply a positive response to the second. Some existing facilities may not offer the desired reliability or performance, and replacement might be more effective in the long run.

5.2 Specification Methods for Material Selection

Suitable materials for CO_2 flooding commonly are selected based on company specifications, manufacturer recommendations, empirical data (field tests), and experience.

5.2.1 Field Tests. ***Overview.*** The purpose of most CO_2-flood field testing is not to test surface facilities, but to collect important data about the reservoir. However, field testing does offer an opportunity to test materials and gain valuable operating experience, which can lead to substantial savings.

A field test can encompass a single well, a pattern, or a small area of the potential flood. Product performance can be gauged by durability, reliability, corrosion resistance, erosion resistance, and

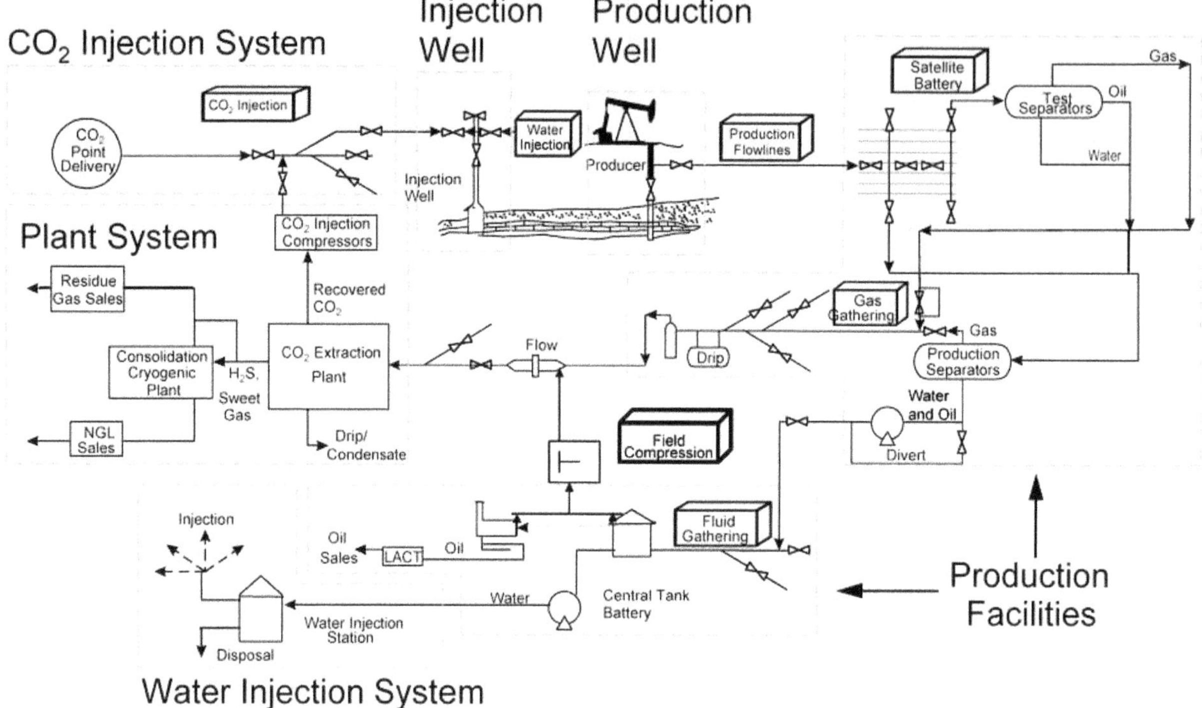

Fig. 5.1—Schematic of a west Texas CO_2 flood. This figure is modified from Fig. 3.16, adding the gas compression leg.

perhaps repair and maintenance costs. A properly designed, installed, and evaluated field test saves money by reducing initial capital requirements, improving system reliability, and lowering future repair and replacement costs. The test also can help establish inventory levels for replacement equipment and repair parts, and it may point out opportunities for standardizing equipment. Standardization not only helps lower purchase costs through larger orders, it also helps prevent confusion about which materials should be used for specific applications.

Field tests should measure material performance in each application—CO_2 injection, water injection, and production—where the material may be used. One might think that only the CO_2 injection system would need evaluation, but nothing could be less true. Many flood design elements are affected by CO_2. Here are a few examples: (1) the presence of CO_2 makes produced water much more corrosive; (2) makeup water from new sources may interact with formation water to create new problems with scale or corrosion; (3) a CO_2 flood may cause paraffins and asphaltenes to precipitate out of the oil, which can cause plugging and emulsion problems; and (4) the potentially dramatic increase in production caused by the flood could cause more formation fines to be entrained in the oil, potentially causing plugging, erosion, and processing problems. Indeed, CO_2 can affect nearly every area of the oil field in one way or another.

The evaluation of material performance depends mostly on visual examination. Material samples should be removed from service and inspected after one, three, and six months. A six-month test should suffice for clear evaluation of most materials.

A visual examination may not tell the entire story, however, and laboratory testing also should be considered. For example, hardness tests and dye-penetrant tests on carbon-steel welds can identify sulfide stress cracking, a dye-penetrant test on stainless steel can find chloride stress cracking, and tensile tests on fiberglass can quantify any loss in strength. How much testing to perform will depend on the cost savings at stake, the time available for testing, and accessibility of resources needed to perform the tests.

The sections below briefly cover piping, valve application, and elastomers, three design choices for which field tests can prove especially valuable.

Piping. Selecting appropriate piping is crucial in a CO_2 flood, and field testing can be an invaluable tool for making the most cost-effective choice.

Materials sometimes are selected for CO_2 applications based on manufacturers' published corrosion resistance, and high-alloy and/or highly specified materials work extremely well in mitigating corrosion-related failures. These products often are more expensive than traditional materials, however, and they can be susceptible to failure mechanisms other than corrosion. For example, austenitic stainless steels such as Grade 316 are susceptible to chloride stress cracking. The best way of evaluating the trade-offs between service capabilities and cost is to field-test each material under operating conditions.

Common piping materials for CO_2 service include fiberglass, stainless steel, cement-lined carbon steel, and internally coated carbon steel, all of which tend to cost more to purchase and install than bare carbon—and which may not always be necessary. We are aware of many fields (Bairoil, Budafa, Crossett, and North Farnsworth) where carbon steel has performed well in three-phase production service (oil, water, gas).

The best way to test pipe materials is to install spool pieces in the service and withdraw them after one, three, and six months for visual inspection and measurements that might indicate changes such as weight loss. When installing different materials in a single pipeline, be aware that galvanic corrosion can adversely affect the test.

Valve Applications. The field test is an ideal opportunity to test valve bodies, seats and trims (internals). How well do the valves perform mechanically? Do they shut off as they are supposed to? Are they easy to open and close? Does the material selected give the corrosion and erosion resistance needed?

While there are some common materials for valve bodies—carbon steel for dry CO_2 injection service and Grade 316 stainless steel in water injection—there can be tremendous variation in the material used, both within the field and between fields. The choice will depend on the corrosivity and erosion potential of the fluids handled.

Materials for valve bodies include standard carbon steel, internally coated carbon steel, Alloy 20, and Grade 316 stainless steel. These and other materials also are used for valve trim. Alloy 20 and nickel aluminum bronze (NiAlBr) are commonly selected materials for corrosion resistance in CO_2 service, although experience has shown that these materials are subject to higher failure rates from erosion and general wear.

The valve seat material can greatly affect valve and system performance, reliability, and total cost. Testing can help with assessing reliability vs. cost for each field's specific operating conditions.

TABLE 5.1—DATA REQUIRED AND DATA SOURCES FOR CO_2 FLOOD FIELD FACILITY DESIGN

Facility	Data Required	Source
CO_2 plant	Location, inlet pressure, maximum capacity, discharge pressure, discharge temperature, discharge composition	Plant design team
Production forecasts	Oil, water, and gas forecasts for individual wells and the total field	Reservoir and production design team
Maps	Topography, well locations, existing facilities locations, lease and unit boundaries, roads	USGS, company sources
Existing facilities	Detailed design criteria, operation and maintenance histories, construction materials, construction and design codes	Operation personnel, facility records, equipment manufacturers
Compositions	Oil, water, and gas (produced and injected) expected through the life of the flood	Reservoir simulation and analog data
Injection pressures	Desired and maximum, surface and bottomhole	Reservoir and production team
Target injection rates	CO_2 and water for each well	Reservoir and production team
Design and construction codes	ASME, API, ASTM, company specifications	Online or library
Environmental and regulatory restrictions	BLM, OSHA, EPA, plus state regulations	Online, library, or agency offices
Environmental conditions	Elevations, pressure and temperature ranges, wind speeds, earthquake or tornado activity, flooding, proximity to populated areas or public facilities	Specifics may be derived from CO_2 release dispersion modeling; U.S. Weather Bureau
Power supplies available	Electricity, fuel gas	Company sources
Sales contract specifications	Oil, gas, NGL, CO_2	Company sources

Elastomers. The performance of elastomers also should be evaluated because failures in these seemingly small components can wreak havoc with operations and cause costly downtime and repairs. Critical locations of elastomers within CO_2 flood surface facilities include spectacle blinds, orifice fittings, and downhole packers.

Elastomers can fail because of swelling, hardening, or cracking, all of which are the direct result of exposure to CO_2. Carbon dioxide diffuses into elastomeric compounds under pressure and temperature. Repeated pressurizing and depressurizing of an elastomer can cause a phenomenon called explosive decompression, which shows up as blistering, fracturing, or cracking. There is no single solution to this problem, which is affected by three factors: the rate of decompression, the permeability of the elastomer to CO_2, and the strength of the elastomer. Because of the difficulty of controlling the rate of decompression, and because most commonly used elastomers have some permeability to CO_2, the most common way of reducing susceptibility to explosive decompression is to increase the hardness (strength) of the elastomer.

The most common elastomers in CO_2 service are peroxide- or sulfur-cured Buna-N (nitrile) and various grades of Viton. Nitrile is a copolymer of butadiene and acrylonitrile monomers and may be compounded for a wide range of service temperatures, from −55 to 125°C. It is the most widely used elastomer in the seal industry and has excellent resistance to petroleum products. For CO_2 service, a durameter rating of 90 or greater usually is specified.

Viton is a trade name for a fluorinated monomer. Fluorocarbon compounds become hard below −20°C but can provide high temperature (225°C) capability. While these compounds have excellent chemical and heat resistance and good mechanical and compression properties, they normally are more expensive than other elastomers. The use of both nitrile and Viton products must be considered carefully if they will be exposed to H_2S because some nitrile and Viton elastomers will degrade in the presence of H_2S.

In addition to the three elastomers discussed above, epichlorohydrine, EPDM (a terpolymer of ethylene and propylene), and Kalrez have been used successfully in CO_2 floods.

To date, there is no industry consensus on which elastomer is best for CO_2 service, but we offer the following guidelines: (1) keep the rate of decompression as low as possible; (2) choose the least permeable elastomer; and (3) select harder (higher strength) compounds because these elastomers tend to resist explosive decompression.

5.2.2 Other Sources of Information. If field testing materials is not possible, the design engineer must turn to other data sources, including analogous CO_2 floods, laboratory testing, and industry publications.

Analogous CO_2 Flood Field Data. Nearby CO_2 floods can be an excellent source of design, operating, and maintenance information. In selecting analogous floods, choose those with process conditions that closely match the conditions anticipated. There are a multitude of process conditions that potentially could be matched. The more that actually *do* match, the better, but the "short list" of matching conditions for surface facility design includes initial crude oil composition (e.g., API gravity and asphaltene content), reservoir pressure and temperature, and H_2S content. In general, any field that provides a good basis for the scaleup process used in screening (see Chap. 3) is a potential source for facility design information.

Laboratory Test Data. Laboratory testing can provide excellent design information and criteria. Typically, one simulates the anticipated fluid composition and then synthesizes these fluids in the laboratory. Tests are run at different conditions, either to match anticipated operating conditions or to provide an accelerated environment (e.g., twice the anticipated temperature) to shorten test duration.

5.3 Production

Before moving on to rigorous technical design of production systems, three complex issues must be analyzed and resolved: how many stages of separation, whether each stage is two-or three-phase separation, and what the operating pressure will be for each stage of separation. The normal design approach is to address gas processing first, then work backward through the remaining systems.

The design of gas processing facilities normally takes precedence in planning a CO_2 flood for four reasons. First, gas processing facilities require substantial capital outlay—they are by far the single most expensive component of a CO_2 flood except for the purchased CO_2 volume (see Sec. 3.3.2 and Fig. 3.20). Second, compression facilities are expensive to operate, primarily because of power requirements and maintenance. Third, compression is a relatively inefficient process because gas absorbs so much energy as force is applied. Thus the best design is one that minimizes overall compression (the gas processing facility plus the field), taking into account the impact of higher backpressure on oil production rates. Fourth, the inlet and outlet pressures fixed by the design of the gas processing plant determine the operating pressure ranges and limits of field equipment.

5.3.1 Separation Issues. Separation is the process of separating the produced gas (hydrocarbons and CO_2), oil, and water. Issues here include the number of stages of separation to be used, the location for each stage of separation, whether each stage will be two-phase or three-phase separation, and the operating pressure for each stage.

Design of the separation facility begins only after the gas processing facility design is fixed, especially for location, inlet pressure, capacity, and outlet pressure, four parameters that are of primary concern in separation facility design.

Separation Locations, Capacities, and Technologies. Separation facilities typically include a few strategically located central facilities (central batteries), as well as several smaller separation facilities (satellites) located throughout the field. The central batteries provide the primary oil and water separation services, and their number is decided based on distance, location relative to the gas processing facility, lease or unit requirements, operating and capital costs, and forecasts of fluid handling requirements. Satellites generally provide a location for well testing and also may provide a stage of gas separation.

The number and location of separation facilities and what separation technology (two-stage or three-stage) is to be used are key concerns because they deal with the need to monitor production rates during a CO_2 flood. This is more important for a CO_2 flood than for a waterflood because in the reservoir, CO_2 is both more costly and more difficult to control than water. Obtaining good areal and vertical sweep depends on fine-tuning the injection of CO_2 and water, and doing this effectively requires accurate information about what is happening in each area of the flood.

Unfortunately, there is no clear-cut answer as to the appropriate number and location of satellites and central batteries—neither is there a simple procedure to find an answer. Once again, one must develop alternatives and analyze trade-offs. In doing so, these data are needed:

- Gas processing facility inlet pressure.
- Crude oil composition.
- Produced gas composition.
- Crude oil sales specifications [e.g., maximum allowable basic sediment and water (BS&W), gravity penalties, and the Reid vapor pressure].
- Regulatory requirements governing the discharge of produced gases.
- Desired injection water quality (e.g., allowable oil and grease and total suspended solids).
- Required water disposal and/or reinjection specifications (e.g., total dissolved solids, ion limits, oil and grease limits).
- Map of well locations.

Number of Satellites. A common monitoring practice is to measure production rates for individual wells for 12 to 24 hours. Assuming that a production engineer must be able to obtain two to four well tests per well per month, then the size of a test site must be limited to 15 to 20 wells. This criterion can be used to arrive at the fewest number of test sites (satellites).

Having more satellites provides the opportunity to test wells more frequently, which can help fine-tune the CO_2 flood to improve oil production and better manage gas production and injection. However, an increase in the number of satellites also means an increase in the number of vessels and support equipment that must be purchased, installed, operated, and maintained. Three-phase meter technology offers one alternative to the use of pressure vessels for well testing, but the effectiveness of this technology depends upon how well the fluids and the process fit the metering specifications. Only a cost/benefit comparison can determine which approach will be more cost-efficient.

Several options are available for the components at CO_2 flood satellites: They can be the same as those used during waterflood operations, they may be upgraded for CO_2 conditions, or they may be newly designed and constructed to handle the increase in pressure, fluid rates (especially gas), and fluid corrosivity associated with CO_2.

Satellite Separation or Not? A satellite can be designed simply to direct all accumulated fluid to a central facility, or it can include a stage of separation. If all produced fluids are routed to a central facility, operation and maintenance requirements are consolidated within a small geographic area, but this also creates the opportunity for slug flow and requires larger diameter lines. Separating the gas at a satellite eliminates slug flow into the central facility and saves money through the use of smaller diameter liquid collection lines—but it requires separators at each satellite, as well as the installation of more gas-gathering lines. A cost/benefit comparison of several alternatives is the only method of finding the cost-effective solution.

Natural Flow or Compression? If the decision is made to install two- or three-phase separation at the satellites, there are two more alternatives. First, each separator can operate at a pressure higher than the sum of the gas processing facility inlet pressure plus pressure losses in the gas-gathering system. This allows the separated gas to flow naturally to the gas processing facility, without the capital and operating cost of field compression. The downside of this approach is that it increases backpressure on the producing wells, thus reducing total production rates.

The second alternative is to operate the satellite at a low pressure and use field compression to move the gas to the gas processing facility. The downside here is the cost to install and operate field compression.

How Many Stages? The more stages of separation there are (sum of the stages in the satellites plus those in the central facility), the more light-end hydrocarbons that stay in the liquid that is to be sold as oil. Thus, more stages of separation increases both the volume and the value of the oil, but it also increases the amount of vessels and support equipment that must be purchased, installed, operated, and maintained.

As these examples show, there are pros and cons for every alternative, and only by performing cost/benefit analyses can one arrive at a solution that takes into consideration all components: capital investment, revenue generated, operating costs, maintenance costs, reliability, and safety.

5.3.2 Gas-Gathering System. The gas-gathering system consists of the piping that delivers produced gas from the separation and field compression facilities to the gas processing facility inlet. Design of the system requires several sets of data: the maximum anticipated rates from each of the gas sources (separation facilities and compression), gas composition, field topography, regulatory requirements, and industry design specifications.

Multiple design issues are involved, including material selection, sizing for flow rates, design for pressure rating, selection of the route, and the location and design of drip vessels. The final design will depend on the material, installation, and maintenance costs for each alternative investigated.

Materials. A number of materials can be used as piping in the gas-gathering system, including stainless steel, fiberglass, aluminum, internally coated carbon steel, and other corrosion resistant alloys. Polyethylene piping has been used in a few cases because it is inexpensive to install, but its use is rare because it absorbs hydrocarbons and can be used only at low pressures.

Of the above materials, the most common are fiberglass, stainless steel, and internally coated carbon steel, each of which has advantages and disadvantages. The final decision on materials usually is made on the basis of corrosion resistance and cost (initial purchase, installation, and maintenance).

Corrosion resistance is very important because the drop in pressure and temperature that occurs as produced gas flows to the gas processing facility causes water and hydrocarbon condensation. Water in the presence of CO_2 forms carbonic acid, which readily attacks a number of materials.

The design engineer also must be aware of external corrosion resulting from electrolytic attack. Typically, selecting corrosion resistant materials, applying external coatings, and/or using cathodic protection mitigates this problem.

In the sections that follow, we discuss various piping materials. From the standpoint of corrosion resistance and cost, we have found fiberglass to be the most satisfactory material for wet CO_2-rich gas applications.

Aluminum. Aluminum has superb resistance to CO_2 and H_2S corrosion mechanisms—it can provide adequate pressure ratings at high temperatures and it has exceptional cold temperature ($<-20°F$) properties.

On the downside, aluminum requires external corrosion protection, such as an external coating or impressed-current cathodic protection. It also can be susceptible to attack from heavy metal ions, the most common of which is iron, which can originate in carbonic acid attack of any bare carbon steel in the system (e.g., casing, tubing, valves, flowlines, etc.). Trace amounts of other heavy metals, such as mercury, also have been observed in some production streams and could prove troublesome. The produced fluid must be tested for iron and other heavy metals before using aluminum.

For the line diameters usually used in gas-gathering systems, the aluminum joining technique is tungsten-inert-gas (TIG) welding, which is a relatively slow and expensive technique that drives up the total cost. Large-diameter aluminum lines (greater than four inches) cannot be plowed into the ground and require excavation in addition to specialized welding, all of which usually makes them more costly than internally coated carbon steel.

In contrast, small-diameter aluminum lines (less than four inches) can be installed from a coil and may be plowed into place, factors which may reduce costs enough to make small-diameter aluminum lines more attractive than internally coated carbon steel.

Carbon Steel. Carbon steel's primary advantages are cost and high-pressure ratings—and every operator in the oil field is familiar with carbon steel and the welding process. The use of bare carbon steel is not common in wet (water saturated) CO_2 service because as temperature and pressure drop, water condenses and reacts with CO_2 to form carbonic acid.

Carbonic acid is a strong corrosive agent in carbon steel, but using corrosion inhibitors or internal coatings, both of which increase the total cost, can mitigate its attack. The use of internal coatings mandates different joining methods—threaded connections, flange connections, or press-together crimp-type connections—because conventional welding techniques are no longer suitable. Threaded connections are not common in line-pipe applications because of the potential for thread leaks.

Carbon steel also is susceptible to electrolytic attack, but using impressed-current cathodic protection and/or the application of an external coating can reduce this susceptibility. Again, both of these increase the total cost. Although carbon steel is not common in gathering systems, internally coated carbon steel is a common material in above-ground separation facilities.

Stainless Steel. Stainless steel provides exceptional corrosion resistance to carbonic acid and H_2S attack, but is susceptible to chloride stress cracking and electrolytic attack. Because water condensation in the gas-gathering system is almost a given, chloride stress cracking is a serious concern. It can occur internally or externally and depends on chloride concentration and operating temperature. NACE *Stand. MR0175-91*[1] provides guidelines for the application of stainless steel when there is potential for chloride stress cracking.

Like other metals, stainless steels can use impressed-current cathodic protection or external coatings to combat electrolytic attack. Stainless steel can be cost-competitive with internally coated carbon steel and, therefore, is a common material in above-ground separation facilities.

Alloys. Numerous corrosion-resistant alloys are available for use in wet CO_2 service, although these alloys generally are not cost effective for piping applications. Such alloys often are used in specialty applications such as downhole products, wellheads, valves, and vessel internals. Three popular alloys for use in wet CO_2 service are NiAlBr, chrome/molybdenum combinations (9Cr/1Mo or 13Cr/1Mo), and Alloy 20.

Fiberglass. Fiberglass (fiber-reinforced plastic, or FRP) is a popular material for wet gas service. It has excellent corrosion resistance; is relatively inexpensive; and when used by experienced personnel, its installation costs can approach those for carbon steel.

Fiberglass works under a wide range of pressures. The maximum operating pressure decreases as temperature increases, however, so operating temperature is an important design parameter. Most gas-gathering applications operate at temperatures well within fiberglass capabilities, although temperature issues may be encountered in production flowlines and compressor piping applications.

Because fiberglass nearly always is buried to prevent ultraviolet degradation of the resins, installation quality is a major concern. Extreme care must be taken during the joining process (whether each connection is glued, threaded, or flanged) to ensure effective, long-lasting connections.

There also are other installation concerns: The bedding material must be compacted around the fiberglass and, along with the backfill material, must be free of large rocks. Manufacturer specifications must be followed to ensure that allowable bends are not exceeded. Some type of marker (e.g., metallic tape, surface markers) must be installed to locate the line once it is buried. Finally, the maximum allowable burial depth may impose limits and must be checked using the procedure outlined in ASTM *Stand. D 3839-89*.[2]

Gas-Gathering Lines. While materials are being selected for the gas-gathering system, the pipeline route or right-of-way (ROW) must be determined, and it should avoid large elevation changes, if possible. Other factors influencing ROW selection can include waterway and highway crossings, special permitting requirements (historical or archeological sites), distance, accessibility, county line crossings, presence of housing or public roads, and lease or unit boundaries. The length and topography of each line influences other design aspects.

Line Diameter. Once the ROW is established, a computer model generally is used to develop the optimum line diameters based on frictional pressure drop, cost, and the degree of confidence in the gas-rate forecast. A common design approach is to use the American Gas Association (AGA) gas equation3 that follows, in conjunction with pipeline efficiencies.

$$q=C_1 E\left[(p_2^2-p_1^2)\,d^5/\gamma_g L T f \bar{z}\right]^{0.5}, \qquad (5.1a)$$

where: q=flow rate, scf/D; C_1=units constant; E=pipeline efficiency, fraction; p_1=upstream pressure, psia; p_2=downstream pressure, psia; d=pipe diameter, ft; γ_g=gas gravity (air=1.0); T=gas temperature, °R; L=pipe length, ft; f=friction factor; and $\bar{z}$=gas compressibility factor at average pressure ($\bar{p}$); and where $\bar{p}$, in psia, is determined by

$$\bar{p}=0.67\,[p_1+p_2-(p_1\,p_2)/(p_1+p_2)] \quad \text{(5.1b)}$$

In Eq. 5.1a, the friction factor, f, is determined using either of the following relationships, which are for partially turbulent flow and fully turbulent flow, respectively:

$$(1/f)^{0.5}=4\log\,[N_{\mathrm{Re}}/(1/f)^{0.5}] \quad \text{(5.1c)}$$

$$(1/f)^{0.5}=4\log\,(3.7\,d/k)\,, \quad \text{(5.1d)}$$

where N_{Re}=Reynolds number; and k=roughness, in.

The design begins with a trial-and-error approach. The plant inlet pressure becomes fixed during the gas processing facility design. Diameters for each line are selected and input along with the appropriate flow rates, and the model is executed to calculate the upstream pressure for each gathering line. The final sizing selection is made after analyzing the cost for lines and vessels for several different pipe diameters.

Once again, the process involves analyzing alternatives and selecting the best one. The trade-offs are straightforward: Larger diameter piping costs more but provides lower operating pressures that enable the use of vessels that are less expensive because they have lower operating pressures.

The inlet pressures to the gas-gathering system provide the basis for vessel, compressor, and piping design at the separation facilities, as described below.

Design Pressure and the Pressure Relief System. The inlet side of the gas-gathering system is connected to pressure vessels (e.g., separators) or compressors, both of which have some type of pressure relief system to provide protection from overpressurizing. The pressure relief system may comprise pressure relief valves (also known as pressure safety valves or PSV's), rupture disks (also called burst plates), or a combination of these.

The highest pressure in a gas-gathering system generally occurs at the system inlet, as dictated by the pressure relief valve there. Most gas-gathering applications use the pressure relief valve setting as the design basis for piping, adding a safety factor of 10%. This safety factor also should be adjusted to accommodate other conditions, such as piping that runs through a populated area.

The pressure-drop model selected for piping design should provide a summary of pressures throughout the pipe system. This pressure summary gives the maximum pressure that each component of the system must be capable of withstanding.

Once the design pressure is established, the remaining design steps follow conventional techniques that determine the required minimum wall thickness for the specific material. As is done in conventional facilities design, choose the next largest commercially available wall thickness.

Metering. *Meter Type and Metering Points.* There are two notable differences between metering of CO_2 and metering of natural gas. One difference is that the materials used for CO_2 metering usually are constructed out of Grade 316 stainless steel to resist corrosion. The other is that CO_2 requires density adjustments for custody transfer applications.

Turbine meters commonly are used for data collection applications, although vortex meters also are emerging as a cost effective method. Orifice meters—the same used for natural gas production—typically are used for custody transfer measurement.

The selection of metering locations is governed by several factors, including the layout of the CO_2 flood. Metering may be required if the gathering system crosses lease or unit boundaries or if gas production from multiple leases will be commingled. If the CO_2 flood encompasses only one lease, a single total flow meter would suffice.

Allocation metering requires additional meter locations. The purpose of allocation metering is to gather data needed to make spending, staffing, and resource decisions. Because allocation meters assign production to a lease, formation, or individual well, accurate allocation metering is important in effective management of a CO_2 flood. The number and locations of allocation meters may be governed by regulatory agencies, with final decisions influenced by the size of the CO_2 flood.

Allocation meters for gas can be either turbine or orifice meters, whereas for liquids they nearly always are turbine meters. There is no standard in place for allocation meters, such as there is for custody transfer meters.

Orifice Meters. API *Stand. MPMS 14.3*[4] provides extensive specifications for the design of an orifice meter run.

The general orifice meter equation developed by the AGA[5] is:

$$q_b=C'\,(p_{dw}\,p_{sa})^{0.5}, \quad \text{(5.2a)}$$

where q_b=rate of flow at base conditions, ft^3/hr; C'=orifice flow constant; p_{dw}=differential pressure, in. of water at 60°F; p_{sa}=absolute static pressure, psia; and $(p_{dw}\,p_{sa})^{0.5}$=the pressure extension.

In Eq. 5.2a, C' is calculated as

$$C'=F_b\,F_{\mathrm{Re}}\,F_e\,F_{pb}\,F_{tb}\,F_{tf}\,F_g\,F_{pv}\,F_m\,F_l\,F_a, \quad \text{(5.2b)}$$

where F_b=basic orifice factor; F_{Re}=Reynolds number factor; F_e=expansion factor; F_{pb}=pressure base factor; F_{tb}=temperature base factor; F_{tf}=flowing temperature factor; F_g=specific gravity factor; F_{pv}=supercompressibility factor; F_m=manometer factor; F_l=gauge location factor; and F_a=orifice thermal expansion factor. Ikoku[6] provides an excellent reference for the calculation procedure.

Gas Density. The density of CO_2 is the single largest obstacle to accurate gas measurement. Carbon dioxide does not behave like an ideal gas; it undergoes phase changes and its density impacts the basic orifice factor, the Reynolds number, specific gravity factor, and the supercompressibility factor used in orifice-meter flow calculations.

The CO_2 density generally is determined using an in-line densitometer or by using an algorithm that incorporates pressure, temperature, and gas composition. Custody transfer applications generally require densitometers if composition is variable; for allocation measurements or situations in which composition is fairly constant, a correlation typically is used. Densitometers require frequent calibration and are relatively expensive to purchase.

Drip Vessels. *Definition and Function.* The final step in the design of the gas-gathering system is to design and select the locations for drip vessels (drips) that collect liquids that either condense out of the gas stream along the pipeline or are carried over from the upstream process. Drips are crucial to the effective operation of a gas-gathering system because the presence of liquids in the system creates a second phase that causes an unnecessary pressure drop. Under the worst case scenario, liquids can build up in low spots until upstream pressure drives them forward in large volumes, referred to as slugs. Liquid buildup also increases susceptibility to freezing in cold temperatures. It is far better to install too many drips than too few.

A drip vessel is classified as a code pressure vessel if it operates at pressures over 15 psig, and most gas-gathering system pressures exceed this limit. Even if a drip vessel's maximum pressure is under 15 psig, it still may qualify as a code pressure vessel if it exceeds the minimum size specified in ASME *Sec. VIII ASME Boiler and Pressure Vessel Code.*[7] Code vessels must be manufactured and maintained under a specific set of requirements spelled out in Sec. VIII.

Drip Design. In designing a drip, the primary need is to determine the optimum size. Some companies have standardized sizes; however, it is much better to size drips technically to arrive at the right drip size (and cost) for each use. An EOS software package can predict liquid condensation rates reasonably accurately, and the drips can be sized accordingly. The design engineer also should consider the size of potential liquid slugs that could originate from liquid carryover or upsets in upstream vessels. In addition, if a drip must be emptied manually, drip size should be based on the desired service interval. This factor is less important if the drip can be emptied automatically to a lower-pressure facility. As always, one must strike a compromise between ease of operation

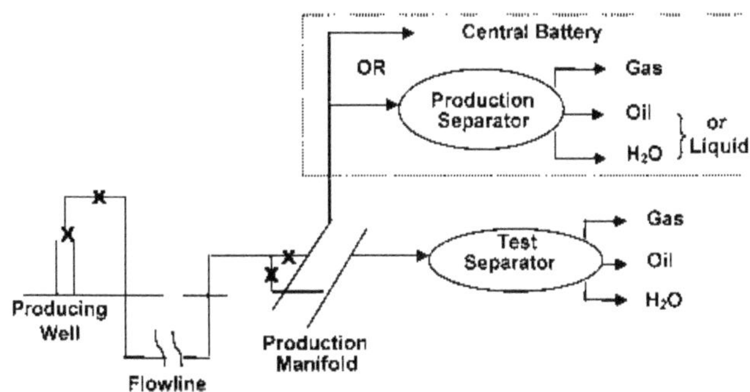

Fig. 5.2—Production satellite schematic.

(larger drips mean less frequent emptying) and initial cost (larger drips cost more to purchase).

Drips should be located at the low spots in the topography of the gas-gathering system route. If the route is relatively flat, EOS software can be used to forecast liquid condensation for drip placement.

Drip vessels usually are constructed of internally coated material because of the potential corrosivity of the condensed liquids. If the drip is classified as a pressure vessel (code vessel), its design must conform to ASME Section VIII as noted above. Drips often are buried beneath or to the side of the gas-gathering system to facilitate the accumulation of liquids. If they are buried, drip vessels also must be externally coated to mitigate electrolytic corrosion. Controls, monitors, and access areas for buried drip vessels should be located above ground.

5.3.3 Production Satellite Design. ***Satellite Batteries.*** The gas-gathering system design completion fixes the maximum design pressure for vessels at the production satellites. **Fig. 5.2** shows a schematic layout for a typical satellite battery, and **Table 5.2** compares alternatives and relative costs (L=low, M=medium, H=high).

Production satellites generally contain a production manifold, a two- or three-phase separator (referred to as a production separator), and a smaller three-phase test separator to provide well test data. Auxiliary equipment at a production satellite may include an instrument air compressor, chemical injection equipment, and automation systems.

The three-phase test separator cycles from one individual well to the next, periodically measuring production rates for each. Production from all the other wells served by the satellite may go directly to a central battery for phase separation, or it may be separated at the satellite by a two- or three-phase production separator. Three-phase separators (gas, oil, and water) are useful for acquiring data on the performance of that specific group of wells, data that are important in effective flood management.

If the satellite is used for three-phase separation, the water and oil phases can either be kept separate or be recombined before being sent on to a central tank battery. Fluids separated at satellites may be recombined, which would reduce capital costs by eliminating the need for two pipelines.

The central battery provides the final cleanup. Oil is separated and processed to meet pipeline sales specifications; water is separated and cleaned to meet injection-water quality guidelines; and CO_2 gas is processed to remove any liquids present, then compressed and sent to the gas processing facility.

If a satellite is used for a stage of separation, the gas dump valves provide the inlet gas feed either to a field compressor or to the gas-gathering system.

Production Flowlines. Production flowlines connect individual producing wells to a production satellite. In most cases, new production flowlines will be needed to handle the higher rates and pressures that CO_2 floods generate, as well as to mitigate corrosion, improve safety, and achieve acceptable reliability. The main design considerations for production flowlines are capacity, pressure rating, and corrosion resistance.

Multiphase Flow. The size for CO_2 flood production flowlines is dictated by multiphase flow considerations. The addition of any gas to a predominately liquid system complicates the flow and the pressure drop calculations. Multiphase flow causes frictional pressure drops that significantly exceed those found in single-phase flow.

Multiphase flow calculations introduce a modification to the basic flow equation to account for fluids and fluid-flow properties in terms of two phases rather than one. The various flow correlations in use differ from each other in how they define two-phase properties and in the range of properties over which they are valid. Most define a flow regime and efficiency loss (sometimes referred to as a holdup).

The general form of the multiphase flow equation[3] is:

$$(dp/dL)=[(f_{tp}\,\rho_{tp}\,v_{tp}^2)/(2\,g_c\,d)]+[(g\,\rho_{tp}\sin\theta)/g_c]$$
$$+\{[(\rho_{tp}\,v_{tp})/g_c]\,(d\,v_{tp}/d\,L)\}, \quad (5.3)$$

where (dp/dL)=pressure drop per unit length; p=pressure; d=pipe diameter, ft; L=length; f_{tp}=friction factor, two-phase; ρ_{tp}=density, two-phase; v_{tp}=velocity, two-phase; g=gravitational acceleration,; g_c=gravitational constant; θ=angle of inclination,; $[(f_{tp}\,\rho_{tp}\,v_{tp}^2)/(2\,g_c\,d)]$=frictional component;

TABLE 5.2—ALTERNATIVES AND RELATIVE COSTS FOR SEPARATOR DESIGN

Production Separator	Test Separator Fluids	Capital Cost	Operator Cost	Downstream Piping	Issues
None	Recombine and send to central battery	L	H	Three-phase collection system	Slugging, higher frictional pressure drop
Two-phase	Gas to gas processing facility; combined liquid to central battery	M	M	Gas collection and liquid collection	
Three-phase	Gas to gas processing facility; water to central battery or water injection station; oil to central battery	H	L	Gas collection, oil collection, and water collection	High capital requirements

$[(g\ \rho_{tp}\ \sin\theta)/g_c]$=elevation component; and $\{[(\rho_{tp}\ v_{tp})/g_c]\ (d\,v_{tp}/d\,L)\}$=acceleration component.

Computer software packages for multiphase flow designs offer a number of equations and correlations to calculate pressure drops. One common design technique uses the Beggs and Brill correlation[8] modified to use Colebrook[9] (Moody diagram) friction factors.

As was the case with the gas-gathering system design, the optimum flowline diameter is determined by trial-and-error.

Pressure Considerations. The pressure rating for production flowlines is an important issue. While normal operating pressures may be relatively low, flowlines easily can be exposed to significantly higher pressures when unexpected conditions occur. In fact, producing wells in a CO_2 flood may be able to flow after gas breakthrough, when gas flow increases significantly.

The possibility of flowing conditions means, of course, that high pressures can be transmitted to surface facilities. If flowlines are not designed adequately, serious mechanical failures can result. Production flowlines must be able to handle the maximum anticipated shut-in surface pressure.

At first glance, this might not seem like a significant issue, but producing wells may be capable of flowing at hundreds of psia, and if flow is blocked in (e.g., valve closed or line plugged), the pressure can build quickly to even higher levels. The maximum possible pressure is the pressure of the reservoir minus the hydrostatic pressure created by a full column of produced gas. Production flowlines must be designed to handle this maximum pressure, or there must be a built-in capability to automatically isolate the flowline from the well.

Another safeguard is to have in place is a method of leak detection. A common technique is to monitor the temperature and/or pressure, to detect changes from normal conditions.

Corrosion Resistance. A final design issue is corrosion mitigation. Oil, water, and carbon dioxide are produced fluids, and water and CO_2 together create the potential for carbonic acid attack. Bare carbon steel has been used effectively in some fields; however, its use can be risky, and field-testing is the only means of determining how it will perform. Most production flowlines in CO_2 floods use fiberglass.

Production Manifold. The production manifold is an arrangement of valves and piping that directs fluid flow from individual wells to a specific process (test or separation). Production manifolds usually are constructed in a modular fashion that uses flange connections to facilitate expansion or contraction as the CO_2 flood matures.

Valves. Valving in the manifold can range from a series of manually operated valves to a single automated valve with an actuator that, depending upon its position, directs produced fluid either into a line feeding the production separator or into another line feeding the test separator. Usually only one well is tested at a time, so the remaining wells coming into the production satellite are directed to the satellite production separator or downstream to the central battery.

Valving must have excellent corrosion and erosion resistance, and must not induce an unnecessary pressure drop. Typical materials used in these valves include Alloy 20, Grade 316 stainless steel (depending on chloride stress cracking potential), Monels, chrome/molybdenum combinations, and other less-common corrosion-resistant alloys, such as Inconels. Valve selection depends on variables such as temperature, pressure, and the composition of the produced fluids.

Piping. The piping used in production manifolds usually is internally coated carbon steel or Grade 316 stainless steel. Fiberglass has been used, but it does not provide adequate safety during unplanned occurrences, such as large liquid slugs that could cause pipe-burst conditions. Safety is paramount in a CO_2 flood, so material selection processes should take into account all risks associated with potential failures.

Pressure Design. The manifold must be designed to accommodate the highest probable pressure. Because producing wells in a CO_2 flood may be able to flow instead of having to be pumped, the manifold must be rated to handle the maximum shut-in surface pressure. The alternative is to use a method of pressure relief.

Separators/Vessels. Vessel design in a CO_2 flood operation is extremely important. Design issues include high gas production rates, adequate pressure rating, sufficient relief systems, separation quality, corrosion resistance, and whether to use two- or three-phase separation.

Cost-effective vessel design depends on the accuracy of the forecasts for oil, water, and gas production rates. Inaccuracies can be accommodated using a conservative design approach, although this is an expensive method of dealing with low-quality information.

Separation Quality—Retention Time. Achieving high quality separation depends on there being adequate retention time (residence time) in the separation vessel. Residence time is a function of the fluid flow rate and the size of the vessel.

The best way to determine retention time (also called residence time) is to conduct a static emulsion breakout test at the anticipated operating temperature and pressure, measuring the relative quality of each fluid over time. The test also can be conducted in the presence of separation aids such as demulsifiers, or at an increased temperature. Though demulsifiers may enable you to reduce capital costs by installing smaller vessels, the demulsifier cost will be an ongoing expense.

Another effective method for determining retention time is by analogy to other CO_2 floods. It is best to run the comparison with a flood whose operating conditions closely match the one you are designing.

Once the retention time has been determined, the size of the vessel components (e.g., oil and water buckets, nozzles) is calculated exactly, just as it is for conventional operations.

Retention time can be used in the following equation to determine the vessel size:

$$t=(V_l\,/\,q_l)\ 1{,}440, \quad \text{(5.4)}$$

where t=design retention time, min; V_l=required separator liquid section volume, bbl; and q_l=liquid throughput, B/D.

Table 5.3 shows the API *Spec 12J*[10] retention time recommendations for gas/oil separation. These guidelines do not account for production problems such as emulsions, foaming, solids, or the presence of water.

Retention times in a CO_2 flood are adversely affected by high gas rates, dissolved CO_2 in the produced liquids, low operating temperatures, emulsions, surging flow rates, and the presence of paraffins and asphaltenes in the produced fluid.

Vessel Corrosion. Of the other pressure-vessel modifications that must be considered in a CO_2 flood; the foremost is the potential for corrosion. The vessel shell, heads, and internals will be in contact with carbonic acid, and possibly with H_2S and sulfate-reducing bacteria, as well.

Two methods commonly are used to prevent corrosion attack: The typical method is to coat the vessel shell, head, and nozzles internally with a coating appropriate for all of the fluids and temperatures to which the vessel will be exposed. Manufacturer data often are used to determine material suitability, and laboratory testing helps, as well. NACE *Standard RP-01-78*[11] and NACE *Standard RP-01-81*[12] provide excellent guidelines for surface preparation and internal coating selection, respectively.

In the second method, sacrificial anodes are used inside the vessel along with internal coatings. Sacrificial anodes have been used in conventional vessel applications, but their use in CO_2 flooding

TABLE 5.3—RETENTION TIME RECOMMENDATION FOR GAS/OIL SEPARATION

Oil Relative Density	Minutes
Below 0.85	1
0.85 to 0.93	1 to 2
0.93 to 1.00	2 to 4

After Ref. 10.

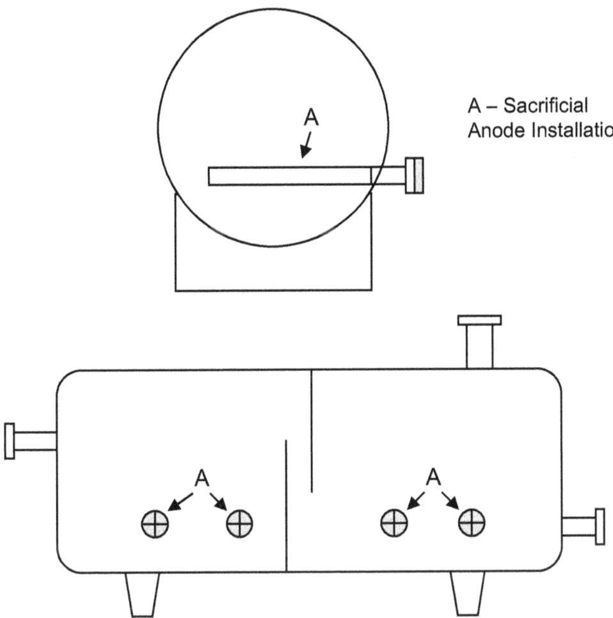

Fig. 5.3—Sacrificial anode installation in a pressure vessel. The top drawing shows the view from the end of the vessel; the bottom shows a side view. The anode is connected at the nozzle.

is even more important, considering the potential corrosivity of the fluids. The decision to use sacrificial anodes must be made early in the design process so that appropriate nozzles can be included during vessel manufacturing.

Sacrificial anodes are designed using the same procedure used for conventional applications. **Fig. 5.3** shows a common sacrificial anode installation.

Vessel internals (mist extractors, weirs, baffles, and vortex breakers) also must be able to resist potential corrosion. Most are made of corrosion-resistant materials such as fiberglass, stainless steel, Monels, and coated carbon steel. Because internal components can be subjected to high stress levels, the metal products are the most common.

Design for Redundancy. One valuable design consideration in a CO_2 flood is redundancy, or spare equipment. No vessel, regardless of application, can be expected be 100% available during its life. This design approach is especially important in a CO_2 flood because of the potential loss of revenue associated with the high oil production rates created by the flood.

If the satellite has both separation and well-test functions, redundancy can be created by designing the test separator to handle most, if not all, of the production separator's capacity. In addition to selecting the appropriate vessel size for this, the piping system connecting the production manifold and the test separator must have adequate capacity. Also, it must be possible to isolate the two vessels through valving. The use of spectacle blinds is one technique to ensure that the vessel being worked on is isolated from both pressure and produced fluid.

Auxiliary Systems. The auxiliary systems in a production satellite—actuators, chemical injection systems, and automation systems—are critical to efficient operation. Actuators require energy to operate. Typically, this will be compressed air, fuel gas, electricity, or other compressed gases, but which one of these will depend on what process is being maintained and, therefore, what hazards might result from an actuator failure caused by an interruption in the energy supply.

Actuators for Dump Valves. Dump valve operation generally is considered to be hazardous service because if a dump valve actuator fails, the vessel either will relieve to atmosphere or will become overpressurized and possibly burst. For this reason, dump valve actuators must have a reliable energy source. Typically, compressed air is used.

Most air compressors contain surge bottles that store energy for use in the event of an electrical interruption, the potential for which is high over the life of the flood. (Electrical actuation rarely is used for this reason.) Bottles of compressed gas, such as air or nitrogen, can be used as emergency backup for highly critical service, although they are not a cost effective primary energy source.

In some cases, dump valve actuators use another high-pressure gas, such as CO_2 or fuel gas. One important consideration in using CO_2 is that the CO_2 must be dehydrated before use to eliminate carbonic acid and its potential corrosivity. Another is that actuator control gas is vented off at the actuator, and in a confined area such as a building, venting CO_2 would be unsafe.

There are similar issues with using fuel gas as an actuator energy source: fuel gas is relatively expensive and also can create a potentially explosive atmosphere. One way to handle this would be to install piping and vent the CO_2 or fuel gas outside the building.

Typically, the cost to dehydrate and vent CO_2 or to purchase fuel gas dictates the use of compressed air.

Chemical Injection Systems. Chemical injection systems are an integral part of production satellites. Chemicals can include demulsifiers, solid control agents, scale inhibitors, and water clarifiers. Demulsifiers and solid control agents (also called interface control chemicals) typically are used to improve separation in the production separator, test separator, and downstream vessels. Because pressure drop and gas breaking out of solution tend to create an ideal environment for scale, scale inhibitors may be injected to prevent downstream scale deposition. Water clarifiers may be added to help achieve water quality targets for reinjection or disposal. Typically, all these chemicals are injected at the production manifold to allow adequate mixing time.

Automation. Automated systems commonly are used to monitor vessel operation and to warn of the existence of adverse conditions. Automation also can be used to control well testing and to control and monitor chemical injection. The degree of automation can range from data acquisition, to monitoring, to on-site control (automatic shutdown or divert), to remote control.

Within the industry, opinions differ about how much and what to automate. Larger operators tend to be proautomation, while small operators do not.

We support the use of automation in CO_2 floods for several reasons. First, CO_2 is so expensive that we believe automation will easily pay for itself through better CO_2 control. Second, conditions in a CO_2 flood (high pressures and flow rates, corrosive fluids, dangerous gas) are such that the risk of catastrophic failure is much higher than in primary or waterflood operations. Because automation makes possible a much safer operation through automatic shut-in capabilities, in an overall risk-management approach to a CO_2 flood, one should have little trouble justifying automation. Third, automation makes sense for well testing because individual well production rates are so much higher than they are in a waterflood. When wells perform at high levels, they should be monitored closely. One of the best ways of doing that is to test them frequently, which is only practical through automation. In summary, while CO_2 automation adds capital and operating costs, we believe it pays for itself many times over.

5.3.4 Liquid Gathering: Satellites to Central Battery. The piping system used to deliver produced liquids from the production satellite to downstream processes at the central battery commonly is referred to as the liquid-gathering or liquid-collection system. The design of this piping system is quite similar to that of the gas-gathering system: The pressure rating, material selection, capacity, and route must be established.

The pressure rating of the line should at least match the maximum allowable working pressure of the upstream vessel or the pump discharge pressure. If the topography of the system creates higher pressures because of hydrostatic effects, this must be taken into account, as well. The density associated with liquids is significant, and hydrostatic pressures can be high.

Fluid Phases and Number of Lines. If a production satellite performs only test separation, then the liquid-gathering system will be designed to accommodate three-phase flow. However, if the upstream satellite also provides a stage of separation, then the design of the liquid-gathering system will be driven by the previous decisions regarding two-phase vs. three-phase separation. For two-phase separation, the mixture of oil and water is piped through a single line to downstream processes. For three-phase separation, the liquids can be recombined and transported through a single line (to reduce capital costs) or kept segregated by using separate lines.

Multiphase Considerations. It is important to remember that the liquids dumped into the gathering lines from the satellite vessels are in equilibrium at the vessel's operating conditions. Additional pressure drop (such as that experienced across the dump valve) will cause the evolution of gas from the liquid. Thus, two- or three-phase flow may result, rather than the single- or two-phase flow expected. See Sec. 5.3.3 for multiphase flow calculations.

The selection of materials and the determination of ROW's for the line follow the same process outlined for the gas-gathering system (see Sec. 5.3.2). Again, fiberglass is the most commonly used material for liquid-gathering systems because gas evolution creates the potential for carbonic acid attack.

5.3.5 Central Battery Design. The central production battery typically consists of separators, tanks, compressors, a LACT unit, a flare or vent stack, piping, valves, and numerous auxiliary systems, as shown in **Fig. 5.4.**

The central production battery is the heart of CO_2 field facilities, providing the final separation and cleanup, of individual fluids before sale or shipment to another process. All liquids are transported to this destination, and processes here are important to sales and other downstream operations. For this reason, redundant systems are common.

Separators. The separators in production batteries typically are designed for two distinct goals: bulk separation and final cleanup.

At the knockout (the first stage of separation in this facility), the bulk of the gas is taken off the produced fluids. The nearly all the water is removed here, as well, leaving a relatively water-free volume of oil.

Additional separation steps clean each stream further. The treater (heater/treater) removes gas and breaks any remaining emulsion so that virtually all water can be removed. Other vessels may clarify the water, capturing some additional oil. The fluids then are sent downstream for accumulation. Water is collected in tanks, and oil, which now should be sale quality, is sent to a storage or surge vessel before custody transfer.

Separator Design Process. The design process for pressure vessels, or separators, for the battery application is the same as that of the production satellites (see Sec. 5.3.3); however, the pressure requirements for central battery vessels are lower, as are the temperatures of the produced fluids. Because of lower pressures, less energy is available to move fluids, which makes frictional pressure analysis particularly important here.

The lower operating temperature of the central battery can lead to separation quality problems. With additional retention time, chemicals and/or applied heat can mitigate these problems. The lower temperature also may affect the decision about which internal coating to use.

As we discussed earlier, a common design practice for facilities that are critical to the bottom line is vessel redundancy. The presence of redundant systems also implies that adequate valving or other isolation equipment is in place.

Two- or Three-Phase Separation. The type of separators chosen at the production battery depends on what type was selected for the satellites. If two-phase separation was chosen for the satellites, then a single line delivers fluids to the battery and the separators at the battery must be three-phase. If three-phase separation was chosen for the satellites, then two separate lines typically will deliver the water and oil phases, and the separators at the battery need to be two-phase.

Water cleanup can be accomplished with two-phase separation if the downstream water tanks are designed with oil recovery capabilities (oil skimming). The oil stream requires three-phase separation. Three-phase separation at the battery also is required if the satellites provide only a well-testing function.

Three-phase separation at the satellites usually leads to better oil quality at the battery because of reduced volumes of fluid transported, lower gas breakout rates, and reduced surges. Two-phase satellite separation offers the advantage of lower cost vessels because fewer nozzles and vessel internals are required.

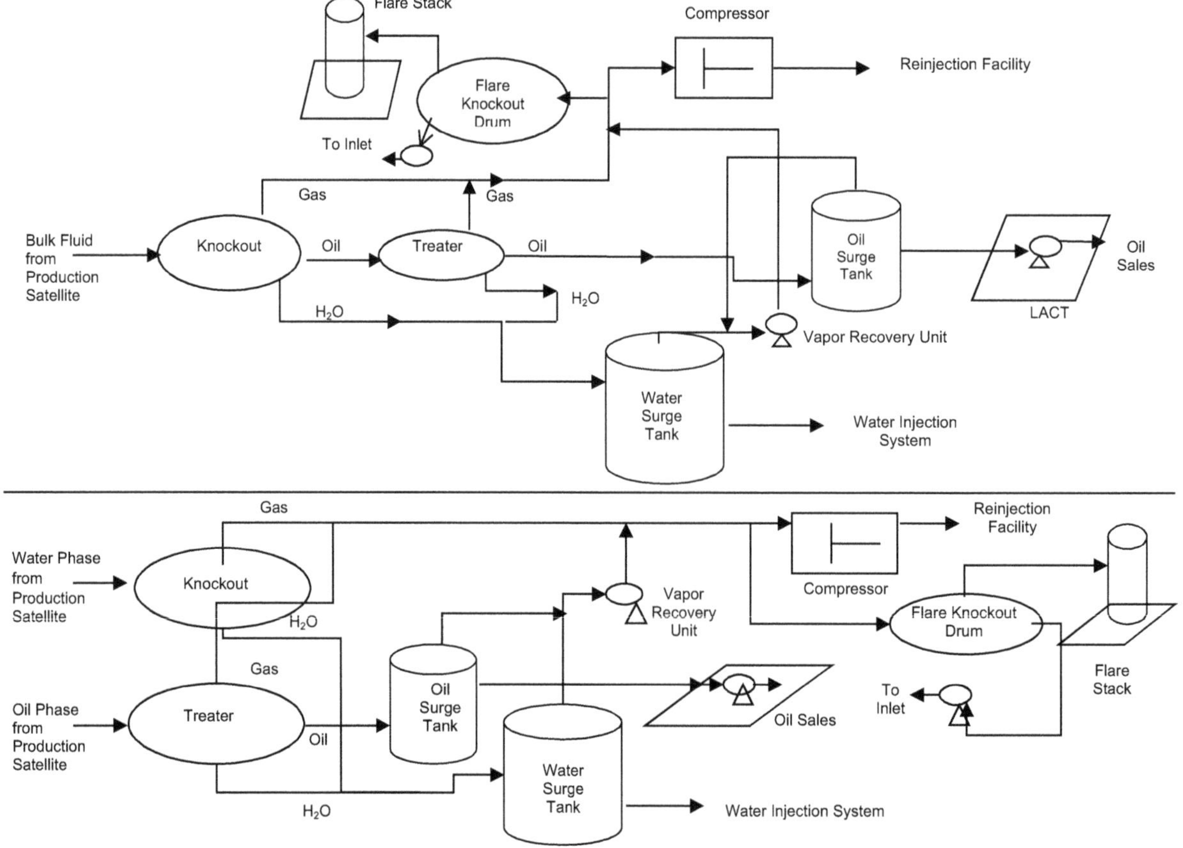

Fig. 5.4—Central battery configuration. The central battery provides final separation and cleanup to get fluids ready for sale or shipment to another process.

Piping. Most production battery piping is metal; nonmetal piping has greater potential for burst failure. Internally coated carbon steel commonly is used for liquid lines, and Grade 316 stainless steel for gas lines.

Internally coated carbon steel usually is joined through flanges, a technique that requires that the system be designed and prefabricated before installation. Stainless steel can be joined by welding or through flanged connections.

Valving used within the production battery must be corrosion and erosion resistant, as outlined in the production satellite section (see Sec.5.3.3).

Tanks. *Tank Uses.* Tanks perform a number of roles at production batteries. "Bad oil" or "slop" tanks collect "non-spec" oil and other fluids from the field. These fluids usually are pumped back to the inlet of the battery and reprocessed. In CO_2 flooding, the volume of non-spec oil can increase because asphaltenes and paraffins tend to precipitate out and remain suspended in the oil, forcing it outside the range of sale specifications.

Another common tank use is to provide surge capacity and suction pressure for waterflood injection pumps. The location of these tanks depends on the location of the injection pump station.

Water cleanup also may use tanks. While this separation process takes place naturally under gravity, the quality and speed of separation can be improved with the use of vortex nozzles and/or the addition of clarifying chemicals. Depending on the volume of separated oil and the amount of solids dropping out, it might not be feasible to carry out the water cleanup function in a surge tank, and special tanks may be required upstream from the surge capacity.

Other tank applications include sump collection, drip liquids collection, and product storage.

Internal Tank Corrosion. To provide a reliable tank system, the design engineer must counteract corrosion. Internal corrosion, in particular, can be severe. Most tanks operate at or near atmospheric pressure, which facilitates gas breakout as liquids enter the tank, and leads to the development of a blanket of corrosive gas above the liquid. Corrosion attack can be mitigated by the use of an internal coating or by using a noncorrosive blanket of nitrogen gas or fuel gas.

Liquid entering a tank still contains small amounts of CO_2 that can create small carbonic acid corrosion cells. This type of corrosion attack usually is stopped by the use of an internal coating.

The accumulation of solids (sludge) in the bottom of a tank also presents challenges. Sludge is a prime environment for the growth of bacteria that can attack the tank bottom and cause severe corrosion problems. This situation can be prevented by using an internal coating, a fiberglass bottom, and/or sacrificial anodes.

Vapor Recovery Unit (VRU). *VRU Compression Options.* The gas vapor liberated as pressure decreases during the separation process must be handled in one of three ways: compressing the gas to a sufficient pressure to allow it to enter the gas-gathering system; flaring; or venting to the atmosphere. Although regulatory permits can be obtained for flaring and venting, most operators install a VRU because of the value of the gas (CO_2, NGL's, and hydrocarbon gas) and because of regulatory considerations.

Recovering all gas vapors may require multiple compressors operating at different pressures, including a compressor that will recover tank vapors at a pressure slightly above atmospheric pressure. Compressor design is discussed in Sec. 5.3.6.

Compression Piping. Gas vapors coming from the separators and tanks will be saturated with water at each vessel's operating temperature and pressure. As pressure drops and the gas cools en route to the compressor, additional liquids will condense out of the gas stream. The presence of these liquids affects material selection and compression design.

As might be expected, corrosion is a key concern in compression piping. The corrosion resistance of internally coated carbon steel and Grade 316 stainless steel make them the two most popular materials for the application. If either material is to be buried, external coatings also should be used, along with insulating flanges to mitigate electrolytic corrosion.

Flare System. The flare system directs gas from the separation facility to a location where it can be safely and efficiently burned. A typical flare system consists of piping, valves, a flare knockout drum, and a flare stack. **Fig. 5.5** illustrates a typical configuration.

Why Flare? Gas vapors must be handled safely during periods when there are downstream problems or when field compression is not fully operational. The alternatives to flaring are venting and shutting in production. The value loss associated with shutting in production makes it highly undesirable, and venting typically is not permitted because of high concentrations of H_2S in the gas. Permits usually are required to flare gas. The most common permit is an "emergency use only" one that allows flaring in the event of a process upset or interruption.

Flare System Inlet. Typically, the flare system ties into the piping before the first-stage compressor suction(s). A control valve that senses pressure in the compressor inlet piping usually controls this tie-in: If the pressure climbs past a pre-established setpoint, the valve opens and allows gas to flow to the flare. The control valve usually uses compressed air actuation because of the critical service requirement. Selection of the pressure setpoint for the flare control valve is crucial. It must be high enough to allow high-rate gas flow through the entire flare system, yet low enough to prevent the operation of relief valves and rupture disks on upstream vessels.

System Capacity and Flare Stack Sizing. The first, and perhaps most important, step in the design of a flare system is to determine the design gas rate. API *RP 521*[13] provides a list of potential gas sources that a flare system should be sized to handle. Some operators have installed emergency shutdown (ESD) systems to mechanically restrict the amount of gas flow that the flare system must handle. We will discuss this later in this chapter.

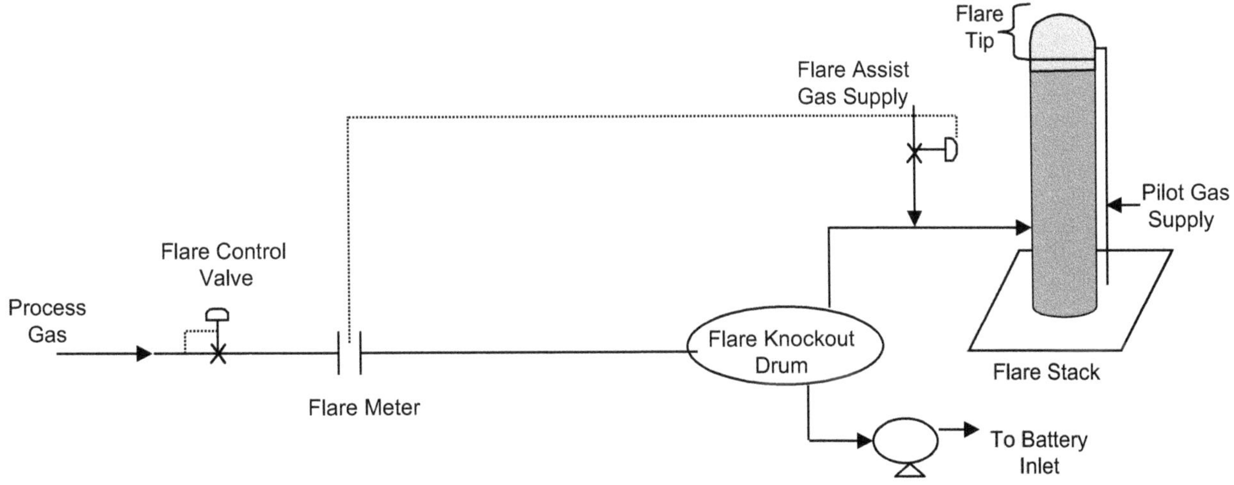

Fig. 5.5—Flare system configuration.

Even though a flare is intended for emergency use only, it may have to be quite large. If a liquid dump valve in a production separator fails in the open position, all gas handled by the satellite flows to the production battery and ultimately to the flare system to provide vessel relief. The gas rate can be quite high.

One common cause of flaring is compressor downtime. If field compression at a battery becomes unavailable, the flare system must be able to handle the rate normally being processed through the compression equipment.

Flare stack sizing is very important. A flare stack whose diameter is too small can create noise pollution problems or might not be sufficient to handle the upstream pressure. A flare stack that is too large can make it difficult to keep the flare burning on a windy day if gas velocities are low.

Liquid Handling. Flare systems are designed primarily for gas; however, flare system design must also account for the presence of liquids. Liquids originate from condensation, carryover, and vessel upsets, and they can cause severe problems that range from the system's reduced ability to relieve pressure to the potential for releasing burning fluids out the top of the stack. It is virtually impossible to keep liquids out of the flare system. A knockout drum must be installed to collect and remove liquids safely.

A flare knockout drum essentially functions as drip (see Sec. 5.3.2) and is installed upstream of the flare stack. The knockout system collects liquids and pumps them back into the upstream process.

The design of the knockout drum must prevent liquid carryover or re-entrainment in the gas stream. Liquid levels should be kept to an absolute minimum in the knockout drum, and level indicators must be highly reliable to ensure that excess liquids do not build up. If the drum is located above ground, the design must include a means to prevent liquids from freezing in cold weather.

Flare Control Valve and Piping Design. The pressure setpoint for the flare control valve should be established concurrent with the flare system piping design. The piping design and material selection process is very close to that for gas-gathering systems (see Sec. 5.3.2), though there are a few differences. One difference is that the operating pressure of the flare system piping usually is much lower, which means both that it does not take as much wall thickness to withstand operating pressures and that little frictional pressure drop can be tolerated. Another difference is that the flare system piping must be sloped toward the flare knockout drum so that liquids in the piping will drain naturally into the drum for removal. A third difference is that, because the flare system is open to the atmosphere, oxygen can accelerate corrosion. Once again, internally coated carbon steel and fiberglass are generally the materials of choice.

Flare Stack Design. The flare itself is comprised of the foundation, guying, stack, tip, and igniter. The presence of H_2S in many CO_2 floods requires the use of a vertical flare system to minimize the potential for hazardous accumulation of H_2S or sulfur dioxide (SO_2) at ground level. Flare systems that handle H_2S also must have a continuous pilot flame to reduce the possibility that the gas will not be burned.

The design of the flare stack is similar to that of conventional operations. When CO_2 is present in high concentrations, it will have an extremely low heating value or heat of combustion, as referenced in API *RP 521*.[13] Heating values less than 200 Btu/scf are common in existing CO_2 floods. Because a gas normally must have a heating value of at least 300 Btu/scf to burn, another energy source—fuel gas, propane, or butane—must be added.

The equipment used to add this external energy is the flare assist system, which is designed to add higher-Btu gas automatically whenever the flare system becomes active. The flare assist system often is controlled off flow rate through the flare system.

The low heating value of CO_2 also affects calculations of stack height and diameter. Normally, stack design is derived from API *RP 521*[13] specifications; however, the low heating value of CO_2 often yields a short flare stack using these calculations, and short stacks are not recommended if there is a possibility of H_2S presence (which is likely to occur if the CO_2 flood follows waterflooding). Common industry practice is to use a tall flare stack to improve dispersion and minimize the potential for hazardous concentrations of H_2S or SO_2 at ground level.

Flare Tip and Igniter. Numerous flare tip designs are available commercially. Options include flame retention rings, windshields, mixing chambers, connections for pilot lines, and smokeless designs.

Smokeless designs commonly use steam to dampen emissions and are used for flare gases that contain significant amounts of heavier hydrocarbons (above C_3).

Pilot connections and mixing chambers are used in flare systems where flame dependability is critical. If the flare gas has a low heating value and if it contains H_2S, a continuously burning pilot is required. Flame retention rings help protect the flare and pilot from the wind. All flare-tip materials must be resistant to high temperatures.

There are two common styles of flare igniter: flame front and electrical sparking systems. In the flame-front design, a thermocouple monitors temperature at the flare tip; if the temperature drops, gas is ignited in the piping at ground level and transported through a line to ignite the pilot or the flare gas. The sparking design relies on an electric charge that bridges a gap between electrodes suspended above the pilot or flare gas. The sparking igniter can be set up using a timer, so that it sparks every few seconds, which ensures that the pilot or flare remains ignited.

5.3.6 Field Compression. As mentioned in our discussion on the separation process (see Sec. 5.3.1), compression facilities must be installed (or existing facilities modified) to recover gas and boost it from low-pressure sources to the gas processing facility.

The compression facility includes compressors, interstage coolers, scrubbers, piping, valves, and auxiliary support systems such as lube oil, instrument air, and automation. The design of field compression for a CO_2 flood is similar to the design for conventional gas compression; however, the properties of CO_2 have a significant impact on compressor design and operation, as well as on the design and performance of the interstage coolers.

Compressors—Number and Location. Compressor downtime is the most common cause of flaring. A thorough reliability and redundancy analysis must be undertaken to address downtime.

One of the first decisions in the design phase is the number and location of compressor stations. If a single central battery is used, the logical solution is one compression station at that battery. Multiple batteries require analysis of economic alternatives, which include having a single, centrally located compression facility connected to the batteries by piping; locating compression facilities at each battery; or some combination of these two. A single compression facility costs less to build and operate, but the cost of piping increases. Perhaps the most important component of the compressor design calculation is the normal operating gas rate. The best way to determine this is to use process simulation software that models flash separation of produced gas from liquids. Accurate liquid rate forecasts are required to develop accurate gas rate forecasts.

Prime Movers. The required input for compressor design includes the available suction pressure and temperature, the required discharge pressure and temperature, and the composition and properties of the gas stream to be compressed.

Most CO_2 compressors are reciprocating, although centrifugal units can be effective. Reciprocating units tend to be the first choice because they are more available, operators are more familiar with them, and they generally are more efficient than centrifugal units. Reciprocating compressors can be driven by alternating current (AC) motors, direct current (DC) motors, or gas engines, depending on the details of the compression process and the available power supplies. Natural gas can be a cost-effective source if it is available onsite.

Compressor driver decisions should be made only after a thorough analysis of purchase, installation, maintenance, and power costs. Gas engines are maintenance-intensive. Direct current motors require more maintenance than their AC counterparts but have natural variable-speed characteristics. Alternating current motors operate at a fixed speeds but can provide variable-speed capability if they are equipped with variable-speed drives (VSD's). Note, however, that VSD's are quite expensive.

The most efficient way to handle gas-rate variation is through variable-speed capability (natural or drive); another common but inefficient way is to use a gas bypass or recycle line. This approach increases flexibility and is much less expensive than a VSD, but adds incremental operating cost because of its inefficiency: Any gas that gets recycled has been compressed, then depressurized in the recycle line, then recompressed.

Finally, gas-rate fluctuation can be handled using variable clearance pockets, which simply increase or decrease the volume of a cylinder to change the capacity of the compressor. Clearances can be adjusted automatically or manually.

The most common approach is reciprocating compression with variable clearance pockets and a gas recycle line. Alternating current motors and gas engines are the most common compression drivers.

Compressor Design and Horsepower Determination. The first step in compressor design is to determine the number of stages of compression, using the following compression ratio equation:[6]

$$F_c=(p_{df}/p_s)^{(1/n)}, \quad (5.5)$$

where F_c=compression ratio; p_{df}=final discharge pressure, psia; p_s=suction pressure, psia; and n=number of stages of compression.

The calculation is iterative. Increasing values (integers) for n are substituted into Eq. 5.5 until the desired compression ratio is achieved. Design compression ratios usually range from three to six per stage.

In determining compressor horsepower and discharge temperature for CO_2, the conventional compressor equations apply. For multiple-stage compressors, calculations are made for each stage; horsepower requirements for each stage are totaled to determine the required driver size.

Cylinder discharge temperatures are important in terms of cylinder temperature limitations and also in the design of interstage cooling.

Ideal Isentropic Horsepower Calculation. For reciprocating compressors, the ideal isentropic horsepower can be calculated as follows:

$$-w=\{C_{p/v}/(C_{p/v}-1)\}(3.027\,p_b/T_b)$$

$$T_s\{[(p_d/p_s)^{z_s}]^{(C_{p/v}-1)/C_{p/v}}-1\}, \quad (5.6a)$$

where w=adiabatic work of compression, hp/MMscf/D ($-w$ denotes work output); $C_{p/v}$=ratio of specific heats (C_p/C_v) at suction conditions; p_b=base pressure, psia; T_b=base temperature, °R; T_s=suction temperature, °R; p_s=suction pressure, psia; p_d=discharge pressure, psia; z_s=compressibility factor at suction conditions; C_p=specific heat at constant pressure, Btu/lbm-°R; and C_v=specific heat at constant volume, Btu/lbm-°R. Base conditions refers to the temperature and pressure required to specify 1.0 scf.

In Eq. 5.6a, the variable $C_{p/v}$ can be determined with Kay's mixing rule,[6]

$$C_{p/v}=\sum(y_i M_i C_{pi})/[\sum(y_i M_i C_{pi})-1.986], \quad (5.6b)$$

where y_i=mole fraction for each component; M_i=molecular weight for each component, C_{pi} is specific heat at constant pressure for each component, Btu/lbm-°R; and C_p for pure CO_2 at 60°F and 14.696 psia=0.199 Btu/lbm-°F.

Discharge Temperature Calculation. The discharge temperature can be calculated as

$$T_d/T_s=(p_d/p_s)^{[z_s(C_{p/v}-1)/C_{p/v}]}, \quad (5.7)$$

where T_d=discharge temperature, °R.

Required Duty of Interstage Coolers Calculation. The required duty of interstage coolers can be calculated as follows:

$$dH_g=n_m C_{pa}\,dT, \quad (5.8)$$

where dH_g=required duty (change in enthalpy), Btu; n_m=number of moles, lbm mol; C_{pa}=constant pressure molal specific heat at cooler operating pressure and average cooler temperature, Btu/(lbm mol-°F); and dT=temperature drop across cooler, °F.

Brake Horsepower. Brake horsepower, or the actual horsepower required at the prime mover, is determined by dividing the ideal horsepower by the compression efficiency. Compression efficiency accounts for losses of energy stemming from thermodynamic and mechanical causes, and the in modern compressors it typically ranges from 75 to 85% of isentropic compression. Compression efficiency also is a function of compression ratio; it peaks at compression ratios of 3.5 to 4.5, after which it begins to decrease.[6]

Coolers. When multiple-stage compressors are used, the gas must be cooled between stages, which is done by heat exchangers commonly referred to as interstage coolers. Shell-and-tube heat exchangers and aerial (also called fin fan) coolers are the most common.

Shell-and-tube exchangers work by transferring heat from a hot fluid through a conductive material (multiple-tube walls) to a low-temperature fluid. These exchangers usually are more efficient, but they require a sustainable source of low-temperature fluid. Fin fan coolers work by circulating air across a set of tubes. They are less efficient and depend on ambient temperatures.

Interstage coolers usually are included by the vendor as part of a compression package. Cooler design is relatively easy; it requires knowledge of gas composition, operating pressure and temperature for each stage; maximum allowable temperatures for the compressor cylinder; and environmental conditions such as elevation and ambient temperature ranges. Most process simulation software can design these heat exchangers.

The choice of tube material for interstage coolers is critical: Any condensed liquids present will contain water and create the potential for carbonic acid attack. All the austenitic stainless steels (e.g., Grade 316 stainless) are susceptible to chloride stress cracking, and chlorides commonly are present in water. The potential for chloride stress cracking is a function of chloride concentration and temperature; the metal is more susceptible at the high temperatures present in interstage cooling applications. Before selecting an austenitic stainless, it is wise to analyze the chloride content of the condensed liquids in the laboratory, using a synthetic gas to match process conditions. Although condensation yields liquids with low total dissolved solids, at high temperatures it only takes chloride concentrations of a few parts per million to cause chloride stress cracking.

If austenitic stainless steel is not acceptable, alternatives include carbon steel and corrosion resistant metals such as Monel and titanium alloys. Carbon steel can provide adequate service life if the design includes sufficient corrosion allowance (additional metal thickness of 1/16 to 1/8 in.) and if a corrosion inhibitor is added to the gas stream. Corrosion resistant alloys such as Monels and titaniums offer exceptional performance but are extremely expensive.

Scrubbers. Gas leaving an upstream vessel is saturated with liquids at the upstream vessel's operating pressure and temperature. These liquids (water and hydrocarbons) condense out of the gas stream as it cools en route to the field compression facility. A two-phase separator, often called a scrubber, must be installed upstream from the compressor suction to catch both condensed liquids and those from process upsets.

Because gas velocity and volume dominates the smaller liquid volumes, scrubbers typically are vertical separators. Most scrubbers are internally coated and use corrosion resistant materials for internal components. Depending on operating pressure, scrubber liquids either are dumped directly to a low-pressure tank or pumped back into the process.

Scrubbers also must be used between stages of compression to capture and remove liquids that condense during interstage cooling. Accurate prediction of liquid condensation rates requires calculation of mass flow rate, which entails the use of phase envelopes (from a simulator or the GPSA Gas Processors Suppliers Association Engineering Data Book) and process data, including pressure, temperature, gas composition, and gas flow rate.

5.3.7 Emergency Shutdown (ESD) System. No matter how well the planning and design phases are carried out, emergency situations will arise during a CO_2 flood. Good design practice anticipates such events and strives to prevent their occurrence and/or mitigate their effects by including an ESD system. An ESD system should be designed to handle process upsets, to accommodate electrical and process shutdowns, and to manage unexpected failures or releases safely. Though they are not mandatory, we strongly recommend that ESD systems be included in all CO_2 floods.

The scope and complexity of the ESD system installed is dictated by such factors as the complexity of the process, environmental conditions (e.g., offshore, near a river, probable tornadoes) facility staffing, and remoteness and anticipated response time.

Fail-Close Valves. The fail-close valve (ESD valve) is common in ESD systems and is designed to stop flow at its locations should a failure occur. This valve commonly is used on vessel inlets (to stop process flow into the vessel), gathering and distribution systems (to stop flow in the event of a line failure), and wellheads (to shut off produced fluids in the event of a downstream failure). Some operators have considered the use of a downhole ESD valve (subsurface safety valve) to prevent releases to atmosphere in the event of a wellhead failure.

The ESD valve can be designed to close on virtually any parameter, including pressure, temperature, liquid level, and flow rate. It normally remains open and only activates when the monitored operating condition falls outside the allowable range. Valve actuation usually is pneumatic, although electrical actuation with high-pressure nitrogen bottles as a backup also has been used. Either way, the actuator must be designed so that the valve fails closed if actuation energy is lost.

The operation of an ESD valve can cause a release of produced fluids in the upstream process—e.g., when an activated ESD valve causes pressure in an upstream vessel to increase until the vessel's relief valve opens. Because of this, it is important to consider where and how releases should take place when installing ESD valves.

It also is wise to design the release location in a way that prevents damage or large cleanup costs. It should be constructed so that liquids can be contained in a tank or overflow storage pit.

Remote ESD Devices. Remote shutdown capabilities may be desirable to prevent sustained discharges, to manage major upsets, or to manage events that impact a large area of the system or field facilities. In these cases, valve actuators either are hard-wired to a central location or they are equipped with telemetry capabilities to receive a signal initiating the shutdown process. The primary advantage of remote capability is that a large area can be shut down quickly and safely.

One common remote ESD application is the ability to shut down an entire process, such as a field-compression or water injection facility. Typically these systemwide shutdowns are accomplished by interfacing with equipment controllers or the main power supply. Shutdown generally is an automatic process based on operating conditions, although most installations include provisions for manual initiation.

Critical ESD systems (those that protect the public), generally are monitored by a host computer using a direct or telemetry communication link. Telemetry is the most common link because of the large distances involved. The host computer checks to make sure that actuation energy is available and that ESD devices are operating properly. If the host senses a problem at an ESD site, it can initiate an automatic shutdown or activate a visual and/or audible alarm to call for human intervention. The host computer also provides an audit trail, 24-hour monitoring and quick response for emergencies.

Gas-Release Dispersion Models. Computer simulators using gas dispersion models predict the travel (velocity, direction, and height) and concentration of compounds resulting from a gas release to the atmosphere, and they commonly are used in designing critical ESD systems. Input parameters for a dispersion model include pressure, temperature, gas composition, direction of the release (vertical or horizontal), and environmental conditions such as elevation, wind velocity, and ambient temperature. The model predicts the direction and speed of the vapor cloud and, more importantly, it predicts the dispersion of gases and provides concentrations along the path. The results from the dispersion model can be used to locate ESD devices (e.g., valves) to minimize the chance that hazardous gas concentrations will reach a public location.

5.4 Injection

Because most current CO_2 floods use the WAG process, there are two injection systems: water and CO_2. We will begin with the water injection system.

Most CO_2 floods are implemented after some period of waterflooding, so a system of water injection facilities already is in place. Plans for the water side of the CO_2 flood should begin with an analysis of these facilities to see if any can or should be used. The analysis also should investigate modifications to existing facilities that can enable their use, which can have a favorable impact on capital investment. Critical factors in determining whether they can be used or modified include the maximum pump discharge pressure, desired injection rates, pressure rating of the flowlines, and maintenance and operating history of the facilities.

5.4.1 Repressurization. Sometimes the reservoir pressure is below the thermodynamic MMP for CO_2, and if miscible CO_2 displacement is desired, the reservoir pressure must be raised above the thermodynamic MMP before injecting CO_2.

The three general methods for repressurizing are to increase the injection rate until it is above the total fluid production rate, to restrict production, and to both increase injection rate and reduce the production rate. Repressurization must proceed relatively slowly to prevent fracture development in the reservoir. The reservoir team should provide a repressurization plan that will underlie the design of the facility. Jarrell and Stein outlined a method for accelerating reservoir pressure growth using Hearn and Hall plots, a technique used successfully in a Slaughter field CO_2 flood.[14]

Water Supply: Compatibility and Corrosion. Because the goal is to inject more fluid than is produced, repressurization may require an additional supply of water. The facility engineer must analyze the required capacity and delivery pressure of the water supply to ensure that it will be adequate. The makeup water also must be analyzed for compatibility with the produced fluid and for corrosion potential.

Compatibility with produced water is an important factor in selecting the makeup water supply. For instance, mixing a water high in calcium with a water high in sulfates can cause precipitation of calcium sulfate scale. An easy test for potential scale precipitation is to mix the desired waters in various ratios and visually inspect the mixture for color change or solids precipitation. Note, however, that this test does not guarantee water compatibility because scale precipitation also depends on pressure and temperature. In addition to physical testing, we advise calculating the scaling tendencies across the range of pressures and temperatures likely to be encountered in the flood operation.

Transporting additional water to the field requires piping that is designed for the appropriate pressure and capacity. The corrosion and scaling tendency of the makeup water can affect the selection of piping materials. Shallow aquifers often have high oxygen concentrations that can render the water extremely corrosive, and it may be more economical to combat this with corrosion resistant materials (e.g., HDPE or fiberglass) than with chemicals.

Regardless of how you manage oxygen in the makeup water piping, oxygen-accelerated corrosion must be addressed before mixing the makeup water with produced water for reinjection. The most common technique is to use an oxygen scavenger, although these chemicals tend to be both toxic (an important factor when surface discharge is a possibility) and expensive. Alternatives are to use corrosion inhibitors, to relocate the oxygen-scavenger treatment point further downstream so that any emergency situation discharges untreated water, to find another source of makeup water, or to live with higher corrosion rates.

Balancing Makeup and Injection Rates. Once the makeup water supply is identified and its means of delivery to the field determined, the facility engineer must add the appropriate amount of makeup water to the produced water to achieve injection target volumes. One method for doing this is to use the level of produced water in a surge tank to control the feed of makeup water. There also must be a way of adjusting upstream water supply, which can be done using variable-speed controllers on pumps, using a large surge tank, by turning multiple wells on or off on demand, or by continuously discharging excess fresh water.

Although the primary need for makeup water occurs during repressurization, makeup water may also be needed at certain times during the CO_2 flood, e.g., when reservoir pressure increases are necessary. Water may also be continuously needed if there are off-structure fluid losses.

In addition to the routine flood uses, the makeup water supply also can be useful for fresh water dilution, fire systems, cooling systems, and perhaps even as a potable water supply.

5.4.2 Water Injection Facilities. Two critical design parameters must be analyzed to determine whether existing water-injection facilities can be used for a WAG CO_2 flood. The first is injection pressure: The injection pressure required for a CO_2 flood usually is greater than the pressure required during waterflooding, and this new pressure may affect the usability of existing flowlines and waterflood pumps. While it may be possible to use existing equipment in its entirety, some modifications may be required.

The second design parameter is the required injection rate, which varies during the course of the flood. During repressurization, the injection rate may be greater than the capacity of the existing waterflood system. When CO_2 injection begins, the rate should decline; and as the flood matures, the rate will increase. Because the need for increased injection rate is relatively short-term, it often can be met by installing temporary additional pump capacity and by eliminating large pressure drops in the system.

Water injection facilities include surge tanks (discussed above), waterflood pumps, piping, measurement and control devices, and a water disposal system. As with production facilities, the size, number, and location of new injection equipment depends on the existing facilities that can be used.

Injection Pumps. The water level in surge tanks usually provides the hydrostatic head (suction pressure) for the injection pumps. Suction pressure requirements are unique to each type of pump, and adequate suction pressure must be provided to prevent cavitation. If the surge tank water level is insufficient, a booster pump will be necessary.

Centralized vs. Decentralized Pump Locations. It would be unusual to install a new water injection pump station solely for starting a CO_2 flood because, in most cases, existing pumps can be modified to meet injection requirements.

If injection capacity is to be increased through the installation of additional pumps, the design engineer must determine the optimal facility site. There are two types of pumping facilities, centralized and decentralized, each with advantages and disadvantages. Ultimately, the choice between them is made by weighing the economics of each. Installation costs for a centralized location may be lower, but the pumps may have to operate at higher pressures to overcome higher frictional pressure drop, thereby increasing operating costs. In addition, a central facility requires that all produced water be transported to one location, and piping costs for this will drive up the total cost. A decentralized system has an advantage in that it can allow a portion of total water injection to continue in the event of a system failure that impacts one pump station.

Pump Types. The two pump types commonly used for water injection are centrifugal and reciprocating, and each has advantages and disadvantages. Reciprocating pumps maintain a fixed discharge rate at all pressures up to the maximum capacity of the pump, while the rate and pressure produced by a centrifugal pump vary inversely: As discharge pressure increases, the discharge rate decreases. In some cases, booster pumps provide an economic means of boosting rates and pressures from existing equipment.

Piping and Valves. Waterflood pumps discharge water into an injection trunk line, a large-diameter line designed for long-distance transport with minimal frictional pressure losses. Design concerns here are selection of piping material, arrangement of water injection piping, and materials used for waterflood valves.

Piping Material. Piping design for a waterflood is identical to design for conventional facilities: The route, material, pressure rating, and capacity must be determined. Most piping design equations handle single-phase water flow extremely well.

Because of pressure requirements, waterflood piping usually is externally coated carbon steel. Internally coated and cement-lined carbon steel have been used where the water is extremely corrosive. Fiberglass also has been used effectively in water injection service.

Water Injection Piping Arrangement. The piping that takes water from the trunkline to the injection facilities can be arranged in one of two ways. The first is a tree configuration that uses a trunkline with lateral lines branching off to each injection well. Well control, automation, and measurement devices are located at each well. In the second configuration, a trunkline delivers injection water to a central manifold (referred to as an injection header), from which individual flowlines run to each injection well. In this configuration, all the well control, automation, and measurement equipment is located at the injection header.

Using injection headers facilitates automation of the water injection process because this arrangement requires only a single remote telemetry unit (RTU) at the header rather than one at each injection well. Note, however, that if each well has its own RTU, waterflood automation can be combined with CO_2 automation. Once again, you must weigh alternatives.

Valves. Valves are used throughout the water injection system to isolate flow and control flow rate. The materials selected for waterflood valves should be based on the corrosivity of water, solids content (erosion problems), dissolved CO_2 content, water composition (specifically the chloride content), and the potential for a sour operating environment (sulfide stress cracking). Serious consideration should be given to using isolation valves at selected points in the distribution system trunklines; these valves can allow downstream maintenance and inspection to take place without a costly, complete system shutdown.

Measurement and Control. *Control Parameters—Rate vs. Pressure.* Two variables—injection rate and pressure—are used to control water injection. Normally, control is based on whichever criteria, pressure or rate, is reached first, although there may be specific operations in which there is only one control parameter. For example., wells injecting on a vacuum would be rate controlled.

Rate control would be advantageous when accuracy is critical in injecting a certain volume of water for each volume of CO_2 injected (maintaining an exact WAG ratio), when the water injection system pressure fluctuates frequently, or when the water supply is limited.

Pressure control would be the first choice when formation parting pressures are substantially lower than the pressure capabilities of the system and you want to preclude the possibility of initiating a fracture.

Measurement and Control Devices. Water injection measurement and control can be accomplished manually or with automated equipment. Injection rate usually is measured with a turbine meter, and in some CO_2 floods, the same turbine meter is used for both water and CO_2 injection. Turbine meters can be automated to collect data at regular intervals.

Pressure can be measured with a simple pressure gauge or, if automated control is desired, a pressure transducer can be used. The transducer converts pressure to an electric signal, which is sent to a local microprocessor (or host computer) as a control variable. A filter screen should be installed upstream of each meter to protect it from plugging or damage by foreign materials in the piping or injection fluid.

Rate and pressure are controlled by adjusting a choke or ball valve. A choke is commonly used because it is specifically designed to handle large pressure drops with minimal erosion. Ball valves have been used effectively when the required pressure drops are not large.

The valve either is opened or closed, manually or automatically, as the control parameter dictates. Automation offers the distinct advantage of frequent and rapid control—a computer can monitor and adjust a valve every few seconds. On the downside, automated controls require additional upfront capital investment, as well as operators with specialized skills.

Water Disposal System. During the life of a CO_2 flood, produced water rates may exceed the water injection requirement, which means that it will become necessary to dispose of excess water. Disposal also will be required if the water injection system goes down for any reason. Water can be disposed of several ways: disposal wells, injection in a nearby waterflood, hauling to a water disposal site, storage, or surface discharge. The choice depends on environmental regulations and total cost.

If the disposal system involves the use of disposal wells within the field, it is feasible to tie the disposal system into the injection control function. This allows disposal wells to be opened up when pressure limits or water surge-tank levels rise above a threshold value.

5.4.3 CO_2 Distribution System.

The CO_2 distribution system delivers CO_2 to individual injection wells from the pipeline supply and the gas processing facility. This system is separate and independent of the water injection system and includes its own piping, valves, measurement, and control facilities.

Piping and Valves. *Piping Arrangement and Materials.* Components of a CO_2 distribution system include trunklines, laterals, and individual well injection flowlines. Trunklines transport CO_2 from the source (e.g., pipeline or gas processing facility) to and through the field. Laterals branch off to deliver CO_2 to several injection wells. Injection flowlines branch off the laterals to deliver CO_2 to each well.

A common material for CO_2 piping is carbon steel, which is excellent as long as free water is not present; that is, as long as the CO_2 has been dehydrated to prevent free water from condensing out of the injection stream when pressure and temperature drop en route to injection wells. Other materials used for the CO_2 distribution piping are stainless steel and fiberglass.

Carbon steel piping has one significant disadvantage: Most carbon steel is rated only to $-20°F$, which can be insufficient for aboveground installations in cold climates. Note that the temperature of pipe does not depend solely on weather—its temperature can drop significantly below ambient temperature when CO_2 is throttled through a valve or if a line leaks. Carbon steel may have to be ordered to a cold-temperature specification if operation below $-20°F$ is anticipated.

The governing specification for carbon steel in cold temperature service is ASTM *Spec. A333/A333M-88a*,[15] which uses a Charpy V-notch test to establish cold-temperature suitability. The test measures the ability of a material to absorb energy, which reveals how susceptible the material will be to potentially catastrophic brittle failure.

Standard stainless steel produced by normal manufacturing processes all is rated for service at low temperatures.

Fiberglass pipe poses a different set of potential problems: It might not withstand high pressures, it might have limited applicability in cold-temperature service, and its pipe length may change significantly as temperature changes. Given these caveats, one must obtain manufacturer recommendations before using fiberglass for this application.

Handling H_2S. Hydrogen sulfide often is present in the CO_2 reinjection stream but is only problematic if free water is present because the combination of free water and H_2S can cause sulfide stress cracking. The design engineer must design the system to prevent the presence of free water or to operate in sour conditions.

Free water can originate from insufficient dehydration, excess pressure or temperature drop, or plant upsets. The WAG process itself creates another source of free water in the CO_2 distribution system: If the supply pressure drops while the injection well is on a CO_2 cycle (injecting CO_2), the injection well has the potential to backflow water from the formation if the check valve fails to hold.

The CO_2 distribution and gas processing systems should be designed to prevent free water from occurring. However, design provisions for a sour environment should also be considered, including ones for H_2S removal, dehydration, and/or mechanisms to prevent backflow conditions. If downstream components are not designed to handle sour conditions, the design engineer must further enhance system integrity by including automated controls, alarms, and even backup systems that can prevent sour conditions from occurring.

Another alternative is to design downstream components to handle the combination of H_2S and water, which may necessitate the use of different materials and welding procedures. Carbon steel still can be used, as long as sulfide stress cracking and corrosion are adequately addressed.

Blowdown Capability. To perform maintenance on the CO_2 distribution system (normal or emergency), one must be able to depressurize the system safely and effectively. This is accomplished through a blowdown station, which simply is a vertical vent pipe that has an isolation (blowdown) valve and is welded to the main distribution line. To depressurize the system, the blowdown valve is opened, and CO_2 vents to the atmosphere.

Blowdown pipe and valve sizing is important, as is the location of blowdown stations. The extreme noise, large temperature drop, and presence of CO_2 during blowdown operations all can cause safety and health problems for both operating personnel and the public (see Chap. 9 for recommended measures for avoiding damage to hearing). Pipes and valves must be sized to minimize these problems and to prevent the formation of pipe-blocking hydrates that can be generated by a large pressure drop. Hydrates can be prevented by large, nonrestrictive valves or by adding heat.

Importance of CO_2 Phases. Piping in CO_2 distribution systems must be designed taking CO_2 phase changes into account because the frictional pressure drop of dense-phase CO_2 (which behaves more like a liquid) exceeds that of supercritical CO_2, which behaves more like a gas. Dense-phase CO_2 can generate much larger hydrostatic pressures than supercritical CO_2.

Phase changes within the distribution system are even more troublesome. A two-phase mixture of liquid-like and gas-like CO_2 generates a larger frictional pressure drop than either phase alone.

Because of the complexities of phase behavior and the number of alternative arrangements for piping, a pipeline simulator model generally is used to design the CO_2 distribution system. Because heat transfer and pressure changes can cause CO_2 to change phase, the simulator model should include a thermodynamic package to account for heat losses through the piping.

Incremental pressure drop caused by multiphase flow must be accommodated by using larger-diameter piping or adding horsepower to the compression system. The alternative to this expense is to prevent multiphase mixtures in the distribution system, which can be done by installing cooling capabilities at the compressor discharge, using insulation on piping to prevent heat gain, increasing discharge pressure to overcome density changes, and/or selecting pipe diameters and a layout that will minimize pressure drop.

Valves. Valves in CO_2 distribution systems usually are carbon steel and may need to be specified for cold temperature service. If the valve is rated for cold temperatures, the bolts holding it in place also must be cold-temperature rated as outlined in ASTM *Spec. A320/A320M-00a*.[16]

The most common uses of valves in the CO_2 distribution system are for isolation and control at laterals and well sites. Measurement and control devices usually have isolation valves to facilitate maintenance. Other valve functions include emergency shutdown service, bleeders, blowdown service, and check valves to prevent backflow.

Metering and Control. The relatively high cost of CO_2 makes metering and control an important issue. The level of automation can range from none, to monitoring, to data collection and storage, to full control. The use of automation commonly reflects the operator's overall philosophy on benefits.

The costs for metering and control depend on the design of the distribution and injection system. A trunkline/flowline arrangement (see Sec. 5.4.2) may result in lower capital requirements because the piping costs less; however, the cost of instrumentation at every well will increase capital and operating expenses higher than they would be in a configuration with all control at a central manifold.

Metering and Control Skids. It is beneficial to place all metering and control equipment on skids, and, if environmental conditions warrant, enclose the skids in small utility buildings.

A skid arrangement, while not required, offers two distinct advantages. The first is that it confines metering and control to a single location, which improves maintenance and repair efficiency. Second, a skid is mobile so that the equipment is easy to move during a phased CO_2 flood implementation.

A CO_2 metering and control skid **(see Fig. 5.6)** consists of a choke, metering device, a pressure gauge or transducer, a temperature gauge or transducer, a screen to protect the metering device, isolation valves, and an enclosure, if warranted. The skid is set away from the wellhead to minimize interference with downhole operations. A flange-end steel line connects the wellhead to the skid.

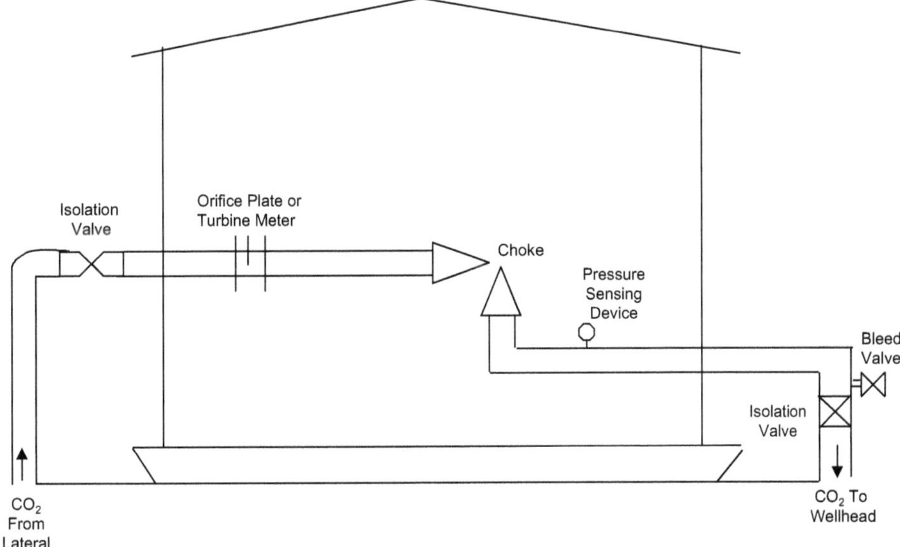

Fig. 5.6—Carbon dioxide injection skid. The use of skids simplifies the task of relocating metering and control equipment as the CO_2 flood is implemented in different parts of the field.

Pressure Control. CO_2 injection can be controlled on rate or pressure or both. Pressure control, using a pressure gauge or transducer and a choke, probably is the most common. A manual system requires a pressure gauge downstream of the choke; the operator simply adjusts the choke to meet the desired injection pressure. An automated system uses a pressure transducer, a pressure setpoint, and an actuator that adjusts the choke valve automatically to keep pressure at the desired level. Automated systems require an on-site or centrally located microprocessor.

Rate Control. Rate control requires some type of measurement device and a choke, and it also can be manual or automatic. The two most common measurement devices are turbine meters and orifice meters.

Most CO_2 floods inject dense-phase CO_2, which means that some form of density compensation must be included for accurate rate measurement. Density compensation can be accomplished using an algorithm or an in-line densitometer. Densitometers are more accurate but quite expensive to purchase and maintain.

Density algorithms are based on an EOS, and can be expressed either as an equation or a table. These algorithms require measures of pressure and temperature, plus a measured or assumed gas composition. Because of their complexity, they are used best in conjunction with an automated system.

Because orifice meters tend to be more accurate than turbine meters, and for that reason they commonly are used for custody transfer. They are more expensive to purchase, but once installed need less maintenance. Orifice meter calculations require measurements of gas composition, differential pressure across the orifice plate, and flowing temperature, as well as knowledge of the meter run configuration. If orifice meters are used for CO_2 injection metering, the orifice plates should be sized to operate as close to the middle of the differential pressure range as possible.

Amoco tested turbine meters for supercritical CO_2 measurement and found excessive bearing wear and failures in the meters. The CO_2 apparently was stripping the bearing lubrication, which accelerated frictional wear and led to gradually reduced accuracy of measurement.* Forced lubrication was tried unsuccessfully. However, at least two operators, Exxon and Chevron, used turbine meters for supercritical CO_2 measurement and experienced adequate measurement accuracy with lower operating costs than with orifice meters. Turbine meters have been used at the Rangely Weber Sand Unit, Wickett Unit, Cordona Lake Unit, and Means Units.

In general, orifice meters are more expensive to install but easier and less expensive to maintain than turbine meters. Orifice meters require only periodic inspections of the orifice plate (and replacement of the elastomer O-rings), while turbine meters tend to wear out bearings and turbine blades more frequently. Once again the decision comes down to a trade-off between capital and operating costs.

Control Devices. In most cases, chokes or ball valves are used to adjust flow rate to meet the rate or pressure setpoint. A choke differs from most other valves in that it is designed to operate with large pressure drops and its construction materials are highly resistant to erosion and cavitation. In most cases, a 90°-angle choke is installed as the piping rises vertically out of the ground, and rate measurement takes place on the horizontal portion of the pipe. Typical chokes have a carbon-steel body with internals made mostly of carbon steel. The disks that restrict flow usually are ceramic.

Ball valves can be used when pressure drops are lower. Ball valves should be sized to operate within a 20 to 80% open range to provide smoother and more consistent control of the injection pressure and/or rate. Slotted-ball control valves with ring-joint flanged ends have been used successfully in CO_2 floods in the Slaughter and Wasson fields. This valve type can be ordered with several port sizes, and care should be taken to ensure the proper operating range.

5.5 Gas Processing

The gas processing facility receives produced gas at a relatively low pressure, processes the gas to remove various components and contaminants, and then discharges the gas at a pressure high enough to be injected into the formation. In its simplest form, this facility includes inlet liquid separation, gas compression, and gas dehydration. Additional equipment enables recovery of NGL's and removal of H_2S.

Gas processing facilities are virtually a necessity; the alternatives are to flare the gas (not only environmentally unacceptable, but also a major waste of money); to reinject the gas into a disposal horizon that will never be produced again; and to sell it to someone else. The last alternative can be an attractive option if a buyer can be found and the cost for transportation is not prohibitive. However, by far the most common approach for managing produced gas in a CO_2 flood is to reinject it into the CO_2-flooded horizon to reduce long-term purchase volumes and improve flood economics.

The design of the gas processing facility should be completed early in the overall CO_2 flood design process because the inlet pressure to this facility impacts many design aspects of other field facilities. For example, the gas processing facility inlet pressure is required in order to design the gas collection system, field compression facilities, and the vessels at the central battery and satellites.

*Personal communication with Gary Simmons, Amoco (1993).

The inlet pressure at the gas processing facility also affects the processing rate of the reservoir because it establishes the minimum backpressure that a producing well encounters. This affects the well's production rate, which, in turn, affects the injection rates at the offset injection wells.

Clearly, then, an integrated team approach is crucial to the successful design of a CO_2 flood. Selection of the inlet pressure at the gas processing facility is not a simple process and requires analysis of several alternatives. A low inlet pressure increases the processing rate of the CO_2 flood and lowers the cost of compression, piping, and vessels in the field, but it increases compression costs at the gas processing facility. A high inlet pressure has the opposite effect—it lowers the processing rate and raises the costs of compression, piping, and vessels, but lowers the compression costs. Obviously, the design of the gas processing facility affects both initial capital requirements and ongoing operating costs for the CO_2 flood, and it is important to have as much input as possible.

The desired purity of injected CO_2 also must be considered in the design of a gas processing facility. Purity definitely will have an impact on the density of the injectant and on the thermodynamic MMP in the reservoir (see Sec. 2.1.6). Density and the thermodynamic MMP, in turn, dictate the required discharge pressure of the gas processing facility, and that will affect the amount of compression required.

5.5.1 Basic Process Selection. Gas processing facilities have three basic functions: liquids separation (to prevent liquids from damaging the compressors); gas compression (to raise the pressure to a level suitable for injection); and gas dehydration (to prevent liquid condensation at any time during the process and to avoid the possibility of corrosion). In a CO_2 flood, the critical processes that should be considered are CO_2 dehydration, NGL recovery, and H_2S removal. The following subsections briefly describe typical systems used for these processes.

Dehydration. Four basic chemical products are used to dehydrate gas: glycol, calcium chloride, glycerol, and solid desiccants (e.g., mole sieve and silica gel). Triethylene glycol (TEG) systems **(see Fig. 5.7)** and mole sieve systems are the most common.

NGL Recovery. In some cases, it is economically attractive to recover hydrocarbons in the form of NGL's from the produced gas stream. The value of the NGL's must be weighed against the loss of value that results from the reduced density of the injectant and from the increase in thermodynamic MMP caused by the removal of hydrocarbons from the CO_2. Three methods enable recovery of NGL's: refrigeration, CO_2 fractionation, and membrane separation.

Refrigeration (using ammonia or propane as the refrigerant) is the simplest and least expensive process. The injection gas stream is cooled and hydrocarbon liquids condense out and fall to the bottom of the vessel for collection. The remaining vapor rises and is directed to the reinjection process. Although refrigeration is simple and low-cost, it has a distinct disadvantage—it cannot recover as much hydrocarbon product as other processes can, and attempts at deeper recovery only lead to substantial CO_2 contamination of the NGL product.

Carbon dioxide fractionation or membrane separation can be used to obtain deeper hydrocarbon recoveries, if economically justified. The recovery of uncontaminated hydrocarbons (especially lighter products) from a CO_2-rich gas stream is difficult because CO_2 and ethane form an azeotrope, a constant-boiling-point mixture. This property prevents ethane recovery without significant CO_2 contamination.

One means of separating CO_2, ethane, and other hydrocarbons is the Ryan-Holmes process shown in **Fig. 5.8.**

The Ryan-Holmes process separates CO_2 and ethane by introducing an additive into the production stream and performing an extractive distillation. The additive typically is generated by the Ryan-Holmes process and is comprised of hydrocarbons.

Membrane separation also can be used to remove hydrocarbons from a CO_2-rich gas stream **(Fig. 5.9).**

Membrane system design involves two key parameters—pore size and solubility of the membrane material. The membrane simply is a section of fibers through which the gas passes. Some molecules are too large to pass, and other molecules are absorbed by the fibers. NGL's and hydrocarbon gases are separated in the first membrane stage. A second stage can be added to obtain higher purity CO_2, if desired.

It is important to note that the physical and thermodynamic properties of the reinjection gas are changed when NGL's are removed, and these changes must be accounted for in designing the gas processing facility.

5.5.2 H_2S Removal. ***Sour Gas Reinjection.*** The produced gas stream often contains H_2S, which either can be removed or can be reinjected along with the CO_2. If H_2S is left in the CO_2, the injection gas stream becomes "sour," extremely corrosive and deadly. Sour gas reinjection can be used effectively and economically as long as the proper materials are used, dehydration is effective and reliable, and adequate safety systems are in place.

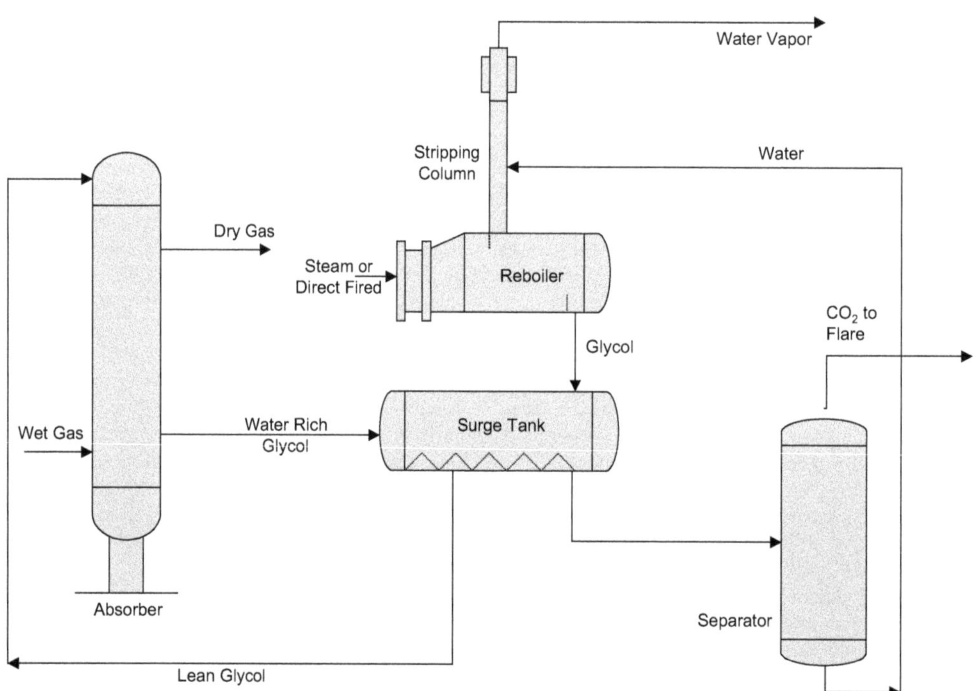

Fig. 5.7—Triethylene glycol dehydration system. This is the most common dehydration process in oil field operations.

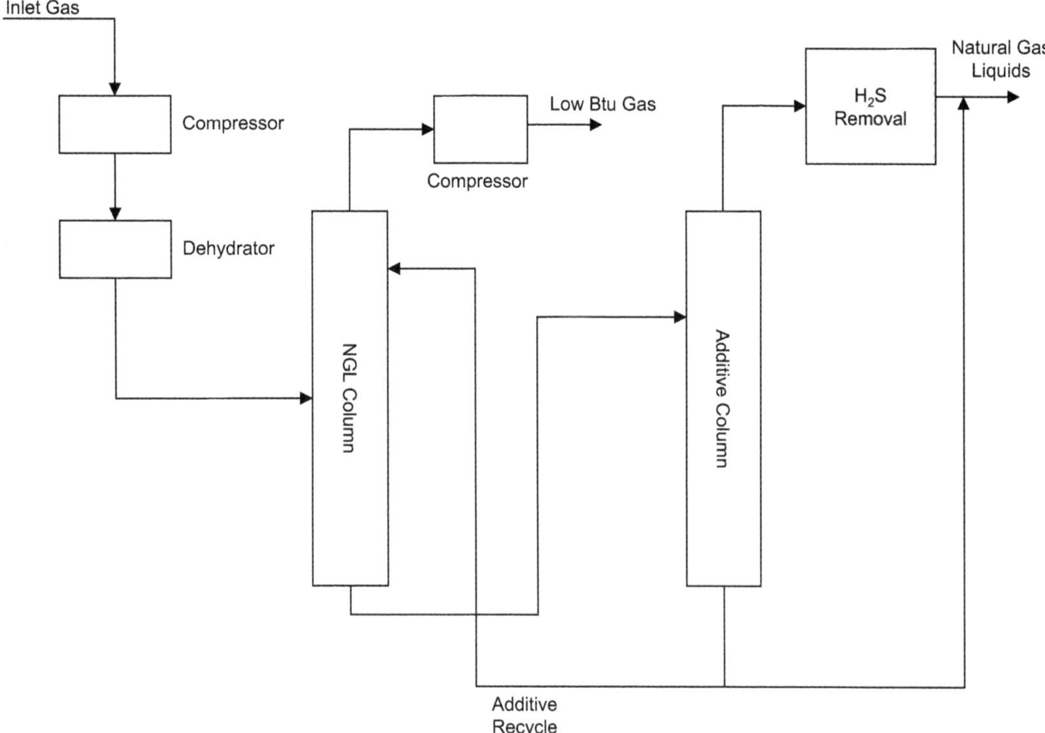

Fig. 5.8—Two-column Ryan-Holmes process. In this process, a hydrocarbon-based additive stream is added to a distillation column to break the azeotrope and allow deeper NGL recoveries.

Although the removal of H_2S adds to the initial capital requirement and to the ongoing operating costs of the gas processing facility, it has benefits: lower costs for downstream materials; lower costs for some safety systems; and lower risks associated with a gas release. In addition, H_2S removal may be necessary to meet product specifications for NGL or CO_2 sales.

There are four basic methods to remove H_2S: recirculating solvent systems, direct REDOX (reduction-oxidation) conversion systems, regenerative adsorption systems, and throw-away solid and liquid scavengers.[17] The REDOX and scavenger systems are direct conversion systems that create a sulfur product that can be disposed of directly; the other systems—recirculating solvent and regenerative adsorption—generate an acid gas that must be further treated, usually by a Claus sulfur-recovery process.

Recirculating Solvent Systems. Recirculating solvent systems can be used on the front end of the plant to treat the inlet gas stream, and they can be used to treat hydrocarbon product streams such as NGL's. The solvent contacts the gas stream and strips out the H_2S. The solvent is then cleaned up to remove the H_2S, and recycled back through the process. Amine systems **(see Fig. 5.10)** are the most commonly used recirculating solvent systems, and have the advantage of being relatively simple and having a high degree of operator familiarity.

REDOX System. The REDOX system uses a solvent (there are many choices of solvent) to convert H_2S directly to elemental sulfur, as shown in **Fig. 5.11.**

REDOX systems have the advantage of being highly selective in removing H_2S and of having a high conversion rate of H_2S to sul-

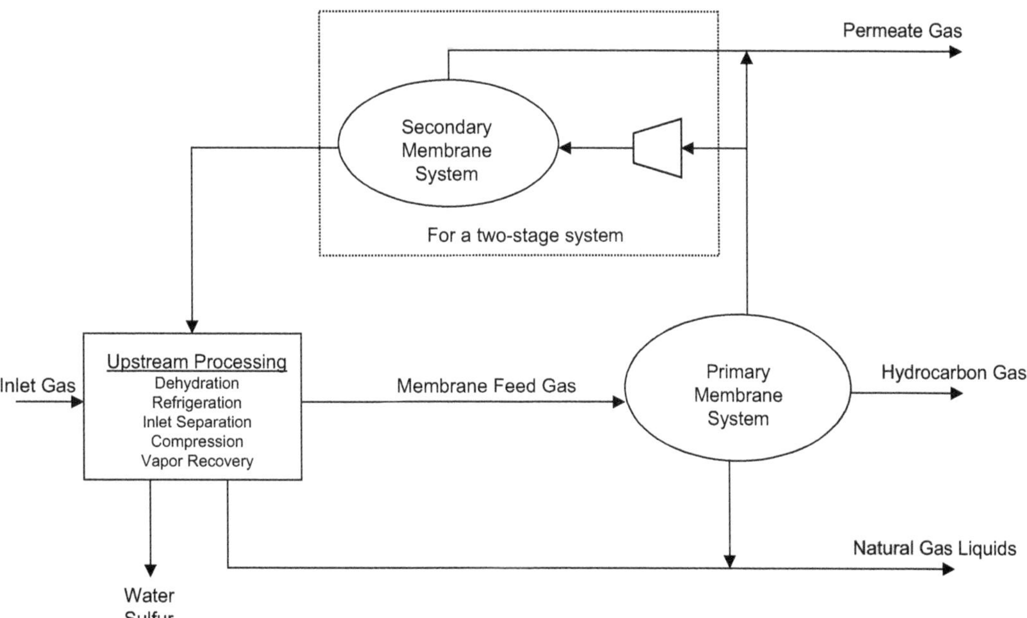

Fig. 5.9—Membrane system with recycle. Engineered fibers absorb or block certain molecules while allowing others to pass through, enabling separation of hydrocarbons and CO_2.

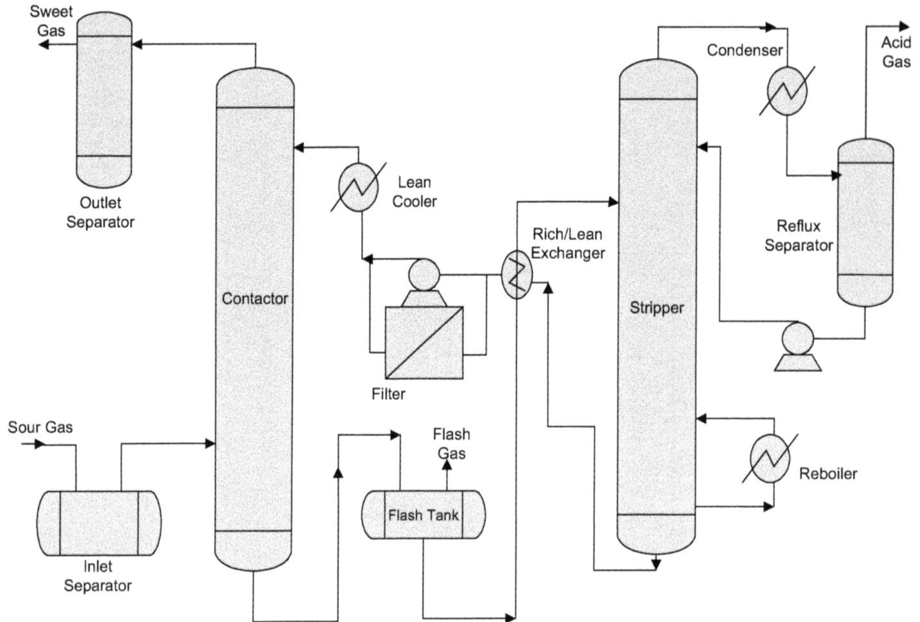

Fig. 5.10—Amine solvent system for H_2S removal. As gas flows upward in the contactor tower, amine solvent flows downward and removes H_2S from the gas. The contaminated amine then is sent to the stripper column, where the acid gas is removed. The lean amine is sent back to the contactor.

fur. The primary disadvantages include the complex chemistry and the high chemical costs that accrue because of solvent degradation.

Regenerative Adsorption Systems. Because regenerative adsorption systems can be expensive to install, they generally are used for smaller gas volumes and for treating final products. The mole-sieve system is one example of the regenerative adsorption process. The H_2S-contaminated gas passes through a vessel containing the mole sieve. The H_2S is adsorbed onto the mole sieve, which can be regenerated through a chemical, temperature, or mechanical process. A mole sieve can be regenerated a limited number of times before needing replacement.

Solid and Liquid Scavenger Systems. Like regenerative adsorption systems, the solid and liquid scavenger systems generally are limited to treating small volumes of gas or product when a continuous process cannot be justified. This is a convenient approach for batch treatment. **Fig. 5.12** shows a simple diagram for the most common H_2S scavenger processes.

5.5.3 Sulfur Recovery and Disposal. The recirculating solvent systems create an acid-gas stream that can be injected into a disposal well, flared, or processed into elemental sulfur. The most common process used to convert the acid gas to elemental sulfur is the Claus process, diagrammed in **Fig. 5.13.**

The sulfur created by direct conversion (REDOX) systems or by a sulfur recovery process such as the Claus process must be disposed of properly. Although it may be a saleable product in the agriculture industry, it is more commonly sent to a landfill, where it must meet the landfill regulatory requirements. In general, sulfur is not considered hazardous waste.

Nomenclature

C' = orifice flow constant, dimensionless

C_1 = units constant, dimensionless

C_p = specific heat at constant pressure, Btu/lbm-°R

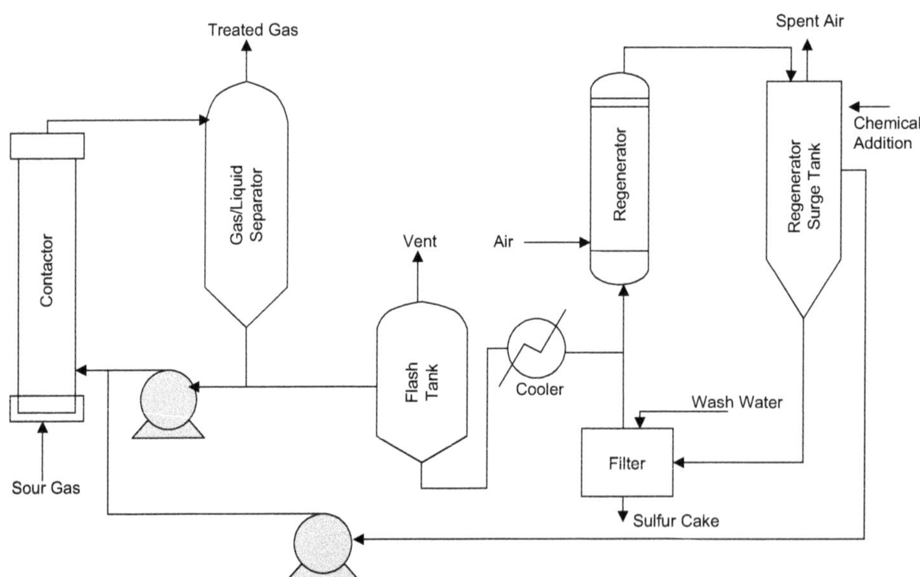

Fig. 5.11—Direct conversion (REDOX) system. The sour gas contacts the treatment chemical in the contactor and flows to the separator. Here, the purified CO_2 leaves the top of the tower, while the chemical and sulfur compounds collect at the bottom. The rest of the process completes the reduction/oxidation of H_2S to elemental sulfur and recirculates the chemical, sending it back to the contactor.

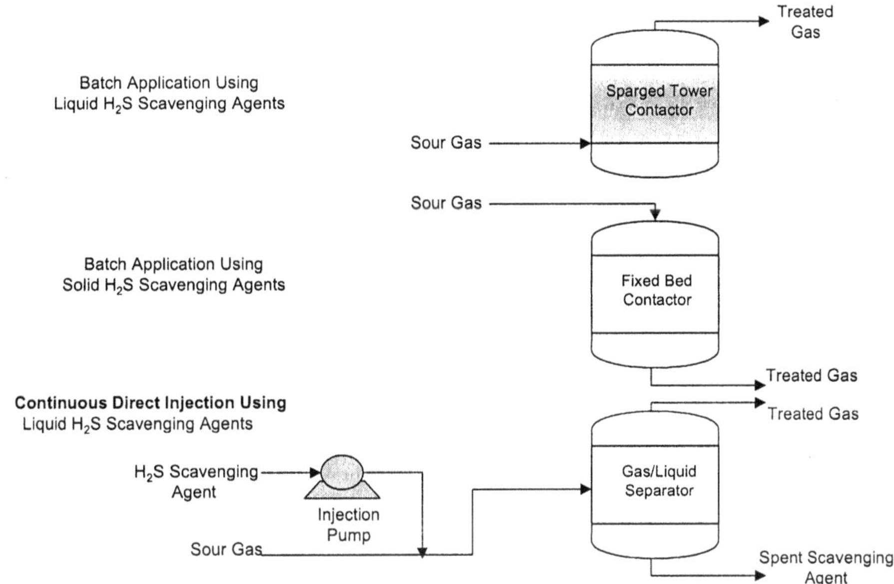

Fig. 5.12—Solid and liquid scavenger systems. There are three types: sparged towers, fixed bed contactors, and gas/liquid separators. The first two generally are used in batch operations, while the third can be used continuously.

C_{pa} = constant pressure molal specific heat at cooler operating pressure and average cooler temperature, Btu/(lbm mol-°F)
C_{pi} = specific heat at constant pressure for each component, using Kay's mixing rule, Btu/lbm-°R
$C_{p/v}$ = ratio of specific heats (Cp/Cv) at suction conditions
C_v = specific heat at constant volume, Btu/lbm-°R
d = pipe diameter, L, ft
dH_g = required duty (change in enthalpy), Btu
dT = temperature drop across a cooler, T, °F
E = pipeline efficiency, fraction
f = friction factor, dimensionless
f_{tp} = friction factor, two-phase, dimensionless
F_a = orifice thermal expansion factor, dimensionless
F_b = basic orifice factor, dimensionless
F_c = compression ratio, dimensionless
F_e = expansion factor, dimensionless
F_g = specific gravity factor, dimensionless
F_l = gauge location factor, dimensionless
F_m = manometer factor, dimensionless
F_{pb} = pressure base factor, dimensionless
F_{pv} = supercompressibility factor, dimensionless
F_{Re} = Reynolds number factor, dimensionless
F_{tb} = temperature base factor, dimensionless
F_{tf} = flowing temperature factor, dimensionless
g = gravitational acceleration, L/t^2
g_c = gravitational constant, dimensionless
k = roughness, L, in.
L = pipe length, L, ft
M_i = molecular weight of each component, using Kay's mixing rule, lbm/lbm-mol
n = number of stages of compression
n_m = number of moles, m, lbm mol
N_{Re} = Reynolds number, dimensionless
p = pressure, m/Lt^2

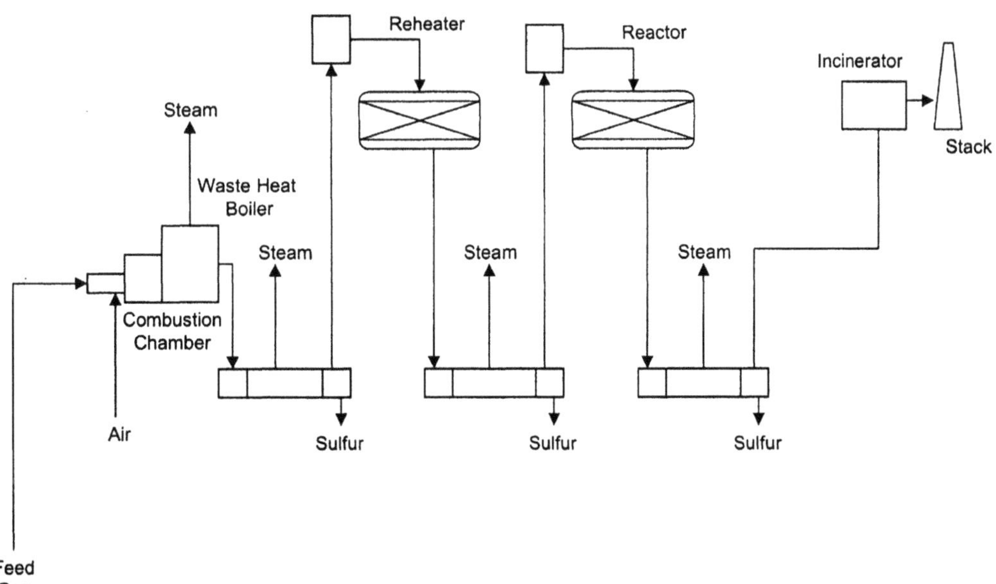

Fig. 5.13—The Claus process. This is the staple of sulfur recovery. The feed gas is the acid-gas stream from solvent recovery systems, which may contain up to 25 to 30% H_2S.

$\overline{p}$ = average pressure, m/Lt^2, psia
p_b = base pressure, m/Lt^2, psia
p_d = discharge pressure, m/Lt^2, psia
p_{df} = final discharge pressure, m/Lt^2, psia
p_{dw} = differential pressure, m/Lt^2, in. of water at 60°F
p_{sa} = absolute static pressure, m/Lt^2, psia
p_s = suction pressure, m/Lt^2, psia
p_1 = upstream pressure, m/Lt^2, psia
p_2 = downstream pressure, m/Lt^2, psia
q = flow rate, L^3/t, scf/D
q_b = rate of flow at base conditions, L^3/t, ft^3/hr
q_l = liquid throughput, L^3/t, B/D
t = design retention time, t, min
T = gas temperature, T, °R
T_b = base temperature, T, °R
T_d = discharge temperature, T, °R
T_s = suction temperature, T, °R
v_{tp} = velocity, two-phase, L/t, ft/sec
V_l = required separator liquid section volume, L^3, bbl
w = adiabatic work of compression, hp/MMscf/D
y_i = mole fraction for each component, using Kay's mixing rule, dimensionless
$\overline{z}$ = gas compressibility factor at average pressure, dimensionless
z_s = compressibility factor at suction conditions
ρ_{tp} = density, two-phase, m/L^3
θ = angle of inclination, degrees
γ_g = gas gravity (air=1.0), dimensionless

References

1. *Std. MR0175-91, Standard Material Requirements Sulfide Stress Cracking Resistant–Metallic Materials for Oilfield Equipment,* NACE, Houston (January 1991).
2. *Std. D 3839-89, Standard Practice for Underground Installation of Fiberglass (Glass-Fiber Reinforced Thermosetting Resin) Pipe,* ASTM, West Conshohocken, Pennsylvania (April 1989).
3. *Multiphase Network Simulation Model Users Guide, Version 1.0,* Scientific Software-Intercomp, Engelwood, Colorado (1991).
4. *Std. MPMS 14.3, Orifice Metering of Natural Gas & Other Hydrocarbon Related Fluids,* API, Washington, D.C.
5. *Orifice Metering of Natural Gas,* Gas Measurement Committee Report No. 3, American Gas Association, New York City (January 1956).
6. Ikoku, C.U.: *Natural Gas Production Engineering,* John Wiley & Sons, New York City (1984).
7. "Unfired Pressure Vessels," Section VIII ASME Boiler and Pressure Vessel Code, ASME, New York City, 1974.
8. Beggs, H.D. and Brill, J.P.: "A Study of Two-Phase Flow in Inclined Pipes," *Trans.,* AIME (1973) **255,** 607.
9. Moody, L.F.: "Friction Factors for Pipe Flow," *Trans.,* ASME (1944) **66,** 671.
10. *Spec. 12J, Oil and Gas Separators,* seventh edition, API, Dallas (October 1989).
11. *Std. RP-01-78, Design, Fabrication, and Surface Finish of Metal Tanks and Vessels to be Lined for Chemical Immersion Service,* NACE, Houston (December 1977).
12. *Std. RP-01-81, Liquid Applied Internal Protective Linings and Coatings for Oil Field Production Equipment,* NACE, Houston (June 1981).
13. *RP 521, Guide for Pressure-Relieving and Depressuring Systems,* second edition, API, Dallas (1982).
14. Jarrell, P.M. and Stein, M.H.: "Maximizing Injection in Wells Recently Converted To Injection Using Hearn and Hall Plots," paper SPE 21724 presented at the 1991 SPE Production Operations Symposium, Oklahoma City, Oklahoma, 7–9 April.
15. *Spec. A333/A333M-88a, Standard Specification for Seamless and Welded Steel Pipe for Low-Temperature Service,* ASTM, West Conshohocken, Pennsylvania (September 1988).
16. *Spec. A320/A320M-00a, Standard Specification for Alloy/Steel Bolting Materials for Low-Temperature Service,* ASTM, West Conshohocken, Pennsylvania (2000).
17. CO_2 Facilities and Plants, CEED CO_2 Flood Short Course No. 7, U. of Texas of the Permian Basin Center for Energy and Economic Diversification, Midland, Texas (1998).

SI Metric Conversion Factors

bbl × 1.589 873	E − 01	= m^3
Btu × 1.055 056	E + 00	= kJ
Btu/lbm-°F × 2.3885	E − 04	= J/Kg-K
°C °C + 273.15		= K
°F (°F − 32)/1.8		= °C
°F °F + 459.67		= °R
ft × 3.048*	E − 01	= m
ft^3 × 2.831 685	E − 02	= m^3
hp × 1.341	E + 00	= kW
in. × 2.54*	E + 00	= cm
lbf × 4.448 222	E + 00	= N
lbm × 4.535 924	E − 01	= kg
lbm mol × 4.535 924	E − 01	= kmol
min × 6.0*	E − 01	= s
psi × 6.894 757*	E + 00	= kPa
°R × °R/1.8		= K

*Conversion factor is exact.

Chapter 6
Well Design

This chapter discusses components of equipment design for producing wells and injection wells, from the wellhead to the bottom of the wellbore. We also discuss the preparatory well work required to put the wellbore in optimum condition for producing or injecting the gassier fluids involved in CO_2 floods. Sec. 6.1 covers producing well design, specifically wellhead design, lift equipment, and flowing wells. Sec. 6.2 covers injection well design, specifically injection wellhead design, issues with switching CO_2 and water cycles, tubing and packer assemblies, and wellbore modifications.

In a CO_2 flood, the principal well design factors that differ from primary or secondary recovery include pressure, corrosion potential, and production and injection rates. The costs for WAG injection well equipment usually will be much higher than for producing wells because of higher pressures and the high potential for corrosion. Production well equipment costs can exceed WAG injection costs if significant changes in artificial-lift equipment must be made later in the flood life. In any case, the total cost of well equipment usually is only a small fraction of the total project investment.

6.1 Producing Well: Wellhead to Bottomhole

6.1.1 Wellhead Design. Design factors to evaluate for the producing wellhead components of a CO_2 flood are metallurgy, valve configuration and type, pressure ratings, and elastomers and seals (in valves and stuffing box, etc.).

The extent of changes in wellhead equipment depends on the degree of change anticipated in the physical and chemical operating conditions. Some operators have made substantial changes before flood startup, while others, particularly in recent years, have made few or no upfront changes.[1] In the latter case, making adjustments to wellhead components on an as-needed basis makes sense, especially if substantial changes in operating conditions are expected in only a few areas rather than fieldwide. Such was the case in Slaughter field,[2] where changes were made as CO_2 breakthrough occurred. **Fig. 6.1** illustrates the Slaughter field changes, which included installation of higher pressure valves and new elastomers and seals. In all cases, the wellhead should be inspected to ensure that all low-pressure fittings are replaced to match the maximum expected shut-in pressure.

Corrosion can affect carbon steel wellhead master valves severely in just a few months.[3] For each field, it is important to anticipate the probable changes in operating conditions that can lead to high corrosivity, such as the amount of CO_2 in the produced gas, higher operating pressures, the amount and composition of produced water, salt concentration, and changes in the composition of the oil. See Sec. 8.3.3 for additional information on corrosion control.

When configuring valves, it is wise to build in an ability to replace a faulty casing valve that has high pressure below it without having to kill the well. Some floods, such as Cedar Lake Unit, have installed sacrificial casing valves, which essentially are two casing valves in tandem. With this configuration, the well can be blown down using the upper valve; if this valve washes out from repeated blowdowns, the lower casing valve remains in good condition. The well still can be shut in without concern about having a washed-out valve during blowdown. A well without this type of valve configuration must be killed to replace a casing valve.

The sacrificial valve configuration also was used in Denver Unit[4] **(Fig. 6.2).**

Several operators use backpressure valves **(Fig. 6.3)** to control the efficiency of rod-pumping wells with high gas/liquid ratios (GLR's). The increased pressure that these valves impose on the fluids in the tubing forces most of the free gas into solution and reduces gas-lock situations. In some wells of their Slaughter and Cedar Lake Units, Amoco has installed backpressure valves when CO_2 has broken through.*

Backpressure valves use spring force on a piston to hold the desired backpressure, which can be set from 50 to 1,500 psia. The valves are available with either flanged or threaded connection ends, and special options include a carbide-steel ball and seat with a peroxide-cured Buna-N or Viton seal for use in highly corrosive environments. The valves typically are installed on the fluid collection line at the wellhead, as shown in **Fig. 6.4.**

If a backpressure valve is not needed (or is unavailable), high-pressure double-pack stuffing boxes also can perform a similar function,[5] as shown in **Fig. 6.5.**

There are two ways to power ESP's. One way is a to use a pigtail power connection, in which the cable plugs into the side of the wellhead. The other is to use a lower-pressure wellhead, in which the cable passes through packing. We recommend that only pigtail connections be used in a CO_2 flood because they can contain the high pressures that would result should the well have to be shut in. Although pigtail connections are more expensive, their expense is justified in light of the risk of leaks and failure from uncontrolled releases that have occurred in through-packing connections.

6.1.2 Lift Equipment: Surface and Downhole. Changes and additions to lift equipment involved in preparing for CO_2 flood operations include those made to rod metallurgy, pump metallurgy,

*Perry M. Jarrell, personal notes from discussion sessions at the CO_2 Forum sponsored by the Petroleum Industry Alliance of the CEED, 19 May 1994.

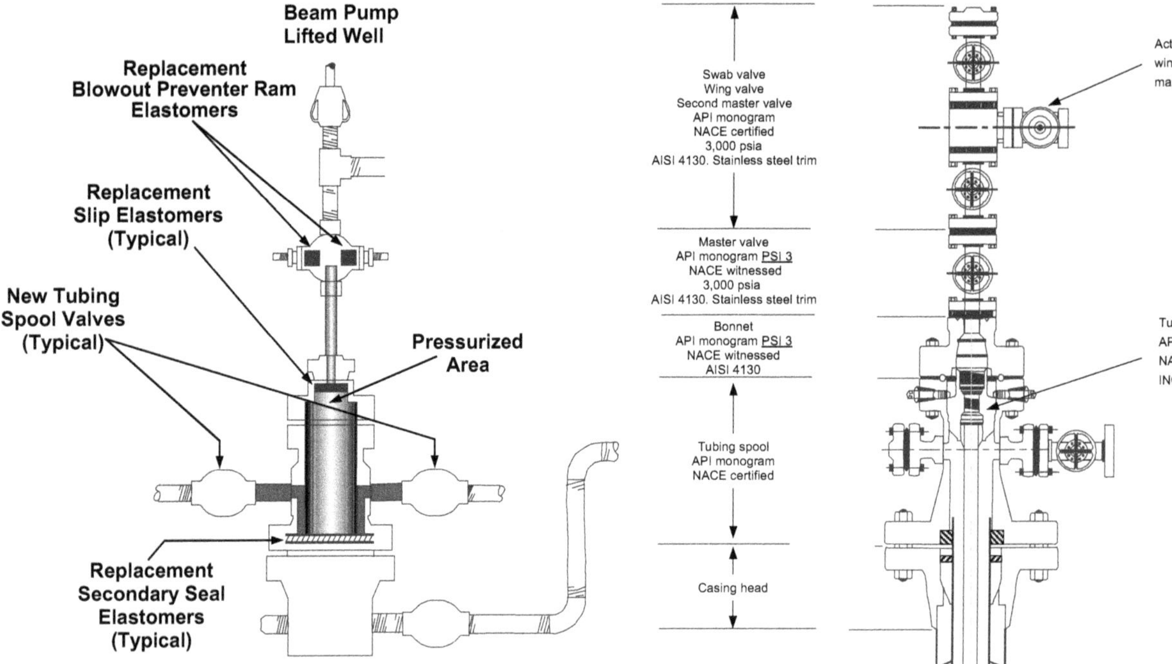

Fig. 6.1—Production wellhead modifications. This diagram is based on the Slaughter field (after Ref. 2).

Fig. 6.2—Production wellhead design cross section. This design is typical of the Denver Unit, with double valves and a double-pack stuffing box (after Ref. 3).

stuffing box seals, gas separation equipment, tubing coatings, and coating types.

The most significant changes in lift operations during a CO_2 flood are in total fluid production and in the increases in gas production, which can be dramatic after CO_2 breakthrough. Total fluid production usually decreases early in the flood, when the miscible oil bank is forming, and then increases as the oil bank breaks through at the producing wells. Sometimes total fluid production also fluctuates with each passing WAG cycle. If water is used as the chase injection fluid during the later stages of the project, then as water injection is being increased, water production increases.

Designing the necessary modifications to lift equipment at the surface and downhole requires an understanding of the magnitude and timing of the changes that will occur. Many operators choose to modify lift equipment on an as-needed basis, which spreads the costs over time and enables them to design for specific problems encountered.

After the first cycle of CO_2 injection in a WAG project, the decline of total production mentioned above may allow one to save money by downsizing lift equipment until response increases later in the project. Such downsizing could include changing from ESP to beam; from large beam to small beam (e.g., from 640 to 320); or from flowing to beam. Downsizing usually provides power cost savings and increased pump efficiency. Fieldwide equipment swapping can reduce equipment acquisition costs.

Downhole Separators. The principles and practices of downhole hydrocarbon gas and liquid separation are fundamentally the same for a CO_2 flood as they are for conventional primary and secondary production. As always, one should strive for the most cost-effective downhole gas-separation solutions.

For rod-pumped wells, one of the best solutions is the "natural gas anchor," in which the pump intake is placed below the deepest fluid entry point to provide the largest possible gas separation space. Sometimes in openhole completions it is not possible to install the pump intake below the deepest fluid/gas entry point, and in those cases, other gas-separation equipment should be used. Some of the most effective gas separation equipment used in CO_2 floods are "poor-boy" assemblies,[6] cup-type and packer-type separators. Which separator is appropriate depends on the vertical location of major gas inflow (top or bottom of pay zone), the allowed fluid velocities, and the downhole pumping pressure.

Poor-boy gas separators work best in low-volume wells in which fluid velocities between the gas dip-tube and the mud anchor are less than 0.5 ft/sec. For $2\frac{7}{8}$-in. tubing with a $1\frac{1}{4}$-in. stinger, the maximum total fluid rate possible is 175 B/D (including the gas volume). Assuming that 30% of the produced volume is gas, the maximum fluid rate in a $2\frac{7}{8}$-in. poor-boy separator is 125 B/D.

If the production rate is too high to be handled effectively by a poor-boy assembly, a cup-type separator would be the next choice. Unfortunately, these separators often cannot be used because the inside diameter of the casing often is too small to accommodate the cup-shaped extrusions on the separator. This is especially true in west Texas. If downhole gas separation equipment cannot be effectively sized, it probably should not be installed because an improperly sized downhole separator actually can restrict flow.

For ESP-pumped wells, the most effective choice for liquid/gas separation is rotary gas separators, which are widely used and work very well.

Rods, Liners, and Pumps. If gas interference problems (excessive gas flow into the wellbore) cannot be effectively managed, then the subsequent gas influx into the downhole pump must be managed. Bottom-discharge pumps, "gas-busters," "charged" pumps, and jet pumps are types of rod pumps that work best in gassy pumping conditions.

Spray metal polish-rods without liners work very well in beam-lifted CO_2 production. It is best to use rods that do not require liners because standard rubber liners may not withstand the pressures associated with shut-in producing wells in a CO_2 flood. If a liner is used, it should be sized so that it will be above the BOP at the top of the pump stroke, so that the BOP can close properly.

Exxon used a long-stroke pump unit successfully to handle both high- and low-gas-rate production. Long-stroke units help increase the compression ratio by allowing more fluid to fill the pump barrel with each stroke, which, in turn, helps prevent gas locking. However, higher-stress rods are required for this type of pump unit, and these rods can become brittle in a high H_2S environment.*

In the Denver Unit, 3% chromium-steel sucker-rod strings were field tested. They lasted for three years of CO_2 service, showing no evidence of pitting or wear on the rod bodies,[7] and only minor wear on the rod couplings. No further field tests were performed because conversion from pumped to flowing was planned for 80% of the wells and

*Perry M. Jarrell, personal notes from discussion sessions at the CO_2 Forum sponsored by the Petroleum Industry Alliance of the CEED, 19 May 1994.

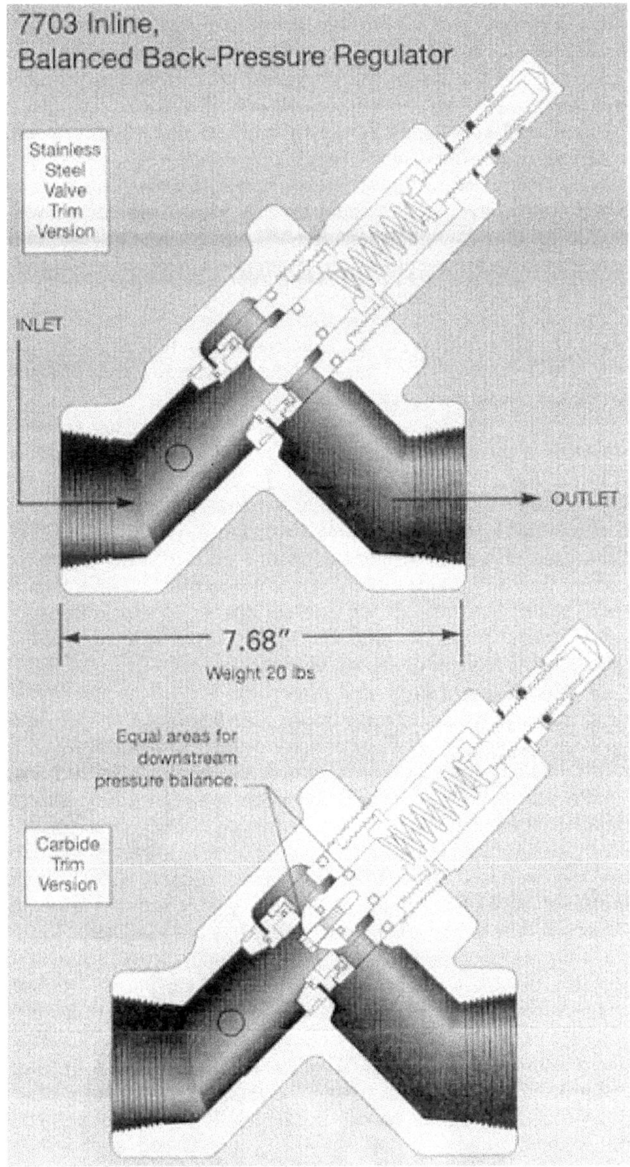

Fig. 6.3—Backpressure valve schematic. These valves can be set for backpressures ranging from 50 to 1,500 psia (from Taylor Valve Technology Inc.).

Fig. 6.4—Location of the backpressure valve. These valves typically are installed on the fluid collection line at the wellhead. (Photo from Taylor Valve Technology Inc.)

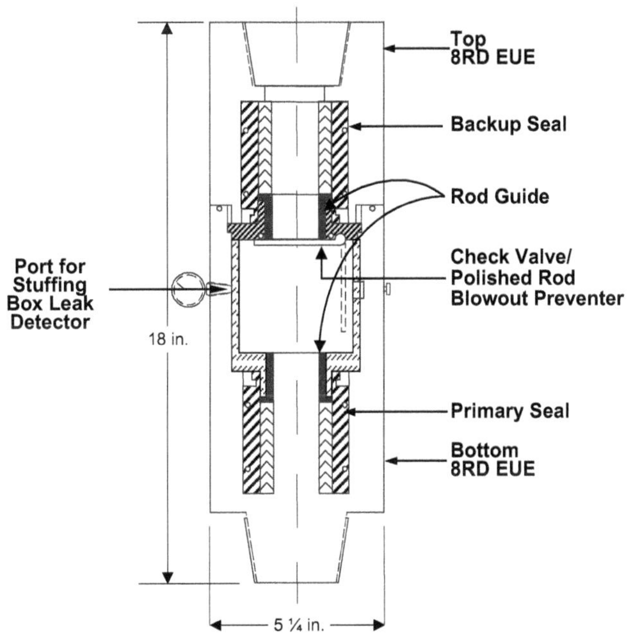

Fig. 6.5—Cross section schematic of a double-pack stuffing box (after Ref. 6). This is an alternative when a backpressure valve is not needed or is unavailable.

long-term rod-corrosion resistance would not be required. Arco uses grade C and grade D rods with about 0.04 failures/yr per well.*

In many cases, corrosion is controlled by injecting corrosion inhibitor downhole, either continuously or on a batch basis. To get the corrosion inhibitor downhole, Arco used bottom holddown pumps with bottom discharge valves and some top holddown pumps. After acid jobs, they used brass-nickel-carbide pumps with bronze-chromium barrels to aid in corrosion prevention.* In some cases, chrome-alloy pumps can cause problems; for example, at Cedar Lake, chrome flakes from brass-chrome pump barrels caused the inhibitor injection pump to stick. A switch to brass-nickel-carbon barrels solved the problem.

Electrical submersible pump motors and seals should be coated with a minimum 12-mil Monel or stainless steel to provide sufficient corrosion protection. Electrical submersible pump cables should be encased with galvanized armor; without it, the cable may fail if CO_2 penetrates the insulation and splits it.

6.1.3 Flowing Wells. As the flood matures, increased gas production may cause the total production from artificially lifted wells to reach a flowing GLR threshold that enables the well to start flowing. Because converting wells from pumped to flowing can lead to significant cost savings for a CO_2 flood, operators should estimate the number, location, and timing of producers that could reach the flowing GLR threshold. These estimates can be made using the predictions from the CO_2 reservoir simulator, in combination with commercially available producing system computer models that provide both the inflow performance relationships for each well and the estimated flowing requirements for tubing size, GLR's, and surface chokes in the producing system.

If there are plans to re-equip a significant number of wells for flowing operations using packers, be sure to include the cost and timing for packers and pump conversions in the overall design plan and cost estimate. Another item that should be included in the cost estimate is operating benefits, which include power cost savings, elimination of rod and pump failures, and equipment maintenance expense reduction (no pumps to gas-lock and no more backside corrosion inhibition treatment programs). Another benefit provided by flowing conditions is the improved recovery in heterogeneous

*Perry M. Jarrell, personal notes from discussion sessions at the CO_2 Forum sponsored by the Petroleum Industry Alliance of the CEED, 19 May 1994.

reservoirs that results from more equitable processing of layers. The increased backpressure on the formation under flowing conditions reduces gas-cycling spikes and promotes more even sweep. **Table 6.1** shows a more complete list of benefits and costs.

The plan for the well producing process to be used (artificial-lift, flowing, etc.) should consider the timing of floodbank response. An unpublished economic analysis performed on flowing and artificial-lift wells in Wasson field arrived at several conclusions:

- A larger drop in oil rate occurred when wells were converted to flowing status before reaching peak tertiary oil response than when they were converted after reaching peak response.
- Wells converted after peak response showed smaller overall oil-rate losses.
- The economics of converting artificial-lift wells to flowing were influenced most by savings in operating costs, such as power, well repair (equipment failures), and CO_2 processing (from lower gas production).
- Conversions caused small losses of oil production rate (around 1 to 3 B/D) for several months because of the higher bottomhole wellface pressure, *but the losses in production were not great enough to negate the operating cost benefits.*
- Production rates eventually increased and no reserve loss occurred.
- For the conversions of more than 40 wells from artificial-lift to flowing in Wasson ODC in 1991, the annual savings amounted to about $800,000, primarily from reduced electrical, pulling, and gas-processing costs.

"Flumping." In designing a lift process, also consider the well inflow performance relationship (IPR) and vertical lift performance to assess the possibility of simultaneously producing up the tubing by way of rod-pump and the up the casing-tubing annulus by way of natural lift. This condition of simultaneous flowing and pumping is commonly known as "flumping." Under certain conditions, a flumping well can produce significantly more fluids than if it were producing by flowing or rod-pumping alone.

One key to flumping success relates to the effectiveness of downhole gas separation. With an efficient downhole separator, the GLR in the casing-tubing annulus will be higher than in the natural formation and the tubing GLR, thereby promoting flow in the annulus while pumping continues in the tubing. If it appears that flumping will work, expect to upgrade the corrosion control program to accommodate the increased annular flow rates.

Wellhead Equipment for Flowing Wells. Wellheads for flowing wells are similar to those for water injection wells. To ensure safe operations, wellhead fittings below the slips should be rated for higher-than-expected pressures. For many flowing wells in the Slaughter and Wasson fields, this translates to 2,000 psia or 3,000 psia wellhead fittings.

Flowing wells should have two similarly rated high-pressure wing valves for well control. They can be either gate valves or ball valves constructed of highly corrosion resistant material, such as 316 stainless steel or aluminum bronze. Master valves, as well, should be similarly high-pressure rated and constructed, and disk trim should be 316 stainless steel or ceramic. Check valves used on flowing wells also should be constructed of 316 stainless steel or aluminum bronze. All wellhead fittings above the master valve should be either plastic coated, stainless steel or aluminum bronze. All NACE requirements must be maintained for high-pressure service in flowing wells if H_2S is produced.

The most important piece of equipment on the flowing well wellhead is the choke, which should be rated for pressures higher than the maximum shut-in pressure of the flowing well. Choke valves should be constructed of 316 stainless steel with tungsten or ceramic disks to prevent corrosion-induced valve failure. The valves also should be designed to operate from 20 to 80% open for normal flow to provide more consistent backpressure and, thus, a more stable flow rate. If the cooling effects of expanding CO_2 cause freezing problems, an in-line heater may need to be added or the well may have to be shut in until the ice melts.

Tubing Size and Corrosion Protection. The tubing size should be determined by calculating specific flowing conditions for each well. In rod-pumped wells, as gas production increases to the GLR threshold that allows the well to flow, it may be necessary to install smaller tubing during the conversion from rod-pumped to flowing. Smaller tubing actually can allow increased tubing volume velocity by containing the gas enough to force it to lift the fluid, thus causing increased total fluid production. If total fluid volume increases beyond a certain point, tubing size may have to be increased later to accommodate it.

If the well operations design indicates that corrosion may be a problem, flowing-well tubing should be coated with plastic, Duoline, or Fluoroline. Yellow band tubing can be used to a depth of 8,000 ft and blue band to 5,000 ft. "Yellow band" is an inspection classification denoting about zero to 15% bodywall reduction for used tubing. "Blue band" denotes about 16 to 30% bodywall reduction.

All tubing, whether new or used, should be bullet-nosed (the pin connection is not a flat edge but a rounded one) so that there will be no sharp edges to break the continuity of the coating

TABLE 6.1—BENEFITS AND COSTS OF CONVERTING ARTIFICIAL LIFT WELLS TO FLOWING WELLS

Benefits	Costs
Lower energy costs (no beam unit or ESP unit to power)	One-time cost to re-equip with downhole packer, surface choke, pulling costs, possible tubing coating
Lower maintenance costs (fewer moving parts, no downhole pump parts to replace or unstick)	Possible paraffin scraping required in tubing
Increased production if downhole pump was acting as a choke	Decreased production because flowing conditions add backpressure to wellface inflow if well was pumped-off before change
Ability to redeploy artificial lift equipment to other wells or fields to save costs there	Too little flow during water cycles and too much flow during gas cycles in high mobility patterns sensitive to gas/water injection bank progression
Reduced potential for casing corrosion (because gases are kept off the annulus)	
Better control of "problem pattern" gas production with choke	
Improved recovery in heterogeneous reservoirs through more equitable layer processing (reduced gas cycling spikes from increased backpressure on the formation)	
Reduces CO_2 processing costs because CO_2 is not recycled as rapidly	

process at the pin end. Using bullet-nosed, coated pipe reduces the possibility of exposing bare steel when the tubing connections are made up, thus helping to prevent a common source of corrosion failure.

In its Slaughter floods, Amoco found that a modified thick-film epoxy was an effective plastic coating for all new and used flowing-well tubing internals. The thick coating was more pliable than other coatings that had been used, and was compatible with CO_2. Another benefit of a thick coating is that it can cover small pits in used tubing.

Tubing couplings should be made from the same material as the tubing to eliminate potential for galvanic corrosion. Full-size couplings should be used whenever possible because they provide extra metal for corrosion resistance. The only exception to using full-size couplings applies to dual-zone production wells, which require special clearance couplings to allow the outside diameter of two sets of couplings to fit within the inside diameter of the existing casing. Ryton is a very effective coating for couplings that require internal coating.

Paraffins and Hydrates. Some flowing wells in Shell's South Crossett Unit developed severe paraffin and hydrate problems because of wellhead tubing cooling effects[8]—temperatures would drop as low as 14 to 30°F. When a downhole electric impedance heating system was installed, oil production increased by about 10%. The heating system applied 40 kW of power uniformly over the 2,000-ft tubing span and maintained the wellhead tubing temperature at 45°F.

Downhole Equipment. In their Denver Unit, Shell used the same downhole equipment as is normally used in flowing wells during primary production, except that Teflon pipe dope was used for connections, and nitrile packer rubbers with varying durometer hardness ratings were used to get a better seal.[5] For the best seal with the lowest risk of extrusion, they used a stacked multidurometer 90/70/90 packer seal.

Bottomhole tools for a flowing production well can be very similar to those used in a CO_2 injection well. The main difference is that the production well should have two perforated subs and an additional profile nipple below the packer **(Fig. 6.6).**

The additional bottom profile nipple should be located below the perforated subs, while the top profile nipple should be located below the on/off tool but above the packer. The purpose and location of the two profile nipples is an important design aspect. The primary function of the top profile nipple is to allow the tubing to be pulled or unlatched from the packer without pulling the packer. The bottom profile nipple is designed so that pressure tools can set and collect data without blocking the flow. This nipple also should be designed to allow other common-sized tools to pass through it. The nipples, subs, and the on/off tool should be constructed of highly corrosion resistant material such as 316 stainless steel.

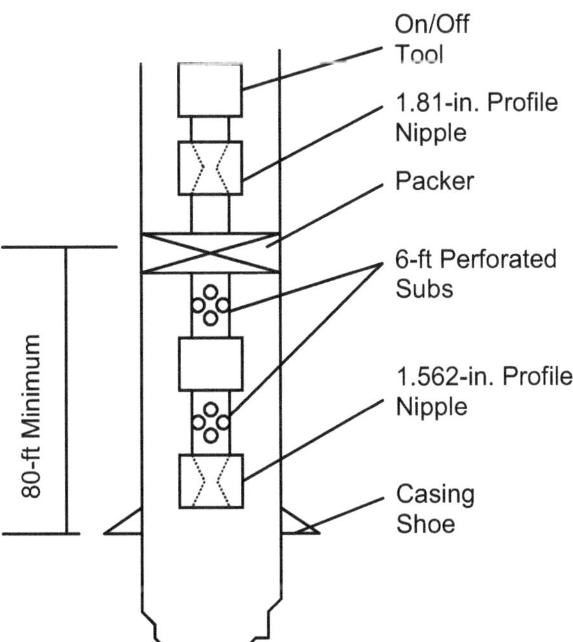

Fig. 6.6—Downhole configuration for flowing wells. The top profile nipple allows the tubing to be pulled without pulling the packer. The bottom nipple facilitates use of pressure tools and other tools.

The on/off tool and the top profile nipple provide a good means of well control, and they allow corrosion inhibitors to be circulated up the annulus. The corrosion control benefits realized when using these devices outweighs the increased cost. These tools also offer an environmental benefit in that they help prevent tubing pulling jobs and thus eliminate the associated atmospheric gas emissions. Finally, these devices allow the tubing to be replaced without pulling the packer; in older wells it can be difficult to get a packer to reset successfully.

Installing two perforated subs below the packer allows near-normal flow up the wellbore when pressure tools are landed in the lower profile nipple. To prevent corrosion, consider using subs that either are plastic-coated both internally and externally or are made of 316 stainless steel to prevent corrosion.

For packers, consider neutral-set, one-piece, through-conduit mandrels with bottom rubbers, which work very well for flowing-well service. Typically, this type of packer also performs better than tension-set packers when well-control problems arise. The mandrels can be plastic-coated or they can be made of 316 stainless steel, an effective but costly alternative.

Plunger Lifting. As a flowing well becomes gassier and ceases to flow efficiently, plunger lifting may be the next logical lift process. Plunger lifting has been effective in some fields, such as those where the GLR was too high for rod-pumping and in flowing wells with lower GLR's that are prone to paraffin problems. Although plunger lifting has been used successfully in several CO_2 floods, its use is not widespread.

Most plunger-lift systems in the oil and gas industry are used to improve productivity in natural gas fields with wells that are burdened by liquid inflow. Gas fields usually are equipped with the surface facilities needed to handle the production and pressure swings associated with plunger-lift systems, e.g., on-site gas/liquid separators that allow the gas stream to flow from the well-site down a collection line to sales.

Oil fields under CO_2 flooding, however, have surface facilities that may not adequately handle pressure and production swings. In oilfield CO_2 applications, excessive wellhead pressures can accumulate as a result of friction-induced pressure drop in long gathering lines running to centralized separation facilities. The pressure drop stems not only from the length of the flowline, but also from the increased pressure required to mobilize multiphase fluids (oil, water, and gas) rather than just gas.

Another potential problem is that oil production facilities such as separator vessels usually are not designed to handle the large swings in gas production that are associated with plunger lifting. Though the vessels may be designed to accumulate the gas, pressure may build up because of flowline size constriction, and this increase in pressure may further reduce well productivity. Gas measurement devices also must be designed to handle the large swings in gas production during the plunger cycles of loading and unloading.

6.2 Injection Well: Wellhead to Bottomhole

The design of a CO_2 injection well is very similar to that of a water injection well. Most downhole equipment will be virtually the same; however, the wellhead for a CO_2 injection well will differ significantly from that of a water injection well.

6.2.1 Wellhead Design. A typical wellhead configuration for a WAG injection well is shown in **Fig. 6.7.** Notable features include double valves on the CO_2 injection line, line blinds, and steel ring-joint flanged components.

Metallurgy. Wellhead valves and flanges usually must be upgraded for higher pressure ratings and corrosion resistance. Of all the equipment in a WAG flood, the wellhead assembly will have the highest potential for corrosion because this is where CO_2 and

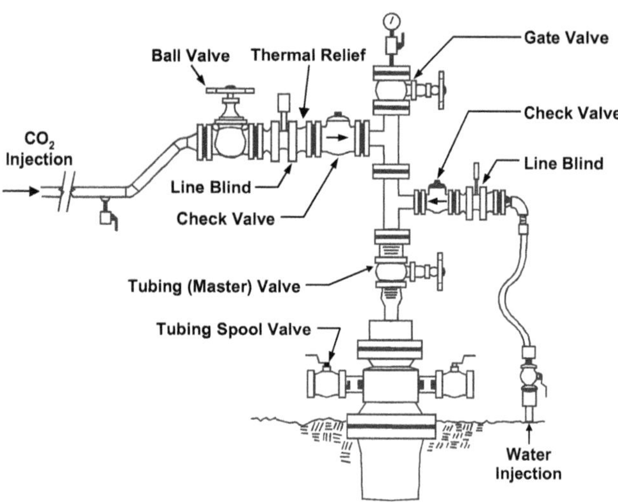

Fig. 6.7—Carbon dioxide injection wellhead configuration. The wellhead typically includes additional valves and upgraded metallurgy.

water meet most often and are most likely to form carbonic acid. Because of the potential for carbonic acid creation when water and CO_2 are switched, one must seriously consider more resistant metallurgy than bare carbon steel. The common metal used in WAG injection wellhead valves and fittings is 316 stainless steel,[2,4,5,9,10] although aluminum bronze also has been used in some cases.

If the WAG design does not include frequent switches between CO_2 and water, it may be possible to use less expensive metallurgy than stainless steel. For example, Northeast Purdy Springer injection wells were installed with stainless steel wellheads, but the operator realized later that stainless was not necessary because there were few switches from CO_2 to water.[9,11] In general, though, most valves and trim should be 316 stainless steel or aluminum bronze.

For some valves and fittings, grade 304 stainless steel may suffice. This grade has virtually the same mechanical, physical, and fabrication characteristics as the more expensive grade 316; however, it does not provide equivalent corrosion resistance, particularly in chloride environments or where pitting corrosion tends to occur. In these cases, grade 316 is a necessity. Aluminum bronze has roughly the same corrosion resistance as 316 stainless, but it does not wear as well in erosional environments.

The requirements for piping are less rigorous because piping often is bare carbon steel on the CO_2 side and plastic-coated carbon steel on the water side. Also, valves that are exposed only to CO_2 can be made of carbon steel bodies with stainless steel trim. Grade 304 stainless steel is adequate for valves exposed to CO_2, even if the gas is saturated with water, but not for valves exposed to brine water mixed with CO_2.

Valves. Injection wells commonly are equipped with two valves for well control, an outside valve that is put to regular use and an inside valve that is reserved for safety. They usually are gate or ball valves and are constructed of 316 stainless steel or aluminum bronze. Master valves commonly are gate valves constructed of 316 stainless steel or aluminum bronze. Aluminum bronze master valves were installed at Wertz Tensleep.[12] Successful disk trim construction has been 316 stainless steel or ceramic. The 316 stainless steel is preferred because ceramic disks are more susceptible to cracking or breaking from handle overtorquing. Additionally, the check valves used on the CO_2 and water side of injection wells should be constructed of 316 stainless steel, as aluminum bronze exhibits poor wear resistance, and check valves are crucial for safe operation.

At Amoco's Slaughter and Wasson floods, the preferred head fitting above the master valve is either stainless steel or aluminum bronze, and all the fittings are ring-joint flange type, per API wellhead specs. Below the tubing valve, however, other connection types are common, including raised-face flanges, threads, and even some mechanically joined connections.

We recommend a positive shutoff on all CO_2 wells to ensure that no leaks occur and to prevent CO_2 from inadvertently flowing back into the water injection system. The most effective positive shutoff is a line blind, also known as a spectacle blind. The line blind can be made of either stainless steel or aluminum bronze. It should be tapped and equipped with a bleed valve in the CO_2 supply side of the blind to facilitate bleeding the CO_2 when rotating the blind. The bleed valve should be a ball valve of sufficient size; needle valves and ball valves sized too small can freeze up and prevent effective bleedoff, creating a safety hazard. Shorter bleedoff time is another benefit of using ball valves rather than needle valves.

Many different wellhead valve configurations can be developed, as long as two important rules are followed: First, the CO_2 supply-side check valve should be placed downstream of the line blind; if not, CO_2 can be trapped between the check valve and the blind. If atmospheric temperature increases, the trapped CO_2 will expand and cause high pressure, possibly to a point beyond the pressure limit of the materials. Second, the CO_2 supply-side valves and fittings should be ring-joint flange type to prevent CO_2 from permeating the fitting; screw-type fittings are more likely to have problems. Thermal relief valves are required between all block valves and between block and choke valves.

The water-side valves and fittings exposed to CO_2 also should be a ring-joint flange type, while those not exposed to CO_2 can be a flanged or screw type. Some floods have used threaded turbine meters and have not seen any problems stemming from CO_2 permeating the threads.

Seals. For seals that are always contained and not exposed to the atmosphere during routine operations, a 90 to 95 durometer Buna-N or Viton elastomer works well. At Cedar Lake it was found that the 90 to 95 durometer seals performed better if they had been peroxide cured.

For seals regularly exposed to the atmosphere, such as the seal on the line blind and the top bonnet seal of the simplex orifice fitting, we recommend Teflon. Although Teflon seals generally are not reusable, they do not absorb CO_2 and are less likely to fail.

Safety Features. Annular pressure monitors help detect leaks in packers and tubing, which is important for taking quick corrective action. To prevent dangerous high pressure buildup on surface equipment, CO_2 injection must be stopped as soon as leaks occur. Rupture disks and pop-off valves can be used to relieve built-up pressure.

For safety considerations, wellhead fittings below the slips should be pressure-rated to the highest expected bottomhole CO_2 injection pressure *at a minimum*. For Cedar Lake wells, wellhead fittings were rated the same as the wellhead (2,000 psia) because of cost considerations, even though initial design called for higher-pressure-rated fittings. These wells have performed acceptably.

6.2.2 Switching CO_2/Water Cycles. For WAG injection projects, knowing the frequency of switches from CO_2 to water and vice versa will help determine the most appropriate process and equipment for switching.

Although some units, such as Salt Creek, have automated switching, in most cases, switching is performed manually (without automation) for safety reasons. Spectacle-blind flanges should be used for all switches. Manual switching significantly reduces the potential for accidental CO_2 injection into the water-injection line, which would send CO_2 back to the water station. High-pressure CO_2 flowing back to a lower-pressure water station has obvious safety consequences. While the switching process can be automated, it is a costly investment because redundant valves must be installed as a safety measure to reduce CO_2 valve leaks into the water line.

Some operations, such as Rangely,[13] Bairoil, and Wasson ODC, experimented with simultaneous CO_2/water injection to reduce gas cycling problems and increase recovery. The wellhead configurations for simultaneous injection were not modified, although the upstream control system and some measurement systems were changed.

6.2.3 Tubing and Packer Assembly. A typical downhole configuration for a Permian Basin CO_2 injection well is illustrated in **Fig. 6.8.**[14] This configuration includes a double-grip type packer, an on/off tool with a profile nipple, and a downhole shutoff valve. If a packer must be pulled for remedial work, the well can be shut in either at the packer using the shutoff valve, or at the profile nipple. This allows for circulation of a kill fluid above the packer before the packer is pulled. Though some operators have experienced minor safety issues (e.g., small leaks) with tension-set packers, they generally are used successfully if the tubing is slacked off to place it in neutral or in compression. Cooling from CO_2 injection can cause tubing to contract, which can shear the packer if the tubing is taut.

Neutral-set, one-piece, through-conduit mandrels with bottom rubbers are the most widely used for CO_2 injection service. The mandrels are typically either plastic-coated or 316 stainless steel. Fiberglass subs, about 4 ft long, can be installed below the mandrel-related packers to decrease potential turbulent flow that could cause erosional wear.

Materials. Unocal had success in their Dollarhide Unit injection-well packer-casing wall seal with a polyepichlorhydrin rubber, which resists CO_2 degradation at a hardness as low as 70 to 80 durometer.[15]

Corrosion resistant alloys (CRA's) have had mixed results in service. On the downside, several Maljamar wells equipped with CRA-wetted parts on packers and on/off tools failed from corrosion after 6 to 13 months.[16] The CRA's that were used—9 chrome-1 moly and 13 chrome—were replaced with stainless steel, which performed satisfactorily. On the upside, Wertz Tensleep successfully used 9 chrome-1 moly material for wetted metal parts of packers.[12] In the Willard Unit, Incoloy 925 has worked very well on wetted packer parts and K-monel has worked well for the profile nipples.

Well Control. Almost all CO_2 injection wells are equipped with on/off tools and profile nipples. The on/off tool and profile nipple provide a means of well control and allow treated fluid (such as corrosion inhibitors) to be circulated in the annulus. Typically, the benefits of reduced corrosion outweigh the incremental costs of these tools. Note that the on/off tool can be used only with a neutral-set packer. For these tools, most corrosion-resistant alloys perform well where less than 1% H_2S is present in the gas stream. When H_2S is present, it causes pitting and under-deposit corrosion.

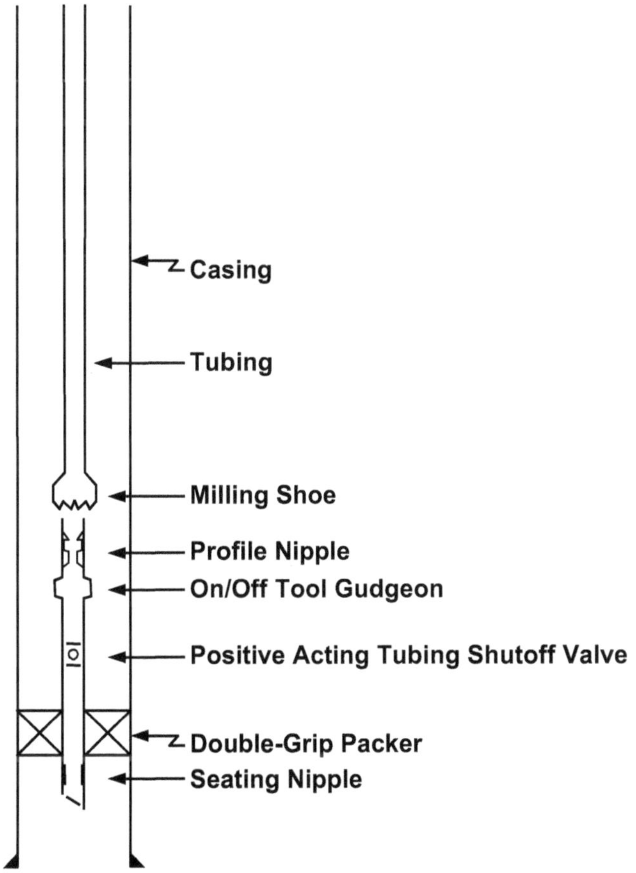

Fig. 6.8—Downhole completion equipment for a typical CO_2 injection well in the Permian Basin.

Tubing Options. The optimal tubing type and size should be determined on a lease-by-lease basis because several options have proved effective. The most widely used kinds of tubing for WAG injection include bare carbon steel, plastic-coated steel, fiberglass-lined steel, and cement-lined steel. Amoco found that a modified thick-film epoxy was effective in all new and used tubing internals in its Slaughter floods. The thick coating was more pliable than other coatings that had been used, was CO_2 compatible, and covered small pits in used tubing.

Arco did not have success with plastic coating in west Texas, and changed to cement-lined tubing. However, they did have success with Tuboscope's TK7 and TK7A plastic coating in the East Velma West Block Sims Sand Unit in Oklahoma.

Texaco burned out plastic coating before replacing it with cement-lined tubing (Mabee Unit).* In the Denver Unit, failures in tubing plastic coating (a thin-film modified phenolic coating) were caused by periodic downhole wireline tool surveys.[7] A change to thick-film modified epoxy/phenolic coating reduced the failure rate.

Many of the WAG injection wells at Bairoil are equipped with bare carbon steel. Though some corrosion is expected, the life of the tubing is expected to match the life of the WAG injection scheme, so that longer life (and more expensive) alternatives were not required.

In projects where plastic coatings have failed, operators have switched to fiberglass-lined and cement-lined tubing. Cement-lined tubing was used heavily in Salt Creek, Wasson ODC, and Anton Irish Clearfork in the early 1990's. The dolomitized (or flouridized) cement provides the most effective coating to protect against acid. Though the operators of Salt Creek have had success with fiberglass-lined tubing, a good number of wells later were equipped with cement-lined tubing because of the lower cost of the cement product.

The Rangely Unit started with fiberglass and internally plastic-coated steel tubing and has changed to using cement-lined tubing. The fiberglass tubing experienced stress-related failures because of the large temperature swings, and the plastic-coated tubing failed because the coating was being cut away by necessary, repeated wireline runs.[17]

Bare steel tubing is acceptable for continuous CO_2 injection (nonWAG injection) and is used in the Denver Unit[5] and Bairoil. The Ford Geraldine Unit also uses bare tubing in injection WAG service without experiencing corrosion problems.**

At the Bennett Ranch Unit, cement-lined tubing in all injection wells was replaced with fiberglass-lined tubing for the 1995 startup of CO_2 injection,[4] because the cement lining was very old. And most recently, cement-lined and fiberglass-lined tubing is not being installed in the Permian Basin because these linings have proved to be too brittle, have cracked too easily, and have had generally poor quality control. It appears that plastic linings are once again gaining favor as the technology improves and operators progress along the learning curve.

Couplings. Tubing collar leaks seem to be one of most widespread problems associated with introducing WAG injection into previous water injection wells.[2,5,10] An effective solution has been to modify tubing connections.

To eliminate any potential galvanic corrosion, couplings should be made out of the same material as the tubing. Ryton (a rubbery plastic or polymer with high coating durability in the joint-thread makeup area) is a very effective coating for couplings that require internal coating.

For their API eight-round tubing connections, Unocal used a high graphite, nonmetallic thread lubricant composed of graphite particles suspended in a lithium-based grease. The graphite particles deformed to make a better seal than metallic particles could,

---

*Perry M. Jarrell, personal notes from discussion sessions at the CO_2 Forum sponsored by the Petroleum Industry Alliance of the CEED, 19 May 1994.

**Personal communication with W.F. Flanders, Transpetco Engineering (1998).

mainly because of the crystal-lattice shearing nature of deformation. Nonmetallic Teflon pipe dope also works well on eight-round connections.

6.2.4 Wellbore Modifications. One of the most significant areas of preCO_2 flood preparation is improving injection conformance. While some out-of-pay injection can be tolerated in most waterflooding operations, the relatively high cost of CO_2 compared to water demands that CO_2 injection be confined to oil-bearing pay. The conformance improvement work that many operators have performed before initiating CO_2 injection includes mechanical zone isolation, plugbacks, liner installations, chemical injection (polymers, surfactants, etc.), and gas cap shutoffs. Several of these techniques were applied[18] successfully in the Rangely Weber Unit.

"Puddle-packing" open holes was successful at Maljamar,[19] reducing out-of-pay injection as much as 20%.[16] Puddle-packing is the process of placing a permeable-resin fill material across voids in the open hole caused by nitroglycerin stimulations (also called "shots"), followed by underreaming the excess material to allow installation of a liner. This process works where cementing a liner across the open hole would not work because the resulting 20-in. thick cement sheath behind the pipe could not be effectively penetrated with perforations.

Higher surface injection pressures mean that some state regulatory commissions (such as the Texas Railroad Commission) require higher pressure casing tests, which may find some casing leaks that were not detected while the well was in water-injection service. Testing wells at higher pressures some time before converting to CO_2 injection will help identify the scope of potential casing repair work required.

The casing just below a packer is likely to become corroded during a WAG flood because it will be an area of frequent CO_2 and water mixing. This can be a significant problem if the operator is depending on this casing to isolate gas caps or other transmissive nonpay zones. Some operators have run casing liners made of CRA's or fiberglass to solve this problem.

In an effort to improve recovery by improving sweep, operators of the Sundown Unit found that cementing fiberglass liners across the open hole with graphite collars was effective in isolating zones and improving conformance.[10]

References

1. Grigg, R.B. and Schechter, D.S.: "State of the Industry in CO_2 Floods," paper SPE 38849 presented at the 1997 SPE Annual Technical Conference and Exhibition, San Antonio, Texas, 5–8 October.
2. Linn, L.R.: "CO_2 Injection and Production Field Facilities Design Evaluation and Considerations," paper SPE 16830 presented at the 1987 SPE Annual Technical Conference and Exhibition, Dallas, 27–30 September.
3. Baskin, G.M. and Spriggs, D.M.: "Wellbore Operations in CO_2 Flooding—A Case History," NACE Corrosion 1985, Boston, Massachusetts (1985) **34,** 1.
4. Hsu, C.F. *et al.*: "Design and Implementation of a Grass-Roots CO_2 Project for the Bennett Ranch Unit," paper SPE 35188 presented at the 1996 SPE Permian Basin Oil and Gas Recovery Conference, Midland, Texas, 27–29 March.
5. Fleming, E.A., Brown, L.M., and Cook, R.L.: "Overview of Production Engineering Aspects of Operating the Denver Unit CO_2 Flood," paper SPE 24157 presented at the 1992 SPE/DOE Symposium on Enhanced Oil Recovery, Tulsa, 22–24 April.
6. Clegg, J.D.: "Another Look at Gas Anchors," *Proc.*, 36th Annual Southwestern Petroleum Short Course, Lubbock, Texas (1989) 293.
7. Iken, G.: "Denver Unit CO_2 Flood: Material Aspects of the Design and Operation of Injection and Production Facilities," *Energy Progress* (1987) **7,** 193.
8. Baud, W. and Eastlund, B.J.: "Electric tubing heater improves well production in CO_2 flood," *Oil & Gas J.* (1994) **92,** No. 16, 60.
9. Brinlee, L.D. and Brandt, J.A.: "Planning and Development of the Northeast Purdy Springer CO_2 Miscible Project," paper SPE 11163 presented at the 1982 SPE Annual Technical Conference and Exhibition, New Orleans, 26–29 September.
10. Folger, L.K. and Guillot, S.N.: "A Case Study of the Development of the Sundown Slaughter Unit CO_2 Flood Hockley County, Texas," paper SPE 35189 presented at the 1996 SPE Permian Basin Oil & Gas Recovery Conference, Midland, Texas, 27–29 March.
11. Fox, M.J. *et al.*: "Review of CO_2 Flood, Springer "A" Sand, Northeast Purdy Unit, Garvin County, Oklahoma," *SPERE* (November 1988) 1161.
12. Kleinstelber, S.W.: "The Wertz Tensleep CO_2 Flood: Design and Initial Performance," *JPT* (May 1990) 630.
13. Robie, D.R., Roedell, J.W., and Wackowski, R.K.: "Field Trial of Simultaneous Injection of CO_2 and Water, Rangely Weber Sand Unit, Colorado," paper SPE 29521 presented at the 1995 SPE Production Operations Symposium, Oklahoma City, Oklahoma, 2–4 April.
14. Spriggs, D. and Tompkins, C.R.: "Equipment Considerations in Carbon Dioxide Flooding," *Proc.,* Thirty-Second Annual Southwestern Petroleum Short Course, Lubbock, Texas (1985) 518.
15. Collier, T.S.: "State of the Art Installation of CO_2 Injection Equipment: A Case Study," paper SPE 17294 presented at the 1988 SPE Permian Basin Oil and Gas Recovery Conference, Midland, Texas, 10–11 March.
16. Bowser, P.D. *et al.*: "Innovative Techniques for Converting Old Waterflood Injectors to State-of-the-Art CO_2 Injectors," paper SPE 18976 presented at the 1989 SPE Joint Rocky Mountain Regional/Low Permeability Reservoirs Symposium and Exhibition, Denver, 6–8 March.
17. Wackowski, R.K. and Masoner, L.O.: "Rangely Weber Sand Unit CO_2 Project Update: Operating History," paper SPE 27755 presented at the 1994 SPE/DOE Symposium on Improved Oil Recovery, Tulsa, 17–20 April.
18. Fullbright, G.D. *et al.*: "Evolution of Conformance Improvement Efforts in a Major CO_2 WAG Injection Project," paper SPE 35361 presented at the 1996 SPE/DOE Symposium on Improved Recovery, Tulsa, 21–24 April.
19. Darr, C.D.K., Brown, E.K., and Murphey, J.R.: "A Method To Convert Multiple-Shot Section Openhole Completions Into Cased-Hole Completions With Zonal Isolation," paper SPE 15009 presented at the 1986 Permian Basin Oil and Gas Recovery Conference, Midland, Texas, 13–14 March.

SI Metric Conversion Factors

bbl ×	1.589 873	E − 01	= m^3
°F	(°F − 32)/1.8		= °C
°F	(°F + 459.67)/1.8		= K
ft ×	3.048*	E − 01	= m
in. ×	2.54*	E + 00	= cm
kW ×	1.0*	E + 00	= kW
mil ×	2.54*	E − 05	= m
psi ×	6.894 757	E + 00	= kPa
psia	6.894 757	E + 00	= kPa

*Conversion factor is exact.

Chapter 7
Implementation: How Can the Design Become Reality?

Once it is clear that a CO_2 flood can be economically successful, the most important ingredients in that success are good planning and well-managed implementation. This chapter reviews all of the critical elements of these areas, including:

- Planning—Sec. 7.1 covers timing and staging of startup and gives a startup checklist.
- Preinjection data gathering—Sec. 7.2 presents a comprehensive list of information needed for a CO_2 flood.
- Permits, contracts, and agreements—Sec. 7.3 discusses key issues having to do with partners, service companies, regulatory bodies, and government incentives.
- Monitoring—Sec. 7.4 covers general monitoring considerations, monitoring prerequisites, components of the monitoring program, setting expectations, and running the monitoring program.

7.1 Planning

7.1.1 Timing and Staging of Startup. Theoretically, a CO_2 flood can be implemented at any time after a field's discovery. In practice, however, a CO_2 flood normally is installed as a tertiary recovery mechanism following the injection of water for secondary recovery, although several successful CO_2 floods (Hanford, Aneth, and Hansford Marmaton) have used CO_2 as the secondary injectant.[1] **Fig 7.1** illustrates how the injection of CO_2 at four points during the life of the field might increase oil production. The solid line shows typical oil production during primary and secondary waterflood recovery, and the dotted lines indicate potential improved oil production from a CO_2 flood.

Type A floods are the secondary CO_2 floods mentioned above (Hanford, Aneth, and Hansford Marmaton[2]). Type B floods are implemented at or near the waterflood production peak; for example, the South Wasson Clearfork Unit's CO_2 flood was started only five years after waterflood initiation.[3] Many more CO_2 floods, including the Wasson Denver Unit[4] and Wertz,[5] are Type C floods that were started after waterflood decline had been established. Also common are Type D floods, in which the operator waits until the field is near its waterflood economic limit before beginning CO_2 injection (Wasson Bennett Ranch Unit). A production profile for a vertical flood (discussed in Chap. 2) is not shown. In general, the ultimate incremental oil production should be approximately the same for all types if reserve losses from economic abandonment are ignored.*

*Personal communication with R.L. Koinis, Shell Western E&P Inc. (1997).

Timing. In deciding when to implement the CO_2 flood, several economic and reservoir-related factors must be weighed, including oil price, time value of money, financial risk, allocation of operating costs, reservoir pressures, and waterflood feasibility.

Oil Price. The oil price normally is the most significant factor in profitability. An operator will not implement a flood unless the oil price is forecast to be high enough during the oil response period for the flood to return an adequate profit.

Time Value of Money. The earlier a flood is implemented, the sooner oil will be produced and so the higher the expected net present value of profits will be.

Financial Risk. Using CO_2 very early in the recovery process poses additional financial risks, most of which relate to the purchase and handling of CO_2. For example, in 2000, the capital and CO_2 purchase commitments for a 10 million bbl OOIP oil field located near the Denver City Hub** easily could have totaled more than $4 million. The monetary return on such a significant investment may be poor if the reservoir does not perform as expected because of inefficient use of injected CO_2.

By operating a waterflood before starting a CO_2 flood, the operator gains useful information about how the reservoir responds to injection. Reservoir models developed during a waterflood provide the basic geologic and reservoir information needed to assess the reservoir's probable response to CO_2 injection, information that greatly reduces the risk of technical project failure.

Playing it safe can be overdone, however; the operator need not wait until the economic limit of the waterflood has been reached. Very little additional information is gained by deferring a CO_2 flood to point D on Fig. 7.1, and the result is unnecessary delay in capturing profitable reserves.

Allocation of Operating Costs. If the CO_2 flood is implemented while waterflooding is still underway, operating costs should be allocated to each of the two processes, and only the incremental costs should be assigned to the CO_2 flood. If secondary waterflood operations are allowed to continue to their economic limit, all the operating costs of subsequent CO_2 flooding must be borne by the CO_2 EOR production. Thus, CO_2 flooding is less profitable after the waterflood economic limit.

**The Denver City Hub is the location near Denver City, Texas, where four major CO_2 transportation pipelines connect and where CO_2 prices often are quoted.

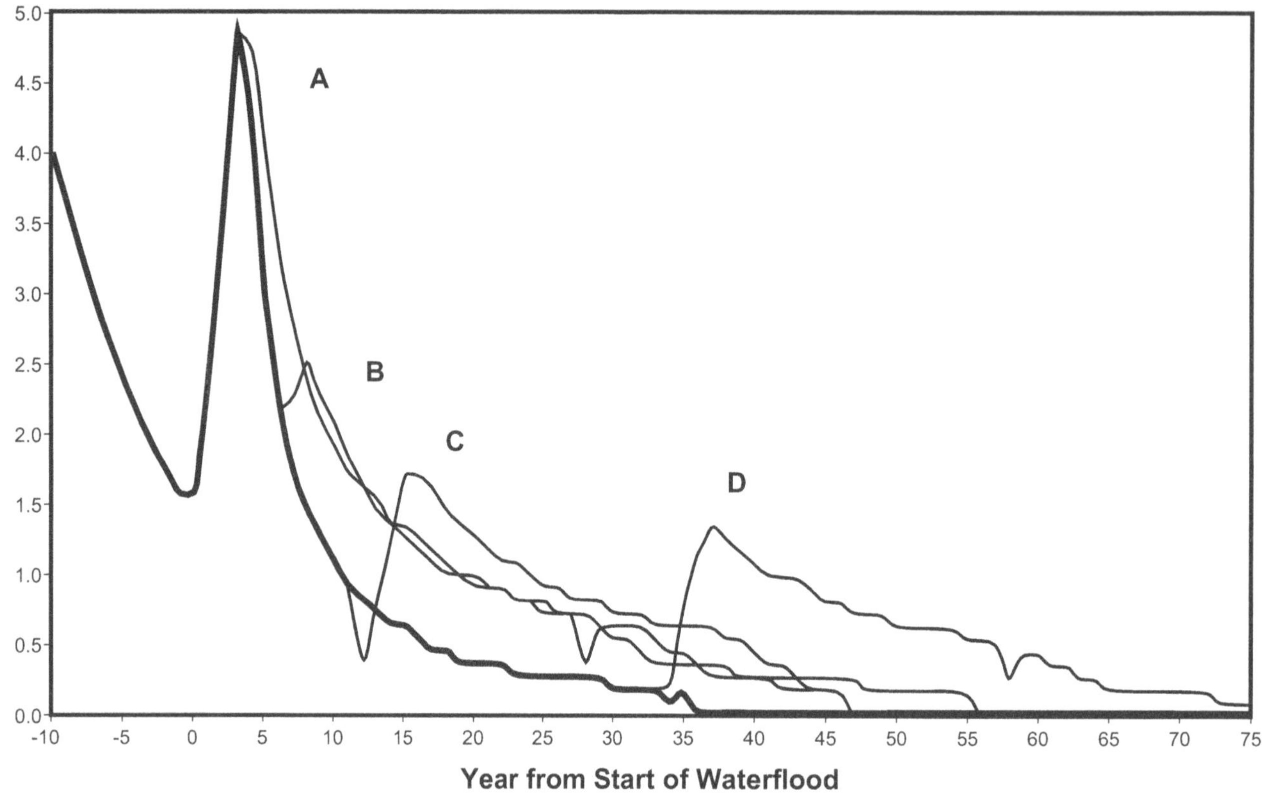

Fig. 7.1—Enhanced oil recovery and waterflood production forecasts. The baseline is a waterflood implemented in year zero. The CO_2 floods A, B, C, and D are implemented in years 0, 6, 13, and 33, respectively.

Reservoir Pressures. Delaying a CO_2 flood until a few years after the waterflood has started to gain more injection response data not only reduces technical risk but may also make the CO_2 flood more financially successful. In many cases, during primary production the reservoir pressure drops to below the MMP. This pressure reduction renders a CO_2 flood less effective until enough CO_2 has been injected to raise reservoir pressure to the MMP, potentially an expensive proposition. For most fields (e.g., Rangely[6]), a waterflood is an inexpensive and commonly used means of increasing reservoir pressure to the point where CO_2 and oil become miscible. In certain cases where water injection was thought to be problematical (North Cross[7], Hansford Marmaton[2]), CO_2 was used as a secondary injectant to increase reservoir pressure.

Waterflood Feasibility. The above discussion is predicated upon the feasibility of waterflooding in addition to CO_2 flooding. However, there are some formations—for example, the Delaware Sands and the Devonian formation, both in west Texas—that often cannot be waterflooded economically. The Devonian formation may have such small pore throats and such low permeability to water that gas injection is the only way to inject at significant rates.[7] Relative permeability curves of Delaware Sand fields reveal that waterflooding is sometimes uneconomical because the residual oil saturation is high.

When a waterflood is not feasible because of very low permeability, the operator must implement the CO_2 flood immediately after the primary recovery period or after hydrocarbon gas injection. The North Cross Unit, which was the second commercial CO_2 flood, is an example of this. After almost 25 years of operation, CO_2 flooding at the North Cross Unit is expected to recover more than 45% of the OOIP.*

Staging. The operator must make decisions not only about timing of the CO_2 flood, but also about whether and how the flood will be staged. This topic also is discussed in Sec. 3.3.1.

In reservoirs with many injectors, CO_2 floods often are implemented in stages. Although staging delays oil response and exacts a toll on the present value of the oil recovered, typically the benefits outweigh the costs—financial risk is reduced because all of the capital is not committed upfront. Once the flood proves itself economically, additional expansions may be justified and uncertainty reduced.[8] An additional benefit is that the infrastructure can be used more efficiently when injection and production are leveled over the course of the project.

Staging can be accomplished through combinations of three basic approaches: starting in a limited area of the field and expanding to other areas (Denver Unit,[4] Rangely[6]); starting with one production zone and expanding vertically to other production zones in the same field (Wasson Denver Unit Transition Zone); and gradually increasing the size of the CO_2 slug injected.

At the beginning of most projects, operators do not commit to purchase the entire CO_2 volume that may ultimately be required but only make an initial commitment that is large enough to recover sufficient oil to yield an economic rate of return. Many operators design and contract for an initial 30 to 40% HCPV slug, even while anticipating a larger ultimate slug. If the flood justifies it, the CO_2 purchase obligation can be increased.

Staging offers several benefits:

- The initial response in the limited area verifies the technical viability of the project before all the investments are made.
- It is less expensive to "learn as you go" on a smaller project.
- The CO_2 handling facilities are more economical because less capital is required and more of the capacity can be used fully over the life of the project.
- Carbon dioxide purchases are reduced because more CO_2 can be recovered from initial injection areas and recycled into areas that begin injection later.
- Full-time labor needs are minimized because of steadier development, injection, and production operations.
- The operator has flexibility to accelerate or decelerate development as economic conditions change.[9,10]

The relative risks and rewards of implementing a CO_2 project in one, two, and four stages are illustrated in **Fig. 7.2,** which presents cumulative net present value cash flow curves for a theoretical project in the San Andres formation of the Permian Basin of west Texas.

Fig. 7.2 presents cash flow in a dimensionless form: The y-axis represents the cumulative present value of the project at any partic-

*Personal communication with J.F. Simmons, Altura Energy Ltd. (1997).

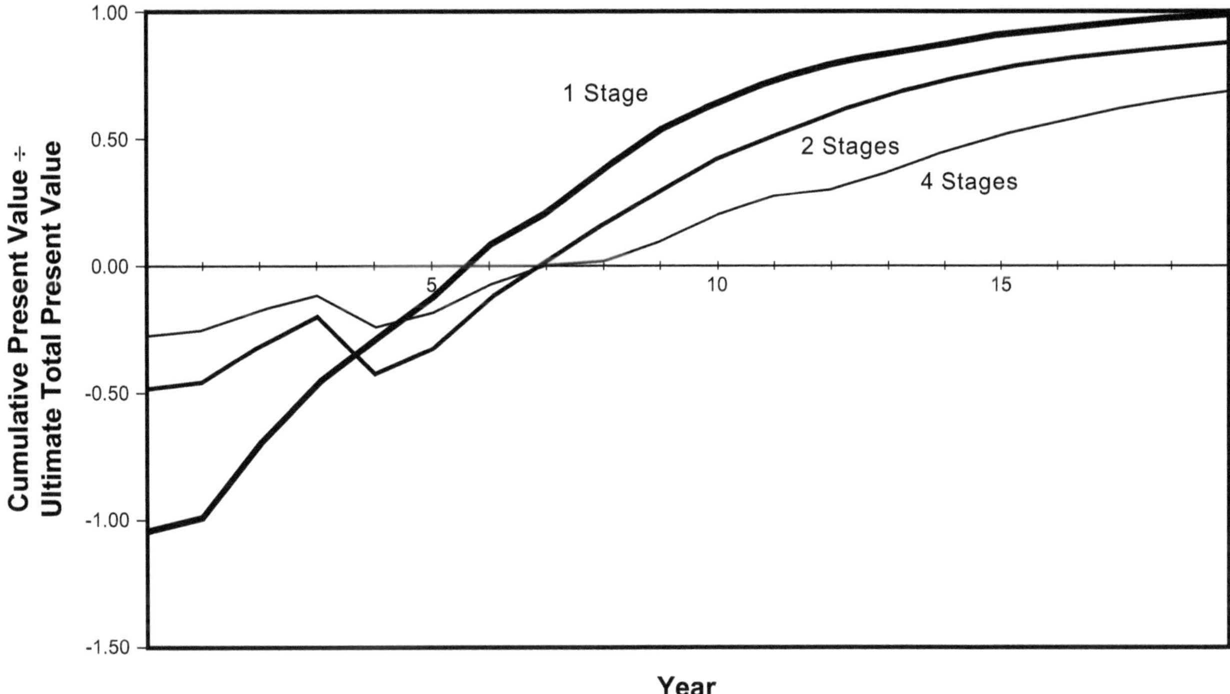

Fig. 7.2—Effects of staging on cumulative net present value. In this dimensionless format, the y-axis is cumulative present value divided by the ultimate total present value of the flood for a one-stage project. Conducting a CO_2 flood in several stages has the benefit of lowering upfront costs, although, in this example, it also reduces the total present value of the project.

ular time, divided by the ultimate total present value of the flood for a one-stage project. Negative cash flow in the early part of the flood stems from the costs of CO_2 purchases and installation of the facilities for the CO_2 flood and gas handling. As production pays out these costs, cumulative cash flow becomes positive. This technique makes it possible to compare different fields and approaches in one graph.

In the example shown, stages are implemented four years apart. In the two-stage and four-stage cases, the costs of CO_2 recycle compressor purchases occur in year zero and year four, respectively.

As Fig. 7.2 shows, implementing the project in one stage makes it possible to realize the maximum reward in the shortest amount of time. In this example, the time value of money outweighs the effects of using smaller compression capacity more efficiently.

Staging the project alters the relationship between risk and reward. In Fig. 7.2, risk is defined as the maximum cumulative present-value negative cash flow while reward is defined as the ultimate present-value of the cash flow. The risk/reward ratio for the one-stage project is 1.05:1.00, for the two-stage project is 0.49:0.89, and for the four-stage project is 0.28:0.70. Stated another way, compared to the one-stage project, the four-stage project trades a 30% reduction in present value profit for a 72% reduction in risk (defined as the maximum cumulative present-value negative cash flow).

7.1.2 Startup Checklist. Implementing a CO_2 flood involves coordinating many technical and operational disciplines, both within the operator's company and between it and other parties, including CO_2 suppliers, facility construction firms, and compression rental companies.

To keep the project on schedule, an organized implementation plan is essential. Before the implementation plan is put in place, the reservoir study should be complete, the CO_2 injection plan should be defined for the project area, and flood costs (both capital and operating) should be estimated carefully. Once the flood is found to be economically viable, an implementation plan similar to the one shown in **Fig. 7.3** can be developed.

The plan shown in Fig. 7.3 is based on the experience of several operators in the Permian Basin, and it is designed so the project can be implemented within 12 months. Shorter completion times are possible if some of the steps are overlapped, but this increases risk because the operator commits funds before obtaining all major approvals and contractual arrangements. In this example, most of the detailed design work is not started until the authorization for expenditures (AFE) is approved internally by the operator and is sent out to the partners. Also, recycle compression is not scheduled for installation until 6 to 12 months after injection is initiated.

Implementing the Project Management Structure. The first task in CO_2-flood implementation is defining the structure of the implementation team: who will be in charge, who will work for whom, and what other organizations will provide help getting the project done.[11] Project management techniques (with a designated project manager) have worked well in planning, designing, implementing, and starting new projects. Defining clear expectations for each team member ensures that the implementation moves forward smoothly. Not much should transpire until the implementation team is in place to define the scope, schedule, and projected cash flows of the project.

Usually the bulk of the implementation team personnel is provided by the facilities, production, reservoir engineering disciplines, along with production operations and construction groups. Geologists and petrophysical engineers often are included on the team as subject-matter experts. (After project implementation is complete, the same skills will be needed for continued operations, but the relative importance of each discipline may change, with increased emphasis on reservoir engineering and reduced emphasis on construction management).

The team should note that cash flows are affected by more than just expenditures for new equipment and supplies of CO_2. Implementing a CO_2 flood may temporarily reduce oil production because of well work, conversions, and/or facility modifications. In addition, reduced injection, which may occur after CO_2 flooding is started, may cause a temporary reduction in oil production. These reductions in cash flow should be considered when preparing operating targets.

Obtaining Partner/Supplier Agreements. One of the first tasks for the implementation team is to make certain that the initiation of the CO_2 flood is acceptable under the current operating or joint venture agreement, and that there has been no troublesome change in scope. A CO_2 flood may require an extraordinary approval process if the parties are under a joint venture agreement that contains a change of scope provision.

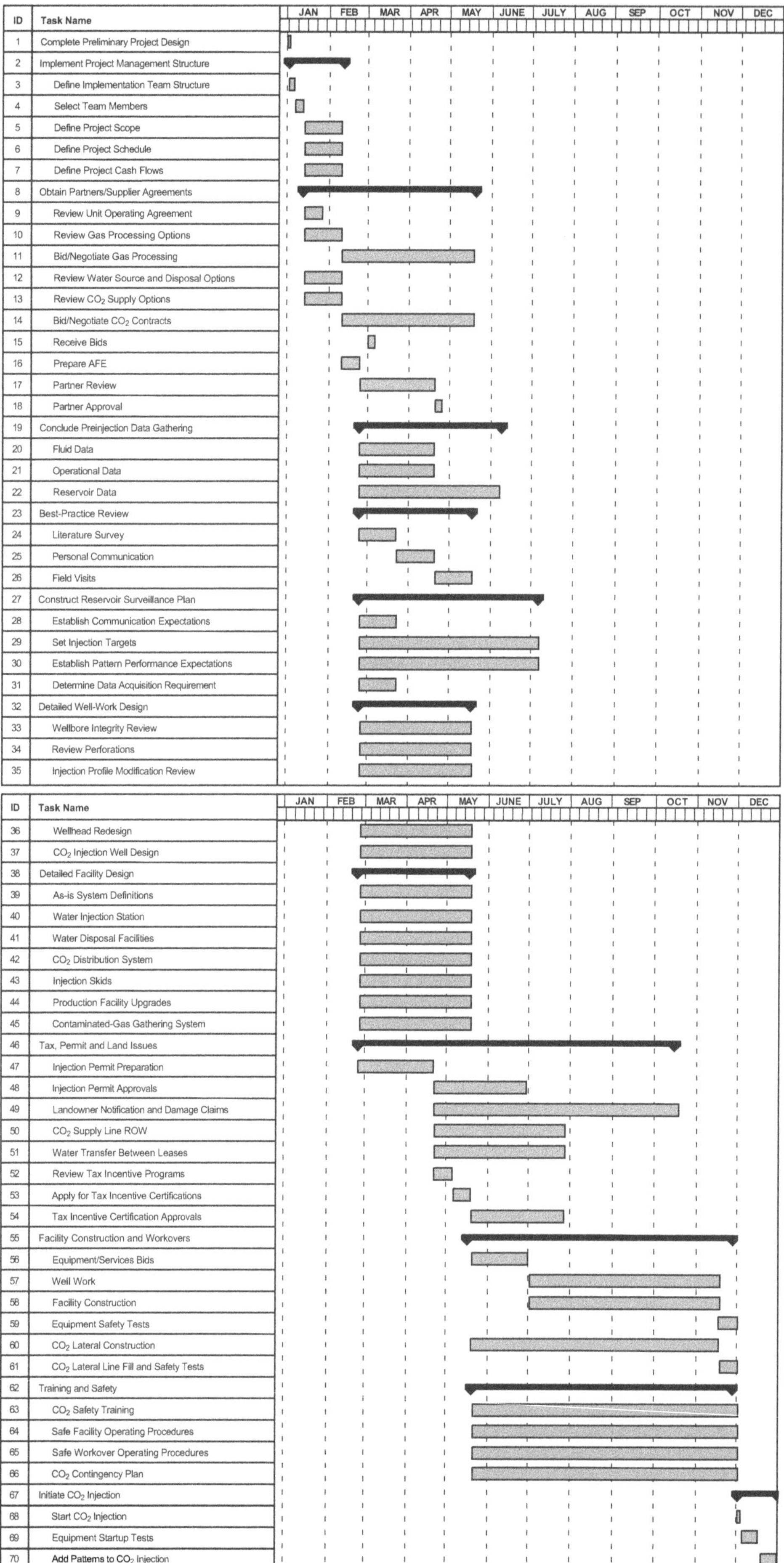

Fig. 7.3—Sample CO_2 flood implementation schedule. A good plan is crucial; it is not hard for a project to fall a year behind schedule, one day at a time.

After reviewing the joint venture agreement, the team explores gas processing options. The choices include reinjecting the gas on-site, sending the CO_2 to a gas plant that can handle CO_2-contaminated gas and enable the economic recovery of NGL's, and building a gas plant specifically for the CO_2 project. The decision to build a plant should be thought of as a completely separate investment—it may well become a bigger project than the CO_2 flood itself.

Next, the operator makes arrangements to process the produced gas. An operator who plans to reinject a project's own produced CO_2 must contact compression companies and make the lease/buy decision for compression equipment. There are several possible options: simple reinjection using existing equipment, purchasing compressors and providing in-house maintenance, purchasing compressors and signing a maintenance contract from a compression operating company, and renting the compressors. An operator who plans to ship a project's gas to an outside plant bids and negotiates for both this service and for the transportation of the gas from the CO_2 project to the processing plant (and back).

Regardless of the specific gas processing decision made, the bidding and negotiating process easily can take several months. While this is underway, the team reviews water source and disposal options. In many cases, once a CO_2 flood starts, not all the produced water can be reinjected into the zone of interest. In the early part of the CO_2 flood, operating conditions often change from using makeup water to disposing of unneeded water. Disposal options include drilling disposal wells, signing a contract with a water disposal facility, and signing an agreement with an adjacent unit that needs water.

While dealing with the gas processing questions, the team reviews CO_2 supply options, which include in-kind supply from joint venture owners (who can supply their own CO_2) and the purchase of CO_2 from a supplier (for those joint venture owners who do not have a source of CO_2). The team also analyzes the market for CO_2 and decides whether they want to arrange for short-term or long-term supplies. As with the bid and negotiation process for gas processing, the process for CO_2 supply easily can take several months by the time negotiations are completed.

Once the operator has the agreements and knows the cost of CO_2 supply and gas processing, the final AFE is prepared and sent out for partner review. It is not necessary to wait for the gas processing and CO_2 supply agreements to be finalized before completing economic evaluations and sending out the AFE, as long as the operator has strong sense of what the price of commodities, equipment, and services will be.

Partner review will take about two months if there are no significant disagreements on the technical or business issues. While the partners conduct their reviews, a host of other activities get underway: gathering preinjection data, reviewing best practices, building a reservoir monitoring (surveillance) plan, preparing detailed well-work procedures, and designing the facilities to be constructed. However, no permits should be submitted to governmental bodies until the partners have approved the project, and no construction should begin without a firm commitment for the CO_2 supply.

Data Gathering, Best-Practice Review, Well Monitoring Plans. Gathering the preinjection data probably will take three to four months to complete, depending on how much was gathered before the decision was made to go ahead with the CO_2 flood. Preinjection data allow the operator to establish baselines for future operations (see Sec. 7.2). The cost of collecting the data may be included in the preflood AFE.

A best-practice review of CO_2 operations and implementation at this time can prevent many questions and headaches later. This review should include a literature survey, personal communications with other knowledgeable people in the industry, and, if possible, field visits to other CO_2 projects.

The reservoir monitoring plan also is constructed now. The first part of this plan is to establish the requirements for data acquisition (rates, temperatures, pressures, concentrations) and the communication of the various disciplines involved. The second part is to establish injection targets and pattern performance expectations. Completing the monitoring plan may take longer than the Gantt chart shows, depending on the status of preinjection data gathering. For example, frequently, injection targets cannot be set without first running step-rate tests to determine the fracture pressure of the reservoir. If these tests are necessary and have not yet been conducted, they should be run now.

Detailed Well-Work Design. Each well is reviewed to determine its suitability for a CO_2 flood. The review looks at well history, wellbore integrity, perforation status, and injection profiles. In older wells, wellbore integrity may be compromised by wellheads and casing that is not adequate to contain the higher pressures that accompany a CO_2 flood. The condition at some wells, especially injectors, may not meet regulatory requirements. Both CO_2 injection wells and production wellheads may need redesign and/or replacement. A review of plugged wells also is advisable (if not required by regulatory bodies) because inadequately plugged wells may become a problem.

The perforation review ensures that CO_2 is injected into the same zones that feed the producing wells. In many waterfloods, when the field nears the end of its waterflood life, perforations are plugged or total depths are plugged back to solve water breakthrough problems. However, injecting CO_2 into the field causes previously immobile oil to become mobile and turns watered-out zones into oil producers once again. Hence, it is important to make sure that these potential pay zones are open at both the producing wells and the injectors. Given the history of many fields, assumptions made about plugged zones may well be inaccurate, and it is worth the effort to conduct a well-by-well review.

While conducting the perforation review, the operator examines injection profiles, injection logs, and temperature logs to determine whether any of the perforations or cement squeezes need alteration. Because of the expense of CO_2, injection losses and inefficiencies that could be tolerated in a waterflood cannot be tolerated in a CO_2 flood. While moderate deviations from the desired injection profile distribution as determined by the calculated permeability-thickness product (kh) distribution (which may not be very accurate) may be tolerated, gross deviations can lead to CO_2 losses that may make the flood not viable.

Detailed Facility Design. While subsurface professionals labor over reservoirs and wellbores, facility engineers start the detailed facilities design stage. Chap. 5 provides a more detailed discussion of facilities design.

Injection Facilities. If they are in reasonable condition, existing waterflood facilities normally can be modified for use in a CO_2 flood. Depending on the size of the project and the extent of the modifications needed (metering, separation, etc.), the detailed facility design work may be completed in as little as several months.

Because most CO_2 floods alternate water and CO_2 injection, the water injection station should be examined to determine whether modifications are needed to cope with higher pressures, and if facilities must be added to connect the station with a water disposal system (which also may need to be constructed). The produced-water tanks must have adequate capacity to allow dissolved CO_2 to escape before reinjection of the water.

In all cases, the facility engineer must design the CO_2 injection distribution system within the field. This system includes the piping that transports CO_2 from the supply terminus to the injection wells, the CO_2 injection headers, and the CO_2 metering and control system. The facility engineer also must make provision for mixing recovered and recycled CO_2 with purchased CO_2.

Production Facilities. The field's production facilities normally need upgrading because under a CO_2 flood, the field may produce more oil and gas than it has for a long time or, perhaps, ever. The task of upgrading production facilities includes designing the contaminated-gas gathering system. The existing gathering system may not have the throughput capacity to handle the volume of CO_2 that will be produced. In these cases, based on the economics of the system, the team decides whether to expand the gathering system or change the WAG ratio and inject more water and less CO_2 into the wells.

Facility engineers also work with other team members to choose the optimum operating pressure for the gas-gathering system. The higher the system pressure, the less compression that is needed to reinject the gas into the reservoir. However, increased pressure in the gathering system also increases backpressure on the wells, which reduces production. The design trade-off is between costs and production.

Tax, Permit, and Land Issues. Issues related to taxes, permits, and land can take more than half a year.

Preparation of injection permits normally takes about two months, and approval in the U.S. may take 2 to 3 months, depending on the workload at the regulating authority.

Preparing tax abatement applications also is time consuming. Before submitting the application, the operator should review and thoroughly understand the applicable tax incentive programs. The U.S. federal incentives allow for tax credits on qualified expenditures. State incentives generally allow for reduced severance taxes. Simply researching the statutes can take about a month. Writing the justification often takes another month. Tax abatement approvals can take two to three months.

Landowner notifications and damage claims also are a time-consuming issue, particularly in cases where new facilities and wells must be sited and land purchased. Running CO_2 lines across agricultural land may cause crop damage during the growing season, and provision for damage claims must be established. Use of government land may be complicated by strict regulations. Right-of-way for the CO_2 transportation lines must also be obtained, although in most cases, the CO_2 transporter does this.

In cases where the operator does not have enough water disposal capacity, water transfer agreements may be negotiated between leases, or other means of disposal may be arranged.

In Fig. 7.3, we assume that most of the tax, permit, and land issues are not initiated until the AFE is approved by partners; however, the overall time of the project can be shortened by starting work on these issues before the AFE is approved.

Facilities Construction and Workovers. In the example shown, workovers and construction of new facilities do not begin until the CO_2 and gas-processing contracts are successfully bid and negotiated.

The facilities construction and workover phase includes bidding for equipment and services, awarding the bids, doing the proper well work, building the needed facilities, and performing safety tests. All these activities can take seven months or longer, depending upon the size of the project.

While the operator is building production facilities and working over wells, the CO_2 supplier may be installing, filling, and testing the CO_2 distribution system. It is unlikely, however, that the supplier will begin any construction until the CO_2 contract is signed.

Training and Safety. Once construction begins, training and safety must be addressed. Even experienced oilfield personnel accustomed to working with H_2S must be trained on safe practices in a CO_2 environment.

Training can take place at any time while facilities are being constructed. Spreading it out over a long period of time helps reduce the stresses and demands on operating personnel, who must continue to operate the field while at the same time being involved with new construction.

In addition to safety training for operating personnel, documentation covering safe working practices in a CO_2 flood must be compiled and made available to operating personnel.

Finally, a contingency plan must be developed to cover accidental release of CO_2 to the environment. The contingency plan must provide not only for the safety of operators and contractors, but also for the safety of the general public. It must also include notification procedures for the police and appropriate emergency personnel. Texas Railroad Commission Rule 36 defines the requirements for a contingency plan very well; however, the specific regulations for the jurisdiction in which the CO_2 flood resides should be always be consulted.

Start CO_2 Injection. After all the facilities and workover procedures are completed and the training and safety training is done, CO_2 can be injected and equipment startup tests may be run as they would be in any oilfield operation. As the flood progresses and the operator gains experience, additional CO_2 injection patterns can be added, as warranted.

7.2 Preinjection Data Gathering

Before initiating CO_2 injection, certain information should be gathered about the field to fine-tune operating parameters for the flood and to establish a baseline for future operations and problem solving. Much of this information may be irretrievable or difficult to obtain once the flood starts.

A CO_2 flood will forever change the nature of both the produced oil and the oil that remains in the reservoir because CO_2 tends to extract the lighter hydrocarbons and leave the heavier ends in the ground. The gas composition also will change, and to determine whether injected CO_2 has broken through and is being recovered, it may be important to know the amount of CO_2 naturally present in produced gas.

Before initiating CO_2 injection, operators should consider gathering data on oil and gas composition, water quality, MMP, reservoir pressure and temperature, formation parting pressure, vertical distribution of injected fluids, asphaltene flocculation susceptibility, and reservoir fluid saturation. Normally, critical design data will have been collected before the project AFE is complete.

Oil and Gas Composition. Oil composition tests should include an extended PVT analysis to determine the relative composition of C_1 to C_{30+} hydrocarbons (lumping all C_{30+} molecules together as a single category), and possibly to compile a pressure-composition (P-X) diagram. These data make it possible to estimate MMP and to lump oil components into hydrocarbon pseudocomponents should EOS reservoir modeling be used. These tests typically are carried out during the evaluation phase of the project. Chap. 4 discusses more fully the use of EOS models.

Gas composition data are important for evaluating the feasibility and requirements of gas processing/NGL recovery and reinjection. These tests also provide baseline CO_2 content, which will be useful in determining the field's response to CO_2 injection and the occurrence of early CO_2 breakthrough. The H_2S content also is important. If H_2S is in the gas, it also is in the oil and water. As the GOR increases during the course of the CO_2 flood, the gas strips the H_2S out of the oil and water, such that the H_2S entering a processing plant can increase significantly. Obtaining a baseline measurement helps facilitate subsequent predictions.

Water Analysis. Water analysis provides information about the minerals and contaminants in the water. It is a good idea to identify scaling tendencies before startup because CO_2 injection will increase mineral dissolution in the formation and enhance conditions that cause scale to form. If the water quality is poor, remedial action can be taken, such as increasing settling time in the separation tanks or installing filters. In most cases, water analysis is performed at some point before the CO_2 flood.

MMP, Reservoir Pressure, and Formation Parting Pressure. The MMP for the CO_2/reservoir oil mixture may be determined through slim-tube experiments using both pure CO_2 and hydrocarbon-contaminated CO_2. This test, which also is normally carried out during the design phase, determines the pressure required to operate a miscible flood and also helps evaluate the effects of reinjecting hydrocarbon-contaminated CO_2. See Chap. 2 for more information on slim-tube tests and other methods for determining the MMP.

Reservoir pressure should be mapped. Low-pressure areas may need additional water injection before the CO_2 flood is effective, and high-pressure areas may need to be depressurized. Operating at higher-than-desired pressures wastes energy, and requires the purchase of greater amounts of CO_2 because at higher pressures CO_2 occupies a smaller volume in the reservoir. When reducing pressure, be careful not to go below the MMP because lost production may occur (see Chap. 2).

The formation parting pressure may set the upper limit for injection pressure. Most operators desire to recover oil as quickly as possible, and high injection rates increase oil production because injection drives production. Although some waterfloods operate profitably by injecting above fracture pressure, CO_2 flooding in this manner is risky. Normally, water is inexpensive and some loss of injected water out-of-zone can be tolerated, but CO_2 is much more expensive, and out-of-zone CO_2 losses are among the important causes of economic failure. The formation parting pressure can be estimated using standard step-rate tests.

Injection Profiles. Injection profile logs measure the vertical distribution of injected fluid. While these logs do not reveal where the fluid flows after it leaves the wellbore area, they do identify near-wellbore problems that should be remedied before starting the CO_2 flood. Modifying vertical distribution for injected fluids before injecting CO_2 increases profits two ways: more CO_2 is injected in the places where it is wanted, and waterflood workover operations generally are less expensive because of lower wellbore pressures.

Asphaltene Fouling. Asphaltene fouling can cause problems, and if these problems occur under CO_2 flood conditions, treatment with a chemical such as xylene is required. The susceptibility of the oil/CO_2 mixture to asphaltene flocculation can be estimated by a precipitation test in which a sample of the oil is titrated with varying amounts of solvent (CO_2) until asphaltene flocculation occurs. Note, however, that a precipitation test is qualitative. The investigator should compare the test results to those for another oil where the extent of field problems with asphaltene flocculation have been determined.

Reservoir Fluid Saturation. In reservoir fluid saturation tests, compensated neutron logs and pulsed neutron logs are used in a time-lapse logging technique to help identify breakthrough zones for which remedial work is needed.[12,13] The compensated neutron log measures the porosity of the formation that is saturated with hydrogen-rich fluids such as oil or water. A compensated neutron log might read 20% porosity before the CO_2 flood and 15% after it is underway, with the difference indicating the presence of CO_2. (Analysis of shut-in wells may be difficult because the near-wellbore saturations may be affected by crossflow). Without a baseline measurement there is no way to quantitatively estimate the amount of CO_2 or brine in the formation.

The pulsed neutron log measures thermal absorption cross section. High energy gamma rays are pulsed into the formation. Chlorine, present in most formation waters, is the most common absorber of the resultant neutron scatter, and water saturations can be determined from the rate at which these neutrons return to the tool. Differences between preCO_2 flood and in-progress readings indicate changes in water saturation. Often such changes are assumed to result from the replacement of brine in the formation with CO_2, but water saturation changes also could be caused by increased saturations of oil and/or hydrocarbon gas near the wellbore or by changes in water salinity.

Although the compensated neutron technique may be preferred to the pulsed neutron technique because the CO_2 saturation is more directly inferred, using the pulsed neutron tool sometimes is more advantageous in cased hole. The compensated neutron tool can be run in cased hole, but it is more affected by distortions caused by the thickness and quality of the casing cement. Also, the pulsed neutron tool has a smaller diameter (1.6975 in.) and can be run inside tubing. The well then may be logged while the well is flowing, thus eliminating the cross flow problem endemic to logging of producing wells while they are shut in.

An operator may use both these logging techniques (measurement of apparent porosity decreases or water saturation changes) in shut-in observation wells or in operating wells. The monitoring of CO_2 fronts can be especially critical in gravity-dominated floods because recompletion downward and away from the CO_2 front is an important part of these projects.[14]

7.3 Permits, Contracts, and Agreements

Before initiating a CO_2 flood, operators must obtain contracts, file permits, and certify their project with four groups:

- Partners and working-interest owners, who approve the injection project;
- Service companies, including CO_2 suppliers, transporters, and gas processors;
- Governmental bodies that regulate oilfield injection, freshwater usage, and air pollution; and
- National and state taxing authorities that have enacted legislation providing incentives for EOR projects.

7.3.1 Partners. The partner-approval process has three phases: justification of the project; preparation of a standard AFE; and review of the unit operating agreement and/or the joint venture agreement. The latter is particularly important because some of these agreements provide for higher-than-normal voting approval if there is a significant change in project scope. A transition from waterflood or primary production to a CO_2 flood can be considered a significant change, so the operator must review these agreements to be sure that the proper approvals from its partners are obtained. Operators who neglect this task can experience delays.

Partner approval can be obtained either before or after an agreement for CO_2 supply and transportation is bid and negotiated. Each partner is responsible for its own working-interest percentage of the CO_2 supply, and those who own a CO_2 supply or have already negotiated a blanket contract applicable to the project may desire to contribute their share of the CO_2 in-kind. If the majority of the partners already have secured a supply of CO_2, or if the CO_2 supply situation is well-known, the operator may propose the project to the partners before bidding the CO_2 agreement.

7.3.2 Service Company Agreements. ***CO_2 Suppliers and Transporters.*** In most cases, in securing a supply of CO_2, the operator contacts its partners and negotiates a "majority purchase contract" for all the owners of the unit. This process has several phases: determining the amount of CO_2 required to complete the entire project, based on the technical report; drawing up a plan for CO_2 purchases in order to maximize investment return while limiting risk; and polling the working-interest owners to find out whether they plan to supply their CO_2 in-kind or would prefer to be a party to the majority purchase contract.

The majority purchase contract covers the volume of CO_2 to be delivered to the project, not including the CO_2 that will be supplied in-kind. Although the operator negotiates this contract, all working-interest owners become parties to it. If one of the companies that commits to supply CO_2 in-kind is unable to fulfill its obligation, CO_2 typically is purchased for that partner under this contract.

The process of obtaining a majority purchase contract starts when the operator sends out a bid. The bid includes the required delivery schedule (daily contract quantity), total volume (total contract quantity), delivery point, pressure, and CO_2 quality. Suppliers and transporters typically respond to the bid with widely varying terms and conditions.

One of the more important aspects of a contract is take-or-pay provisions, which give an operator good reason to minimize exposure in case of subpar flood performance. Under these provisions, rather than committing initially to the entire volume of CO_2 that a project ultimately will require, an operator commits only to a lesser volume that will be sufficient to yield a reasonable rate of return and profit. If the project is successful, the operator arranges for additional volumes.

The topics generally covered in a CO_2 purchase and sale contract are shown in **Table 7.1.** Note that CO_2 purchase and transportation can be obtained in one contract or separate contracts, but, in general, the contracts for both will contain the same topics. Some key points about typical CO_2 purchase and sale contracts follow:

Definitions. The key definition in a CO_2 purchase and sale contract is the unit of measurement to be used for CO_2. In the U.S., CO_2 typically is priced in terms of 1,000 ft^3 (Mcf), where ft^3 is defined as the amount of CO_2 necessary to fill 1.0 ft^3 of space at base temperature and pressure. Permian Basin suppliers use a base pressure of 14.65 psia and a base temperature of 60°F, while sales from the LaBarge source field in the Rocky Mountain area are measured using a base pressure of 14.73 psia. Other countries measure CO_2 in terms of m^3 or tonne.

Other terms, such as operator, working-interest owner, CO_2, day, month, annual period, injection field, source field, and transportation pipeline, usually are defined in this section, as well.

Quantities. A typical CO_2 purchase and sale contract includes a limit on the total volume that is purchased under that contract, as well as a schedule of deliveries. A percentage of scheduled deliveries is designated as a "take-or-pay quantity," an amount of CO_2 the purchaser is required to pay for whether or not it actually is taken.

Some contracts offer provisions in which a purchaser who pays for quantities not taken has rights to accept those volumes later

TABLE 7.1—ANNOTATED CONTENTS LIST FOR A CO_2 PURCHASE AND SALE CONTRACT

Article	Title	Includes
I	Definitions	
II	Scope of Contract	Limits obligations of seller and buyer
III	Quantities	Daily contract quantities, total contract quantities, take-or-pay quantities, and/or makeup and banked quantities
IV	Prices	Product price, including cost reimbursements, if any
V	Delivery Point and Pressures	Title passage, indemnifications, delivery pressures
VI	Measurement	Who will carry out measurements, how it will be done, and what equipment will be used
VII	Measuring Equipment and Testing	Maintenance and testing of measurement equipment, plus procedures to be used in case of meter failure
VIII	Quality	Product specifications
IX	Accounting	Payments, failure to pay, auditing, refunds
X	Force Majeure	Addresses events or effects that cannot be reasonably anticipated or controlled
XI	Notices	How to formally contact buyer and seller
XII	Successors and Assigns	Terms binding successors to the contract
XIII	Term	The period of time the contract is in effect
XIV	Governmental Regulation	The laws, rules, regulations, and proclamations applicable to the contract
XV	Miscellaneous	

without paying again. In some contracts, purchased quantities that exceed the daily contract quantity may be "banked" to prevent paying take-or-pay penalties in the future.

Pricing. Carbon dioxide is priced several ways. The price may remain fixed, regardless of changing conditions; it may escalate according to a schedule contained in the contract; or it may be tied to the price the operator receives for its produced hydrocarbons, to the producer price index, or to overall oil or gas prices. All these pricing schemes are in use at the time of publication.

Delivery Point and Pressures. The point of custody transfer must be defined clearly. In some cases, the supplier or purchaser may incur unexpected third-party meter or pump charges. For example, if custody is transferred immediately upstream of a third-party facility (e.g., a meter), the third party may charge the customer a fee that the customer did not plan for. The delivery pressure also should be sufficient to enter the purchaser's facility.

Measurement. Most custody transfers of large volumes of CO_2 are measured through orifice meters, which generally are constructed and installed in accordance with the standards embodied in Chap. 14 of the API Manual of Petroleum Measurement Standards.[15]

An orifice meter measures the flowing pressure and the pressure drop across an orifice. When these readings are combined with temperature (as determined by an online thermometer accurate to $\pm 0.1°F$), density (as determined by an online densitometer or by sampling the composition of the CO_2 and using EOS calculations), and the atmospheric pressure (often based on 14.65 psia at sea level, corrected to actual elevation using the procedure given in the Chap. 14 of the *API Manual of Petroleum Measurement Standards*[15]), the volumetric flow rate can be calculated by a flow computer. In general, if meters are well maintained, their measurement accuracy will be within 1%.

Quality. Standard pipeline quality specifications for CO_2 transported to the Permian Basin of west Texas are given in **Table 7.2.** The major CO_2 source fields supplying the Permian Basin market (McElmo Dome, Sheep Mountain, and Bravo Dome) at the time of publication produce CO_2 that contains only 1 to 2% impurities. The CO_2 shipped from the processing plant at the Rangely field to the Rocky Mountain market is of similar quality.

Gas Processing. If NGL's are recovered, gas processing is carried out at a liquids recovery plant. Such plants may separate only the NGL's from the gas, or they may also separate natural gas from the CO_2. If neither NGL nor natural gas recovery is deemed profitable, the operator reinjects the entire produced mixture of CO_2 and natural gas. If simple reinjection is decided on, the operator buys or rents compression to accomplish the task.

A CO_2 recycle contract with an NGL processing plant where NGL's are recovered and methane is separated from CO_2 looks very similar to the CO_2 purchase and sale contract already discussed. The fee or price that the gas processor charges may be fixed or tied to natural gas or NGL prices. Moreover, the agreement may state that the plant will keep a portion or all of the NGL's or the natural gas in return for processing. Normally, the agreement also provides for the return of the CO_2 or the CO_2 /natural gas mixture to the operator at a specified pressure.

If the operator and its partners decide to reinject the entire produced gas stream instead of processing it, and if they decide to rent compression, that rental contract generally includes a simple monthly rental fee that is either fixed or inflated according to a producer price index, and another fee that is tied to energy usage (either the cost of electrical power or natural gas), if the operator does not supply the utilities.

7.3.3 Permits and Regulations. There are four types of permits that may be needed: injection, water use, air, and CO_2 transportation. Obtaining the necessary permits should be begun early in the planning and implementation process. If an operator is unfamiliar with permitting procedures, the operator should contact the appropriate regulating body for help.

Injection Permits. Injection permits for CO_2 are necessary because in most cases, the operator has not included CO_2 on the initial permit as a permitted injection fluid, along with brine or

TABLE 7.2—STANDARD PIPELINE SPECIFICATIONS FOR TRANSPORTATION OF PERMIAN BASIN CO_2	
Carbon dioxide	Containing at least 95.0 mol% carbon dioxide
Water	Containing no free water and not more than 30.0 lbm of water in the vapor phase per MMcf of carbon dioxide
Hydrogen sulfide	Containing by weight not more than 20 parts hydrogen sulfide per million parts carbon dioxide
Temperature	Not exceeding a temperature of 120°F
Nitrogen	Containing not more than 4.0 mol% nitrogen
Hydrocarbons	Containing not more than 5.0 mol% hydrocarbons
Oxygen	Containing by weight not more than 10 parts oxygen per million parts carbon dioxide

fresh water. Another consideration in getting new injection permits is that CO_2 is less dense than water. In order to reach the same bottomhole pressure, which is key to maintaining throughput, a higher surface pressure is necessary. Where permits have been based on water injection at a given maximum surface pressure, to maintain the same bottomhole injection pressure, the operator must determine the maximum wellhead injection pressure for the less dense CO_2.

The injection permit process is time consuming and should be undertaken as early as possible. Some information that U.S. regulatory bodies may require from the operator are:

- State or U.S. Environmental Protection Agency (EPA) forms that contain project and individual well information such as depth of injection, surface injection pressure, and injection rate.
- An "Area of Review," which investigates the wellbore integrity of all wells within a certain radius of each injector (usually one mile).
- Old injection permits.
- A determination of the depth to which usable fresh water must be protected in the injection area (which may be carried out by a regulatory body such as a natural resources conservation commission).
- A list of the owners of the surface land.
- A copy of the electric log for each well.
- An "affidavit of publication," in which a local newspaper attests that public notice of the proposed permit was given in its publication.

It may take several weeks or months to assemble the needed information and provide time for public comment, and recent experience indicates that at least two months should be allowed for the approval process at the regulating body, assuming the application is in order. Contacting the regulatory body to determine its current job load will help the operator plan its business.

There are several practices that can simplify CO_2 injection permit processes:

- Permit the maximum reasonable rates and pressures, to cover all anticipated conditions. Repermitting should be minimized.
- Permit all wells in a field for the same pressures and rates. Having identical permits for all wells simplifies compliance monitoring because it is easier for the operating personnel to remember the limits. It may not be possible to get identical pressures and rates in every jurisdiction because some agencies require a step-rate test for each well to set the maximum surface injection pressure or because the fracture parting pressure varies greatly in the field. In those cases, permitting different pressures may be unavoidable. Wherever a specific step-rate test is not required for each individual well, however, and where fracture parting pressures are consistent, the operator should determine the maximum surface pressures for both CO_2 and water at which the reservoir will not be fractured and permit those rates for each well.
- Permit the entire unitized interval. Oil response to the initial injection program may indicate that expanding the initial injection interval is profitable. Take advantage of any means to minimize or eliminate repermitting.
- Put permit information in a visual format such as a permit map and display it on a wall on-site. A permit map **(Fig. 7.4)**[11] provides a quick check to see that every injector has a permit. Color-coding can designate special limits.

Water-Use Permits. Normally, initiation of CO_2 injection reduces or eliminates the need to inject fresh water as a makeup volume in the oil field. Where regulating bodies are operating under a "use it or lose it" philosophy, existing freshwater injection permits may be invalidated when freshwater injection ceases. In such cases, the operator is forced to reapply for freshwater injection permits at the tail end of the CO_2 project, when additional volumes of injectant may be needed to replace reservoir fluids that are not reinjected.

Air Permits. There are several factors to consider when evaluating the need for air permits. First, the CO_2 flood may require the installation of recycle compressors, and if these compressors are gas-fired, they are a point source of air pollution. Using electric compressors eliminates this particular need to review air permit requirements. A second point source for emissions at most compression facilities is the glycol dehydrator's reboiler. Nonpoint source emissions from valves and mechanical connections also may require permitting in some critical nonattainment areas. In

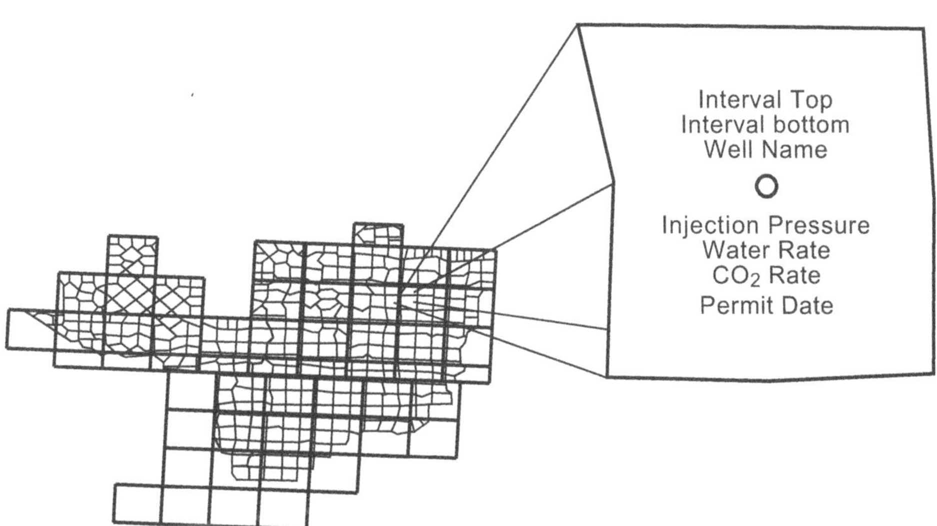

Fig. 7.4—Permit map for a west Texas CO_2 flood (after Ref. 11). Displaying a map of this kind on a wall provides a quick check to see that every injector has a permit.

some cases, the operator may be required to replace mechanical connections with welded ones.

CO_2 Transportation Permits. Operators of CO_2 floods must determine whether or not they will own or operate the CO_2 pipeline lateral that brings the CO_2 to their fields from the CO_2 supplier's mainline. Operation of a CO_2 pipeline in the U.S. must follow the CFR *49—Part 195, Hazardous Liquids Pipeline Safety Rules.*[16] This regulation also is known as DOT 195 because the U.S. Dept. of Transportation (U.S. DOT) administers the regulations.

DOT 195 applies to both interstate and intrastate transportation, from the outlet flange of the source field to the inlet of the injection system at the use field. Among the provisions in the regulation are drug testing, operator qualification and training, operations manuals, a one-call system, a damage prevention program, and air patrols. There is a learning curve associated with operating a pipeline facility in compliance with these regulations, and oilfield operators should seriously consider whether they want to expand their operations into the pipeline business.

7.3.4 Government Incentives. This section addresses tax incentives offered by the U.S. federal and state governments. At this time, significant statutory tax incentives for CO_2 EOR projects are offered only within the United States. (The Canadian provincial government of Saskatchewan did approve new incentives for support of the Weyburn project, which included foregoing the 7% sales tax on the purchase of CO_2 and lowering the effective royalty rate.[17])

U.S. Federal EOR Tax Incentives. The EOR tax credit is a 15% credit defined in Section 43 (c) (2) (A) of the Internal Revenue Code of 1986. Although a tax credit can be repealed at any time, this one has been in place for more than a decade, and it is a significant factor in promoting CO_2 flooding and, therefore, bringing its economic benefits to the public.

Costs that qualify for the EOR credit include the tangible and intangible costs associated with installing the flood, CO_2 purchase costs, and operating costs associated with CO_2 injection. Injection into the project area must have begun on or after 1 January 1991, and the project area may be either an entirely new flood or an expansion of an existing flood. The tax credit phases out as oil prices rise above $28/bbl (indexed to inflation with a base year of 1991).

When the 15% credit is taken, the remaining 85% of the qualifying costs are expensed or depreciated normally. In addition, the credit cannot be taken during years in which an owner is in an alternative minimum tax situation, although it can be banked in these years and claimed in the future.

To receive the credit, the operator first submits to the U.S. Internal Revenue Service (IRS) a petroleum engineer's statement that more than an insignificant volume of oil will be produced because of CO_2 injection, and the owners then claim a credit on their tax returns. The petroleum engineer's statement should be that of a registered professional engineer and should identify and describe the project, using maps that define the project area. It also should define the tertiary recovery method, give the background on the field, provide some details about project design, show production history, and estimate the amount of oil the project will recover.

Because the engineer's statement must follow strict guidelines, the engineer should read the IRS code carefully. Appendix E gives an example of an engineer's statement found in the public record.

U.S. State EOR Tax Incentives. At the time of this writing, nine states offer EOR incentives; however, like the U.S. EOR tax credit, these incentives can be repealed. These states include Arkansas, Colorado, Mississippi, Montana, New Mexico, North Dakota, Texas, Oklahoma, and Wyoming. Because most state EOR plans are similar, and because CO_2 flooding is most active in Texas, we will use Texas' Statewide Rule 50 to discuss state plans here. The most common difference between the Texas statutes and the others is that most states qualify only the incremental oil produced for the severance tax reduction, but in Texas, all the oil in the project area qualifies for the tax reduction.

Under Rule 50, there is a 50% reduction in the severance tax for projects that begin CO_2 injection after January 1, 1989. Projects may be entirely new or may be areal or vertical expansions to an existing CO_2 flood. For example, in 1994, the Texas Railroad Commission (TRRC) approved a vertical expansion of the Denver Unit CO_2 flood.

As mentioned earlier, in Texas, the reduction in tax applies to *all* the oil produced from the project area—not just the incremental volume. The severance tax reduction lasts for a period of 10 years from the time that the TRRC certifies that the project is producing incremental oil.

Obtaining the severance tax incentive in Texas is a two-step process. First, the operator must submit an H-12 form and supporting documentation, designating the project area and establishing the production decline curve. After the EOR project has commenced, the operator submits an H-13 form and supporting documentation, showing that positive oil production response above the previously established decline curve has occurred. The H-13 submission must be made within five years of the approval of the H-12 application.

For vertical expansions such as the Denver Unit Transition Zone project, the oil from the expansion project is commingled with oil from a nonqualifying project. In this case, the TRRC determined that oil production that is greater than the established decline curve will be taxed at half the normal rate.

There is one caveat: Do not begin injection of CO_2 until the H-12 form is approved. The Texas statute specifically states that for the oil to qualify for severance tax reduction, the H-12 form must be approved (and not simply submitted) before beginning injection. Careful reading of every state's statutes is always prudent.

7.4 Monitoring a CO_2 Flood

The expense of CO_2 makes a flood monitoring program vital. The improved reservoir management that will result from a CO_2 flood monitoring plan also may provide a substantial increase in production over that of waterflood operations. **Fig. 7.5** illustrates this phenomenon at the Means San Andres Unit.[8]

In general, good waterflood practices are good CO_2 flood practices, although there are some additional concerns unique to the CO_2 process because CO_2 costs more than water; CO_2 is more mobile than water in displacing oil; and, when a CO_2 flood follows a waterflood, the largest portion of the reservoir's oil already may have been removed by the waterflood. Because of these factors, the margin for error in a CO_2 flood is small, and a good monitoring program is necessary to optimize performance.[11]

Regardless of the care with which the initial design work is undertaken, actual performance never exactly follows the initial forecast. Because response is not 100% predictable, it is most important that a monitoring plan be in place *before* the flood commences. This way, deviations from the expected performance can be detected as early as possible and corrective measures taken. Early flood results also may indicate opportunities to enhance production that were not anticipated at the start of the project.

7.4.1 General Considerations for Monitoring. Unlike some other oil and gas production activities, monitoring work is difficult and tedious—and it lasts the life of the flood. The task of collecting, organizing, and analyzing large amounts of data is time consuming and often is postponed when unforeseen additional work arises. It is all too easy to forget that injecting CO_2 is the most costly part of the EOR process, and injection management is the most important part of making a CO_2 flood profitable.

While most operators use some type of monitoring for their waterflood operations, monitoring must be much more scrupulously applied to CO_2 flooding to ensure that maximum use is made of purchased and recycled volumes. A successful monitoring or surveillance process includes the elements shown in **Fig. 7.6.**

As Fig. 7.6 shows, there are well defined tasks that must be accomplished in a good monitoring program:

- Assemble the monitoring team and appoint a team leader.
- Gather field data.
- Define an operating philosophy.
- Set expectations.
- Collect data, and identify and analyze variances between actual performance and expectations.
- Take corrective actions and revise expectations, as needed.

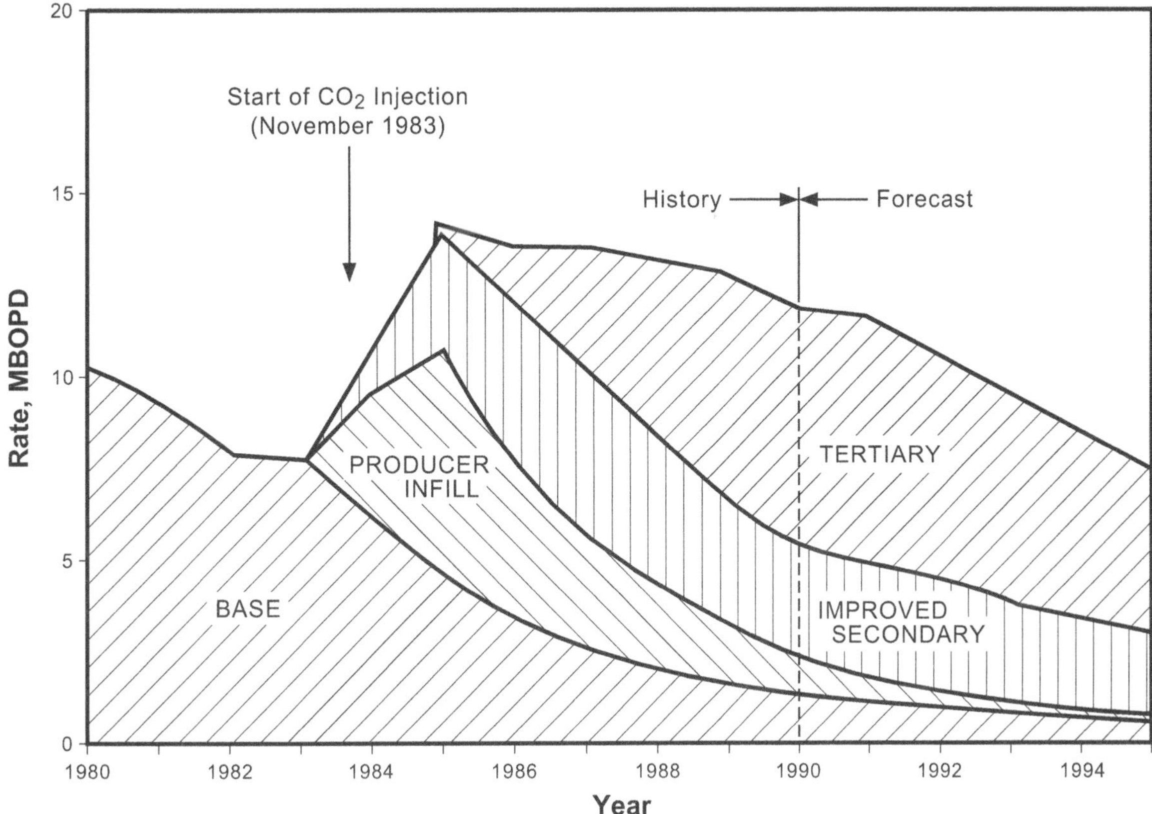

Fig. 7.5—Response components of the Means San Andres Unit CO_2 injection project area (after Ref. 8). Good monitoring and reservoir management can improve production from both CO_2 and waterflood operations.

Because the flood monitoring process addresses the needs of both operations and engineering, it requires participation from both groups throughout the life of the project. Surveillance must be tied to field activities because if data gathered through monitoring do not play an immediate role in optimizing field activities, the value of the effort may be lost.

It is also necessary to establish a monitoring plan for equipment performance, but because this task essentially is like that of waterflood operations, we do not discuss it in this monograph.

7.4.2 Monitoring Prerequisites. Before any monitoring activities begin, the operator should designate who is responsible for establishing an operational philosophy, setting expectations, collecting background data, and monitoring the flood's progress. Thakur[18] gives five keys to a successful waterflood monitoring plan, which also apply to CO_2 flooding:

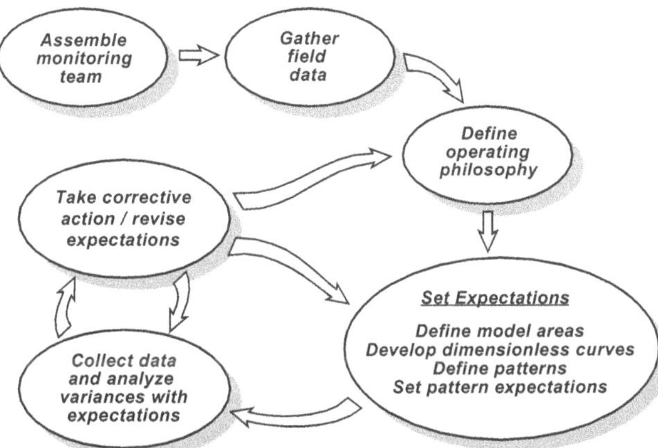

Fig. 7.6—Carbon dioxide flood monitoring process. Injection management is the most important part of making a CO_2 flood profitable.

- It should involve all functional groups.
- It must be flexible.
- It must have management support.
- It must have the support of field personnel.
- It must require periodic review meetings in the field offices.

7.4.3 Components of the Monitoring Program. ***The Monitoring Team.*** The monitoring team should have an assigned leader and should include operations personnel (both lease operators and operations foremen), plus production, reservoir, geological, and petrophysical engineers, as well as the facilities engineer, when necessary. The team should meet on a regular basis—monthly to quarterly, depending on the flood and the stage it is in.

The Field Data. The monitoring team begins by compiling and organizing the data required for everyday reservoir management. This should take place well before the flood starts. After startup, many decisions must be made quickly, and these decisions can be made more efficiently if the data already are organized. Background data should include:[19]

- Maps, including geological maps of the fields, structure maps, and isopach maps.
- Individual well data, including mechanical well sketches and a petrophysical summary for each well (pay intervals, pay thickness, formation tops, and an evaluated openhole log over the entire unitized interval).
- Geological data, including rock properties and core data.
- PVT and other fluid analyses for the oil, gas, and water.
- Reservoir pressure data.

The Operating Philosophy. The operating philosophy determines what work will be carried out as part of monitoring, and what the practical objectives are of that work. Here is a useful philosophy compiled from several sources:[11,20,21]

- *Account for all injected and produced fluids.* The most rudimentary troubleshooting is impossible without this information.
- *Complete injectors/producers in all floodable pay.* Sometimes zones plugged during previous waterflood operations are good CO_2 flood candidates.

• *Maintain reservoir pressure above MMP (miscible floods).* The ability of CO_2 to reduce the residual oil saturation is reduced when the pressure of the displacement front drops below the MMP.

• *Maximize injection below fracture pressure to maximize production.* The reservoir should be processed as quickly as possible. Care must be taken, however, not to exceed the formation fracture pressure while injecting because fracturing can create loss zones that waste CO_2 and severely impact project economics.

• *Pump off producers.* Maximum throughput is the objective. In low-permeability reservoirs, processing the reservoir as fast as possible often is required to achieve adequate financial returns. On the other hand, several reasons to maintain a fluid level above the downhole pumps do exist: (1) there are theoretical advantages to maintaining backpressure to improve miscible displacement; (2) increased pressure in the wellbore may keep gas in solution and increase pump efficiency; and (3) decreased pressure drop across the reservoir can decrease asphaltene and/or paraffin deposition.

• *Maintain good water quality.* Good water quality is a prerequisite for maintaining maximum injectivity because waterborne sediments can plug formations. The need for filtration is highly dependent on the reservoir's characteristics. A high-permeability, well-consolidated sandstone reservoir with large pore throats may be adequately filtered using no more than water tanks sized to provide adequate settling time. Other reservoirs may require installation of filters along with periodic water sampling and measurement.

• *Obtain good vertical distribution of injected fluids.* Attempt to flood all the pay. Good practice includes profile control to ensure the best vertical distribution practical. One way of accomplishing this is by using alternating cycles of CO_2 and water according to a predefined WAG ratio (see below).

• *Balance flood patterns.* Attempt to flood-out the patterns evenly so that wells and hydrocarbon reserves are not abandoned prematurely. Pattern balancing is difficult and tends to conflict with maximizing injection and/or production, so practical compromises are necessary. Assigning targets that restrict production in some wells may be considered in cases when irregular patterns or directional permeability issues exist. Streamline or other types of modeling may be used to determine pattern allocation factors.

• *Monitor WAG ratio and cycles.* Because CO_2 flows more easily that oil, it tends to break through more porous formations and leave oil behind. Using alternating cycles of CO_2 and water (which flows less easily than oil) helps maintain a more uniform CO_2 front and prevents CO_2 from being wasted. WAG ratios and CO_2 slug size are part of the design of the flood, and actual performance in the field normally dictates necessary changes in design values.

• *Limit gas production to the recycle capacity.* There is a limited amount of compression available at any one time. Limiting CO_2 injection to control gas production prevents having to shut in oil production.

• *Maintain the stability of the CO_2/oil interface.* This is a concern in gravity-stable vertical CO_2 floods or in horizontal floods where gravity override may occur.

• *Maintain a continuous optimization program.* Oil prices, tax incentives, surface processing bottlenecks, and other conditions change, and the entire process demands continuous flood management optimization.

7.4.4 Setting Expectations. Setting expectations is an important part of an effective monitoring program. **Fig. 7.7**[19] diagrams the process. The approach is to:

• Collect and analyze basic data.

• Define prototype areas so that a few well-defined areas can serve as models for the entire field, which greatly simplifies forecasting and planning.

• Develop dimensionless curves by plotting output from the reservoir simulation of the prototype areas in dimensionless form (as described in Chap. 3 and below).

• Define injection patterns.

• Set performance expectations for each pattern using pattern injection rates, OOIP, and the dimensionless recovery curves to forecast response.

Collect and Analyze Basic Data. The process shown in Fig. 7.7 starts with gaining a geological understanding of what the rock is like and where the oil is; gathering field history data that can be used to construct injection patterns; and devising a development plan. These activities normally are accomplished during the design phase of the project.

The importance of understanding the geology and petrophysics of the field cannot be overstated.

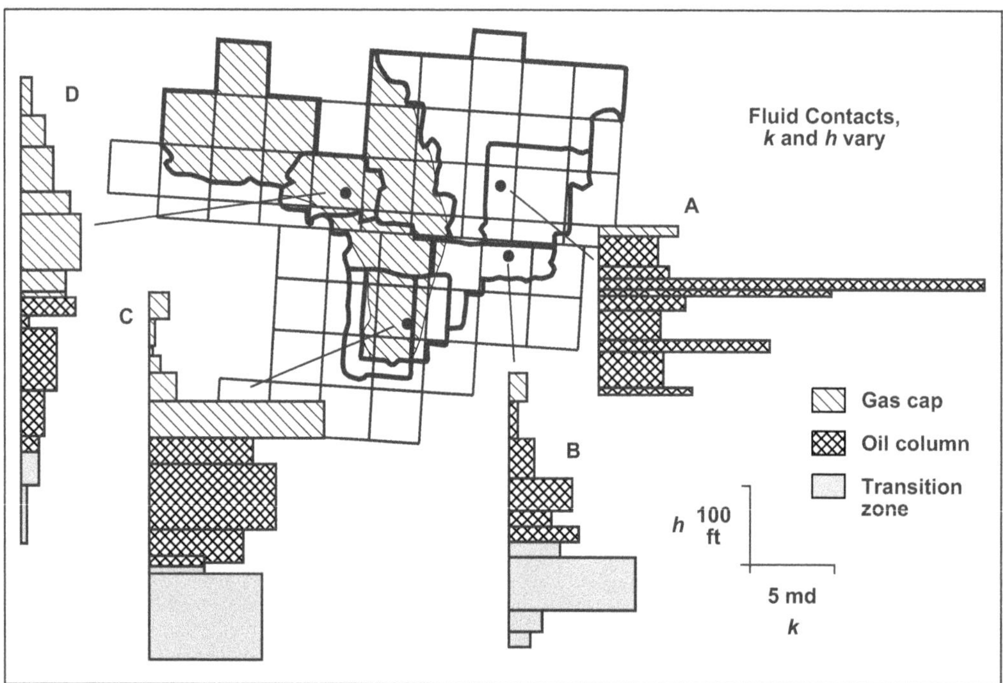

Fig. 7.7—Setting expectations for a CO_2 flood. A large Permian Basin field is divided into 11 prototype areas. Within each prototype area, both the geology and development history must be similar. The length of the horizontal bar represents permeability (*k*), and the hatching indicates water, oil, or gas. Between prototype areas, the permeability, thickness (*h*), and fluid contacts will vary (after Ref. 19).

Define Prototype Areas. A prototype area represents a geologically similar part of the field. Fig. 7.7[19] shows four prototype areas within a Permian Basin field that had a total of 11 prototype areas. The stacked boxes represent permeability vs. depth, and the cross-hatching denotes water, oil, or gas. For example, Prototype A shows a thief zone, a narrow formation of great permeability through which CO_2 can flow too rapidly, and the B prototype shows the transition zone between oil and water.

Fig. 7.8[19] shows a completed prototype simulation model for one area. The nine-spot design of injector and producers allows one-fourth of the area to be used in the model, making it easier to run. The design shown here was based on history data from the previous waterflood.

As simulation hardware, software, and technique improve, engineers and geoscientists will undertake more full-field simulations, which may be thought of as the ultimate way of breaking a field into prototype areas. The techniques discussed in the following sections apply to both prototype and full-field studies.

Develop Dimensionless Curves. The output from the simulation model for each prototype unit may be presented as a dimensionless curve. On that curve, the vertical axis is the produced oil, expressed as a fraction of OOIP, and the horizontal axis is the cumulative EOR injection (including both water and CO_2 in the WAG cycle), expressed as a fraction of HCPV. The shape of the dimensionless curve depends on the WAG scheme used in the simulation.

EOR production for a CO_2 flood is defined as the oil production that is greater than that which would have been produced by continued waterflood operations. The forecasted EOR production is easy to determine: simply subtract the forecasts of continued waterflood operations from those of the CO_2 flood.

The use of dimensionless values in the curve shown in **Fig. 7.9**[19] is what allows the prototype model to be scaled easily to different patterns and changing injection rates.

Define Injection Patterns. Injection and production are allocated to patterns by geometry, by streamline modeling, or by other, less rigorous methods. As a matter of personal preference, some engineers allocate flow streams to producers, while others allocate to injectors. This discussion uses injector-centered patterns because the field studied had inverted nine-spot patterns.

Set Expectations for Each Pattern. The next step is to set expectations for each pattern, and, later, to monitor the results continuously and reset expectations as the flood progresses. The dimensionless curves described above (with cumulative oil recovery as a fraction of OOIP on the y-axis and cumulative CO_2 and water injection as a fraction of HCPV on the x-axis) are critical to this work.

At the beginning of a CO_2 project, simulator projections for total cumulative oil production can be combined with an oil-production regression fit for waterflood alone to separate out that component of the EOR which is caused entirely by CO_2. This information can be used to monitor the effectiveness of CO_2 injection. Including both water and CO_2 on the x-axis of the dimensionless graphs accounts for the continuation of water injection during the WAG process and rationalizes the curves.

As the project progresses (and this topic could also fall under the section on monitoring below), actual oil production can be compared with simulator predictions—again using a waterflood regression fit to correct for the effect of water—to monitor flood effectiveness.

For example, Fig. 7.9 shows the simulator's forecast of cumulative oil recovery as a result of both CO_2 and water injection, and it shows the actual results achieved. **Fig. 7.10** shows the same actual production data as Fig. 7.9, plus *the waterflood regression fit*, which is the projection of cumulative waterflood recovery if CO_2 were not being used.

The solid line curving gently to the right is the waterflood regression fit. To determine the EOR production resulting from CO_2, the waterflood regression fit must be subtracted from total cumulative oil production. If this is not done, the CO_2 flood may appear far more effective than it actually is; for example, CO_2 may be injected at faster rates than water was during the waterflood, and this higher throughput rate increases production regardless of whether or not miscible benefits are being realized. In Fig. 7.10, the open-diamond line represents EOR produced strictly as a result of CO_2 injection and shows that this flood is quite successful thus far.

Note that all of these data are handled in dimensionless form, i.e., injection is stated as a fraction of HCPV and recovery as a fraction of OOIP. Making these dimensionless calculations instead of expressing waterflood decline as BOPD on a semilog scale helps distinguish between problems associated with injection targets, formation damage, or mechanical difficulties and those associated with CO_2 process difficulties.

7.4.5 Running the Monitoring Program. The ongoing flood monitoring program consists of collecting data, identifying and analyzing variances, and taking corrective action. These activities are the main body of the monitoring program diagrammed in Fig. 7.6.

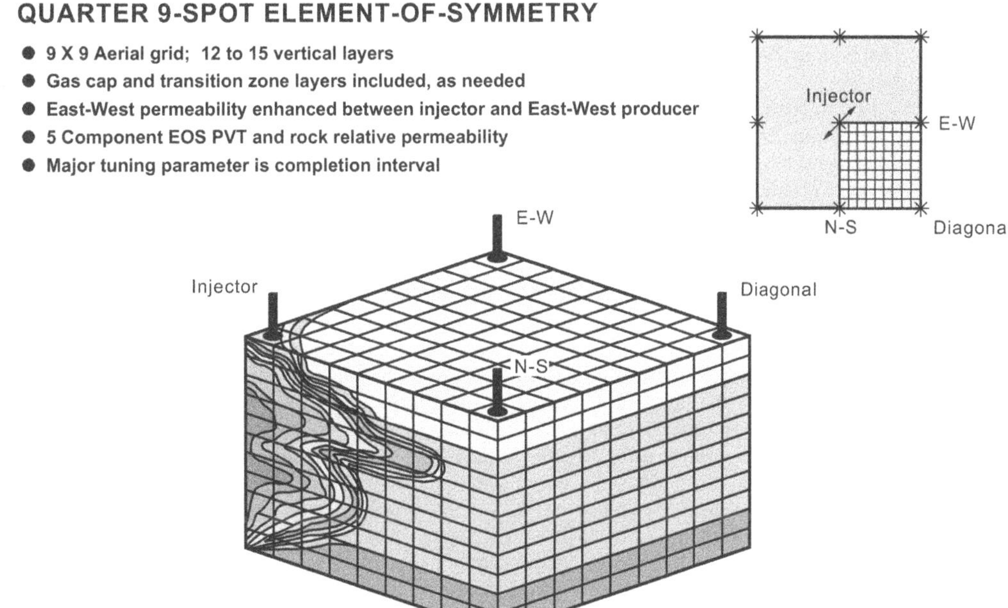

Fig. 7.8—A completed prototype simulation model for one area. One-quarter of a nine-spot element-of-symmetry model can be used to develop expectations for a large field (after Ref. 19).

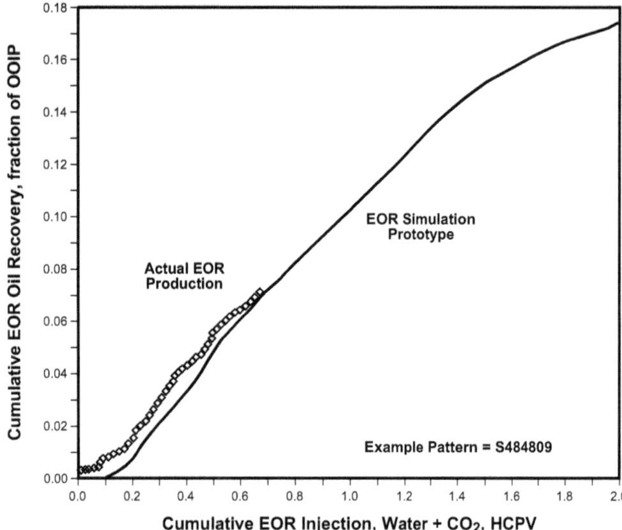

Fig. 7.9—Simulated dimensionless EOR production compared with actual EOR production. The one-quarter nine-spot pattern illustrated in Fig. 7.8 was used to develop the simulated forecast. Here the actual results (open diamonds) verify the simulator's projections (after Ref. 19).

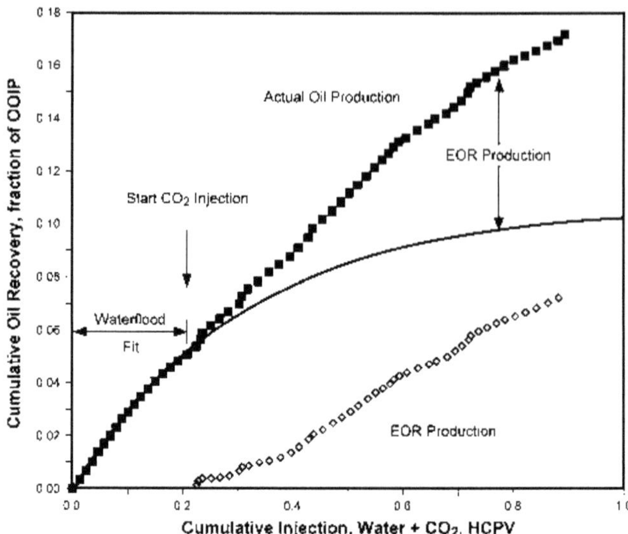

Fig. 7.10—Using dimensionless cumulatives to set and monitor expectations, based on the actual EOR production shown in Fig 7.9 (after Ref. 19). The solid line curving gently to the right is the projected recovery for the waterflood alone, and the solid diamonds represent total cumulative oil production. Carbon dioxide EOR production (the open diamonds) is obtained by subtracting the projected waterflood recovery from total oil production.

Clear lines of communication and good working relationships between field and office staffs are absolutely essential. The following sections present one approach to monitoring which relates to our earlier discussion on the operating philosophy[11] (see Sec. 7.4.3).

Meter Injected and Produced Fluids. All injection wells should have meters. The greatest project expense is purchase of CO_2, and one must accurately measure its injection into each well. Surface pressure measurements should be made downstream of any chokes so that the best available data will be available to estimate bottomhole pressures and ensure that they remain below the formation parting pressure.[21]

It is not always necessary to measure the CO_2 content of the produced gas; in some cases, the GOR is low enough that CO_2 sensors are not needed, and CO_2 breakthrough can be detected simply by noting a rise in the produced GOR.

Compositional analysis of the recycle gas should be performed on a regular basis because minor fluctuations in gas composition can make significant changes in the gas compressibility factor, especially when a gas is near its critical point.[2] An inaccurate estimate of the compressibility factor will lead to inaccurate flow measurement. Producing wells can be production-tested anywhere from once every quarter to four times a month.[2,22] Most operators test their wells on a monthly basis.

Conduct Regular Cross Section Reviews. These reviews help assure that all the floodable pay is open and that injectors and producers are completed in correlating zones.

Measure Injection Profiles Periodically. Injection profiles reveal where the injectant is flowing. If the injectant is flowing into a zone where it is not wanted, the situation sometimes can be corrected by squeezing off the zone with cement or injecting gel polymers into the formation. Some operators run one survey per year per injector.[23]

Measure Pressures Periodically. Reservoir pressure data are taken twice a year at the Ford Geraldine Unit;[24] followup tests are conducted at the Means field once a year;[21] and bottomhole pressures are monitored quarterly at the Hansford Marmaton Unit.[2] In the Marmaton Unit, streamline modeling is used to allocate production and injection to each pattern and fractional pattern. The bottomhole pressure determined from periodic measurements not only helps make sure that reservoir pressure remains above the MMP, it also helps verify that the injection requirements derived from the streamline models are accurate. When changes in bottomhole pressure do not conform to the pressure that would be expected based on allocated production and injection, the streamline model can be fine-tuned to represent the reservoir more closely. This, in turn, helps determine injection requirements more accurately.

Conduct Step-Rate Tests. These tests determine the maximum bottomhole pressure at which injection can occur without fracturing the reservoir, and operating at the maximum pressure maximizes throughput. Step-rate tests performed at the Ford Geraldine Unit[23] are used to derive maximum injection pressure targets for each well. Continued step-rate testing may show increased parting pressures as the reservoir pressure increases; in other words, the reservoir is less likely to crack as reservoir pressures go up over time.

Monitor Fluid Levels. A fluid-level control program for pumping wells is necessary to make sure that wells are pumped off, a condition necessary to maximize fluid production. Fluid levels are measured monthly for pumping producers at the Ford Geraldine Unit, and tubing pressures for flowing wells are reported twice each month.[24]

Balance Injection Patterns. A pattern balancing program helps improve the advance of the flood front from multiple injection wells toward producing wells. For example, at the Means San Andres Unit, a balancing program measures the IWR's for every injection pattern in the CO_2 project. The program allocates production injection to each pattern on the basis of each well's location and the volumes injected and withdrawn. Fluid level monitoring allows the producer to make periodic adjustments in injection targets.[21]

Pattern balancing has also been used at the Hansford Marmaton Unit[2] and at the Hanford flood.[22]

Monitor WAG. It is important to keep records of how much CO_2 and water is injected in each cycle so the actual WAG ratio can be compared to the desired CO_2 injection policy. Many operators have monitored and changed WAG ratios and half-cycle sizes to increase profitability.[4,25,26]

Maintain Good Water Quality. Where water quality is an issue, a periodic water sampling and measurement program is a must to assist the operator in performing proper maintenance on the filtration system.

Periodically Compare Performance With Expectations. The progress of the flood must be compared to the original expectations. Common parameters of comparison include oil production rate, CO_2 and water injection rates, CO_2 injected vs. EOR oil produced, and total EOR injection vs. EOR oil produced.

One of the key activities in performance comparison is frequent monitoring of the project on an individual pattern basis.[9] When managing on a pattern basis, it is important to take observed differences in geology into account to make sure that patterns that are aggregated within a prototype actually have similar geology.[27]

Take Corrective Action and Revise Expectations. After all the monitoring data are collected, corrective actions are taken as warranted, and/or the model is updated as needed.

In reviewing our earlier discussion on setting expectations, (see Sec. 7.4.4) one can see how the geologic field plans and field history are used toward this. In pattern analysis (the key element in ongoing monitoring), expectations are plotted against results to fine-tune the flood, uncover field process problems, and refine the model. All this continually unfolding information cycles back into the process to allow further prediction and enable economic success.

Nomenclature

k = permeability, L^2, md
h = thickness, L, ft

References

1. Moritis, G.: "EOR Survey And Analysis," *Oil & Gas J.* (20 April 1998) 49.
2. Flanders, W.A., Stanberry, W.A., and Martinez, M.: "CO_2 Injection Increases Hansford Marmaton Production," *JPT* (January 1990) 68.
3. Burbank, D.E.: "Early CO_2 Flood Experience at the South Wasson Clearfork Unit," paper SPE 24160 presented at the 1992 SPE/DOE Eighth Symposium on Enhanced Oil Recovery, Tulsa, 22–24 April.
4. Tanner, C.S. *et al.*: "Production Performance of the Wasson Denver Unit CO_2 Flood," paper SPE 24156 presented at the 1992 SPE/DOE Eighth Symposium on Enhanced Oil Recovery Symposium, Tulsa, 22–24 April.
5. Kleinstelber, S.W.: "The Wertz Tensleep CO_2 Flood: Design and Initial Performance." *JPT* (May 1990) 630.
6. Masoner, L.O. and Wackowski, R.K: "Rangely Weber Sand Unit CO_2 Project Update," *SPERE* (August 1995) 103.
7. Pontious S.B. and Tham M.J.: "North Cross (Devonian) Unit CO_2 Flood—Review of Flood Performance and Numerical Simulation Model," *Miscible Processes II,* Reprint Series, SPE, Richardson, Texas (1985) 18, 682.
8. Hadlow, R.E.: "Update of Industry Experience With CO_2 Injection," paper SPE 24928 presented at the 1992 SPE Annual Technical Conference and Exhibition, Washington DC, 4–7 October.
9. Poole, E.S.: "Evaluation and Implementation of CO_2 Injection at the Dollarhide Unit," paper SPE 17277 presented at the 1988 SPE Permian Basin Oil and Gas Recovery Conference, Midland, Texas, 10–11 March.
10. Pariani, G.J. *et al.*: "An Approach To Optimize Economics in a West Texas CO_2 Flood," *JPT* (September 1992) 984.
11. Melzer, L.S. *et al.* (eds.): "How to Put Together a CO_2 Flood," CO_2 Flood Short Course No. 5, Center for Energy and Economic Diversification, U. of Texas of the Permian Basin, Midland, Texas (1996).
12. Kittridge, M.G.: "Quantitative CO_2 Flood Monitoring, Denver Unit, Wasson (San Andres) Field," paper SPE 24644 presented at the 1992 SPE Annual Technical Conference and Exhibition, Washington, DC, 4–7 October.
13. Charleson, G.S., Bilhartz, H.L. Jr., and Stalkup, F.I.: "Use of Time-Lapse Logging Techniques in Evaluating the Willard Unit CO_2 Flood Mini-Test," *Miscible Processes II,* SPE Reprint (1985) **18,** 691.
14. Odom, R.: "A Pulsed Neutron Analysis Model for Carbon Dioxide Floods: Application to the Reinecke Field, West Texas," paper SPE 59717 presented at the 2000 SPE Permian Basin Oil and Gas Recovery Conference, Midland, Texas, 21–23 March.
15. API Manual of Petroleum Measurement Standards, Chap. 14.3, Part 1, General Equations and Uncertainty Guidelines—Concentric, Square-Edged Orifice Meters (A.G.A. Report No. 3), third edition, September 1990, Reaffirmed, August 1995. ANSI/API 2530, Part 1, 1991.
16. U.S. Code of Federal Regulation, Title 49 Part 195, Hazardous Liquids Pipeline Safety Rules, National Archives and Records Administration.
17. "Big Canadian miscible CO_2 EOR project, pipeline advance," *Oil & Gas J.* (7 July 1997) 23.
18. Thakur, G.C.: "Implementation of a Reservoir Management Program," paper SPE 20748 presented at the 1990 SPE Annual Technical Conference and Exhibition, New Orleans, 23–26 September.
19. Melzer, L.S. *et al.* (eds.): "How CO_2 Flood Surveillance Helps Assure a Successful EOR Program," CO_2 Flood Short Course No. 4, Center for Energy and Economic Diversification, U. of Texas of the Permian Basin, Midland, Texas (1996).
20. Bangia, V.K., Yau, F.F., and Hendricks, G.R.: "Reservoir Performance of Gravity-Stable Vertical CO_2 Miscible Flood: Wolfcamp Reservoir, Wellman Unit," *SPERE* (November 1993) 261.
21. Magruder, J.B., Stiles, L.H., and Yelverton, T.D.: "Review of the Means San Andres Unit CO_2 Tertiary Project," *JPT* (May 1990) 638.
22. Merritt, M.B. and Groce, J.F.: "A Case History of the Hanford San Andres Miscible CO_2 Project," *JPT* (August 1992) 924.
23. Pittaway, K.R. and Runyan, E.E.: "The Ford Geraldine Unit CO_2 Flood: Operating History," *SPEPE* (August 1990) 333.
24. Pittaway, K.R. and Rosato, R.J.: "The Ford Geraldine Unit CO_2 Flood—Update 1990," *SPERE* (November 1991) 410.
25. Stein, M.H. *et al.*: "Slaughter Estate Unit CO_2 Flood: Comparison Between Pilot and Field-Scale Performance," *JPT* (September 1992) 1026.
26. Harpole, K.J. and Hallenbeck, L.D.: "East Vacuum Grayburn San Andres Unit CO_2 Flood Ten Year Performance Review: Evolution of a Reservoir Management Strategy and Results of WAG Optimization," paper SPE 36710 presented at the 1996 SPE Annual Technical Conference and Exhibition, Denver, 6–9 October.
27. Dane, A.V.: "Performance Review of a Large-Scale CO_2-WAG Enhanced Recovery Project SACROC Unit—Kelly-Snyder Field," *EOR Field Case Histories,* Reprint Series, SPE, Richardson, Texas (1987) **23,** 52.

SI Metric Conversion Factors

bbl ×	1.589 873	E − 01	= m^3
°F	(°F − 32)/1.8		= °C
°F	(°F + 459.67)/1.8		= K
ft ×	3.048*	E − 01	= m
ft^3 ×	2.831 685	E − 02	= m^3
lbm ×	4.535 924	E − 01	= kg
mile ×	1.609 344*	E + 00	= km
psia ×	6.894 757	E + 00	= kPa
tonne ×	1.0*	E + 00	= Mg

*Conversion factor is exact.

Chapter 8
Operations: How Can the CO_2 Flood Process Be Managed Effectively?

In this chapter, we focus on activities specific to CO_2 flooding (omitting those similar to waterflooding). We divide the operation of a CO_2 flood into three topics:

• Reservoir management. Sec. 8.1 covers what to inject; how fast to inject; how much to inject; surveillance methods and tools; tools to manage oil recovery, sweep efficiency and injection; and recommendations for WAG decisions.

• Well management. Sec. 8.2 covers producing method choices and wellbore remedial work, including selection of workovers, chemical treatment, and CO_2 breakthrough.

• Facility management. Sec. 8.3 covers the reinjection plant, separation and metering, corrosion control, and facility optimization.

8.1 Reservoir Management

The overwhelming majority of CO_2 floods have been undertaken in reservoirs where gravity forces were relatively unimportant, leading to the use of pattern injection rather than top-down injection. While continuous injection of CO_2 followed by chase water is an option for pattern CO_2 floods, most alternate between water and CO_2 at some time during the life of the project. This section covers WAG management for pattern CO_2 floods.

8.1.1 What to Inject: Optimizing Half-Cycle Slug Size and WAG Ratio. The implementation team forecasts the scheduled use of CO_2 (before the supply arrangements for CO_2 are finalized) and later fine-tunes the flood design regarding two operating parameters, the half-cycle slug size and the WAG ratio.

A half-cycle is defined as the reservoir volume of CO_2 or water that is injected before switching to the alternate fluid. Half-cycle slug size often is measured in fraction of HCPV injected. Sometimes it is measured in time.

The WAG ratio is a ratio of the reservoir volume of water injected to that of CO_2 injected during a full cycle. Both components of the WAG ratio (water and CO_2) should be expressed as reservoir volume, not stock tank barrels or standard cubic feet. Sometimes the WAG ratio is expressed in terms of its reciprocal, the gas/water ratio (GWR).

As noted by Masoner and Wackowski,[1] WAG parameters are the primary tools to improve the oil cut and CO_2 use, and WAG management is important in controlling and managing operating costs as a flood matures. A sound WAG policy allows the operator to:

• Control sweep efficiency by maintaining a more uniform front through the reservoir.

• Control CO_2 recycle volumes.

• Facilitate management of produced gas and liquid ratios under both flowing and artificial-lift situations.

• Maximize profitability.

Half-Cycle Slug Sizes. Industry WAG policies are widely published.[2–12] Typical WAG ratios vary from 1:2 to 3:1, but are usually about 1:1 **(Table 8.1).** Half-cycle slug sizes reported in publications vary from 0.1 to 6.0% of HCPV, although most tend to be in the 1.0 to 2.0% range. These references reflect the state of industry practice in the late 1980's.

During the 1990's, many operators stopped basing WAG switches on time and started basing them on volume injected. In a volume-based system, the entire HCPV half-cycle slug is injected before the fluid is switched, no matter how long it takes. A volume-based process facilitates injection of the planned WAG ratio and corrects for any variations in rates that may occur during the cycles.

Another recent change, initiated by operators at the Rangely Weber Sand Unit[3] and others, has been reduction of WAG half-cycles. Shorter WAG cycles tend to minimize swings in fluid and gas production.

Controlling gas production is particularly important because of gas processing costs and capacity constraints. Because gas production usually is difficult to forecast accurately, facilities may end up being improperly sized. WAG design may need to be altered to accommodate facility limitations.

Minimizing water production also may be important because an excess of higher density water can cause flowing wells to stop producing. Maintaining either flowing or nonflowing conditions—rather than cycling between the two conditions—makes it easier to optimize artificial lift.

The main drawback of short cycles and small slugs is the additional labor required to handle the increased switching. Switching frequency varies from system to system, and is limited by the degree to which switches are automated. Masoner reports that one week is the minimum practical injection interval for manual changes at Rangely,[1] while some Salt Creek wells are switched more than once a day using automatic controls.

WAG Ratios. WAG ratios often change during the course of a flood, and a ratio that maximizes oil recovery at the beginning of the EOR project may be inappropriate later in the project.[13] In the early 1980's, CO_2 flood design usually called for a specific WAG ratio maintained over the life of the flood, but more recently, tapered WAG designs have been implemented. In this design, the initial WAG ratio is low to accelerate EOR response, but to limit gas production it is increased after CO_2 breakthrough.

TABLE 8.1—1980s INDUSTRY WAG POLICIES

Field	Source	WAG Ratio	CO_2 Half-Cycle
Rangely	Attanucci *et al.*[2]; Masoner and Wackowski[2]	1:1 to 2:1 to 3:1	1.5 to 0.25 to 0.1% (1 to 3 weeks)
South Wasson Clearfork	Burbank[4]	8:1 to 2:1	2 months with 2:1 WAG
South Welch	Hill *et al.*[5]	1.1	2.6 to 8.2% (2 to 6 months)
Twofreds	Kirkpatrick *et al.*[6]	1:5:3	
SACROC	Kane[7]	0.47:1 to 1:1 to 3:1	6.0 to 1.5%
Hanford	Merritt and Groce[8]	1:1	3.0%
Denver Unit	Tanner *et al.*[9]	1:1	
North Ward Estes	Ring and Smith[10]	1:1	1.5% (1 month)
Slaughter	Stein *et al.*[11]; Merchant and Thakur[12]	1:2 to 1:3 to 1:4	1.0 to 1.5 to 2.0%

Where gas processing capacity exists before the CO_2 flood commences, the initial WAG ratio may continue until some point before the maximum gas production limit is reached, after which the WAG is tapered by increasing the amount of water. This is to help prevent gas production from exceeding gas processing capacity.

In projects where gas production facilities must be built or acquired, a tapered WAG design may be used from the start. By gradually increasing the injected water/CO_2 ratio over time, smaller gas processing plants can be installed, and the entire plant can be used for a longer period at higher utilization.

While increasing the WAG ratio helps control gas production, it also has some undesirable effects, especially in strongly water-wet formations, as noted by Holm and O'Brien.[14] Water is detrimental to the oil recovery mechanism for three reasons: it decreases the displacement efficiency the CO_2 solvent front; it prevents some of the solvent phase from reaching the residual oil because of dissolution of the solvent; and in water wet formations, it can trap large amounts of previously CO_2-mobilized oil and/or solvent. Sec. 2.2.2 contains additional information about water blocking in CO_2 floods. We recommend that the amount of water used in the WAG operation be minimized and that it be used only to help control excessive gas production related to channeling or to minimize the cost of injection.

WAG management can have a profound effect on the economics of a flood. Most WAG decisions involve striking a balance between maximizing oil recovery and controlling the operating costs associated with gas production.

8.1.2 How Fast to Inject: Setting Injection Pressure and Rate Targets. The reason for optimizing injection pressure and rate is rather simple: to maximize throughput by injecting as much CO_2 or water as possible without fracturing the reservoir.

Setting Injection Pressure. To set the injection pressures for water and CO_2, one first must obtain an estimate of the formation parting pressure. If a waterflood is underway, step-rate or other formation parting pressure tests probably have already been performed; if not, they should be run before starting the CO_2 flood.

Establishing the appropriate wellhead CO_2 injection pressure is more complex than determining the corresponding water injection pressure. Normal step-rate tests may directly measure the surface pressure for water injection, but because CO_2 is less dense than water and is a compressible fluid, some calculations are required to set the CO_2 wellhead injection pressure.

First, the bottomhole formation parting pressure is estimated from the water injection step rate test. Then, the pressure increase during CO_2 injection between the surface and bottomhole is calculated using one of the methods used in the natural gas industry, such as that of Cullender and Smith.[15] Surface injection pressure then is set at a value less than the bottomhole formation parting pressure minus the wellbore pressure increase.

Another complicating factor is that the density of the injected gas may change over time as the recycled CO_2 picks up contaminants such as nitrogen and hydrocarbon gases. As the composition of the injected gas changes, so does its density—and this requires a concomitant change in the surface injection pressure. **Fig. 8.1** shows well depth vs. bottomhole pressure for two gases, 100% CO_2 and 90% CO_2 plus 10% methane. Minor changes in composition can make significant differences in bottomhole pressures and can cause dramatic changes in injection rates. Fluctuations in wellhead temperature also can affect the corresponding bottomhole pressure, as well as the measured volume of CO_2.

So far, we have discussed injection here as though the optimal goal always were to maximize throughput without injecting above formation parting pressure. As we discussed in Chap. 7, however, there are several reasons why it would be desirable in many cases to restrict injection further:

• Pattern balancing. Processing patterns at the same HCPV rate may maximize oil recovery, but may not maximize economic present value if overall recovery must be slowed to accommodate low-permeability areas in the reservoir. Injection at different HCPV rates can enable faster oil recovery, but it also may lead to premature abandonment of wells because of early CO_2 breakthrough caused by the uneven sweep. A trade-off often is required.

• Equipment limitations. It might not be possible to inject as much CO_2 as the well theoretically could take because either excessive equipment is required or the injection pressures needed are not be economical, or CO_2 is in short supply.

• Preserving stability in a gravity-stable vertical flood. Bangia *et al.*[16] describe this process in detail for the Wellman Unit. Also see Chap. 2, Sec. 2.4.

Note that injection pressure cannot be maintained exactly at a constant value; instead, injection pressure will fluctuate slightly within a deadband or range of control error. Choosing the size of the deadband is an important decision: Deadbands that are too small

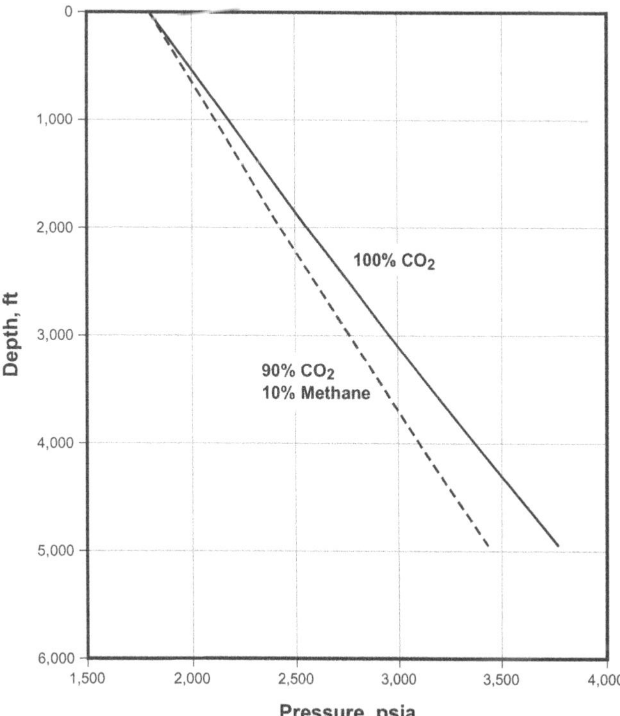

Fig. 8.1—The effect of hydrocarbon gas contamination of CO_2 on static bottomhole pressure. Contamination reduces the density of the injected gas; to compensate, higher injection pressures must be used.

will cause the valve system to open and close repeatedly, wearing out the valves and/or running down the battery (if the control unit is not connected to AC current). If the deadband is too large, the well may not always be injecting at a high enough rate. Generally, operators using SCADA systems set a deadband of about 30 to 40 psi, which may be about 2% of the injection pressure.

Setting Injection Rate. In addition to setting injection pressures, the operator must establish injection rate targets. Rate targets can help identify wells that need stimulation, and they also can be used to limit injection in high-rate wells.

As an initial estimate, one may assume that the CO_2 flood will have the same injectivity as the waterflood. This injectivity estimate, based on reservoir volume injected, then can be modified to account for differences in viscosities and relative permeabilities or to account for experience with similar projects.

In many cases, as discussed in Chap. 2, CO_2 projects experience a loss of injectivity when water is injected as part of the WAG cycle **(Fig. 8.2).**[17] In Fig. 8.2, the first bar in every group of three represents initial waterflood injectivity; the second bar shows the injectivity of CO_2 compared to waterflood injectivity; and the third bar shows water injectivity after the CO_2 portion of the cycle. Because the gas is much less viscous than water, and reduced viscosity typically outweighs relative permeability effects, CO_2 often has greater injectivity than water (at least during the first cycle). In contrast, water injectivity often is reduced after a CO_2 cycle. The six projects surveyed averaged a loss of about 20% of injectivity during the WAG process, and this amount should be expected. Note, however, that SACROC saw a slight injectivity increase.

Several authors have explored the injectivity loss illustrated in Fig. 8.2. Phase behavior, inorganic precipitation, and wettability are among the causes investigated. Patel *et al.*[18] conclude that neither phase behavior effects by themselves nor inorganic precipitates reduced injectivity at the Denver (San Andres) Unit project referenced in Fig. 8.2; instead, the low CO_2 injectivity stemmed from mixed wettability and the resulting relative permeability hysteresis, as shown in **Fig. 8.3.**[19] The hysteresis is caused by the trapped gas effect in the oil-wet pores, as discussed in Chap. 2, Sec. 2.3.4.

Christman and Gorell[19] extend the work of Patel *et al.*[18] and conclude that when viscous forces dominate gravity and capillary forces, tertiary displacement of residual oil by CO_2 can be broken down into the travel of four slugs through the reservoir—waterflood residual, mobile oil bank, CO_2-rich phase, and followup water. Injectivity is controlled by the least mobile slug, which is the oil bank, so the loss of injectivity is not permanent. Once the oil bank is produced, overall mobility in the reservoir increases. As the oil saturation decreases, the relative permeabilities of both oil and water approach their initial values.

Chopra *et al.*[20] also note that trapped CO_2 prevents the water relative permeability from reaching the maximum value obtained in waterflooding. If the water cycle is long enough, or if the process has moved to chase water, the trapped CO_2 may dissolve and water injectivity may exceed the preCO_2 flood value.

Although nearly total recovery of injectivity may not occur until the end of the flood (and probably will not occur if inorganic precipitation occurs), injectivity may increase during a WAG cycle. At the Hanford San Andres CO_2 project, Merritt notes that water injection was 20% lower at the beginning of each WAG cycle than at the end, and CO_2 injectivity typically was 13% lower at the beginning than at the end.[8] This phenomenon may be explained by relative permeability effects: as a fluid is injected into the reservoir, its saturation near the wellbore area increases and, therefore, so does the injectivity. (See also Chap. 2, Secs. 2.3 through 2.5).

8.1.3 How Much To Inject: Ultimate Slug Size. While WAG design has the greatest impact on how *fast* the oil is recovered, the total amount of CO_2 injected is the parameter that most affects the *amount* of oil recovered. In light of this, Hadlow[17] notes another factor that becomes more important as projects age: the use of CO_2 in greater amounts than originally planned.

The more CO_2 you inject, the more oil you recover, but unfortunately the process is subject to the law of diminishing returns. **Fig. 8.4**[17] shows the results of a reservoir simulation study of the Means-San Andres CO_2 project using varying solvent slug sizes and a WAG ratio of 2:1. As the graph shows, the larger the slug size, the greater the ultimate oil recovery—but the incremental oil recovered is less for each incremental increase in slug size.

The optimum slug size for a particular project (or for an individual pattern) depends on economic factors such as oil prices, operating expenses, and CO_2 costs. Fortunately, the ultimate slug size (the total amount of CO_2 injected) need not be determined at the start of the project, but can be decided on later, when the operator has more information. Note, however, that the anticipated ultimate slug size could affect the price of the CO_2 and the selection of the CO_2 source.

8.1.4 Overview of Surveillance Methods and Tools. Appropriate surveillance is the only way to gather the project-management information necessary to maximize financial performance. The primary method of monitoring any enhanced recovery project is to collect data and present them in graphical form. In order for plots and maps to reveal trends and problems, they must be properly constructed. Some effective surveillance can be accomplished using total field data; however, remedial action is undertaken well-by-well or pattern-by-pattern, so that most presentations must make it possible to examine the field on this smaller scale. This section focuses on pattern surveillance because well surveillance in a CO_2 flood is quite similar to that in a waterflood.

Every CO_2 flood should start with plotted expectations for each injection pattern. As the flood progresses, differences between expec-

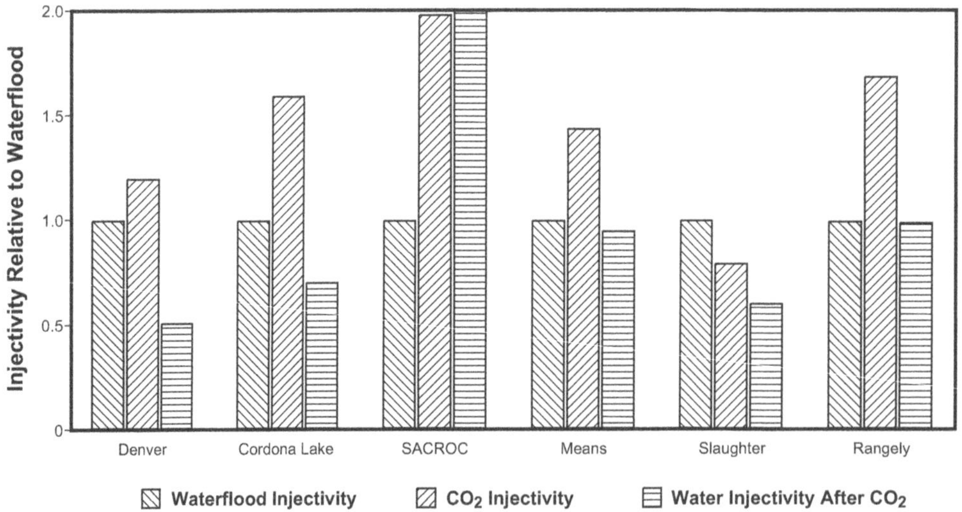

Fig. 8.2—WAG injectivity data (after Ref. 17). Water injectivity can change dramatically after injection of CO_2.

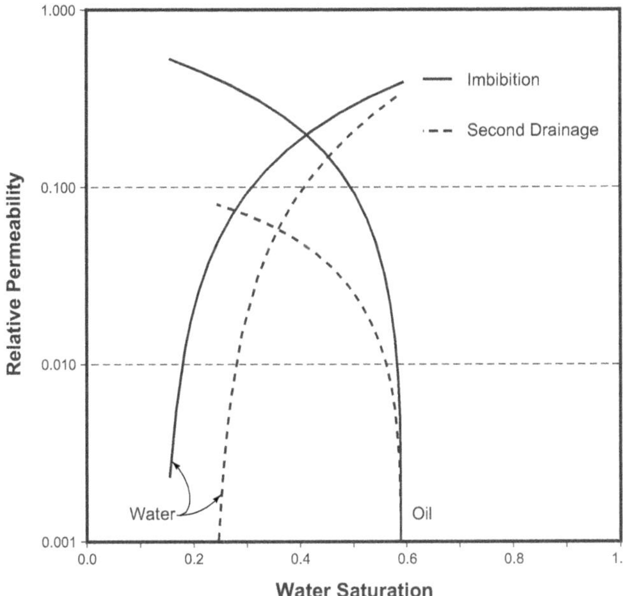

Fig. 8.3—Imbibition and secondary drainage (hysteresis) relative permeabilities for Wasson (Denver Unit) (after Ref. 19). The relative permeabilities of oil and water change during the course of the flood, as shown by the hysteresis curves for both.

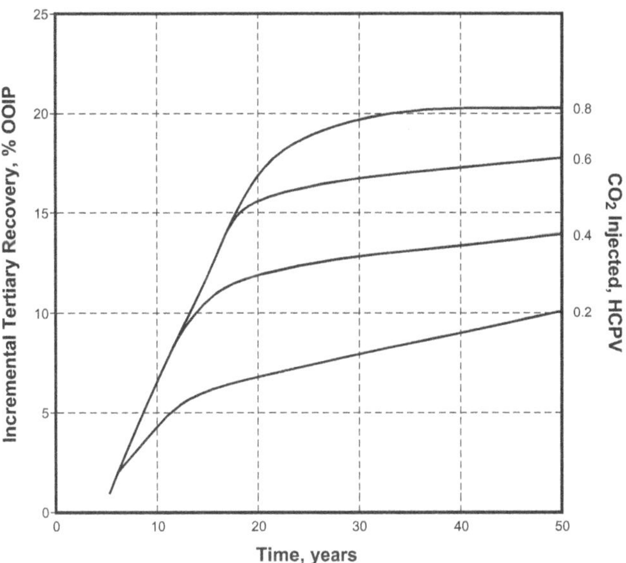

Fig. 8.4—Effect of CO_2 slug size on oil recovery at the Means-San Andres field (after Ref. 17). As the slug size (CO_2 bank) is increased, recovery also increases, but the incremental oil recovered is less for each incremental increase in slug size. Theoretical calculations and a WAG ratio of 2:1 were used.

tations and actual performance can be investigated and the operator can initiate remedial action or revise expectations, accordingly.

The fundamental data on which all surveillance rests are injection and production volumes. Oil, gas, and water production rates and CO_2 and water injection volumes must be metered accurately. All measurements of flood effectiveness depend on these most basic parameters.

After injection and production data are collected, the volumes must be allocated to patterns. Patterns may be either production- or injection-well centered, i.e., injection data are allocated to the production well, or production data to the injection well. If the injection pattern is a line drive or five-spot, arguably it makes little difference which type of well one centers the pattern on. On the other hand, if the injection pattern is an inverted nine-spot, an injection-well centered pattern makes more sense. We prefer to use injection-well centered patterns because much of CO_2 flood surveillance focuses on optimizing injection.

In addition to volume data, reservoir-related parameters such as OOIP and HCPV also must be allocated to each pattern.

Allocation Methods. Several methods are used to allocate data. For areally homogeneous reservoirs with regularized or geometrically symmetrical patterns, simple geometric allocation may suffice. In cases where patterns are irregular, the reservoir is areally heterogeneous, and/or injection and production rates vary significantly, streamtube modeling may improve the surveillance effort because it provides a better estimate of the swept area around each well.

Plotting Options. Many of the surveillance techniques plot EOR rather than total oil. In order to estimate the EOR, the engineer first must establish a baseline for preCO_2 flood decline. Some people use simple decline rates (oil rate vs. time), but this method is not accurate if the injection rates before and after the start of the CO_2 flood are different. For example, if the injection rate is lower during WAG injection than during the waterflood, a decrease that shows up in oil production might be related to the change in injection rate rather than to a process failure. One also may see an increase in oil production and mistakenly forecast additional reserves, when the occurrence might be nothing more than an increase in the production rate of oil that ultimately would have been produced anyway.

A better method for estimating waterflood decline is to tie it to water injection, as illustrated in **Fig. 8.5.**[21] A regression fit over the waterflood history is performed on data plotted on a cumulative basis. The regression fit could be based on log (WOR) vs. cumulative oil production or another standard method to forecast waterflood recovery. The EOR production is calculated by subtracting the total production from the waterflood forecast.

Note the dimensionless nature of Fig. 8.5. By normalizing the volume data, the progress of patterns can be compared even if they are at different stages of recovery or are processing at different rates. Industry practice is to divide oil production by OOIP and to divide injection by initial HCPV. For reservoirs in which oil has displaced all the water except the residual water saturation during migration into the reservoir, initial HCPV approximates movable pore volume (MPV) in a CO_2 flood because residual oil saturation to CO_2 flooding is almost always very low. However, if the reservoir didn't fill completely with oil during the initial saturation phase, the initial HCPV will not be equivalent to the MPV because movable water remains. In such reservoirs, using the MPV instead of the HCPV may facilitate more meaningful comparison of dimensionless plots between projects.

Once patterns are established and volume data are allocated and nondimensionalized, there are a number of simple diagnostic tools that can help identify problems and areas for improvement. These tools are of three types: those that help manage oil recovery and CO_2 use; those that help manage sweep efficiency, CO_2 breakthrough, and CO_2 use; and those that help manage injection.

8.1.5 Tools to Help Manage Oil Recovery and CO_2 Use. The single best measure of pattern performance is shown in **Figs. 8.6**[2] **and 8.7,**[21] in which cumulative EOR is plotted against cumulative injection from the start of the CO_2 flood. This method highlights the effectiveness of the CO_2 injection process and compares the actual EOR to the expected values.

Both Figs. 8.6 and 8.7 are useful plots. Fig. 8.6 shows only cumulative CO_2 injection along the x-axis, which leads to discontinuities during water injection half-cycles but clearly shows how much CO_2 has been injected. Fig. 8.7 is smooth because its x-axis shows cumulative EOR injection for both CO_2 and water. For forecasting purposes, we find that converting the dimensionless cumulative volumes to dimensional rates is easier using the smooth total EOR injection shown in Fig 8.7.

The shape of the curve in Fig. 8.7 depends on the WAG history. Tapered injection increases curvature. For constant WAG schemes, larger water/CO_2 ratios straighten the curve by slowing recovery. At some point, the slope of every curve for every injection pattern will approach zero, indicating that CO_2 injection is losing its effectiveness.

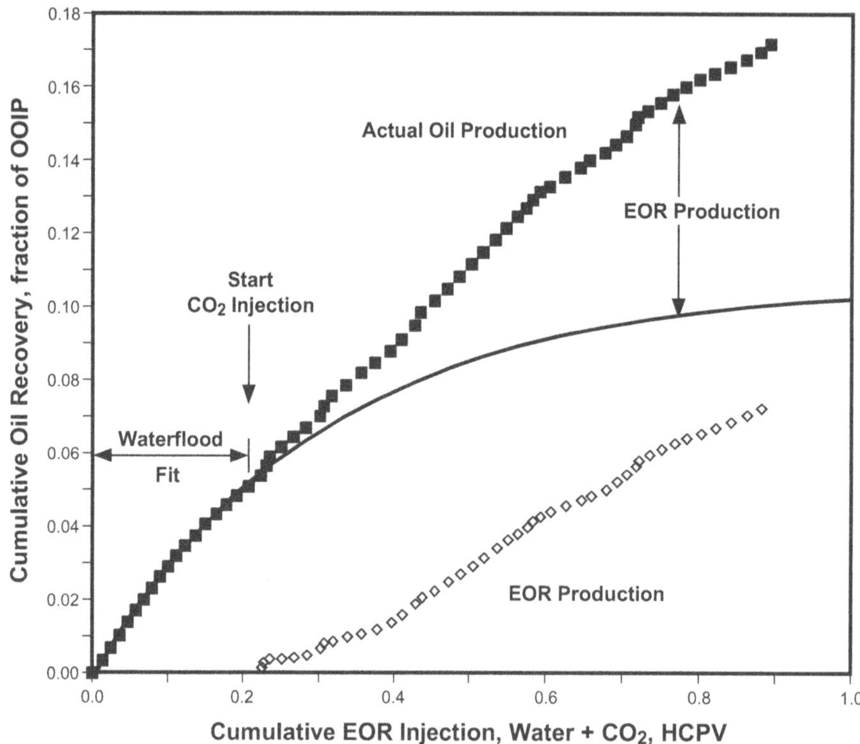

Fig. 8.5—Dimensionless waterflood recovery estimates, using a method that ties the waterflood decline to water injection (after Ref. 21).

Another way to present the data is in a diagnostic pattern performance scatter plot, as shown in **Fig. 8.8.**[21] Here, the projected EOR for a prototype area is plotted as a solid line, and actual individual pattern performances are compared to it.

If the whole reservoir is fairly homogeneous, one expected-performance curve may suffice for the entire field; however, in larger reservoirs or where the geology changes significantly across the reservoir, several prototype curves may be necessary. In Fig. 8.8, patterns A and B fall fairly close to the expected results, while pattern C plots well below expectations. The situation at C and other patterns that fall below forecast should be investigated and remedial action taken as warranted.

Fig. 8.9[2] shows another measure of the efficiency of CO_2 injection. This graph plots the cumulative gross CO_2 utilization ratio against the cumulative CO_2 injection. In this figure, the gross CO_2 utilization ratio is equal to the total amount of CO_2 injected divided by the enhanced oil recovered. (The net CO_2 utilization ratio is defined as the total amount of CO_2 purchased divided by the enhanced oil recovered.) Because of the lag between CO_2 injection and oil production, the plot starts with an infinite utilization ratio. As oil response occurs, the utilization ratio rapidly decreases to a minimum. If the ratio begins to increase, then the process is becoming less efficient. As this happens, the WAG ratio should be tapered as economically warranted. Additional water in the WAG cycles will limit the use of CO_2, while permitting recovery of the remaining oil.

One certain means of determining whether or not a pattern is responding effectively to CO_2 injection is shown in **Fig. 8.10.** The

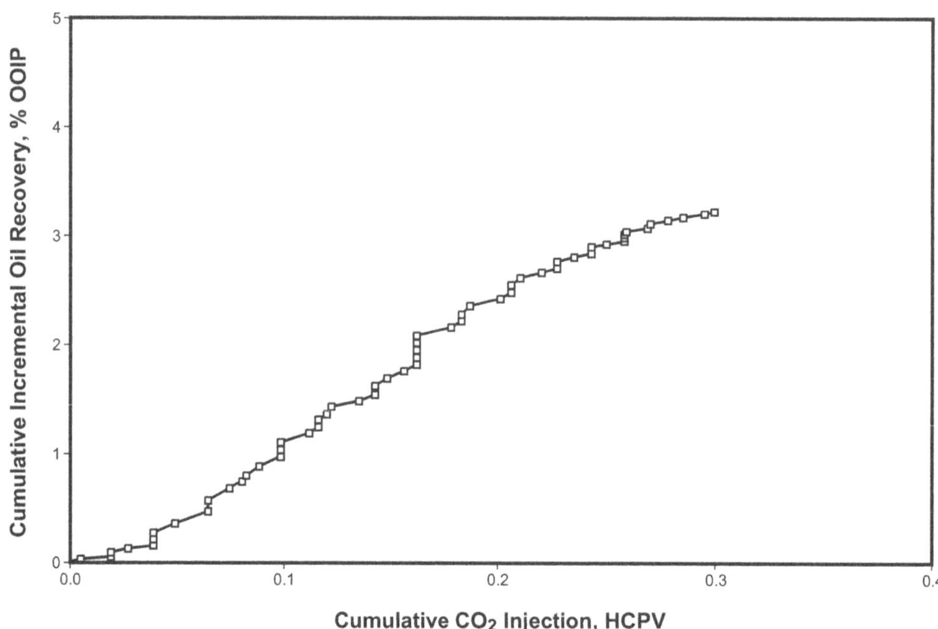

Fig. 8.6—Cumulative incremental oil recovery vs. cumulative CO_2 injected. The discontinuities occur during water half-cycles (after Ref. 2).

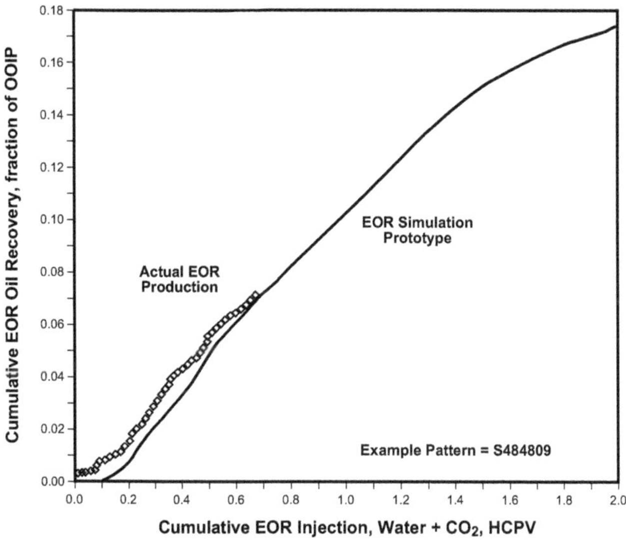

Fig. 8.7—Cumulative incremental oil recovery vs. cumulative HCPV EOR injection (after Ref. 21). This curve differs from that of Fig 8.6 in that it shows both water and CO_2 injected.

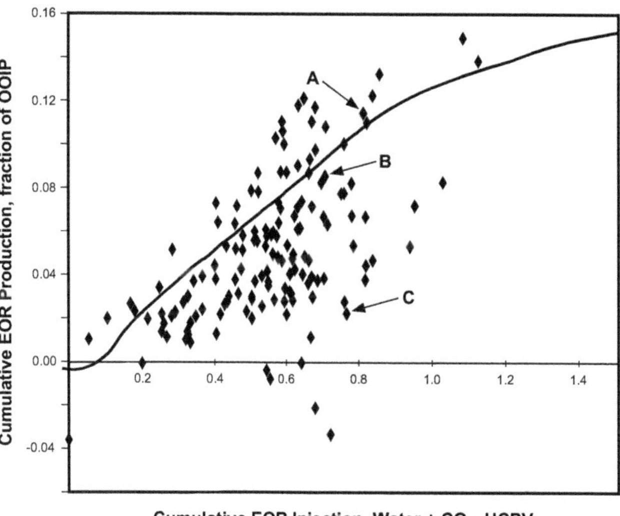

Fig. 8.8—Diagnostic pattern performance scatter plot (after Ref. 21). The solid line is the projected EOR curve and the points represent the performance of individual patterns.

success of the EOR process is apparent when the oil cut reverses its downward trend and begins to increase. (The oil cut is defined in the standard way, which is oil production divided by the sum of oil and water production over a given length of time [STBO/(STBO+STBW)]. If the oil cut shown on this plot does not increase after the CO_2 flood commences, the pattern's performance should be investigated.

Fig. 8.11 illustrates the final diagnostic tool for oil recovery and CO_2 utilization. It shows actual and planned cumulative WAG ratios vs. cumulative EOR injection for an individual pattern. Clearly the operator in this case did not follow the injection plan for the pattern. If the injection plan is not followed, the forecast curves, which normally are based on a specific injection scheme, may not be valid.

8.1.6 Tools To Help Manage Sweep Efficiency, CO_2 Breakthrough, and CO_2 Use. The plots shown in **Figs. 8.12 and 8.13** help assess sweep efficiency. In cases where a pattern is not being repressurized or maintained at a pressure similar to the adjacent patterns, the voidage replacement ratio (reservoir production divided by reservoir injection) can become much greater than 1.0. In such a case, a CO_2 sweep imbalance or an actual loss may be occurring—an expensive problem.

Fig. 8.13 shows CO_2 retention vs. cumulative injection. The CO_2 retention is the fraction of the total CO_2 injection remaining in the reservoir (purchased CO_2/injected CO_2). It is equal to the net CO_2 utilization divided by the gross CO_2 utilization. If all the injected CO_2 were produced, the retention rate would be zero; if none of the CO_2 were recovered, the retention rate would be 1.0.

A high retention rate may be either good or bad. If the voidage matches injection, and the CO_2 is being injected into the target oil zone, a high retention rate indicates that CO_2 is efficiently mobilizing oil and is not breaking through thief zones, while a low retention rate may indicate a breakthrough problem. However, a high retention rate coupled with high voidage replacement ratio could indicate CO_2 losses out of the oil pay zone (or it could indicate pattern allocation errors).

Carbon dioxide breakthrough manifests itself in high GOR's. GOR plots identify likely patterns for WAG tapering, which can be especially important when either the calculated economic limit for CO_2 injection or the facility limit for CO_2 compression is reached.

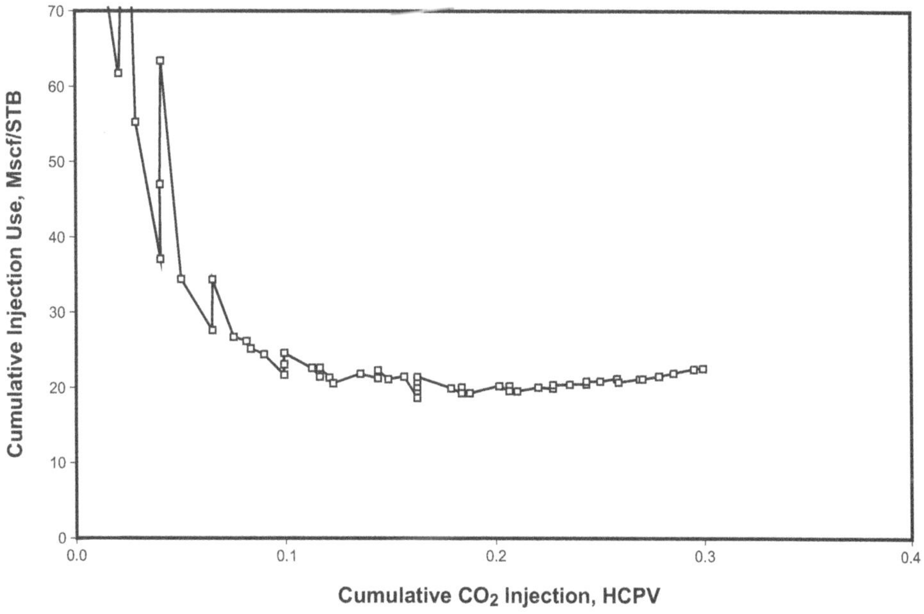

Fig. 8.9—Cumulative gross CO_2 utilization ratio vs. HCPV CO_2 injection (after Ref. 2).

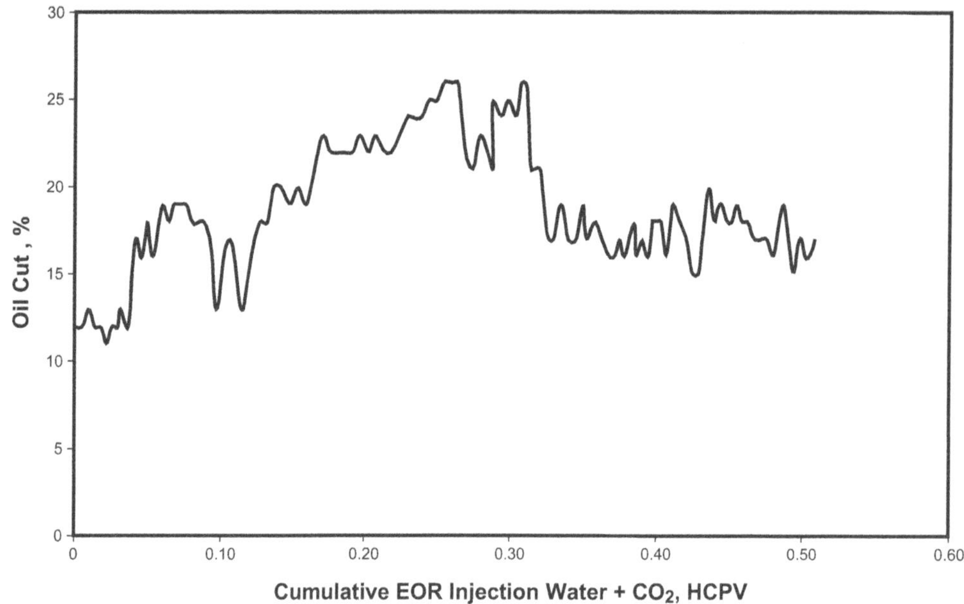

Fig. 8.10—Diagnostic plot of oil cut. The oil cut should increase after the CO_2 flood commences.

Fig. 8.14 shows the incremental GOR (incremental total gas production divided by incremental EOR oil production).

The GLR shown in **Fig. 8.15** also is an important parameter to track and manage. A widely varying GLR can be costly. When the GLR increases, previously pumped wells may begin to flow. When the GLR decreases, flowing wells stop flowing. In either case, a change in artificial lift may be required. If the GLR fluctuates significantly, faster WAG switches should be considered.

8.1.7 Tools To Help Manage Injection. In addition to the plots already presented, a number of other diagnostic tools may be helpful in monitoring a CO_2 flood:

- Comparison of actual and target injection volumes vs. time.
- Comparison of actual and target injection rates vs. time.
- Composition and density of the injected gas stream vs. time. (Impurities in the recycled CO_2 may change the density of the injected gas.)
- Injection profiles.
- Time-lapse logging.

The natural process of CO_2 flooding may cause injection profiles to change. These changes can be dramatic after much of the oil has been produced from a thief zone and a high-velocity CO_2-filled channel has been established.

Time-lapse logging with neutron tools is a useful technique for determining changes in CO_2 saturations.[22,23] (See also Sec. 7.2.) Both compensated neutron and pulsed neutron tools can be used in open and cased holes to detect the presence of CO_2.

Compensated neutron logs measure the presence of hydrogen: Oil and water have almost the same amount of hydrogen per unit volume, while CO_2 has none. If CO_2 has replaced oil or water in the reservoir, its presence will be shown by a decrease in apparent neutron porosity.

Pulsed neutron logs, which normally measure the capture cross section of the reservoir fluid to estimate water saturation, also can be used like compensated neutron logs to determine CO_2 saturation in cases where a small-diameter tool is required.

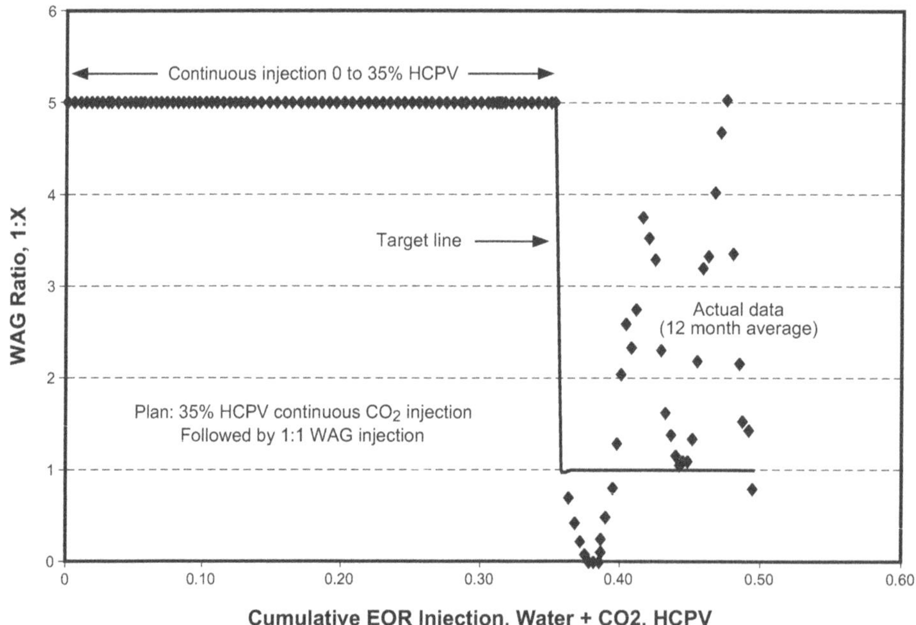

Fig. 8.11—Diagnostic plot of actual vs. target WAG injection ratio. If the WAG injection plan is not followed (as was the case here), the forecast curves for oil recovery may not be valid.

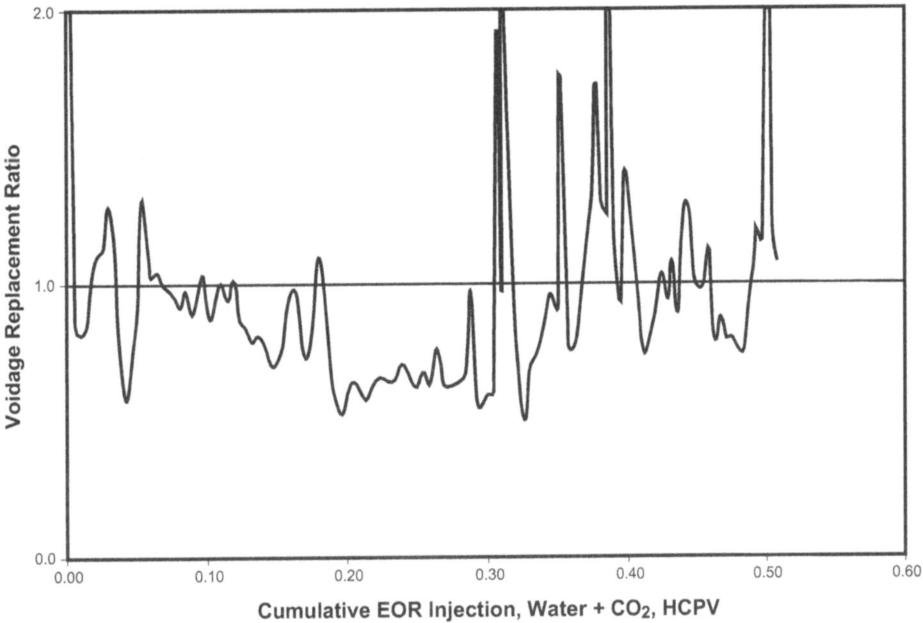

Fig. 8.12—Diagnostic plot of voidage replacement ratio. If the voidage replacement ratio varies significantly from 1.0, the reservoir pressure is changing or sweep imbalance is occurring.

8.1.8 Making WAG Decisions—Recommendations. In conclusion, we recommend the following steps in managing injection and flood performance:

- Initiate the WAG based on the CO_2 flood design.
- Regularly review diagnostic plots to identify problems.
- Attempt to cure problems first by reducing the half-cycle length, then by altering the WAG ratio, if necessary (note that well-by-well changes to the WAG ratio and cycle sizes often are made by trial and error, using reservoir simulation as a guide).
- Maintain the maximum allowable bottomhole injection pressure.
- Alter expectations as required (such as for unexpected reduced injectivity).
- Make certain that wellbore mechanics, formation damage, and/or lift problems are not degrading performance.

While the above techniques are our favorite approaches for monitoring the progress of a CO_2 flood, other methods are available: For example, Masoner *et al.* present a comprehensive methodology used to optimize performance at the Rangely Weber Sand Unit.[24]

8.2 Well Management

8.2.1 Choice of Producing Method. Optimizing the producing method is an important element in maximizing the economic return of a CO_2 project. Many authors report the necessity for changing from one type of artificial lift equipment to another, including Kane at SACROC,[7] Tanner *et al.* at the Denver Unit,[9] Wackowski and Masoner at Rangely,[25] and Pittaway and Runyon at Ford Geraldine.[26]

There are three common producing methods used in CO_2 floods today: rod pumps, ESP's, and natural flow. Early in an EOR project, rod pumps may be in place at the wells; later, as liquid production increases because of increased injection, submersible pumps may be needed in wells with high fluid rates. As CO_2 begins to break through, and production and GLR increase still further, the

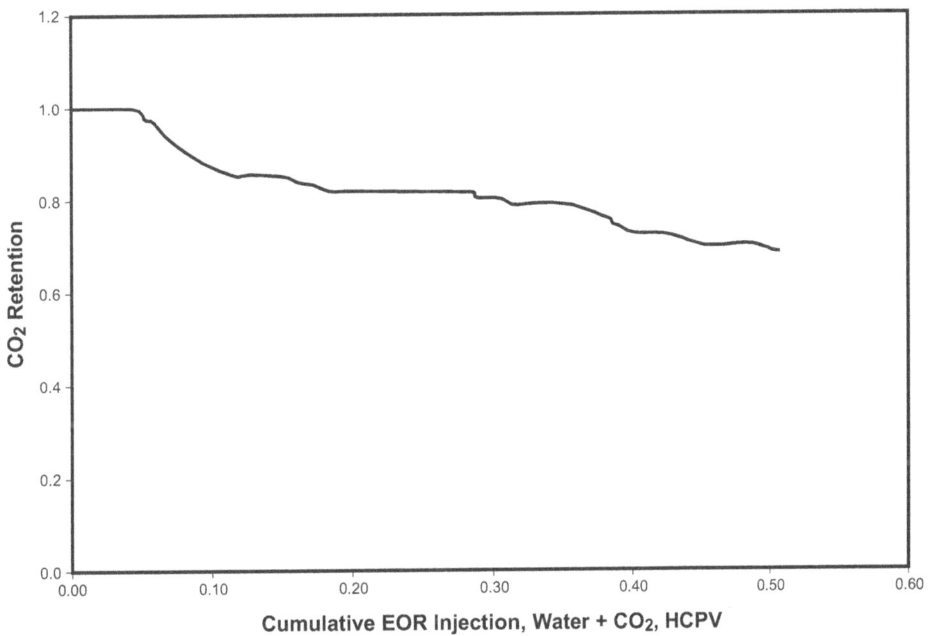

Fig. 8.13—Diagnostic plot of CO_2 retention. High retention may be good or bad, depending on voidage results.

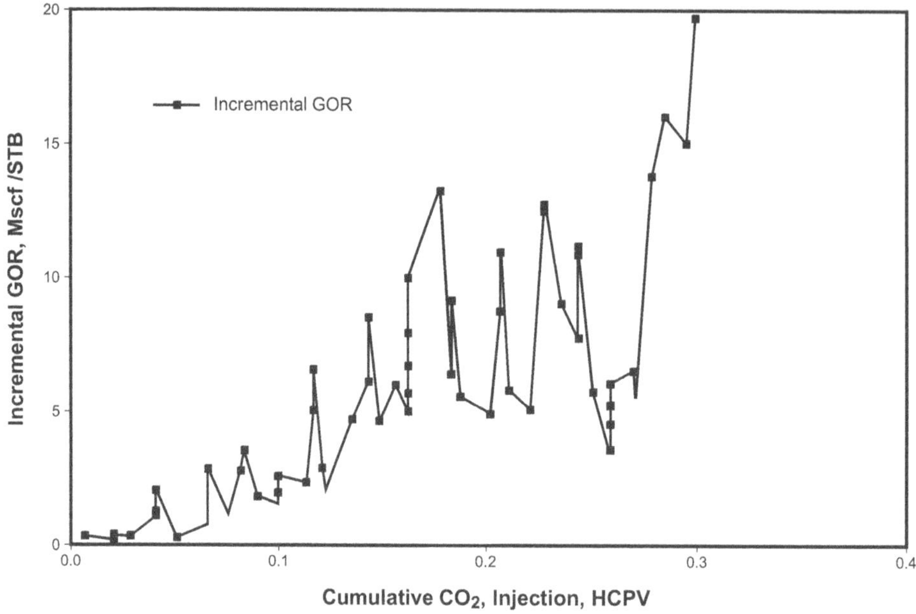

Fig. 8.14—Diagnostic plot of incremental GOR vs. HCPV CO_2 injected (after Ref. 2). The high incremental GOR at the end of this plot indicates that this pattern is a good candidate for WAG tapering.

well may be converted to flowing conditions. Toward the end of the flood, as the WAG is tapered and more water is injected, the wells may require pumping again.

The decision between rod and submersible pumping is driven by the liquid production rate or wellbore conditions, while decision between pumping and flowing is governed by the GLR. Putting the right lift equipment on the well at the right time saves money and optimizes production.

Rod Pumping. For nonflowing conditions and low fluid rates, rod pumps generally are more economical. Another benefit of rod pumping is that the wells are pumped off, which maximizes drawdown on the reservoir and increases throughput. Rod stress limits the maximum fluid production.

A common problem associated with rod pumping in a CO_2 flood is gas locking, which can occur when CO_2 breaks through. Gas locking problems have been solved in several ways, including increasing backpressure by up to 300 psi at the surface (which forces gas back into solution[26]), moving the pump farther downhole (below the point where the gas breaks out of solution), and installing gas anchors.

Electric Submersible Pumping. For wells with a large production volume of fluids, electrical submersible pumping often is chosen. Submersible pumps have greater capacity than rod pumps, but they also have several disadvantages: They consume considerable electricity, have a narrow operating range, and present maintenance challenges—repairs to the pump body are much more expensive than repairs to simple downhole rod pumps.

Like rod pumps, ESP's also can have problems with gas locking, but unlike rod pumps, unshrouded ESP's must operate with a fluid level above the pump because the moving fluid acts as the pump's coolant. Standard remedies for gas locking are similar to those used in rod pumping. They include setting the depth of the pump at the top perforation, installing a gas separator intake or motor shroud and holding backpressure on the well to keep gas in

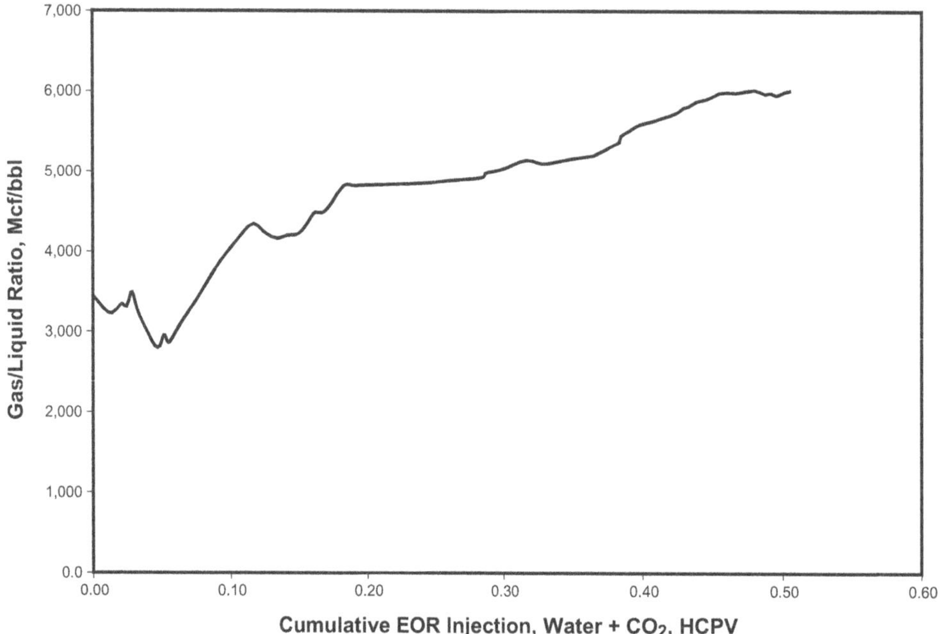

Fig. 8.15—Diagnostic plot of GLR. Flowing conditions were maintained over the life of this well. Plots for other wells show widely varying GLR's that can require costly changes in artificial lift.

solution, and using an auger-type intake. At Rangely, for keeping gas in solution through the pump, an auger intake was more effective than gas separators.[3]

Although gas locking may not occur, gas breakthrough can cause another problem—on/off cycling of the ESP even when there is no fluid to pump. Without fluid to cool the pump, the failure rate increases. Even if spurious cycling does not cause failure, it still reduces ESP efficiency.[3]

Still another problem associated with electric submersible pumping in a CO_2 flood is operation outside of optimal conditions. ESP's have a limited operating range, and, unfortunately, fluid production volumes can change dramatically during the WAG cycle. If WAG injection cannot be managed in a way that steadies a well's productive capacity, then a variable speed drive (VSD) can be installed to adjust the pump speed.

Brownlee and Sugg reported success using VSD's at the East Vacuum Grayburg-San Andres Unit.[27] Not only were VSD's used to fine-tune pumping operations at individual wells, but their use also allowed the ESP's to be moved from well to well.

Masoner and Wackowski[1] were not as satisfied with VSD operations at Rangely. They reported high operating costs for VSD's, and they preferred to install smaller ESP's to handle the reduced volumes that occurred under WAG injection. The move to smaller ESP's led to fewer failures and lower repair costs.

Natural Flow. In some situations, converting the well to flow may be better than using artificial lift. Flowing wells reduce operating costs because they consume no electricity and have no moving parts.

One of the problems associated with flowing wells is that if reservoir conditions change and flow ceases, the that flow cessation may not be apparent. A lack of flow causes no surface indication between well tests, in contrast with rod pumping, where the beam stops moving, or submersible pumping, where a chart records the pump's activity. Masoner and Wackowski[1] observed this phenomenon but reported that even though the flowing performance of Rangely wells was inconsistent, reduced production was acceptable in exchange for reduced operating costs.

On the other hand, pumping a well may draw down the bottomhole pressure more than allowing the well to flow. Fleming notes that while converting wells to flow saves lift expenses, production can be maximized by installing equipment to delay this conversion.[28] She states that wellheads were modified at the Denver Unit by installing electric feed-through (EFT) connections on submersible-pumped wells to contain pressure at the wellhead and prevent gas locking. For rod-pumped wells, high-pressure double-pack stuffing boxes were installed to perform a similar function.

Conversion Decisions. To make the decision about converting from a rod pump to an ESP, the only datum required is an estimation of the total fluid rate, because the calculations to determine the maximum capacities of rod and submersible pumps are straightforward. However, deciding on the best time to switch from pumping a well to flowing it is not so easy. One must predict both the fluid rate and the GLR, and then calculate whether or not flowing conditions are to be expected.

Greene's method, which uses inflow and outflow performance curves, can be used to calculate an envelope of flowing conditions.[29] A graph of this envelope for a typical well in the Denver Unit is shown in **Fig. 8.16.** In this plot, a stable flow region exists at GLR's above 1,000 scf/bbl and between about 150 and 350 BFPD. If operations are planned within this stable flow region, then converting it from artificial lift to flowing would be warranted; if the well would not be operating within the stable region, then artificial lift would be better.

Clearly, it is important to maintain open communications between those who are managing the WAG injection process and those responsible for optimizing lift performance. Ideally, the flood should be managed to keep wells either within the region of stable flow or within the pumping region, because repeated switching between flowing and pumping configurations is expensive.

Recommendations. We recommend the following approach to lift:

• Decide on on a lift strategy, i.e., minimize costs by flowing or maximize production by pumping.

• Manage the WAG cycle in a way that minimizes production swings and maximizes performance.

• Plan for artificial lift changes to be needed as the project matures.

• Be flexible.

8.2.2 Wellbore Remedial Work. ***Selection of Workovers.*** There are several major differences in wellbore remedial work between a waterflood and a CO_2 flood:

• Selection of workovers for producing wells in a CO_2 flood is not as straightforward as in a mature waterflood.

• Chemical treatment can be required to solve problems of scale, paraffin, and asphaltene deposition caused by injection of CO_2 into a reservoir.

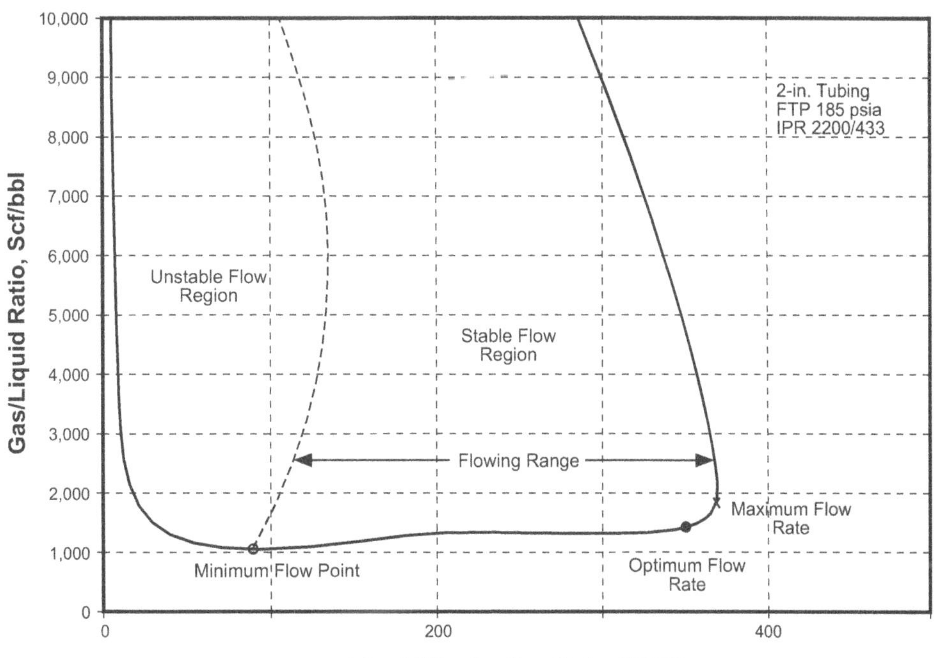

Fig. 8.16—Envelope of flowing conditions for a typical Denver Unit well (after Ref. 29). Conversion from pumping to natural flow is warranted within the stable flow region.

• CO_2 breakthrough is more costly than water breakthrough.

• Safety considerations are more important because production of CO_2 increases surface tubing and casing pressures and makes well control even more important during workover.

These problems are surmountable, and production problems have not been a major factor in CO_2 flooding.[17]

Selecting producing well workovers in a CO_2 flood is not as straightforward as in a mature waterflood. Total fluid production should remain constant or increase gradually in a mature waterflood, and a decrease in fluid production often indicates formation damage (barring mechanical problems or operational changes).

In a CO_2 flood, however, total fluid production measured in reservoir volume initially decreases because relative permeability effects cause a reduction in amount of water that can be injected. Later, after EOR response appears (indicated by an oil-cut increase), the total fluid production levels off—unless gas interference in the production wellbore causes fluid production to decrease further. Before EOR response and stabilization of total fluid production, selecting wells that result in successful workovers is difficult. After EOR response, however, the major factor in selecting producer workover candidates is the same as in a mature waterflood—a reduction in total fluid production.

Nevertheless, one should approach injection-well workovers in a CO_2 flood with caution because a decrease in injection rate may be caused by mobilization of the oil bank or by relative permeability effects, rather than by a problem that can be remedied by a workover. A review of workovers in the Slaughter Estate Unit CO_2 flood for the 1988–1990 period reveals that slightly more than one-half of the workovers were performed on wells without formation damage. In many cases the decreased fluid production that led to the workovers was related to mobilization of the tertiary oil bank, and the workovers were not really needed. Thirty-seven percent of the workovers were performed on wells with calcium sulfate and calcium carbonate scale damage (some of these wells also had damage related to asphaltene and paraffin deposition), and nine percent of the workovers were on wells that had only asphaltene and/or paraffin deposition problems.

Chemical Treatment. Two factors that affect well chemistry in a CO_2 flood are lower pH and lower temperature. As CO_2 mixes with water in the reservoir, it forms carbonic acid and lowers the pH of reservoir fluids, as shown in **Fig. 8.17.**[28] Bottomhole temperatures drop significantly in CO_2 production wells because of the Joule-Thompson cooling effect caused by CO_2 expansion as it enters the wellbore. For example, Fleming reports bottomhole temperature drops of 65°F at the Denver Unit,[28] and at the Levelland Unit pilot, a bottomhole temperature of 36°F was recorded in a well producing at a GOR of over 20 Mcf/STB.

As produced fluids move up the wellbore past choke points, additional temperature drops occur because pressure continues to decrease. This reduced temperature can lead to increased scale, paraffin, and asphaltene deposition; on occasion it even can cause fluids in the wellhead to freeze.

Several authors recommend reviewing scale and paraffin inhibitor performance for a CO_2 flood because the pH of the well may fall as low as 4.7, a level at which many conventional inhibitors are ineffective. According to Shuler, the performance of an inhibitor at a low pH cannot be predicted based on its performance at neutral pH.[30]

Scale Formation. Increased scaling problems in CO_2 floods have been reported by Shuler *et al.*,[30] Brownlee and Sugg,[27] Fleming *et al.*,[28] and Morgenthaler *et al.*[31] The change is related mainly to temperature, the effects of which are shown in **Fig. 8.18.**[31] Lower bottomhole temperature increases gypsum scaling tendencies and decreases calcite scaling tendencies. **Fig. 8.19**[31] shows overall trends in gypsum and calcite scale as determined by Morgenthaler *et al.* Gypsum scaling predominates in the wellbore, while calcite scale is more likely to be found in low-pressure surface equipment.

While both polymer and phosphonate treatments may be used to combat scale, a review of CO_2 flood case studies indicates a preference for phosphonate inhibitors. Schuler *et al.*[30] at the Rangely Weber Sand Unit selects phosphonates over polymer-based scale inhibitors because residuals can be determined more reliably (to determine how often to treat the well), because the chemicals are less expensive, and because phosphonates provide the longest-acting squeezes. Fleming *et al.*[28] also reports that phosphonate treatments have performed better than polymers at the Denver Unit.

Brownlee and Sugg[27] discuss the use of phosphonate inhibitors to combat sulfate scale problems at the East Vacuum Unit. As with the Rangely Unit, phosphonate levels in produced fluids at East Vacuum are monitored to determine timing and necessity for squeeze treatment. Brownlee and Sugg also note that the operator does not rely on squeeze treatments for electric-submersible-pumped wells, but uses surface chemical pumps to inject the phosphonate inhibitor continuously.

Paraffins. Paraffin problems have been reported in various fields.[6,27,28,32,33] These deposits form when the temperature of the crude oil drops below its cloud point, which generally ranges between 60 and 65°F. Conditions favorable to paraffin deposition

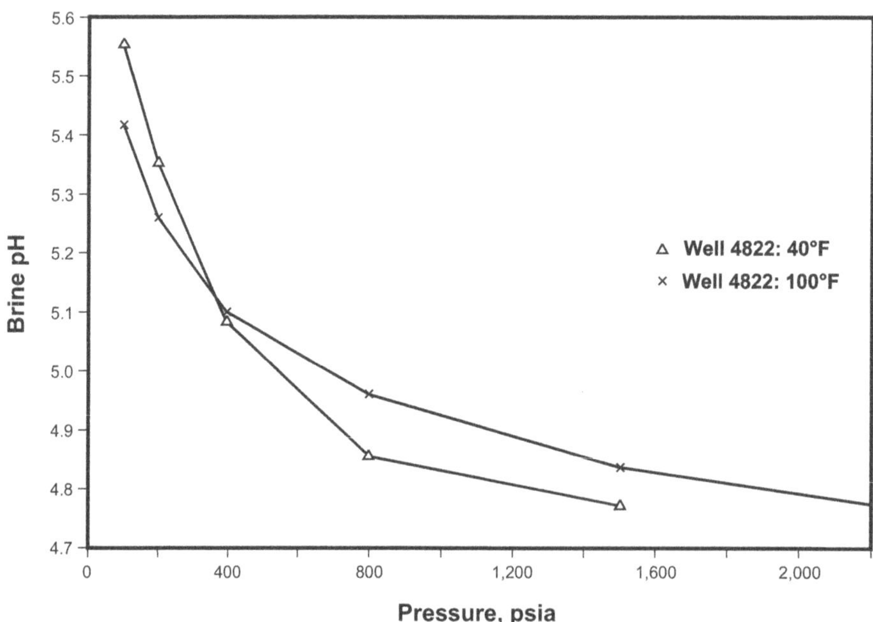

Fig. 8.17—Brine pH decreases as CO_2 pressure (and resultant solubility) increases and as temperature decreases (after Ref. 28). These data are taken from two CO_2 response wells in the Denver Unit.

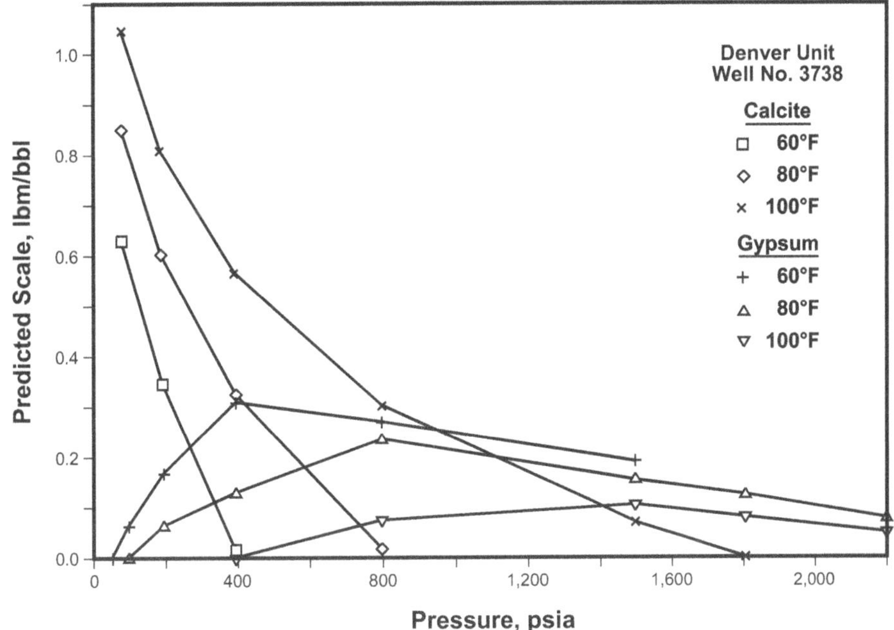

Fig. 8.18—Gypsum and calcite scaling tendencies as a function of pressure and temperature (after Ref. 31). These curves are for Denver Unit Well No. 3738.

possibly are created when CO_2 expands as the reservoir fluids flow through into the wellbore and up the tubing and annulus.

Operators handle paraffin deposition in several ways. To remove existing deposits, some companies use hot oiling or hot water combined with a paraffin solvent,[6] while others pump heavy aromatic solvents downhole.[27] Mechanical cleanouts also are used. In efforts to prevent paraffin deposition, operators may increase backpressure on the wells to keep both CO_2 and light-end hydrocarbons in solution.[28] In some cases, downhole heaters even are used.[33] Crystal modifiers, which raise the cloud point, can be very effective, but are costly to use.* Perhaps because of varied crude compositions and operating conditions, an industry consensus has not formed on how best to handle the problem.

Asphaltenes. Increased asphaltene precipitation has occurred in many CO_2 flood fields.[25,27,28,32,34–38] Normally, asphaltene deposition does not occur during primary or waterflood operation, but often appears soon after CO_2 breakthrough.[38] Asphaltenes can hang up plungers, clog wellheads, and cause plugging in tubulars, chokes, surface, and production lines. At the Little Creek Field, Hansen[32] reported that asphaltene problems were most severe during cold weather, and according to Srivastava,[34] the most important factor governing deposition is the concentration of CO_2 in the oil. He also notes that asphaltene flocculation is relatively insensitive to operating pressure.

Fortunately, there is little evidence of production declines caused by asphaltene deposition in the reservoir—problems seem to be confined to production equipment.[37,38] The most common treatment to remove deposits is the use of solvents, particularly xylene. Other methods of dealing with asphaltene deposition are mechanical cleanouts and the use of hot water to clean flowline plugging.

To prevent asphaltene deposition, Fleming reports that backpressure is successful in keeping light hydrocarbons in solution,[28] which helps prevent asphaltene deposition. As noted earlier, the Denver Unit uses high-pressure wellhead equipment—either high-pressure stuffing boxes for rod pumps or EFT submersible wellheads for ESP's.

Recommendations. Based on the industry experience documented above, we make the following recommendations:

- Gather water chemistry data to establish preCO_2 baselines.
- Be prepared for increased scaling problems.
- Test inhibitors on conditions anticipated during the CO_2 flood.
- Reduce pressure drops where possible.
- Experiment with various chemical solutions.
- Attempt to distinguish between natural injection and production rate declines and those caused by formation damage and wellbore plugging.

CO_2 Breakthrough. Carbon dioxide breakthrough can be a significant problem, reducing sweep efficiency and increasing operating costs. Besides managing WAG ratios and half-cycle slug sizes, the industry has tried many mechanical remedies to reduce unwanted CO_2 production: cement plugbacks and squeezes, liners, selective injection equipment, selective production equipment, gel polymer squeezes, and CO_2 foam injection. Unlike the first four techniques, gel polymer squeezes and CO_2 foam injection work beyond the near wellbore region.

Cement Squeezes and Plugbacks. Cement squeezes and plugbacks are comparatively inexpensive. The squeezes are not always effective because carbonic acid, which forms when CO_2 and water mix, dissolves cement. Plugbacks may work better because a cast-iron bridge plug provides an additional barrier to flow.

Fiberglass Liners. Fiberglass liners can help control CO_2 breakthrough by controlling the injection profile at the wellbore. They have been run into injection wells across pay in fields in Texas and

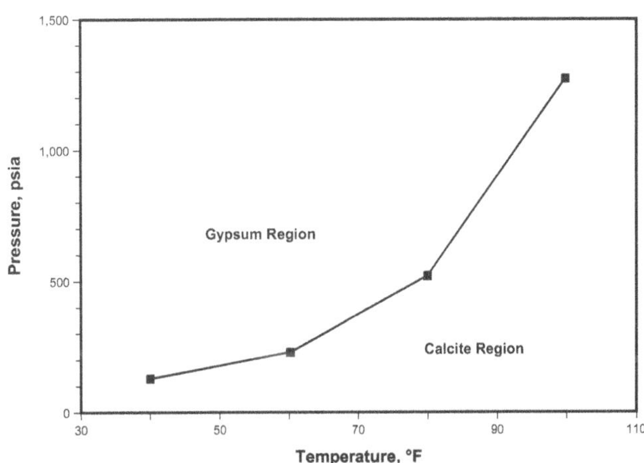

Fig. 8.19—Gypsum vs. calcite scaling (after Ref. 31). The occurrence of calcite or gypsum scale at the Denver Unit depends on both temperature and pressure.

*Personal communication with M.T. Cox, ExxonMobil Production Co. (2000).

New Mexico.[39,40] Injection profile logs for wells that use fiberglass liners demonstrate an improvement in control.

Selective Injection Equipment. Selective injection equipment (SIE) was used at the Rangely Weber Sand Unit.[41] It consisted of several packers and mandrels placed vertically across the Weber interval, with wireline-retrievable chokes placed in the mandrels to control the amount of fluid entering each major pay zone. The operator estimated in 1991 that the installation of SIE increased recovery by more than 6% of the OOIP. Subsequent experience revealed that maintaining full injection coverage over the entire injection interval often required removal of the SIE, which required expensive fishing operations. Further analysis indicated that the operational cost of SIE completions did not support the benefit, and the SIE was reduced to one or two isolation packers.*

Selective Production Equipment. Selective production equipment is used at the Wellman project, a gravity-stable, vertical CO_2 flood.[16] Here, a straddle packer assembly blocks production from gassed-out perforated intervals. The equipment consists of a hydraulically set retrievable packer on the bottom and a mechanically set retrievable packer with an on/off tool at the top **(Fig. 8.20).**[16] The Wellman operator believes that this technique is more reliable than cement squeezing.

Gel Polymer Squeezes. Gel polymer squeezes have been tried in several reservoirs. They worked at the Rangely Weber Unit[1] and at the Wertz Tensleep projects.[42] And at Wertz, Borling[42] reports that acrylamide-polymer/chromium(III)-carboxylate gels improved tertiary oil recovery by 35,000 to 140,000 barrels of oil per pattern.

Foam Injection. The industry has studied CO_2 foam injection for many years because it provides temporary control of injection profiles. Although foam is not established yet as a widespread economically successful practice, several field applications show that the process may have technical merit.[43,44]

At the North Ward Estes field, the foam apparently diverted CO_2 from the thief zone to unswept regions, and incremental recovery may have resulted. Martin *et al.*[44] also report that CO_2 diversion occurred at the East Vacuum Grayburg/San Andres Unit CO_2 foam test. They add that incremental oil was produced and gas cycling was reduced, and they further state that foam injection shows promise of being an economical conformance control method.

Recommendations. Any of the control techniques discussed above may work in a particular application. When evaluating the CO_2 injection control techniques, engineers should plan multiple jobs because some failures are certain to occur. In general, the rules are to expect problems, consider several control techniques, try the less expensive techniques first, and plan multiple jobs so that a technique can be evaluated adequately.

8.3 Facility Management

Facility management for a CO_2 flood is similar to that for waterflood operations. The major differences between the two processes are discussed below, and involve the reinjection plant, separation and metering, corrosion, and facility optimization.

8.3.1 Reinjection Plant. In most ways, CO_2 compression is similar to that of natural gas. This section focuses on operational differences between the two. In general, there are five special factors that must be addressed in CO_2 compression:

- Flowline maintenance.
- Pressure swings at the compressor inlet and discharge side.
- Interactions between CO_2 and the compressor lubricating oils.
- Alterations in dehydrator operation.
- Erratic CO_2 volumes at startup.

Flowline Maintenance. Melzer *et al.*[45] note several flowline maintenance problems in a CO_2 flood, the most important of which relate to low temperatures. Because CO_2 is much more compressible than hydrocarbon gases such as methane, the Joule-Thompson effect is pronounced. When CO_2 expands, its temperature drops dramatically. Because CO_2 expands from the moment it flows up the well with the oil and natural gas, hydrocarbons emerge from the ground much colder than they would be in a waterflood. As the CO_2 escapes from the hydrocarbons during transportation and processing, the temperature of the oil and gas in the flowlines continues to drop, giving rise to two potential problems: paraffins and hydrates.

*Personal communication with L.O. Masoner, Chevron USA Production Co. (2000).

Paraffin is the more common of these two problems. At low temperatures, typically below 60 to 65°F, paraffins precipitate and deposit in flowlines. The deposits either block flow directly or break off in chunks to clog downstream valves. Hydrates also block flowlines. Hydrates are slushy crystals of CO_2 and water that may form when the temperature drops below about 55°F (when pressures are below 5,000 psia). The risk of hydrate formation may be gauged using the plot in **Fig. 8.21.**[46] Hydrates may form in Regions 2 and 5 of the illustration, while ice may form in Region 3.

Blockages occur in locations that allow entrained CO_2 to expand and/or vaporize, for example, in pipe elbows upstream of the compression plant.* Both paraffin and hydrate formation can be reduced or prevented by reducing the pressure drop that occurs both between the wellhead and the separation facilities and between the separation facilities and the CO_2 processing plant.

Once CO_2 breakthrough occurs, leaks become a potentially major problem for several reasons. First, varying amounts of CO_2 produced by the well can create gas surges that cause lines to fail, particularly at points where temporary repair clamps are used. Second, paraffin and hydrate plugs can cause pressure-related failures.

Melzer *et al.*[45] also note that corrosion in small couplings and collars may be a common problem on headers because of leakage of CO_2 and produced water, which can occur anywhere a connection seal is used. Teflon-based threaded components used with Teflon tape appears to be a good preventive measure for this problem.

*Personal communication with R.G. Thompson, Shell CO_2 Co. (1997).

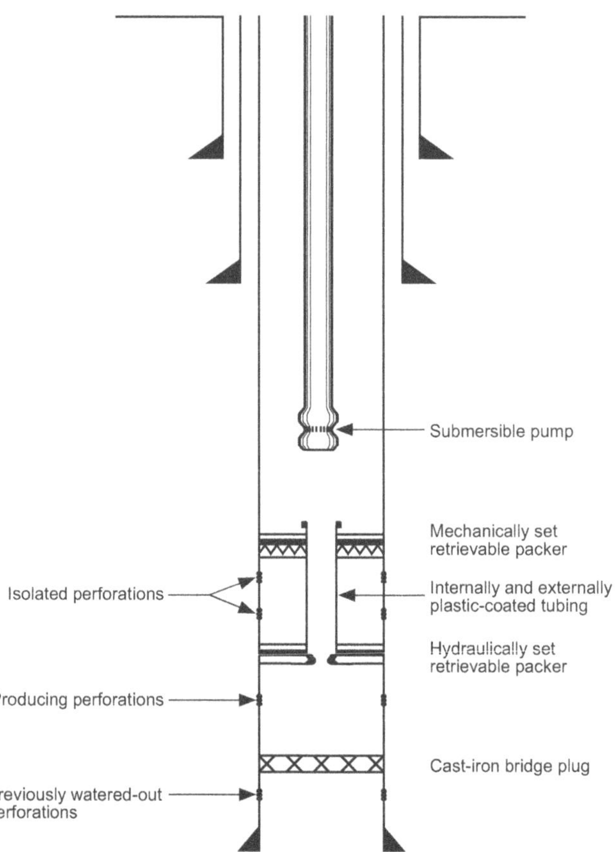

Fig. 8.20—Typical downhole isolation equipment for recompletion of wells at the Wellman Unit (after Ref. 16). This straddle-packer configuration blocks production from gassed-out perforated intervals.

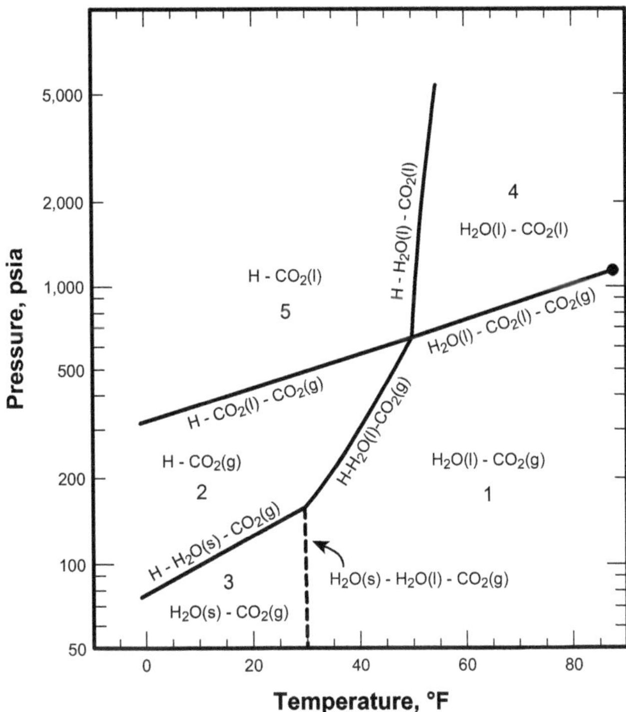

Fig. 8.21—Pressure/temperature projection for CO_2/water systems (after Ref. 46). Hydrates of CO_2 and water may form in Regions 2 and 5, while ice may form in Region 3.

Pressure Swings at the Compressor Inlet or Outlet. Large pressure swings at the compressor inlet stem from the much greater temperature sensitivity of CO_2 compared to that of natural gas. While day/night and seasonal temperature differences have little effect on an aboveground natural gas flowline, they can have a sizable effect on CO_2 flowlines. Heat from sunlight can increase the CO_2 pressure inside the line dramatically, while cooler nighttime temperatures may cause enough loss of pressure that the CO_2 stops flowing altogether. These problems are reported by Skopak and Phillips at the Ford Geraldine Unit[47] and at the Allison (Coal Bed Methane) Unit.*

Outlet pressure swings also may create problems. WAG changes, temperature variation, wellwork, CO_2 purchases, and other factors all may cause discharge pressure fluctuations that would result in reduced injection.

Drastic pressure changes in flowline pressure cause similarly drastic swings in compression ratios at the plant. In the worst case, the compressor's piston rods may become overly stressed and may bend, halting operations. While compressors generally have sensors to shut down operations when stresses rise, it is not uncommon for operational personnel to disable sensors for the sake of convenience.

In other cases, pressure swings may cause a line failure, a diversion of gas to flare stacks, and/or the need to start or stop a compressor. To manage these operational problems, the operator may need to intervene by shutting in or opening producing or injection wells to control the pressure swings.

Any pressure change that causes these operational problems is drastic. The equipment may be designed to move the drastic range outside of typical operations. Solutions may include installing VSD's that automatically adjust compressor speed and laying lines with pressure ratings far in excess of expected maximum pressure. In any case, the ability to divert flow to flare stacks and shut-in wells is required.

Ideally, CO_2 lines should be buried deeply enough so that the problem of temperature swings does not arise. In the Permian Basin of west Texas, burying the line 2.5 to 3.0 ft deep normally prevents serious day/night fluctuations.

*Personal communication with R.T. Bradley, Shell CO_2 Co. (1998).

Interactions Between CO_2 and Compressor Lubricating Oils. Carbon dioxide is used as an EOR injectant because it is soluble in oil, but this advantage becomes a disadvantage at the CO_2 compression plant. First, the dissolution of CO_2 in the compressor's lubricant renders the lubricant less effective and more likely to form deposits on cylinder walls. Second, the oil tends to dissolve in the CO_2 and travel downstream with it. If downstream seals are not compatible with the oil, troublesome and potentially dangerous leaks can occur. For these reasons, it may be advisable to use dual packing systems on CO_2 compressors to keep the lubricant and CO_2 separated, and to use a hydrocarbon or synthetic oil that will be compatible with downstream seals. Compressors at the McElmo Dome CO_2 source field experienced a buildup of hydrocarbon deposits while using hydrocarbon lubrication oil. Such a buildup did not occur while using a synthetic oil.

Be aware, too, that two lubricating systems exist in a gas compressor, one for the cylinders and the other for the crankshaft and piston rods. In CO_2 compression, the crankcase and the cylinder/packing lube systems cannot be segregated completely, and identical or compatible lubricants should be used for both systems.**

Dehydrator Issues. The dehydrator removes water from CO_2 to prevent the formation of carbonic acid, which is extremely corrosive to bare carbon steel.

Ball and Harrell[48] recommend that the operator seek assistance from an experienced CO_2 dehydration equipment supplier when designing and operating the equipment. They summarize several issues that affect dehydrator operation in CO_2 compression:

- Carbon dioxide contains more water vapor at saturation than does natural gas, which means that larger equipment is needed to remove a greater volume of water.
- Because CO_2 is soluble in glycol, the load on the glycol regenerator normally is higher when dehydrating a CO_2 stream than when dehydrating a natural gas stream. When glycol pressure is reduced in the regenerator, the dissolved CO_2 is released, cooling the glycol much more than does the release of natural gas. Dehydrator operation in a CO_2 flood will entail higher energy costs than normally experienced when processing natural gas.
- More glycol is lost in CO_2 recycling than in natural gas processing because CO_2 is more soluble in glycol than is natural gas.
- Minor general pitting corrosion is likely to occur.

Erratic CO_2 Volumes. One final difference between natural gas and CO_2 compression in an EOR field is that the operator may need to deal with problems associated with compressing small erratic volumes of CO_2 during the startup phase of the flood. Generally, the time required for gas production to stabilize is about 2 years. (Even after this point, erratic volumes can occur, especially if the WAG cycles are not managed closely.)

During the early period of erratic CO_2 volumes, a means of handling large slugs of gas that exceed compression capabilities must be found. There are several possibilities. One option is venting. Another is to install compression in small increments. A third choice is to install compressors with VSD's that can handle a range of operating conditions. There are disadvantages with each of these methods, however—venting may not be permitted, and installing compression in small increments or purchasing VSD's may increase operating costs. A fourth and final option is to curtail production.

8.3.2 Separation and Metering. Separation and metering of produced fluids in a CO_2 flood is not significantly different from separating and metering fluids in a waterflood. **Fig. 8.22** shows a typical CO_2 flood operation.

Reservoir fluids in a CO_2 flood—oil, hydrocarbon gas, water, and CO_2—are produced at the wellhead assembly in much the same manner as they are in a typical waterflood. The major difference is that in a CO_2 flood, gas flows up the annulus of producing wells in a greater amount and at higher velocity. This gas is mostly composed of CO_2, and when recombined with the oil and water, it helps move the pumped fluid down the flowline to the production satellite.

**Personal communication with R.G. Thompson, Shell CO_2 Co. (1997).

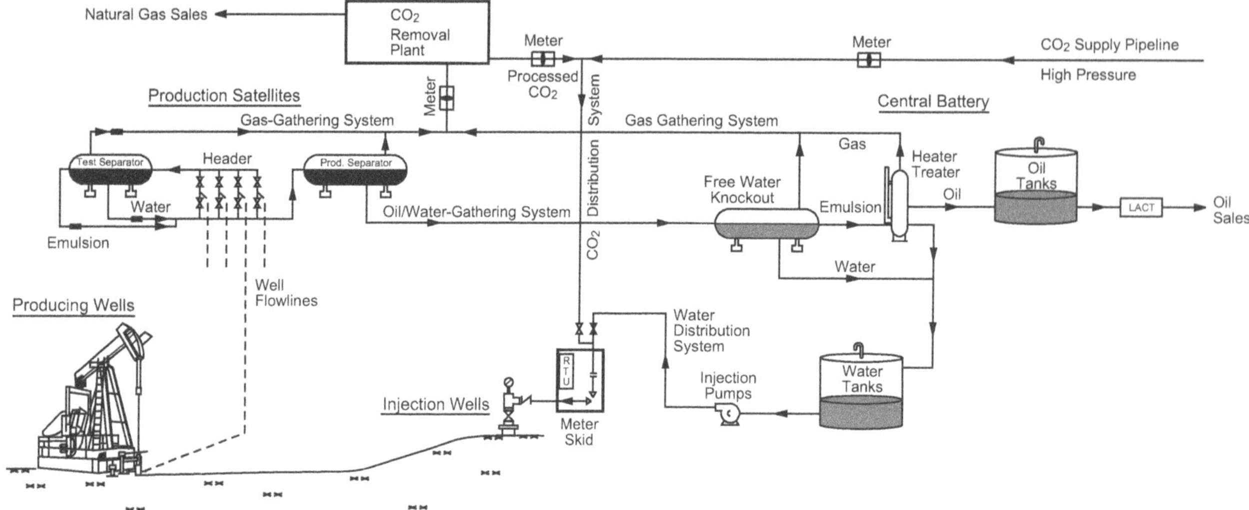

Fig. 8.22—Typical CO_2 flood operation (after Ref. 45).

The production satellite often consists of two units: a three-phase test separator and a larger two-phase production separator.

Test Separator. The test separator exists solely for the purpose of measuring the performance of individual wells. The output from each well cycles periodically to the test separator so that all fluids from that well (oil, water, gases) can be separated and metered before being recombined and sent on to the production separator. In some cases, infrared sensors measure the amount of CO_2, water, and hydrocarbons in the gas phase.

The frequency of well tests varies significantly in practice. In smaller fields such as Hanford,[8] or large fields in which conditions change rapidly, such as Salt Creek,[35] wells are tested four to five times a month. For most units in which conditions do not change rapidly, such as the Denver Unit, wells are tested monthly.

Anywhere CO_2 has not yet broken out of a fluid, it is possible for oil foaming and paraffin buildup to occur, and the test separator is no exception. Foaming can be controlled through continuous injection of antifoam chemicals.

Paraffin buildup can be prevented or controlled with a paraffin dispersant, solvent, or inhibitor. If a paraffin solvent is used, the operator should perform a solvency test to choose the most suitable solvent. A dispersant chemical can be continuously injected into the system, or it can be used in conjunction with hot oil or hot water treatments. Existing paraffin deposits can be removed by hot oiling or hot water treatments.

One other problem that can occur, especially in the test vessel, is a gas surge with enough force to blow all of the fluid out of the vessel. If CO_2 breakthrough occurs suddenly and the test separator is not large enough, the gas will run down the liquid flowlines into the production separator and the test results will be incorrect. If a surge of sufficient force hits the production separator and the low level shutdown equipment fails, gas will be sent on to the central battery, where it may cause an upset.

Production Separator. The production separator separates liquids from gases, although in some cases, such as the Rangely Field, three-phase separators are used successfully at satellites.[25]

Under waterflood conditions, liquids from the production separator normally flow directly to the central battery for further separation of oil and water. However, a CO_2 flood changes conditions in two important ways—it dissolves CO_2 in the oil and water and it causes greater volumes of oil and gas to be produced. If the tankage and residence time at the production separator is inadequate for CO_2 to break free of the oil, adding a free gas knockout may be necessary.

Central Battery. After most of the gas has been removed from the fluid, the oil/water emulsion moves to the free water knockout at the central battery where entrained CO_2 continues to break out.

In some cases, the CO_2 that remains in the oil can react with paraffins to create a paraffin interface at the oil/water contact. This interface interferes with the separation process at the battery and causes higher water carryover downstream. Although paraffins can be removed by scheduled cleanouts of the free water knockout, problem prevention through the use of continuously injected emulsion breaker chemicals is preferred.[45] Operators often ask chemical vendors to perform shakeout tests when they are selecting emulsion breakers.

Carbon dioxide also can cause separation difficulties simply as a result of the oil cooling that occurs when entrained CO_2 vaporizes. Bangia reports that at the Wellman field, the surface temperature of the oil may be less than 10°F at wells producing 5 MMcf/D of gas, while the bottomhole temperature is 150°F. Such low temperatures not only may cause hydrates to form at well restrictions (see Sec. 8.3.1) but also may make separating the oil and water very difficult. The Wellman operator finally installed indirect heaters at the batteries to solve the separation problem.[49]

Downstream of the free water knockout, all facilities should operate exactly as they would in a waterflood. Oil moves on to the oil tanks LACT unit for sale, and water moves to the skim tanks and then to the injection system. Any gas that continues to break out at those tanks is caught by a vapor recovery system.

Metering. Metering of CO_2 is a vital part of the surveillance plan. Metering occurs at several points:

• At the CO_2 custody transfer meter for CO_2 entering the site. The custody transfer measuring device usually is an orifice meter that measures purchased CO_2 with an accuracy within 1%.

• At the CO_2 recycle meter at the compressor. This meter often is an orifice meter similar to the custody transfer meter. The recycle meter determines the amount of CO_2 returned to the field for reinjection. (If the gas is transported to a processing plant, a custody transfer meter is installed at the plant inlet, in addition to the recycle meter at the plant tailgate.)

• At the CO_2 allocation meters at each wellhead injection skid to determine how much CO_2 is injected at each well. These may be turbine, wedge-type, or orifice meters.

• At test meters at the test separator to measure the amount of oil, water, natural gas, and CO_2 that is produced by each well. These normally are turbine meters.

The sum of the allocation meter reading should be compared to the sum of the recycle and purchase volumes on a regular basis to see that both totals are trending similarly. If the totals start to diverge, maintenance personnel should check the meters to determine the cause of the change. Certain allocation meters may need repair or replacement—or you may even find that one of the custody transfer meters is not performing properly.

Because all flood monitoring depends on accurate data, the importance of accurate metering cannot be overstated. Note that the gas temperature can make a significant difference on turbine

TABLE 8.2—FAILURES/YEAR AT FORD GERALDINE DURING CORROSION CONTROL PROGRAM	
Rod parts (pumping wells)	0.44
Subsurface pumps	0.10
Tubing leaks	0.05
Total failure frequency (pumps and flowlines)	0.41

After Ref. 26.

accuracy. At Rangely, injection targets normally are 10% higher than average during the warm weather months.*

Standardization is one of the factors that should be taken into account in planning an approach to metering. If all equipment is standardized, then measurement and maintenance should be less of a problem than if different systems are used at random.[45]

Most operators use turbine meters to measure the gas production rate, although wedge meters are installed parallel with turbine meters at test facilities in the Salt Creek field. Gas is directed to the turbine meters when the production rate is very low (less than 250 Mcf/D). The wedge meters have a wider gas measurement range than do turbine meters, and can better handle varying production. They also require less maintenance.**

8.3.3 Corrosion Control. Corrosion has been a concern in CO_2 flooding from its inception. It is well known that when CO_2 dissolves in water, a small fraction hydrolyzes to form carbonic acid. (The remainder exists as physically dissolved carbon dioxide.) Carbonic acid dissociates to form bicarbonate ions, carbonate ions, and hydrogen ions.

Carbonic acid is quite corrosive to most carbon steels. In 1965, in an early CO_2 test at the Mead Strawn field, corrosion coupon test results indicated that corrosion rates in CO_2 injection wells were nil during CO_2 injection, but increased to 4 to 60 mil/yr during the early stages of carbonated water injection.[50] When a corrosion inhibitor and a bactericide were added to the water injection system, the corrosion rate dropped to 1 to 6 mil/yr. It is important to note that in this test, corrosion was significant in producing wells outside the CO_2 pilot area; so while corrosion may have been exacerbated by CO_2 injection, it was present before the start of the CO_2 flood.

The Little Creek pilot in Mississippi also encountered significant corrosion.[32] Hansen reports that despite using plastic-coated tubing and monthly batch inhibitor treatments, corrosion still caused producing well problems. On the other hand, he noted that there were no corrosion problems in surface facilities that were protected by inhibitor treatments or in the CO_2 injection well in which pure CO_2 was injected. Corrosion was not a problem at the gas recycling equipment, either, because the recycled gas stream was dehydrated effectively. In the full-scale project, the operator used continuous inhibition for the wells and facilities and encountered virtually no problems.†

While the early pilots noted significant corrosion problems, several CO_2 floods were reported as operating with almost no corrosion. These include the Ford Geraldine Unit,[51] where the operator started an extensive and carefully monitored corrosion control program, the result of which was that total failures actually decreased after CO_2 injection began. The program included batch and squeeze treating for the wells, and continuous monitoring of the iron count in the produced water. When the iron count rose above a predetermined level (40 ppm), the treatment was repeated.‡ **Table 8.2** shows failure rates per year per well.

At the Twofreds field,[6] the operator also controlled corrosion with chemical inhibition, treating 35 pumping wells weekly and 10 flowing wells every two months with an oil-soluble blend of a dimertrimer acid and imidazoline. For the 45 wells, there were no corrosion-related failures for a period of seven years, from the start of the program until the data were published.

At the Wellman field,[49] Bangia *et al.* report that yearly ultrasonic testing of all lines and vessels identified almost no corrosion after eight years of operation. They did note that casing inspection logs revealed moderate to severe corrosion of the production casing below the pump intake, which is a section of pipe that could not be treated regularly for corrosion.

The Denver Unit also saw higher corrosion-related failures of injection tubing in the WAG injection area than in the continuous CO_2 injection area.* The Rangely Weber Sand Unit also experienced corrosion problems.** On the other hand, authors such as Fleming *et al.*[28] note that corrosion problems can be handled effectively and do not cause problems that have severe economic impact.

Based on the published record, operators normally are successful at controlling corrosion when they have in place carefully implemented corrosion control programs.

8.3.4 Facility Optimization. As a field progresses through the life of the CO_2 flood, conditions change dramatically. **Fig. 8.23** represents the changing amounts of oil, water, and CO_2 production forecasted for a west Texas flood during the expected 20 to 30 years of its EOR life. In general, CO_2 begins breaking through a year or so after startup. At the same time, the volume of water produced declines significantly. Although water production declines, there probably is more than enough water for WAG injection, and some means of water disposal may need to be found. If a tapered WAG cycle is used, the need for water will increase gradually, and when the use of CO_2 is no longer economical, water disposal needs will turn to water supply requirements.

Dealing with gas- and water-handling systems is the major challenge of facilities optimization. Fortunate operators are able to implement the flood plan as designed, and the capital investments required for new gas- and water-handling systems will be timed to coincide with actual field needs. In most cases, however, operators must respond to events as they unfold.

CO_2 Handling Considerations. The breakthrough of CO_2 has two important implications: (1) CO_2 processing facilities must be available—or else wells producing breakthrough CO_2 must be shut in, and (2) natural gas sales to existing gas plants or pipelines will be curtailed as CO_2 breakthrough at individual wells causes the produced gas to deviate from pipeline specifications.

In many cases, operators construct a second, parallel gas-gathering system. The original system continues to transport sales gas while the new line collects CO_2-contaminated gas that does not meet pipeline specifications and transports this gas to the CO_2 dehydration and recompression facility. As CO_2 contaminates the produced gas stream, wells are switched from the sales system to the reinjection system.

Parallel systems may operate at different pressures—if the CO_2-contaminated gas is gathered at a higher pressure, this can reduce recompression costs.

Water-Handling Considerations. As water production decreases over time, the operator can reduce operating costs by consolidating water plants at larger fields. This has been done both at the Rangely field[25] and at the Denver Unit.

Another opportunity for optimization is filtering the water before reinjecting it. Produced water invariably contains particulates that, when reinjected, can plug the formation and require major well workovers to restore injectivity. Changing filters at the surface water facility is a nuisance, but it may prevent considerable expense in downhole cleanout workovers. Operators that report water filtering include the Hansford Unit,[52] Twofreds,[6] and Rangely.[25] Those who filter their injection water reduce injection problems.

References

1. Masoner, L.O. and Wackowski, R.K.: "Rangely Weber Sand Unit CO_2 Project Update: Decisions and Issues Facing a Maturing EOR Project,"

*Personal communication with R.K. Wackowski, Chevron USA Production Co. (2000).
**Personal communication with M.K. Moshell, Mobil Exploration and Producing U.S. Inc. (1998).
†Personal communication with R.T. Bradley, Shell CO_2 Co. (1998).
‡Personal communication with W.F. Flanders, Transpetco Engineering (1998).

*Personal communication with S.G. Hendrickson, Shell Western E&P Inc. (1998).
**Personal communication with L.O. Masoner, Chevron USA Production Co. (2000).

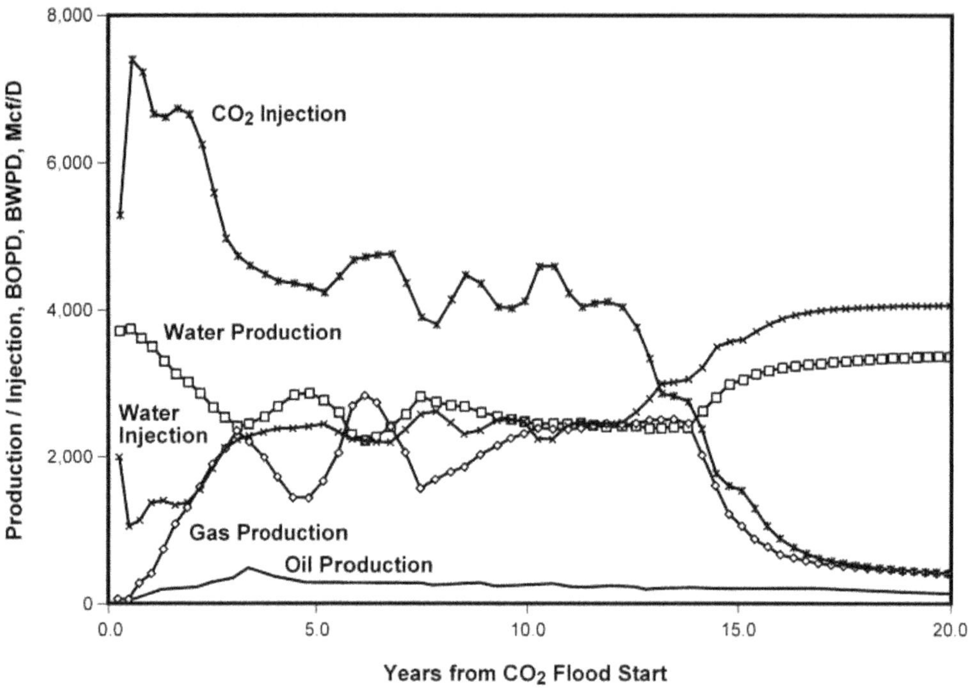

Fig. 8.23—Production and injection forecasts for a CO_2 project area.

paper SPE 27756 presented at the 1994 SPE/DOE Improved Oil Recovery Symposium, Tulsa, 17–20 April.

2. Attanucci, V. *et al.*: "WAG Process Optimization in the Rangely CO_2 Miscible Flood," paper SPE 26622 presented at the 1993 SPE Annual Technical Conference and Exhibition, Houston, 3–6 October.
3. Masoner, L.O. and Wackowski, R.K.: "Rangely Weber Sand Unit CO_2 Project Update," *SPERE* (August 1995) 203.
4. Burbank, D.E.: "Early CO_2 Flood Experience at the South Wasson Clearfork Unit," paper SPE 24160 presented the 1992 SPE/DOE Enhanced Oil Recovery Symposium, Tulsa, 22–24 April.
5. Hill, W.J. *et al.*: "CO_2 Operating Plan, South Welch Unit, Dawson County, Texas," paper SPE 27676 presented at the 1994 SPE Permian Basin Oil and Gas Recovery Conference, Midland, Texas, 16–18 March.
6. Kirkpatrick, R.K., Flanders, W.A., and Depauw, R.M.: "Performance of the Twofreds CO_2 Injection Project," *EOR Field Case Histories,* Reprint Series, SPE, Richardson, Texas (1987) **23,** 95.
7. Kane, A.V.: "Performance Review of a Large-Scale CO_2-WAG Enhanced Recovery Project SACROC Unit—Kelly-Snyder Field," *EOR Field Case Histories,* Reprint Series, SPE, Richardson, Texas (1987) **23,** 52.
8. Merritt, M.B. and Groce, J.F.: " A Case History of the Hanford San Andres Miscible CO_2 Project," *JPT* (August 1992) 924.
9. Tanner, C.S. *et al.*: "Production Performance of the Wasson Denver Unit CO_2 Flood," paper SPE 24156 presented the 1992 SPE/DOE Enhanced Oil Recovery Symposium, Tulsa, 22–24 April.
10. Ring, J.N. and Smith, D.J.: "An Overview of the North Ward Estes CO_2 Flood," paper SPE 30729 presented at the 1995 SPE Annual Technical Conference and Exhibition, Dallas, 22–25 October.
11. Stein, M.H. *et al.*: "Slaughter Estate Unit CO_2 Flood: Comparison Between Pilot and Field-Scale Performance," *JPT* (September 1992) 1026.
12. Merchant, D.H. and Thakur, S.C.: "Reservoir Management in Tertiary CO_2 Floods," paper SPE 26624 presented at the 1993 SPE Annual Technical Conference and Exhibition, Houston, 3–6 October
13. Pariani, G.J. *et al.*: "An Approach to Optimize Economics in a West Texas CO_2 Flood," *JPT* (September 1992) 984.
14. Holm, L.W. and O'Brien, L.J.: "Factors to Consider When Designing a CO_2 Flood," paper SPE 14105 presented at the 1986 SPE International Meeting on Petroleum Engineering, Beijing, 17–20 May.
15. Cullender, M.H. and Smith, R.V.: "Practical Solution of Gas-Flow Equations for Wells and Pipelines With Large Temperature Gradients," *Trans.,* AIME (1965) **207,** 281.
16. Bangia, V.K., Yau, F.F., and Hendricks, G.R.: "Reservoir Performance of a Gravity-Stable, Vertical CO_2 Miscible Flood: Wolfcamp Reef Reservoir, Wellman Unit," *SPERE* (November 1993) 261.
17. Hadlow, R.E.: "Update of Industry Experience with CO_2 Injection," paper SPE 24928 presented at the 1992 SPE Annual Technical Conference and Exhibition, Washington DC, 4–7 October.
18. Patel, P.D., Christman, P.G., and Gardner, J.W.: "Investigation of Unexpectedly Low Field-Observed Fluid Mobilities During Some CO_2 Tertiary Floods," *SPERE* (November 1987) 507.
19. Christman P.G. and Gorell, S.B.: "Comparison of Laboratory-and-Field Observed CO_2 Tertiary Injectivity," *JPT* (February 1990) 226.
20. Chopra, A.K., Stein, M.H., and Dismuke, C.T.: "Prediction of Performance of Miscible-Gas Pilots," *JPT* (December 1990) 1564.
21. Melzer, L.S. *et al.* (eds.): "How CO_2 Flood Surveillance helps assure a successful EOR Program," CO_2 Flood Short Course No. 4, Center for Energy and Economic Diversification, U. of Texas of the Permian Basin, Midland, Texas (1996).
22. Kittridge, M.G.: "Quantitative CO_2 Flood Monitoring, Denver Unit, Wasson (San Andres) Field," paper SPE 24644 presented at the 1992 SPE Annual Technical Conference and Exhibition, Washington, DC, 4–7 October.
23. Charleson, G.S., Bilhartz, H.L. Jr., and Stalkup, F.I.: "Use of Time-Lapse Logging Techniques in Evaluating the Willard Unit CO_2 Flood Mini-Test," *Miscible Processes II,* SPE Reprint (1985) **18,** 691.
24. Masoner, L.O., Abidi, H.R., and Hild, G.P.: paper SPE 35363 "Diagnosing CO_2 Flood Performance Using Actual Performance Data," presented at the 1996 SPE/DOE Tenth Symposium on Improved Oil Recovery, Tulsa, 21–24 April.
25. Wackowski, R.K. and Masoner, L.O.: "Rangely Weber Sand Unit CO_2 Project Update: Operating History," paper SPE 27755 presented at the 1994 SPE/DOE Improved Oil Recovery Symposium, Tulsa, 17–20 April.
26. Pittaway, K.R. and Runyan, E.E.: "The Fort Geraldine Unit CO_2 Flood: Operating History," paper SPE 17278 presented at the 1988 SPE/DOE Enhanced Oil Recovery Symposium, Tulsa, 17–20 April.
27. Brownlee, M.H. and Sugg, L.A.: "East Vacuum Grayburg-San Andres Unit CO_2 Injection Project: Development and Results to Date," paper SPE 16721 presented at the 1987 SPE Annual Technical Conference and Exhibition, Dallas, 27–30 September.
28. Fleming, E.A., Brown, L.M., and Cook, R.I.: "Overview of Production Engineering Aspects of Operating the Denver Unit CO_2 Flood," paper SPE 24157 presented at the 1992 SPE/DOE Eighth Symposium on Enhanced Oil Recovery, Tulsa, 22–24 April.
29. Greene, W.R.: "Analyzing the Flowing Performance of Oil Wells: Denver Unit CO_2 Flood," paper SPE 19725 presented at the 1989 SPE Annual Technical Conference and Exhibition, San Antonio, Texas, 8–11 October.

30. Shuler, P.J., Freitas, E.A., and Bowker, K.A.: "Selection and Application of $BaSO_4$ Scale Inhibitors for a CO_2 Flood, Rangely Weber Sand Unit, Colorado," *SPEPE* (August 1991) 259.
31. Morgenthaler, L.N. *et al.*: "Scale Prediction and Control in the Denver Unit CO_2 Flood," paper SPE 26603 presented at the 1993 SPE Annual Technical Conference and Exhibition, Houston, 3–6 October.
32. Hansen, P.W.: "A CO_2 Teritary Recovery Pilot Little Creek Field, Mississippi," *EOR Field Case Histories,* Reprint Series, SPE, Richardson, Texas (1987) **23,** 75.
33. Baud, W. and Eastlund, B.: "Electric Tubing Heater Improves Well Production in CO_2 Flood," *Oil & Gas J.* (18 April 1994) 60.
34. Srivastava, V.A. *et al.*: "Quantification of Asphaltene Flocculation during Miscible CO_2 Flooding in the Weyburn Reservoir," *J. Cdn. Pet. Tech.* (October 1995) 31.
35. Moshell, M.K.: "The Salt Creek CO_2 Flood: A Status Report After One Year of Operation," Southwestern Petroleum Assn. Short Course, Southwestern Petroleum Short Course Assn., Lubbock, Texas (1995).
36. Kleinstelber, S.W.: "The Wertz Tensleep CO_2 Flood: Design and Initial Performance," *JPT* (May 1990) 630.
37. Beliveau, D., Payne, D.A., and Mundry, M.: "Waterflood and CO_2 Flood of the Fractured Midale Field," *JPT* (September 1993) 881.
38. Hervey, J.R. and Iakovakis, A.C.: "Performance Review of a Miscible CO_2 Tertiary Project: Rangely Weber Sand Unit, Colorado," *SPERE* (May 1991) 163.
39. Bowser, P.D. *et al.*: "Innovative Techniques for Converting Old Waterflood Injectors to State-of-the-Art CO_2 Injectors," paper SPE 18976 presented at the 1989 SPE Joint Rocky Mountain Regional/Low Permeability Reservoirs Symposium and Exhibition, Denver, 6–8 March.
40. Lacy, R.D. and Tally, C.L.: "A Unique Method for Installing Slimline Fiberglass Liners in a CO_2 Miscible Flood," paper SPE 20124 presented at the 1990 SPE Permian Basin Oil and Gas Recovery Conference, Midland, Texas, 8–9 March.
41. Franks, A.P.: "The Use of Selective Injection Equipment in the Rangely Weber Sand Unit," paper SPE 21649 presented at the 1991 SPE Production Operations Symposium, Oklahoma City, Oklahoma, 7–9 April.
42. Borling, D.C.: "Injection Conformance Control Case Histories Using Gels at the Wertz Field CO_2 Tertiary Flood in Wyoming," paper SPE 27825 presented at the 1994 SPE/DOE Improved Oil Recovery Symposium, Tulsa, 17–20 April.
43. Chou, S.I. *et al.*: "CO_2 Foam Field Trial at North Ward-Estes," paper SPE 24643 presented at the 1992 SPE Annual Technical Conference and Exhibition, Washington, DC, 4–7 October.
44. Martin, F.D., Stevens, J.E., and Harpole, K.J.: "CO_2 Foam Field Verification Pilot Test EVGSAU: Phase IIIB—Project Operations and Performance Review," paper SPE 27786 presented at the 1994 SPE/DOE Symposium on Improved Oil Recovery, Tulsa, 17–20 April.
45. Melzer, L.S. *et al.* (eds.): "Equipping and Day-to-Day Operations of a CO_2 Flood," CO_2 Flood Short Course No. 3, Center for Energy and Economic Diversification, U. of Texas of the Permian Basin, Midland, Texas, (1995).
46. Song, K.Y. and Kobayashi, R.: "Water Content of CO_2 in Equilibrium with Liquid Water and/or Hydrates," *SPEFE* (December 1987) 500.
47. Skopak, J.E. and Phillips, L.A.: "Design and Operation of a High Pressure CO_2 Dehydration Facility," Southwestern Petroleum Assn. Short Course, Southwestern Petroleum Short Course Assn., Lubbock, Texas (1985).
48. Ball, B. and Harrell, B.: "Understanding CO_2 Dehydration with Conventional TEG (Triethylene Glycol) Systems in CO_2 EOR Projects," Southwestern Petroleum Assn. Short Course, Southwestern Petroleum Short Course Assn., Lubbock, Texas (1985).
49. Bangia, V.K., Yau, F.F., and Hendricks, G.R.: "Reservoir Performance of a Gravity-Stable Vertical CO_2 Miscible Flood: Wolfcamp Reef Reservoir, Wellman Unit," paper SPE 22898 presented at the 1991 SPE Annual Technical Conference and Exhibition, Dallas, 6–9 October.
50. Holm, L.W. and O'Brien, L.J.: "Carbon Dioxide Test at the Mead-Strawn Field," *EOR Field Case Histories,* Reprint Series, SPE, Richardson, Texas (1987) **23,** 83.
51. Pittaway, K.R. and Rosato, R.J.: "The Ford Geraldine Unit CO_2 Flood—Update 1990," *SPERE* (November 1991) 410.
52. Flanders, W.A., Stanberry, W.A., and Martinez, M.: "CO_2 Injection Increases Hansford Marmaton Production," *JPT* (January 1990) 68.

SI Metric Conversion Factors

bbl × 1.589 873	E − 01	= m^3
°F (°F − 32)/1.8		= °C
°F (°F + 459.67)/1.8		= K
ft × 3.048*	E − 01	= m
ft3 × 2.831 685	E − 02	= m^3
in. × 2.54*	E + 00	= cm
lbm × 4.535 924	E − 01	= kg
mil × 2.54*	E − 05	= m
psi × 6.894 757	E + 00	= kPa
psia × 6.894 757	E + 00	= kPa

*Conversion factor is exact.

Chapter 9

CO_2 Flood Environmental, Health, and Safety Planning

Overall, the environmental advantages of CO_2 injection outweigh the disadvantages. Although CO_2 is a greenhouse gas and venting should be minimized, large-scale releases to the atmosphere normally do not occur because produced CO_2 is reinjected into the reservoir. Using CO_2 also offers the environmental benefit of reducing the need for fresh water as a makeup injectant. In areas of the world where fresh water is both scarce and in demand for agricultural and other uses, this may be an important consideration. The use of CO_2 does add some complexity to preparing and implementing an environmental, health, and safety (EHS) plan. This chapter outlines the major EHS considerations for CO_2 EOR, including:

- The CO_2 EHS plan—Sec. 9.1 covers key elements of this plan, focusing on the factors that make a CO_2 flood more complex than primary or secondary production.
- The dangers of CO_2—Sec. 9.2 reviews atmospheric hazards, noise levels, frostbite, hydrates and ice plugs, and high pressures.

9.1 The CO_2 EHS Plan

A written EHS plan for a CO_2 flood should be prepared in advance and kept updated for each facility and operation, as it would be for any oil recovery operation. The plan should include leak monitoring and reporting, safety equipment usage, safe operating procedures, and emergency response. The emergency response plan must incorporate risk assessment, including dispersion modeling, maps of facilities and residences within the release and dispersion range, and contact numbers for local police and other emergency response agencies.

The EHS plan also should contain a training program to educate employees, contractors, and the general public about the hazards associated with CO_2. To ensure that all personnel are knowledgeable and proficient in their jobs at all times, regardless of turnover, training must be carried out in a planned manner and reported periodically to senior management responsible for overall EHS issues. Management and operating personnel share responsibility for ensuring compliance with the plan, but upper management involvement and commitment will be the driving force in making sure that the plan is, and remains, a good one.

As a normal constituent of the atmosphere, CO_2 is considered harmless. It is nonflammable, colorless, tasteless and odorless and is 1.5 times heavier than air. (At 32°F and 1 atm, the specific gas gravity of CO_2 is 1.5240.) As used in EOR, however, CO_2 can pose serious safety concerns—asphyxiation, atmospheric hazard control, noise level (during pressure relief), frostbite, hydrates/ice plugs, and high pressures. The work required to make sure that a CO_2 flood meets EHS standards presents an ideal opportunity to reassess (and potentially upgrade) all EHS systems related to the field.

9.2 The Dangers of CO_2

At normal conditions, the atmospheric concentration of CO_2 is 0.03% (300 ppm), a nontoxic amount. Most people with normal cardiovascular, pulmonary-respiratory, and neurological functions tolerate exposure of up to 0.5 to 1.5% CO_2 for one to several hours without harm. Higher concentrations or exposures of longer duration are hazardous and may cause asphyxiation if the concentration of oxygen in the air is reduced to below the 16% required to sustain human life **(Fig. 9.1).** The current U.S. standard for the maximum allowable concentration of CO_2 in the air for eight continuous hours of exposure is 0.5%, while 3.0% is the maximum concentration that operating personnel can be exposed to for a short period of time.

As Fig. 9.1 shows, when people are exposed to increasing concentrations of CO_2, the effects become more pronounced. At concentrations up to 1.5%, there are no noticeable physical consequences for one hour or more. Noticeable effects occur above this level, namely changes in respiration and blood pH that lead to increased heart rate, discomfort, nausea, and unconsciousness. At concentrations above 20.0%, death can occur in 20 to 30 minutes.[1]

People exposed to CO_2 who are still conscious and alert should be taken into fresh air and kept under observation. Unconscious or disoriented persons should be moved to fresh air and be treated with a respirator or receive cardiopulmonary resuscitation (CPR), as warranted. First aid always should be followed by a professional medical examination.

9.2.1 Atmospheric Hazard Control. Venting of CO_2 should be minimized, although for safety and other operational reasons, it sometimes cannot be avoided. Adequate ventilation *must* be provided when CO_2 is discharged into the air, and CO_2 vents should be located at high elevations and in areas where maximum dispersion can occur.

Because of its high density, released CO_2 will flow to low elevations and collect there, especially under stagnant airflow conditions. High concentrations can persist in open pits, tanks, and buildings. For this reason, operators should install monitors wherever CO_2 might concentrate and should regularly check and calibrate the equipment. It is also a good idea to have portable monitors, because CO_2 could collect in so many places that

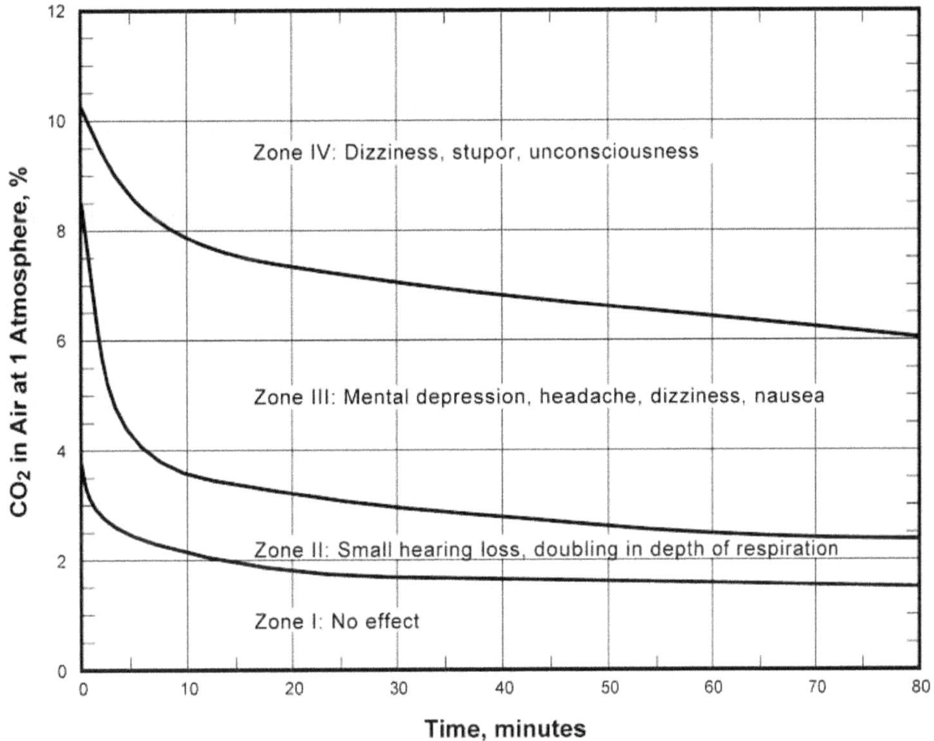

Fig. 9.1—Effects of CO_2 exposure on humans (after Ref. 1).

installation of a fixed monitor at every location would not be economically feasible.

Fixed CO_2 monitors should sound an alarm and/or turn on safety equipment such as ventilation fans when activated. They also may activate emergency shut-down procedures if the concentration is particularly high.

At least three types of CO_2 monitors are available: tube reaction (similar to a Drager H_2S monitor), thermal conductivity, and nondispersive infrared.

Personnel should take precautions whenever they enter areas where air is stagnant, such as sewers, wells, and closed-off rooms. If the presence of CO_2 is suspected, entry into such areas should be carefully planned according to written operating procedures that should include preventing additional CO_2 from entering the area, clearing the area by forced ventilation, and testing and continuously monitoring the atmosphere to detect an oxygen deficiency.

Where it is not feasible to install ventilation to create a safe breathing atmosphere in an area, personnel entering that area should be trained and proficient in the use of appropriate respiratory equipment (refer to ANSI Standard Z88.Z1969, Practices for Respiratory Protection[2] or equivalent). Currently, the only equipment suitable for this use is airline respirators or self-contained breathing apparatus (SCBA). Most operators use SCBA. Gas canisters suitable for general use are not available at this time.[3]

9.2.2 Noise Levels. High noise levels can result whenever pressure is relieved, such as when vessels are evacuated for maintenance or when CO_2 is bled from a wellhead before switching to water injection. Hearing protection *must* be worn whenever the noise exceeds 90 dB for an extended period of time (United States Dept. of Labor, Occupational Safety and Health Standards 29 CFR Part 1910[4]).

9.2.3 Frostbite. Frostbite (freeze burn) is a serious injury that can result from contact with cold surfaces, solid CO_2 (dry ice), or escaping liquid CO_2. Any CO_2 pressure drop can cause a hazardously cold condition, and frost is not uncommon on wellheads and flowlines where a large amount of CO_2 is being produced.

When containment pressures are released accidentally and a small amount of CO_2 turns to gas, the temperature of the remaining liquid immediately drops to approximately −101°F, nearly the temperature of solid dry ice.[5] Personnel should avoid entering a CO_2 vapor cloud not only because of the high concentration of CO_2, but also because of the danger of frostbite. Precautions such as wearing gloves and eye goggles must be taken to prevent coming in contact with very cold surfaces and vapors.

Should frostbite occur, the most important element of treatment is speed. The longer a body part is frozen, the more likely it is to be destroyed. Suggested treatments for frostbite are found in many publications.[6]

9.2.4 Hydrates/Ice Plugs. Hydrates, or ice plugs, can form in the piping of facilities and flowlines, especially at pipe bends, depressions, and locations downstream of restriction devices.

It is not necessary to have subfreezing temperatures for hydrates to form. For example, wet CO_2 forms a solid hydrate, $CO_2 \cdot 8H_2O$, at elevated operating pressures, even at temperatures up to 55°F (see Sec. 8.3.1). Using glycols, alcohols, and other freeze depressants in the line or facility can prevent or reduce hydrate formation. Installing heat tracing on pipes also is effective.

The first indication of an ice plug will be an abnormally low pressure on the downstream side of the blockage. If an ice plug is suspected, the section of the line in which the plug is located should be isolated as soon as practical. Pressure should be maintained as close to equal as possible on both sides of the plug to prevent it from being dislodged by a pressure differential and then damaging equipment downstream. Applying heat to the pipe exterior in the form of heating pads or hot air blowers can melt the plug.

9.2.5 High Pressures. ***Key Safety Considerations.*** A principal source of danger in a CO_2 facility is the high pressure (generally above 1,100 psia) at which CO_2 is transported and injected. High pressure is particularly dangerous with CO_2 because of CO_2's high coefficient of thermal expansion—a small change in temperature can cause a large change in pressure.

To prevent overpressurizing the system (which could cause emergency releases or burst pipes), the CO_2 must never be trapped or blocked. Following these rules helps maintain proper pressures:

- Do not close more than one flow valve at a time without adequate venting.
- Prevent formation of ice or hydrate plugs.
- Do not pressurize the system above the rating for the weakest part of the system.

• Do not bury piping less than 2.5 ft below surface, a depth at which constant temperature can be maintained (at least in moderate latitudes).[3]

Thermal Relief Requirements. Thermal relief requirements must be considered for piping segments aboveground where CO_2 may be trapped and subjected to significant temperature changes. The requirement for thermal relief is dictated by the fact that CO_2 is more thermally expansive than are hydrocarbon gases: When a sample of pure CO_2 at 60°F and 2,000 psia is heated to only 80°F, the pressure increases to approximately 3,300 psia. In contrast, pure methane would increase from 2,000 psia to only about 2,180 psia. Emergency and thermal relief systems also should be computer modeled to check for the proper dispersion of any CO_2 gas that may be released.

Wellhead Considerations. Well problems often accompany CO_2 injection simply because CO_2 may cause wells to flow more than they have in the past, increasing the potential for wellhead and tubing failures. Additional failure potential also exists because of the presence of carbonic acid, which exacerbates corrosion. In the event of a catastrophic failure, the release of CO_2 and wellbore and reservoir fluids is possible.

Two CO_2 floods in the Wasson field provide examples of how equipment systems can be designed to prevent or minimize accidental releases of fluids into the environment. At the Wasson Willard CO_2 flood,[7] stuffing-box leak detectors are installed on all wells. These leak detectors and existing high/low pressure switches are connected to pump off controllers (POC). Most wells in the field also are equipped with a high-performance butterfly safety shutdown valve to which the POC is connected. Pressure switches operate these valves through electrohydraulic actuators.

The Wasson Denver Unit[1] uses a similar system in which a vibration detector on the beam unit shuts down the pump in the event of a rod or wrist-pin failure. In addition, injector valves are programmed to close upon detection of low tubing pressure, low injection line pressure, high casing pressure, and other potentially dangerous conditions.

The status of this safety equipment at both floods is continuously monitored to enable quick detection of leaks and blocked flowlines in wells where casing integrity may be compromised.

Protection of Near-Surface Waters. In some cases, operators have installed casing pressure relief valves to protect shallow freshwater zones. These valves ensure that in high casing pressures, the result is a surface release rather than an underground blowout. Although any release is costly and unacceptable, a surface release in an uninhabited area may be preferable to a release into a freshwater aquifer.

Workovers. Increased well pressures make workovers more difficult. If well-kill operations are required, sodium chloride brine may not be adequate, which would mandate the use of heavier-weight fluids. The use of heavier fluids in injection well workovers may be eliminated by shutting in the well to let it stabilize, or by switching to water injection before the workover.[1] If possible, workovers should be done through tubing to avoid pulling equipment out of the well.

References

1. Fleming, E.A., Brown, L.M., and Cook, R.I.: "Overview of Production Engineering Aspects of Operating the Denver Unit CO_2 Flood," paper SPE/DOE 24157 presented at the 1992 SPE/DOE Enhanced Oil Recovery Symposium, Tulsa, 22–24 April.
2. American National Standard Institute ANSI Standard Z88.2-1992, Respiratory Protection, Washington, DC.
3. Boone, S.A. III: "CO_2 and HOS Safety Regulations and Recommended Practices," Southwestern Petroleum Assn. Short Course, presented at the 1985 Southwestern Petroleum Short Course Assn. Meeting, Lubbock, Texas, 23–25 April.
4. U.S. Code of Federal Regulations, Title 29 Part 1910, Occupational Safety and Health Standards (2001).
5. Sides, W.S. III: "Injection Safety Valve Solutions for CO_2 WAG Cycle Wells," paper SPE 24000 presented at the 1992 SPE Permian Basin Oil & Gas Recovery Conference, Midland, Texas, 18–20 March.
6. American Red Cross Standard First Aid, American Red Cross, Washington, DC (1988).
7. D'Souza, D.F. and Barnes, M.: "Willard CO_2 Flood-Well Site Automation System," Southwestern Petroleum Assn. Short Course, presented at the 1995 Southwestern Petroleum Assn. Meeting, Lubbock, Texas, 19–20 April.

SI Metric Conversion Factors

atm ×	1.013 250*	E + 02	= kPa
°F	(°F − 32)/1.8		= °C
°F	(°F + 459.67)/1.8		= K
psia ×	6.894 757	E + 00	= kPa

*Conversion factor is exact.

Appendix A
Worldwide CO_2 Floods

A.1 Active Floods

Table A.1—Basic Data for Active Floods
Table A.2—Reservoir Data for Active Floods
Table A.3—Production and Project Data for Active Floods

A.2 Inactive Floods

Table A.4—Basic Data for Inactive Floods
Table A.5—Reservoir Data for Inactive Floods
Table A.6—Production and Project Data for Inactive Floods

Abbreviations used in **Tables A.1 through A.6** are listed in **Table A.7.**

Nomenclature

D = depth, L, ft
k = permeability, L^2, md
T = temperature, T, °F
ϕ = porosity, %
γ_o = oil gravity, °API
μ_o = oil viscosity, m/Lt, cp

References

1. Moritis, G.: "EOR Oil Production Up Slightly," *Oil & Gas J.* (1998) **96,** No. 16, 49
2. Moritis, G.: "CO_2 and HC injection lead EOR production increase," *Oil & Gas J.* (1990) **88,** No. 17, 49.
3. Taber, J.J.: "Current Status of the Use of CO_2 for Enhanced Oil Recovery," U.S. Dept. of Energy Argonne National Laboratory Recovery and Use of Waste CO_2 in EOR Workshop, Denver (1988), 65–99.
4. Cox, B. and Schubert, J.: *1986 EOR Project Sourcebook,* Pasha Publications, Inc., Arlington, Virginia (1986) 101–167.
5. Brock, W.R. and Bryan, L.A.: "Summary Results of CO_2 EOR Field Tests, 1972–1987," SPE paper 18977 presented at the 1989 SPE Joint Rocky Mountain Region/Low Permeability Reservoirs Symposium and Exhibition, Denver, 6–8 March.

SI Metric Conversion Factors

acre ×	4.046 859	E + 03	= m^2
°API	141.5/(131.5 + °API)		= g/cm^3
bbl ×	1.589 873	E − 01	= m^3
cp ×	1.0*	E − 03	= Pa·s
°F	(°F − 32)/1.8		= °C
°F	(°F + 459.67)/1.8		= K
ft ×	3.048*	E − 01	= m

*Conversion factor is exact.

TABLE A.1—BASIC DATA FOR ACTIVE FLOODS

No.	Field	Country	Location	Flood Type	Operator	County
1	Joffre Viking	Canada	Alberta	Miscible	Vikor	
2	Joffre Viking	Canada	Alberta	Miscible	Vikor	
3	Joffre Viking	Canada	Alberta	Miscible	Vikor	
4	Midale	Canada	Saskatchewan	Miscible	Shell Canada	
5	Area 2102	Trinidad	Forest Reserve	Immiscible	Trintoc	
6	Area 2121	Trinidad	Forest Reserve	Immiscible	Trintoc	
7	Area 2124	Trinidad	Forest Reserve	Immiscible	Trintoc	
8	Oropouche	Trinidad	Forest Reserve	Immiscible	Trintoc	
9	Bati Raman	Turkey	Batman	Immiscible	TPAO	
10	Adair San Andres Unit	U.S.	Texas	Miscible	Amerada Hess	Gaines
11	Alvord South Field	U.S.	Texas	Miscible	Mitchell Energy	Wise
12	Aneth/McElmo Creek Unit	U.S.	Utah	Miscible	Mobil	San Juan
13	Anton Irish	U.S.	Texas	Miscible	Altura	Hale
14	Bennett Ranch	U.S.	Texas	Miscible	Altura	Yoakum
15	Cedar Lake	U.S.	Texas	Miscible	Altura	Gaines
16	Cordona Lake	U.S.	Texas	Miscible	Exxon	Crane
17	Crossett (Mid Cross-Devonian Unit)	U.S.	Texas	Miscible	Altura	Crane and Upton
18	Crossett (N. Cross)	U.S.	Texas	Miscible	Altura	Crane and Upton
19	Crossett (S. Cross)	U.S.	Texas	Miscible	Altura	Crockett
20	Dollarhide	U.S.	Texas	Miscible	Spirit Energy	Andrews
21	Dollarhide (Clearfork)	U.S.	Texas	Miscible	Spirit Energy	Andrews
22	Dollarhide, North	U.S.	Texas	Miscible	Oxy USA	Andrews
23	East Ford	U.S.	Texas	Miscible	Orla Petco	Reeves
24	East Penwell (SA) Unit	U.S.	Texas	Miscible	Fina	Ector
25	El Mar	U.S.	Texas	Miscible	Burlington	Loving
26	Ford Geraldine Unit	U.S.	Texas	Miscible	Conoco	Reeves and Culberson
27	Gmk South	U.S.	Texas	Miscible	Mobil	Gaines
28	Goldsmith	U.S.	Texas	Miscible	Chevron	Ector
29	Hanford	U.S.	Texas	Miscible	Fasken	Gaines
30	Hanford East	U.S.	Texas	Miscible	Fasken	Gaines
31	Hansford Marmaton	U.S.	Texas	Miscible (Immiscible)	Stanberry Oil	Hansford
32	Huntley, East	U.S.	Texas	Miscible	Southwest Royalty	Garza
33	Huntley, South	U.S.	Texas	Miscible	Southwest Royalty	Garza
34	Little Creek Field	U.S.	Mississippi	Miscible	Shell	Lincoln and Pike
35	Lost Soldier	U.S.	Wyoming	Miscible	Amoco	Sweetwater
36	Lost Soldier	U.S.	Wyoming	Miscible	Amoco	Sweetwater

TABLE A.1 (continued)—BASIC DATA FOR ACTIVE FLOODS						
No.	Field	Country	Location	Flood Type	Operator	County
37	Lost Soldier	U.S.	Wyoming	Miscible	Amoco	Sweetwater
38	Mabee	U.S.	Texas	Miscible	Texaco	Andrews/Martin
39	Maljamar	U.S.	New Mexico	Miscible	Conoco	Lea, Eddy
40	Means (San Andres)	U.S.	Texas	Miscible	Exxon	Andrews
41	North Cowden	U.S.	Texas	Miscible	Altura	Ector
42	Northeast Purdy	U.S.	Oklahoma	Miscible	Oxy USA	Garvin
43	Olive	U.S.	Mississippi	Miscible	Shell	Pike
44	Paradis	U.S.	Louisiana	Miscible	Texaco	St. Charles
45	Paradis	U.S.	Louisiana	Miscible	Texaco/MVP	St. Charles
46	Postle	U.S.	Oklahoma	Miscible	Mobil	Texas
47	Rangely Weber Sand	U.S.	Colorado	Miscible	Chevron	Rio Blanco
48	Reinecke	U.S.	Texas	Miscible	Spirit Energy	Borden
49	Sable	U.S.	Texas	Miscible	Whiting	Yoakum
50	Sacroc Unit	U.S.	Texas	Miscible	Pennzoil	Scurry
51	Salt Creek	U.S.	Texas	Miscible	Mobil	Kent
52	Seminole Unit	U.S.	Texas	Miscible	Amerada Hess	Gaines
53	Seminole Unit ROZ Pilot	U.S.	Texas	Miscible	Amerada Hess	Gaines
54	Sho-Vel-Tum	U.S.	Oklahoma	Miscible	Henry Petroleum	
55	Slaughter (CMU)	U.S.	Texas	Miscible	Altura	Hockley
56	Slaughter (Coons)	U.S.	Texas	Miscible	Exxon	Hockley
57	Slaughter (E. Mallet)	U.S.	Texas	Miscible	Mobil	Hockley
58	Slaughter (FU)	U.S.	Texas	Miscible	Altura	Hockley
59	Slaughter (SEU)	U.S.	Texas	Miscible	Altura	Hockley
60	Slaughter (Sundown)	U.S.	Texas	Miscible	Texaco	Hockley
61	South Cowden	U.S.	Texas	Miscible	Phillips	Lea
62	Vacuum	U.S.	New Mexico	Miscible	Phillips	Lea
63	Vacuum	U.S.	New Mexico	Miscible	Texaco	Lea
64	Wasson (Cornell Unit)	U.S.	Texas	Miscible	Exxon	Yoakum
65	Wasson (Denver)	U.S.	Texas	Miscible	Altura	Yoakum and Gaines
66	Wasson (ODC)	U.S.	Texas	Miscible	Altura	Yoakum
67	Wasson (South)	U.S.	Texas	Miscible	Altura	Gaines
68	Wasson (Willard)	U.S.	Texas	Miscible	Arco	Yoakum
69	Wasson (Mahoney)	U.S.	Texas	Miscible	Mobil	Yoakum
70	Welch, South	U.S.	Texas	Miscible	Oxy USA	Ector
71	Welch, West	U.S.	Texas	Miscible	Oxy USA	Ector
72	Wellman	U.S.	Texas	Miscible	Weiser	Terry
73	Wertz	U.S.	Wyoming	Miscible	Amoco	Carbon, Sweetwater
74	West Brahaney Unit	U.S.	Texas	Miscible	Fina	Yoakum
75	West Mallalieu	U.S.	Mississippi	Miscible	Shell	Lincoln

TABLE A.2—RESERVOIR DATA FOR ACTIVE FLOODS

No.	Start Date	Size (acre)	No. of Producers	No. of Injectors	Pay Zone	Formation Type	ϕ (%)	k (md)	D (ft)	γ_o (°API)	μ_o (cp)
1	Oct. 1991	1,280	5	2	Viking	S	13.0	500.0	5,000	42.0	1.14
2	Aug. 1985	480	6	2	Viking	S	13.0	500.0	5,140	42.0	1.14
3	Jun. 1988	480	4	1	Viking	S	13.0	500.0	5,075	42.0	1.14
4	1992	2,560	60	8	Mississippi Midale	LS/Dolo	10.0 to 35.0	1.0 to 100.0	4,500	27.0	3.00
5	Jun. 1976	58	6	3	Forest Sands	S	32.0	175.0	3,000	19.0	16.00
6	Jan. 1974	29	3	2	Forest Sands	S	30.0	150.0	2,600	17.0	32.00
7	Jan. 1986	184	1	1	Forest Sands	S	31.0	334.0	3,850	25.0	6.00
8	Jun. 1990	175	4	3	Retrench	S	30.0	36.0	2,400	29.0	5.00
9	Mar. 1986	10,709	145	41	Garzan	LS	18.0	58.0	4,265	13.0	592.00
10	Nov. 1997	5,338	82	63	San Andres	Dolo	15.0	8.0	4,852	35.0	1.00
11	1980	2,291	11	1	Caddo	Congt	12.8	55.0	5,700	44.0	0.39
12	Feb. 1985	13,440	143	120	Ismay Desert Creek	LS	14.0	5.0	5,600	41.0	0.50
13	Apr. 1997	1,600	82	40	Clearfork	Dolo	7.0	5.0	5,900	28.0	2.70
14	Jun. 1995	160	15	7	San Andres	Dolo	10.0	7.0	5,200	33.0	1.20
15	Aug. 1994	2,500	175	166	San Andres	Dolo	14.0	5.0	4,700	32.0	2.40
16	Dec. 1985	2,084	30	20	Devonian	Tripol	22.0	4.0	5,500	40.0	0.50
17	Jul. 1997	1,326	12	6	Devonian	Tripol	18.0	2.0	5,400	42.0	0.38
18	Apr. 1972	1,155	25	12	Devonian	Tripol	22.0	5.0	5,300	44.0	0.42
19	Jun. 1988	1,200	22	10	Devonian	Tripol	21.0	4.0	5,200	43.0	0.60
20	May 1985	6,183	83	66	Devonian	Dolo/ Tripol	13.5	9.0	8,000	40.0	0.44
21	Nov. 1995	160	21	4	Clearfork	Dolo	11.5	3.5	6,500	40.0	0.44
22	Nov. 1997	1,280	22	4	Devonian	S	22.0	3.0 to 4.0	8,000	40.0	0.60
23	Jul. 1995	1,953	12	8	Delaware	S	23.0	64.0	2,680	40.0	0.77
24	May 1996	540	34	13	San Andres	Dolo	10.0	4.0	4,000	34.0	2.00
25	Apr. 1994	6,000	69	56	Delaware	S	21.8	23.7	4,500	40.5	
26	Feb. 1981	3,850	91	69	Delaware	S	23.0	64.0	2,680	40.0	1.40
27	1982	1,143	24	24	San Andres	Dolo	9.8	3.0	5,400	30.0	2.57
28	Dec. 1996	330	16	9	San Andres	Dolo	11.6	32.0	4,200		
29	Jul. 1986	1,120	28	26	San Andres	Dolo	10.5	4.0	5,500	32.0	1.38
30	Mar. 1997	340	7	4	San Andres	Dolo	10.0	4.0	5,500	32.0	1.38
31	Jun. 1980	2,010	9	10	Marmaton	S	18.1	48.0	6,500	44.0	1.56
32	Jan. 1994	700	38	15	San Andres	Dolo	16.9	5.6	3,100	37.0	2.10

TABLE A.2 (continued)—RESERVOIR DATA FOR ACTIVE FLOODS

No.	Start Date	Size (acre)	No. of Producers	No. of Injectors	Pay Zone	Formation Type	ϕ (%)	k (md)	D (ft)	γ_o (°API)	μ_o (cp)
33	Jan. 1994	560	31	8	San Andres	Dolo	15.1	6.1	3,400	37.0	2.60
34	Dec. 1985	6,200	19	19	Lower Tuscaloosa	S	23.0	33.0	10,640	39.0	0.40
35	May 1989	1,345	54	60	Tensleep	S	9.9	31.0	5,000	35.0	1.30
36	May 1989	790	20	41	Darwin-Madison	S/LS/Dolo	10.3	4.0	5,400	35.0	1.40
37	Jun. 1996	120	5	4	Cambrian	S	7.0	10.0	7,000	35.0	
38	Jan. 1992	12,824	390	85	San Andres	Dolo	9.0	4.0	4,700	32.0	2.30
39	Jan. 1989	1,760	94	22	Grayburg-San Andres	Dolo/S	10.8	5.0	3,650 to 4,200	37.0	1.00
40	Nov. 1983	8,500	484	284	San Andres	Dolo	9.0	20.0	4,300	29.0	6.00
41	Feb. 1995	200	30	18	San Andres	Dolo	10.0	2.0 to 5.0	4,200	34.0	1.46
42	Sep. 1982	3,400	75	44	Springer	S	13.0	44.0	9,400	38.0	1.20
43	Oct. 1987	1,280	8	4	Lower Tuscaloosa	S	26.0	50.0	10,500	39.0	0.34
44	Feb. 1982	347	20	1	Lower 9,000 Ft	S	26.0	770.0	10,400	37.0	0.50
45	May 1989	298	4	2	#16 Sand	S	24.0	245.0	9,950	38.0	35.00
46	Nov. 1995	11,000	140	110	Morrow	S	16.0	30.0	6,100	36.0	1.30
47	Oct. 1986	15,000	204	200	Weber Ss	S	12.0	10.0	6,000	35.0	1.70
48	Nov. 1998	700	25	5	Cisco Canyon Reef	Dolo/LS	10.5	150.0	6,800	43.5	2.00 to 3.00
49	Mar. 1984	825	33	32	San Andres	Dolo	8.4	2.0	5,200	32.0	1.46
50	Jan. 1972	49,900	325	57	Canyon Reef	LS	3.9	19.0	6,700	41.0	0.35
51	Oct. 1993	12,000	85	48	Canyon	LS	20.0	12.0	6,300	39.2	0.95
52	Mar. 1983	15,699	408	160	San Andres	Dolo	12.0	1.3 to 12.3	5,300	35.0	1.07
53	Jul. 1996	500	15	10	San Andres	Dolo	12.0	1.3 to 12.3	5,300	35.0	1.07
54	Sep. 1982	1,100	60	40	Sims	S	16.0	37.0	6,200	25.0	3.30
55	1984	6,412	175	134	San Andres	Dolo/LS	10.8	2.0	4,900	31.0	1.40
56	May 1985	569	24	11	San Andres	Dolo	12.5	6.0	4,900	32.0	1.30
57	Jun. 1989	2,495	84	47	San Andres	Dolo	10.3	3.0	5,000	32.0	1.60
58	Dec. 1984	1,600	59	52	San Andres	Dolo/LS	10.0	4.0	4,950	31.0	1.40
59	Dec. 1984	5,700	185	161	San Andres	Dolo/LS	12.0	5.0	4,950	31.0	1.40
60	Jan. 1994	8,685	280	187	San Andres	Dolo	11.0	6.0	4,950	33.0	1.40
61	Feb. 1981	4,900	192	100	San Andres	Dolo	11.7	11.0	4,500	38.0	1.00
62	Feb. 1981	4,900	192	100	San Andres	Dolo	11.7	11.0	4,500	38.0	1.00
63	Jul. 1997	2,240	95	85	San Andres	Dolo	12.0	22.0	4,550	38.0	1.00
64	Jul. 1985	1,923	62	50	San Andres	Dolo	8.6	2.0	4,500	33.0	1.00

TABLE A.2 (continued)—RESERVOIR DATA FOR ACTIVE FLOODS

No.	Start Date	Size (acre)	No. of Producers	No. of Injectors	Pay Zone	Formation Type	ϕ (%)	k (md)	D (ft)	γ_o (°API)	μ_o (cp)
65	Apr. 1983	27,848	735	365	San Andres	Dolo	12.0	8.0	5,200	33.0	1.24
66	Nov. 1984	7,800	293	290	San Andres	Dolo/LS	9.0	5.0	5,100	32.0	1.30
67	Jan. 1986	4,960	105	70	Clearfork	Dolo	6.0	2.0	6,700	35.0	1.00
68	Jan. 1986	8,000	282	226	San Andres	Dolo	9.0	1.0	5,100	32.0	2.01
69	Oct. 1985	640	30	26	San Andres	Dolo	13.0	6.2	5,100	33.0	0.97
70	Apr. 1996	900	38	19	San Andres	Dolo	11.0	4.0	4,550	34.0	2.10
71	Oct. 1997	640	30	13	San Andres	Dolo	11.0	4.0	4,900	34.0	2.10
72	Jul. 1983	1,400	14	9	Wolfcamp	LS	9.2	100.0	9,800	43.5	0.54
73	Oct. 1986	1,400	28	41	Tensleep	S	10.0	20.0	6,000	35.0	1.16
74	Jun. 1996	400	15	8	San Andres	Dolo	10.0	2.0	5,300	33.0	2.00
75	Nov. 1986	5,760	6	2	Lower Tuscaloosa	S	25.0	20.0	10,365	38.0	0.50

TABLE A.3—PRODUCTION AND PROJECT DATA FOR ACTIVE FLOODS

No.	T (°F)	Previous Production	Starting Saturation (%)	Ending Saturation (%)	Project Maturity	Total Production (B/D)	Enhanced Production (B/D)	Project Evaluation	Profit	Project Scope	Reference Source
1	133	WF	38.0	28.0	HF			Succ		FW	1
2	133	WF	38.0	27.0	NC			Succ		P(EL)	1
3	133	WF	36.0		HF			Succ		P(EL)	1
4	145	WF	50.0		JS	750	250	Prom		LW	1
5	120	Prim	56.0		HF	270	270	Succ	Yes	RW	1
6	120	Prim	60.0		HF	50	50	Prom		RW	1
7	130	WF	44.0		HF	77	77	Prom		RW	1
8	120	Prim	53.0	48.0	HF	88	88	Prom		P	1
9	129		78.0		NC	14,000	13,500	Succ	Yes	FW	1
10	98	WF			JS	1,800		TETT		RW	1
11	154	WF	60.0	52.0	NC	60	60	Prom	No	FW	1
12	125	Prim/WF	50.0		JS	7,000	3,500	Succ	Yes	LW	1
13	115	Prim/WF			JS	5,000	1,000	TETT		FW	1
14	105	WF	55.0	37.0	HF	3,200	100	TETT		P	1
15	103	WF			HF	6,365	1,744	TETT	Yes	LW	1
16	101	WF			HF	1,100	400	Prom	Yes	LW	1
17	105	Prim/GI	60.0	20.0	JS	50	0	TETT		FW	1
18	106	GI	49.0	21.0	HF	1,345	1,345	Succ	Yes	FW	1
19	104	GI	43.0	24.0	HF	1,570	1,500	Prom		EL	1
20	122	Prim/WF	35.0	22.0	JS	2,350	900		Yes	FW	1
21		Prim/WF			JS	355		TETT		P	1
22	123	WF			JS	2000		TETT		RW	1
23	83	Prim	49.0	36.0	JS	60	60	TETT		FW	1
24	86	WF	55.0	40.0	JS	575	100	Prom	Yes	RW	1
25	97	Prim/WF	40.3		JS	630	630	TETT		RW	1
26	83	Prim/WF	41.0	35.0	NC	600	600	Succ	No	FW	1
27	101	WF	55.0	28.0	JS	1,300	400	Succ	Yes	LW	1
28		WF			JS			TETT		P	1
29	104	Prim	60.7	18.7	HF	600	600	Succ		LW	1
30	104	WF		18.7	JS	110	50	Succ		LW	1
31	142	Prim	43.0		HF	440	440	Succ	Yes	FW	1
32	100	WF	45.0	31.0	JS	450	130	TETT	TETT	FW	1
33	105	WF	45.0	34.0	JS	1,300	150	TETT	TETT	FW	1
34	248	WF	15.0	2.0	NC	850	850	Prom		FW	1
35	178	WF			NC	4,500	4,050	Succ	Yes	FW	1
36	181	WF			NC	1,600	1,100	Succ	Yes	FW	1
37		WF			JS	1,350	1,080	Succ	Yes	FW	1
38	104	WF	36.0	10.0	JS	6,500	5,500	Prom	No	FW	1
39	95	Prim/GI/WF	55.6	40.0	NC	1,400	100	TETT		EUL	1
40	97	WF			HF	10,700	7,200	Succ	Yes	FW	1
41	91	WF			JS	17,600	839	TETT	Yes	LW	1
42	148	WF	46.0	40.0	HF	2,600	1,950	Succ		FW	1

TABLE A.3 (continued)—PRODUCTION AND PROJECT DATA FOR ACTIVE FLOODS

No.	*T* (°F)	Previous Production	Starting Saturation (%)	Ending Saturation (%)	Project Maturity	Total Production (B/D)	Enhanced Production (B/D)	Project Evaluation	Profit	Project Scope	Reference Source
43	250	WF	17.0	2.0	JS	320	320	Succ	Yes	FW	1
44	205	Prim	62.0	48.0	HF	100	100	Prom	No	RW	1
45	190	Prim	44.0	24.0	HF	270	270	Prom		P	1
46	147	WF	37.0	25.0	JS	5,800	4,600	TETT		EL	1
47	160	WF	38.0	29.0	JS	23,881	13,881	Succ		FW	1
48	140	WF	32.0 to 36.0	4.0 to 6.0	JS	1,900		TETT		RW	1
49	107	WF			NC	520	300	Succ	Yes	FW	1
50	130	Prim/WF	63.3	46.8	NC	9,000	9,000	Succ	Yes	FW	1
51	125	WF	89.0	15.0	JS	26,000	12,000	Succ	Yes	EL	1
52	140	WF			HF	32,000	30,000	Succ	Yes	FW	1
53	140	Prim			JS	1,000	500	TETT		P	1
54	115	WF	59.0	42.0	HF	1,700	1,700	Succ	Yes	LW	1
55	105	WF			HF	4,400	2,600	Succ	Yes	FW	1
56	110	WF			HF	700	550	Succ	Yes	LW	1
57	107	WF	45.0	8.0	JS	6,800	2,000	Succ	Yes	LW	1
58	105	WF			HF	2,500	1,600	Succ	Yes	FW	1
59	105	WF			HF	7,400	4,200	Succ	Yes	FW	1
60	105	WF	41.0	25.0	JS	6,700	4,400	TETT	Yes	LW	1
61	101	Prim	70.0	50.0	JS	500	250	Succ	Yes	FW	1
62	101	Prim	70.0	50.0	HF	7,800	4,700	Succ	Yes	FW	1
63	101	WF	36.0	15.0	JS	3,900		TETT	Yes	FW	1
64	106	WF			HF	1,600	1,100	Prom	Prom	LW	1
65	105	WF	51.0	30.5	HF	36,600	30,700	Succ	Yes	FW	1
66	110	WF			HF	13,313	9,300	Succ	Yes	FW	1
67	105	WF	60.0		JS	5,000	1,000	Prom		FW	1
68	110	WF	55.5	41.0	JS	6,000	4,900	Succ	Yes	FW	1
69	110	WF	54.4	39.2	HF	1,700	300	Succ	Yes	LW	1
70	98	WF	50.0	38.0	JS	1,900	950	Prom	Yes	RW	1
71	98	WF			JS	240		TETT		RW	1
72	151	WF	35.0	10.0	HF	1,400	1,400	Succ	Yes	FW	1
73	163	WF			NC	1,000	800	Succ	Yes	FW	1
74	108	WF	66.0	50.0	JS	79		TETT		RW	1
75	245	Prim	15.0	1.0	NC	225	225	Disc		P(EUL)	1

TABLE A.4—BASIC DATA FOR INACTIVE FLOODS

No.	Field	Country	Location	Flood Type	Operator	County
1	Midale	Canada	Saskatchewan	Miscible	Shell Canada	
2	Budafa	Hungary		Miscible-Partially	MOL RT NKFV	
3	Nagylengyel	Hungary		Immiscible (Gas Cap)	MOL RT NKFV	
4	Nagylengyel	Hungary		Immiscible (Gas Cap)	MOL RT NKFV	
5	Bay St. Elaine	U.S.	Louisiana	Miscible	Texaco	Terrebonne
6	Bayou Sale	U.S.	Louisiana	Cyclic-Immiscible	Texaco	St. Mary
7	Bridger Lake	U.S.	Utah	Miscible	Phillips	Summit
8	Cote Blanche Bay West	U.S.	Louisiana	Cyclic-Immiscible	Texaco	St. Mary
9	Dillinger Ranch	U.S.	Wyoming	Miscible	Tenneco	Campbell
10	East Coyote-Hualde Dome Units	U.S.	California	Immiscible	Unocal	Orange
11	East Velma	U.S.	Oklahoma	Miscible	Arco	Stephens
12	Farnsworth, North	U.S.	Texas	Miscible	Stanberry Oil	Ochiltree
13	Garber	U.S.	Oklahoma	Miscible	Arco	Garfield
14	Granny's Creek	U.S.	West Virginia	Miscible	Columbia Gas Transportation	Clay
15	Greater Aneth	U.S.	Utah	Miscible	Superior	San Juan
16	Heidelberg	U.S.	Mississippi	Immiscible	Chevron	Jasper
17	Kane	U.S.	Pennsylvania		Pennzoil	Elk
18	Kurten	U.S.	Texas	Miscible	Chevron	Brazos
19	Lafitte	U.S.	Louisiana	Cyclic-Immiscible	Texaco	Jefferson
20	Lake Barre	U.S.	Louisiana	Cyclic-Immiscible	Texaco	Terrebonne
21	Levelland	U.S.	Texas	Miscible	Amoco	Hockley
22	Levelland (minipilot)	U.S.	Texas	Miscible	Amoco	Hockley
23	Lick Creek	U.S.	Arkansas	Immiscible	Phillips	Bradley-Union
24	Little Creek Pilot	U.S.	Mississippi	Miscible	Shell	Lincoln and Pike
25	Little Knife	U.S.	North Dakota	Miscible	Gulf	Billings
26	Magnet Whithers BH&S State Tracts	U.S.	Texas	Cyclic-Immiscible	Texaco	Wharton
27	Magnet Whithers Pierce Estate B&C	U.S.	Texas	Immiscible	Texaco	Wharton
28	Maljamar	U.S.	New Mexico	Miscible	Conoco	Lea
29	Manvel	U.S.	Texas	Cyclic-Immiscible	Texaco	Brazoria
30	Manvel	U.S.	Texas	Immiscible	Texaco	Brazoria
31	McElroy	U.S.	Texas	Miscible	Southland Royalty	Upton
32	Mead Strawn	U.S.	Texas	Miscible	Union	Jones
33	North Coles Levee	U.S.	California	Miscible	Arco	Arco
34	North Ward Estes	U.S.	Texas	Miscible	Chevron	Ward
35	North Cowden Unit	U.S.	Texas	Miscible	Amoco	Ector
36	Northeast Purdy (Garber)	U.S.	Oklahoma	Miscible	Cities Service	Garvin
37	Paradis	U.S.	Louisiana	Immiscible	Texaco	St. Charles
38	Pewitt Ranch	U.S.	Texas	Immiscible	Exxon	Titus
39	Pickett Ridge	U.S.	Texas	Cyclic-Immiscible	Texaco	Wharton

TABLE A.4 (continued)—BASIC DATA FOR INACTIVE FLOODS

No.	Field	Country	Location	Flood Type	Operator	County
40	Pierce Ranch	U.S.	Texas	Immiscible	Texaco	Wharton
41	Pittsburg	U.S.	Texas	Immiscible	Chevron	Camp
42	Pittsburg	U.S.	Texas	Miscible	Chevron	Camp
43	Plymouth	U.S.	Texas	Immiscible	Texaco	San Patricio
44	Port Neches	U.S.	Texas		Texaco	Orange
45	Rankin	U.S.	Texas	Miscible	Petromac Inc.	Harris
46	Richardson	U.S.	West Virginia	Miscible	Pennzoil	Calhoun, Roane
47	Rose City North	U.S.	Texas	Miscible	George R. Brown	Orange
48	Rose City South	U.S.	Texas	Miscible	George R. Brown	Orange
49	Rock Creek	U.S.	Wyoming	Miscible	Pennzoil	Roane
50	S. Bishop Ranch (9,200 ft)	U.S.	Wyoming	Miscible	Grace Petroleum	Campbell
51	S. Bishop Ranch (9,400 ft)	U.S.	Wyoming	Miscible	Grace Petroleum	Campbell
52	Sho-Vel-Tum	U.S.	Oklahoma	Immiscible	Texaco	Stevens
53	South Marsh Island Block 6	U.S.	Louisiana	Immiscible	Texaco	Offshore
54	Talco	U.S.	Texas	Immiscible	Texaco	Franklin
55	Thompson	U.S.	Texas	Cyclic-Immiscible	Texaco	Fort Bend
56	University Waddell	U.S.	Texas	Miscible	Chevron	Crane
57	Weeks Island	U.S.	Louisiana	Miscible	Shell	New Iberia
58	Weeks Island	U.S.	Louisiana	Immiscible	Shell	New Iberia
59	Weeks Island	U.S.	Louisiana	Immiscible	Shell	New Iberia
60	Welch	U.S.	Texas	Miscible	Oxy (Cities Service)	Dawson
61	West Columbia	U.S.	Texas	Cyclic-Immiscible	Texaco	Brazoria
62	West Delta Block 109	U.S.	Louisana	Immiscible	Texaco	Offshore
63	West Delta Block 110	U.S.	Louisiana	Immiscible	Texaco	Offshore
64	West Delta Block 111	U.S.	Louisiana	Immiscible	Texaco	Offshore
65	West Delta Block 112	U.S.	Louisiana	Immiscible	Texaco	Offshore
66	West Sussex Unit	U.S.	Wyoming	Miscible	Conoco	Johnson
67	Wilmington	U.S.	California	Immiscible	UPR (Champlin)	Los Angeles
68	Withers North	U.S.	Texas	Immiscible	Texaco	Wharton
69	Withers North	U.S.	Texas	Cyclic-Immiscible	Texaco	Wharton

TABLE A.5—RESERVOIR DATA FOR INACTIVE FLOODS

No.	Start Date	Size (acre)	No. of Producers	No. of Injectors	Pay Zone	Formation Type	ϕ (%)	k (md)	D (ft)	γ_o (°API)	μ_o (cp)
1	Oct. 1985	5	3	4	Midale	LS/Dolo	16.0	3.0	4,600	28.0	3.00
2	1981	1,013	77	60	Zaia Kerettye	S	22.0	110.0	3,000	42.0	1.00
3	Oct. 1988	2,890	104	10	I-IV Rudistic	LS	1.1	1,000.0	6,700	18.0	22.00
4	Sep. 1980	200	5	1	Triassic	Dolo	2.5	1,000.0	7,400	16.0	92.00
5	Jan. 1981	9	2	1	8,000-Foot	S	32.9	1,480.0	7,400	36.0	0.67
6	May 1984	564	5	2	St. Mary	S	31.0	500.0	10,000	34.0	0.40
7	Apr. 1970	3,800	7	1	Dakata	S	12.8	79.0	15,600	40.0	0.36
8	Mar. 1984	55	3	3	14 Sand	S	29.0	322.0	8,000	32.0	1.30
9	Oct. 1980	600	20	10	Minnelusa Sand	S	13.0	10.0 to 100.0	9,000	37.0	0.86
10	Jun. 1982	766	92	21	1,2,3, Anaheim	S	26.0	400.0	3,300 to 460	23.0	12.00
11	1983				Sims	S	17.0	70.0		26.0	2.50
12	Jun. 1980	1,431	9	5	Marmaton B	LS	11.5	140.0	6,400	44.0	1.61
13	Oct. 1981	80	9	4	Crews	S	19.0	12.0	1,900	44.0	1.00
14	Jun. 1976	7	4	1	Pocono Big Injun	S	16.0	7.0	2,000	45.0	3.14
15	1982	13,357	140	21	Aneth	LS	10.0		5,750	42.0	0.47
16	Dec. 1983	40	1		Eutaw	S	25.0	74.0	5,060	20.0	15.00
17	Nov. 1991	41	9	4	Kane	S	12.0				
18	Aug. 1981	672	5	4	Woodbine	S	12.0	0.0	8,300	38.0	0.40
19	Aug. 1984	271	5	2	8,900-Foot	S	27.0	250.0	8,900	34.0	0.70
20	Mar. 1984	1,164	12	4	Upper M	S	25.0	139.0	13,000	33.0	0.40
21	Mar. 1973	13	2	6	San Andres	Dolo	11.5	4.0	4,900	30.0	2.30
22	Aug. 1978	1.5	1	4	San Andres	Dolo	11.8	4.0	4,900	30.0	2.30
23	Jan. 1986	1,640	39	13	Meakin	S	30.3	1,200.0	1,100 to 170	17.0	160.00
24	Feb. 1974	31	3	1	Tuscaloosa	S	23.0	33.0	10,700	39.0	0.40
25	Jan. 1981	5	4	1	Madison Canyon Zone D	LS/Dolo	18.0	22.0	9,800	43.0	0.20
26	Oct. 1983	1,224			Magnet Withers	S	23.0	1,700.0	5,500	26.0	2.30
27	Jul. 1983	500				S	23.0	1,700.0	5,500	26.0	2.30
28	May 1983	5	1	4	San Andres	Dolo	10.0	11.2	4,050	36.0	0.80
29	Nov. 1983	128			Oligocene	S	30.0	1,000.0	5,000	26.7	7.20
30	Oct. 1982	43			Oakville	S	30.0	400.0	4,000	25.0	4.40
31	Feb. 1981	640	38	20	San Andres	Dolo	11.6	2.0	3,850	31.0	2.30
32	Dec. 1964	43	3	4	Strawn	S	11.0	12.0	4,475	41.0	1.30
33	Jun. 1981	70	8	3	Stevens	S	19.5	9.0	9,000	36.0	0.45
34	Mar. 1989	3,840	190	194	Yates	S	16.0	37.0	2,600	35.0	1.40
35	Feb. 1979	12	2	6	Grayburg	Dolo/LS	10.0	5.0	4,300	34.0	1.67
36	Sep. 1982	8,320	106	102	Springer	S	13.0	44.0	9,400	38.0	1.20
37	Mar. 1984	110	2	2	Main Pay	S	28.0	1,033.0	10,200	38.0	0.50
38	Jun. 1983		1		Paluxy	S	24.0	1,000.0 to 1,500.0	4,500	19.0	30.00

TABLE A.5 (continued)—RESERVOIR DATA FOR INACTIVE FLOODS

No.	Start Date	Size (acre)	No. of Producers	No. of Injectors	Pay Zone	Formation Type	ϕ (%)	k (md)	D (ft)	γ_o (°API)	μ_o (cp)
39	Aug. 1983	726			Pickett Ridge	S	30.0	1,200.0	4,600	25.0	2.50
40	Jan. 1983	480			Pierce Ranch	S	31.8	534.0	4,900	24.4	4.57
41	Jun. 1985	43	4	1	Pittsburg	Limey S	110	2.0	8,000	41.0	
42	Nov. 1983	120	3		Sub-Clarksville	S	23.0	460.0	3,800	14.0	2,200.00
43	Oct. 1983	380			Main Greta	S	310	350.0	4,650	23.3	3.19
44	Jul. 1993	235	5	4	Marginulina	S	30.0	750.0	5,800	38.0	2.40
45	Jan. 1981	80	6	1	Yogua	S	27.0	300.0	7,900	37.0	0.60
46	Aug. 1986	78	8	4	Berea	S	14.0	40.0	2,300	46.0	3.00
47	Apr. 1981	800	3	5	Hackberry	S	37.0	4,500.0	8,200	37.0	2.00
48	Jan. 1983	900	8	5	Hackberry	S	37.0	4,500.0	8,200	37.0	2.00
49	Nov. 1976	20	2	6	Big Injun	S	22.0	20.0	2,000	43.0	3.20
50	Jan. 1982	640	5	2	Minnelusa	S	16.0	50.0	9,200	35.0	1.14
51	Jan. 1982	1,280	5	4	Minnelusa	S	15.0	150.0	9,400	34.0	
52	May 1983	120	2		Deese	S	13.0	100.0	5,530	24.0	18.00
53	Jun. 1985	82	2	1	Rob E-S	S	29.6	323.0	11,200	34.0	0.30
54	May 1982	240			Paluxy	S	25.0	388.0	3,785	23.0	25.00
55	Jan. 1983	100			Frio	S	27.0	100.0 to 1,000.0	5,100	25.2	2.70
56	May 1983	920	50	13	Devonian	Dolo	12.0	14.0	8,500	43.0	0.45
57	1979	8	2	1	S Res B	S	26.0	1,800.0	12,760	32.0	0.50
58	Jan. 1988	480	18	6	S Res A	S	23.0	150.0 to 4,000.0	1,400	33.0	0.42
59	Jan. 1988	980	8	2	R Res A	S	23.0	100.0 to 1,500.0	13,200	34.0	0.41
60	Feb. 1982	2,675	129	132	San Andres	LS	9.3	9.0	4,890	34.0	2.15
61	Jun. 1983	33			PTSD	S	30.0	560.0	2,600	30.0	8.00
62	Jun. 1985	48	1	1	10,200-Foot	S	27.0	205.0	10,000	37.3	0.26
63	Jun. 1985	74	1	1	10,800-Foot	S	30.0	1,900.0	10,500	36.5	0.28
64	Jun. 1985	75	1	1	13,100-Foot	S	29.7	68.0	11,325	37.0	0.30
65	Jun. 1985	78	1	1	12,500-Foot	S	27.0	1,032.0	12,000	34.1	0.25
66	Dec. 1982	10	3	1	Shannon	S	19.5	121.0	3,040	38.0	1.70
67	Mar. 1981	41	3	4	Tar	S	24.0	465.0	2,500	14.0	283.00
68	May 1983	768			C-Sand	S	25.0	400.0	5,320	25.3	2.90
69	Mar. 1983	454			Withers N.	S	25.0	1,050.0	5,250	25.7	2.45

TABLE A.6—PRODUCTION AND PROJECT DATA FOR INACTIVE FLOODS

No.	T (°F)	Previous Production	Starting Saturation (%)	Ending Saturation (%)	Project Maturity	Total Production (B/D)	Enhanced Production (B/D)	Project Evaluation	Profit	Project Scope	Reference Source
1	150	Prim/WF	35.0		C					P	1–4
2	147	Prim			NC	400	400	Succ			1–4
3	237	Prim			JS	3,700	3700	Succ			1–4
4	248	Prim			NC	60	60	Succ			1–4
5	170	Prim	20.0	5.0	NC	7	7	Disc	No		1–4
6	194	Prim	50.0	45.0	JS	2,000	30	Succ	Yes		1–4
7	225	Prim	70.0	61.0	NC	650	250	Succ	Yes	FW	1–4
8	184	WF	50.0	45.0	JS	80	10	Prom	Yes		1–4
9	230	Prim/WF			HF	120	60				1–4
10	130	WF			Term	1,600		Disc	No		1–4
11		Prim/WF			JS						1–4
12	131	WF	57.0		NC	200	200	Succ	Yes	FW	1–4
13	100	WF	30.0	16.0	NC	35		Succ	Yes	P	1–4
14	75	Prim/WF			Term						1–4
15	135	Prim/WF	43.0								1–4
16	150	Prim			HF						1–4
17											1–4
18	230	Prim	40.0		JS	160	120				1–4
19	185	Prim	50.0	45.0	JS	400	30	Succ	Yes		1–4
20	236	WF	50.0	45.0	JS	650	50	Succ	Yes		1–4
21	105	WF	74.0		HF	47	21	TETT			1–4
22	105	WF	43.0		HF	9	6	TETT			1–4
23	118	Prim	55.0	46.0	NC	450	400	Succ	Yes	LW	1–4
24	248	Prim/WF	21.0		Term			Succ			1–4
25	240	Prim									1–4
26	154	GI	35.0	31.0	JS	563	5	Succ	Yes		1–4
27	154	GI	35.0	34.0	JS	200	17	Succ	Yes		1–4
28	90									P	5
29	149	Prim	45.0	20.0	JS	550	30	Succ	Yes		1–4
30	149	Prim	42.0 to 65.0	Variable	JS	127					1–4
31	86	WF			Term						
32	135	Prim	39.0		Term			Succ			1–4
33	235	Prim	34.0	25.8	HF	150		Disc	No	P	1–4
34	83	WF	25.0	10.0	JS	3,800	2,675	TETT			1–4
35	94	WF	46.0		JS	22		TETT		P	1–4
36	148	WF	46.0	40.0	NC	3,500	950	TETT	TETT	LW	1–4
37	195	WF	50.0	45.0	JS	250	15	Succ	Yes		1–4
38	160	Prim			JS						1–4
39	138	Prim	29.0	28.0	HF	112			No		1–4

TABLE A.6 (continued)—PRODUCTION AND PROJECT DATA FOR INACTIVE FLOODS

No.	T (°F)	Previous Production	Starting Saturation (%)	Ending Saturation (%)	Project Maturity	Total Production (B/D)	Enhanced Production (B/D)	Project Evaluation	Profit	Project Scope	Reference Source
40	155	Prim			JS						1–4
41	205	WF			NC	237	105	Succ	Yes		1–4
42	120	Prim			JS						1–4
43	150	Prim	31.5	20.0	Term	150		No			1–4
44	165	WF	30.0		JS	225	225	TETT	No	RW	1–4
45	192	WF	55.0	25.0	JS	80	80	Prom	No		1–4
46	80	WF	48.0		Term	80	40	Prom	No	P	1–4
47	180	WF	50.0	35.0	NC	160	160	Prom		FW	1–4
48	180	WF	50.0	35.0	HF	800	800	Prom		FW	1–4
49	73	Prim			NC						1–4
50	220	WF									1–4
51	180	WF			JS						1–4
52	129	Prim	49.0	40.0	Term	12	8	Prom	Yes		1–4
53	208	Prim	50.0	45.0	JS	220	100	Succ	Yes		1–4
54	147	Prim	50.0	45.0	Term	462		Disc	No		1–4
55	120	Prim	45.0 to 65.0	Variable	JS	93	3	Prom	Yes		1–4
56	140	WF	71.0		JS	2,000	70	TETT	Yes		1–4
57	225	WF	22.0	2.0	NC	160	160	Succ	No	P	1–4
58	242	Prim	28.0	5.0	JS			TETT	Yes	RW	1–4
59	241	Prim	28.0	5.0	JS			TETT	Yes	RW	1–4
60	96	WF	30.0	18.0	HF	3,100	300	Succ	No	LW	1–4
61	116	Prim	36.0 to 48.0	10.0 to 20.0	JS	290	9	Prom	Yes		1–4
62	194	Prim	50.0	45.0	JS	3,000	10	Disc	Yes		1–4
63	200	Prim	50.0	45.0	JS	2,200	300	Succ	Yes		1–4
64	210	Prim	50.0	45.0	JS	500	10	Succ	Yes		1–4
65	218		50.0	45.0	JS	1,600		Disc	Yes		1–4
66	96				Term	79	79			P	1–4
67	123	WF	51.0	30.0	JS	375	375	TETT	TETT	P	1–4
68	147	WF	32.0	30.0	Term	98	4	Disc	No		1–4
69	145	Prim	35.0	32.0	HF	300	4	Prom	Yes		1–4

TABLE A.7—ABBREVIATIONS FOR APPENDIX A TABLES

Formation Type	Project Maturity	Previous Production	Project Evaluation	Project Scope
S = sandstone	JS = just started	Prim = primary	TETT = too early to tell	P = pilot
LS = limestone	HF = half-finished	WF = waterflood	Prom = promising	FW = fieldwide
Dolo = dolomite	NC = nearing completion	GI = gas injection	Succ = successful	LW = leasewide
Congl = conglomerate	C = completed		Disc = discouraging	RW = reservoirwide
Tripol = tripolite	Term = terminated			EL = expansion likely
				EUL = expansion unlikely

Appendix B
Single-Well Cyclic CO_2 Injection Treatment

Single-well cyclic CO_2 injection treatment, also known as "CO_2 huff 'n' puff," is considerably different from CO_2 flooding. In a CO_2 flood, a slug of CO_2 is propagated through the reservoir from multiple injection wells to multiple producing wells. In a cyclic CO_2 injection treatment, CO_2 is injected into a single well, which then is shut in for a time and produced back. The process may be repeated several times.

Carbon dioxide floods involve many wells, can require substantial financial investment, and can yield significant increased oil recovery. In comparison, the cyclic CO_2 injection treatment involves one well at a time, requires much less investment, and yields a much smaller increase in oil recovery.

The single-well cyclic CO_2 injection treatment is similar to steam soak treatments used in heavy-oil reservoirs. Instead of oil viscosity being reduced by steam heat, viscosity is reduced by CO_2 solubility in the crude oil. Carbon dioxide also swells the oil.

B.1 Procedure

There are four steps in a cyclic CO_2 injection treatment:

- Pressure-test the casing and/or tubing to ensure that the pressurized CO_2 will go to the intended destination.
- Inject the CO_2 for a period of from one day to several weeks, depending on the flow capacity of the well. Volumes have ranged from 2 to 50 MMscf of CO_2.
- Shut in the well for a soak period that usually lasts two to four weeks, although soak times as long as half a year have been used. During this time, the pressurized CO_2 dissolves into the oil, swells its volume, and reduces its viscosity as the reservoir pressure and CO_2 saturations redistribute. The oil in the reservoir is energized in much the same way that water in a soda bottle is energized by pressurized CO_2.
- Produce the well again. If the well produces all CO_2 for the first 24 hours, shut it in again for another week, and repeat the process until the well produces oil.[1]

The oil rate after a single-well cyclic CO_2 injection treatment can show up to a 10-fold increase[1] over the preCO_2 oil rate, although a three-fold increase is more common. In most cases, after the well is brought back into production, the oil rate declines rapidly back to the base oil-rate trend, and can even cross below the preCO_2 oil rate trend line, as shown in **Fig. B.1.** The time required for the oil rate to return to the previous oil-rate trend after a single treatment takes from one month to as many 100 months.[2]

Single-well cyclic CO_2 injection treatment has been repeated successfully on the same well,[1,3–5] although each repeated cycle generally yields less oil per Mscf of CO_2 injected than the cycle preceding it.[6] After three cycles, the procedure generally is no longer effective.

B.2 Mechanisms

Cyclic CO_2 injection treatment usually takes place below the thermodynamic MMP of the reservoir, which means that CO_2 is immiscible with the crude oil; however, the injection of CO_2 sometimes can increase reservoir pressure enough to vaporize some crude oil components.[6]

The three major mechanisms by which the cyclic CO_2 injection treatment works are:

- Swelling the oil—Oil swelling by dissolved CO_2 allows oil saturation to increase, which increases the oil's relative permeability and its mobility.
- Reducing oil viscosity—Carbon dioxide can reduce the viscosity of heavy oils by more than a factor of 10.[4] For a light oil, Simpson[7] reported a viscosity reduction from 2.8 cp to 0.5 cp.
- Blocking water—The injected CO_2 increases gas saturation, as well as oil saturation, which decreases water saturation and water relative permeability. In effect, the CO_2 pushes water away from the well and acts as a water block. This can reduce lifting costs and improve the economics of the cyclic CO_2 injection treatment, although it typically is a short-term effect. Simpson[7] found that cyclic CO_2 injection treatment also can reduce water coning, although coning eventually returned when the oil production rate returned to the previous oil-rate trend. If there is a strong natural water drive, the water can return during the soak period, displacing and dissolving the CO_2.[7]

A strong water drive also promotes a gas relative permeability hysteresis effect.[4] As incoming water decreases gas saturation, the gas phase follows a gas relative permeability hysteresis curve like the one shown in **Fig. B.2.** Gas flowing on the relative permeability hysteresis curve has reduced mobility (because of reduced gas relative permeability), which makes the next cycle of CO_2 injection less effective in increasing the reservoir pressure.

Field experience[1,2,8] and parametric model studies[4,6,9] have shown that the most important process variables in the performance of a cyclic CO_2 injection well treatment are:

- Injection volume—Increasing the CO_2 injection volume increases both the initial oil rate **(Fig. B.3)** and the oil recovery **(Fig. B.4)** because more oil is contacted.
- Containment—Containment of CO_2 is critical to ensure that the CO_2 contacts a large volume of oil. The process will be ineffi-

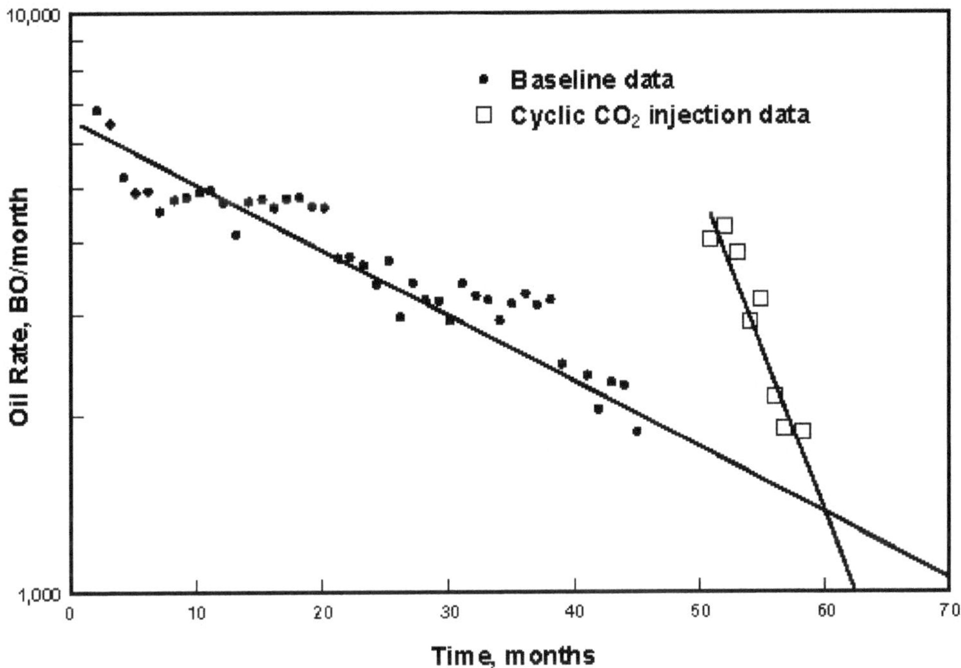

Fig. B.1—Decline curve analysis of a single-well CO_2 cyclic injection treatment (after Ref. 2). Although the oil rate improves dramatically after a CO_2 injection treatment, it also then declines rapidly, sometimes below the base oil-rate trend.

cient if CO_2 is lost through casing leaks, tubing leaks, gas-lift valves, or through fractures or high permeability channels. Increasing the soak time beyond four weeks has not been found to have a strong impact on incremental oil recovery.[2,5,6]

• Bottomhole injection pressure—A higher CO_2 bottomhole injection pressure leads to increased reservoir pressure, which will increase oil production rates when the well is brought back on-stream.

Operators have been successful with two[1,5] and three[3,4] cycles.

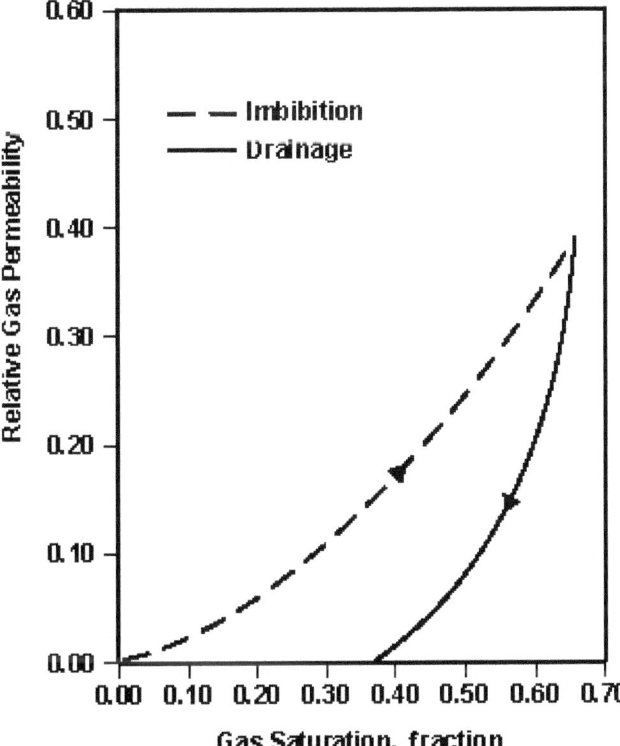

Fig. B.2—Carbon dioxide hysteresis curve (after Ref. 4). The solid line represents the hysteresis phenomenon created when water moves back into the reservoir after the first cycle of CO_2 cyclic treatment. Air permeability was used as the base for relative permeability data.

However, a model study[6] indicated that after two cycles, the decreased remaining oil saturation causes inefficient CO_2 use.

B.3 Applications and Design Considerations

Single-well cyclic CO_2 injection treatments have been applied successfully to both heavy[3,4,10] and light oil reservoirs.[1,2,5,7,8] Many of the heavy oil applications have been in California, while many of the successful light oil applications have been in Kentucky, Louisiana, and Texas.

With heavy-oil reservoirs, the reduction in oil viscosity and oil swelling have been the main reasons that the cyclic CO_2 injection treatments are economical. In light-oil reservoirs, additional factors have been important. Single-well cyclic CO_2 injection treatments in Kentucky were successful (compared to existing operations) because waterflooding was unsuccessful,[2,8] the low reservoir pressure yielded a favorable CO_2 utilization factor, and the fields were close to their economic limits.

Haskin and Alston[5] found that reservoirs with poor continuity between wells responded well to cyclic CO_2 injection treatments. Several successful Louisiana projects were adjacent to ongoing CO_2 floods, which reduced the investment costs.

In one economically successful program, produced natural gas was combusted, and the flue gas (87% N_2, 13% CO_2) was used in a small fieldwide cyclic treatment program.[11] The field had 51 wells spread over 225 acres, and it was treated over a 10-year period. The cumulative incremental oil recovery was 43,000 STB.

Reservoir simulation can be used to predict the performance of a cyclic CO_2 injection treatment.[4,6] The time required for this exercise may not be warranted because of the relatively small investment costs for cyclic CO_2 treatments, and the data required for reservoir simulation may not be available for an economically marginal field. The success of a cyclic CO_2 injection treatment is not as sensitive to reservoir heterogeneities as is CO_2 flooding,[12] so a less detailed reservoir description suffices when using simulators.

Cyclic CO_2 injection treatments also can be designed using correlations and performance from other projects. Field results from the early treatments in the target reservoir then can be used to improve subsequent treatments.

Empirical correlations have been developed to predict incremental oil recovery from cyclic CO_2 injection treatments,[5,9] but most of these correlations—and most of the early published work about cyclic CO_2 injection treatments—do not account for the negative effect of lost oil production during the periods of CO_2 injec-

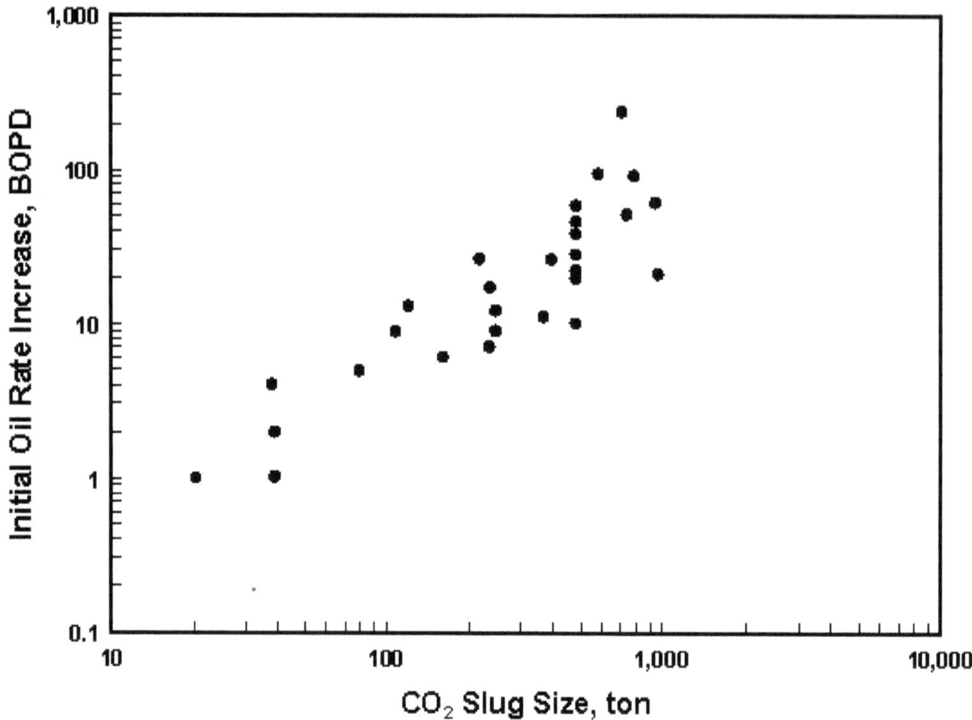

Fig. B.3—Initial increase in oil rate as a function of slug size (after Ref. 2). The larger the slug of CO_2 injected during the cyclic injection treatment, the greater the initial oil rate when the well is brought back into production.

tion and soaking. As a result, the correlations yield only approximate results.[13]

Fig. B.5 shows how a cyclic CO_2 injection treatment affects total cumulative incremental oil recovery. In this figure, the single-hatched area is only slightly greater than the double-hatched area. In some ways, the cyclic CO_2 injection treatment may be likened to damming a river for a while and then blowing up the dam. It may look as though the process has increased the amount of water flowing, but the effect lasts only until the artificial lake is emptied. Cyclic CO_2 injection treatments are not entirely analogous to this situation, though, because the reservoir is energized by the high-pressure CO_2 and because oil swelling and viscosity reduction also contribute to increased oil rates and recovery.

B.4 Results

Incremental cumulative oil recoveries for single-well cyclic CO_2 injection treatments have ranged from 0 to 15,000 STB. However, most of the reported incremental oil recoveries do not include the lost oil production discussed above. Although many of the treatments still would show positive incremental oil recovery if the periods of lost oil production were included in the analysis, it is clear that some of treatments that were thought to be successful were not.

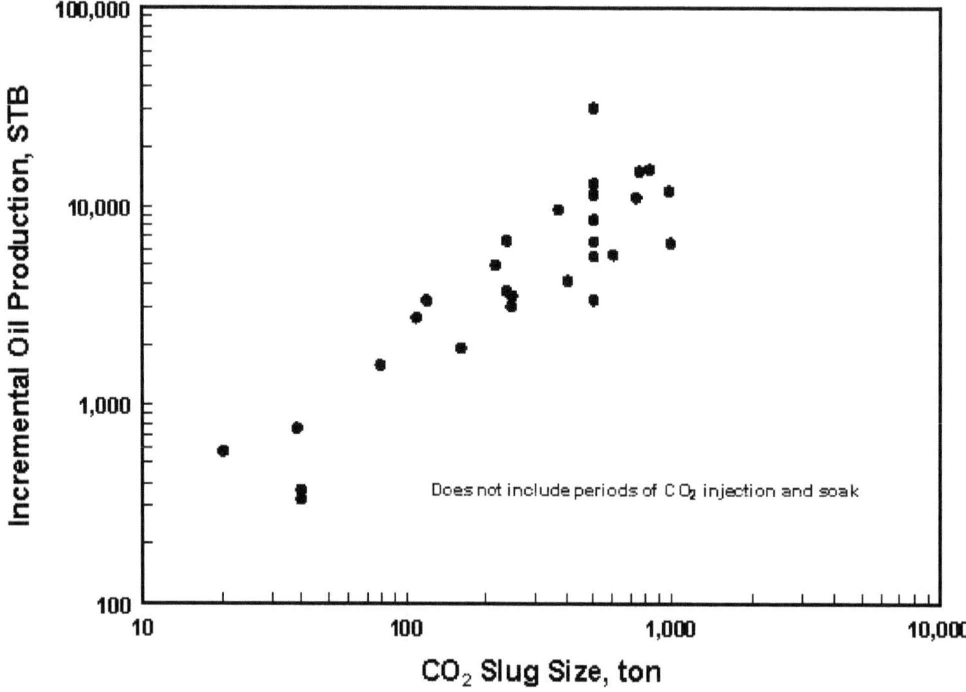

Fig. B.4—Incremental oil recovery as a function of slug size (after Ref. 2). The larger the slug of CO_2 injected during a cyclic injection treatment, the greater the total incremental oil recovered when the well is brought back into production.

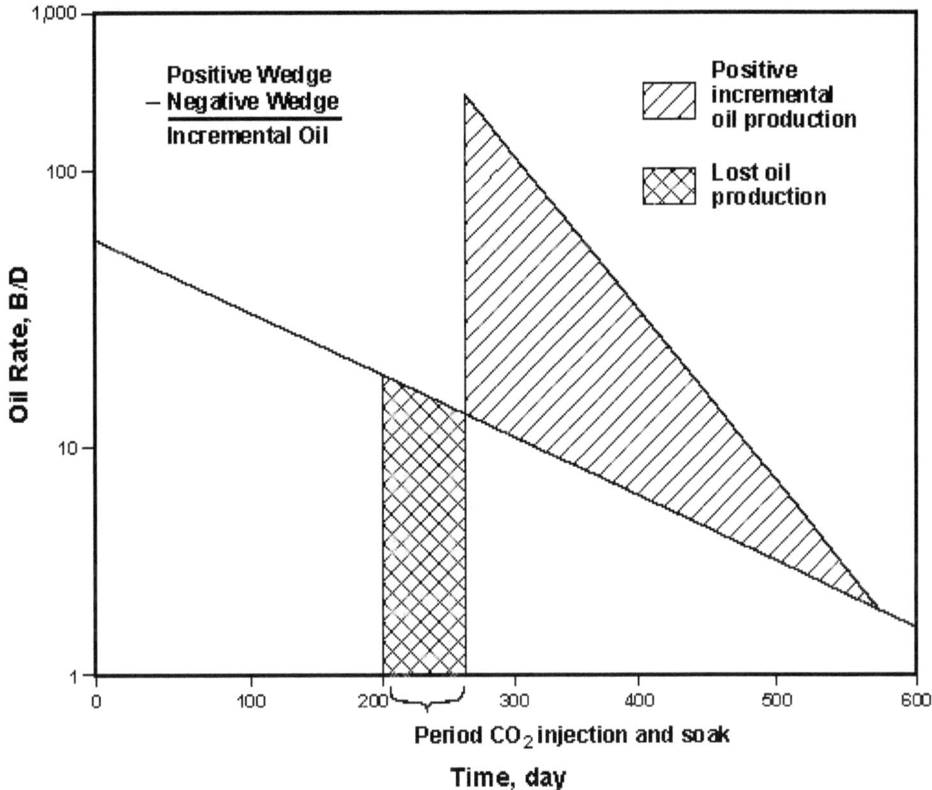

Fig. B.5—Illustration of incremental oil recovery from a cyclic CO_2 injection treatment. The positive effect of the cyclic CO_2 treatment on incremental oil recovery must be weighed against the lost oil production during CO_2 injection and soaking. Early literature often ignored this factor.

Averaging the Haskins and Alston[5] data from single-well cyclic CO_2 injection treatments in Texas Gulf Coast reservoirs gives the results shown in **Table B.1.** These data may be useful in planning future treatments.

TABLE B.1—AVERAGE SINGLE-WELL CYCLIC CO_2 INJECTION RESULTS FROM TEXAS GULF COAST RESERVOIRS

Lost oil production during the soak period	917 STB
Ratio of peak oil rate to base oil rate	3:1
Average CO_2 slug size	4.7 MMscf
Resultant CO_2 utilization factor	5.2 Mscf/STB

After Haskin and Alston.[5]

References

1. Palmer, F.S., Landry, R.W., and Bou-Mikael, S.: "Design and Implementation of Immiscible Carbon Dioxide Displacement Projects (CO_2 Huff-Puff) in South Louisiana," paper SPE 15497 presented at the 1986 SPE Annual Technical Conference and Exhibition, New Orleans, 5–8 October.
2. Thomas, G.A. and Monger-McClure, T.G.: "Feasibility of Cyclic CO_2 Injection for Light-Oil Recovery," *SPERE* (May 1991) 179.
3. Patton, J.T. *et al.*: "Carbon Dioxide Well Stimulation: Part 2—Design of Aminoil's North Bolsa Strip Project," *JPT* (August 1982) 1805.
4. Denoyelle, L.C. and Lemonnier, P.: "Simulation of CO_2 Huff 'n' Puff Using Relative Permeability Hysteresis," paper SPE 16710 presented at the 1987 Annual Technical Conference and Exhibition, Dallas, 27–30 September.
5. Haskin, H.K. and Alston, R.B.: "An Evaluation of CO_2 Huff 'n' Puff Tests in Texas," *JPT* (February 1989) 177.
6. Hsu, H.H. and Brugman, R.J.: "CO_2 Huff-Puff Simulation Using a Compositional Reservoir Simulator," paper SPE 15503 presented at the 1986 Annual Technical Conference and Exhibition, New Orleans, 5–8 October.
7. Simpson, M.R.: "The CO_2 Huff 'n' Puff Process in a Bottomwater Drive Reservoir," *JPT* (July 1988) 887.
8. Monger, T.G. and Coma, J.M.: "A Laboratory and Field Evaluation of the CO_2 Huff 'n' Puff Process for Light-Oil Recovery," *SPERE* (November 1988) 1168.
9. Patton, J.T., Coats, K.H., and Spence, K.: "Carbon Dioxide Well Stimulation: Part 1—A Parametric Study," *JPT* (August 1982) 1798.
10. Olenick, S. *et al.*: "Cyclic CO_2 Injection for Heavy-Oil Recovery in Halfmoon Field: Laboratory Evaluation and Pilot Performance," paper SPE 24645 presented at the 1992 Annual Technical Conference and Exhibition, Washington, DC, 4–7 October.
11. Johnson, H.R., Schmidt, L.D., and Thrash, L.D.: "A Flue Gas Huff 'n' Puff Process for Oil Recovery From Shallow Formations," paper SPE 20269 presented at the SPE/DOE 1990 Symposium on Enhanced Oil Recovery, Tulsa, 22–25 April.
12. Wehner, S.C., Prieditis, J., and Cole, R.M.: "CO_2 Huff 'n' Puff: Initial Results From a Waterflooded SSC Reservoir," paper SPE 35223 presented at the SPE/DOE 1996 Symposium on Improved Oil Recovery, Tulsa, 21–24 April.
13. Emanuel, A.S. *et al.*: "Analytic and Numerical Model Studies of Cyclic CO_2 Injection Projects," paper SPE 22934 presented at the 1991 Annual Technical Conference and Exhibition, Dallas, 6–9 October.

SI Metric Conversion Factors

acre × 4.046 856	E + 03	= m^2
bbl × 1.589 873	E − 01	= m^3
cp × 1.0*	E − 03	= Pa·s
ft^3 × 2.831 658	E − 02	= m^3
ton × 9.071 847	E − 01	= Mg

*Conversion factor is exact.

Appendix C
Case Histories

This appendix lists published papers that relate to the following fields, organized alphabetically.

Alvord South

Craig, F.F. III: "Field Use of Halogen Compounds to Trace Injected CO_2," paper SPE 14309 presented at the 1985 Annual Technical Conference and Exhibition, Las Vegas, Nevada, 22–25 September.

Hopson, S.: "Mitchell's EOR Project Uses CO_2 From Exhaust Gas," *Proc.,* Thirty-Fourth Oklahoma University Gas Conditioning Conference, Norman, Oklahoma (1984) D1–D9.

Bati Raman

Karaoguz, D. *et al.*: "Performance of a Heavy-Oil Recovery Process by an Immiscible CO_2 Application, Bati Raman Field," paper SPE 18002 presented at the 1989 SPE Middle East Oil Technical Conference and Exhibition, Manama, Bahrain, 11–14 March.

Issever, K., Pamir, N., and Tirek, A.: "Performance of a Heavy-Oil Field Under Carbon Dioxide Injection, Bati Raman, Turkey," paper 20883 presented at the 1990 Europec, The Hague, 22–24 October.

Connor, T.E.: "Application of CO_2 EOR Technology in Turkey," *Proc.,* EOR Using CO_2: Technol, ECON Use Conference, Houston (1984) 1–18.

Bay St. Elaine

Palmer, F.S., Nute, A.J., and Peterson, R.L.: "Implementation of a Gravity-Stable, Miscible CO_2 Flood in the 8,000 Foot Sand, Bay St. Elaine Field," *JPT* (January 1984) 101.

McNally, R.: "Texaco EOR Projects," *Pet. Eng. Intl.* (1988) **60,** No. 11, 34.

Budafa

Dank, V. *et al.*: "Geological Characteristics and Production Possibilities of Pure CO_2 and High CO_2 Content Gas Deposits in Hungary," Szkfi and Ifp int CO_2 Enhanced Oil Recovery, Budapest, Hungary (1983) 457–475.

Papay, J. *et al.*: "Basic Concepts of Production Regulation for the Case of Producing From the Zala-Kerettye Series Using CO_2," *Proc.,* Third AGIP SPA ET AL Improved Oil Recovery Conference, Rome (1985) 247–254.

Udvardi, G. *et al.*: "Processing Facilities for Enhanced Oil Recovery in Hungary," *Proc.,* Worldwide Gas Processing, Sixty-Eighth GPA Annual Convention, Phoenix, Arizona (1990) 127–134.

Cedar Creek—(Injectivity Test)

Good, P.A. and Downer, D.G.: "Cedar Creek Anticline Carbon Dioxide Injectivity Test: Design, Implementation, and Analysis," paper SPE 17326 presented at the 1988 SPE/DOE Enhanced Oil Recovery Symposium, Tulsa, 17–20 April.

Crossett (North Cross)

Frey, R.P.: "Operating Practices in the North Cross CO_2 Flood," *Proc.,* Twenty-Second Annual Southwestern Petroleum Short Course, Lubbock, Texas, 159–164.

Mizenko, G.J.: "North Cross (Devonian) Unit CO_2 Flood: Status Report," paper SPE 24210 presented at the 1992 SPE/DOE Symposium on Enhanced Oil Recovery, Tulsa, 22–24 April.

Pontious, S.B. and Tham, M.J.: "North Cross (Devonian) Unit CO_2 Flood—Review of Flood Performance and Numerical Simulation Model," *JPT* (December 1978) 1706.

Dollarhide

"Dipping into Dollarhide's Reserves," *Southwest Oil World* (1989) **36,** 16.

Collier, T.S.: "Successful Installation of CO_2 Injection Equipment: A Case Study," Thirty-Fifth Annual Southwestern Petroleum Short Course, Lubbock, Texas (1988) 328–337.

Poole, E.S.: "Evaluation and Implementation of CO_2 Injection at the Dollarhide Devonian Unit," paper SPE 17277 presented at the 1988 SPE Permian Basin Oil and Gas Recovery Conference, Midland, Texas, 10–11 March.

Ford Geraldine

Lee, K.H. and El-Saleh, M.M.: "A Full-Field Numerical Modeling Study for the Ford Geraldine Unit CO_2 Flood," paper SPE 20227 presented at the 1990 SPE/DOE Symposium on Enhanced Oil Recovery, Tulsa, 22–25 April.

Phillips, L.A., McPherson, J.L., and Leibrecht, R.J.: "CO_2 Flood: Design and Initial Operations, Ford Geraldine (Delaware Sand) Unit," paper SPE 12197 presented at the 1983 SPE Annual Technical Conference and Exhibition, San Francisco, 5–8 October.

Pittaway, K.R. and Runyan, E.E.: "The Ford Geraldine Unit CO_2 Flood: Operating History," *SPEPE* (August 1990) 3330.

Pittaway, K.R. and Rosato, R.J.: "The Ford Geraldine Unit CO_2 Flood—Update 1990," *SPERE* (November 1991) 410.

Garber

Kumar, R. and Eibeck, J.N.: "CO_2 Flooding a Waterflooded Shallow Pennsylvanian Sand in Oklahoma—A Case History," paper SPE 12668 presented at the 1984 SPE/DOE Symposium on Enhanced Oil Recovery, Tulsa, 15–18 April.

MacAllister, D.J.: "The Garber CO_2 Pilot: Successful CO_2 Flooding of a Shallow, Watered-Out Reservoir," *Proc.,* Sixth Kansas U. *et al.* Tertiary Oil Recovery Conf. (1985) Wichita, Kansas.

Granny's Creek

Connor, W.D.: "Granny's Creek CO_2 Injection Project, Clay Co., West Virginia," *Proc.,* ERDA Symposium on Enhanced Oil, Gas Recovery, and Improved Drilling Methods, Tulsa (1977), 30 August–1 September.

Connor, W.D.: "Granny's Creek CO_2 Injection Project," CO_2 Corrosion in Oil and Gas Production—Selected Papers, Contract No. EF-76-C-05-5302, 1978, U.S. DOE, Washington, DC (1984).

Watts, R.J., *et al.*: "CO_2 Injection for Tertiary Oil Recovery, Granny's Creek Field, Clay County, West Virginia," paper SPE 10693 presented at the 1982 Joint SPE/DOE Enhanced Oil Recovery Symposium, Tulsa, 4–7 April.

Hanford

Merritt, M.B. and Groce, J.F.: "A Case History of the Hanford San Andres Miscible CO_2 Project," *JPT* (August 1992) 924.

Hansford Marmaton

Flanders, W.A., Stanberry, W.A., and Martinez, M: "CO_2 Injection Increases Hansford Marmaton Production," *JPT* (January 1990) 68.

Joffre Viking

Ilsley, D.B, MacIntrye, K.J., and Stephenson, D.J.: "Joffre Viking Tertiary Carbon Dioxide Enhanced Oil Recovery Project," *Proc.,* Fourth Annual Advances in Petroleum Technology Conference, Calgary (1983).

"Tertiary Oil Recovery Project Makes Canadian Oil History," *Energy Processing/Canada* (1985) **78,** No. 1, 22.

Kelly Snyder (SACROC)

Dicharry, R.M., Perryman, T.L., and Ronquille, J.D.: "Evaluation and Design of a CO_2 Miscible Flood Project—SACROC Unit, Kelly-Snyder Field," *JPT* (November 1973) 1309.

Graue, P.J. and Blevins, T.R.: "SACROC Tertiary CO_2 Pilot Project," paper SPE 7090 presented at the 1978 SPE Symposium on Improved Methods for Oil Recovery, Tulsa, 16–19 April.

Kane, A.V.: "Performance Review of a Large-Scale CO_2-WAG Enhanced Recovery Project, SACROC Unit—Kelly-Snyder Field," *JPT* (February 1979) 217.

Langston, M.V., Hoadley, S.F., and Young, D.N.: "Definite CO_2 Flooding Response in the SACROC Unit," paper SPE 17321 presented at the 1988 SPE/DOE Enhanced Oil Recovery Symposium, Tulsa, 17–20 April.

Newton, L.E. Jr. and McClay, R.A.: "Corrosion and Operation Problems, CO_2 Project, Sacroc Unit," paper 6391 presented at the 1977 Permian Basin Oil and Gas Recovery Conference, Midland, Texas, 10–11 March.

Patterson, K.W.: "Downhole Corrosion Encountered in the CO_2 Flood at the SACROC Unit," CO_2 Corrosion in Oil and Gas Production—Selected Papers, Southwest Petroleum Short Course, Texas Tech U., Lubbock, Texas (1984).

Thomas, G.: "SACROC: Grande Dame of CO_2 EOR," *Well Service* (1985) **20,** 22.

Levelland

Graham, B.D. *et al.*: "Design and Implementation of a Levelland Unit CO_2 Tertiary Pilot," paper SPE 8831 presented at the 1980 Joint SPE/DOE Symposium on Enhanced Oil Recovery, Tulsa, 20–23 April.

Henry, R.L. *et al.*: "Utilization of Composition Observation Wells in a West Texas CO_2 Pilot Flood," paper SPE 9786 presented at the 1981 Joint SPE/DOE Enhanced Oil Recovery Symposium, Tulsa, 5–8 April.

Macon, R.B.: "Design and Operation of the Levelland Unit CO_2 Injection Facility," paper SPE 8410 presented at the 1979 Annual Fall Technical Conference and Exhibition, Las Vegas, Nevada, 23–26 September.

Macon, R.B.: "Amoco CO_2 Facility Proves Efficient," *Oil & Gas J.* (7 April 1980) **78,** 80.

Lick Creek

Reid, T.B. and Robinson, H.J.: "Lick Creek Meakin Sand Unit Immiscible CO_2/Waterflood Project," *JPT* (September 1981) 1723.

Little Creek

Hansen, P.W.: "A CO_2 Tertiary Recovery Pilot Little Creek Field, Mississippi," paper SPE 6747 presented at the 1977 SPE Annual Technical Conference and Exhibition, Denver, 9–12 October.

Little Knife

Desch, J.B. *et al.*: "Enhanced Oil Recovery by CO_2 Miscible Displacement in the Little Knife Field, Billings County, North Dakota," *JPT* (September 1984) 1592.

Lost Soldier

Jenson, J.R. *et al.*: "Bairoil CO_2 Flexible Project Management in an Uncertain Environment," *Proc.,* Sixty-Eighth GPA Annual Convention, New Gas Processing Projects, San Antonio, Texas (1989) 221.

McElmo Creek

Shaw, E.C.: "A Simple Technique To Forecast CO_2 Flood Performance," paper 23975 presented at the 1992 Permian Basin Oil and Gas Recovery Conference, Midland, Texas, 18–20 March.

Mabee

Roper, M.K. Jr. *et al.*: "Interpretation of a CO_2 WAG Injectivity Test in the San Andres Formation Using a Compositional Simulator," paper SPE 24163 presented at the 1992 SPE/DOE Eighth Symposium on Enhanced Oil Recovery, Tulsa, 22–24 April.

Maljamar

Pittaway, K.R. *et al.*: "The Maljamar CO_2 Pilot: Review and Results," *JPT* (October 1987) 1256.

Mead Strawn

Holm, L.W. and O'Brien, L.J.: "Carbon Dioxide Test at the Mead-Strawn Field," *JPT* (April 1971) 431.

Means

Magruder, J.B., Stiles, L.H., and Yelverton, T.D.: "A Review of the Means San Andres Unit CO_2 Tertiary Project," *JPT* (May 1990) 638.

Stiles, L.H. *et al.*: "Design and Operation of a CO_2 Tertiary Pilot: Means San Andres Unit," paper SPE 11987 presented at the 1983 Annual Technical Conference and Exhibition, San Francisco, 5–8 October.

Stiles, L.H. and Magruder, J.B.: "Reservoir Management in the Means San Andres Unit," *JPT* (April 1992) 469.

Midale

Beliveau, D.A.: "Midale CO_2 Flood Pilot," *Proc.,* 1986 Ann. CIM Petroleum Soc. Tech. Mtg., Calgary, Vol 1.

Beliveau, D. and Payne, D.A.: "Analysis of a Tertiary CO_2 Flood Pilot in a Naturally Fractured Reservoir," paper SPE 22947 presented at the 1991 Annual Technical Conference and Exhibition, Dallas, 6–9 October.

Beliveau, D., Payne, D.A., and Mundry, M.: "Waterflood and CO_2 Flood of the Fractured Midale Field," *JPT* (September 1993) 881.

Northeast Purdy

Brinlee, L.D. and Brandt, J.A.: "Planning and Development of the Northeast Purdy Springer CO_2 Miscible Project," paper SPE 11163 presented at the 1982 Annual Technical Conference and Exhibition, New Orleans, 26–29 September.

Fox, M. J. *et al.*: "Review of CO_2 Flood, Springer 'A' Sand, Northeast Purdy Unit, Garvin County, Oklahoma," *SPERE* (November 1988) 1161.

North Coles Levee

MacAllister, D.J.: "Evaluation of CO_2 Flood Performance: North Coles Levee CO_2 Pilot, Kern County, California," *JPT* (February 1989) 185.

North Ward

Chou, S.I. *et al.*: "CO_2 Foam Field Trial at North Ward-Estes," paper SPE 24643 presented at the 1992 SPE Annual Technical Conference and Exhibition, Washington, DC, 4–7 October.

Power, M.T., Leicht, M.A., and Barnett, K.L.: "Converting Wells in a Mature West Texas Field for CO_2 Injection," paper SPE 20702 presented at the 1990 Annual Technical Conference and Exhibition, New Orleans, 23–26 September.

Ring, J.N. and Smith, D.J.: "An Overview of the North Ward Estes CO_2 Flood," paper SPE 30729 presented at the 1995 SPE Annual Technical Conference and Exhibition, Dallas, 22–25 October.

Winzinger, R. *et al*: "Design of a Major CO_2 Flood, North Ward Estes Field, Ward County, Texas," *SPERE* (February 1991) 11.

Paradis

Bears, D.A. *et al.*: "Paradis CO_2 Flood Gathering, Injection, and Production Systems," *JPT* (August 1984) 1312.

Rangely

Attanucci, V. *et al.*: "WAG Process Optimization in the Rangely CO_2 Miscible Flood," paper SPE 26622 presented at the 1993 SPE Annual Technical Conference and Exhibition, Houston, 3–6 October.

Franks, A.P.: "The Use of Selective Injection Equipment in the Rangely Weber Sand Unit," paper SPE 21649 presented at the 1991 SPE Production Operations Symposium, Oklahoma City, Oklahoma, 7–9 April.

Hervey, J.R. and Iakovakis, A.C.: "Performance Review of a Miscible CO_2 Tertiary Project: Rangely Weber Sand Unit, Colorado," *SPERE* (May 1991) 163.

Jonas, T.M., Chou, S.I., and Vasicek, S.L.: "Evaluation of a CO_2 Foam Field Trial: Rangely Weber Sand Unit," paper SPE 20468 presented at the 1990 Annual Technical Conference and Exhibition, New Orleans, 23–26 September.

"Chevron Rides the Rangely Field," *Western Oil World* (1989) **16,** 26.

Masoner, L.O., Abidi, H.R., and Hild, G.P.: "Diagnosing CO_2 Flood Performance Using Actual Performance Data," paper SPE 35363 presented at the 1996 SPE/DOE Symposium on Improved Oil Recovery, Tulsa, 21–24 April.

Robie, D.R. Jr., Roedell, J.W., and Wackowski, R.K.: "Field Trial of Simultaneous Injection of CO_2 and Water, Rangely Weber Sand Unit, Colorado," paper SPE 29521 presented at the 1995 Production Operations Symposium, Oklahoma City, Oklahoma, 2–4 April.

Shuler, P.J., Freitas, E.A., and Bowker, K.A.: "Selection and Application of $BaSO_4$ Scale Inhibitors for a CO_2 Flood, Rangely Weber Sand Unit, Colorado," *SPEPE* (August 1991) 259.

Wackowski, R.K. and Masoner, L.O.: "Rangely Weber Sand Unit CO_2 Project Update: Operating History," paper SPE 27755 presented at the 1994 SPE/DOE Symposium on Improved Oil Recovery, Tulsa, 17–20 April.

Rock Creek

Brummert, A.C. *et al.*: "Rock Creek Oil Field CO_2 Pilot Tests, Roane County, West Virginia," *JPT* (March 1988) 339.

Brummert, A.C. *et al.*: "Economic Evaluation of a CO_2 EOR Flood at the Rock Creek Field, Roane County, West Virginia," *SPERE* (August 1988) 829.

Heller, J.P., Boone, D.A., and Watts, R.J.: "Testing CO_2-Foam for Mobility Control at Rock Creek," paper 14519 presented at the 1985 SPE Eastern Regional Meeting, Morgantown, West Virginia, 6–8 November.

Watts, R.J.: "CO_2 Injection Must Wait For Economics," *Northeast Oil World* (1987) **7,** 13.

San Filippo, G.P. and Guckert, L.G.S.: "Development of a Pilot Carbon Dioxide Flood in the Rock Creek-Bin Injun Field, Roane Co., West Virginia," *Proc.,* U.S. DOE Symposium on Enhanced Oil and Gas Recovery and Improved Drilling Methods, Tulsa (1977).

Rose City North

Baskin, G.M. and Spriggs, D.M.: "Wellbore Operations in CO_2 Flooding—A Case History," NACE Corrosion 85, Boston, Massachusetts (1985) 1.

Sable

Westerman, G.W.: "Sable Unit Automatic CO_2 Injection System," Thirty-Second Annual Southwestern Petroleum Short Course, Lubbock, Texas (1985) 535.

Salt Creek (Texas)

Moshell, M.K.: "The Salt Creek CO_2 Flood: A Status Report After One Year of Operation," presented at the 1995 Annual Southwestern Petroleum Short Course Assn. Meeting, Lubbock, Texas, 19–20 April.

Slaughter (SEU)

Adams, G.H. and Rowe, H.G.: "Slaughter Estate Unit CO_2 Pilot—Surface and Downhole Equipment Construction and Operation in the Presence of H_2S," *JPT* (June 1981) 1065.

Ader, J.C. and Stein, M.H.: "Slaughter Estate Unit Tertiary Miscible Gas Pilot Reservoir Description," *JPT* (May 1984) 837.

Folger, L.K. and Guillot, S.N.: "A Case Study of the Development of the Sundown Slaughter Unit CO_2 Flood Hockley County, Texas," paper SPE 35189 presented at the 1996 SPE Permian Basin Oil and Gas Recovery Conference, Midland, Texas, 27–29 March.

Jarrell, P.M. and Stein, M.H.: "Maximizing Injection in Wells Recently Converted To Injection Using Hearn and Hall Plots," paper SPE 21724 presented at the 1991 SPE Production Operations Symposium, Oklahoma City, Oklahoma, 7–9 April.

Linn, L.R.: "CO_2 Injection and Production Field Facilities Design Evaluation and Considerations," paper SPE 16830 presented at the 1987 Annual Technical Conference and Exhibition, Dallas, 27–30 September.

Pariani, G.J. *et al.*: "An Approach To Optimize Economics in a West Texas CO_2 Flood," paper SPE 22022 presented at the 1991 SPE Hydrocarbon Economics and Evaluation Symposium, Dallas, 11–12 April.

Rowe, H.G., York, S.D., and Ader, J.C.: "Slaughter Estate Unit Tertiary Pilot Performance," *JPT* (March 1982) 613.

Stein, M.H. *et al.*: "Slaughter Estate Unit CO_2 Flood—A Comparison Between Pilot and Field-Scale Performance," *JPT* (September 1992) 1026.

White, J.T. and Benoit, R.L.: "A CO_2 Injection Measurement and Control System," paper SPE 14288 presented at the 1985 Annual Technical Conference and Exhibition, Las Vegas, Nevada, 22–25 September.

Timbalier Bay

Moore, J.S.: "Design, Installation, and Early Operation of the Timbalier Bay S-2B(RA)SU Gravity-Stable, Miscible CO_2-Injection Project," *SPEPE* (September 1986) 369.

Twofreds

Kirkpatrick, R.K., Flanders, W.A., and DePauw, R.M.: "Performance of the Twofreds CO_2 Injection Project," paper SPE 14439 presented at the 1985 Annual Technical Conference and Exhibition, Las Vegas, Nevada, 22–25 September.

Thrash, J.C.: "Twofreds Field A Tertiary Oil Recovery Project," paper SPE 8382 presented at the 1979 SPE Annual Technical Conference and Exhibition, Las Vegas, Nevada, 23–26 September.

Vacuum, East

Brownlee, M.H. and Sugg, L.A.: "East Vacuum Grayburg-San Andres Unit CO_2 Injection Project: Development and Results to Date," paper SPE 16721 presented at the 1987 Annual Technical Conference and Exhibition, Dallas, 27–30 September.

Harpole, K.J. and Hallenbeck, L.D.: "East Vacuum Grayburg San Andres Unit CO_2 Flood Ten Year Performance Review—Evolution of a Reservoir Management Strategy and Results of WAG Optimization," paper SPE 36710 presented at the 1996 SPE Annual Technical Conference and Exhibition, Denver, 6–9 October.

Martin, F.D. *et al.*: "CO_2-Foam Field Verification Pilot Test at EVGSAU Injection Project Phase I: Project Planning and Initial Results," paper SPE 24176 presented at the 1992 SPE/DOE Symposium on Enhanced Oil Recovery, Tulsa, 22–24 April.

Stevens, J.E. and Martin, F.D.: "CO_2 Foam Field Verification Pilot Test at EVGSAU: Phase IIIB—Project Operations and Performance Review," paper SPE 27786 presented at the 1994 SPE/DOE Symposium on Improved Oil Recovery, Tulsa, 17–20 April.

Wasson (Bennett Ranch)

Hsu, C.F. *et al.*: "Design and Implementation of a Grass-Roots CO_2 Project for the Bennett Ranch Unit," paper SPE 35188 presented at the 1996 SPE Permian Basin Oil and Gas Recovery Conference, Midland, Texas, 27–29 March.

Wasson (Cornell)

Todd, M.R., Cobb, W.M., and McCarter, E.D.: "CO_2 Flood Performance Evaluation for the Cornell Unit, Wasson San Andres Field," *JPT* (October 1982) 2271.

Wasson (Denver)

Bremer, C.: "Report on the Denver Unit CO_2 Flood, Wasson Field, Texas," presented at the 1982 Texas Petroleum Recovery Councils Conference on Enhanced Oil and Gas Recovery Field Operations in Texas, Austin, Texas, 16 November.

Fleming, E.A., Brown, L.M., and Cook, R.L.: "Overview of Production Engineering Aspects of Operating the Denver Unit CO_2 Flood," paper SPE 24157 presented at the 1992 SPE/DOE Symposium on Enhanced Oil Recovery, Tulsa, 22–24 April.

Greene, W.R.: "Analyzing the Flowing Performance of Oil Wells: Denver Unit CO_2 Flood," paper SPE 19725 presented at the 1989 Annual Technical Conference and Exhibition, San Antonio, Texas, 8–11 October.

Iken, G.: "Denver Unit CO_2 Flood: Material Aspects of the Design and Operation of Injection and Production Facilities," *Energy Progress* (1987) **7**, No. 4, 193.

Kittridge, M.G.: "Quantitative CO_2 Flood Monitoring, Denver Unit, Wasson (San Andres) Field," paper SPE 24644 presented at the 1992 Annual Technical Conference and Exhibition, Washington, DC, 4–7 October.

Patel, P.D., Christman, P.G., and Gardner J.W.: "Investigation of Unexpectedly Low Field-Observed Fluid Mobilities During Some CO_2 Tertiary Floods," *SPERE* (November 1987) 507.

Tanner, C.S. *et al.*: "Production Performance of the Wasson Denver Unit CO_2 Flood," paper SPE 24156 presented at the 1992 SPE/DOE Symposium on Enhanced Oil Recovery, Tulsa, 22–24 April.

Barnard, B.: "Shell Moves to Enhanced Recovery in Wasson Field," *Well Servicing* (1982) **22,** No. 1, 14.

Wasson (ODC)

Bullock, G.W., Wood, T.B., and Konecki, M.L.: "A Brief History of the Wasson EOR Project," *SPEPE* (August 1990) 338.

Wasson (Roberts)—(Planned Flood)

Hindi, R., Cheng, C.T., and Wang, B.: "CO_2 Miscible Flood Simulation Study, Roberts Unit, Wasson Field, Yoakum County, Texas," paper SPE 24185 presented at the 1992 SPE/DOE Symposium on Enhanced Oil Recovery, Tulsa, 22–24 April.

Wasson (South)

Burbank, D.E.: "Early CO_2 Flood Experience at the South Wasson Clearfork Unit," paper SPE 24160 presented at the 1992 SPE/DOE Symposium on Enhanced Oil Recovery, Tulsa, 22–24 April.

Wasson (Willard)

Bilhartz, H.L. Jr.: "Case History—A Pressure Core Hole," paper SPE 6389 presented at 1977 SPE Permian Basin Oil and Gas Recovery Conference, Midland, Texas, 10–11 March.

Bilhartz, H.L. and Charlson, G.S.: "Coring for In-Situ Saturation in the Willard Unit CO_2 Flood Mini-Test," paper SPE 7050 presented at the 1978 SPE Symposium on Improved Methods for Oil Recovery, Tulsa, 16–18 April.

Bilhartz, H.L. Jr., Charlson, G.S., and Stalkup, F.I.: "Use of Time-Lapse Logging Techniques in Evaluation the Willard Unit CO_2 Flood Mini-Test," paper SPE 7049 presented at the 1978 SPE Symposium on Improved Methods for Oil Recovery, Tulsa, 16–18 April.

Johnston, J.W.: "A Review of the Willard (San Andres) Unit CO_2 Injection Project," paper SPE 6388 presented at the 1977 Permian Basin Oil and Gas Recovery Conference of SPE of AIME, Midland, Texas, 10–11 March.

Price, B.C. and Gregg, F.L.: "CO_2/EOR: From Source to Resource," *Oil & Gas J.* (27 August 1983) **81,** 116.

Weeks Island

Cole, E.L., Ferrell, H.H., and Miller, D.E.: "An Evaluation of the Weeks Island "S" Sand Reservoir B Gravity Stable CO_2 Displacement Project, Iberia Parish, Louisiana," U.S. DOE Fossil Energy Rep. No. DOE/BC/10830-11 (DE89000727), U.S. DOE, Washington, DC (1989).

Johnston, J.R.: "Weeks Island Gravity Stable CO_2 Pilot," paper SPE 17351 presented at the 1988 SPE/DOE Enhanced Oil Recovery Symposium, Tulsa, 17–20 April.

Johnston, J.R. and Perry, G.E.: "Weeks Island Gravity Stable CO_2 Pilot: Final Report," U.S. DOE Fossil Energy Rep. No. DOE/MC/112004-6 (DE89000719), U.S. DOE, Washington, DC (1989).

Perry, G.E.: "Weeks Island "S" Sand Reservoir B Gravity Stable Miscible CO_2 Displacement, Iberia Parish, Louisiana," paper SPE 10695 presented at the 1982 SPE/DOE Joint Symposium on Enhanced Oil Recovery, Tulsa, 4–7 April.

Welch (South)

Hill, W.J. *et al.*: "CO_2 Operating Plan, South Welch Unit, Dawson County, Texas," paper SPE 27676 presented at the 1994 SPE Permian Basin Oil and Gas Recovery Conference, Midland, Texas, 16–18 March.

Wellman

Bangia, V.K., Yau, F.F., and Hendricks, G.R.: "Reservoir Performance of a Gravity-Stable, Vertical CO_2 Miscible Flood: Wolfcamp Reef Reservoir, Wellman Unit," paper SPE 22898 presented at the 1991 Annual Technical Conference and Exhibition, Dallas, 6–9 October.

Wertz

Borling, D.C.: "Injection Conformance Control Case Histories Using Gels at the Wertz Field CO_2 Tertiary Flood in Wyoming," paper SPE 27825 presented at the 1994 SPE/DOE Improved Oil Recovery Symposium, Tulsa, 17–20 April.

Kleinstelber, S.W.: "The Wertz Tensleep CO_2 Flood: Design and Initial Performance," *JPT* (May 1990) 6300.

West Sussex

Dauben, D.L.: "A Review of the West Sussex Unit CO_2 Flood Project," U.S. DOE Fossil Energy Rep. No. DOE/BC/10830-7 (DE98000705), U.S. DOE, Washington, DC (1988).

Hoiland, R.C., Joyner, H.D., and Stadler, J.L.: "Case History of a Successful Rocky Mountain Pilot CO_2 Flood," paper SPE 14939 presented at the 1986 SPE/DOE Symposium on Enhanced Oil Recovery, Tulsa, 20–23 April.

Weyburn

"Big Canadian miscible CO_2 EOR project, pipeline advance," *Oil & Gas J.* (7 July 1997).

Srivastava, V.A. *et al.*: "Quantification of Asphaltene Flocculation during Miscible CO_2 Flooding in the Weyburn Reservoir," *J. Cdn. Pet. Tech.* (October 1995) 31.

Wilmington

Hlozek, R.J. and Jones, R.D.: "Long Beach Features Unique CO_2 Injection Facilities," *Oil & Gas J.* (6 June 1983) **81,** 55.

Spivak, A., Garrison, W.H., and Nguyen, J.P.: "A Review of an Immiscible CO_2 Project, Tar Zone, Fault Block V, Wilmington Field, California," *SPERE* (May 1990) 1550.

Various Fields

Beeler, P.F.: "West Virginia CO_2 Oil Recovery Project Interim Report," *Proc.,* ERDA Symposium on Enhanced Oil, Gas Recovery, and Improved Drilling Methods, Tulsa (1977).

Brock, W.R. and Bryan, L.A.: "Summary Results of CO_2 EOR Field Tests, 1972–1987," paper SPE 18977 presented at the 1989 SPE Joint Rocky Mountain Regional/Low Permeability Reservoirs Symposium and Exhibition, Denver, 6–8 March.

Byars, H.G. and Galbraith, J.M.: "An Integrated Corrosion Control and Monitoring Program," International Corrosion Forum, Anaheim, California (1983) 50/1–50/18.

Cardenas, R.L. *et al.*: "Laboratory Design of a Gravity-Stable Miscible CO_2 Process," *JPT* (January 1984) 1110.

Crockett, D.H.: "Field Use of Halogen Compounds to Trace Injected CO_2," *Proc.,* 22nd Annual Southwestern Petroleum Short Course, Lubbock, Texas (1975) 159.

Hadlow, R.E.: "Update of Industry Experience with CO_2 Injection," paper SPE 24928 presented at the 1992 SPE Annual Technical Conference and Exhibition, Washington, DC, 4–7 October.

Appendix D

Adjustments to Water Relative Permeability Curves to Account for CO_2 Solubility

The adjustments to water relative permeability (k_{rw}) described in this appendix take into account the solubility of CO_2 in water. These adjustments are required as input to the simulator whenever water cycles are large enough to dissolve sufficient CO_2 around injection wells to affect water injectivity. Reservoir simulator studies for small water cycles (on the order of 2% HCPV) indicate that adjustments are not necessary in these cases.

The k_{rw} should be adjusted for chase water injection and for WAG projects that use large water cycles. Field experience in the Levelland field of west Texas has shown that water injectivity increased significantly during chase water injection after all the WAG CO_2 cycles had been completed.

Standard oil/water relative permeability measurements limit the maximum saturation of water in the reservoir to $(1.0-S_{or})$, where S_{or} is the residual oil saturation to water. During a CO_2 flood, the CO_2 reduces the S_{or} and fills the vacated space. If CO_2 were not soluble in water, the k_{rw} curve would not have to be adjusted; however, the water injected after a CO_2 cycle can dissolve residual CO_2, and the maximum water saturation after a CO_2 cycle can exceed $(1.0-S_{or})$. Understanding this effect is important in calculating the water injection rate for larger water cycles and for chase water following CO_2.

There are two basic types of adjustments, one for when there is no water hysteresis curve and one for when there is a water hysteresis curve.

If there is no water hysteresis curve (certainly the case for water-wet rock and some mixed-wet rock), the only adjustment required is to extrapolate the data for the original k_{rw} curve to a water saturation of 100%. The k_{rw} at 100% water is obtained by dividing the measured water permeability at 100% water by the oil permeability at connate water saturation (S_{wc}). In absence of permeability measured at 100% water, the k_{rw} can be assumed to be 1.0 at 100% water. This latter value is a lower limit based on the assumption that the pores occupied by the connate water have negligible flow capacity. Actual values will be slightly greater than 1.0. If all the CO_2 is dissolved, the maximum achievable water saturation will be slightly less than 100% because of the presence of a residual oil saturation to CO_2.

If there is a k_{rw} hysteresis curve, however, the hysteresis curve must be adjusted using a second set of calculations, discussed in Sec. D.2.

D.1 Extrapolation of the Data for the Original Water Relative Permeability Curve

For a simulator to accurately reflect the effect of CO_2 solubility in water, the k_{rw} for increasing water saturation must be extrapolated to the k_{rw} at 100% water. The adjustments are demonstrated in **Fig. D.1** with an example based on k_{rw} data from west Texas San Andres carbonate rock. Fig. D.1 illustrates the initial k_{rw} curve and the extrapolation that accounts for the dissolution of CO_2. This section uses a k_{rw} of 1.2 at 100% water, based on west Texas San Andres core data.

Fig. D.1 shows that before the CO_2 flood, the rock has a maximum k_{rw} of 0.45 at a water saturation (S_w) of 0.56 volume fraction. After CO_2 has reduced the S_{or}, the saturation and relative permeability of water can increase along the dashed line.

If the k_{rw} data have the proper curvature to be described by Eqs. D.1a and D.1b, then the equation can be used to generate the relative permeability values at water saturations greater than $(1.0-S_{or})$.

$$k_{rw}=k_{rw1}[(S_w-S_{wc})/(1.0-S_{wc})]^a, \quad \text{(D.1a)}$$

where k_{rw}=water relative permeability; k_{rw1}=water relative permeability at 100% water; S_w=water saturation relative to oil; S_{wc}=connate water saturation; and a=a variable exponent that is tied to k_{rw} at (1.0 - S_{or}) by

$$a=\mathrm{Ln}(k_{rw2}/k_{rw1})/\mathrm{Ln}[(1.0-S_{or}-S_{wc})/(1.0-S_{wc})], \quad \text{(D.1b)}$$

where k_{rw2}=the water relative permeability at $(1.0-S_{or})$ and S_{or}=the residual oil saturation to water.

D.2 Adjustments for Water Hysteresis

The next adjustment is to the water hysteresis curve. This adjustment is necessary only for intermediate-wet to oil-wet rock if k_{rwh} (water relative permeability for decreasing water saturation, or water relative permeability hysteresis) differs more than the experiment error in measurement from k_{rw}. Unfortunately, there is no extrapolation approach that can be used in this case.

In Fig. D.1, the original k_{rw} curve starts at an S_w of 0.005 volume fraction with zero relative permeability, and increases to 0.45 at an S_w of 0.56 volume fraction. The original water hysteresis relative permeability curve starts at a value of 0.45 at 0.56

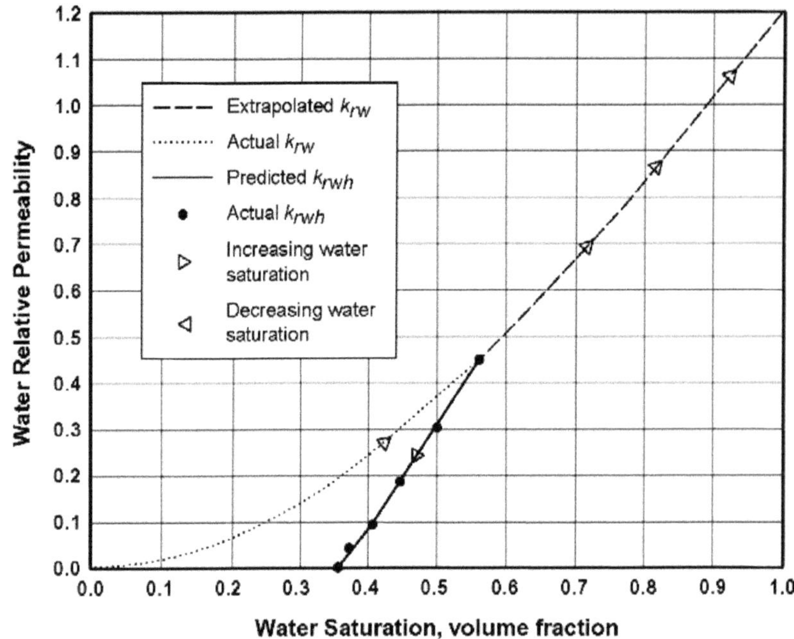

Fig. D.1—Extrapolation of data for a standard k_{rw} curve. This plot shows laboratory-measured k_{rw} that has been extrapolated to a value of 1.2 at 100% water saturation to account for CO_2 solubility effects in water.

volume fraction and moves downward. These data were measured in the laboratory.

Once CO_2 has displaced all the residual oil, the k_{rw} curve can continue up to 100% water. The water hysteresis curve then would start decreasing from a water saturation of 100%. To account for this, the hysteresis curve shown in Fig. D.1 must shift to the right, as shown in **Fig. D.2.**

The new k_{rwh} curve in Fig. D.2 is generated using a method similar to that developed by Land.[1] Beginning at a k_{rw} of 1.2 and 100% water, the k_{rwh} curve decreases with decreasing water saturation. When the value of k_{rwh} equals the value of k_{rw} at $(1.0-S_{or})$, the new k_{rwh} curve will be parallel to the original k_{rw} curve, but shifted to the right.

The parallel behavior agrees with experimental data[2] and theoretical studies.[1,3] The parallel shape also is consistent with the performance of reservoir simulators that generate parallel hysteresis relative permeability curves at lower saturations. Fig. D.1 shows that Land's hysteresis prediction technique[1] (represented by the solid line for decreasing water saturation) closely matches the measured values for hysteresis k_{rw} (the symbols).

Going a step further and using Land's method[1] to generate the displaced water relative hysteresis curve for intermediate-wet to oil-wet rock requires a water trapping constant, C, calculated as

$$C=1.0/(S_{wr}-S_{wc})-1.0/(S_{wi}-S_{wc}), \quad \text{(D.2)}$$

where S_{wr}=residual water saturation to oil displacement (water saturation at which k_{rwh}=0), and S_{wi}=water saturation at which the hysteresis behavior begins.

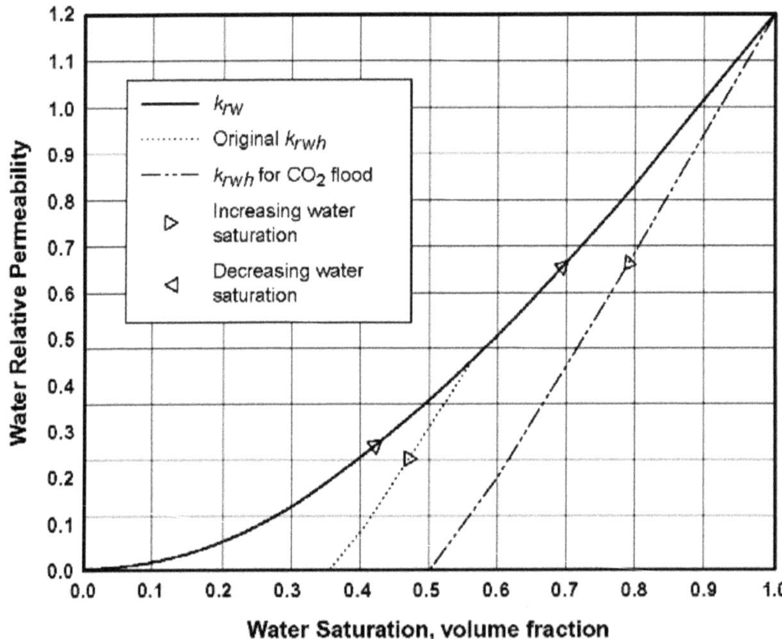

Fig. D.2—Water relative permeability curve applicable to CO_2 flooding. The dashed line is the k_{rw} hysteresis curve consistent with extrapolation of the data for the original k_{rw} curve. This curve also accounts for the effect of CO_2 solubility in water. The hysteresis curve begins at 100% water and is parallel to the original k_{rw} curve when the k_{rw} is less than or equal to the maximum k_{rw} before extrapolation (0.46 in this example).

For the standard water/oil relative permeability data measured in the lab, S_{wi} equals $(1.0-S_{or})$. For the hysteresis curve, the S_{wr} for decreasing water saturation is calculated by rearranging Eq. D.2:

$$S_{wr}=(S_{wi}-S_{wc})/[C\,(S_{wi}-S_{wc})+1]+S_{wc}. \quad \text{(D.3)}$$

Table D.1 shows how Eq. D.3 was used repeatedly at different S_{wi} values to calculate the corresponding S_{wr} values and to apply the parallel shift in S_w to generate the new k_{rwh} curve shown in Fig. D.2. The results calculated using the equations in this Appendix are input to the simulator to account for relative permeability changes caused by CO_2 solubility during large water cycles. Note that the two k_{rwh} curves are parallel for k_{rw} values of less than 0.45, which is the value at S_w equal to $(1.0-S_{or})$.

Nomenclature

a = exponent used in Eq. D.1b and defined by Eq. D.1b, dimensionless
C = water trapping constant calculated using Eq. D.2, dimensionless
k_{rw} = water relative permeability, dimensionless
k_{rw1} = water relative permeability at 100% water, dimensionless
k_{rw2} = water relative permeability at $1.0-S_{or}$, dimensionless
k_{rwh} = hysteresis water relative permeability, dimensionless
S_{or} = residual oil saturation to water, fraction
S_w = water saturation, fraction
S_{wc} = connate water saturation, fraction
S_{wh} = water saturation in Table D.1 that corresponds to the water hysteresis, fraction
S_{wi} = water saturation at which hysteresis behavior begins, fraction
S_{wr} = residual water saturation to oil displacement on hysteresis relative permeability curve (water saturation at which $k_{rwh}=0$), fraction

References

1. Land, C.S.: "Calculation of Imbibition Relative Permeability for Two-Phase and Three-Phase Flow from Rock Properties," *Trans.*, AIME (1968) **243,** 149
2. Geffen, T.M *et al.*: "Experimental Investigation of Factors Affecting Laboratory Relative Permeability Measurements," *Trans.*, AIME (1951) **192,** 99.
3. Carlson, F.M.: "Simulation of Relative Permeability Hysteresis to the Nonwetting Phase," paper SPE 10157 presented at the 1981 SPE Annual Technical Conference and Exhibition, San Antonio, Texas, 5–7 October.

TABLE D.1—ADJUSTMENTS TO WATER RELATIVE PERMEABILITY HYSTERESIS DATA

New S_{wi} (fraction)	= 1.000
S_{wc} (fraction)	= 0.005
Eq. D.3. S_{wr} (fraction)	= 0.500
Eq. D.2. C	= 1.015

Original Values				Values Calculated Using Eq. D.3		
Col. 1 S_w (fraction)	Col. 2 k_{rw} (dimensionless)	Col. 3 S_{wi} (fraction)	Col. 4 S_{wr} (fraction)	Col. 5 S_{wr} Shift (fraction)	Col. 6 S_{wh} (fraction)	Col. 7 kr_{wh} (dimensionless)
0.005	0.0000	0.005	0.005	0.495	0.500	0.0000
0.100	0.0105	0.100	0.092	0.408	0.508	0.0105
0.150	0.0370	0.150	0.131	0.369	0.519	0.0370
0.200	0.0600	0.200	0.168	0.332	0.532	0.0600
0.250	0.0900	0.250	0.201	0.299	0.549	0.0900
0.300	0.1350	0.300	0.232	0.268	0.568	0.1350
0.350	0.1900	0.350	0.261	0.239	0.589	0.1900
0.400	0.2400	0.400	0.287	0.213	0.613	0.2400
0.450	0.3000	0.450	0.312	0.188	0.638	0.3000
0.500	0.3600	0.500	0.334	0.166	0.666	0.3600
0.560	0.4500	0.560	0.360	0.140	0.700	0.4500
0.650	0.5755	0.650	0.395	0.105	0.755	0.5755
0.750	0.7348	0.750	0.429	0.071	0.821	0.7348
0.850	0.9097	0.850	0.460	0.040	0.890	0.9097
1.000	1.2000	1.000	0.500	0.000	1.000	1.2000

Spreadsheet Steps

Col.	Source
1	Input from relative permeability data
2	Input from relative permeability data.
3	Equals input S_{wi}.
4	Eq. D.3 with S_{wi} from Col. 3.
5	Original S_{wr} at top of spreadsheet minus the S_{wr} calculated in Col. 4.
6	S_{wh} is the water saturation corresponding to the water hysteresis in Col. 7. S_{wh} is calculated by adding S_{wi} from Col. 3 to the S_{wr} shift in Col. 5.
7	Equals input k_{rw}.

Appendix E
Sample Self-Certification Document for Obtaining EOR Tax Credits

A document patterned on the following Engineer's Statement may serve as self-certification for obtaining the U.S. federal EOR tax credit. The format and content were valid at the time of publication in this monograph. The document reproduced here is a copy of the Engineer's Statement filed by Mobil Producing Texas & New Mexico Inc. at the Oklahoma Corporate Commission in support of an application for severance tax reduction for the Postle CO_2 flood.

ENGINEER'S STATEMENT

TABLE OF CONTENTS

I. Identification of Project
II. Property Ownership
III. Description of Project
  A. Tertiary Recovery Method
  B. Background
  C. Project Design
  D. Increase in Recovery
IV. Engineer's Statement

TABLE OF EXHIBITS

1. Property Description
2. Lease Map
3. Oklahoma Commission Order No. 67195 establishing the Hough Morrow A Unit
4. Production History (1968–90)
5. Production Forecast

I. IDENTIFICATION OF PROJECT

The Hough Morrow A Unit Project is a new Carbon Dioxide Miscible Tertiary Recovery Project in the Postle Field located in Texas County, Oklahoma. The project covers 3,360 acres in the eastern and central portion of the Postle Field. A description of the property and a map showing the location of the unit are shown in Exhibits 1 and 2. Also attached (Exhibit 3) is the Corporation Commission of Oklahoma Order No. 67195 granting approval of the application to form the Hough Morrow A Unit.

II. PROPERTY OWNERSHIP

Mobil Oil Corporation, the Operator of the Hough Morrow A Unit, owns 89.31 percent working interest, Mobil Producing Texas & New Mexico Inc. owns 2.72 percent working interest and Mobil Exploration and Producing North America Inc. owns 1.16 percent working interest. The remaining interest is owned by Anadarko Petroleum Corporation, Oxy USA Inc., Mrs. M.O. Rife Jr. and Hoover Capital Company.

III. DESCRIPTION OF THE PROJECT

A. Tertiary Recovery Method

The Hough Morrow A Unit Project involves a tertiary recovery method as defined in section 43(c) (2) (A) (i) of the Internal Revenue Code of 1986 and Proposed Treasury Regulation section 1.43-2(e) (2), specifically, Carbon Dioxide Miscible fluid displacement. The first injection of carbon dioxide (CO_2) under the Hough Morrow A Unit Project is planned for December 1995.

B. Background

The Hough Morrow A Unit Project is a 3,360 acre waterflood project in the Postle Field. The field was discovered in 1958 by Republic Natural Gas. The Hough Morrow A Unit began waterflood operations in 1967 with the unitization of the working interest owners and royalty interest. Currently, the unit is developed on an irregular pattern with 8 active producers and 14 active injectors. The Unit oil production is approximately 360 barrels per day from the Morrow A sandstone formation at an average depth of 6,150 feet.

C. Project Design

The Carbon Dioxide flood project for the Hough Morrow A Unit is designed to inject a slug of CO_2 equal in size to 35% of the reservoir's original hydrocarbon pore volume or 52 billion standard cubic feet of CO_2. In accordance with sound engineering practices, the CO_2 will be injected alternately with equally sized slugs of water. The alternating of slugs of CO_2 and water will serve to improve sweep efficiency.

D. Increase in Recovery

The application of the CO_2 miscible process in the Hough Morrow A Unit is expected to result in more than an insignificant increase in the amount of crude oil that ultimately will be recovered from the Morrow A reservoir. With the tertiary method, it is estimated that 7,935,000 barrels of additional oil will be recovered due to the injection of carbon dioxide.

The additional reserve recovery is a result of the CO_2 mobilizing the remaining oil contained in the Morrow A reservoir pay intervals currently open to production under secondary waterflood operations.

E. Production History and Forecast

The production history of the unit since 1968 is shown in Exhibit 4. A forecast of oil production without and with the tertiary recovery project is shown in Exhibit 5. The volumes are in thousands of barrels of oil per year for the period 1996 through year 2014. Also shown are the reserves recovered over the life of the project.

IV. ENGINEER'S STATEMENT

The undersigned believes that the Hough Morrow A Unit Project is a "qualified enhanced oil recovery project" within the meaning of section 43(c) (2) (A) of the Internal Revenue Code of 1986, in that (i) the project involves the application (in accordance with sound engineering principles) of one or more tertiary recovery methods (as defined in section 43(c) (2) (A) (i) of the Internal Revenue Code and Proposed Treasury Regulation section 1.43-2(e) (2)) which can reasonably be expected to result in more than an insignificant increase in the amount of crude oil that will ultimately be recovered, (ii) the project is located within the United States, and (iii) the first injection of liquids, gases, or other matter commenced after December 31, 1990.

Under the penalties of perjury, I declare that I have examined this certification, including all attachments and statements, and to the best of my knowledge and belief, it is true, correct, and complete.

(the seal, signature and certification number of the Registered Professional Engineer)

SI Metric Conversion Factors

acre × 4.046 856	E + 03 = m^2
acre × 4.046 856	E − 01 = ha
bbl × 1.589 873	E − 01 = m^3
ft × 3.048*	E − 01 = m
ft^3 × 2.831 685	E − 02 = m^3

*Conversion factor is exact.

Appendix F
CO_2 Properties

This appendix provides quick access to key data on CO_2 physical and thermodynamic properties necessary for basic calculations and estimations in the design and operation of pure CO_2 flow in pipeline, compression, surface facility, and well systems.

It is imperative that one know the CO_2 physical state for each segment of flow through a system. Slight deviations in pressure and/or temperature from the triple point will cause CO_2 to exist as a solid, liquid, or vapor **(Fig. F.1).** At pressures and temperatures above the critical point (1,071 psia and 87.8°F), CO_2 is not a true vapor or liquid, but exists instead as a dense vapor phase.

Fig. F.2 shows the very wide range of CO_2 densities that can exist around 1,000 psia. At 120°F, CO_2 will act like a gas with a density of about 11 lbm/ft^3, and at 80°F it is more liquid-like, with a density of about 44 lbm/ft^3.

Fig. F.3 shows the pressure/enthalpy diagram for CO_2, which is useful for compressor and heat-exchanger calculations. Other important data such as viscosity, compression factors, and volume factors are listed in **Table F.1** for temperature ranges of 0 to 360°F and pressure ranges of 100 to 10,000 psia. These data are also available in the additional media files that accompany this monograph.

References

1. *Engineering Data Book*, FPS Version, Vol. II, Gas Processors Suppliers Association (1998).

SI Metric Conversion Factors

bbl × 1.589 873	E − 01	= m^3
Btu × 1.055 056	E + 00	= kJ
Btu/lbm-°F × 4.186 8*	E + 00	= kJ/(kg•K)
Btu/lbm-°R × 4.186 8*	E + 00	= kJ/(kg•K)
°F (°F − 32)/1.8		= °C
°F (°F + 459.67)/1.8		= K
ft × 3.048*	E − 01	= m
ft^3 × 2.831 685	E − 02	= m^3
lbm × 4.535 924	E − 01	= kg
psia × 6.894 757		= kPa
°R × 5/9		= K

*Conversion factor is exact.

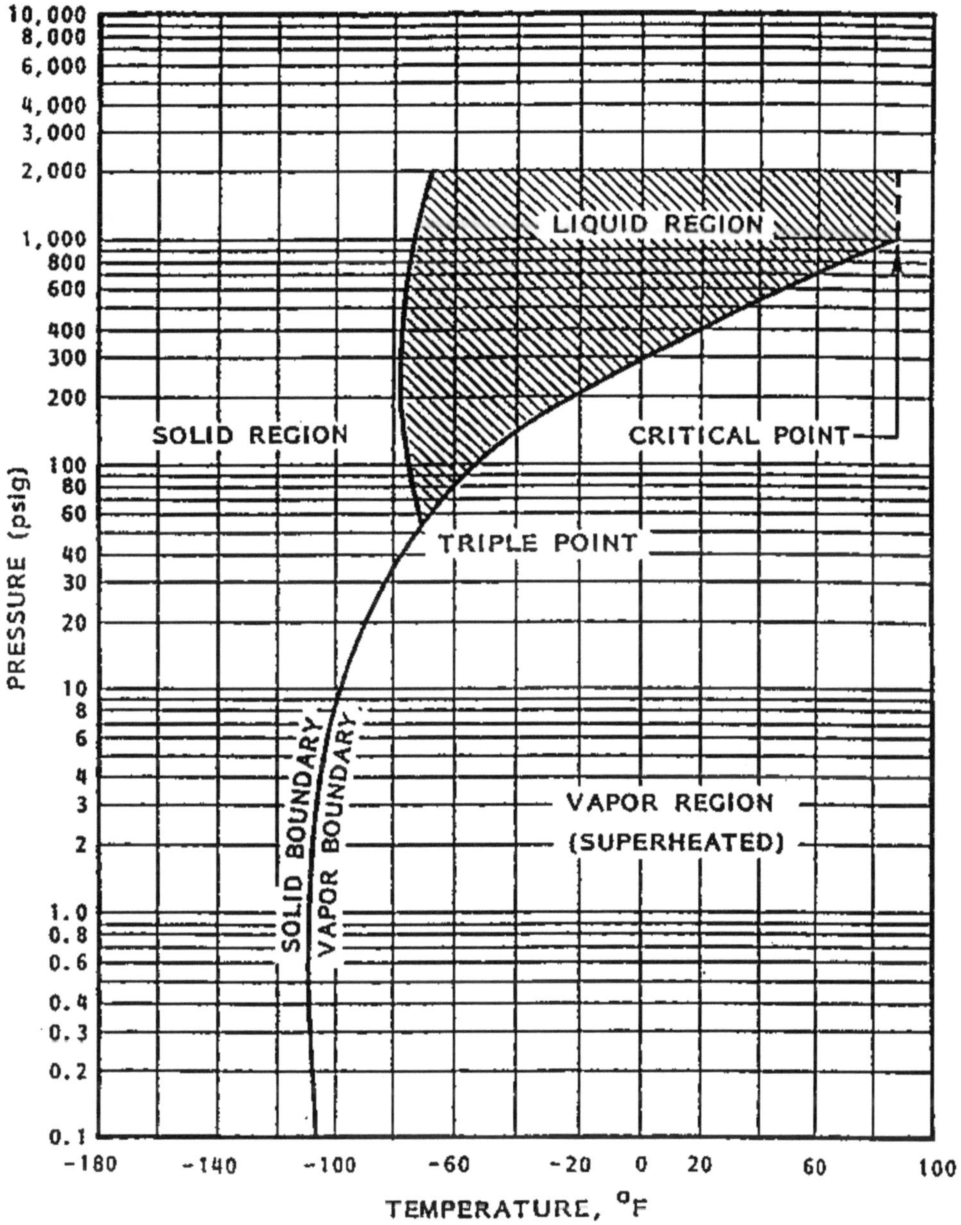

10,000
8,000
6,000
4,000
3,000
2,000
1,000
800
600
400
300
200
100
80
60
40
30
20
10
8
6
4
3
2
1.0
0.8
0.6
0.4
0.3
0.2
0.1
PRESSURE (psig)
LIQUID REGION
SOLID REGION
CRITICAL POINT
TRIPLE POINT
SOLID BOUNDARY
VAPOR BOUNDARY
VAPOR REGION
(SUPERHEATED)
-180
-140
-100
-60
-20
0
20
60
100
TEMPERATURE, °F

Fig. F.1—CO_2 phase equilibrium diagram.

Temperatures, °F
0
10
20
30
40
50
60
70
80
90
100
110
120
80
70
60
50
40
30
20
10
-
Density, lbm/ft³
-
1,000
2,000
3,000
4,000
Pressure, psia

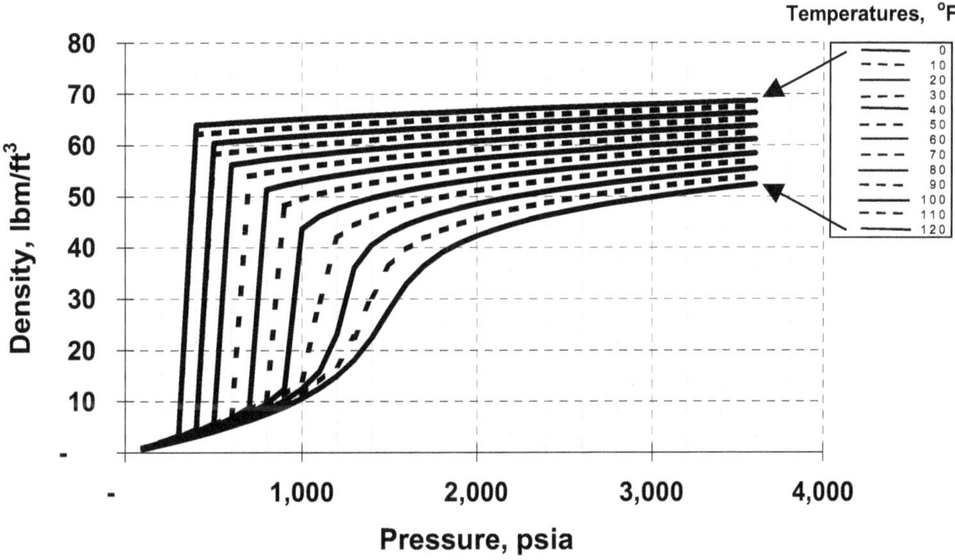

Fig. F.2—CO_2 densities at various temperatures, °F (from Table F.1).

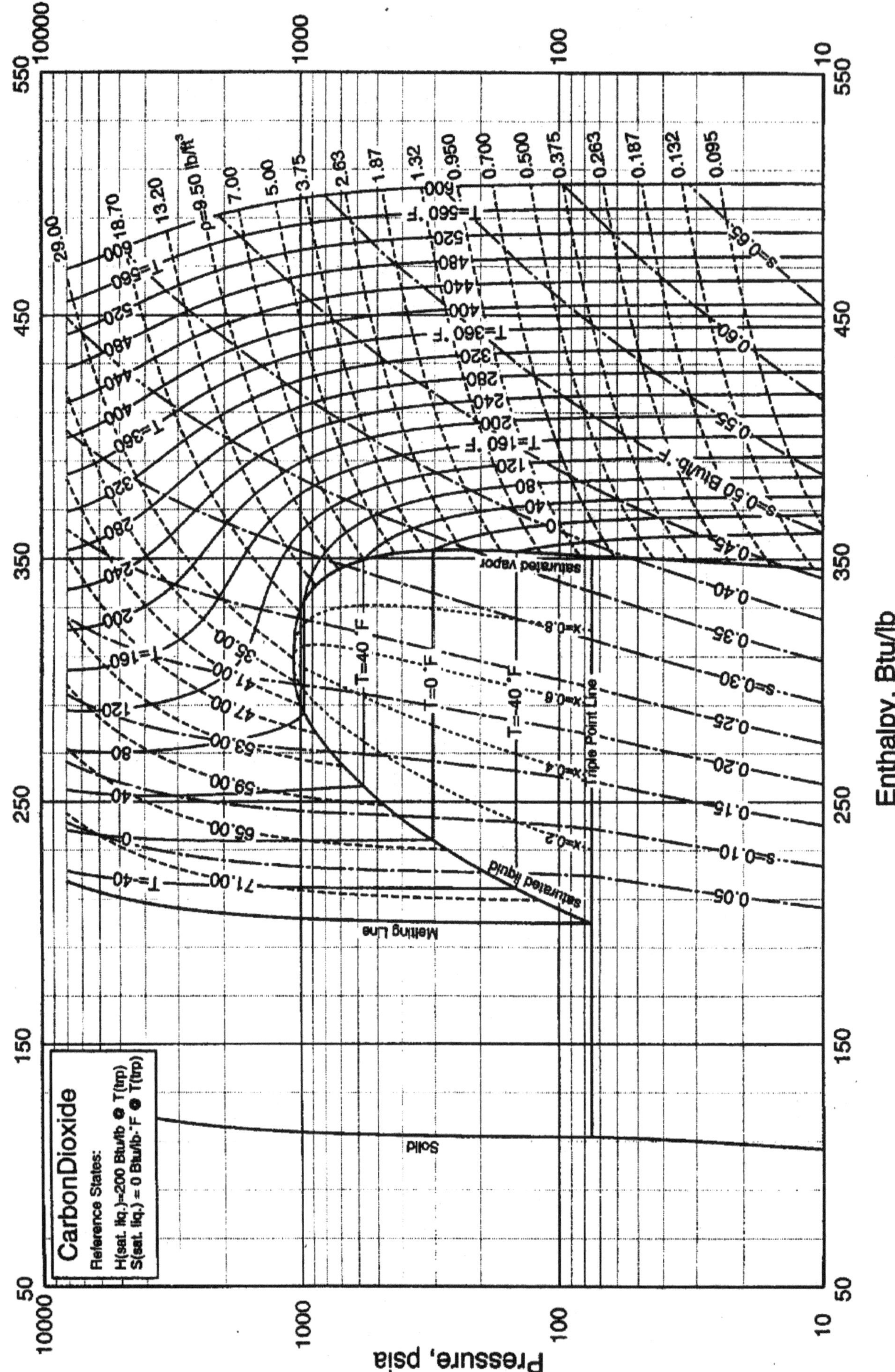

Fig. F.3—Pressure/enthalpy diagram for CO_2 (from Ref. 1).

TABLE F.1—CO_2 PROPERTIES AT VARIOUS TEMPERATURES, °F (TABLE ALSO AVAILABLE ONLINE)											
Temperature (°F)	Pressure (psia)	Density (lbm/ft^3)	Compressibility Factor	Heat Capacity (Btu/lbm°F)	Heat Capacity Ratio (C_p/C_v)	Sonic Velocity (ft/s)	Enthalpy (Btu/lbm)	Entropy [Btu/(lbm°F)]	Viscosity (cp)	Phase	Gas Formation Volume Factor (res bbl/Mcf)
0	14.696	0.13222	0.99158	0.19362	1.3199	820.93	202.34	0.6234	0.012832	V	156.316676
0	100	0.949	0.94009	0.21404	1.3865	796.35	198.79	0.53136	0.012901	V	21.779411
0	200	2.0454	0.87234	0.24922	1.502	763.49	194.05	0.49252	0.013034	V	10.104911
0	300	3.3804	0.79175	0.31264	1.7088	723.6	188.25	0.46459	0.013254	V	6.114255
0	305.74	3.468	0.78653	0.31802	1.7262	720.99	187.86	0.46308	0.013271	V	5.959910
0	305.74	63.757	0.042782	0.52381	2.3484	2264	68.514	0.20344	0.13432	L	0.324179
0	400	63.954	0.055799	0.5192	2.327	2288.2	68.448	0.2027	0.13569	L	0.323179
0	500	64.157	0.069529	0.51466	2.3059	2313	68.385	0.20194	0.13711	L	0.322161
0	600	64.354	0.083179	0.51042	2.286	2337.1	68.331	0.20119	0.13852	L	0.321173
0	700	64.545	0.096755	0.50645	2.2674	2360.4	68.284	0.20047	0.1399	L	0.320223
0	800	64.731	0.11026	0.50273	2.2498	2383.1	68.244	0.19976	0.14126	L	0.319304
0	900	64.912	0.1237	0.49923	2.2331	2405.2	68.211	0.19906	0.1426	L	0.318423
0	1000	65.089	0.13707	0.49593	2.2174	2426.7	68.183	0.19838	0.14393	L	0.317555
0	1100	65.261	0.15038	0.4928	2.2024	2447.8	68.162	0.19772	0.14523	D	0.316719
0	1200	65.429	0.16362	0.48985	2.1882	2468.3	68.145	0.19707	0.14652	D	0.315887
0	1300	65.594	0.17681	0.48704	2.1746	2488.5	68.134	0.19643	0.1478	D	0.315094
0	1400	65.755	0.18995	0.48437	2.1617	2508.2	68.128	0.1958	0.14906	D	0.314332
0	1500	65.912	0.20303	0.48183	2.1494	2527.6	68.126	0.19519	0.15031	D	0.313578
0	1600	66.067	0.21606	0.4794	2.1376	2546.6	68.129	0.19458	0.15154	D	0.312846
0	1700	66.218	0.22904	0.47708	2.1262	2565.3	68.135	0.19399	0.15277	D	0.312133
0	1800	66.366	0.24197	0.47487	2.1154	2583.6	68.146	0.1934	0.15398	D	0.311434
0	1900	66.511	0.25486	0.47275	2.105	2601.7	68.16	0.19283	0.15518	D	0.310760
0	2000	66.654	0.26769	0.47071	2.0949	2619.5	68.178	0.19226	0.15636	D	0.310084
0	2100	66.794	0.28049	0.46876	2.0853	2636.9	68.199	0.1917	0.15754	D	0.309439
0	2200	66.932	0.29324	0.46688	2.076	2654.2	68.224	0.19115	0.15871	D	0.308800
0	2300	67.068	0.30595	0.46508	2.067	2671.1	68.252	0.19061	0.15987	D	0.308176
0	2400	67.201	0.31862	0.46334	2.0584	2687.9	68.282	0.19008	0.16102	D	0.307566
0	2500	67.332	0.33125	0.46167	2.05	2704.4	68.316	0.18955	0.16216	D	0.306968
0	2600	67.46	0.34384	0.46005	2.042	2720.7	68.352	0.18903	0.16329	D	0.306380
0	2700	67.587	0.3564	0.4585	2.0342	2736.7	68.391	0.18852	0.16442	D	0.305809
0	2800	67.712	0.36892	0.45699	2.0266	2752.6	68.432	0.18802	0.16553	D	0.305247
0	2900	67.835	0.3814	0.45554	2.0193	2768.3	68.476	0.18752	0.16664	D	0.304691
0	3000	67.956	0.39385	0.45413	2.0122	2783.8	68.522	0.18703	0.16774	D	0.304149
0	3100	68.076	0.40626	0.45277	2.0053	2799.1	68.571	0.18654	0.16884	D	0.303612
0	3200	68.193	0.41864	0.45145	1.9986	2814.2	68.622	0.18606	0.16993	D	0.303087
0	3300	68.31	0.43099	0.45018	1.9921	2829.1	68.675	0.18558	0.17101	D	0.302573
0	3400	68.424	0.44331	0.44894	1.9858	2843.9	68.73	0.18511	0.17208	D	0.302068
0	3500	68.537	0.4556	0.44774	1.9797	2858.5	68.786	0.18465	0.17315	D	0.301573
0	3600	68.649	0.46785	0.44657	1.9737	2873	68.845	0.18419	0.17421	D	0.301079
0	3700	68.759	0.48008	0.44544	1.9679	2887.3	68.906	0.18373	0.17527	D	0.300600
0	3800	68.867	0.49227	0.44434	1.9622	2901.5	68.969	0.18329	0.17632	D	0.300121
0	3900	68.975	0.50444	0.44327	1.9567	2915.5	69.033	0.18284	0.17736	D	0.299655
0	4000	69.081	0.51658	0.44223	1.9513	2929.4	69.099	0.1824	0.1784	D	0.299195
0	4100	69.185	0.5287	0.44122	1.9461	2943.2	69.166	0.18196	0.17944	D	0.298746
0	4200	69.289	0.54078	0.44023	1.941	2956.8	69.236	0.18153	0.18047	D	0.298296
0	4300	69.391	0.55284	0.43927	1.936	2970.3	69.306	0.18111	0.18149	D	0.297857

TABLE F.1 (continued)—CO_2 PROPERTIES AT VARIOUS TEMPERATURES, °F											
Temperature (°F)	Pressure (psia)	Density (lbm/ft³)	Compressibility Factor	Heat Capacity (Btu/lbm°F)	Heat Capacity Ratio (C_p/C_v)	Sonic Velocity (ft/s)	Enthalpy (Btu/lbm)	Entropy [Btu/(lbm°F)]	Viscosity (cp)	Phase	Gas Formation Volume Factor (res bbl/Mcf)
0	4400	69.492	0.56488	0.43834	1.9311	2983.7	69.378	0.18068	0.18251	D	0.297427
0	4500	69.592	0.57689	0.43743	1.9263	2996.9	69.452	0.18026	0.18353	D	0.297001
0	4600	69.691	0.58887	0.43654	1.9217	3010	69.527	0.17985	0.18454	D	0.296578
0	4700	69.789	0.60083	0.43568	1.9171	3023	69.604	0.17944	0.18554	D	0.296163
0	4800	69.885	0.61276	0.43483	1.9127	3035.9	69.682	0.17903	0.18654	D	0.295751
0	4900	69.981	0.62467	0.43401	1.9083	3048.7	69.761	0.17863	0.18754	D	0.295346
0	5000	70.076	0.63656	0.4332	1.904	3061.4	69.841	0.17823	0.18853	D	0.294948
0	5100	70.169	0.64843	0.43242	1.8999	3074	69.923	0.17783	0.18952	D	0.294557
0	5200	70.262	0.66027	0.43165	1.8958	3086.4	70.006	0.17744	0.19051	D	0.294168
0	5300	70.354	0.67209	0.4309	1.8918	3098.8	70.09	0.17705	0.19149	D	0.293784
0	5400	70.444	0.68389	0.43017	1.8878	3111.1	70.175	0.17666	0.19246	D	0.293406
0	5500	70.534	0.69566	0.42945	1.884	3123.3	70.261	0.17628	0.19344	D	0.293029
0	5600	70.623	0.70742	0.42875	1.8802	3135.3	70.349	0.1759	0.19441	D	0.292662
0	5700	70.712	0.71915	0.42806	1.8765	3147.3	70.437	0.17552	0.19537	D	0.292295
0	5800	70.799	0.73087	0.42739	1.8729	3159.2	70.527	0.17514	0.19634	D	0.291937
0	5900	70.885	0.74256	0.42673	1.8693	3171	70.617	0.17477	0.1973	D	0.291579
0	6000	70.971	0.75423	0.42609	1.8658	3182.7	70.709	0.1744	0.19825	D	0.291225
0	6100	71.056	0.76589	0.42546	1.8624	3194.3	70.802	0.17404	0.1992	D	0.290880
0	6200	71.14	0.77752	0.42484	1.859	3205.9	70.895	0.17367	0.20015	D	0.290534
0	6300	71.224	0.78914	0.42424	1.8557	3217.3	70.989	0.17331	0.2011	D	0.290195
0	6400	71.306	0.80073	0.42365	1.8524	3228.7	71.085	0.17296	0.20204	D	0.289856
0	6500	71.388	0.81231	0.42307	1.8492	3240	71.181	0.1726	0.20298	D	0.289524
0	6600	71.47	0.82387	0.4225	1.8461	3251.2	71.278	0.17225	0.20392	D	0.289195
0	6700	71.55	0.83541	0.42194	1.843	3262.4	71.376	0.1719	0.20486	D	0.288869
0	6800	71.63	0.84694	0.42139	1.84	3273.4	71.475	0.17155	0.20579	D	0.288550
0	6900	71.709	0.85844	0.42086	1.837	3284.4	71.574	0.1712	0.20672	D	0.288229
0	7000	71.788	0.86993	0.42033	1.834	3295.3	71.675	0.17086	0.20764	D	0.287914
0	7100	71.866	0.8814	0.41982	1.8311	3306.2	71.776	0.17052	0.20857	D	0.287602
0	7200	71.943	0.89285	0.41931	1.8283	3316.9	71.878	0.17018	0.20949	D	0.287291
0	7300	72.02	0.90429	0.41881	1.8254	3327.6	71.98	0.16984	0.21041	D	0.286987
0	7400	72.096	0.91571	0.41832	1.8227	3338.3	72.084	0.16951	0.21132	D	0.286684
0	7500	72.171	0.92711	0.41784	1.8199	3348.8	72.188	0.16918	0.21224	D	0.286383
0	7600	72.246	0.9385	0.41737	1.8173	3359.3	72.293	0.16885	0.21315	D	0.286087
0	7700	72.321	0.94987	0.41691	1.8146	3369.7	72.398	0.16852	0.21406	D	0.285792
0	7800	72.394	0.96123	0.41645	1.812	3380.1	72.505	0.1682	0.21496	D	0.285502
0	7900	72.468	0.97257	0.41601	1.8094	3390.4	72.612	0.16787	0.21587	D	0.285214
0	8000	72.54	0.98389	0.41557	1.8069	3400.6	72.719	0.16755	0.21677	D	0.284927
0	8100	72.612	0.9952	0.41514	1.8044	3410.8	72.827	0.16723	0.21767	D	0.284644
0	8200	72.684	1.0065	0.41471	1.8019	3420.9	72.936	0.16691	0.21856	D	0.284365
0	8300	72.755	1.0178	0.4143	1.7995	3431	73.045	0.1666	0.21946	D	0.284093
0	8400	72.826	1.029	0.41389	1.7971	3441	73.155	0.16628	0.22035	D	0.283800
0	8500	72.896	1.0403	0.41348	1.7947	3450.9	73.266	0.16597	0.22124	D	0.283541
0	8600	72.965	1.0515	0.41309	1.7924	3460.8	73.377	0.16566	0.22213	D	0.283261
0	8700	73.034	1.0627	0.4127	1.7901	3470.6	73.489	0.16535	0.22302	D	0.282988
0	8800	73.103	1.074	0.41231	1.7878	3480.4	73.601	0.16504	0.2239	D	0.282747
0	8900	73.171	1.0851	0.41193	1.7856	3490.1	73.714	0.16474	0.22478	D	0.282460

TABLE F.1 (continued)—CO_2 PROPERTIES AT VARIOUS TEMPERATURES, °F											
Temperature (°F)	Pressure (psia)	Density (lbm/ft^3)	Compressibility Factor	Heat Capacity (Btu/lbm°F)	Heat Capacity Ratio (C_p/C_v)	Sonic Velocity (ft/s)	Enthalpy (Btu/lbm)	Entropy [Btu/(lbm°F)]	Viscosity (cp)	Phase	Gas Formation Volume Factor (res bbl/Mcf)
0	9000	73.239	1.0963	0.41156	1.7833	3499.8	73.828	0.16444	0.22567	D	0.282204
0	9100	73.306	1.1075	0.4112	1.7811	3509.4	73.942	0.16413	0.22654	D	0.281955
0	9200	73.373	1.1186	0.41084	1.779	3518.9	74.056	0.16383	0.22742	D	0.281685
0	9300	73.439	1.1298	0.41048	1.7768	3528.4	74.171	0.16354	0.2283	D	0.281446
0	9400	73.505	1.1409	0.41013	1.7747	3537.8	74.287	0.16324	0.22917	D	0.281188
0	9500	73.571	1.152	0.40979	1.7726	3547.2	74.403	0.16294	0.23004	D	0.280935
0	9600	73.636	1.1631	0.40945	1.7706	3556.6	74.52	0.16265	0.23091	D	0.280687
0	9700	73.7	1.1742	0.40912	1.7685	3565.9	74.637	0.16236	0.23178	D	0.280445
0	9800	73.765	1.1853	0.40879	1.7665	3575.1	74.754	0.16207	0.23265	D	0.280207
0	9900	73.828	1.1963	0.40847	1.7645	3584.3	74.872	0.16178	0.23351	D	0.279951
0	10000	73.892	1.2074	0.40815	1.7626	3593.4	74.99	0.16149	0.23437	D	0.279723
10	14.696	0.12933	0.99216	0.19485	1.3161	829.1	204.28	0.62758	0.013107	V	159.810727
10	100	0.92451	0.94445	0.21333	1.377	806.16	200.93	0.53595	0.013175	V	22.356424
10	200	1.9788	0.88252	0.24378	1.4784	776.01	196.51	0.49783	0.0133	V	10.445228
10	300	3.2306	0.81082	0.29401	1.6447	740.47	191.27	0.47111	0.013502	V	6.397740
10	360.41	4.1418	0.7598	0.34698	1.8166	714.68	187.45	0.45645	0.013686	V	4.990290
10	360.41	62.069	0.0507	0.5425	2.4345	2115.7	73.787	0.21445	0.12249	L	0.332993
10	400	62.168	0.05618	0.53982	2.4229	2128	73.737	0.21409	0.1231	L	0.332464
10	500	62.411	0.069952	0.53346	2.3951	2157.9	73.621	0.21321	0.1246	L	0.331172
10	600	62.645	0.083629	0.52765	2.3694	2186.5	73.517	0.21236	0.12607	L	0.329935
10	700	62.87	0.097217	0.5223	2.3454	2213.8	73.424	0.21153	0.12751	L	0.328751
10	800	63.089	0.11072	0.51735	2.323	2240	73.341	0.21073	0.12892	L	0.327612
10	900	63.3	0.12415	0.51275	2.302	2265.3	73.268	0.20995	0.1303	L	0.326533
10	1000	63.504	0.1375	0.50846	2.2824	2289.8	73.203	0.20919	0.13166	L	0.325481
10	1100	63.703	0.15077	0.50445	2.2638	2313.5	73.146	0.20845	0.133	D	0.324448
10	1200	63.896	0.16398	0.50069	2.2463	2336.6	73.096	0.20772	0.13432	D	0.323469
10	1300	64.084	0.17713	0.49715	2.2298	2359.1	73.053	0.20701	0.13562	D	0.322531
10	1400	64.268	0.19021	0.49382	2.2141	2381	73.017	0.20632	0.1369	D	0.321609
10	1500	64.446	0.20323	0.49067	2.1992	2402.3	72.987	0.20565	0.13817	D	0.320715
10	1600	64.621	0.21619	0.48768	2.185	2423.2	72.963	0.20498	0.13941	D	0.319845
10	1700	64.791	0.2291	0.48485	2.1715	2443.7	72.944	0.20433	0.14065	D	0.319006
10	1800	64.958	0.24195	0.48215	2.1587	2463.7	72.93	0.2037	0.14187	D	0.318183
10	1900	65.121	0.25476	0.47959	2.1464	2483.3	72.921	0.20307	0.14308	D	0.317396
10	2000	65.281	0.26751	0.47714	2.1346	2502.6	72.916	0.20246	0.14427	D	0.316616
10	2100	65.437	0.28021	0.47481	2.1233	2521.5	72.916	0.20185	0.14545	D	0.315855
10	2200	65.59	0.29287	0.47258	2.1125	2540.1	72.92	0.20126	0.14662	D	0.315120
10	2300	65.74	0.30548	0.47044	2.1021	2558.4	72.928	0.20068	0.14778	D	0.314397
10	2400	65.888	0.31805	0.46839	2.0921	2576.3	72.94	0.2001	0.14893	D	0.313695
10	2500	66.032	0.33058	0.46643	2.0825	2594	72.955	0.19954	0.15007	D	0.313011
10	2600	66.175	0.34306	0.46454	2.0733	2611.5	72.974	0.19898	0.1512	D	0.312335
10	2700	66.314	0.35551	0.46272	2.0644	2628.6	72.996	0.19843	0.15232	D	0.311682
10	2800	66.451	0.36791	0.46098	2.0557	2645.6	73.022	0.19789	0.15343	D	0.311033
10	2900	66.586	0.38028	0.45929	2.0474	2662.2	73.05	0.19736	0.15453	D	0.310405
10	3000	66.719	0.39261	0.45767	2.0394	2678.7	73.081	0.19684	0.15563	D	0.309787
10	3100	66.85	0.4049	0.45611	2.0316	2695	73.115	0.19632	0.15671	D	0.309179
10	3200	66.978	0.41716	0.45459	2.0241	2711	73.152	0.19581	0.15779	D	0.308586

TABLE F.1 (continued)—CO_2 PROPERTIES AT VARIOUS TEMPERATURES, °F											
Temperature (°F)	Pressure (psia)	Density (lbm/ft^3)	Compressibility Factor	Heat Capacity (Btu/lbm°F)	Heat Capacity Ratio (C_p/C_v)	Sonic Velocity (ft/s)	Enthalpy (Btu/lbm)	Entropy [Btu/(lbm°F)]	Viscosity (cp)	Phase	Gas Formation Volume Factor (res bbl/Mcf)
10	3300	67.105	0.42939	0.45313	2.0168	2726.8	73.191	0.1953	0.15886	D	0.308008
10	3400	67.23	0.44158	0.45172	2.0097	2742.5	73.232	0.1948	0.15992	D	0.307435
10	3500	67.353	0.45374	0.45036	2.0029	2757.9	73.277	0.19431	0.16098	D	0.306876
10	3600	67.474	0.46586	0.44903	1.9962	2773.2	73.323	0.19383	0.16203	D	0.306321
10	3700	67.593	0.47796	0.44775	1.9897	2788.3	73.372	0.19335	0.16308	D	0.305783
10	3800	67.711	0.49002	0.44651	1.9834	2803.2	73.422	0.19287	0.16411	D	0.305249
10	3900	67.827	0.50206	0.4453	1.9773	2818	73.475	0.1924	0.16515	D	0.304729
10	4000	67.942	0.51406	0.44413	1.9714	2832.6	73.53	0.19194	0.16617	D	0.304213
10	4100	68.055	0.52604	0.443	1.9656	2847	73.587	0.19148	0.16719	D	0.303709
10	4200	68.166	0.53798	0.44189	1.96	2861.3	73.646	0.19103	0.16821	D	0.303208
10	4300	68.277	0.5499	0.44082	1.9545	2875.4	73.706	0.19058	0.16922	D	0.302718
10	4400	68.385	0.5618	0.43978	1.9491	2889.4	73.769	0.19013	0.17022	D	0.302240
10	4500	68.493	0.57366	0.43876	1.9439	2903.3	73.833	0.18969	0.17122	D	0.301763
10	4600	68.599	0.5855	0.43778	1.9388	2917	73.899	0.18926	0.17221	D	0.301295
10	4700	68.704	0.59732	0.43682	1.9338	2930.6	73.966	0.18883	0.1732	D	0.300838
10	4800	68.808	0.60911	0.43588	1.929	2944.1	74.035	0.1884	0.17419	D	0.300385
10	4900	68.91	0.62087	0.43497	1.9242	2957.5	74.105	0.18798	0.17517	D	0.299936
10	5000	69.012	0.63261	0.43408	1.9196	2970.7	74.177	0.18756	0.17614	D	0.299495
10	5100	69.112	0.64433	0.43321	1.915	2983.8	74.251	0.18714	0.17711	D	0.299062
10	5200	69.211	0.65602	0.43237	1.9106	2996.8	74.326	0.18673	0.17808	D	0.298633
10	5300	69.309	0.66769	0.43154	1.9063	3009.7	74.402	0.18633	0.17904	D	0.298210
10	5400	69.406	0.67934	0.43073	1.902	3022.4	74.479	0.18592	0.18	D	0.297795
10	5500	69.502	0.69097	0.42995	1.8979	3035.1	74.558	0.18552	0.18096	D	0.297386
10	5600	69.597	0.70257	0.42918	1.8938	3047.6	74.638	0.18513	0.18191	D	0.296978
10	5700	69.691	0.71415	0.42843	1.8898	3060.1	74.72	0.18473	0.18285	D	0.296577
10	5800	69.784	0.72571	0.42769	1.8859	3072.4	74.802	0.18434	0.1838	D	0.296182
10	5900	69.876	0.73725	0.42698	1.8821	3084.7	74.886	0.18396	0.18474	D	0.295792
10	6000	69.967	0.74877	0.42627	1.8783	3096.8	74.971	0.18357	0.18568	D	0.295407
10	6100	70.057	0.76027	0.42559	1.8746	3108.9	75.057	0.18319	0.18661	D	0.295027
10	6200	70.147	0.77175	0.42492	1.871	3120.9	75.144	0.18282	0.18754	D	0.294651
10	6300	70.235	0.7832	0.42426	1.8674	3132.7	75.232	0.18244	0.18846	D	0.294276
10	6400	70.323	0.79464	0.42362	1.8639	3144.5	75.321	0.18207	0.18939	D	0.293910
10	6500	70.41	0.80606	0.42299	1.8605	3156.2	75.411	0.1817	0.19031	D	0.293547
10	6600	70.496	0.81746	0.42237	1.8572	3167.8	75.502	0.18134	0.19123	D	0.293188
10	6700	70.582	0.82885	0.42177	1.8539	3179.3	75.594	0.18098	0.19214	D	0.292836
10	6800	70.666	0.84021	0.42117	1.8506	3190.8	75.687	0.18062	0.19305	D	0.292484
10	6900	70.75	0.85156	0.42059	1.8474	3202.1	75.781	0.18026	0.19396	D	0.292139
10	7000	70.833	0.86288	0.42002	1.8443	3213.4	75.876	0.1799	0.19486	D	0.291794
10	7100	70.916	0.87419	0.41947	1.8412	3224.6	75.972	0.17955	0.19577	D	0.291455
10	7200	70.997	0.88549	0.41892	1.8381	3235.7	76.069	0.1792	0.19667	D	0.291122
10	7300	71.078	0.89676	0.41838	1.8352	3246.8	76.166	0.17885	0.19757	D	0.290788
10	7400	71.159	0.90802	0.41786	1.8322	3257.8	76.265	0.17851	0.19846	D	0.290460
10	7500	71.238	0.91926	0.41734	1.8293	3268.7	76.364	0.17817	0.19935	D	0.290135
10	7600	71.317	0.93049	0.41683	1.8265	3279.5	76.464	0.17783	0.20024	D	0.289815
10	7700	71.396	0.94169	0.41634	1.8237	3290.2	76.565	0.17749	0.20113	D	0.289495
10	7800	71.473	0.95289	0.41585	1.8209	3300.9	76.666	0.17715	0.20201	D	0.289182

TABLE F.1 (continued)—CO_2 PROPERTIES AT VARIOUS TEMPERATURES, °F

Temperature (°F)	Pressure (psia)	Density (lbm/ft^3)	Compressibility Factor	Heat Capacity (Btu/lbm°F)	Heat Capacity Ratio (C_p/C_v)	Sonic Velocity (ft/s)	Enthalpy (Btu/lbm)	Entropy [Btu/(lbm°F)]	Viscosity (cp)	Phase	Gas Formation Volume Factor (res bbl/Mcf)
10	7900	71.55	0.96406	0.41537	1.8182	3311.5	76.768	0.17682	0.2029	D	0.288869
10	8000	71.627	0.97522	0.4149	1.8155	3322.1	76.871	0.17649	0.20378	D	0.288560
10	8100	71.703	0.98637	0.41444	1.8129	3332.5	76.975	0.17616	0.20465	D	0.288256
10	8200	71.778	0.9975	0.41398	1.8103	3343	77.079	0.17583	0.20553	D	0.287954
10	8300	71.853	1.0086	0.41354	1.8077	3353.3	77.185	0.1755	0.2064	D	0.287650
10	8400	71.927	1.0197	0.4131	1.8052	3363.6	77.29	0.17518	0.20727	D	0.287353
10	8500	72.001	1.0308	0.41267	1.8027	3373.8	77.397	0.17486	0.20814	D	0.287064
10	8600	72.074	1.0419	0.41224	1.8002	3384	77.504	0.17454	0.20901	D	0.286781
10	8700	72.146	1.0529	0.41183	1.7978	3394.1	77.612	0.17422	0.20988	D	0.286478
10	8800	72.219	1.064	0.41142	1.7954	3404.1	77.72	0.17391	0.21074	D	0.286208
10	8900	72.29	1.075	0.41102	1.793	3414.1	77.829	0.1736	0.2116	D	0.285918
10	9000	72.361	1.086	0.41062	1.7907	3424	77.939	0.17328	0.21246	D	0.285635
10	9100	72.432	1.097	0.41023	1.7884	3433.9	78.049	0.17297	0.21331	D	0.285357
10	9200	72.502	1.108	0.40985	1.7861	3443.7	78.16	0.17267	0.21417	D	0.285086
10	9300	72.571	1.1189	0.40947	1.7839	3453.4	78.271	0.17236	0.21502	D	0.284795
10	9400	72.64	1.1299	0.4091	1.7817	3463.1	78.383	0.17205	0.21587	D	0.284535
10	9500	72.709	1.1408	0.40873	1.7795	3472.7	78.495	0.17175	0.21672	D	0.284256
10	9600	72.777	1.1518	0.40838	1.7773	3482.3	78.609	0.17145	0.21757	D	0.284007
10	9700	72.845	1.1627	0.40802	1.7752	3491.9	78.722	0.17115	0.21842	D	0.283739
10	9800	72.912	1.1736	0.40767	1.7731	3501.3	78.836	0.17085	0.21926	D	0.283477
10	9900	72.979	1.1845	0.40733	1.771	3510.8	78.951	0.17056	0.2201	D	0.283220
10	10000	73.045	1.1954	0.40699	1.7689	3520.1	79.066	0.17026	0.22094	D	0.282968
20	14.696	0.12657	0.99269	0.1961	1.3125	837.17	206.23	0.63169	0.013382	V	163.300531
20	100	0.90148	0.94839	0.2129	1.3683	815.71	203.06	0.54044	0.013447	V	22.927677
20	200	1.9179	0.89153	0.23957	1.4584	787.93	198.93	0.50292	0.013567	V	10.776533
20	300	3.1011	0.82708	0.28056	1.5966	755.99	194.14	0.47715	0.013753	V	6.664988
20	400	4.5589	0.75013	0.35666	1.8475	717.04	188.26	0.45461	0.014055	V	4.533667
20	421.91	4.9373	0.73058	0.38411	1.9356	706.95	186.73	0.44963	0.014146	V	4.186210
20	421.91	60.263	0.059856	0.56685	2.5357	1958.8	79.228	0.22552	0.11156	L	0.342974
20	500	60.497	0.07066	0.55953	2.507	1988.6	79.078	0.2247	0.11282	L	0.341646
20	600	60.784	0.084393	0.55108	2.4729	2024.1	78.904	0.2237	0.11439	L	0.340039
20	700	61.057	0.098017	0.54349	2.4414	2057.4	78.747	0.22274	0.11592	L	0.338514
20	800	61.319	0.11154	0.53661	2.4123	2088.8	78.606	0.22182	0.1174	L	0.337065
20	900	61.57	0.12497	0.53034	2.3852	2118.6	78.479	0.22093	0.11885	L	0.335688
20	1000	61.811	0.13832	0.52458	2.3601	2147	78.364	0.22006	0.12027	L	0.334394
20	1100	62.044	0.15158	0.51928	2.3367	2174.3	78.261	0.21922	0.12166	D	0.333137
20	1200	62.27	0.16476	0.51437	2.3147	2200.6	78.169	0.21841	0.12302	D	0.331928
20	1300	62.488	0.17787	0.5098	2.2941	2225.9	78.086	0.21762	0.12436	D	0.330775
20	1400	62.699	0.1909	0.50554	2.2748	2250.5	78.012	0.21685	0.12567	D	0.329648
20	1500	62.904	0.20387	0.50156	2.2565	2274.3	77.946	0.21609	0.12697	D	0.328575
20	1600	63.103	0.21678	0.49782	2.2393	2297.5	77.888	0.21536	0.12824	D	0.327546
20	1700	63.297	0.22962	0.4943	2.223	2320.1	77.838	0.21464	0.12949	D	0.326538
20	1800	63.486	0.2424	0.49098	2.2076	2342.1	77.794	0.21394	0.13073	D	0.325562
20	1900	63.67	0.25513	0.48784	2.1929	2363.5	77.757	0.21326	0.13195	D	0.324624
20	2000	63.85	0.2678	0.48487	2.1789	2384.6	77.725	0.21259	0.13316	D	0.323708
20	2100	64.025	0.28042	0.48205	2.1656	2405.1	77.699	0.21193	0.13435	D	0.322822

TABLE F.1 (continued)—CO_2 PROPERTIES AT VARIOUS TEMPERATURES, °F											
Temperature (°F)	Pressure (psia)	Density (lbm/ft³)	Compressibility Factor	Heat Capacity (Btu/lbm°F)	Heat Capacity Ratio (C_p/C_v)	Sonic Velocity (ft/s)	Enthalpy (Btu/lbm)	Entropy [Btu/(lbm°F)]	Viscosity (cp)	Phase	Gas Formation Volume Factor (res bbl/Mcf)
20	2200	64.197	0.29299	0.47937	2.1529	2425.3	77.679	0.21129	0.13553	D	0.321961
20	2300	64.365	0.30551	0.47681	2.1407	2445	77.664	0.21065	0.13669	D	0.321122
20	2400	64.529	0.31798	0.47438	2.1291	2464.4	77.653	0.21003	0.13785	D	0.320303
20	2500	64.689	0.33041	0.47205	2.118	2483.4	77.647	0.20942	0.13899	D	0.319511
20	2600	64.847	0.34279	0.46983	2.1073	2502.1	77.645	0.20882	0.14012	D	0.318734
20	2700	65.001	0.35513	0.4677	2.097	2520.5	77.648	0.20823	0.14124	D	0.317978
20	2800	65.153	0.36742	0.46566	2.0872	2538.6	77.654	0.20765	0.14234	D	0.317233
20	2900	65.301	0.37968	0.46371	2.0777	2556.4	77.664	0.20708	0.14344	D	0.316514
20	3000	65.447	0.39189	0.46183	2.0685	2574	77.678	0.20652	0.14453	D	0.315803
20	3100	65.591	0.40407	0.46002	2.0597	2591.2	77.695	0.20597	0.14562	D	0.315114
20	3200	65.732	0.41621	0.45828	2.0512	2608.3	77.715	0.20542	0.14669	D	0.314438
20	3300	65.87	0.42832	0.45661	2.0429	2625	77.739	0.20488	0.14775	D	0.313782
20	3400	66.006	0.44039	0.45499	2.035	2641.6	77.766	0.20435	0.14881	D	0.313135
20	3500	66.141	0.45242	0.45343	2.0273	2657.9	77.795	0.20383	0.14986	D	0.312498
20	3600	66.272	0.46442	0.45193	2.0199	2674.1	77.827	0.20332	0.1509	D	0.311876
20	3700	66.402	0.47639	0.45047	2.0126	2690	77.862	0.20281	0.15193	D	0.311268
20	3800	66.53	0.48832	0.44907	2.0056	2705.7	77.9	0.2023	0.15296	D	0.310666
20	3900	66.656	0.50022	0.44771	1.9988	2721.2	77.94	0.20181	0.15398	D	0.310077
20	4000	66.78	0.5121	0.44639	1.9923	2736.6	77.983	0.20132	0.15499	D	0.309505
20	4100	66.903	0.52394	0.44511	1.9859	2751.8	78.027	0.20083	0.156	D	0.308938
20	4200	67.023	0.53575	0.44388	1.9796	2766.8	78.074	0.20036	0.157	D	0.308380
20	4300	67.142	0.54754	0.44268	1.9736	2781.6	78.124	0.19988	0.158	D	0.307837
20	4400	67.26	0.55929	0.44151	1.9677	2796.3	78.175	0.19942	0.15899	D	0.307296
20	4500	67.376	0.57102	0.44038	1.962	2810.8	78.228	0.19895	0.15997	D	0.306769
20	4600	67.49	0.58272	0.43928	1.9564	2825.1	78.284	0.1985	0.16095	D	0.306249
20	4700	67.603	0.59439	0.43822	1.9509	2839.4	78.341	0.19804	0.16193	D	0.305736
20	4800	67.714	0.60604	0.43718	1.9456	2853.4	78.4	0.1976	0.1629	D	0.305234
20	4900	67.824	0.61766	0.43617	1.9404	2867.4	78.461	0.19715	0.16386	D	0.304738
20	5000	67.933	0.62926	0.43519	1.9354	2881.2	78.524	0.19672	0.16482	D	0.304252
20	5100	68.04	0.64083	0.43423	1.9305	2894.8	78.588	0.19628	0.16578	D	0.303771
20	5200	68.146	0.65238	0.4333	1.9256	2908.4	78.654	0.19585	0.16673	D	0.303299
20	5300	68.251	0.66391	0.43239	1.9209	2921.8	78.721	0.19543	0.16767	D	0.302835
20	5400	68.355	0.67541	0.43151	1.9163	2935	78.79	0.19501	0.16862	D	0.302376
20	5500	68.457	0.68688	0.43064	1.9118	2948.2	78.861	0.19459	0.16955	D	0.301920
20	5600	68.559	0.69834	0.4298	1.9074	2961.2	78.933	0.19418	0.17049	D	0.301475
20	5700	68.659	0.70977	0.42898	1.9031	2974.2	79.007	0.19377	0.17142	D	0.301034
20	5800	68.758	0.72118	0.42818	1.8989	2987	79.081	0.19336	0.17234	D	0.300600
20	5900	68.856	0.73257	0.4274	1.8948	2999.7	79.158	0.19296	0.17326	D	0.300172
20	6000	68.953	0.74394	0.42663	1.8908	3012.3	79.235	0.19256	0.17418	D	0.299750
20	6100	69.049	0.75529	0.42588	1.8868	3024.8	79.314	0.19216	0.1751	D	0.299335
20	6200	69.144	0.76661	0.42515	1.8829	3037.2	79.394	0.19177	0.17601	D	0.298921
20	6300	69.238	0.77792	0.42444	1.8791	3049.5	79.475	0.19138	0.17692	D	0.298516
20	6400	69.331	0.78921	0.42374	1.8754	3061.7	79.558	0.191	0.17782	D	0.298116
20	6500	69.424	0.80047	0.42306	1.8717	3073.8	79.641	0.19062	0.17872	D	0.297718
20	6600	69.515	0.81172	0.42239	1.8681	3085.8	79.726	0.19024	0.17962	D	0.297328
20	6700	69.605	0.82295	0.42174	1.8646	3097.7	79.812	0.18986	0.18052	D	0.296942

TABLE F.1 (continued)—CO_2 PROPERTIES AT VARIOUS TEMPERATURES, °F											
Temperature (°F)	Pressure (psia)	Density (lbm/ft^3)	Compressibility Factor	Heat Capacity (Btu/lbm°F)	Heat Capacity Ratio (C_p/C_v)	Sonic Velocity (ft/s)	Enthalpy (Btu/lbm)	Entropy [Btu/(lbm°F)]	Viscosity (cp)	Phase	Gas Formation Volume Factor (res bbl/Mcf)
20	6800	69.695	0.83416	0.4211	1.8611	3109.5	79.899	0.18949	0.18141	D	0.296561
20	6900	69.784	0.84535	0.42047	1.8577	3121.3	79.987	0.18912	0.1823	D	0.296183
20	7000	69.872	0.85652	0.41986	1.8544	3132.9	80.076	0.18875	0.18318	D	0.295810
20	7100	69.959	0.86767	0.41926	1.8511	3144.5	80.166	0.18839	0.18407	D	0.295440
20	7200	70.045	0.87881	0.41867	1.8479	3155.9	80.257	0.18802	0.18495	D	0.295077
20	7300	70.131	0.88993	0.41809	1.8447	3167.3	80.349	0.18766	0.18583	D	0.294718
20	7400	70.216	0.90103	0.41753	1.8416	3178.6	80.441	0.18731	0.1867	D	0.294361
20	7500	70.3	0.91211	0.41697	1.8385	3189.9	80.535	0.18695	0.18757	D	0.294008
20	7600	70.383	0.92318	0.41643	1.8355	3201	80.63	0.1866	0.18844	D	0.293661
20	7700	70.466	0.93423	0.41589	1.8325	3212.1	80.726	0.18625	0.18931	D	0.293316
20	7800	70.548	0.94526	0.41537	1.8296	3223.1	80.822	0.18591	0.19017	D	0.292974
20	7900	70.629	0.95628	0.41486	1.8267	3234	80.919	0.18556	0.19104	D	0.292638
20	8000	70.709	0.96728	0.41435	1.8239	3244.9	81.017	0.18522	0.1919	D	0.292304
20	8100	70.789	0.97827	0.41386	1.8211	3255.7	81.116	0.18488	0.19275	D	0.291976
20	8200	70.869	0.98924	0.41337	1.8183	3266.4	81.216	0.18455	0.19361	D	0.291649
20	8300	70.947	1.0002	0.4129	1.8156	3277	81.317	0.18421	0.19446	D	0.291328
20	8400	71.025	1.0111	0.41243	1.813	3287.6	81.418	0.18388	0.19531	D	0.290997
20	8500	71.103	1.0221	0.41197	1.8103	3298.1	81.52	0.18355	0.19616	D	0.290702
20	8600	71.18	1.033	0.41152	1.8077	3308.5	81.623	0.18322	0.19701	D	0.290386
20	8700	71.256	1.0439	0.41107	1.8052	3318.9	81.726	0.18289	0.19785	D	0.290077
20	8800	71.331	1.0547	0.41064	1.8027	3329.2	81.83	0.18257	0.19869	D	0.289747
20	8900	71.406	1.0656	0.41021	1.8002	3339.5	81.935	0.18225	0.19953	D	0.289452
20	9000	71.481	1.0764	0.40979	1.7977	3349.6	82.041	0.18193	0.20037	D	0.289137
20	9100	71.555	1.0873	0.40938	1.7953	3359.8	82.147	0.18161	0.20121	D	0.288856
20	9200	71.628	1.0981	0.40897	1.7929	3369.8	82.254	0.18129	0.20204	D	0.288554
20	9300	71.701	1.1089	0.40857	1.7906	3379.8	82.361	0.18098	0.20287	D	0.288259
20	9400	71.774	1.1197	0.40818	1.7883	3389.8	82.469	0.18066	0.20371	D	0.287970
20	9500	71.845	1.1305	0.40779	1.786	3399.6	82.578	0.18035	0.20453	D	0.287687
20	9600	71.917	1.1413	0.40741	1.7837	3409.5	82.687	0.18004	0.20536	D	0.287410
20	9700	71.988	1.152	0.40703	1.7815	3419.2	82.797	0.17974	0.20619	D	0.287114
20	9800	72.058	1.1627	0.40666	1.7793	3428.9	82.908	0.17943	0.20701	D	0.286823
20	9900	72.128	1.1735	0.4063	1.7771	3438.6	83.019	0.17913	0.20783	D	0.286564
20	10000	72.197	1.1842	0.40594	1.7749	3448.2	83.131	0.17882	0.20865	D	0.286285
30	14.696	0.12392	0.99317	0.19737	1.309	845.13	208.2	0.63575	0.013656	V	166.785574
30	100	0.87976	0.95195	0.21271	1.3604	825.03	205.19	0.54483	0.013719	V	23.493524
30	200	1.862	0.89956	0.23631	1.441	799.33	201.31	0.50782	0.013833	V	11.100286
30	300	2.9869	0.84116	0.27059	1.5586	770.41	196.9	0.48284	0.014007	V	6.919766
30	400	4.33	0.77366	0.32742	1.7512	736.49	191.67	0.46164	0.014276	V	4.773360
30	490.77	5.8847	0.69845	0.43368	2.0982	697.71	185.64	0.44251	0.014671	V	3.512298
30	490.77	58.31	0.070488	0.59966	2.6609	1793.8	84.881	0.23673	0.10136	L	0.354463
30	500	58.345	0.071771	0.59831	2.6563	1798.4	84.853	0.23662	0.10152	L	0.354253
30	600	58.712	0.085586	0.58485	2.6089	1845.5	84.573	0.2354	0.10327	L	0.352035
30	700	59.056	0.099269	0.57323	2.5656	1887.9	84.322	0.23425	0.10494	L	0.349985
30	800	59.38	0.11283	0.56304	2.526	1926.7	84.097	0.23315	0.10654	L	0.348072
30	900	59.687	0.12628	0.554	2.4898	1962.8	83.894	0.2321	0.10809	L	0.346279
30	1000	59.98	0.13963	0.5459	2.4565	1996.7	83.711	0.23109	0.1096	L	0.344598

TABLE F.1 (continued)—CO_2 PROPERTIES AT VARIOUS TEMPERATURES, °F

Temperature (°F)	Pressure (psia)	Density (lbm/ft³)	Compressibility Factor	Heat Capacity (Btu/lbm°F)	Heat Capacity Ratio (C_p/C_v)	Sonic Velocity (ft/s)	Enthalpy (Btu/lbm)	Entropy [Btu/(lbm°F)]	Viscosity (cp)	Phase	Gas Formation Volume Factor (res bbl/Mcf)
30	1100	60.259	0.15288	0.53859	2.4258	2028.7	83.546	0.23013	0.11106	D	0.342998
30	1200	60.526	0.16604	0.53194	2.3974	2059.1	83.396	0.22919	0.11249	D	0.341480
30	1300	60.783	0.17912	0.52586	2.3711	2088.2	83.261	0.22829	0.11388	D	0.340044
30	1400	61.03	0.19212	0.52027	2.3466	2116.1	83.138	0.22742	0.11524	D	0.338671
30	1500	61.269	0.20504	0.51511	2.3238	2143	83.027	0.22658	0.11658	D	0.337351
30	1600	61.499	0.21789	0.51032	2.3024	2168.9	82.927	0.22576	0.11789	D	0.336087
30	1700	61.723	0.23067	0.50586	2.2823	2194	82.837	0.22496	0.11918	D	0.334870
30	1800	61.939	0.24338	0.5017	2.2635	2218.4	82.756	0.22418	0.12044	D	0.333692
30	1900	62.149	0.25604	0.4978	2.2457	2242.1	82.683	0.22342	0.12169	D	0.332574
30	2000	62.353	0.26863	0.49414	2.2289	2265.1	82.619	0.22269	0.12291	D	0.331481
30	2100	62.552	0.28116	0.49069	2.2129	2287.6	82.562	0.22196	0.12412	D	0.330421
30	2200	62.745	0.29365	0.48743	2.1978	2309.5	82.512	0.22126	0.12531	D	0.329413
30	2300	62.934	0.30607	0.48435	2.1835	2331	82.468	0.22057	0.12649	D	0.328418
30	2400	63.117	0.31845	0.48143	2.1698	2352	82.431	0.21989	0.12765	D	0.327464
30	2500	63.297	0.33078	0.47866	2.1568	2372.5	82.4	0.21923	0.1288	D	0.326538
30	2600	63.473	0.34306	0.47602	2.1444	2392.6	82.374	0.21858	0.12994	D	0.325635
30	2700	63.644	0.35529	0.47351	2.1325	2412.4	82.353	0.21794	0.13106	D	0.324753
30	2800	63.812	0.36748	0.47112	2.1211	2431.7	82.337	0.21732	0.13217	D	0.323899
30	2900	63.977	0.37963	0.46883	2.1102	2450.8	82.326	0.2167	0.13327	D	0.323070
30	3000	64.138	0.39173	0.46664	2.0997	2469.5	82.32	0.2161	0.13436	D	0.322255
30	3100	64.296	0.40379	0.46455	2.0897	2487.9	82.317	0.2155	0.13544	D	0.321461
30	3200	64.451	0.41582	0.46254	2.08	2506	82.319	0.21492	0.13651	D	0.320693
30	3300	64.603	0.4278	0.46061	2.0707	2523.8	82.325	0.21435	0.13757	D	0.319934
30	3400	64.752	0.43975	0.45876	2.0617	2541.4	82.334	0.21378	0.13862	D	0.319199
30	3500	64.899	0.45166	0.45698	2.0531	2558.6	82.347	0.21322	0.13966	D	0.318477
30	3600	65.043	0.46354	0.45526	2.0447	2575.7	82.363	0.21267	0.14069	D	0.317774
30	3700	65.184	0.47538	0.45361	2.0366	2592.5	82.383	0.21213	0.14172	D	0.317083
30	3800	65.323	0.48719	0.45202	2.0288	2609.1	82.405	0.2116	0.14274	D	0.316409
30	3900	65.46	0.49896	0.45048	2.0213	2625.4	82.431	0.21107	0.14375	D	0.315744
30	4000	65.595	0.5107	0.449	2.0139	2641.6	82.459	0.21055	0.14476	D	0.315094
30	4100	65.728	0.52241	0.44756	2.0068	2657.5	82.491	0.21004	0.14575	D	0.314457
30	4200	65.858	0.53409	0.44618	2	2673.3	82.524	0.20954	0.14674	D	0.313833
30	4300	65.987	0.54574	0.44483	1.9933	2688.8	82.561	0.20904	0.14773	D	0.313221
30	4400	66.114	0.55736	0.44353	1.9868	2704.2	82.6	0.20855	0.14871	D	0.312620
30	4500	66.239	0.56896	0.44227	1.9805	2719.4	82.641	0.20806	0.14968	D	0.312035
30	4600	66.362	0.58052	0.44105	1.9744	2734.4	82.685	0.20758	0.15065	D	0.311454
30	4700	66.484	0.59206	0.43987	1.9684	2749.3	82.731	0.2071	0.15161	D	0.310887
30	4800	66.603	0.60357	0.43872	1.9626	2763.9	82.779	0.20663	0.15256	D	0.310328
30	4900	66.722	0.61505	0.4376	1.9569	2778.5	82.83	0.20617	0.15351	D	0.309776
30	5000	66.838	0.6265	0.43652	1.9514	2792.9	82.882	0.20571	0.15446	D	0.309232
30	5100	66.954	0.63794	0.43546	1.9461	2807.1	82.936	0.20525	0.1554	D	0.308705
30	5200	67.067	0.64934	0.43444	1.9408	2821.2	82.992	0.2048	0.15634	D	0.308179
30	5300	67.18	0.66072	0.43344	1.9357	2835.1	83.05	0.20436	0.15727	D	0.307663
30	5400	67.29	0.67208	0.43247	1.9308	2849	83.11	0.20392	0.15819	D	0.307158
30	5500	67.4	0.68341	0.43152	1.9259	2862.6	83.172	0.20348	0.15912	D	0.306657
30	5600	67.508	0.69472	0.4306	1.9211	2876.2	83.235	0.20305	0.16003	D	0.306165

TABLE F.1 (continued)—CO_2 PROPERTIES AT VARIOUS TEMPERATURES, °F											
Temperature (°F)	Pressure (psia)	Density (lbm/ft^3)	Compressibility Factor	Heat Capacity (Btu/lbm°F)	Heat Capacity Ratio (C_P/C_V)	Sonic Velocity (ft/s)	Enthalpy (Btu/lbm)	Entropy [Btu/(lbm°F)]	Viscosity (cp)	Phase	Gas Formation Volume Factor (res bbl/Mcf)
30	5700	67.615	0.70601	0.4297	1.9165	2889.6	83.3	0.20262	0.16095	D	0.305682
30	5800	67.721	0.71727	0.42883	1.9119	2902.9	83.366	0.2022	0.16186	D	0.305203
30	5900	67.825	0.72852	0.42798	1.9075	2916.1	83.434	0.20178	0.16276	D	0.304736
30	6000	67.929	0.73974	0.42714	1.9032	2929.1	83.504	0.20137	0.16366	D	0.304272
30	6100	68.031	0.75094	0.42633	1.8989	2942.1	83.575	0.20096	0.16456	D	0.303815
30	6200	68.132	0.76211	0.42554	1.8947	2954.9	83.647	0.20055	0.16546	D	0.303361
30	6300	68.232	0.77327	0.42476	1.8907	2967.6	83.721	0.20014	0.16635	D	0.302918
30	6400	68.331	0.78441	0.42401	1.8867	2980.3	83.796	0.19974	0.16723	D	0.302480
30	6500	68.429	0.79553	0.42327	1.8828	2992.8	83.873	0.19935	0.16812	D	0.302049
30	6600	68.526	0.80662	0.42255	1.8789	3005.2	83.951	0.19895	0.169	D	0.301619
30	6700	68.622	0.8177	0.42184	1.8751	3017.5	84.03	0.19856	0.16987	D	0.301199
30	6800	68.717	0.82876	0.42115	1.8715	3029.7	84.11	0.19818	0.17075	D	0.300783
30	6900	68.811	0.8398	0.42048	1.8678	3041.8	84.191	0.19779	0.17162	D	0.300373
30	7000	68.904	0.85082	0.41982	1.8643	3053.8	84.274	0.19741	0.17249	D	0.299967
30	7100	68.996	0.86182	0.41917	1.8608	3065.7	84.358	0.19704	0.17335	D	0.299566
30	7200	69.087	0.8728	0.41854	1.8574	3077.6	84.443	0.19666	0.17421	D	0.299169
30	7300	69.178	0.88377	0.41792	1.854	3089.3	84.529	0.19629	0.17507	D	0.298779
30	7400	69.267	0.89471	0.41731	1.8507	3101	84.616	0.19592	0.17593	D	0.298390
30	7500	69.356	0.90564	0.41671	1.8474	3112.5	84.704	0.19556	0.17678	D	0.298008
30	7600	69.444	0.91656	0.41613	1.8442	3124	84.793	0.19519	0.17763	D	0.297633
30	7700	69.531	0.92745	0.41556	1.8411	3135.4	84.883	0.19483	0.17848	D	0.297258
30	7800	69.617	0.93833	0.415	1.838	3146.8	84.974	0.19447	0.17933	D	0.296890
30	7900	69.703	0.9492	0.41445	1.8349	3158	85.066	0.19412	0.18017	D	0.296527
30	8000	69.788	0.96004	0.41392	1.832	3169.2	85.159	0.19377	0.18101	D	0.296165
30	8100	69.872	0.97087	0.41339	1.829	3180.2	85.253	0.19342	0.18185	D	0.295808
30	8200	69.955	0.98169	0.41287	1.8261	3191.2	85.347	0.19307	0.18268	D	0.295457
30	8300	70.038	0.99248	0.41236	1.8233	3202.2	85.443	0.19272	0.18352	D	0.295106
30	8400	70.12	1.0033	0.41186	1.8204	3213	85.539	0.19238	0.18435	D	0.294772
30	8500	70.202	1.014	0.41137	1.8177	3223.8	85.636	0.19204	0.18518	D	0.294410
30	8600	70.282	1.0248	0.41089	1.815	3234.5	85.735	0.1917	0.18601	D	0.294086
30	8700	70.362	1.0355	0.41042	1.8123	3245.2	85.833	0.19137	0.18683	D	0.293741
30	8800	70.442	1.0462	0.40996	1.8096	3255.8	85.933	0.19103	0.18765	D	0.293404
30	8900	70.52	1.057	0.4095	1.807	3266.3	86.034	0.1907	0.18847	D	0.293102
30	9000	70.599	1.0676	0.40906	1.8044	3276.7	86.135	0.19037	0.18929	D	0.292752
30	9100	70.676	1.0783	0.40862	1.8019	3287.1	86.237	0.19005	0.19011	D	0.292437
30	9200	70.753	1.089	0.40819	1.7994	3297.4	86.339	0.18972	0.19092	D	0.292129
30	9300	70.83	1.0996	0.40776	1.7969	3307.7	86.443	0.1894	0.19174	D	0.291800
30	9400	70.905	1.1103	0.40735	1.7945	3317.8	86.547	0.18908	0.19255	D	0.291505
30	9500	70.981	1.1209	0.40694	1.7921	3328	86.652	0.18876	0.19335	D	0.291191
30	9600	71.055	1.1315	0.40653	1.7897	3338	86.757	0.18844	0.19416	D	0.290882
30	9700	71.129	1.1421	0.40614	1.7874	3348	86.863	0.18812	0.19497	D	0.290580
30	9800	71.203	1.1527	0.40575	1.7851	3358	86.97	0.18781	0.19577	D	0.290285
30	9900	71.276	1.1632	0.40537	1.7828	3367.9	87.077	0.1875	0.19657	D	0.289970
30	10000	71.349	1.1738	0.40499	1.7806	3377.7	87.185	0.18719	0.19737	D	0.289686
40	14.696	0.12139	0.99361	0.19864	1.3056	852.99	210.18	0.63976	0.013929	V	170.267054
40	100	0.85923	0.95519	0.2127	1.3531	834.13	207.32	0.54913	0.013991	V	24.054901

TABLE F.1 (continued)—CO_2 PROPERTIES AT VARIOUS TEMPERATURES, °F

Temperature (°F)	Pressure (psia)	Density (lbm/ft³)	Compressibility Factor	Heat Capacity (Btu/lbm°F)	Heat Capacity Ratio (C_p/C_v)	Sonic Velocity (ft/s)	Enthalpy (Btu/lbm)	Entropy [Btu/(lbm°F)]	Viscosity (cp)	Phase	Gas Formation Volume Factor (res bbl/Mcf)
40	200	1.8102	0.90677	0.23377	1.4258	810.28	203.66	0.51257	0.014099	V	11.417761
40	300	2.8847	0.85353	0.26301	1.5275	783.92	199.56	0.48823	0.014263	V	7.164920
40	400	4.1376	0.79343	0.30771	1.6823	753.89	194.84	0.46805	0.014507	V	4.995310
40	500	5.6814	0.72229	0.38871	1.9571	717.81	189.08	0.44887	0.014888	V	3.637939
40	567.58	7.0281	0.66281	0.50377	2.3314	686.8	184.09	0.43491	0.015292	V	2.940870
40	567.58	56.165	0.08294	0.64603	2.8319	1622.6	90.809	0.24822	0.091751	L	0.368002
40	600	56.324	0.08743	0.63852	2.8085	1643	90.666	0.24772	0.0924	L	0.366964
40	700	56.783	0.10118	0.61835	2.7418	1700	90.262	0.24625	0.094315	L	0.364008
40	800	57.206	0.11478	0.60165	2.6826	1750.1	89.907	0.24489	0.096119	L	0.361318
40	900	57.597	0.12825	0.5875	2.6298	1795.3	89.591	0.24361	0.097834	L	0.358863
40	1000	57.963	0.1416	0.57529	2.5825	1836.6	89.309	0.24241	0.099476	L	0.356596
40	1100	58.308	0.15483	0.56462	2.5399	1875	89.055	0.24126	0.10105	D	0.354467
40	1200	58.633	0.16797	0.55517	2.5013	1910.9	88.826	0.24017	0.10258	D	0.352504
40	1300	58.943	0.18101	0.54672	2.4662	1944.7	88.619	0.23912	0.10405	D	0.350649
40	1400	59.238	0.19397	0.53912	2.4341	1976.9	88.431	0.23812	0.10549	D	0.348916
40	1500	59.521	0.20684	0.53222	2.4045	2007.5	88.26	0.23716	0.10689	D	0.347262
40	1600	59.791	0.21962	0.52592	2.3772	2036.9	88.105	0.23622	0.10825	D	0.345673
40	1700	60.052	0.23234	0.52014	2.3519	2065.1	87.964	0.23532	0.10958	D	0.344183
40	1800	60.303	0.24498	0.51481	2.3283	2092.3	87.836	0.23445	0.11089	D	0.342746
40	1900	60.545	0.25756	0.50987	2.3064	2118.6	87.72	0.23361	0.11217	D	0.341380
40	2000	60.779	0.27007	0.50528	2.2858	2144	87.614	0.23278	0.11342	D	0.340064
40	2100	61.006	0.28252	0.501	2.2665	2168.7	87.519	0.23198	0.11466	D	0.338800
40	2200	61.226	0.29491	0.497	2.2483	2192.7	87.433	0.23121	0.11587	D	0.337583
40	2300	61.439	0.30724	0.49324	2.2312	2216.1	87.355	0.23045	0.11707	D	0.336406
40	2400	61.647	0.31952	0.48971	2.215	2238.9	87.286	0.2297	0.11825	D	0.335275
40	2500	61.849	0.33175	0.48637	2.1996	2261.1	87.224	0.22898	0.11941	D	0.334183
40	2600	62.046	0.34392	0.48323	2.185	2282.9	87.169	0.22827	0.12055	D	0.333118
40	2700	62.237	0.35605	0.48024	2.1711	2304.1	87.121	0.22758	0.12169	D	0.332094
40	2800	62.425	0.36813	0.47742	2.1579	2324.9	87.079	0.2269	0.1228	D	0.331098
40	2900	62.607	0.38016	0.47473	2.1453	2345.3	87.043	0.22624	0.12391	D	0.330128
40	3000	62.786	0.39215	0.47217	2.1333	2365.3	87.013	0.22559	0.125	D	0.329189
40	3100	62.961	0.4041	0.46973	2.1218	2385	86.988	0.22495	0.12608	D	0.328277
40	3200	63.132	0.41601	0.4674	2.1107	2404.2	86.968	0.22432	0.12715	D	0.327392
40	3300	63.299	0.42787	0.46518	2.1001	2423.2	86.953	0.2237	0.12821	D	0.326521
40	3400	63.463	0.4397	0.46305	2.09	2441.8	86.943	0.2231	0.12926	D	0.325680
40	3500	63.624	0.45149	0.46101	2.0802	2460.1	86.936	0.2225	0.1303	D	0.324858
40	3600	63.782	0.46324	0.45905	2.0708	2478.1	86.934	0.22191	0.13133	D	0.324054
40	3700	63.937	0.47495	0.45717	2.0617	2495.9	86.936	0.22134	0.13235	D	0.323266
40	3800	64.089	0.48663	0.45536	2.053	2513.4	86.942	0.22077	0.13336	D	0.322500
40	3900	64.238	0.49828	0.45362	2.0445	2530.6	86.951	0.22021	0.13437	D	0.321753
40	4000	64.384	0.50989	0.45195	2.0364	2547.6	86.964	0.21966	0.13536	D	0.321019
40	4100	64.529	0.52147	0.45034	2.0285	2564.3	86.98	0.21912	0.13635	D	0.320302
40	4200	64.67	0.53302	0.44878	2.0209	2580.9	86.999	0.21858	0.13733	D	0.319601
40	4300	64.81	0.54454	0.44728	2.0135	2597.2	87.021	0.21806	0.13831	D	0.318915
40	4400	64.947	0.55602	0.44582	2.0064	2613.3	87.047	0.21753	0.13928	D	0.318238
40	4500	65.082	0.56748	0.44442	1.9994	2629.2	87.075	0.21702	0.14024	D	0.317579

TABLE F.1 (continued)—CO_2 PROPERTIES AT VARIOUS TEMPERATURES, °F

Temperature (°F)	Pressure (psia)	Density (lbm/ft³)	Compressibility Factor	Heat Capacity (Btu/lbm°F)	Heat Capacity Ratio (C_p/C_v)	Sonic Velocity (ft/s)	Enthalpy (Btu/lbm)	Entropy [Btu/(lbm°F)]	Viscosity (cp)	Phase	Gas Formation Volume Factor (res bbl/Mcf)
40	4600	65.215	0.57891	0.44306	1.9927	2644.9	87.106	0.21651	0.14119	D	0.316933
40	4700	65.346	0.59031	0.44175	1.9862	2660.4	87.139	0.21601	0.14214	D	0.316298
40	4800	65.475	0.60168	0.44048	1.9798	2675.7	87.175	0.21552	0.14309	D	0.315674
40	4900	65.602	0.61302	0.43924	1.9737	2690.9	87.214	0.21503	0.14403	D	0.315059
40	5000	65.728	0.62434	0.43805	1.9677	2705.9	87.255	0.21455	0.14496	D	0.314460
40	5100	65.851	0.63563	0.43688	1.9618	2720.7	87.298	0.21407	0.14588	D	0.313869
40	5200	65.973	0.6469	0.43576	1.9561	2735.4	87.343	0.2136	0.14681	D	0.313291
40	5300	66.094	0.65814	0.43466	1.9506	2749.9	87.391	0.21313	0.14772	D	0.312720
40	5400	66.212	0.66935	0.4336	1.9452	2764.2	87.44	0.21267	0.14864	D	0.312157
40	5500	66.329	0.68054	0.43256	1.9399	2778.4	87.492	0.21222	0.14954	D	0.311605
40	5600	66.445	0.69171	0.43156	1.9348	2792.5	87.546	0.21177	0.15045	D	0.311064
40	5700	66.559	0.70285	0.43058	1.9298	2806.4	87.601	0.21132	0.15135	D	0.310529
40	5800	66.672	0.71398	0.42963	1.9249	2820.2	87.658	0.21088	0.15224	D	0.310007
40	5900	66.784	0.72507	0.4287	1.9201	2833.9	87.718	0.21044	0.15313	D	0.309487
40	6000	66.894	0.73615	0.42779	1.9155	2847.4	87.778	0.21001	0.15402	D	0.308979
40	6100	67.003	0.7472	0.42691	1.9109	2860.8	87.841	0.20958	0.1549	D	0.308476
40	6200	67.11	0.75823	0.42605	1.9064	2874.1	87.905	0.20916	0.15578	D	0.307980
40	6300	67.216	0.76925	0.42521	1.9021	2887.2	87.971	0.20874	0.15665	D	0.307497
40	6400	67.322	0.78024	0.4244	1.8978	2900.3	88.038	0.20832	0.15752	D	0.307017
40	6500	67.426	0.79121	0.4236	1.8936	2913.2	88.107	0.20791	0.15839	D	0.306544
40	6600	67.528	0.80215	0.42282	1.8895	2926	88.177	0.2075	0.15925	D	0.306073
40	6700	67.63	0.81308	0.42206	1.8855	2938.7	88.249	0.20709	0.16011	D	0.305613
40	6800	67.731	0.82399	0.42132	1.8816	2951.3	88.322	0.20669	0.16097	D	0.305159
40	6900	67.83	0.83488	0.42059	1.8777	2963.8	88.397	0.20629	0.16182	D	0.304711
40	7000	67.929	0.84576	0.41988	1.8739	2976.2	88.472	0.2059	0.16267	D	0.304273
40	7100	68.026	0.85661	0.41919	1.8702	2988.5	88.549	0.20551	0.16352	D	0.303836
40	7200	68.123	0.86744	0.41851	1.8666	3000.7	88.628	0.20512	0.16436	D	0.303404
40	7300	68.218	0.87826	0.41784	1.863	3012.8	88.707	0.20474	0.16521	D	0.302980
40	7400	68.313	0.88906	0.41719	1.8595	3024.8	88.788	0.20436	0.16604	D	0.302561
40	7500	68.407	0.89984	0.41656	1.856	3036.7	88.87	0.20398	0.16688	D	0.302147
40	7600	68.499	0.9106	0.41503	1.8527	3048.5	88.953	0.2036	0.16771	D	0.301737
40	7700	68.591	0.92134	0.41532	1.8493	3060.2	89.037	0.20323	0.16854	D	0.301330
40	7800	68.682	0.93207	0.41473	1.8461	3071.9	89.122	0.20286	0.16937	D	0.300932
40	7900	68.773	0.94278	0.41414	1.8429	3083.4	89.209	0.20249	0.1702	D	0.300536
40	8000	68.862	0.95348	0.41357	1.8397	3094.9	89.296	0.20213	0.17102	D	0.300148
40	8100	68.951	0.96415	0.41301	1.8366	3106.3	89.384	0.20177	0.17184	D	0.299760
40	8200	69.038	0.97482	0.41245	1.8335	3117.6	89.474	0.20141	0.17266	D	0.299381
40	8300	69.125	0.98546	0.41191	1.8305	3128.8	89.564	0.20106	0.17347	D	0.299002
40	8400	69.212	0.99609	0.41138	1.8276	3140	89.655	0.2007	0.17428	D	0.298630
40	8500	69.297	1.0067	0.41086	1.8247	3151	89.748	0.20035	0.17509	D	0.298260
40	8600	69.382	1.0173	0.41036	1.8218	3162	89.841	0.2	0.1759	D	0.297896
40	8700	69.466	1.0279	0.40985	1.819	3172.9	89.935	0.19966	0.17671	D	0.297540
40	8800	69.549	1.0385	0.40936	1.8162	3183.8	90.03	0.19932	0.17751	D	0.297192
40	8900	69.632	1.049	0.40888	1.8135	3194.6	90.125	0.19897	0.17831	D	0.296824
40	9000	69.714	1.0596	0.40841	1.8108	3205.3	90.222	0.19864	0.17911	D	0.296492
40	9100	69.795	1.0701	0.40794	1.8081	3215.9	90.319	0.1983	0.17991	D	0.296140

TABLE F.1 (continued)—CO_2 PROPERTIES AT VARIOUS TEMPERATURES, °F

Temperature (°F)	Pressure (psia)	Density (lbm/ft^3)	Compressibility Factor	Heat Capacity (Btu/lbm°F)	Heat Capacity Ratio (C_p/C_v)	Sonic Velocity (ft/s)	Enthalpy (Btu/lbm)	Entropy [Btu/(lbm°F)]	Viscosity (cp)	Phase	Gas Formation Volume Factor (res bbl/Mcf)
40	9200	69.876	1.0806	0.40749	1.8055	3226.5	90.418	0.19797	0.18071	D	0.295795
40	9300	69.956	1.0911	0.40704	1.8029	3237	90.517	0.19763	0.1815	D	0.295458
40	9400	70.036	1.1016	0.4066	1.8004	3247.4	90.616	0.1973	0.18229	D	0.295128
40	9500	70.114	1.112	0.40617	1.7979	3257.7	90.717	0.19698	0.18308	D	0.294778
40	9600	70.193	1.1225	0.40574	1.7954	3268	90.818	0.19665	0.18387	D	0.294462
40	9700	70.27	1.1329	0.40532	1.793	3278.3	90.92	0.19633	0.18466	D	0.294126
40	9800	70.347	1.1434	0.40491	1.7905	3288.5	91.023	0.19601	0.18544	D	0.293823
40	9900	70.424	1.1538	0.40451	1.7882	3298.6	91.127	0.19569	0.18622	D	0.293501
40	10000	70.499	1.1642	0.40411	1.7858	3308.6	91.231	0.19537	0.187	D	0.293185
50	14.696	0.11896	0.99401	0.19991	1.3023	860.76	212.17	0.64371	0.014201	V	173.744561
50	100	0.83977	0.95815	0.21285	1.3464	843.04	209.44	0.55335	0.014261	V	24.612352
50	200	1.7621	0.91327	0.23181	1.4122	820.85	205.98	0.51719	0.014366	V	11.729751
50	300	2.7923	0.86448	0.25714	1.5013	796.68	202.16	0.49338	0.014519	V	7.402072
50	400	3.9715	0.81041	0.29356	1.6297	769.78	197.84	0.474	0.014744	V	5.204325
50	500	5.3748	0.74852	0.35231	1.8347	738.77	192.77	0.45618	0.015077	V	3.845502
50	600	7.1714	0.6732	0.47047	2.2349	700.42	186.38	0.43777	0.015611	V	2.882123
50	652.99	8.4375	0.62271	0.61133	2.6903	673.91	181.93	0.42655	0.016059	V	2.449621
50	652.99	53.758	0.097736	0.71644	3.0931	1446.7	97.111	0.26013	0.082557	L	0.384475
50	700	54.069	0.10417	0.69735	3.0338	1483.4	96.798	0.2592	0.083656	L	0.382265
50	800	54.668	0.11775	0.6646	2.9256	1551.7	96.209	0.25738	0.085828	L	0.378086
50	900	55.201	0.13119	0.63918	2.8361	1610.6	95.704	0.25572	0.087827	L	0.374436
50	1000	55.685	0.1445	0.61868	2.7604	1662.8	95.263	0.2542	0.089695	L	0.371182
50	1100	56.129	0.15769	0.60167	2.6953	1710.1	94.875	0.25279	0.091457	D	0.368240
50	1200	56.541	0.17077	0.58724	2.6386	1753.4	94.529	0.25147	0.093132	D	0.365553
50	1300	56.926	0.18375	0.57481	2.5887	1793.6	94.219	0.25022	0.094736	D	0.363081
50	1400	57.288	0.19664	0.56393	2.5441	1831.3	93.94	0.24904	0.096277	D	0.360797
50	1500	57.63	0.20943	0.55432	2.5042	1866.7	93.688	0.24791	0.097766	D	0.358647
50	1600	57.954	0.22214	0.54573	2.468	1900.4	93.459	0.24683	0.099207	D	0.356637
50	1700	58.263	0.23477	0.53801	2.4351	1932.4	93.251	0.2458	0.10061	D	0.354743
50	1800	58.559	0.24733	0.531	2.4049	1963.1	93.062	0.24481	0.10197	D	0.352959
50	1900	58.842	0.25981	0.52461	2.3771	1992.5	92.89	0.24385	0.1033	D	0.351254
50	2000	59.114	0.27223	0.51876	2.3514	2020.8	92.732	0.24292	0.1046	D	0.349644
50	2100	59.376	0.28458	0.51336	2.3276	2048.2	92.589	0.24203	0.10587	D	0.348101
50	2200	59.628	0.29687	0.50836	2.3054	2074.6	92.458	0.24116	0.10712	D	0.346628
50	2300	59.872	0.3091	0.50372	2.2846	2100.2	92.339	0.24032	0.10834	D	0.345216
50	2400	60.109	0.32127	0.4994	2.2652	2125.1	92.23	0.2395	0.10954	D	0.343858
50	2500	60.337	0.33339	0.49535	2.2469	2149.3	92.131	0.23871	0.11072	D	0.342556
50	2600	60.559	0.34545	0.49156	2.2296	2172.8	92.042	0.23793	0.11189	D	0.341296
50	2700	60.775	0.35746	0.48799	2.2133	2195.8	91.961	0.23717	0.11303	D	0.340082
50	2800	60.985	0.36943	0.48463	2.1979	2218.2	91.889	0.23643	0.11416	D	0.338917
50	2900	61.189	0.38134	0.48145	2.1832	2240.1	91.824	0.23571	0.11527	D	0.337780
50	3000	61.388	0.39322	0.47845	2.1693	2261.5	91.766	0.235	0.11637	D	0.336693
50	3100	61.582	0.40504	0.4756	2.156	2282.5	91.714	0.23431	0.11746	D	0.335626
50	3200	61.772	0.41683	0.47289	2.1434	2303	91.669	0.23363	0.11853	D	0.334602
50	3300	61.957	0.42857	0.47032	2.1313	2323.2	91.63	0.23297	0.11959	D	0.333601
50	3400	62.138	0.44027	0.46786	2.1197	2342.9	91.597	0.23232	0.12064	D	0.332629

TABLE F.1 (continued)—CO_2 PROPERTIES AT VARIOUS TEMPERATURES, °F											
Temperature (°F)	Pressure (psia)	Density (lbm/ft^3)	Compressibility Factor	Heat Capacity (Btu/lbm°F)	Heat Capacity Ratio (C_p/C_v)	Sonic Velocity (ft/s)	Enthalpy (Btu/lbm)	Entropy [Btu/(lbm°F)]	Viscosity (cp)	Phase	Gas Formation Volume Factor (res bbl/Mcf)
50	3500	62.315	0.45193	0.46552	2.1087	2362.4	91.569	0.23168	0.12168	D	0.331683
50	3600	62.488	0.46355	0.46328	2.0981	2381.4	91.546	0.23105	0.12271	D	0.330761
50	3700	62.658	0.47514	0.46114	2.0879	2400.2	91.527	0.23044	0.12373	D	0.329867
50	3800	62.824	0.48669	0.45909	2.0781	2418.7	91.514	0.22983	0.12474	D	0.328994
50	3900	62.987	0.49821	0.45712	2.0687	2436.8	91.504	0.22923	0.12574	D	0.328146
50	4000	63.147	0.50969	0.45523	2.0596	2454.7	91.499	0.22865	0.12673	D	0.327315
50	4100	63.304	0.52113	0.45342	2.0508	2472.3	91.498	0.22807	0.12771	D	0.326499
50	4200	63.458	0.53255	0.45167	2.0424	2489.6	91.501	0.2275	0.12869	D	0.325710
50	4300	63.609	0.54393	0.44999	2.0342	2506.8	91.507	0.22694	0.12965	D	0.324933
50	4400	63.758	0.55528	0.44837	2.0263	2523.6	91.517	0.22639	0.13061	D	0.324175
50	4500	63.904	0.5666	0.44681	2.0187	2540.3	91.531	0.22585	0.13157	D	0.323433
50	4600	64.048	0.5779	0.4453	2.0113	2556.7	91.547	0.22531	0.13251	D	0.322712
50	4700	64.189	0.58916	0.44384	2.0042	2572.9	91.567	0.22479	0.13345	D	0.321999
50	4800	64.328	0.60039	0.44243	1.9972	2588.9	91.59	0.22427	0.13438	D	0.321301
50	4900	64.465	0.6116	0.44107	1.9905	2604.7	91.615	0.22375	0.13531	D	0.320620
50	5000	64.6	0.62278	0.43975	1.984	2620.3	91.643	0.22324	0.13623	D	0.319952
50	5100	64.733	0.63393	0.43848	1.9776	2635.7	91.674	0.22274	0.13715	D	0.319294
50	5200	64.864	0.64505	0.43724	1.9714	2651	91.708	0.22225	0.13806	D	0.318647
50	5300	64.993	0.65615	0.43604	1.9654	2666.1	91.744	0.22176	0.13896	D	0.318014
50	5400	65.12	0.66723	0.43487	1.9596	2681	91.783	0.22128	0.13986	D	0.317396
50	5500	65.246	0.67828	0.43375	1.9539	2695.7	91.823	0.2208	0.14075	D	0.316786
50	5600	65.369	0.6893	0.43265	1.9484	2710.3	91.867	0.22033	0.14164	D	0.316184
50	5700	65.491	0.7003	0.43158	1.943	2724.8	91.912	0.21986	0.14253	D	0.315594
50	5800	65.612	0.71128	0.43055	1.9377	2739	91.959	0.2194	0.14341	D	0.315016
50	5900	65.731	0.72223	0.42954	1.9326	2753.2	92.009	0.21894	0.14428	D	0.314444
50	6000	65.848	0.73317	0.42856	1.9276	2767.2	92.06	0.21849	0.14516	D	0.313887
50	6100	65.964	0.74408	0.4276	1.9227	2781	92.113	0.21805	0.14602	D	0.313335
50	6200	66.078	0.75496	0.42667	1.9179	2794.8	92.169	0.2176	0.14689	D	0.312789
50	6300	66.191	0.76583	0.42577	1.9132	2808.4	92.226	0.21717	0.14774	D	0.312256
50	6400	66.303	0.77668	0.42489	1.9087	2821.8	92.285	0.21673	0.1486	D	0.311732
50	6500	66.414	0.7875	0.42403	1.9042	2835.2	92.345	0.2163	0.14945	D	0.311212
50	6600	66.523	0.7983	0.42319	1.8998	2848.4	92.407	0.21588	0.1503	D	0.310700
50	6700	66.631	0.80909	0.42237	1.8956	2861.5	92.471	0.21546	0.15114	D	0.310200
50	6800	66.737	0.81985	0.42157	1.8914	2874.5	92.536	0.21504	0.15198	D	0.309703
50	6900	66.843	0.8306	0.42079	1.8873	2887.4	92.603	0.21463	0.15282	D	0.309216
50	7000	66.947	0.84132	0.42003	1.8833	2900.1	92.672	0.21422	0.15366	D	0.308733
50	7100	67.05	0.85203	0.41929	1.8793	2912.8	92.742	0.21382	0.15449	D	0.308259
50	7200	67.152	0.86271	0.41856	1.8755	2925.3	92.813	0.21342	0.15531	D	0.307788
50	7300	67.253	0.87338	0.41785	1.8717	2937.8	92.886	0.21302	0.15614	D	0.307326
50	7400	67.353	0.88403	0.41715	1.868	2950.1	92.96	0.21262	0.15696	D	0.306870
50	7500	67.452	0.89466	0.41648	1.8643	2962.4	93.035	0.21223	0.15778	D	0.306419
50	7600	67.55	0.90528	0.41581	1.8608	2974.5	93.112	0.21184	0.1586	D	0.305977
50	7700	67.647	0.91588	0.41516	1.8572	2986.6	93.189	0.21146	0.15941	D	0.305540
50	7800	67.743	0.92646	0.41453	1.8538	2998.5	93.269	0.21108	0.16022	D	0.305107
50	7900	67.838	0.93702	0.4139	1.8504	3010.4	93.349	0.2107	0.16103	D	0.304678
50	8000	67.932	0.94756	0.41329	1.8471	3022.2	93.43	0.21032	0.16183	D	0.304254

TABLE F.1 (continued)—CO_2 PROPERTIES AT VARIOUS TEMPERATURES, °F											
Temperature (°F)	Pressure (psia)	Density (lbm/ft^3)	Compressibility Factor	Heat Capacity (Btu/lbm°F)	Heat Capacity Ratio (C_p/C_v)	Sonic Velocity (ft/s)	Enthalpy (Btu/lbm)	Entropy [Btu/(lbm°F)]	Viscosity (cp)	Phase	Gas Formation Volume Factor (res bbl/Mcf)
50	8100	68.025	0.95809	0.4127	1.8438	3033.9	93.513	0.20995	0.16263	D	0.303837
50	8200	68.118	0.96861	0.41211	1.8406	3045.5	93.597	0.20958	0.16343	D	0.303427
50	8300	68.209	0.9791	0.41154	1.8375	3057	93.681	0.20921	0.16423	D	0.303018
50	8400	68.3	0.98958	0.41098	1.8343	3068.4	93.767	0.20885	0.16503	D	0.302616
50	8500	68.39	1	0.41043	1.8313	3079.8	93.854	0.20849	0.16582	D	0.302204
50	8600	68.479	1.0105	0.40989	1.8283	3091	93.942	0.20813	0.16661	D	0.301827
50	8700	68.567	1.0209	0.40936	1.8253	3102.2	94.031	0.20778	0.1674	D	0.301428
50	8800	68.655	1.0314	0.40884	1.8224	3113.3	94.121	0.20742	0.16818	D	0.301068
50	8900	68.741	1.0418	0.40833	1.8196	3124.4	94.211	0.20707	0.16897	D	0.300687
50	9000	68.827	1.0521	0.40783	1.8167	3135.3	94.303	0.20672	0.16975	D	0.300285
50	9100	68.913	1.0625	0.40734	1.8139	3146.2	94.396	0.20638	0.17053	D	0.299921
50	9200	68.997	1.0729	0.40686	1.8112	3157	94.489	0.20603	0.17131	D	0.299565
50	9300	69.081	1.0832	0.40639	1.8085	3167.8	94.584	0.20569	0.17208	D	0.299189
50	9400	69.164	1.0936	0.40592	1.8059	3178.4	94.679	0.20535	0.17286	D	0.298848
50	9500	69.247	1.1039	0.40547	1.8032	3189	94.775	0.20502	0.17363	D	0.298487
50	9600	69.329	1.1142	0.40502	1.8006	3199.6	94.872	0.20468	0.1744	D	0.298134
50	9700	69.41	1.1245	0.40458	1.7981	3210	94.97	0.20435	0.17517	D	0.297788
50	9800	69.49	1.1347	0.40415	1.7956	3220.4	95.068	0.20402	0.17593	D	0.297423
50	9900	69.57	1.145	0.40372	1.7931	3230.8	95.168	0.20369	0.1767	D	0.297091
50	10000	69.649	1.1553	0.40331	1.7907	3241	95.268	0.20337	0.17746	D	0.296766
60	14.696	0.11663	0.99439	0.20119	1.2992	868.44	214.18	0.6476	0.014473	V	177.221247
60	100	0.8213	0.96085	0.21313	1.3401	851.77	211.57	0.55749	0.014531	V	25.165976
60	200	1.7171	0.91915	0.23029	1.4001	831.07	208.29	0.52168	0.014631	V	12.036898
60	300	2.7079	0.87427	0.25253	1.479	808.81	204.71	0.49833	0.014777	V	7.632776
60	400	3.8251	0.82522	0.283	1.5878	784.49	200.72	0.4796	0.014984	V	5.403410
60	500	5.1215	0.77042	0.3283	1.7492	757.27	196.17	0.46278	0.015283	V	4.035671
60	600	6.6988	0.70682	0.40545	2.0198	725.55	190.73	0.44622	0.01573	V	3.085430
60	700	8.8107	0.62697	0.57987	2.6092	685.08	183.61	0.42785	0.016473	V	2.345886
60	747.75	10.237	0.57641	0.79841	3.3088	658.44	178.89	0.41697	0.017068	V	2.018986
60	747.75	50.964	0.11579	0.837	3.5442	1264.5	103.96	0.27279	0.073528	L	0.405577
60	800	51.468	0.12266	0.793	3.3995	1315.7	103.4	0.27135	0.075053	L	0.401579
60	900	52.295	0.13581	0.73298	3.1938	1398.6	102.51	0.26895	0.077645	L	0.395228
60	1000	52.998	0.1489	0.69111	3.044	1468.3	101.78	0.26686	0.079938	L	0.389989
60	1100	53.613	0.16191	0.65973	2.9278	1528.9	101.16	0.265	0.082023	D	0.385513
60	1200	54.164	0.17483	0.63507	2.8339	1583	100.62	0.26331	0.083951	D	0.381586
60	1300	54.665	0.18767	0.61502	2.7558	1632.1	100.15	0.26175	0.085756	D	0.378103
60	1400	55.126	0.20042	0.59829	2.6894	1677.2	99.741	0.26031	0.087463	D	0.374948
60	1500	55.553	0.21308	0.58407	2.632	1719	99.372	0.25895	0.089088	D	0.372057
60	1600	55.952	0.22566	0.57177	2.5817	1758.2	99.04	0.25768	0.090644	D	0.369397
60	1700	56.328	0.23817	0.56101	2.537	1795.1	98.741	0.25647	0.09214	D	0.366941
60	1800	56.683	0.2506	0.55147	2.497	1830.1	98.47	0.25531	0.093585	D	0.364642
60	1900	57.02	0.26296	0.54296	2.4609	1863.4	98.224	0.25421	0.094985	D	0.362489
60	2000	57.34	0.27525	0.53528	2.4281	1895.1	97.999	0.25316	0.096344	D	0.360459
60	2100	57.646	0.28748	0.52832	2.3981	1925.6	97.795	0.25214	0.097667	D	0.358547
60	2200	57.94	0.29964	0.52197	2.3705	1954.9	97.607	0.25117	0.098958	D	0.356726
60	2300	58.222	0.31174	0.51615	2.345	1983.1	97.436	0.25022	0.10022	D	0.354996

TABLE F.1 (continued)—CO_2 PROPERTIES AT VARIOUS TEMPERATURES, °F											
Temperature (°F)	Pressure (psia)	Density (lbm/ft^3)	Compressibility Factor	Heat Capacity (Btu/lbm°F)	Heat Capacity Ratio (C_p/C_v)	Sonic Velocity (ft/s)	Enthalpy (Btu/lbm)	Entropy [Btu/(lbm°F)]	Viscosity (cp)	Phase	Gas Formation Volume Factor (res bbl/Mcf)
60	2400	58.493	0.32379	0.51078	2.3213	2010.4	97.279	0.24931	0.10145	D	0.353354
60	2500	58.754	0.33578	0.50581	2.2992	2036.8	97.136	0.24843	0.10266	D	0.351782
60	2600	59.007	0.34772	0.5012	2.2786	2062.4	97.005	0.24757	0.10385	D	0.350279
60	2700	59.251	0.35961	0.49689	2.2593	2087.3	96.885	0.24674	0.10502	D	0.348840
60	2800	59.487	0.37144	0.49287	2.2411	2111.5	96.775	0.24593	0.10616	D	0.347447
60	2900	59.717	0.38323	0.4891	2.224	2135.1	96.676	0.24514	0.10729	D	0.346114
60	3000	59.939	0.39497	0.48555	2.2078	2158.1	96.585	0.24437	0.1084	D	0.344827
60	3100	60.156	0.40667	0.4822	2.1925	2180.5	96.503	0.24362	0.1095	D	0.343589
60	3200	60.367	0.41832	0.47904	2.178	2202.5	96.428	0.24288	0.11058	D	0.342387
60	3300	60.572	0.42993	0.47605	2.1642	2223.9	96.361	0.24216	0.11165	D	0.341226
60	3400	60.772	0.4415	0.47322	2.151	2245	96.302	0.24146	0.1127	D	0.340103
60	3500	60.968	0.45303	0.47052	2.1385	2265.6	96.248	0.24077	0.11374	D	0.339014
60	3600	61.159	0.46452	0.46796	2.1265	2285.8	96.201	0.2401	0.11477	D	0.337956
60	3700	61.345	0.47597	0.46552	2.115	2305.6	96.16	0.23944	0.11579	D	0.336927
60	3800	61.527	0.48738	0.46319	2.104	2325.1	96.125	0.23879	0.1168	D	0.335925
60	3900	61.706	0.49876	0.46096	2.0935	2344.2	96.095	0.23815	0.11779	D	0.334954
60	4000	61.88	0.51011	0.45883	2.0834	2363	96.069	0.23753	0.11878	D	0.334012
60	4100	62.051	0.52142	0.45679	2.0737	2381.5	96.049	0.23691	0.11976	D	0.333090
60	4200	62.219	0.5327	0.45483	2.0643	2399.7	96.033	0.23631	0.12073	D	0.332194
60	4300	62.384	0.54394	0.45295	2.0553	2417.7	96.022	0.23572	0.12169	D	0.331315
60	4400	62.545	0.55516	0.45114	2.0466	2435.3	96.015	0.23513	0.12264	D	0.330464
60	4500	62.703	0.56634	0.4494	2.0382	2452.7	96.012	0.23456	0.12359	D	0.329627
60	4600	62.859	0.57749	0.44773	2.0301	2469.9	96.012	0.23399	0.12452	D	0.328810
60	4700	63.012	0.58861	0.44612	2.0223	2486.8	96.017	0.23343	0.12545	D	0.328011
60	4800	63.162	0.59971	0.44456	2.0147	2503.5	96.024	0.23288	0.12638	D	0.327234
60	4900	63.31	0.61077	0.44306	2.0074	2520	96.036	0.23234	0.12729	D	0.326467
60	5000	63.455	0.62181	0.44161	2.0002	2536.2	96.05	0.23181	0.12821	D	0.325721
60	5100	63.598	0.63282	0.44021	1.9933	2552.3	96.068	0.23128	0.12911	D	0.324989
60	5200	63.739	0.64381	0.43886	1.9867	2568.1	96.088	0.23076	0.13001	D	0.324274
60	5300	63.877	0.65477	0.43754	1.9802	2583.8	96.112	0.23025	0.1309	D	0.323572
60	5400	64.013	0.6657	0.43627	1.9738	2599.3	96.138	0.22974	0.13179	D	0.322881
60	5500	64.148	0.67661	0.43504	1.9677	2614.6	96.167	0.22924	0.13267	D	0.322200
60	5600	64.28	0.68749	0.43385	1.9617	2629.7	96.199	0.22875	0.13355	D	0.321541
60	5700	64.411	0.69835	0.43269	1.9559	2644.7	96.233	0.22826	0.13442	D	0.320890
60	5800	64.539	0.70918	0.43157	1.9503	2659.4	96.27	0.22778	0.13529	D	0.320248
60	5900	64.666	0.72	0.43047	1.9448	2674.1	96.309	0.2273	0.13615	D	0.319623
60	6000	64.791	0.73079	0.42941	1.9394	2688.6	96.35	0.22683	0.13701	D	0.319006
60	6100	64.915	0.74155	0.42838	1.9342	2702.9	96.393	0.22636	0.13786	D	0.318397
60	6200	65.037	0.7523	0.42738	1.9291	2717.1	96.439	0.2259	0.13871	D	0.317803
60	6300	65.157	0.76302	0.4264	1.9241	2731.1	96.487	0.22545	0.13955	D	0.317215
60	6400	65.276	0.77372	0.42545	1.9192	2745	96.536	0.22499	0.14039	D	0.316637
60	6500	65.393	0.7844	0.42453	1.9144	2758.8	96.588	0.22455	0.14123	D	0.316069
60	6600	65.509	0.79506	0.42363	1.9098	2772.4	96.641	0.22411	0.14206	D	0.315511
60	6700	65.623	0.8057	0.42275	1.9053	2785.9	96.697	0.22367	0.14289	D	0.314961
60	6800	65.736	0.81632	0.42189	1.9008	2799.3	96.754	0.22324	0.14372	D	0.314420
60	6900	65.848	0.82692	0.42106	1.8965	2812.5	96.813	0.22281	0.14454	D	0.313886

TABLE F.1 (continued)—CO_2 PROPERTIES AT VARIOUS TEMPERATURES, °F												
Temperature (°F)	Pressure (psia)	Density (lbm/ft³)	Compressibility Factor	Heat Capacity (Btu/lbm°F)	Heat Capacity Ratio (C_p/C_v)	Sonic Velocity (ft/s)	Enthalpy (Btu/lbm)	Entropy [Btu/(lbm°F)]	Viscosity (cp)	Phase	Gas Formation Volume Factor (res bbl/Mcf)
60	7000	65.959	0.8375	0.42025	1.8922	2825.7	96.873	0.22239	0.14536	D	0.313361
60	7100	66.068	0.84806	0.41945	1.888	2838.7	96.935	0.22197	0.14617	D	0.312843
60	7200	66.176	0.8586	0.41868	1.884	2851.6	96.999	0.22155	0.14699	D	0.312332
60	7300	66.282	0.86912	0.41792	1.88	2864.4	97.064	0.22114	0.14779	D	0.311828
60	7400	66.388	0.87963	0.41718	1.876	2877.1	97.131	0.22073	0.1486	D	0.311334
60	7500	66.492	0.89011	0.41646	1.8722	2889.7	97.2	0.22032	0.1494	D	0.310843
60	7600	66.596	0.90058	0.41575	1.8684	2902.1	97.269	0.21992	0.1502	D	0.310361
60	7700	66.698	0.91103	0.41506	1.8647	2914.5	97.341	0.21952	0.151	D	0.309885
60	7800	66.799	0.92147	0.41438	1.8611	2926.8	97.413	0.21913	0.15179	D	0.309417
60	7900	66.899	0.93188	0.41372	1.8575	2939	97.487	0.21874	0.15258	D	0.308952
60	8000	66.998	0.94228	0.41308	1.8541	2951	97.562	0.21835	0.15337	D	0.308495
60	8100	67.097	0.95267	0.41244	1.8506	2963	97.639	0.21797	0.15416	D	0.308046
60	8200	67.194	0.96303	0.41182	1.8472	2974.9	97.716	0.21759	0.15494	D	0.307598
60	8300	67.29	0.97338	0.41122	1.8439	2986.7	97.795	0.21721	0.15572	D	0.307158
60	8400	67.385	0.98372	0.41062	1.8407	2998.4	97.875	0.21683	0.1565	D	0.306726
60	8500	67.48	0.99403	0.41004	1.8375	3010.1	97.956	0.21646	0.15728	D	0.306294
60	8600	67.573	1.0043	0.40947	1.8343	3021.6	98.039	0.21609	0.15805	D	0.305860
60	8700	67.666	1.0146	0.40891	1.8312	3033.1	98.122	0.21572	0.15882	D	0.305446
60	8800	67.758	1.0249	0.40837	1.8282	3044.4	98.207	0.21536	0.15959	D	0.305040
60	8900	67.849	1.0352	0.40783	1.8252	3055.7	98.292	0.215	0.16036	D	0.304644
60	9000	67.939	1.0454	0.4073	1.8222	3066.9	98.379	0.21464	0.16112	D	0.304227
60	9100	68.028	1.0556	0.40679	1.8193	3078.1	98.466	0.21429	0.16188	D	0.303820
60	9200	68.117	1.0658	0.40628	1.8165	3089.1	98.555	0.21393	0.16265	D	0.303421
60	9300	68.205	1.076	0.40578	1.8137	3100.1	98.645	0.21358	0.1634	D	0.303031
60	9400	68.292	1.0862	0.4053	1.8109	3111	98.735	0.21324	0.16416	D	0.302650
60	9500	68.378	1.0964	0.40482	1.8082	3121.8	98.827	0.21289	0.16491	D	0.302276
60	9600	68.464	1.1065	0.40435	1.8055	3132.6	98.919	0.21255	0.16567	D	0.301883
60	9700	68.548	1.1167	0.40389	1.8028	3143.3	99.012	0.21221	0.16642	D	0.301525
60	9800	68.633	1.1268	0.40343	1.8002	3153.9	99.106	0.21187	0.16717	D	0.301147
60	9900	68.716	1.1369	0.40299	1.7976	3164.5	99.201	0.21153	0.16791	D	0.300777
60	10000	68.799	1.147	0.40255	1.7951	3174.9	99.297	0.2112	0.16866	D	0.300415
70	14.696	0.11439	0.99473	0.20247	1.2961	876.04	216.2	0.65145	0.014743	V	180.693273
70	100	0.80372	0.96333	0.21352	1.3342	860.34	213.71	0.56155	0.0148	V	25.716449
70	200	1.6749	0.92451	0.22915	1.3892	840.99	210.59	0.52605	0.014896	V	12.340067
70	300	2.6303	0.88306	0.24887	1.4596	820.4	207.22	0.50311	0.015035	V	7.857871
70	400	3.6943	0.83831	0.27488	1.5534	798.25	203.51	0.48491	0.015228	V	5.594748
70	500	4.9054	0.78918	0.31129	1.6851	774.03	199.36	0.46887	0.0155	V	4.213490
70	600	6.3297	0.73391	0.36708	1.8856	746.85	194.58	0.45355	0.015888	V	3.265333
70	700	8.1016	0.66896	0.46698	2.2378	714.91	188.77	0.43769	0.016476	V	2.551162
70	800	10.59	0.58489	0.72049	3.0936	673.03	180.86	0.41897	0.017498	V	1.951732
70	852.82	12.689	0.52037	1.2053	4.6183	639.01	174.41	0.40518	0.018538	V	1.628886
70	852.82	47.517	0.13896	1.0895	4.4697	1065.5	111.74	0.28685	0.064234	L	0.434979
70	900	48.289	0.1443	0.9766	4.0829	1130.9	110.81	0.28476	0.066192	L	0.428016
70	1000	49.55	0.15625	0.84283	3.6116	1238	109.33	0.28126	0.069556	L	0.417115
70	1100	50.529	0.16855	0.76658	3.3357	1322.6	108.23	0.27847	0.072319	D	0.409046
70	1200	51.342	0.18096	0.7156	3.147	1393.9	107.34	0.2761	0.074721	D	0.402566

TABLE F.1 (continued)—CO_2 PROPERTIES AT VARIOUS TEMPERATURES, °F											
Temperature (°F)	Pressure (psia)	Density (lbm/ft^3)	Compressibility Factor	Heat Capacity (Btu/lbm°F)	Heat Capacity Ratio (C_p/C_v)	Sonic Velocity (ft/s)	Enthalpy (Btu/lbm)	Entropy [Btu/(lbm°F)]	Viscosity (cp)	Phase	Gas Formation Volume Factor (res bbl/Mcf)
70	1300	52.043	0.1934	0.67842	3.0065	1456.2	106.6	0.27403	0.076878	D	0.397145
70	1400	52.663	0.20582	0.64977	2.8962	1511.9	105.96	0.27216	0.078856	D	0.392460
70	1500	53.222	0.21821	0.62683	2.8065	1562.5	105.41	0.27047	0.080696	D	0.388346
70	1600	53.732	0.23055	0.60794	2.7314	1609	104.93	0.2689	0.082426	D	0.384664
70	1700	54.202	0.24283	0.59204	2.6674	1652.1	104.5	0.26744	0.084066	D	0.381320
70	1800	54.64	0.25506	0.57842	2.6119	1692.5	104.11	0.26607	0.085629	D	0.378273
70	1900	55.049	0.26723	0.56659	2.5631	1730.6	103.77	0.26478	0.087128	D	0.375463
70	2000	55.434	0.27934	0.55618	2.5198	1766.6	103.45	0.26355	0.088572	D	0.372854
70	2100	55.798	0.29139	0.54694	2.481	1800.8	103.17	0.26238	0.089967	D	0.370417
70	2200	56.143	0.30339	0.53866	2.4458	1833.5	102.91	0.26127	0.091319	D	0.368141
70	2300	56.473	0.31533	0.53118	2.4139	1864.8	102.67	0.2602	0.092632	D	0.365993
70	2400	56.787	0.32722	0.52439	2.3846	1894.8	102.45	0.25917	0.093912	D	0.363969
70	2500	57.088	0.33905	0.51819	2.3577	1923.8	102.25	0.25818	0.09516	D	0.362043
70	2600	57.378	0.35084	0.51249	2.3328	1951.7	102.07	0.25723	0.09638	D	0.360223
70	2700	57.656	0.36257	0.50723	2.3097	1978.8	101.9	0.2563	0.097574	D	0.358479
70	2800	57.925	0.37426	0.50236	2.2882	2005	101.75	0.25541	0.098745	D	0.356822
70	2900	58.184	0.3859	0.49784	2.268	2030.4	101.61	0.25454	0.099894	D	0.355233
70	3000	58.435	0.39749	0.49361	2.2491	2055.1	101.48	0.2537	0.10102	D	0.353705
70	3100	58.678	0.40904	0.48966	2.2314	2079.2	101.36	0.25288	0.10213	D	0.352241
70	3200	58.913	0.42055	0.48595	2.2146	2102.7	101.25	0.25208	0.10322	D	0.350836
70	3300	59.142	0.43201	0.48246	2.1987	2125.6	101.15	0.2513	0.1043	D	0.349475
70	3400	59.364	0.44344	0.47917	2.1837	2148	101.06	0.25054	0.10536	D	0.348170
70	3500	59.581	0.45482	0.47607	2.1695	2169.8	100.98	0.24979	0.10641	D	0.346902
70	3600	59.791	0.46617	0.47312	2.1559	2191.2	100.91	0.24907	0.10744	D	0.345683
70	3700	59.997	0.47748	0.47033	2.143	2212.2	100.84	0.24836	0.10846	D	0.344500
70	3800	60.197	0.48875	0.46768	2.1307	2232.8	100.78	0.24766	0.10947	D	0.343352
70	3900	60.393	0.49998	0.46515	2.1189	2252.9	100.72	0.24698	0.11047	D	0.342235
70	4000	60.584	0.51119	0.46275	2.1077	2272.7	100.68	0.24631	0.11145	D	0.341160
70	4100	60.771	0.52235	0.46045	2.0969	2292.2	100.63	0.24565	0.11243	D	0.340105
70	4200	60.954	0.53349	0.45825	2.0865	2311.3	100.6	0.24501	0.11339	D	0.339088
70	4300	61.133	0.54459	0.45615	2.0765	2330.1	100.57	0.24438	0.11435	D	0.338094
70	4400	61.308	0.55566	0.45414	2.067	2348.6	100.54	0.24376	0.1153	D	0.337126
70	4500	61.48	0.5667	0.4522	2.0578	2366.8	100.52	0.24315	0.11624	D	0.336184
70	4600	61.649	0.57771	0.45035	2.0489	2384.7	100.5	0.24255	0.11717	D	0.335265
70	4700	61.814	0.58869	0.44857	2.0403	2402.3	100.49	0.24196	0.11809	D	0.334368
70	4800	61.976	0.59964	0.44685	2.0321	2419.7	100.48	0.24138	0.11901	D	0.333492
70	4900	62.136	0.61056	0.4452	2.0241	2436.9	100.48	0.24081	0.11991	D	0.332635
70	5000	62.292	0.62146	0.44361	2.0163	2453.8	100.48	0.24024	0.12082	D	0.331802
70	5100	62.446	0.63233	0.44207	2.0089	2470.5	100.48	0.23969	0.12171	D	0.330986
70	5200	62.597	0.64317	0.44059	2.0016	2487	100.49	0.23914	0.1226	D	0.330186
70	5300	62.746	0.65398	0.43916	1.9946	2503.2	100.5	0.2386	0.12348	D	0.329401
70	5400	62.892	0.66477	0.43777	1.9878	2519.3	100.51	0.23807	0.12436	D	0.328635
70	5500	63.036	0.67554	0.43643	1.9812	2535.1	100.52	0.23754	0.12523	D	0.327887
70	5600	63.178	0.68628	0.43514	1.9748	2550.8	100.54	0.23703	0.12609	D	0.327152
70	5700	63.318	0.69699	0.43388	1.9686	2566.2	100.57	0.23652	0.12695	D	0.326428
70	5800	63.455	0.70768	0.43266	1.9625	2581.5	100.59	0.23601	0.12781	D	0.325720

TABLE F.1 (continued)—CO_2 PROPERTIES AT VARIOUS TEMPERATURES, °F											
Temperature (°F)	Pressure (psia)	Density (lbm/ft³)	Compressibility Factor	Heat Capacity (Btu/lbm°F)	Heat Capacity Ratio (C_p/C_v)	Sonic Velocity (ft/s)	Enthalpy (Btu/lbm)	Entropy [Btu/(lbm°F)]	Viscosity (cp)	Phase	Gas Formation Volume Factor (res bbl/Mcf)
70	5900	63.59	0.71835	0.43148	1.9566	2596.7	100.62	0.23551	0.12866	D	0.325027
70	6000	63.724	0.729	0.43034	1.9509	2611.6	100.65	0.23502	0.1295	D	0.324349
70	6100	63.856	0.73962	0.42923	1.9453	2626.4	100.68	0.23454	0.13034	D	0.323679
70	6200	63.985	0.75022	0.42815	1.9398	2641.1	100.72	0.23405	0.13118	D	0.323023
70	6300	64.113	0.7608	0.4271	1.9345	2655.5	100.75	0.23358	0.13201	D	0.322378
70	6400	64.24	0.77136	0.42608	1.9293	2669.9	100.79	0.23311	0.13284	D	0.321746
70	6500	64.364	0.78189	0.42509	1.9243	2684	100.84	0.23265	0.13366	D	0.321121
70	6600	64.487	0.79241	0.42412	1.9193	2698.1	100.88	0.23219	0.13448	D	0.320510
70	6700	64.608	0.8029	0.42318	1.9145	2712	100.93	0.23173	0.13529	D	0.319906
70	6800	64.728	0.81338	0.42227	1.9098	2725.8	100.97	0.23128	0.1361	D	0.319316
70	6900	64.847	0.82383	0.42138	1.9052	2739.4	101.02	0.23084	0.13691	D	0.318731
70	7000	64.964	0.83427	0.42051	1.9007	2752.9	101.08	0.2304	0.13771	D	0.318159
70	7100	65.079	0.84469	0.41966	1.8963	2766.3	101.13	0.22996	0.13851	D	0.317596
70	7200	65.193	0.85508	0.41884	1.892	2779.6	101.19	0.22953	0.13931	D	0.317037
70	7300	65.306	0.86546	0.41803	1.8878	2792.7	101.24	0.2291	0.14011	D	0.316490
70	7400	65.417	0.87582	0.41725	1.8836	2805.7	101.3	0.22868	0.1409	D	0.315951
70	7500	65.528	0.88616	0.41648	1.8796	2818.6	101.36	0.22826	0.14168	D	0.315418
70	7600	65.637	0.89649	0.41573	1.8756	2831.4	101.43	0.22785	0.14247	D	0.314896
70	7700	65.744	0.9068	0.415	1.8718	2844.1	101.49	0.22743	0.14325	D	0.314381
70	7800	65.851	0.91708	0.41428	1.8679	2856.7	101.56	0.22703	0.14403	D	0.313869
70	7900	65.957	0.92736	0.41358	1.8642	2869.2	101.62	0.22662	0.1448	D	0.313370
70	8000	66.061	0.93761	0.4129	1.8605	2881.6	101.69	0.22622	0.14557	D	0.312873
70	8100	66.164	0.94785	0.41223	1.8569	2893.8	101.76	0.22583	0.14634	D	0.312385
70	8200	66.266	0.95807	0.41158	1.8534	2906	101.83	0.22543	0.14711	D	0.311903
70	8300	66.368	0.96828	0.41094	1.8499	2918.1	101.91	0.22504	0.14788	D	0.311429
70	8400	66.468	0.97847	0.41031	1.8465	2930.1	101.98	0.22466	0.14864	D	0.310960
70	8500	66.567	0.98864	0.4097	1.8432	2942	102.06	0.22427	0.1494	D	0.310495
70	8600	66.665	0.9988	0.4091	1.8399	2953.8	102.13	0.22389	0.15016	D	0.310039
70	8700	66.762	1.0089	0.40851	1.8367	2965.5	102.21	0.22351	0.15091	D	0.309574
70	8800	66.859	1.0191	0.40793	1.8335	2977.1	102.29	0.22314	0.15166	D	0.309151
70	8900	66.954	1.0292	0.40737	1.8304	2988.7	102.37	0.22277	0.15241	D	0.308706
70	9000	67.048	1.0393	0.40682	1.8273	3000.1	102.45	0.2224	0.15316	D	0.308272
70	9100	67.142	1.0494	0.40628	1.8243	3011.5	102.53	0.22204	0.15391	D	0.307847
70	9200	67.235	1.0594	0.40574	1.8213	3022.8	102.62	0.22167	0.15465	D	0.307403
70	9300	67.327	1.0695	0.40522	1.8184	3034	102.7	0.22131	0.15539	D	0.306997
70	9400	67.418	1.0795	0.40471	1.8155	3045.2	102.79	0.22095	0.15613	D	0.306571
70	9500	67.508	1.0895	0.40421	1.8126	3056.2	102.87	0.2206	0.15687	D	0.306154
70	9600	67.598	1.0996	0.40372	1.8098	3067.2	102.96	0.22025	0.15761	D	0.305773
70	9700	67.686	1.1096	0.40324	1.8071	3078.1	103.05	0.2199	0.15834	D	0.305373
70	9800	67.774	1.1195	0.40276	1.8044	3089	103.14	0.21955	0.15907	D	0.304954
70	9900	67.862	1.1295	0.4023	1.8017	3099.7	103.23	0.21921	0.1598	D	0.304570
70	10000	67.948	1.1395	0.40184	1.799	3110.4	103.32	0.21886	0.16053	D	0.304194
80	14.696	0.11223	0.99505	0.20374	1.2932	883.56	218.23	0.65525	0.015013	V	184.163930
80	100	0.78697	0.9656	0.214	1.3288	868.75	215.84	0.56555	0.015068	V	26.263710
80	200	1.6353	0.92939	0.2283	1.3793	850.63	212.88	0.53033	0.015161	V	12.639410
80	300	2.5586	0.89101	0.24595	1.4426	831.54	209.69	0.50773	0.015293	V	8.078303

TABLE F.1 (continued)—CO_2 PROPERTIES AT VARIOUS TEMPERATURES, °F											
Temperature (°F)	Pressure (psia)	Density (lbm/ft^3)	Compressibility Factor	Heat Capacity (Btu/lbm°F)	Heat Capacity Ratio (C_p/C_v)	Sonic Velocity (ft/s)	Enthalpy (Btu/lbm)	Entropy [Btu/(lbm°F)]	Viscosity (cp)	Phase	Gas Formation Volume Factor (res bbl/Mcf)
80	400	3.5762	0.84996	0.26851	1.5245	811.25	206.23	0.48999	0.015474	V	5.779594
80	500	4.7167	0.80554	0.29866	1.6347	789.45	202.41	0.47457	0.015724	V	4.382036
80	600	6.0259	0.75663	0.34158	1.7916	765.65	198.11	0.46017	0.016069	V	3.429976
80	700	7.5845	0.70134	0.4089	2.0355	739	193.12	0.44583	0.016562	V	2.725144
80	800	9.5612	0.63582	0.53392	2.479	707.69	186.98	0.43041	0.017317	V	2.161738
80	900	12.452	0.54925	0.87774	3.6428	666.14	178.39	0.41132	0.018682	V	1.659917
80	969.57	16.587	0.4442	2.7259	9.1985	609.46	166.88	0.3883	0.021173	V	1.246115
80	969.57	42.618	0.17288	2.0049	7.5324	815.91	121.52	0.30424	0.053524	L	0.484981
80	1000	43.873	0.1732	1.5044	5.9305	904.55	119.93	0.30105	0.056082	L	0.471093
80	1100	46.227	0.18082	1.0385	4.3251	1067.4	116.98	0.29484	0.061279	D	0.447108
80	1200	47.703	0.19116	0.8789	3.7552	1175.4	115.2	0.2908	0.064837	D	0.433286
80	1300	48.822	0.20234	0.79136	3.4356	1260.6	113.88	0.28766	0.067706	D	0.423348
80	1400	49.738	0.21389	0.73397	3.2216	1332.5	112.84	0.28503	0.070176	D	0.415548
80	1500	50.52	0.22562	0.69262	3.0645	1395.4	111.98	0.28276	0.072378	D	0.409115
80	1600	51.206	0.23744	0.66103	2.9426	1451.7	111.26	0.28073	0.074384	D	0.403639
80	1700	51.82	0.24929	0.63591	2.8442	1502.9	110.62	0.2789	0.076242	D	0.398855
80	1800	52.377	0.26115	0.61535	2.7626	1550.1	110.07	0.27721	0.077982	D	0.394617
80	1900	52.888	0.27299	0.59813	2.6935	1594	109.58	0.27565	0.079625	D	0.390798
80	2000	53.361	0.28481	0.58346	2.6339	1635	109.14	0.2742	0.081187	D	0.387333
80	2100	53.803	0.2966	0.57076	2.5819	1673.8	108.75	0.27283	0.082682	D	0.384159
80	2200	54.217	0.30835	0.55964	2.5359	1710.4	108.39	0.27153	0.084118	D	0.381224
80	2300	54.607	0.32006	0.5498	2.4949	1745.3	108.07	0.2703	0.085503	D	0.378497
80	2400	54.977	0.33173	0.54102	2.458	1778.5	107.78	0.26913	0.086843	D	0.375952
80	2500	55.328	0.34336	0.53312	2.4245	1810.4	107.51	0.26801	0.088142	D	0.373567
80	2600	55.663	0.35495	0.52597	2.3939	1841	107.26	0.26694	0.089407	D	0.371324
80	2700	55.983	0.36649	0.51945	2.3659	1870.4	107.04	0.2659	0.090639	D	0.369196
80	2800	56.29	0.37799	0.51347	2.3401	1898.9	106.83	0.26491	0.091842	D	0.367182
80	2900	56.585	0.38945	0.50798	2.3161	1926.3	106.64	0.26394	0.093018	D	0.365269
80	3000	56.868	0.40087	0.5029	2.2939	1953	106.46	0.26301	0.09417	D	0.363447
80	3100	57.142	0.41225	0.49818	2.2731	1978.8	106.3	0.26211	0.0953	D	0.361708
80	3200	57.407	0.42359	0.4938	2.2536	2003.9	106.15	0.26124	0.096408	D	0.360043
80	3300	57.662	0.43489	0.4897	2.2354	2028.4	106.01	0.26039	0.097498	D	0.358440
80	3400	57.91	0.44615	0.48586	2.2182	2052.2	105.89	0.25956	0.09857	D	0.356912
80	3500	58.151	0.45737	0.48225	2.2019	2075.4	105.77	0.25875	0.099626	D	0.355434
80	3600	58.384	0.46856	0.47886	2.1866	2098.1	105.67	0.25797	0.10067	D	0.354015
80	3700	58.611	0.47971	0.47565	2.172	2120.3	105.57	0.2572	0.10169	D	0.352643
80	3800	58.832	0.49082	0.47262	2.1581	2142	105.48	0.25645	0.1027	D	0.351316
80	3900	59.047	0.50191	0.46975	2.145	2163.2	105.4	0.25572	0.1037	D	0.350042
80	4000	59.257	0.51295	0.46703	2.1324	2184.1	105.33	0.255	0.10469	D	0.348798
80	4100	59.461	0.52397	0.46443	2.1204	2204.5	105.26	0.2543	0.10566	D	0.347601
80	4200	59.661	0.53495	0.46197	2.109	2224.5	105.2	0.25362	0.10663	D	0.346436
80	4300	59.856	0.5459	0.45961	2.098	2244.2	105.15	0.25294	0.10758	D	0.345305
80	4400	60.047	0.55682	0.45737	2.0874	2263.5	105.1	0.25228	0.10852	D	0.344208
80	4500	60.234	0.56771	0.45522	2.0773	2282.5	105.06	0.25163	0.10946	D	0.343141
80	4600	60.417	0.57857	0.45316	2.0676	2301.2	105.02	0.251	0.11038	D	0.342103
80	4700	60.596	0.5894	0.45119	2.0583	2319.6	104.99	0.25037	0.1113	D	0.341092

TABLE F.1 (continued)—CO_2 PROPERTIES AT VARIOUS TEMPERATURES, °F											
Temperature (°F)	Pressure (psia)	Density (lbm/ft^3)	Compressibility Factor	Heat Capacity (Btu/lbm°F)	Heat Capacity Ratio (C_p/C_v)	Sonic Velocity (ft/s)	Enthalpy (Btu/lbm)	Entropy [Btu/(lbm°F)]	Viscosity (cp)	Phase	Gas Formation Volume Factor (res bbl/Mcf)
80	4800	60.771	0.6002	0.4493	2.0493	2337.8	104.96	0.24976	0.11221	D	0.340105
80	4900	60.943	0.61098	0.44748	2.0406	2355.6	104.94	0.24915	0.11311	D	0.339148
80	5000	61.112	0.62173	0.44573	2.0322	2373.2	104.92	0.24856	0.114	D	0.338213
80	5100	61.278	0.63245	0.44405	2.0241	2390.5	104.91	0.24797	0.11489	D	0.337299
80	5200	61.44	0.64314	0.44243	2.0163	2407.6	104.9	0.2474	0.11577	D	0.336404
80	5300	61.6	0.65381	0.44087	2.0087	2424.4	104.9	0.24683	0.11664	D	0.335532
80	5400	61.757	0.66445	0.43936	2.0014	2441.1	104.89	0.24627	0.11751	D	0.334678
80	5500	61.911	0.67507	0.43791	1.9943	2457.5	104.9	0.24572	0.11837	D	0.333845
80	5600	62.063	0.68566	0.4365	1.9874	2473.7	104.9	0.24518	0.11922	D	0.333027
80	5700	62.212	0.69623	0.43514	1.9808	2489.7	104.91	0.24464	0.12007	D	0.332228
80	5800	62.359	0.70678	0.43383	1.9743	2505.5	104.92	0.24411	0.12091	D	0.331448
80	5900	62.504	0.7173	0.43256	1.968	2521.1	104.94	0.24359	0.12175	D	0.330680
80	6000	62.646	0.7278	0.43132	1.9619	2536.5	104.96	0.24308	0.12258	D	0.329928
80	6100	62.786	0.73828	0.43013	1.9559	2551.8	104.98	0.24257	0.12341	D	0.329193
80	6200	62.924	0.74873	0.42897	1.9501	2566.9	105	0.24207	0.12424	D	0.328467
80	6300	63.06	0.75917	0.42784	1.9445	2581.8	105.03	0.24157	0.12505	D	0.327761
80	6400	63.195	0.76958	0.42675	1.939	2596.5	105.06	0.24109	0.12587	D	0.327064
80	6500	63.327	0.77997	0.42569	1.9336	2611.1	105.09	0.2406	0.12668	D	0.326380
80	6600	63.457	0.79034	0.42466	1.9284	2625.6	105.12	0.24012	0.12748	D	0.325708
80	6700	63.586	0.80069	0.42366	1.9233	2639.9	105.16	0.23965	0.12828	D	0.325049
80	6800	63.713	0.81103	0.42268	1.9183	2654	105.2	0.23918	0.12908	D	0.324404
80	6900	63.839	0.82134	0.42173	1.9134	2668	105.24	0.23872	0.12987	D	0.323767
80	7000	63.962	0.83163	0.42081	1.9087	2681.9	105.28	0.23827	0.13066	D	0.323140
80	7100	64.084	0.8419	0.41991	1.904	2695.7	105.33	0.23781	0.13145	D	0.322523
80	7200	64.205	0.85215	0.41903	1.8995	2709.3	105.38	0.23737	0.13223	D	0.321916
80	7300	64.324	0.86239	0.41818	1.8951	2722.8	105.43	0.23692	0.13301	D	0.321321
80	7400	64.442	0.87261	0.41734	1.8907	2736.1	105.48	0.23649	0.13379	D	0.320736
80	7500	64.558	0.8828	0.41653	1.8865	2749.3	105.53	0.23605	0.13456	D	0.320155
80	7600	64.673	0.89298	0.41574	1.8823	2762.5	105.58	0.23562	0.13533	D	0.319585
80	7700	64.787	0.90315	0.41496	1.8782	2775.5	105.64	0.2352	0.1361	D	0.319027
80	7800	64.899	0.91329	0.41421	1.8743	2788.4	105.7	0.23478	0.13686	D	0.318473
80	7900	65.01	0.92342	0.41347	1.8703	2801.1	105.76	0.23436	0.13762	D	0.317930
80	8000	65.12	0.93353	0.41275	1.8665	2813.8	105.82	0.23394	0.13838	D	0.317393
80	8100	65.229	0.94363	0.41204	1.8627	2826.4	105.88	0.23353	0.13913	D	0.316866
80	8200	65.336	0.95371	0.41136	1.8591	2838.8	105.95	0.23313	0.13988	D	0.316345
80	8300	65.442	0.96377	0.41068	1.8554	2851.2	106.01	0.23273	0.14063	D	0.315831
80	8400	65.548	0.97382	0.41002	1.8519	2863.4	106.08	0.23233	0.14138	D	0.315325
80	8500	65.652	0.98385	0.40938	1.8484	2875.6	106.15	0.23193	0.14212	D	0.314825
80	8600	65.755	0.99386	0.40875	1.845	2887.7	106.22	0.23154	0.14287	D	0.314330
80	8700	65.857	1.0039	0.40813	1.8416	2899.6	106.29	0.23115	0.1436	D	0.313856
80	8800	65.958	1.0138	0.40753	1.8383	2911.5	106.37	0.23077	0.14434	D	0.313349
80	8900	66.058	1.0238	0.40694	1.835	2923.3	106.44	0.23038	0.14508	D	0.312884
80	9000	66.157	1.0338	0.40636	1.8318	2935	106.52	0.23001	0.14581	D	0.312430
80	9100	66.255	1.0437	0.40579	1.8287	2946.6	106.59	0.22963	0.14654	D	0.311956
80	9200	66.352	1.0536	0.40523	1.8256	2958.2	106.67	0.22926	0.14727	D	0.311492
80	9300	66.448	1.0635	0.40469	1.8225	2969.6	106.75	0.22889	0.14799	D	0.311038

TABLE F.1 (continued)—CO_2 PROPERTIES AT VARIOUS TEMPERATURES, °F											
Temperature (°F)	Pressure (psia)	Density (lbm/ft³)	Compressibility Factor	Heat Capacity (Btu/lbm°F)	Heat Capacity Ratio (C_p/C_v)	Sonic Velocity (ft/s)	Enthalpy (Btu/lbm)	Entropy [Btu/(lbm°F)]	Viscosity (cp)	Phase	Gas Formation Volume Factor (res bbl/Mcf)
80	9400	66.543	1.0734	0.40416	1.8195	2981	106.83	0.22852	0.14872	D	0.310594
80	9500	66.638	1.0833	0.40363	1.8166	2992.2	106.91	0.22815	0.14944	D	0.310159
80	9600	66.731	1.0932	0.40312	1.8137	3003.4	106.99	0.22779	0.15016	D	0.309733
80	9700	66.824	1.103	0.40261	1.8108	3014.6	107.08	0.22743	0.15088	D	0.309288
80	9800	66.916	1.1129	0.40212	1.808	3025.6	107.16	0.22708	0.15159	D	0.308879
80	9900	67.007	1.1227	0.40163	1.8053	3036.6	107.25	0.22672	0.15231	D	0.308452
80	10000	67.098	1.1325	0.40116	1.8025	3047.5	107.33	0.22637	0.15302	D	0.308033
90	14.696	0.11016	0.99534	0.20501	1.2904	890.99	220.27	0.659	0.015281	V	187.631126
90	100	0.77098	0.9677	0.21456	1.3236	877.02	217.99	0.56949	0.015335	D	26.808549
90	200	1.5978	0.93386	0.22769	1.3702	860.03	215.16	0.53452	0.015425	D	12.935534
90	300	2.4918	0.89822	0.2436	1.4276	842.28	212.14	0.51223	0.015551	D	8.294573
90	400	3.4684	0.86042	0.26342	1.4999	823.61	208.89	0.49487	0.015722	D	5.959133
90	500	4.5494	0.81996	0.28895	1.5939	803.83	205.35	0.47996	0.015953	D	4.543131
90	600	5.7675	0.77615	0.32338	1.7212	782.68	201.43	0.46626	0.016264	D	3.583662
90	700	7.1751	0.72786	0.37291	1.9039	759.75	197.02	0.45299	0.016693	D	2.880596
90	800	8.8656	0.67323	0.45147	2.1911	734.29	191.86	0.43938	0.017304	D	2.331342
90	900	11.036	0.60842	0.59912	2.7194	704.78	185.49	0.42437	0.018243	D	1.872809
90	1000	14.264	0.52303	1.0036	4.1019	666.45	176.54	0.40537	0.019964	D	1.448969
90	1100	28.23	0.29071	29.72	76.332	524.55	145.76	0.34743	0.031977	D	0.732150
90	1200	41.913	0.21361	1.4657	5.8064	889.39	126.09	0.31078	0.052356	D	0.493143
90	1300	44.406	0.21841	1.0531	4.3873	1032.5	122.88	0.30416	0.057399	D	0.465438
90	1400	46.013	0.227	0.89211	3.8143	1134.7	120.88	0.29978	0.060951	D	0.449190
90	1500	47.232	0.23694	0.80189	3.4834	1216.7	119.41	0.29638	0.063826	D	0.437602
90	1600	48.227	0.24752	0.74243	3.2601	1286.5	118.24	0.29355	0.0663	D	0.428571
90	1700	49.075	0.25845	0.69951	3.0959	1348	117.28	0.29112	0.068502	D	0.421173
90	1800	49.816	0.26958	0.66669	2.9683	1403.5	116.47	0.28895	0.070507	D	0.414904
90	1900	50.478	0.28082	0.64059	2.8655	1454.3	115.77	0.287	0.07236	D	0.409456
90	2000	51.077	0.29214	0.61921	2.7804	1501.3	115.15	0.28522	0.074091	D	0.404663
90	2100	51.624	0.30349	0.60132	2.7084	1545.1	114.6	0.28357	0.075724	D	0.400366
90	2200	52.13	0.31486	0.58607	2.6465	1586.3	114.12	0.28204	0.077275	D	0.396486
90	2300	52.601	0.32622	0.57288	2.5926	1625.2	113.68	0.2806	0.078756	D	0.392930
90	2400	53.041	0.33758	0.56134	2.5449	1662	113.29	0.27924	0.080176	D	0.389671
90	2500	53.456	0.34892	0.55115	2.5025	1697.1	112.93	0.27796	0.081545	D	0.386650
90	2600	53.847	0.36024	0.54205	2.4644	1730.6	112.6	0.27674	0.082867	D	0.383841
90	2700	54.219	0.37153	0.53387	2.43	1762.7	112.3	0.27557	0.084149	D	0.381209
90	2800	54.573	0.38279	0.52648	2.3986	1793.6	112.03	0.27445	0.085394	D	0.378735
90	2900	54.91	0.39403	0.51975	2.3698	1823.3	111.77	0.27338	0.086606	D	0.376412
90	3000	55.234	0.40523	0.51359	2.3433	1852	111.54	0.27234	0.087789	D	0.374208
90	3100	55.544	0.4164	0.50793	2.3188	1879.8	111.33	0.27135	0.088945	D	0.372119
90	3200	55.842	0.42753	0.5027	2.2961	1906.6	111.13	0.27038	0.090076	D	0.370126
90	3300	56.129	0.43864	0.49786	2.2749	1932.7	110.95	0.26945	0.091184	D	0.368237
90	3400	56.407	0.44971	0.49336	2.255	1958	110.78	0.26855	0.092271	D	0.366426
90	3500	56.675	0.46074	0.48915	2.2364	1982.7	110.63	0.26767	0.093339	D	0.364687
90	3600	56.934	0.47175	0.48522	2.2189	2006.7	110.49	0.26682	0.094389	D	0.363030
90	3700	57.185	0.48272	0.48153	2.2024	2030.2	110.35	0.26599	0.095423	D	0.361432
90	3800	57.429	0.49366	0.47806	2.1868	2053.1	110.23	0.26518	0.09644	D	0.359896

TABLE F.1 (continued)—CO_2 PROPERTIES AT VARIOUS TEMPERATURES, °F											
Temperature (°F)	Pressure (psia)	Density (lbm/ft³)	Compressibility Factor	Heat Capacity (Btu/lbm°F)	Heat Capacity Ratio (C_p/C_v)	Sonic Velocity (ft/s)	Enthalpy (Btu/lbm)	Entropy [Btu/(lbm°F)]	Viscosity (cp)	Phase	Gas Formation Volume Factor (res bbl/Mcf)
90	3900	57.666	0.50457	0.47478	2.172	2075.4	110.12	0.26439	0.097443	D	0.358418
90	4000	57.897	0.51545	0.47169	2.158	2097.3	110.02	0.26362	0.098433	D	0.356993
90	4100	58.121	0.5263	0.46877	2.1446	2118.7	109.92	0.26287	0.099409	D	0.355617
90	4200	58.34	0.53712	0.46599	2.1319	2139.7	109.84	0.26213	0.10037	D	0.354287
90	4300	58.553	0.5479	0.46335	2.1198	2160.3	109.76	0.26142	0.10133	D	0.352992
90	4400	58.761	0.55866	0.46084	2.1082	2180.5	109.69	0.26071	0.10227	D	0.351745
90	4500	58.964	0.56939	0.45845	2.0971	2200.3	109.62	0.26002	0.1032	D	0.350534
90	4600	59.162	0.58009	0.45617	2.0864	2219.8	109.57	0.25935	0.10412	D	0.349358
90	4700	59.356	0.59076	0.45399	2.0762	2239	109.51	0.25868	0.10503	D	0.348214
90	4800	59.546	0.60141	0.4519	2.0664	2257.8	109.47	0.25803	0.10593	D	0.347106
90	4900	59.732	0.61203	0.4499	2.057	2276.3	109.43	0.25739	0.10683	D	0.346026
90	5000	59.914	0.62262	0.44798	2.0479	2294.6	109.39	0.25676	0.10772	D	0.344973
90	5100	60.092	0.63319	0.44614	2.0392	2312.5	109.36	0.25615	0.10859	D	0.343951
90	5200	60.267	0.64373	0.44437	2.0307	2330.2	109.33	0.25554	0.10947	D	0.342952
90	5300	60.439	0.65425	0.44267	2.0226	2347.7	109.31	0.25494	0.11033	D	0.341980
90	5400	60.607	0.66474	0.44103	2.0147	2364.9	109.3	0.25435	0.11119	D	0.341028
90	5500	60.773	0.67521	0.43946	2.0071	2381.8	109.28	0.25378	0.11204	D	0.340102
90	5600	60.935	0.68565	0.43793	1.9997	2398.6	109.27	0.25321	0.11288	D	0.339193
90	5700	61.095	0.69607	0.43647	1.9926	2415.1	109.27	0.25264	0.11372	D	0.338307
90	5800	61.252	0.70647	0.43505	1.9856	2431.4	109.27	0.25209	0.11455	D	0.337441
90	5900	61.406	0.71684	0.43367	1.9789	2447.5	109.27	0.25155	0.11538	D	0.336591
90	6000	61.558	0.72719	0.43235	1.9724	2463.4	109.28	0.25101	0.1162	D	0.335760
90	6100	61.707	0.73752	0.43106	1.9661	2479.1	109.28	0.25048	0.11702	D	0.334947
90	6200	61.854	0.74783	0.42982	1.9599	2494.6	109.3	0.24995	0.11783	D	0.334152
90	6300	61.999	0.75812	0.42862	1.9539	2510	109.31	0.24944	0.11864	D	0.333373
90	6400	62.142	0.76838	0.42745	1.9481	2525.1	109.33	0.24893	0.11944	D	0.332605
90	6500	62.282	0.77863	0.42632	1.9424	2540.2	109.35	0.24842	0.12023	D	0.331857
90	6600	62.421	0.78886	0.42522	1.9369	2555	109.37	0.24793	0.12103	D	0.331122
90	6700	62.557	0.79906	0.42415	1.9315	2569.7	109.4	0.24743	0.12182	D	0.330398
90	6800	62.692	0.80925	0.42311	1.9262	2584.2	109.43	0.24695	0.1226	D	0.329690
90	6900	62.824	0.81941	0.4221	1.9211	2598.6	109.46	0.24647	0.12338	D	0.328992
90	7000	62.955	0.82956	0.42112	1.9161	2612.8	109.49	0.24599	0.12416	D	0.328309
90	7100	63.084	0.83969	0.42017	1.9112	2626.9	109.53	0.24553	0.12493	D	0.327637
90	7200	63.212	0.8498	0.41924	1.9065	2640.9	109.57	0.24506	0.1257	D	0.326977
90	7300	63.338	0.85989	0.41833	1.9018	2654.7	109.61	0.2446	0.12646	D	0.326327
90	7400	63.462	0.86996	0.41745	1.8973	2668.4	109.65	0.24415	0.12722	D	0.325687
90	7500	63.585	0.88001	0.4166	1.8928	2681.9	109.69	0.2437	0.12798	D	0.325057
90	7600	63.706	0.89005	0.41576	1.8885	2695.3	109.74	0.24325	0.12874	D	0.324439
90	7700	63.826	0.90007	0.41494	1.8842	2708.6	109.79	0.24281	0.12949	D	0.323831
90	7800	63.944	0.91007	0.41415	1.88	2721.8	109.84	0.24238	0.13024	D	0.323231
90	7900	64.061	0.92006	0.41337	1.8759	2734.9	109.89	0.24195	0.13098	D	0.322643
90	8000	64.176	0.93003	0.41261	1.8719	2747.8	109.95	0.24152	0.13173	D	0.322062
90	8100	64.291	0.93998	0.41187	1.868	2760.7	110	0.2411	0.13247	D	0.321489
90	8200	64.403	0.94992	0.41115	1.8641	2773.4	110.06	0.24068	0.1332	D	0.320927
90	8300	64.515	0.95984	0.41044	1.8604	2786	110.12	0.24026	0.13394	D	0.320371
90	8400	64.626	0.96974	0.40975	1.8567	2798.6	110.18	0.23985	0.13467	D	0.319822

TABLE F.1 (continued)—CO_2 PROPERTIES AT VARIOUS TEMPERATURES, °F											
Temperature (°F)	Pressure (psia)	Density (lbm/ft³)	Compressibility Factor	Heat Capacity (Btu/lbm°F)	Heat Capacity Ratio (C_p/C_v)	Sonic Velocity (ft/s)	Enthalpy (Btu/lbm)	Entropy [Btu/(lbm°F)]	Viscosity (cp)	Phase	Gas Formation Volume Factor (res bbl/Mcf)
90	8500	64.735	0.97963	0.40908	1.853	2811	110.24	0.23945	0.1354	D	0.319283
90	8600	64.843	0.9895	0.40841	1.8495	2823.3	110.31	0.23904	0.13613	D	0.318750
90	8700	64.95	0.99936	0.40777	1.846	2835.5	110.37	0.23864	0.13685	D	0.318226
90	8800	65.056	1.0092	0.40714	1.8425	2847.6	110.44	0.23825	0.13757	D	0.317707
90	8900	65.16	1.019	0.40652	1.8392	2859.7	110.51	0.23785	0.13829	D	0.317188
90	9000	65.264	1.0288	0.40591	1.8358	2871.6	110.58	0.23746	0.13901	D	0.316680
90	9100	65.367	1.0386	0.40532	1.8326	2883.4	110.65	0.23708	0.13972	D	0.316184
90	9200	65.468	1.0484	0.40474	1.8294	2895.2	110.72	0.23669	0.14044	D	0.315698
90	9300	65.569	1.0582	0.40417	1.8262	2906.9	110.79	0.23631	0.14115	D	0.315223
90	9400	65.669	1.068	0.40362	1.8231	2918.4	110.87	0.23593	0.14185	D	0.314757
90	9500	65.767	1.0777	0.40307	1.8201	2929.9	110.94	0.23556	0.14256	D	0.314273
90	9600	65.865	1.0874	0.40253	1.8171	2941.4	111.02	0.23519	0.14327	D	0.313798
90	9700	65.962	1.0971	0.40201	1.8141	2952.7	111.1	0.23482	0.14397	D	0.313334
90	9800	66.058	1.1068	0.4015	1.8112	2963.9	111.18	0.23445	0.14467	D	0.312878
90	9900	66.153	1.1165	0.40099	1.8083	2975.1	111.26	0.23409	0.14537	D	0.312432
90	10000	66.247	1.1262	0.4005	1.8055	2986.2	111.34	0.23373	0.14606	D	0.311995
100	14.696	0.10816	0.99561	0.20627	1.2877	898.36	222.33	0.66271	0.015549	V	191.096473
100	100	0.75569	0.96963	0.21518	1.3188	885.17	220.14	0.57336	0.015601	D	27.350710
100	200	1.5624	0.93796	0.22729	1.3619	869.21	217.43	0.53862	0.015688	D	13.228691
100	300	2.4295	0.90479	0.24172	1.4141	852.66	214.56	0.5166	0.015809	D	8.507248
100	400	3.3695	0.86986	0.2593	1.4786	835.42	211.5	0.49958	0.015971	D	6.134115
100	500	4.3992	0.83282	0.2813	1.56	817.38	208.2	0.4851	0.016186	D	4.698332
100	600	5.5427	0.7932	0.30976	1.6661	798.38	204.6	0.47197	0.01647	D	3.729014
100	700	6.8358	0.75034	0.34828	1.81	778.24	200.62	0.45948	0.01685	D	3.023588
100	800	8.336	0.70321	0.40375	2.0165	756.66	196.12	0.44706	0.017369	D	2.479463
100	900	10.144	0.65009	0.49133	2.3396	733.17	190.88	0.43409	0.018104	D	2.037481
100	1000	12.468	0.58769	0.65221	2.9221	706.82	184.44	0.41963	0.019227	D	1.657719
100	1100	15.864	0.50808	1.048	4.3043	675.15	175.63	0.40152	0.021228	D	1.302873
100	1200	22.968	0.38283	2.9938	10.577	632.11	159.64	0.37117	0.02682	D	0.899886
100	1300	36.109	0.26381	2.258	8.4469	746.91	137.64	0.33074	0.042661	D	0.572414
100	1400	40.549	0.25299	1.3057	5.3044	909.26	131.48	0.31889	0.050034	D	0.509727
100	1500	42.902	0.25619	1.021	4.306	1024.5	128.37	0.31254	0.054498	D	0.481763
100	1600	44.535	0.26325	0.88153	3.7965	1114.5	126.29	0.30806	0.057864	D	0.464099
100	1700	45.802	0.27197	0.79728	3.4799	1189.6	124.73	0.30454	0.060647	D	0.451268
100	1800	46.845	0.28155	0.74004	3.2606	1255.1	123.48	0.30159	0.063062	D	0.441210
100	1900	47.737	0.29164	0.69811	3.0976	1313.7	122.44	0.29904	0.065222	D	0.432968
100	2000	48.519	0.30204	0.66579	2.9705	1367.2	121.56	0.29679	0.067191	D	0.425988
100	2100	49.217	0.31265	0.63994	2.8677	1416.6	120.8	0.29475	0.069014	D	0.419954
100	2200	49.848	0.32339	0.61868	2.7825	1462.5	120.14	0.29289	0.070718	D	0.414635
100	2300	50.426	0.33422	0.60082	2.7103	1505.7	119.55	0.29117	0.072326	D	0.409890
100	2400	50.958	0.3451	0.58555	2.6481	1546.3	119.02	0.28958	0.073851	D	0.405598
100	2500	51.454	0.35602	0.57233	2.5938	1584.9	118.54	0.28808	0.075308	D	0.401695
100	2600	51.917	0.36696	0.56073	2.5459	1621.5	118.11	0.28667	0.076705	D	0.398114
100	2700	52.352	0.3779	0.55047	2.5032	1656.4	117.72	0.28534	0.078049	D	0.394799
100	2800	52.763	0.38885	0.5413	2.4648	1689.9	117.36	0.28407	0.079348	D	0.391730
100	2900	53.153	0.39978	0.53306	2.43	1721.9	117.04	0.28286	0.080606	D	0.388853

TABLE F.1 (continued)—CO_2 PROPERTIES AT VARIOUS TEMPERATURES, °F											
Temperature (°F)	Pressure (psia)	Density (lbm/ft^3)	Compressibility Factor	Heat Capacity (Btu/lbm°F)	Heat Capacity Ratio (C_p/C_v)	Sonic Velocity (ft/s)	Enthalpy (Btu/lbm)	Entropy [Btu/(lbm°F)]	Viscosity (cp)	Phase	Gas Formation Volume Factor (res bbl/Mcf)
100	3000	53.523	0.4107	0.52559	2.3983	1752.8	116.74	0.28171	0.081828	D	0.386159
100	3100	53.877	0.42161	0.5188	2.3693	1782.5	116.46	0.2806	0.083017	D	0.383629
100	3200	54.215	0.4325	0.51258	2.3425	1811.2	116.21	0.27953	0.084176	D	0.381240
100	3300	54.539	0.44336	0.50686	2.3178	1839	115.97	0.27851	0.085309	D	0.378970
100	3400	54.85	0.4542	0.50158	2.2949	1865.9	115.76	0.27751	0.086416	D	0.376817
100	3500	55.15	0.46502	0.49668	2.2735	1892.1	115.56	0.27656	0.087501	D	0.374771
100	3600	55.439	0.47581	0.49213	2.2534	1917.5	115.37	0.27563	0.088565	D	0.372815
100	3700	55.718	0.48658	0.48788	2.2347	1942.2	115.2	0.27473	0.08961	D	0.370950
100	3800	55.988	0.49732	0.48391	2.217	1966.3	115.04	0.27385	0.090637	D	0.369160
100	3900	56.25	0.50804	0.48018	2.2004	1989.8	114.9	0.273	0.091647	D	0.367448
100	4000	56.503	0.51872	0.47668	2.1847	2012.8	114.76	0.27217	0.092641	D	0.365793
100	4100	56.75	0.52939	0.47338	2.1698	2035.2	114.64	0.27136	0.093621	D	0.364212
100	4200	56.989	0.54002	0.47026	2.1556	2057.2	114.52	0.27057	0.094588	D	0.362680
100	4300	57.222	0.55063	0.46731	2.1422	2078.7	114.41	0.2698	0.095541	D	0.361205
100	4400	57.448	0.56121	0.46451	2.1294	2099.7	114.32	0.26905	0.096482	D	0.359779
100	4500	57.669	0.57177	0.46185	2.1172	2120.4	114.23	0.26832	0.097412	D	0.358403
100	4600	57.885	0.5823	0.45932	2.1055	2140.6	114.14	0.2676	0.098331	D	0.357068
100	4700	58.095	0.5928	0.45691	2.0944	2160.5	114.07	0.26689	0.099239	D	0.355773
100	4800	58.301	0.60328	0.45461	2.0837	2180.1	114	0.2662	0.10014	D	0.354520
100	4900	58.502	0.61373	0.45242	2.0734	2199.3	113.94	0.26553	0.10103	D	0.353300
100	5000	58.698	0.62416	0.45032	2.0635	2218.2	113.88	0.26486	0.10191	D	0.352118
100	5100	58.89	0.63457	0.44831	2.0541	2236.8	113.83	0.26421	0.10278	D	0.350972
100	5200	59.078	0.64495	0.44638	2.045	2255.1	113.79	0.26357	0.10364	D	0.349853
100	5300	59.263	0.65531	0.44453	2.0362	2273.1	113.75	0.26294	0.1045	D	0.348765
100	5400	59.444	0.66564	0.44275	2.0277	2290.9	113.71	0.26232	0.10535	D	0.347703
100	5500	59.621	0.67595	0.44104	2.0195	2308.4	113.69	0.26171	0.10619	D	0.346669
100	5600	59.795	0.68624	0.4394	2.0116	2325.6	113.66	0.26111	0.10703	D	0.345661
100	5700	59.966	0.6965	0.43781	2.004	2342.6	113.64	0.26053	0.10786	D	0.344674
100	5800	60.133	0.70675	0.43629	1.9966	2359.4	113.62	0.25995	0.10868	D	0.343717
100	5900	60.298	0.71697	0.43481	1.9894	2376	113.61	0.25937	0.1095	D	0.342777
100	6000	60.46	0.72717	0.43339	1.9825	2392.4	113.6	0.25881	0.11031	D	0.341859
100	6100	60.619	0.73735	0.43201	1.9757	2408.5	113.6	0.25826	0.11111	D	0.340962
100	6200	60.775	0.74751	0.43068	1.9692	2424.5	113.6	0.25771	0.11191	D	0.340085
100	6300	60.929	0.75764	0.4294	1.9629	2440.2	113.6	0.25717	0.11271	D	0.339223
100	6400	61.081	0.76776	0.42815	1.9567	2455.8	113.61	0.25664	0.1135	D	0.338383
100	6500	61.23	0.77786	0.42695	1.9507	2471.2	113.62	0.25612	0.11428	D	0.337560
100	6600	61.377	0.78793	0.42578	1.9449	2486.4	113.63	0.2556	0.11506	D	0.336749
100	6700	61.522	0.79799	0.42464	1.9392	2501.5	113.64	0.25509	0.11584	D	0.335958
100	6800	61.664	0.80803	0.42354	1.9337	2516.4	113.66	0.25458	0.11661	D	0.335182
100	6900	61.805	0.81805	0.42247	1.9283	2531.1	113.68	0.25408	0.11738	D	0.334421
100	7000	61.943	0.82805	0.42143	1.923	2545.7	113.71	0.25359	0.11814	D	0.333673
100	7100	62.08	0.83803	0.42042	1.9179	2560.1	113.73	0.2531	0.1189	D	0.332938
100	7200	62.214	0.848	0.41944	1.9129	2574.4	113.76	0.25262	0.11966	D	0.332220
100	7300	62.347	0.85794	0.41849	1.908	2588.5	113.79	0.25215	0.12041	D	0.331510
100	7400	62.478	0.86787	0.41756	1.9032	2602.5	113.83	0.25168	0.12116	D	0.330815
100	7500	62.607	0.87778	0.41666	1.8986	2616.4	113.86	0.25121	0.1219	D	0.330132

TABLE F.1 (continued)—CO_2 PROPERTIES AT VARIOUS TEMPERATURES, °F											
Temperature (°F)	Pressure (psia)	Density (lbm/ft^3)	Compressibility Factor	Heat Capacity (Btu/lbm°F)	Heat Capacity Ratio (C_p/C_v)	Sonic Velocity (ft/s)	Enthalpy (Btu/lbm)	Entropy [Btu/(lbm°F)]	Viscosity (cp)	Phase	Gas Formation Volume Factor (res bbl/Mcf)
100	7600	62.735	0.88768	0.41577	1.894	2630.1	113.9	0.25075	0.12264	D	0.329462
100	7700	62.861	0.89755	0.41492	1.8896	2643.7	113.94	0.2503	0.12338	D	0.328799
100	7800	62.986	0.90741	0.41408	1.8852	2657.2	113.98	0.24985	0.12412	D	0.328149
100	7900	63.109	0.91725	0.41326	1.8809	2670.5	114.03	0.2494	0.12485	D	0.327509
100	8000	63.23	0.92708	0.41247	1.8768	2683.8	114.07	0.24896	0.12558	D	0.326881
100	8100	63.35	0.93689	0.41169	1.8727	2696.9	114.12	0.24852	0.1263	D	0.326262
100	8200	63.469	0.94669	0.41094	1.8687	2709.9	114.17	0.24809	0.12702	D	0.325654
100	8300	63.586	0.95646	0.4102	1.8648	2722.7	114.22	0.24766	0.12774	D	0.325051
100	8400	63.702	0.96623	0.40947	1.8609	2735.5	114.28	0.24724	0.12846	D	0.324462
100	8500	63.816	0.97597	0.40877	1.8571	2748.2	114.33	0.24682	0.12918	D	0.323877
100	8600	63.93	0.98571	0.40808	1.8534	2760.7	114.39	0.2464	0.12989	D	0.323306
100	8700	64.042	0.99542	0.4074	1.8498	2773.2	114.45	0.24599	0.1306	D	0.322738
100	8800	64.152	1.0051	0.40675	1.8462	2785.6	114.51	0.24558	0.13131	D	0.322173
100	8900	64.262	1.0148	0.4061	1.8427	2797.8	114.57	0.24518	0.13201	D	0.321627
100	9000	64.371	1.0245	0.40547	1.8393	2810	114.63	0.24478	0.13271	D	0.321094
100	9100	64.478	1.0341	0.40485	1.8359	2822	114.7	0.24438	0.13341	D	0.320541
100	9200	64.584	1.0438	0.40425	1.8326	2834	114.76	0.24398	0.13411	D	0.320031
100	9300	64.689	1.0534	0.40366	1.8293	2845.9	114.83	0.24359	0.13481	D	0.319502
100	9400	64.794	1.063	0.40308	1.8261	2857.7	114.9	0.24321	0.1355	D	0.318983
100	9500	64.897	1.0726	0.40251	1.823	2869.4	114.97	0.24282	0.13619	D	0.318476
100	9600	64.999	1.0822	0.40196	1.8199	2881	115.04	0.24244	0.13688	D	0.317979
100	9700	65.1	1.0918	0.40141	1.8168	2892.5	115.12	0.24206	0.13757	D	0.317493
100	9800	65.2	1.1014	0.40088	1.8138	2903.9	115.19	0.24169	0.13825	D	0.317016
100	9900	65.299	1.1109	0.40036	1.8109	2915.3	115.27	0.24132	0.13894	D	0.316521
100	10000	65.397	1.1204	0.39984	1.808	2926.6	115.34	0.24095	0.13962	D	0.316035
110	14.696	0.10623	0.99586	0.20752	1.2851	905.65	224.4	0.66637	0.015815	V	194.559764
110	100	0.74106	0.97142	0.21585	1.3143	893.18	222.29	0.57718	0.015867	D	27.890797
110	200	1.5288	0.94174	0.22705	1.3543	878.19	219.7	0.54264	0.015951	D	13.519322
110	300	2.3711	0.91081	0.24021	1.402	862.74	216.97	0.52087	0.016067	D	8.716867
110	400	3.278	0.87842	0.25595	1.4599	846.77	214.07	0.50415	0.01622	D	6.305160
110	500	4.2629	0.84436	0.27517	1.5314	830.22	210.98	0.49002	0.016422	D	4.848546
110	600	5.3437	0.80829	0.29923	1.6216	813.03	207.64	0.47735	0.016684	D	3.867852
110	700	6.5461	0.76979	0.33035	1.7388	795.1	204.01	0.46548	0.017027	D	3.157389
110	800	7.9076	0.72829	0.37231	1.897	776.35	199.99	0.45391	0.017479	D	2.613775
110	900	9.4872	0.68291	0.43215	2.1218	756.64	195.48	0.44223	0.018089	D	2.178587
110	1000	11.386	0.63227	0.52449	2.4658	735.85	190.25	0.42991	0.018945	D	1.815334
110	1100	13.796	0.574	0.68458	3.0533	713.83	183.92	0.41622	0.020224	D	1.498211
110	1200	17.154	0.50359	1.0152	4.2408	690.99	175.72	0.39969	0.022367	D	1.204896
110	1300	22.594	0.41421	1.8014	6.9842	673.96	163.93	0.37733	0.026742	D	0.914810
110	1400	30.781	0.32742	2.206	8.4572	715.98	149.29	0.3504	0.035547	D	0.671477
110	1500	36.456	0.2962	1.493	6.0453	832.83	140.69	0.33434	0.043437	D	0.566954
110	1600	39.609	0.2908	1.1359	4.7699	941.04	136.27	0.32572	0.048594	D	0.521829
110	1700	41.713	0.29339	0.95722	4.1062	1031	133.44	0.31996	0.052404	D	0.495508
110	1800	43.289	0.29933	0.85033	3.6987	1107.7	131.39	0.31561	0.05548	D	0.477454
110	1900	44.553	0.307	0.77902	3.4225	1174.9	129.81	0.31208	0.058099	D	0.463915
110	2000	45.612	0.31565	0.72789	3.2222	1235.3	128.52	0.3091	0.060405	D	0.453137

TABLE F.1 (continued)—CO_2 PROPERTIES AT VARIOUS TEMPERATURES, °F											
Temperature (°F)	Pressure (psia)	Density (lbm/ft^3)	Compressibility Factor	Heat Capacity (Btu/lbm°F)	Heat Capacity Ratio (C_p/C_v)	Sonic Velocity (ft/s)	Enthalpy (Btu/lbm)	Entropy [Btu/(lbm°F)]	Viscosity (cp)	Phase	Gas Formation Volume Factor (res bbl/Mcf)
110	2100	46.525	0.32493	0.68927	3.0697	1290.4	127.44	0.3065	0.062485	D	0.444247
110	2200	47.33	0.33462	0.65893	2.949	1341.2	126.52	0.30419	0.064392	D	0.436700
110	2300	48.051	0.34458	0.63436	2.8506	1388.6	125.72	0.3021	0.066162	D	0.430146
110	2400	48.705	0.35473	0.61398	2.7684	1433.1	125.01	0.30019	0.06782	D	0.424366
110	2500	49.303	0.36503	0.59674	2.6985	1475	124.38	0.29843	0.069387	D	0.419220
110	2600	49.857	0.37541	0.58194	2.6379	1514.7	123.82	0.29679	0.070876	D	0.414559
110	2700	50.371	0.38587	0.56905	2.5849	1552.4	123.32	0.29525	0.072298	D	0.410328
110	2800	50.852	0.39638	0.55772	2.538	1588.4	122.86	0.2938	0.073662	D	0.406450
110	2900	51.304	0.40692	0.54765	2.496	1622.9	122.44	0.29243	0.074976	D	0.402870
110	3000	51.731	0.41747	0.53865	2.4582	1655.9	122.06	0.29113	0.076246	D	0.399538
110	3100	52.136	0.42804	0.53052	2.4238	1687.6	121.71	0.28989	0.077475	D	0.396439
110	3200	52.52	0.43861	0.52316	2.3925	1718.2	121.39	0.2887	0.078669	D	0.393534
110	3300	52.887	0.44918	0.51644	2.3637	1747.7	121.09	0.28757	0.079831	D	0.390805
110	3400	53.238	0.45974	0.51028	2.3372	1776.3	120.82	0.28647	0.080963	D	0.388228
110	3500	53.574	0.4703	0.50462	2.3126	1803.9	120.56	0.28542	0.082069	D	0.385799
110	3600	53.897	0.48084	0.49938	2.2898	1830.7	120.33	0.28441	0.083151	D	0.383488
110	3700	54.208	0.49136	0.49452	2.2685	1856.7	120.11	0.28342	0.084211	D	0.381287
110	3800	54.507	0.50187	0.49	2.2485	1882	119.91	0.28247	0.08525	D	0.379194
110	3900	54.796	0.51236	0.48578	2.2298	1906.6	119.73	0.28155	0.086269	D	0.377194
110	4000	55.076	0.52283	0.48183	2.2122	1930.7	119.55	0.28066	0.087272	D	0.375279
110	4100	55.346	0.53328	0.47812	2.1956	1954.1	119.39	0.27979	0.088257	D	0.373444
110	4200	55.609	0.54371	0.47463	2.1799	1977	119.24	0.27894	0.089228	D	0.371682
110	4300	55.863	0.55412	0.47135	2.165	1999.4	119.11	0.27812	0.090184	D	0.369989
110	4400	56.111	0.5645	0.46824	2.1509	2021.3	118.98	0.27731	0.091126	D	0.368354
110	4500	56.351	0.57487	0.4653	2.1375	2042.8	118.86	0.27653	0.092056	D	0.366785
110	4600	56.585	0.58521	0.46251	2.1247	2063.8	118.75	0.27576	0.092973	D	0.365265
110	4700	56.814	0.59553	0.45986	2.1125	2084.5	118.65	0.27501	0.09388	D	0.363797
110	4800	57.036	0.60583	0.45734	2.1008	2104.7	118.56	0.27428	0.094775	D	0.362379
110	4900	57.253	0.61611	0.45494	2.0897	2124.6	118.48	0.27356	0.095661	D	0.361007
110	5000	57.465	0.62636	0.45265	2.079	2144.2	118.4	0.27286	0.096537	D	0.359673
110	5100	57.672	0.6366	0.45047	2.0688	2163.4	118.33	0.27217	0.097404	D	0.358385
110	5200	57.875	0.64681	0.44838	2.0589	2182.3	118.26	0.27149	0.098262	D	0.357131
110	5300	58.073	0.657	0.44637	2.0495	2200.9	118.2	0.27083	0.099112	D	0.355913
110	5400	58.267	0.66716	0.44445	2.0404	2219.2	118.15	0.27018	0.099954	D	0.354724
110	5500	58.457	0.67731	0.44261	2.0316	2237.2	118.1	0.26954	0.10079	D	0.353573
110	5600	58.643	0.68744	0.44084	2.0231	2254.9	118.06	0.26891	0.10162	D	0.352453
110	5700	58.826	0.69754	0.43914	2.015	2272.5	118.03	0.26829	0.10244	D	0.351357
110	5800	59.005	0.70762	0.4375	2.0071	2289.7	117.99	0.26768	0.10325	D	0.350289
110	5900	59.18	0.71769	0.43592	1.9994	2306.7	117.97	0.26709	0.10406	D	0.349252
110	6000	59.353	0.72773	0.4344	1.9921	2323.5	117.94	0.2665	0.10486	D	0.348235
110	6100	59.522	0.73775	0.43293	1.9849	2340.1	117.92	0.26592	0.10565	D	0.347243
110	6200	59.689	0.74775	0.43152	1.978	2356.5	117.91	0.26535	0.10644	D	0.346273
110	6300	59.852	0.75774	0.43015	1.9713	2372.7	117.9	0.26478	0.10723	D	0.345329
110	6400	60.013	0.7677	0.42882	1.9648	2388.6	117.89	0.26423	0.10801	D	0.344402
110	6500	60.171	0.77765	0.42754	1.9584	2404.4	117.89	0.26368	0.10878	D	0.343498
110	6600	60.327	0.78757	0.4263	1.9523	2420	117.89	0.26314	0.10955	D	0.342609

TABLE F.1 (continued)—CO_2 PROPERTIES AT VARIOUS TEMPERATURES, °F											
Temperature (°F)	Pressure (psia)	Density (lbm/ft^3)	Compressibility Factor	Heat Capacity (Btu/lbm°F)	Heat Capacity Ratio (C_p/C_v)	Sonic Velocity (ft/s)	Enthalpy (Btu/lbm)	Entropy [Btu/(lbm°F)]	Viscosity (cp)	Phase	Gas Formation Volume Factor (res bbl/Mcf)
110	6700	60.48	0.79748	0.4251	1.9463	2435.4	117.89	0.26261	0.11031	D	0.341742
110	6800	60.631	0.80737	0.42394	1.9405	2450.7	117.9	0.26209	0.11107	D	0.340893
110	6900	60.78	0.81724	0.42281	1.9348	2465.7	117.91	0.26157	0.11183	D	0.340059
110	7000	60.926	0.82709	0.42171	1.9293	2480.6	117.92	0.26106	0.11258	D	0.339241
110	7100	61.07	0.83693	0.42065	1.9239	2495.4	117.94	0.26055	0.11333	D	0.338442
110	7200	61.212	0.84674	0.41962	1.9187	2510	117.96	0.26005	0.11407	D	0.337654
110	7300	61.353	0.85654	0.41861	1.9136	2524.4	117.98	0.25956	0.11481	D	0.336883
110	7400	61.491	0.86633	0.41764	1.9086	2538.7	118	0.25907	0.11554	D	0.336129
110	7500	61.627	0.87609	0.41669	1.9037	2552.9	118.03	0.25859	0.11627	D	0.335383
110	7600	61.761	0.88584	0.41576	1.899	2566.9	118.06	0.25811	0.117	D	0.334654
110	7700	61.894	0.89557	0.41486	1.8943	2580.7	118.09	0.25764	0.11773	D	0.333936
110	7800	62.025	0.90529	0.41399	1.8898	2594.5	118.12	0.25718	0.11845	D	0.333232
110	7900	62.154	0.91499	0.41314	1.8854	2608.1	118.16	0.25672	0.11917	D	0.332539
110	8000	62.282	0.92467	0.4123	1.881	2621.6	118.2	0.25626	0.11988	D	0.331857
110	8100	62.408	0.93434	0.41149	1.8768	2635	118.24	0.25581	0.12059	D	0.331187
110	8200	62.532	0.94399	0.4107	1.8726	2648.2	118.28	0.25537	0.1213	D	0.330527
110	8300	62.655	0.95363	0.40993	1.8685	2661.3	118.32	0.25492	0.12201	D	0.329880
110	8400	62.777	0.96325	0.40918	1.8646	2674.3	118.37	0.25449	0.12271	D	0.329241
110	8500	62.897	0.97286	0.40844	1.8606	2687.2	118.42	0.25405	0.12341	D	0.328613
110	8600	63.016	0.98245	0.40773	1.8568	2700	118.47	0.25363	0.12411	D	0.327994
110	8700	63.133	0.99203	0.40702	1.853	2712.7	118.52	0.2532	0.12481	D	0.327385
110	8800	63.249	1.0016	0.40634	1.8494	2725.3	118.57	0.25278	0.1255	D	0.326788
110	8900	63.364	1.0111	0.40567	1.8457	2737.8	118.63	0.25237	0.12619	D	0.326180
110	9000	63.477	1.0207	0.40501	1.8422	2750.1	118.69	0.25195	0.12688	D	0.325619
110	9100	63.589	1.0302	0.40437	1.8387	2762.4	118.74	0.25154	0.12757	D	0.325038
110	9200	63.7	1.0397	0.40375	1.8353	2774.6	118.8	0.25114	0.12825	D	0.324470
110	9300	63.81	1.0492	0.40313	1.8319	2786.7	118.87	0.25074	0.12893	D	0.323914
110	9400	63.919	1.0587	0.40253	1.8286	2798.6	118.93	0.25034	0.12961	D	0.323369
110	9500	64.027	1.0681	0.40195	1.8254	2810.5	118.99	0.24995	0.13029	D	0.322806
110	9600	64.133	1.0776	0.40137	1.8222	2822.3	119.06	0.24955	0.13096	D	0.322285
110	9700	64.239	1.087	0.40081	1.819	2834.1	119.13	0.24917	0.13163	D	0.321745
110	9800	64.343	1.0964	0.40026	1.816	2845.7	119.2	0.24878	0.13231	D	0.321216
110	9900	64.446	1.1059	0.39972	1.8129	2857.2	119.27	0.2484	0.13297	D	0.320726
110	10000	64.549	1.1152	0.39919	1.8099	2868.7	119.34	0.24802	0.13364	D	0.320189
120	14.696	0.10438	0.9961	0.20877	1.2825	912.88	226.48	0.66999	0.016081	V	198.022782
120	100	0.72704	0.97308	0.21657	1.31	901.09	224.45	0.58094	0.016131	D	28.428890
120	200	1.4969	0.94522	0.22696	1.3472	886.98	221.97	0.54659	0.016213	D	13.807475
120	300	2.3162	0.91632	0.23902	1.3911	872.53	219.37	0.52504	0.016324	D	8.923542
120	400	3.1931	0.88623	0.25321	1.4434	857.7	216.62	0.50858	0.01647	D	6.472884
120	500	4.1383	0.85478	0.27019	1.5067	842.47	213.7	0.49477	0.01666	D	4.994542
120	600	5.1655	0.82175	0.29089	1.5847	826.81	210.59	0.48249	0.016904	D	4.001288
120	700	6.2934	0.78689	0.31675	1.6826	810.7	207.24	0.4711	0.017216	D	3.284183
120	800	7.5479	0.74983	0.34998	1.809	794.13	203.6	0.46019	0.017618	D	2.738320
120	900	8.9663	0.71012	0.39429	1.9774	777.12	199.59	0.4494	0.018144	D	2.305157
120	1000	10.605	0.6671	0.45615	2.2117	759.75	195.12	0.4384	0.018843	D	1.948957
120	1100	12.555	0.61983	0.54782	2.5562	742.21	190	0.4268	0.019805	D	1.646233

TABLE F.1 (continued)—CO_2 PROPERTIES AT VARIOUS TEMPERATURES, °F

Temperature (°F)	Pressure (psia)	Density (lbm/ft^3)	Compressibility Factor	Heat Capacity (Btu/lbm°F)	Heat Capacity Ratio (C_p/C_v)	Sonic Velocity (ft/s)	Enthalpy (Btu/lbm)	Entropy [Btu/(lbm°F)]	Viscosity (cp)	Phase	Gas Formation Volume Factor (res bbl/Mcf)
120	1200	14.973	0.567	0.69407	3.0997	725.14	183.98	0.41408	0.021193	D	1.380426
120	1300	18.141	0.50697	0.94397	4.0166	710.71	176.62	0.39943	0.023343	D	1.139332
120	1400	22.497	0.44026	1.3399	5.4572	706.55	167.47	0.38206	0.026918	D	0.918740
120	1500	28.018	0.37875	1.6234	6.5231	733.94	157.31	0.36326	0.032522	D	0.737688
120	1600	32.998	0.34304	1.4316	5.8721	803.85	149.25	0.34832	0.038689	D	0.626377
120	1700	36.493	0.32957	1.1703	4.9192	889.3	144.06	0.33844	0.043739	D	0.566383
120	1800	38.948	0.32696	0.99594	4.2643	970.15	140.61	0.33164	0.047696	D	0.530681
120	1900	40.795	0.32949	0.88394	3.8358	1043.3	138.11	0.32652	0.050929	D	0.506641
120	2000	42.268	0.33475	0.807	3.5372	1109.3	136.18	0.32243	0.053681	D	0.488992
120	2100	43.489	0.34162	0.7509	3.3174	1169.5	134.63	0.31901	0.056094	D	0.475264
120	2200	44.534	0.34949	0.70818	3.1488	1224.9	133.35	0.31607	0.058257	D	0.464113
120	2300	45.446	0.35804	0.67454	3.0153	1276.2	132.26	0.31348	0.06023	D	0.454794
120	2400	46.258	0.36706	0.64735	2.9067	1324.2	131.31	0.31116	0.062051	D	0.446825
120	2500	46.988	0.3764	0.62488	2.8165	1369.2	130.49	0.30905	0.063751	D	0.439867
120	2600	47.654	0.38599	0.60596	2.7401	1411.8	129.76	0.30712	0.065349	D	0.433725
120	2700	48.265	0.39576	0.5898	2.6743	1452.2	129.11	0.30533	0.066863	D	0.428232
120	2800	48.831	0.40566	0.5758	2.617	1490.6	128.52	0.30366	0.068305	D	0.423268
120	2900	49.359	0.41566	0.56354	2.5665	1527.2	127.99	0.3021	0.069684	D	0.418747
120	3000	49.853	0.42573	0.55269	2.5215	1562.3	127.51	0.30062	0.071009	D	0.414595
120	3100	50.317	0.43586	0.54303	2.4811	1595.9	127.07	0.29923	0.072286	D	0.410768
120	3200	50.756	0.44603	0.53434	2.4445	1628.2	126.67	0.2979	0.07352	D	0.407217
120	3300	51.173	0.45622	0.52649	2.4112	1659.4	126.3	0.29664	0.074716	D	0.403898
120	3400	51.569	0.46644	0.51935	2.3807	1689.4	125.96	0.29543	0.075878	D	0.400800
120	3500	51.946	0.47667	0.51282	2.3527	1718.5	125.65	0.29427	0.077009	D	0.397888
120	3600	52.307	0.4869	0.50683	2.3268	1746.6	125.36	0.29316	0.078112	D	0.395138
120	3700	52.653	0.49714	0.5013	2.3027	1773.9	125.09	0.29209	0.079189	D	0.392544
120	3800	52.986	0.50737	0.49619	2.2803	1800.4	124.84	0.29105	0.080243	D	0.390079
120	3900	53.306	0.5176	0.49144	2.2594	1826.1	124.61	0.29005	0.081275	D	0.387740
120	4000	53.614	0.52782	0.48701	2.2398	1851.2	124.4	0.28909	0.082287	D	0.385511
120	4100	53.912	0.53803	0.48287	2.2213	1875.6	124.2	0.28815	0.08328	D	0.383384
120	4200	54.199	0.54822	0.479	2.204	1899.5	124.01	0.28724	0.084257	D	0.381344
120	4300	54.478	0.55841	0.47536	2.1876	1922.7	123.84	0.28635	0.085217	D	0.379399
120	4400	54.748	0.56858	0.47193	2.1721	1945.5	123.68	0.28549	0.086162	D	0.377529
120	4500	55.01	0.57873	0.4687	2.1574	1967.7	123.53	0.28465	0.087093	D	0.375729
120	4600	55.264	0.58887	0.46565	2.1434	1989.5	123.39	0.28384	0.088011	D	0.374001
120	4700	55.512	0.59899	0.46275	2.1302	2010.9	123.27	0.28304	0.088917	D	0.372334
120	4800	55.752	0.60909	0.46001	2.1175	2031.8	123.15	0.28226	0.089811	D	0.370725
120	4900	55.987	0.61917	0.4574	2.1054	2052.4	123.04	0.2815	0.090693	D	0.369169
120	5000	56.216	0.62924	0.45492	2.0939	2072.5	122.94	0.28075	0.091566	D	0.367670
120	5100	56.439	0.63929	0.45255	2.0829	2092.3	122.84	0.28003	0.092428	D	0.366218
120	5200	56.657	0.64931	0.45029	2.0723	2111.8	122.76	0.27931	0.093281	D	0.364804
120	5300	56.869	0.65932	0.44814	2.0621	2130.9	122.68	0.27861	0.094125	D	0.363439
120	5400	57.078	0.66932	0.44607	2.0524	2149.8	122.6	0.27793	0.09496	D	0.362119
120	5500	57.281	0.67929	0.4441	2.043	2168.3	122.54	0.27725	0.095788	D	0.360831
120	5600	57.48	0.68924	0.4422	2.034	2186.6	122.48	0.27659	0.096607	D	0.359579
120	5700	57.675	0.69918	0.44038	2.0253	2204.5	122.42	0.27594	0.09742	D	0.358365

TABLE F.1 (continued)—CO_2 PROPERTIES AT VARIOUS TEMPERATURES, °F											
Temperature (°F)	Pressure (psia)	Density (lbm/ft^3)	Compressibility Factor	Heat Capacity (Btu/lbm°F)	Heat Capacity Ratio (C_p/C_v)	Sonic Velocity (ft/s)	Enthalpy (Btu/lbm)	Entropy [Btu/(lbm°F)]	Viscosity (cp)	Phase	Gas Formation Volume Factor (res bbl/Mcf)
120	5800	57.866	0.70909	0.43863	2.0169	2222.3	122.37	0.27531	0.098225	D	0.357178
120	5900	58.054	0.71899	0.43695	2.0088	2239.7	122.33	0.27468	0.099023	D	0.356026
120	6000	58.238	0.72887	0.43533	2.0009	2257	122.29	0.27406	0.099815	D	0.354903
120	6100	58.418	0.73873	0.43378	1.9934	2273.9	122.26	0.27346	0.1006	D	0.353808
120	6200	58.595	0.74857	0.43227	1.9861	2290.7	122.23	0.27286	0.10138	D	0.352738
120	6300	58.769	0.7584	0.43082	1.979	2307.3	122.2	0.27227	0.10215	D	0.351697
120	6400	58.939	0.76821	0.42942	1.9721	2323.6	122.18	0.2717	0.10292	D	0.350680
120	6500	59.107	0.77799	0.42807	1.9654	2339.8	122.17	0.27113	0.10369	D	0.349681
120	6600	59.272	0.78776	0.42676	1.959	2355.7	122.15	0.27056	0.10444	D	0.348708
120	6700	59.434	0.79752	0.4255	1.9527	2371.5	122.15	0.27001	0.1052	D	0.347759
120	6800	59.594	0.80725	0.42427	1.9466	2387	122.14	0.26947	0.10595	D	0.346825
120	6900	59.751	0.81697	0.42308	1.9407	2402.4	122.14	0.26893	0.10669	D	0.345914
120	7000	59.905	0.82667	0.42193	1.9349	2417.6	122.14	0.2684	0.10743	D	0.345021
120	7100	60.058	0.83636	0.42082	1.9293	2432.7	122.14	0.26787	0.10816	D	0.344149
120	7200	60.208	0.84603	0.41973	1.9238	2447.6	122.15	0.26735	0.1089	D	0.343293
120	7300	60.355	0.85568	0.41868	1.9185	2462.3	122.16	0.26684	0.10962	D	0.342452
120	7400	60.501	0.86531	0.41766	1.9133	2476.9	122.18	0.26634	0.11034	D	0.341626
120	7500	60.644	0.87493	0.41667	1.9082	2491.3	122.19	0.26584	0.11106	D	0.340819
120	7600	60.786	0.88454	0.4157	1.9033	2505.6	122.21	0.26535	0.11178	D	0.340028
120	7700	60.925	0.89412	0.41477	1.8984	2519.8	122.24	0.26486	0.11249	D	0.339247
120	7800	61.063	0.90369	0.41385	1.8937	2533.8	122.26	0.26438	0.1132	D	0.338483
120	7900	61.198	0.91325	0.41296	1.8891	2547.6	122.29	0.26391	0.11391	D	0.337733
120	8000	61.332	0.92279	0.4121	1.8846	2561.4	122.32	0.26344	0.11461	D	0.336996
120	8100	61.465	0.93231	0.41125	1.8802	2575	122.35	0.26297	0.11531	D	0.336269
120	8200	61.595	0.94182	0.41043	1.8759	2588.5	122.38	0.26251	0.116	D	0.335556
120	8300	61.724	0.95132	0.40963	1.8717	2601.8	122.42	0.26206	0.1167	D	0.334857
120	8400	61.851	0.9608	0.40885	1.8676	2615.1	122.46	0.26161	0.11739	D	0.334168
120	8500	61.977	0.97026	0.40808	1.8635	2628.2	122.5	0.26116	0.11807	D	0.333488
120	8600	62.101	0.97971	0.40734	1.8596	2641.2	122.54	0.26072	0.11876	D	0.332821
120	8700	62.224	0.98915	0.40661	1.8557	2654.1	122.59	0.26028	0.11944	D	0.332165
120	8800	62.346	0.99857	0.4059	1.8519	2666.9	122.63	0.25985	0.12012	D	0.331518
120	8900	62.465	1.008	0.40521	1.8481	2679.6	122.68	0.25942	0.1208	D	0.330889
120	9000	62.584	1.0174	0.40453	1.8445	2692.1	122.73	0.259	0.12147	D	0.330264
120	9100	62.701	1.0268	0.40387	1.8409	2704.6	122.79	0.25858	0.12214	D	0.329652
120	9200	62.817	1.0361	0.40322	1.8374	2717	122.84	0.25816	0.12281	D	0.329022
120	9300	62.932	1.0455	0.40259	1.8339	2729.2	122.9	0.25775	0.12348	D	0.328437
120	9400	63.045	1.0548	0.40197	1.8305	2741.4	122.95	0.25734	0.12415	D	0.327834
120	9500	63.158	1.0641	0.40136	1.8272	2753.5	123.01	0.25694	0.12481	D	0.327243
120	9600	63.269	1.0735	0.40077	1.8239	2765.5	123.07	0.25653	0.12547	D	0.326695
120	9700	63.379	1.0828	0.40019	1.8207	2777.4	123.13	0.25614	0.12613	D	0.326128
120	9800	63.487	1.092	0.39962	1.8175	2789.2	123.2	0.25574	0.12678	D	0.325543
120	9900	63.595	1.1013	0.39906	1.8144	2800.9	123.26	0.25535	0.12744	D	0.324999
120	10000	63.702	1.1106	0.39852	1.8113	2812.5	123.33	0.25496	0.12809	D	0.324466
130	14.696	0.10258	0.99632	0.21001	1.28	920.04	228.57	0.67357	0.016345	V	201.483402
130	100	0.71358	0.97462	0.21733	1.3059	908.88	226.62	0.58465	0.016395	D	28.965090
130	200	1.4665	0.94844	0.227	1.3406	895.59	224.24	0.55047	0.016474	D	14.093519

TABLE F.1 (continued)—CO_2 PROPERTIES AT VARIOUS TEMPERATURES, °F

Temperature (°F)	Pressure (psia)	Density (lbm/ft^3)	Compressibility Factor	Heat Capacity (Btu/lbm°F)	Heat Capacity Ratio (C_p/C_v)	Sonic Velocity (ft/s)	Enthalpy (Btu/lbm)	Entropy [Btu/(lbm°F)]	Viscosity (cp)	Phase	Gas Formation Volume Factor (res bbl/Mcf)
130	300	2.2644	0.9214	0.2381	1.3811	882.06	221.75	0.52912	0.016581	D	9.127809
130	400	3.1139	0.89337	0.25098	1.4287	868.26	219.14	0.51289	0.016721	D	6.637598
130	500	4.0236	0.86424	0.26612	1.4853	854.19	216.38	0.49935	0.0169	D	5.136933
130	600	5.0041	0.83387	0.28419	1.5535	839.86	213.46	0.4874	0.017127	D	4.130348
130	700	6.0696	0.80208	0.30613	1.6369	825.27	210.35	0.47643	0.017415	D	3.405330
130	800	7.2382	0.76866	0.33333	1.7409	810.47	207.01	0.46603	0.017778	D	2.855511
130	900	8.5351	0.73335	0.36789	1.8733	795.52	203.4	0.4559	0.018241	D	2.421633
130	1000	9.9948	0.69583	0.41308	2.0464	780.56	199.45	0.44581	0.018836	D	2.067963
130	1100	11.667	0.65569	0.47424	2.2797	765.85	195.08	0.43549	0.019614	D	1.771517
130	1200	13.626	0.61247	0.56013	2.6051	751.87	190.18	0.42468	0.020658	D	1.516852
130	1300	15.981	0.56572	0.68479	3.0727	739.71	184.58	0.41305	0.022101	D	1.293295
130	1400	18.886	0.51553	0.86458	3.7397	731.89	178.11	0.40026	0.024168	D	1.094373
130	1500	22.474	0.46417	1.0885	4.5631	733.9	170.76	0.38627	0.027162	D	0.919656
130	1600	26.603	0.41828	1.255	5.1814	755.24	163.1	0.37201	0.031228	D	0.776939
130	1700	30.61	0.38624	1.2364	5.1271	801.49	156.36	0.35948	0.035849	D	0.675224
130	1800	33.906	0.3692	1.1121	4.68	864.73	151.24	0.34982	0.040193	D	0.609577
130	1900	36.462	0.3624	0.98715	4.2167	932.43	147.49	0.34256	0.043937	D	0.566858
130	2000	38.466	0.3616	0.8893	3.8475	998.19	144.67	0.33695	0.047128	D	0.537326
130	2100	40.087	0.36433	0.81643	3.5692	1060.1	142.47	0.33242	0.049893	D	0.515603
130	2200	41.439	0.36922	0.76135	3.3567	1117.9	140.69	0.32864	0.052336	D	0.498772
130	2300	42.595	0.37553	0.71841	3.1895	1171.9	139.22	0.32539	0.054533	D	0.485240
130	2400	43.604	0.38279	0.68401	3.0544	1222.5	137.97	0.32254	0.056537	D	0.474012
130	2500	44.498	0.39073	0.65585	2.9429	1270	136.89	0.32	0.058386	D	0.464490
130	2600	45.301	0.39915	0.63236	2.8493	1314.9	135.95	0.3177	0.060108	D	0.456249
130	2700	46.03	0.40794	0.61249	2.7695	1357.4	135.12	0.31561	0.061726	D	0.449027
130	2800	46.698	0.417	0.59546	2.7006	1397.8	134.38	0.31368	0.063254	D	0.442606
130	2900	47.314	0.42627	0.58069	2.6404	1436.3	133.71	0.31188	0.064706	D	0.436844
130	3000	47.886	0.4357	0.56776	2.5873	1473.1	133.12	0.3102	0.066093	D	0.431624
130	3100	48.421	0.44525	0.55633	2.5401	1508.3	132.57	0.30863	0.067423	D	0.426856
130	3200	48.922	0.4549	0.54615	2.4977	1542.2	132.07	0.30714	0.068702	D	0.422479
130	3300	49.395	0.46463	0.53702	2.4594	1574.8	131.62	0.30573	0.069937	D	0.418440
130	3400	49.842	0.47441	0.52878	2.4246	1606.2	131.2	0.30439	0.071132	D	0.414681
130	3500	50.267	0.48424	0.5213	2.3927	1636.5	130.82	0.30312	0.072291	D	0.411180
130	3600	50.671	0.49411	0.51447	2.3635	1665.9	130.47	0.30189	0.073418	D	0.407907
130	3700	51.056	0.50399	0.50822	2.3365	1694.3	130.14	0.30072	0.074516	D	0.404818
130	3800	51.425	0.5139	0.50246	2.3115	1721.8	129.84	0.29959	0.075587	D	0.401915
130	3900	51.779	0.52382	0.49713	2.2882	1748.6	129.55	0.29851	0.076633	D	0.399169
130	4000	52.119	0.53375	0.4922	2.2665	1774.6	129.29	0.29746	0.077657	D	0.396568
130	4100	52.447	0.54368	0.48761	2.2462	1800	129.05	0.29645	0.07866	D	0.394093
130	4200	52.762	0.55361	0.48332	2.2271	1824.7	128.82	0.29547	0.079643	D	0.391737
130	4300	53.067	0.56354	0.47932	2.2092	1848.8	128.61	0.29452	0.080609	D	0.389490
130	4400	53.361	0.57346	0.47556	2.1923	1872.3	128.42	0.2936	0.081559	D	0.387338
130	4500	53.646	0.58337	0.47202	2.1763	1895.3	128.24	0.2927	0.082492	D	0.385275
130	4600	53.923	0.59328	0.46869	2.1611	1917.8	128.07	0.29183	0.083412	D	0.383302
130	4700	54.191	0.60318	0.46554	2.1468	1939.9	127.91	0.29098	0.084317	D	0.381407
130	4800	54.451	0.61306	0.46257	2.1331	1961.5	127.76	0.29015	0.08521	D	0.379578

TABLE F.1 (continued)—CO_2 PROPERTIES AT VARIOUS TEMPERATURES, °F

Temperature (°F)	Pressure (psia)	Density (lbm/ft^3)	Compressibility Factor	Heat Capacity (Btu/lbm°F)	Heat Capacity Ratio (C_p/C_v)	Sonic Velocity (ft/s)	Enthalpy (Btu/lbm)	Entropy [Btu/(lbm°F)]	Viscosity (cp)	Phase	Gas Formation Volume Factor (res bbl/Mcf)
130	4900	54.705	0.62294	0.45975	2.1201	1982.6	127.62	0.28934	0.086091	D	0.377824
130	5000	54.951	0.6328	0.45707	2.1077	2003.4	127.5	0.28855	0.086961	D	0.376128
130	5100	55.192	0.64265	0.45453	2.0959	2023.8	127.38	0.28778	0.087819	D	0.374493
130	5200	55.426	0.65248	0.4521	2.0846	2043.8	127.27	0.28703	0.088668	D	0.372909
130	5300	55.654	0.6623	0.44979	2.0737	2063.4	127.17	0.28629	0.089507	D	0.371380
130	5400	55.877	0.6721	0.44758	2.0633	2082.8	127.07	0.28557	0.090337	D	0.369896
130	5500	56.095	0.68189	0.44547	2.0534	2101.8	126.99	0.28486	0.091158	D	0.368461
130	5600	56.308	0.69166	0.44345	2.0438	2120.5	126.91	0.28417	0.091971	D	0.367066
130	5700	56.516	0.70142	0.44152	2.0345	2139	126.83	0.28349	0.092776	D	0.365715
130	5800	56.72	0.71116	0.43966	2.0257	2157.1	126.77	0.28282	0.093573	D	0.364400
130	5900	56.92	0.72088	0.43788	2.0171	2175	126.7	0.28216	0.094364	D	0.363120
130	6000	57.115	0.73059	0.43617	2.0089	2192.6	126.65	0.28152	0.095147	D	0.361878
130	6100	57.307	0.74028	0.43452	2.0009	2210	126.6	0.28088	0.095924	D	0.360666
130	6200	57.495	0.74996	0.43293	1.9932	2227.2	126.55	0.28026	0.096694	D	0.359489
130	6300	57.679	0.75962	0.43141	1.9857	2244.1	126.51	0.27965	0.097459	D	0.358340
130	6400	57.86	0.76926	0.42993	1.9785	2260.8	126.48	0.27905	0.098218	D	0.357218
130	6500	58.038	0.77889	0.42851	1.9715	2277.3	126.45	0.27845	0.098971	D	0.356125
130	6600	58.213	0.7885	0.42714	1.9648	2293.6	126.42	0.27787	0.099718	D	0.355056
130	6700	58.384	0.7981	0.42581	1.9582	2309.6	126.4	0.27729	0.10046	D	0.354015
130	6800	58.553	0.80767	0.42452	1.9518	2325.5	126.38	0.27672	0.1012	D	0.352992
130	6900	58.719	0.81724	0.42328	1.9456	2341.2	126.37	0.27617	0.10193	D	0.351998
130	7000	58.882	0.82678	0.42208	1.9396	2356.8	126.36	0.27561	0.10266	D	0.351020
130	7100	59.042	0.83632	0.42091	1.9338	2372.1	126.35	0.27507	0.10338	D	0.350069
130	7200	59.2	0.84583	0.41978	1.9281	2387.3	126.35	0.27453	0.1041	D	0.349132
130	7300	59.356	0.85533	0.41868	1.9225	2402.3	126.35	0.27401	0.10482	D	0.348217
130	7400	59.509	0.86482	0.41762	1.9171	2417.1	126.35	0.27348	0.10553	D	0.347323
130	7500	59.66	0.87429	0.41658	1.9119	2431.8	126.36	0.27297	0.10624	D	0.346445
130	7600	59.808	0.88374	0.41558	1.9068	2446.4	126.37	0.27246	0.10694	D	0.345582
130	7700	59.955	0.89318	0.41461	1.9018	2460.8	126.38	0.27196	0.10764	D	0.344737
130	7800	60.099	0.90261	0.41366	1.8969	2475	126.4	0.27146	0.10834	D	0.343910
130	7900	60.242	0.91202	0.41273	1.8921	2489.1	126.42	0.27097	0.10903	D	0.343097
130	8000	60.382	0.92141	0.41183	1.8875	2503.1	126.44	0.27048	0.10972	D	0.342297
130	8100	60.521	0.93079	0.41096	1.8829	2517	126.46	0.27	0.1104	D	0.341512
130	8200	60.658	0.94016	0.41011	1.8785	2530.7	126.49	0.26953	0.11109	D	0.340743
130	8300	60.793	0.94951	0.40928	1.8741	2544.2	126.52	0.26906	0.11177	D	0.339986
130	8400	60.926	0.95885	0.40847	1.8699	2557.7	126.55	0.26859	0.11244	D	0.339243
130	8500	61.058	0.96817	0.40768	1.8657	2571	126.58	0.26814	0.11312	D	0.338511
130	8600	61.188	0.97748	0.40691	1.8616	2584.2	126.61	0.26768	0.11379	D	0.337792
130	8700	61.316	0.98678	0.40616	1.8576	2597.3	126.65	0.26723	0.11446	D	0.337086
130	8800	61.443	0.99606	0.40543	1.8537	2610.3	126.69	0.26679	0.11513	D	0.336389
130	8900	61.568	1.0053	0.40471	1.8499	2623.2	126.73	0.26635	0.11579	D	0.335695
130	9000	61.692	1.0146	0.40401	1.8461	2636	126.78	0.26591	0.11645	D	0.335036
130	9100	61.814	1.0238	0.40333	1.8424	2648.6	126.82	0.26548	0.11711	D	0.334359
130	9200	61.935	1.0331	0.40266	1.8388	2661.2	126.87	0.26505	0.11777	D	0.333729
130	9300	62.055	1.0423	0.40201	1.8353	2673.6	126.92	0.26463	0.11842	D	0.333081
130	9400	62.173	1.0515	0.40137	1.8318	2686	126.97	0.26421	0.11907	D	0.332446

Temperature (°F)	Pressure (psia)	Density (lbm/ft³)	Compressibility Factor	Heat Capacity (Btu/lbm°F)	Heat Capacity Ratio (C_p/C_v)	Sonic Velocity (ft/s)	Enthalpy (Btu/lbm)	Entropy [Btu/(lbm°F)]	Viscosity (cp)	Phase	Gas Formation Volume Factor (res bbl/Mcf)
130	9500	62.29	1.0607	0.40075	1.8283	2698.3	127.02	0.2638	0.11972	D	0.331825
130	9600	62.406	1.0698	0.40014	1.825	2710.4	127.08	0.26338	0.12037	D	0.331185
130	9700	62.52	1.079	0.39954	1.8217	2722.5	127.13	0.26298	0.12101	D	0.330590
130	9800	62.634	1.0882	0.39896	1.8184	2734.4	127.19	0.26257	0.12166	D	0.330006
130	9900	62.746	1.0973	0.39838	1.8153	2746.3	127.25	0.26217	0.1223	D	0.329405
130	10000	62.857	1.1064	0.39782	1.8121	2758.1	127.31	0.26177	0.12294	D	0.328815
140	14.696	0.10085	0.99652	0.21123	1.2777	927.13	230.68	0.67712	0.016609	V	204.941417
140	100	0.70065	0.97605	0.21812	1.3021	916.57	228.8	0.58831	0.016657	D	29.499518
140	200	1.4376	0.95143	0.22714	1.3345	904.05	226.51	0.55429	0.016734	D	14.377710
140	300	2.2154	0.92608	0.23739	1.372	891.35	224.13	0.53311	0.016837	D	9.329752
140	400	3.0397	0.89992	0.24915	1.4154	878.49	221.64	0.51709	0.016971	D	6.799653
140	500	3.9173	0.87287	0.26276	1.4664	865.46	219.03	0.5038	0.017141	D	5.276214
140	600	4.8568	0.84483	0.27871	1.5267	852.29	216.28	0.49213	0.017355	D	4.255601
140	700	5.8687	0.81569	0.29766	1.5988	839	213.37	0.4815	0.017621	D	3.521843
140	800	6.9665	0.78532	0.32048	1.6864	825.66	210.28	0.47152	0.017953	D	2.966877
140	900	8.1675	0.75357	0.34845	1.7941	812.36	206.98	0.46192	0.018367	D	2.530603
140	1000	9.4943	0.72029	0.38337	1.9288	799.24	203.42	0.45249	0.018886	D	2.176959
140	1100	10.977	0.6853	0.42789	2.1004	786.55	199.58	0.44305	0.019544	D	1.882916
140	1200	12.656	0.64844	0.4858	2.3229	774.67	195.38	0.43342	0.020387	D	1.633170
140	1300	14.584	0.6096	0.56237	2.6152	764.28	190.75	0.42343	0.021483	D	1.417243
140	1400	16.828	0.56893	0.66356	2.9981	756.58	185.63	0.41292	0.022935	D	1.228213
140	1500	19.456	0.52724	0.79096	3.4748	753.78	179.98	0.40179	0.024874	D	1.062331
140	1600	22.483	0.48668	0.92817	3.9826	759.46	173.92	0.39021	0.027432	D	0.919319
140	1700	25.778	0.451	1.0311	4.3619	778.26	167.82	0.37875	0.030622	D	0.801808
140	1800	29.034	0.42397	1.0549	4.4501	812.91	162.23	0.36831	0.034208	D	0.711878
140	1900	31.942	0.40679	1.0086	4.2818	860.77	157.57	0.35952	0.037797	D	0.647082
140	2000	34.386	0.39775	0.93711	4.0184	915.58	153.86	0.3524	0.04112	D	0.601067
140	2100	36.41	0.39443	0.86717	3.7568	972.19	150.92	0.34662	0.044102	D	0.567667
140	2200	38.098	0.3949	0.80717	3.5298	1027.9	148.55	0.34185	0.046764	D	0.542509
140	2300	39.529	0.39791	0.75806	3.3423	1081.4	146.61	0.33782	0.049157	D	0.522877
140	2400	40.764	0.40263	0.71817	3.1885	1132.4	144.99	0.33434	0.051328	D	0.507035
140	2500	41.846	0.40857	0.68542	3.0611	1180.8	143.6	0.33128	0.05332	D	0.493934
140	2600	42.807	0.41537	0.65811	2.9539	1226.8	142.4	0.32856	0.055165	D	0.482842
140	2700	43.671	0.42281	0.635	2.8623	1270.5	141.36	0.3261	0.056886	D	0.473287
140	2800	44.455	0.43073	0.61521	2.7831	1312.1	140.43	0.32386	0.058503	D	0.464933
140	2900	45.173	0.43903	0.59808	2.714	1351.8	139.61	0.32179	0.060031	D	0.457551
140	3000	45.834	0.44761	0.58311	2.6532	1389.8	138.87	0.31988	0.061483	D	0.450943
140	3100	46.448	0.45642	0.56993	2.5991	1426.3	138.2	0.3181	0.062869	D	0.444985
140	3200	47.02	0.46541	0.55824	2.5508	1461.3	137.6	0.31643	0.064196	D	0.439571
140	3300	47.556	0.47454	0.5478	2.5074	1495	137.05	0.31486	0.065472	D	0.434612
140	3400	48.061	0.48379	0.53842	2.468	1527.5	136.54	0.31337	0.066703	D	0.430052
140	3500	48.538	0.49313	0.52994	2.4322	1558.8	136.08	0.31196	0.067892	D	0.425830
140	3600	48.989	0.50254	0.52224	2.3995	1589.2	135.65	0.31061	0.069045	D	0.421901
140	3700	49.419	0.51201	0.51522	2.3694	1618.5	135.26	0.30933	0.070165	D	0.418234
140	3800	49.828	0.52153	0.50878	2.3417	1647	134.89	0.3081	0.071254	D	0.414800
140	3900	50.219	0.53109	0.50286	2.316	1674.6	134.55	0.30692	0.072316	D	0.411573

TABLE F.1 (continued)—CO_2 PROPERTIES AT VARIOUS TEMPERATURES, °F											
Temperature (°F)	Pressure (psia)	Density (lbm/ft^3)	Compressibility Factor	Heat Capacity (Btu/lbm°F)	Heat Capacity Ratio (C_p/C_v)	Sonic Velocity (ft/s)	Enthalpy (Btu/lbm)	Entropy [Btu/(lbm°F)]	Viscosity (cp)	Phase	Gas Formation Volume Factor (res bbl/Mcf)
140	4000	50.594	0.54067	0.49739	2.2921	1701.5	134.24	0.30578	0.073353	D	0.408522
140	4100	50.953	0.55028	0.49232	2.2698	1727.7	133.95	0.30469	0.074367	D	0.405642
140	4200	51.299	0.5599	0.48761	2.249	1753.1	133.68	0.30363	0.07536	D	0.402906
140	4300	51.632	0.56954	0.48321	2.2295	1778	133.43	0.30261	0.076332	D	0.400312
140	4400	51.953	0.57918	0.47911	2.2111	1802.2	133.19	0.30162	0.077287	D	0.397836
140	4500	52.263	0.58883	0.47526	2.1938	1825.9	132.97	0.30066	0.078224	D	0.395476
140	4600	52.563	0.59848	0.47164	2.1775	1849	132.77	0.29973	0.079146	D	0.393219
140	4700	52.853	0.60813	0.46824	2.162	1871.7	132.58	0.29883	0.080053	D	0.391058
140	4800	53.135	0.61778	0.46502	2.1474	1893.9	132.4	0.29795	0.080945	D	0.388987
140	4900	53.408	0.62742	0.46199	2.1334	1915.6	132.23	0.29709	0.081825	D	0.386995
140	5000	53.674	0.63706	0.45911	2.1202	1936.9	132.08	0.29626	0.082693	D	0.385082
140	5100	53.932	0.64669	0.45639	2.1076	1957.8	131.93	0.29544	0.083549	D	0.383238
140	5200	54.183	0.65631	0.45379	2.0955	1978.3	131.8	0.29465	0.084394	D	0.381460
140	5300	54.428	0.66592	0.45133	2.084	1998.5	131.67	0.29387	0.085228	D	0.379742
140	5400	54.667	0.67552	0.44898	2.073	2018.3	131.55	0.29311	0.086053	D	0.378083
140	5500	54.9	0.68511	0.44674	2.0625	2037.8	131.45	0.29236	0.086869	D	0.376479
140	5600	55.127	0.69469	0.44459	2.0523	2056.9	131.35	0.29163	0.087675	D	0.374926
140	5700	55.349	0.70426	0.44254	2.0426	2075.8	131.25	0.29092	0.088474	D	0.373423
140	5800	55.567	0.71381	0.44058	2.0333	2094.4	131.17	0.29022	0.089264	D	0.371961
140	5900	55.779	0.72336	0.4387	2.0243	2112.6	131.09	0.28953	0.090047	D	0.370549
140	6000	55.987	0.73288	0.43689	2.0156	2130.7	131.01	0.28886	0.090822	D	0.369168
140	6100	56.191	0.7424	0.43516	2.0073	2148.4	130.95	0.2882	0.091591	D	0.367833
140	6200	56.39	0.7519	0.43349	1.9992	2165.9	130.89	0.28755	0.092352	D	0.366531
140	6300	56.585	0.76139	0.43189	1.9914	2183.2	130.83	0.28691	0.093108	D	0.365266
140	6400	56.777	0.77086	0.43034	1.9839	2200.2	130.78	0.28628	0.093857	D	0.364031
140	6500	56.965	0.78032	0.42885	1.9766	2217	130.74	0.28566	0.094601	D	0.362829
140	6600	57.15	0.78977	0.42741	1.9695	2233.6	130.7	0.28505	0.095338	D	0.361659
140	6700	57.331	0.7992	0.42603	1.9627	2250	130.66	0.28445	0.096071	D	0.360515
140	6800	57.509	0.80862	0.42468	1.9561	2266.2	130.63	0.28387	0.096798	D	0.359400
140	6900	57.684	0.81802	0.42339	1.9497	2282.2	130.6	0.28328	0.09752	D	0.358309
140	7000	57.856	0.82741	0.42213	1.9434	2298	130.58	0.28271	0.098238	D	0.357245
140	7100	58.025	0.83678	0.42092	1.9373	2313.7	130.56	0.28215	0.09895	D	0.356202
140	7200	58.191	0.84615	0.41974	1.9314	2329.1	130.55	0.28159	0.099659	D	0.355188
140	7300	58.355	0.85549	0.4100	1.9257	2344.4	130.54	0.28105	0.10036	D	0.354189
140	7400	58.516	0.86482	0.4175	1.9201	2359.5	130.53	0.2805	0.10106	D	0.353213
140	7500	58.675	0.87414	0.41643	1.9147	2374.4	130.53	0.27997	0.10176	D	0.352259
140	7600	58.831	0.88345	0.41539	1.9094	2389.2	130.53	0.27945	0.10245	D	0.351327
140	7700	58.985	0.89274	0.41438	1.9042	2403.8	130.53	0.27893	0.10314	D	0.350411
140	7800	59.136	0.90201	0.41339	1.8992	2418.3	130.53	0.27841	0.10382	D	0.349510
140	7900	59.286	0.91127	0.41244	1.8943	2432.7	130.54	0.27791	0.1045	D	0.348628
140	8000	59.433	0.92052	0.41151	1.8895	2446.9	130.55	0.2774	0.10518	D	0.347765
140	8100	59.578	0.92976	0.41061	1.8848	2460.9	130.57	0.27691	0.10585	D	0.346919
140	8200	59.721	0.93898	0.40973	1.8802	2474.8	130.59	0.27642	0.10653	D	0.346087
140	8300	59.862	0.94819	0.40887	1.8758	2488.6	130.61	0.27594	0.10719	D	0.345271
140	8400	60.002	0.95738	0.40804	1.8714	2502.3	130.63	0.27546	0.10786	D	0.344467
140	8500	60.139	0.96656	0.40722	1.8671	2515.8	130.65	0.27499	0.10852	D	0.343679

TABLE F.1 (continued)—CO_2 PROPERTIES AT VARIOUS TEMPERATURES, °F											
Temperature (°F)	Pressure (psia)	Density (lbm/ft^3)	Compressibility Factor	Heat Capacity (Btu/lbm°F)	Heat Capacity Ratio (C_p/C_v)	Sonic Velocity (ft/s)	Enthalpy (Btu/lbm)	Entropy [Btu/(lbm°F)]	Viscosity (cp)	Phase	Gas Formation Volume Factor (res bbl/Mcf)
140	8600	60.275	0.97573	0.40643	1.8629	2529.2	130.68	0.27452	0.10918	D	0.342905
140	8700	60.409	0.98489	0.40566	1.8588	2542.5	130.71	0.27406	0.10983	D	0.342146
140	8800	60.542	0.99403	0.4049	1.8548	2555.7	130.74	0.2736	0.11049	D	0.341397
140	8900	60.672	1.0032	0.40417	1.8509	2568.8	130.78	0.27315	0.11114	D	0.340675
140	9000	60.802	1.0123	0.40345	1.8471	2581.7	130.81	0.2727	0.11179	D	0.339946
140	9100	60.929	1.0214	0.40275	1.8433	2594.5	130.85	0.27226	0.11243	D	0.339232
140	9200	61.055	1.0305	0.40206	1.8396	2607.3	130.89	0.27182	0.11308	D	0.338535
140	9300	61.18	1.0395	0.40139	1.8359	2619.9	130.94	0.27138	0.11372	D	0.337819
140	9400	61.303	1.0486	0.40074	1.8324	2632.4	130.98	0.27095	0.11436	D	0.337151
140	9500	61.425	1.0577	0.4001	1.8289	2644.8	131.03	0.27053	0.11499	D	0.336497
140	9600	61.546	1.0667	0.39947	1.8255	2657.1	131.07	0.27011	0.11563	D	0.335826
140	9700	61.665	1.0757	0.39886	1.8221	2669.4	131.12	0.26969	0.11626	D	0.335168
140	9800	61.782	1.0848	0.39826	1.8188	2681.5	131.18	0.26927	0.11689	D	0.334554
140	9900	61.899	1.0938	0.39768	1.8155	2693.5	131.23	0.26886	0.11752	D	0.333922
140	10000	62.014	1.1028	0.3971	1.8123	2705.5	131.28	0.26846	0.11814	D	0.333303
150	14.696	0.099179	0.99671	0.21245	1.2754	934.16	232.8	0.68062	0.016872	V	208.398714
150	100	0.68822	0.97738	0.21893	1.2984	924.17	230.98	0.59193	0.016918	D	30.032315
150	200	1.4099	0.9542	0.22737	1.3288	912.35	228.79	0.55805	0.016994	D	14.660027
150	300	2.1689	0.93041	0.23688	1.3636	900.44	226.5	0.53704	0.017093	D	9.529683
150	400	2.9699	0.90595	0.24765	1.4035	888.43	224.12	0.5212	0.017221	D	6.959365
150	500	3.8185	0.88077	0.25998	1.4496	876.33	221.64	0.50812	0.017383	D	5.412749
150	600	4.7214	0.8548	0.2742	1.5034	864.18	219.04	0.49671	0.017584	D	4.377626
150	700	5.6869	0.82797	0.29078	1.5666	852.02	216.31	0.48637	0.017833	D	3.634477
150	800	6.7248	0.80021	0.3103	1.6416	839.92	213.43	0.47673	0.018139	D	3.073543
150	900	7.8476	0.77143	0.33356	1.7314	827.97	210.38	0.46755	0.018514	D	2.633779
150	1000	9.0708	0.74156	0.36164	1.8402	816.32	207.14	0.45864	0.018976	D	2.278619
150	1100	10.414	0.71053	0.39594	1.9733	805.18	203.69	0.44985	0.019547	D	1.984792
150	1200	11.9	0.6783	0.43834	2.1378	794.87	199.98	0.44104	0.020257	D	1.736864
150	1300	13.56	0.64489	0.49117	2.3421	785.84	195.99	0.4321	0.021146	D	1.524290
150	1400	15.426	0.61046	0.55687	2.5951	778.85	191.68	0.42293	0.022268	D	1.339845
150	1500	17.534	0.57544	0.63673	2.9003	775.03	187.04	0.41348	0.023692	D	1.178784
150	1600	19.902	0.54077	0.72751	3.2443	776.1	182.11	0.40375	0.02549	D	1.038527
150	1700	22.506	0.50809	0.81697	3.5805	784.38	177	0.39394	0.027714	D	0.918369
150	1800	25.248	0.47955	0.88472	3.8338	802.23	171.95	0.38438	0.030341	D	0.818628
150	1900	27.965	0.45701	0.91338	3.9404	830.87	167.25	0.37554	0.033241	D	0.739090
150	2000	30.491	0.44122	0.90224	3.8981	869.14	163.12	0.36773	0.036215	D	0.677876
150	2100	32.731	0.43157	0.8669	3.7647	913.86	159.64	0.36105	0.039091	D	0.631477
150	2200	34.674	0.42679	0.82347	3.5997	961.7	156.74	0.35539	0.04178	D	0.596097
150	2300	36.351	0.4256	0.78054	3.4353	1010.4	154.33	0.35058	0.044258	D	0.568590
150	2400	37.806	0.42701	0.74159	3.2849	1058.6	152.3	0.34643	0.046533	D	0.546704
150	2500	39.08	0.4303	0.70765	3.1528	1105.5	150.58	0.34282	0.048629	D	0.528879
150	2600	40.207	0.43497	0.67852	3.0387	1150.7	149.09	0.33962	0.050569	D	0.514057
150	2700	41.214	0.44066	0.65356	2.9401	1194.2	147.81	0.33676	0.052377	D	0.501493
150	2800	42.124	0.44712	0.63203	2.8544	1235.9	146.67	0.33418	0.054071	D	0.490672
150	2900	42.951	0.45416	0.6133	2.7791	1275.9	145.67	0.33181	0.055667	D	0.481212
150	3000	43.709	0.46167	0.59688	2.7126	1314.3	144.77	0.32964	0.057179	D	0.472863

TABLE F.1 (continued)—CO_2 PROPERTIES AT VARIOUS TEMPERATURES, °F											
Temperature (°F)	Pressure (psia)	Density (lbm/ft^3)	Compressibility Factor	Heat Capacity (Btu/lbm°F)	Heat Capacity Ratio (C_p/C_v)	Sonic Velocity (ft/s)	Enthalpy (Btu/lbm)	Entropy [Btu/(lbm°F)]	Viscosity (cp)	Phase	Gas Formation Volume Factor (res bbl/Mcf)
160	2200	31.433	0.4632	0.79516	3.5094	920.61	164.87	0.36861	0.037651	D	0.657562
160	2300	33.247	0.45783	0.77052	3.4126	961.28	162.11	0.36324	0.040039	D	0.621681
160	2400	34.86	0.45562	0.74246	3.302	1003.5	159.74	0.35854	0.042297	D	0.592901
160	2500	36.293	0.45587	0.7144	3.1908	1046.3	157.7	0.35441	0.044415	D	0.569498
160	2600	37.569	0.458	0.68802	3.0857	1088.7	155.94	0.35076	0.046397	D	0.550153
160	2700	38.713	0.46157	0.66404	2.9897	1130.3	154.4	0.34749	0.048253	D	0.533906
160	2800	39.744	0.46624	0.64261	2.9034	1170.7	153.05	0.34455	0.049998	D	0.520047
160	2900	40.681	0.47177	0.62359	2.8263	1209.9	151.86	0.34188	0.051642	D	0.508070
160	3000	41.537	0.47798	0.6067	2.7574	1247.9	150.79	0.33944	0.053199	D	0.497599
160	3100	42.325	0.48472	0.59165	2.6956	1284.6	149.84	0.33719	0.054679	D	0.488338
160	3200	43.054	0.49188	0.57817	2.6399	1320.1	148.98	0.3351	0.05609	D	0.480065
160	3300	43.731	0.4994	0.56606	2.5895	1354.4	148.2	0.33315	0.057441	D	0.472635
160	3400	44.364	0.50719	0.55511	2.5436	1387.7	147.49	0.33132	0.058737	D	0.465889
160	3500	44.957	0.51522	0.54518	2.5017	1420	146.84	0.3296	0.059984	D	0.459744
160	3600	45.515	0.52345	0.53614	2.4634	1451.2	146.24	0.32798	0.061187	D	0.454113
160	3700	46.042	0.53183	0.52788	2.4281	1481.6	145.69	0.32645	0.06235	D	0.448913
160	3800	40.541	0.54035	0.52031	2.3966	1511.1	145.19	0.32498	0.063477	D	0.444102
160	3900	47.015	0.54898	0.51336	2.3655	1539.8	144.72	0.32359	0.064572	D	0.439626
160	4000	47.466	0.5577	0.50694	2.3377	1567.8	144.29	0.32226	0.065636	D	0.435443
160	4100	47.897	0.5665	0.50101	2.3118	1595	143.89	0.32099	0.066673	D	0.431526
160	4200	48.309	0.57537	0.49551	2.2877	1621.5	143.51	0.31976	0.067684	D	0.427847
160	4300	48.703	0.5843	0.49041	2.2651	1647.4	143.16	0.31859	0.068672	D	0.424383
160	4400	49.082	0.59327	0.48564	2.244	1672.6	142.84	0.31745	0.069638	D	0.421105
160	4500	49.447	0.60228	0.4812	2.2242	1697.3	142.54	0.31636	0.070585	D	0.418001
160	4600	49.798	0.61132	0.47704	2.2056	1721.5	142.26	0.3153	0.071513	D	0.415051
160	4700	50.137	0.62039	0.47313	2.188	1745.1	141.99	0.31427	0.072423	D	0.412247
160	4800	50.464	0.62948	0.46947	2.1715	1768.2	141.74	0.31328	0.073318	D	0.409573
160	4900	50.781	0.63859	0.46601	2.1558	1790.8	141.51	0.31232	0.074197	D	0.407021
160	5000	51.087	0.64771	0.46275	2.1409	1813	141.3	0.31138	0.075062	D	0.404577
160	5100	51.384	0.65684	0.45967	2.1268	1834.8	141.09	0.31047	0.075914	D	0.402236
160	5200	51.673	0.66598	0.45675	2.1134	1856.1	140.9	0.30958	0.076753	D	0.399990
160	5300	51.953	0.67513	0.45399	2.1006	1877.1	140.73	0.30872	0.077581	D	0.397835
160	5400	52.225	0.68428	0.45136	2.0885	1897.7	140.56	0.30788	0.078397	D	0.395759
160	5500	52.491	0.69343	0.44886	2.0768	1917.9	140.4	0.30706	0.079203	D	0.393759
160	5600	52.749	0.70259	0.44648	2.0657	1937.8	140.26	0.30625	0.079999	D	0.391837
160	5700	53	0.71174	0.44421	2.055	1957.4	140.12	0.30547	0.080786	D	0.389976
160	5800	53.246	0.72088	0.44204	2.0448	1976.7	139.99	0.3047	0.081563	D	0.388174
160	5900	53.485	0.73003	0.43997	2.035	1995.6	139.88	0.30395	0.082332	D	0.386438
160	6000	53.719	0.73917	0.43799	2.0256	2014.3	139.76	0.30321	0.083093	D	0.384755
160	6100	53.948	0.7483	0.43609	2.0166	2032.7	139.66	0.30249	0.083846	D	0.383122
160	6200	54.172	0.75743	0.43426	2.0078	2050.8	139.57	0.30178	0.084592	D	0.381542
160	6300	54.39	0.76655	0.43251	1.9994	2068.7	139.48	0.30109	0.085331	D	0.380006
160	6400	54.605	0.77566	0.43083	1.9913	2086.3	139.39	0.30041	0.086063	D	0.378514
160	6500	54.814	0.78477	0.42922	1.9835	2103.7	139.32	0.29974	0.086788	D	0.377068
160	6600	55.02	0.79386	0.42766	1.976	2120.9	139.25	0.29908	0.087508	D	0.375657
160	6700	55.222	0.80295	0.42616	1.9686	2137.8	139.18	0.29843	0.088221	D	0.374287

TABLE F.1 (continued)—CO_2 PROPERTIES AT VARIOUS TEMPERATURES, °F

Temperature (°F)	Pressure (psia)	Density (lbm/ft^3)	Compressibility Factor	Heat Capacity (Btu/lbm°F)	Heat Capacity Ratio (C_p/C_v)	Sonic Velocity (ft/s)	Enthalpy (Btu/lbm)	Entropy [Btu/(lbm°F)]	Viscosity (cp)	Phase	Gas Formation Volume Factor (res bbl/Mcf)
160	6800	55.419	0.81203	0.42471	1.9616	2154.5	139.12	0.2978	0.088929	D	0.372953
160	6900	55.613	0.8211	0.42332	1.9547	2171	139.07	0.29718	0.089632	D	0.371653
160	7000	55.804	0.83015	0.42197	1.9481	2187.3	139.02	0.29656	0.090329	D	0.370382
160	7100	55.99	0.8392	0.42067	1.9416	2203.5	138.98	0.29596	0.091021	D	0.369146
160	7200	56.174	0.84824	0.41941	1.9354	2219.4	138.94	0.29536	0.091709	D	0.367940
160	7300	56.355	0.85727	0.41819	1.9293	2235.1	138.91	0.29477	0.092391	D	0.366763
160	7400	56.532	0.86628	0.41701	1.9234	2250.7	138.88	0.2942	0.09307	D	0.365610
160	7500	56.706	0.87529	0.41587	1.9177	2266.1	138.85	0.29363	0.093743	D	0.364487
160	7600	56.878	0.88428	0.41476	1.9121	2281.3	138.83	0.29306	0.094413	D	0.363385
160	7700	57.047	0.89327	0.41369	1.9066	2296.3	138.81	0.29251	0.095079	D	0.362312
160	7800	57.213	0.90224	0.41265	1.9013	2311.2	138.8	0.29196	0.095741	D	0.361259
160	7900	57.377	0.9112	0.41163	1.8962	2326	138.78	0.29142	0.096399	D	0.360228
160	8000	57.538	0.92015	0.41065	1.8912	2340.5	138.78	0.29089	0.097053	D	0.359219
160	8100	57.697	0.92909	0.4097	1.8862	2355	138.77	0.29037	0.097704	D	0.358232
160	8200	57.853	0.93801	0.40877	1.8815	2369.3	138.77	0.28985	0.098352	D	0.357260
160	8300	58.007	0.94693	0.40787	1.8768	2383.4	138.77	0.28934	0.098996	D	0.356312
160	8400	58.159	0.95584	0.40699	1.8722	2397.5	138.78	0.28883	0.099637	D	0.355383
160	8500	58.309	0.96473	0.40613	1.8678	2411.3	138.79	0.28833	0.10027	D	0.354469
160	8600	58.457	0.97361	0.4053	1.8634	2425.1	138.8	0.28784	0.10091	D	0.353572
160	8700	58.603	0.98248	0.40449	1.8591	2438.7	138.81	0.28735	0.10154	D	0.352692
160	8800	58.747	0.99134	0.4037	1.855	2452.2	138.83	0.28687	0.10217	D	0.351828
160	8900	58.889	1.0002	0.40293	1.8509	2465.6	138.85	0.28639	0.1028	D	0.350984
160	9000	59.029	1.009	0.40217	1.8469	2478.9	138.87	0.28592	0.10342	D	0.350138
160	9100	59.167	1.0178	0.40144	1.843	2492	138.89	0.28545	0.10404	D	0.349311
160	9200	59.304	1.0267	0.40073	1.8391	2505.1	138.92	0.28499	0.10466	D	0.348535
160	9300	59.439	1.0355	0.40003	1.8354	2518	138.95	0.28453	0.10528	D	0.347743
160	9400	59.572	1.0443	0.39934	1.8317	2530.8	138.98	0.28408	0.10589	D	0.346967
160	9500	59.704	1.053	0.39868	1.8281	2543.5	139.01	0.28363	0.1065	D	0.346175
160	9600	59.834	1.0618	0.39803	1.8245	2556.1	139.05	0.28319	0.10711	D	0.345432
160	9700	59.962	1.0706	0.39739	1.821	2568.6	139.09	0.28275	0.10772	D	0.344704
160	9800	60.09	1.0793	0.39677	1.8176	2581	139.13	0.28232	0.10832	D	0.343959
160	9900	60.215	1.0881	0.39616	1.8143	2593.3	139.17	0.28189	0.10893	D	0.343261
160	10000	60.34	1.0968	0.39556	1.811	2605.5	139.21	0.28146	0.10953	D	0.342546
170	14.696	0.095996	0.99705	0.21485	1.271	948.05	237.07	0.68752	0.017393	V	215.308578
170	100	0.66472	0.97978	0.22063	1.2916	939.08	235.38	0.59902	0.017438	D	31.093679
170	200	1.358	0.95917	0.22806	1.3185	928.55	233.34	0.5654	0.01751	D	15.219806
170	300	2.0827	0.93814	0.2363	1.3486	918.03	231.23	0.54467	0.017603	D	9.924073
170	400	2.842	0.91667	0.24547	1.3826	907.51	229.05	0.52915	0.017721	D	7.272715
170	500	3.6396	0.89471	0.25575	1.4211	897.04	226.79	0.51644	0.017869	D	5.678790
170	600	4.48	0.87226	0.26732	1.4648	886.63	224.45	0.50544	0.01805	D	4.613582
170	700	5.368	0.84929	0.28041	1.5147	876.34	222.01	0.49558	0.018269	D	3.850362
170	800	6.3096	0.82576	0.29532	1.572	866.25	219.48	0.48649	0.018534	D	3.275725
170	900	7.3116	0.80168	0.31239	1.638	856.42	216.83	0.47796	0.018851	D	2.826846
170	1000	8.3817	0.77703	0.33205	1.7143	847	214.06	0.4698	0.019231	D	2.465933
170	1100	9.529	0.75182	0.35477	1.803	838.13	211.16	0.46191	0.019686	D	2.169026
170	1200	10.764	0.7261	0.38114	1.9061	830.02	208.12	0.45417	0.020229	D	1.920254

TABLE F.1 (continued)—CO_2 PROPERTIES AT VARIOUS TEMPERATURES, °F											
Temperature (°F)	Pressure (psia)	Density (lbm/ft^3)	Compressibility Factor	Heat Capacity (Btu/lbm°F)	Heat Capacity Ratio (C_p/C_v)	Sonic Velocity (ft/s)	Enthalpy (Btu/lbm)	Entropy [Btu/(lbm°F)]	Viscosity (cp)	Phase	Gas Formation Volume Factor (res bbl/Mcf)
170	1300	12.097	0.69993	0.41176	2.0261	822.95	204.93	0.44653	0.020879	D	1.708657
170	1400	13.539	0.67344	0.44718	2.1649	817.29	201.58	0.43891	0.021657	D	1.526562
170	1500	15.103	0.64684	0.48772	2.3238	813.5	198.08	0.43129	0.022587	D	1.368514
170	1600	16.795	0.62046	0.53305	2.5012	812.23	194.42	0.42365	0.023696	D	1.230658
170	1700	18.614	0.5948	0.58168	2.6913	814.24	190.65	0.41599	0.025008	D	1.110365
170	1800	20.549	0.57049	0.63043	2.8816	820.44	186.82	0.4084	0.02654	D	1.005817
170	1900	22.568	0.54833	0.67447	3.0535	831.69	182.99	0.40095	0.02829	D	0.915866
170	2000	24.621	0.52906	0.70866	3.1871	848.61	179.27	0.39381	0.030232	D	0.839496
170	2100	26.648	0.51325	0.72945	3.2684	871.29	175.77	0.38709	0.032314	D	0.775628
170	2200	28.593	0.50111	0.73622	3.2947	899.24	172.54	0.3809	0.034469	D	0.722860
170	2300	30.413	0.49254	0.7312	3.2739	931.4	169.64	0.37529	0.03663	D	0.679606
170	2400	32.086	0.48716	0.71817	3.2208	966.52	167.06	0.37026	0.038746	D	0.644175
170	2500	33.609	0.48446	0.7008	3.1501	1003.4	164.79	0.36577	0.040785	D	0.614981
170	2600	34.989	0.48397	0.68169	3.0721	1041.3	162.8	0.36174	0.042731	D	0.590729
170	2700	36.239	0.48524	0.66242	2.9932	1079.4	161.05	0.35813	0.044579	D	0.570343
170	2800	37.374	0.48793	0.64388	2.9171	1117.3	159.49	0.35487	0.046331	D	0.553023
170	2900	38.409	0.49174	0.62655	2.8456	1154.6	158.12	0.3519	0.047993	D	0.538122
170	3000	39.356	0.49646	0.61059	2.7795	1191.2	156.89	0.34919	0.049571	D	0.525178
170	3100	40.227	0.50189	0.59604	2.7189	1226.9	155.78	0.3467	0.051073	D	0.513796
170	3200	41.033	0.50791	0.5828	2.6636	1261.6	154.79	0.3444	0.052507	D	0.503710
170	3300	41.78	0.51441	0.57075	2.613	1295.4	153.88	0.34225	0.05388	D	0.494697
170	3400	42.478	0.5213	0.55977	2.5666	1328.3	153.06	0.34025	0.055197	D	0.486578
170	3500	43.13	0.52852	0.54973	2.524	1360.3	152.31	0.33837	0.056463	D	0.479222
170	3600	43.743	0.536	0.54053	2.4848	1391.5	151.63	0.3366	0.057684	D	0.472504
170	3700	44.321	0.54371	0.53207	2.4485	1421.8	150.99	0.33493	0.058864	D	0.466347
170	3800	44.867	0.55161	0.52428	2.4149	1451.3	150.41	0.33335	0.060005	D	0.460672
170	3900	45.384	0.55967	0.51709	2.3838	1480.1	149.87	0.33184	0.061112	D	0.455419
170	4000	45.876	0.56787	0.51043	2.3548	1508.2	149.38	0.33041	0.062188	D	0.450539
170	4100	46.344	0.57618	0.50426	2.3278	1535.6	148.91	0.32903	0.063234	D	0.445983
170	4200	46.791	0.5846	0.49852	2.3026	1562.3	148.48	0.32772	0.064254	D	0.441726
170	4300	47.219	0.5931	0.49318	2.279	1588.4	148.08	0.32646	0.065249	D	0.437727
170	4400	47.629	0.60167	0.48819	2.2569	1613.9	147.71	0.32525	0.06622	D	0.433960
170	4500	48.022	0.6103	0.48353	2.2362	1638.9	147.36	0.32408	0.067171	D	0.430402
170	4600	48.4	0.61890	0.47917	2.2166	1663.3	147.04	0.32295	0.068102	D	0.427034
170	4700	48.765	0.62771	0.47507	2.1982	1687.2	146.73	0.32186	0.069014	D	0.423843
170	4800	49.116	0.63648	0.47122	2.1809	1710.6	146.45	0.32081	0.06991	D	0.420811
170	4900	49.456	0.64528	0.4676	2.1645	1733.6	146.18	0.31979	0.070789	D	0.417922
170	5000	49.784	0.65411	0.46418	2.1489	1756.1	145.93	0.3188	0.071653	D	0.415168
170	5100	50.102	0.66296	0.46095	2.1342	1778.1	145.7	0.31784	0.072504	D	0.412535
170	5200	50.41	0.67183	0.45789	2.1202	1799.8	145.48	0.31691	0.07334	D	0.410015
170	5300	50.708	0.68072	0.455	2.1068	1821	145.27	0.316	0.074165	D	0.407602
170	5400	50.998	0.68962	0.45225	2.0942	1841.9	145.08	0.31511	0.074978	D	0.405284
170	5500	51.28	0.69852	0.44964	2.082	1862.5	144.9	0.31425	0.075779	D	0.403051
170	5600	51.555	0.70744	0.44715	2.0705	1882.6	144.73	0.31341	0.07657	D	0.400908
170	5700	51.822	0.71636	0.44478	2.0594	1902.5	144.57	0.31258	0.077351	D	0.398841
170	5800	52.082	0.72529	0.44253	2.0488	1922	144.42	0.31178	0.078123	D	0.396851

TABLE F.1 (continued)—CO_2 PROPERTIES AT VARIOUS TEMPERATURES, °F												
Temperature (°F)	Pressure (psia)	Density (lbm/ft³)	Compressibility Factor	Heat Capacity (Btu/lbm°F)	Heat Capacity Ratio (C_p/C_v)	Sonic Velocity (ft/s)	Enthalpy (Btu/lbm)	Entropy [Btu/(lbm°F)]	Viscosity (cp)	Phase	Gas Formation Volume Factor (res bbl/Mcf)
170	5900	52.336	0.73421	0.44037	2.0386	1941.3	144.28	0.31099	0.078886	D	0.394922
170	6000	52.584	0.74314	0.43831	2.0289	1960.2	144.15	0.31023	0.07964	D	0.393064
170	6100	52.825	0.75207	0.43633	2.0195	1978.8	144.02	0.30947	0.080386	D	0.391266
170	6200	53.062	0.761	0.43444	2.0105	1997.2	143.91	0.30874	0.081125	D	0.389526
170	6300	53.292	0.76992	0.43263	2.0018	2015.3	143.8	0.30801	0.081856	D	0.387836
170	6400	53.518	0.77884	0.43088	1.9935	2033.2	143.7	0.3073	0.08258	D	0.386200
170	6500	53.739	0.78776	0.42921	1.9854	2050.8	143.61	0.30661	0.083297	D	0.384613
170	6600	53.956	0.79667	0.4276	1.9776	2068.2	143.52	0.30593	0.084008	D	0.383070
170	6700	54.168	0.80557	0.42605	1.9701	2085.4	143.44	0.30526	0.084713	D	0.381568
170	6800	54.376	0.81447	0.42456	1.9628	2102.3	143.37	0.3046	0.085412	D	0.380110
170	6900	54.579	0.82336	0.42312	1.9558	2119	143.3	0.30395	0.086105	D	0.378690
170	7000	54.779	0.83225	0.42173	1.949	2135.5	143.24	0.30331	0.086792	D	0.377311
170	7100	54.975	0.84112	0.42039	1.9424	2151.8	143.18	0.30269	0.087475	D	0.375961
170	7200	55.168	0.84999	0.41909	1.936	2168	143.13	0.30207	0.088153	D	0.374649
170	7300	55.357	0.85885	0.41784	1.9298	2183.9	143.09	0.30147	0.088825	D	0.373369
170	7400	55.543	0.86771	0.41663	1.9237	2199.6	143.04	0.30087	0.089493	D	0.372123
170	7500	55.726	0.87655	0.41546	1.9178	2215.2	143.01	0.30028	0.090157	D	0.370902
170	7600	55.905	0.88538	0.41432	1.9121	2230.6	142.97	0.2997	0.090816	D	0.369709
170	7700	56.082	0.89421	0.41322	1.9066	2245.8	142.94	0.29913	0.091472	D	0.368547
170	7800	56.255	0.90303	0.41215	1.9012	2260.9	142.92	0.29857	0.092123	D	0.367410
170	7900	56.426	0.91183	0.41111	1.8959	2275.8	142.9	0.29801	0.09277	D	0.366294
170	8000	56.595	0.92063	0.41011	1.8908	2290.6	142.88	0.29746	0.093414	D	0.365207
170	8100	56.76	0.92942	0.40913	1.8858	2305.2	142.87	0.29692	0.094054	D	0.364142
170	8200	56.924	0.93819	0.40818	1.8809	2319.6	142.86	0.29639	0.09469	D	0.363095
170	8300	57.084	0.94696	0.40726	1.8762	2333.9	142.85	0.29586	0.095323	D	0.362074
170	8400	57.243	0.95572	0.40636	1.8715	2348.1	142.85	0.29534	0.095953	D	0.361073
170	8500	57.399	0.96446	0.40549	1.867	2362.1	142.85	0.29483	0.096579	D	0.360088
170	8600	57.553	0.9732	0.40464	1.8626	2376	142.85	0.29432	0.097203	D	0.359126
170	8700	57.705	0.98193	0.40381	1.8582	2389.8	142.85	0.29382	0.097824	D	0.358183
170	8800	57.854	0.99064	0.403	1.854	2403.5	142.86	0.29332	0.098441	D	0.357254
170	8900	58.002	0.99935	0.40222	1.8498	2417	142.87	0.29283	0.099056	D	0.356345
170	9000	58.148	1.008	0.40145	1.8458	2430.4	142.89	0.29235	0.099669	D	0.355436
170	9100	58.292	1.0167	0.40071	1.8418	2443.7	142.91	0.29187	0.10028	D	0.354564
170	9200	58.434	1.0254	0.39998	1.8379	2456.8	142.92	0.2914	0.10089	D	0.353711
170	9300	58.574	1.0341	0.39927	1.8341	2469.9	142.95	0.29093	0.10149	D	0.352877
170	9400	58.712	1.0427	0.39857	1.8304	2482.8	142.97	0.29047	0.10209	D	0.352026
170	9500	58.849	1.0514	0.3979	1.8267	2495.7	143	0.29001	0.10269	D	0.351227
170	9600	58.984	1.06	0.39723	1.8231	2508.4	143.03	0.28956	0.10329	D	0.350411
170	9700	59.117	1.0686	0.39659	1.8196	2521	143.06	0.28911	0.10388	D	0.349613
170	9800	59.249	1.0772	0.39596	1.8161	2533.5	143.09	0.28866	0.10448	D	0.348830
170	9900	59.38	1.0858	0.39534	1.8127	2546	143.13	0.28822	0.10507	D	0.348063
170	10000	59.508	1.0944	0.39473	1.8094	2558.3	143.16	0.28779	0.10566	D	0.347312
180	14.696	0.09448	0.99721	0.21603	1.2689	954.91	239.23	0.69091	0.017653	V	218.763066
180	100	0.65361	0.98087	0.2215	1.2884	946.41	237.59	0.6025	0.017697	D	31.622629
180	200	1.3337	0.96141	0.22849	1.3138	936.47	235.62	0.569	0.017766	D	15.497625
180	300	2.0426	0.94161	0.23619	1.342	926.56	233.6	0.54839	0.017857	D	10.118970

TABLE F.1 (continued)—CO_2 PROPERTIES AT VARIOUS TEMPERATURES, °F											
Temperature (°F)	Pressure (psia)	Density (lbm/ft³)	Compressibility Factor	Heat Capacity (Btu/lbm°F)	Heat Capacity Ratio (C_p/C_v)	Sonic Velocity (ft/s)	Enthalpy (Btu/lbm)	Entropy [Btu/(lbm°F)]	Viscosity (cp)	Phase	Gas Formation Volume Factor (res bbl/Mcf)
180	400	2.783	0.92144	0.2447	1.3735	916.71	231.5	0.53302	0.017971	D	7.426661
180	500	3.5581	0.90089	0.25415	1.4089	906.94	229.34	0.52046	0.018112	D	5.808825
180	600	4.3714	0.87995	0.26469	1.4486	897.28	227.11	0.50963	0.018284	D	4.728172
180	700	5.2268	0.8586	0.27647	1.4935	887.79	224.8	0.49996	0.018492	D	3.954389
180	800	6.1289	0.83683	0.28972	1.5443	878.52	222.4	0.4911	0.01874	D	3.372359
180	900	7.0827	0.81465	0.30466	1.602	869.55	219.91	0.48282	0.019035	D	2.918200
180	1000	8.0942	0.79206	0.32158	1.6677	861	217.32	0.47495	0.019384	D	2.553551
180	1100	9.1695	0.76908	0.34077	1.7426	853	214.63	0.46738	0.019797	D	2.254059
180	1200	10.316	0.74577	0.36259	1.828	845.73	211.83	0.46003	0.020285	D	2.003596
180	1300	11.54	0.7222	0.38737	1.9253	839.39	208.92	0.45282	0.020859	D	1.791021
180	1400	12.85	0.69847	0.41539	2.0354	834.28	205.88	0.4457	0.021535	D	1.608445
180	1500	14.252	0.67476	0.4468	2.1591	830.72	202.74	0.43864	0.022329	D	1.450256
180	1600	15.75	0.65128	0.48143	2.2955	829.16	199.48	0.43162	0.023258	D	1.312303
180	1700	17.345	0.62835	0.5186	2.4418	830.1	196.14	0.42463	0.024338	D	1.191624
180	1800	19.03	0.6064	0.55683	2.5924	834.13	192.73	0.41772	0.025584	D	1.086108
180	1900	20.79	0.58589	0.5938	2.738	841.85	189.31	0.41092	0.027	D	0.994143
180	2000	22.6	0.56736	0.62657	2.8673	853.77	185.93	0.4043	0.028578	D	0.914566
180	2100	24.423	0.55126	0.65239	2.9692	870.17	182.67	0.39796	0.030296	D	0.846299
180	2200	26.22	0.53792	0.66945	3.0366	891	179.57	0.39197	0.032118	D	0.788282
180	2300	27.956	0.52745	0.67742	3.0677	915.88	176.69	0.3864	0.034002	D	0.739333
180	2400	29.601	0.51979	0.67732	3.0662	944.14	174.05	0.38127	0.035904	D	0.698238
180	2500	31.139	0.51471	0.67106	3.0395	974.97	171.66	0.37659	0.037787	D	0.663757
180	2600	32.564	0.51188	0.66075	2.9962	1007.6	169.52	0.37234	0.039626	D	0.634719
180	2700	33.876	0.51098	0.64816	2.9434	1041.3	167.61	0.36847	0.041404	D	0.610136
180	2800	35.082	0.51169	0.63456	2.8864	1075.6	165.9	0.36495	0.043114	D	0.589163
180	2900	36.19	0.51373	0.62078	2.8285	1110.1	164.36	0.36174	0.044752	D	0.571115
180	3000	37.212	0.51686	0.60732	2.7718	1144.4	162.98	0.3588	0.04632	D	0.555441
180	3100	38.155	0.52088	0.59448	2.7176	1178.3	161.74	0.35609	0.04782	D	0.541705
180	3200	39.028	0.52565	0.58241	2.6665	1211.7	160.62	0.35358	0.049257	D	0.529582
180	3300	39.84	0.53103	0.57117	2.6187	1244.5	159.6	0.35125	0.050636	D	0.518790
180	3400	40.597	0.53692	0.56074	2.5742	1276.5	158.67	0.34908	0.05196	D	0.509117
180	3500	41.306	0.54323	0.55109	2.5329	1307.8	157.82	0.34705	0.053235	D	0.500383
180	3600	41.971	0.5499	0.54216	2.4945	1338.4	157.04	0.34514	0.054465	D	0.492456
180	3700	42.597	0.55686	0.53387	2.4587	1368.3	156.33	0.34333	0.055653	D	0.485211
180	3800	43.189	0.56408	0.52618	2.4254	1397.5	155.67	0.34163	0.056803	D	0.478568
180	3900	43.749	0.57151	0.51903	2.3943	1426.1	155.06	0.34001	0.057910	D	0.472430
180	4000	44.281	0.57912	0.51237	2.3652	1454	154.49	0.33847	0.059	D	0.466762
180	4100	44.787	0.5869	0.50616	2.338	1481.3	153.97	0.337	0.060053	D	0.461495
180	4200	45.269	0.59481	0.50036	2.3124	1508	153.48	0.33559	0.061077	D	0.456579
180	4300	45.729	0.60284	0.49494	2.2884	1534.1	153.03	0.33425	0.062077	D	0.451981
180	4400	46.17	0.61097	0.48986	2.2658	1559.7	152.6	0.33295	0.063052	D	0.447666
180	4500	46.593	0.61918	0.48509	2.2446	1584.7	152.21	0.33171	0.064005	D	0.443599
180	4600	46.999	0.62748	0.48061	2.2246	1609.3	151.84	0.33051	0.064938	D	0.439773
180	4700	47.389	0.63583	0.4764	2.2056	1633.3	151.49	0.32936	0.065852	D	0.436144
180	4800	47.765	0.64425	0.47244	2.1878	1656.9	151.17	0.32825	0.066747	D	0.432713
180	4900	48.128	0.65272	0.4687	2.1708	1680	150.86	0.32717	0.067626	D	0.429455

TABLE F.1 (continued)—CO_2 PROPERTIES AT VARIOUS TEMPERATURES, °F											
Temperature (°F)	Pressure (psia)	Density (lbm/ft³)	Compressibility Factor	Heat Capacity (Btu/lbm°F)	Heat Capacity Ratio (C_p/C_v)	Sonic Velocity (ft/s)	Enthalpy (Btu/lbm)	Entropy [Btu/(lbm°F)]	Viscosity (cp)	Phase	Gas Formation Volume Factor (res bbl/Mcf)
180	5000	48.478	0.66122	0.46517	2.1548	1702.7	150.58	0.32612	0.068489	D	0.426346
180	5100	48.817	0.66977	0.46184	2.1396	1724.9	150.31	0.32511	0.069338	D	0.423391
180	5200	49.145	0.67835	0.45868	2.1251	1746.8	150.06	0.32413	0.070172	D	0.420569
180	5300	49.463	0.68695	0.45568	2.1114	1768.2	149.82	0.32317	0.070993	D	0.417865
180	5400	49.771	0.69558	0.45284	2.0983	1789.3	149.6	0.32224	0.071802	D	0.415279
180	5500	50.07	0.70422	0.45014	2.0858	1810.1	149.4	0.32134	0.0726	D	0.412793
180	5600	50.361	0.71289	0.44757	2.0738	1830.5	149.2	0.32045	0.073386	D	0.410413
180	5700	50.644	0.72156	0.44512	2.0624	1850.5	149.02	0.31959	0.074162	D	0.408116
180	5800	50.919	0.73025	0.44279	2.0515	1870.3	148.84	0.31875	0.074928	D	0.405910
180	5900	51.187	0.73895	0.44056	2.041	1889.7	148.68	0.31793	0.075685	D	0.403784
180	6000	51.449	0.74766	0.43843	2.031	1908.9	148.53	0.31713	0.076433	D	0.401735
180	6100	51.704	0.75637	0.43639	2.0213	1927.7	148.39	0.31635	0.077172	D	0.399752
180	6200	51.953	0.76508	0.43444	2.0121	1946.3	148.25	0.31558	0.077903	D	0.397834
180	6300	52.196	0.7738	0.43257	2.0032	1964.6	148.13	0.31483	0.078627	D	0.395981
180	6400	52.434	0.78252	0.43077	1.9946	1982.7	148.01	0.31409	0.079343	D	0.394187
180	6500	52.667	0.79124	0.42905	1.9863	2000.5	147.9	0.31337	0.080053	D	0.392447
180	6600	52.894	0.79995	0.4274	1.9783	2018	147.8	0.31266	0.080755	D	0.390756
180	6700	53.117	0.80867	0.4258	1.9706	2035.4	147.7	0.31197	0.081452	D	0.389120
180	6800	53.335	0.81738	0.42427	1.9632	2052.5	147.62	0.31129	0.082142	D	0.387527
180	6900	53.549	0.82609	0.42279	1.956	2069.4	147.53	0.31061	0.082826	D	0.385980
180	7000	53.758	0.8348	0.42137	1.949	2086.1	147.46	0.30996	0.083505	D	0.384477
180	7100	53.964	0.8435	0.41999	1.9422	2102.6	147.39	0.30931	0.084178	D	0.383013
180	7200	54.165	0.85219	0.41866	1.9357	2118.9	147.32	0.30867	0.084846	D	0.381584
180	7300	54.363	0.86089	0.41738	1.9293	2135	147.26	0.30805	0.08551	D	0.380199
180	7400	54.558	0.86957	0.41614	1.9232	2150.9	147.21	0.30743	0.086168	D	0.378843
180	7500	54.748	0.87825	0.41494	1.9172	2166.6	147.16	0.30682	0.086822	D	0.377523
180	7600	54.936	0.88692	0.41378	1.9114	2182.2	147.11	0.30622	0.087471	D	0.376233
180	7700	55.12	0.89558	0.41265	1.9057	2197.5	147.07	0.30564	0.088116	D	0.374973
180	7800	55.302	0.90424	0.41156	1.9002	2212.8	147.04	0.30506	0.088757	D	0.373745
180	7900	55.48	0.91289	0.4105	1.8949	2227.8	147.01	0.30448	0.089394	D	0.372544
180	8000	55.656	0.92153	0.40948	1.8897	2242.7	146.98	0.30392	0.090027	D	0.371369
180	8100	55.828	0.93016	0.40848	1.8846	2257.4	146.96	0.30336	0.090656	D	0.370219
180	8200	55.998	0.93879	0.40751	1.8797	2272	146.94	0.30281	0.091282	D	0.369098
180	8300	56.166	0.9474	0.40657	1.8748	2286.5	146.92	0.30227	0.091904	D	0.367995
180	8400	56.331	0.95601	0.40566	1.8701	2300.8	146.91	0.30174	0.092523	D	0.366919
180	8500	56.493	0.96461	0.40477	1.8655	2315	146.9	0.30121	0.093139	D	0.365864
180	8600	56.653	0.9732	0.40391	1.861	2329	146.89	0.30069	0.093752	D	0.364830
180	8700	56.811	0.98178	0.40307	1.8566	2342.9	146.89	0.30018	0.094361	D	0.363816
180	8800	56.967	0.99035	0.40225	1.8523	2356.7	146.89	0.29967	0.094968	D	0.362821
180	8900	57.12	0.99892	0.40145	1.8481	2370.3	146.89	0.29917	0.095572	D	0.361849
180	9000	57.271	1.0075	0.40067	1.844	2383.8	146.9	0.29867	0.096173	D	0.360902
180	9100	57.421	1.016	0.39992	1.84	2397.2	146.91	0.29818	0.096771	D	0.359947
180	9200	57.568	1.0245	0.39918	1.8361	2410.5	146.92	0.29769	0.097367	D	0.359013
180	9300	57.714	1.0331	0.39846	1.8322	2423.7	146.93	0.29721	0.09796	D	0.358134
180	9400	57.857	1.0416	0.39775	1.8284	2436.7	146.95	0.29674	0.098551	D	0.357240
180	9500	57.999	1.0501	0.39707	1.8247	2449.7	146.97	0.29627	0.099139	D	0.356364

TABLE F.1 (continued)—CO_2 PROPERTIES AT VARIOUS TEMPERATURES, °F											
Temperature (°F)	Pressure (psia)	Density (lbm/ft^3)	Compressibility Factor	Heat Capacity (Btu/lbm°F)	Heat Capacity Ratio (C_p/C_v)	Sonic Velocity (ft/s)	Enthalpy (Btu/lbm)	Entropy [Btu/(lbm°F)]	Viscosity (cp)	Phase	Gas Formation Volume Factor (res bbl/Mcf)
180	9600	58.139	1.0586	0.39639	1.8211	2462.5	146.99	0.29581	0.099725	D	0.355506
180	9700	58.277	1.0671	0.39574	1.8176	2475.3	147.02	0.29535	0.10031	D	0.354666
180	9800	58.414	1.0756	0.3951	1.8141	2487.9	147.05	0.29489	0.10089	D	0.353844
180	9900	58.549	1.084	0.39448	1.8106	2500.4	147.07	0.29444	0.10147	D	0.353005
180	10000	58.682	1.0925	0.39386	1.8073	2512.8	147.11	0.294	0.10205	D	0.352215
190	14.696	0.093012	0.99735	0.2172	1.2668	961.71	241.39	0.69427	0.017911	V	222.214195
190	100	0.64288	0.98188	0.22238	1.2854	953.65	239.81	0.60595	0.017954	D	32.150058
190	200	1.3103	0.9635	0.22897	1.3093	944.26	237.91	0.57254	0.018022	D	15.774118
190	300	2.0043	0.94483	0.23619	1.3358	934.94	235.96	0.55206	0.01811	D	10.312305
190	400	2.7271	0.92588	0.24411	1.3651	925.71	233.95	0.53681	0.01822	D	7.579107
190	500	3.4813	0.90662	0.25283	1.3977	916.58	231.88	0.52439	0.018355	D	5.937158
190	600	4.2697	0.88704	0.26247	1.4341	907.61	229.75	0.51372	0.01852	D	4.840780
190	700	5.0955	0.86716	0.27315	1.4747	898.83	227.55	0.50422	0.018716	D	4.056248
190	800	5.9624	0.84696	0.28502	1.5202	890.29	225.27	0.49556	0.01895	D	3.466540
190	900	6.8741	0.82645	0.29825	1.5712	882.09	222.92	0.48749	0.019226	D	3.006751
190	1000	7.835	0.80566	0.31301	1.6285	874.31	220.5	0.47987	0.019549	D	2.638002
190	1100	8.8497	0.78461	0.32951	1.6928	867.06	217.98	0.47258	0.019928	D	2.335525
190	1200	9.9231	0.76335	0.34796	1.765	860.5	215.38	0.46553	0.020371	D	2.082887
190	1300	11.06	0.74195	0.36855	1.8458	854.79	212.69	0.45867	0.020885	D	1.868765
190	1400	12.265	0.72051	0.39144	1.9358	850.16	209.91	0.45195	0.021484	D	1.685137
190	1500	13.543	0.69914	0.41668	2.0353	846.87	207.05	0.44532	0.022176	D	1.526147
190	1600	14.896	0.67804	0.44417	2.1438	845.21	204.1	0.43878	0.022976	D	1.387582
190	1700	16.323	0.65741	0.47353	2.2598	845.56	201.08	0.43231	0.023893	D	1.266225
190	1800	17.822	0.63753	0.504	2.3803	848.31	198.02	0.42592	0.024939	D	1.159716
190	1900	19.384	0.61872	0.53437	2.5004	853.88	194.94	0.41964	0.026118	D	1.066262
190	2000	20.994	0.60135	0.56302	2.6138	862.66	191.87	0.41351	0.027429	D	0.984511
190	2100	22.631	0.58574	0.58818	2.7134	874.92	188.86	0.40756	0.028864	D	0.913290
190	2200	24.27	0.5722	0.60829	2.793	890.77	185.95	0.40187	0.030404	D	0.851625
190	2300	25.884	0.5609	0.62241	2.8487	910.13	183.18	0.39648	0.032024	D	0.798511
190	2400	27.449	0.55192	0.63034	2.8795	932.67	180.59	0.39142	0.033695	D	0.752988
190	2500	28.946	0.54519	0.63261	2.8873	957.96	178.19	0.38671	0.035387	D	0.714054
190	2600	30.361	0.54056	0.63025	2.8761	985.45	175.98	0.38236	0.037075	D	0.680760
190	2700	31.689	0.53783	0.62446	2.8508	1014.6	173.98	0.37835	0.038738	D	0.652236
190	2800	32.928	0.53677	0.61641	2.8162	1044.9	172.16	0.37466	0.040362	D	0.627702
190	2900	34.08	0.53714	0.60704	2.776	1075.9	170.51	0.37127	0.041939	D	0.606475
190	3000	35.152	0.53872	0.59702	2.7331	1107.4	169.01	0.36815	0.043464	D	0.587984
190	3100	36.148	0.54134	0.58683	2.6895	1138.9	167.65	0.36526	0.044934	D	0.571784
190	3200	37.075	0.54482	0.57678	2.6463	1170.3	166.42	0.36258	0.046351	D	0.557476
190	3300	37.94	0.54904	0.56705	2.6045	1201.5	165.29	0.36009	0.047717	D	0.544770
190	3400	38.749	0.55387	0.55776	2.5645	1232.2	164.27	0.35776	0.049033	D	0.533399
190	3500	39.507	0.55922	0.54897	2.5265	1262.5	163.32	0.35559	0.050304	D	0.523164
190	3600	40.219	0.56501	0.5407	2.4907	1292.2	162.46	0.35354	0.051531	D	0.513898
190	3700	40.891	0.57117	0.53293	2.4569	1321.4	161.66	0.35161	0.052718	D	0.505460
190	3800	41.525	0.57766	0.52564	2.4251	1350	160.93	0.34979	0.053868	D	0.497751
190	3900	42.125	0.58441	0.5188	2.3952	1378	160.25	0.34806	0.054983	D	0.490655
190	4000	42.694	0.5914	0.51239	2.367	1405.5	159.62	0.34642	0.056067	D	0.484111

TABLE F.1 (continued)—CO_2 PROPERTIES AT VARIOUS TEMPERATURES, °F

Temperature (°F)	Pressure (psia)	Density (lbm/ft^3)	Compressibility Factor	Heat Capacity (Btu/lbm°F)	Heat Capacity Ratio (C_p/C_v)	Sonic Velocity (ft/s)	Enthalpy (Btu/lbm)	Entropy [Btu/(lbm°F)]	Viscosity (cp)	Phase	Gas Formation Volume Factor (res bbl/Mcf)
190	4100	43.236	0.59859	0.50636	2.3405	1432.5	159.03	0.34485	0.05712	D	0.478045
190	4200	43.752	0.60596	0.50069	2.3154	1458.9	158.49	0.34336	0.058146	D	0.472409
190	4300	44.245	0.61348	0.49536	2.2918	1484.8	157.98	0.34193	0.059146	D	0.467149
190	4400	44.716	0.62113	0.49034	2.2694	1510.2	157.5	0.34056	0.060122	D	0.462225
190	4500	45.167	0.6289	0.4856	2.2482	1535.1	157.06	0.33924	0.061076	D	0.457607
190	4600	45.6	0.63677	0.48113	2.2282	1559.6	156.65	0.33798	0.062008	D	0.453261
190	4700	46.016	0.64473	0.47692	2.2092	1583.6	156.26	0.33676	0.062921	D	0.449162
190	4800	46.417	0.65276	0.47293	2.1912	1607.2	155.89	0.33558	0.063816	D	0.445283
190	4900	46.803	0.66087	0.46916	2.1741	1630.3	155.55	0.33444	0.064693	D	0.441614
190	5000	47.175	0.66903	0.46558	2.1579	1653	155.23	0.33334	0.065554	D	0.438126
190	5100	47.535	0.67725	0.4622	2.1425	1675.3	154.93	0.33228	0.066399	D	0.434813
190	5200	47.883	0.68551	0.45899	2.1278	1697.3	154.65	0.33125	0.067231	D	0.431652
190	5300	48.22	0.69381	0.45594	2.1138	1718.8	154.38	0.33024	0.068049	D	0.428635
190	5400	48.546	0.70215	0.45304	2.1004	1740.1	154.13	0.32927	0.068854	D	0.425755
190	5500	48.863	0.71052	0.45029	2.0876	1760.9	153.9	0.32832	0.069647	D	0.422997
190	5600	49.17	0.71892	0.44766	2.0755	1781.4	153.68	0.3274	0.070428	D	0.420355
190	5700	49.469	0.72733	0.44516	2.0638	1801.6	153.47	0.3265	0.071199	D	0.417811
190	5800	49.759	0.73577	0.44277	2.0526	1821.5	153.27	0.32562	0.07196	D	0.415372
190	5900	50.042	0.74423	0.44049	2.042	1841.1	153.09	0.32477	0.07271	D	0.413027
190	6000	50.318	0.7527	0.43831	2.0317	1860.4	152.91	0.32393	0.073452	D	0.410766
190	6100	50.586	0.76118	0.43623	2.0219	1879.4	152.75	0.32312	0.074185	D	0.408584
190	6200	50.848	0.76967	0.43423	2.0124	1898.1	152.6	0.32232	0.074909	D	0.406477
190	6300	51.104	0.77817	0.43232	2.0033	1916.6	152.45	0.32154	0.075626	D	0.404443
190	6400	51.354	0.78668	0.43048	1.9945	1934.8	152.32	0.32077	0.076335	D	0.402477
190	6500	51.598	0.79519	0.42872	1.9861	1952.7	152.19	0.32002	0.077036	D	0.400572
190	6600	51.836	0.80371	0.42703	1.9779	1970.5	152.07	0.31929	0.077731	D	0.398730
190	6700	52.07	0.81223	0.4254	1.9701	1987.9	151.96	0.31857	0.078419	D	0.396942
190	6800	52.298	0.82075	0.42383	1.9625	2005.2	151.86	0.31786	0.079101	D	0.395208
190	6900	52.522	0.82927	0.42232	1.9552	2022.2	151.76	0.31717	0.079777	D	0.393523
190	7000	52.741	0.83779	0.42087	1.9481	2039.1	151.67	0.31649	0.080447	D	0.391887
190	7100	52.956	0.84631	0.41946	1.9412	2055.7	151.58	0.31582	0.081111	D	0.390296
190	7200	53.167	0.85483	0.41811	1.9345	2072.1	151.51	0.31516	0.08177	D	0.388750
190	7300	53.374	0.86334	0.4168	1.9281	2088.4	151.43	0.31452	0.082424	D	0.387242
190	7400	53.577	0.87186	0.41554	1.9218	2104.4	151.37	0.31388	0.083073	D	0.385779
190	7500	53.776	0.88036	0.41431	1.9157	2120.3	151.3	0.31325	0.083717	D	0.384346
190	7600	53.972	0.88887	0.41313	1.9098	2135.9	151.25	0.31264	0.084357	D	0.382955
190	7700	54.164	0.89737	0.41199	1.9041	2151.4	151.2	0.31203	0.084992	D	0.381596
190	7800	54.353	0.90586	0.41088	1.8985	2166.8	151.15	0.31143	0.085623	D	0.380268
190	7900	54.539	0.91435	0.4098	1.8931	2182	151.11	0.31085	0.08625	D	0.378973
190	8000	54.722	0.92283	0.40876	1.8878	2197	151.07	0.31027	0.086873	D	0.377707
190	8100	54.901	0.93131	0.40775	1.8827	2211.9	151.04	0.30969	0.087492	D	0.376472
190	8200	55.078	0.93978	0.40676	1.8777	2226.6	151.01	0.30913	0.088108	D	0.375263
190	8300	55.252	0.94824	0.40581	1.8728	2241.1	150.98	0.30857	0.08872	D	0.374079
190	8400	55.424	0.9567	0.40488	1.868	2255.5	150.96	0.30803	0.089329	D	0.372924
190	8500	55.592	0.96515	0.40398	1.8634	2269.8	150.94	0.30748	0.089934	D	0.371791
190	8600	55.759	0.97359	0.40311	1.8588	2284	150.93	0.30695	0.090536	D	0.370682

TABLE F.1 (continued)—CO_2 PROPERTIES AT VARIOUS TEMPERATURES, °F											
Temperature (°F)	Pressure (psia)	Density (lbm/ft^3)	Compressibility Factor	Heat Capacity (Btu/lbm°F)	Heat Capacity Ratio (C_p/C_v)	Sonic Velocity (ft/s)	Enthalpy (Btu/lbm)	Entropy [Btu/(lbm°F)]	Viscosity (cp)	Phase	Gas Formation Volume Factor (res bbl/Mcf)
190	8700	55.923	0.98202	0.40225	1.8544	2298	150.92	0.30642	0.091135	D	0.369594
190	8800	56.084	0.99045	0.40142	1.85	2311.9	150.91	0.3059	0.091731	D	0.368530
190	8900	56.243	0.99887	0.40062	1.8458	2325.6	150.9	0.30539	0.092324	D	0.367487
190	9000	56.4	1.0073	0.39983	1.8417	2339.2	150.9	0.30488	0.092914	D	0.366471
190	9100	56.555	1.0157	0.39906	1.8376	2352.7	150.9	0.30438	0.093501	D	0.365466
190	9200	56.708	1.0241	0.39832	1.8336	2366.1	150.91	0.30388	0.094086	D	0.364484
190	9300	56.859	1.0325	0.39759	1.8297	2379.4	150.92	0.30339	0.094668	D	0.363522
190	9400	57.008	1.0408	0.39688	1.8259	2392.5	150.93	0.3029	0.095248	D	0.362546
190	9500	57.154	1.0492	0.39618	1.8222	2405.6	150.94	0.30242	0.095825	D	0.361625
190	9600	57.299	1.0576	0.3955	1.8185	2418.5	150.95	0.30195	0.0964	D	0.360723
190	9700	57.443	1.0659	0.39484	1.815	2431.3	150.97	0.30148	0.096973	D	0.359806
190	9800	57.584	1.0743	0.3942	1.8114	2444.1	150.99	0.30102	0.097543	D	0.358941
190	9900	57.724	1.0826	0.39357	1.808	2456.7	151.01	0.30056	0.098112	D	0.358060
190	10000	57.862	1.0909	0.39295	1.8046	2469.2	151.04	0.3001	0.098678	D	0.357197
200	14.696	0.09159	0.99749	0.21836	1.2648	968.46	243.57	0.6976	0.018168	V	225.666284
200	100	0.63252	0.98283	0.22328	1.2825	960.82	242.04	0.60935	0.01821	D	32.676511
200	200	1.2878	0.96545	0.2295	1.3051	951.95	240.2	0.57605	0.018276	D	16.049336
200	300	1.9676	0.94785	0.23627	1.3299	943.17	238.32	0.55566	0.018362	D	10.504506
200	400	2.6738	0.93001	0.24367	1.3573	934.51	236.39	0.54053	0.018468	D	7.730096
200	500	3.4085	0.91193	0.25175	1.3875	925.99	234.4	0.52824	0.018599	D	6.063854
200	600	4.1741	0.89361	0.26061	1.4209	917.64	232.36	0.51771	0.018756	D	4.951697
200	700	4.973	0.87505	0.27034	1.4579	909.5	230.26	0.50837	0.018943	D	4.156158
200	800	5.8082	0.85626	0.28106	1.4989	901.64	228.1	0.49988	0.019164	D	3.558549
200	900	6.6826	0.83725	0.29287	1.5444	894.11	225.88	0.492	0.019422	D	3.092929
200	1000	7.5994	0.81804	0.30591	1.5949	887	223.59	0.48459	0.019724	D	2.719768
200	1100	8.5621	0.79867	0.3203	1.6509	880.42	221.23	0.47754	0.020074	D	2.413970
200	1200	9.5741	0.77918	0.33617	1.7129	874.48	218.8	0.47076	0.020479	D	2.158807
200	1300	10.639	0.75964	0.35364	1.7814	869.33	216.3	0.46418	0.020946	D	1.942772
200	1400	11.759	0.74012	0.37279	1.8567	865.14	213.73	0.45778	0.021483	D	1.757646
200	1500	12.938	0.72073	0.39364	1.9389	862.11	211.09	0.45151	0.022098	D	1.597492
200	1600	14.177	0.70161	0.4161	2.0276	860.47	208.39	0.44534	0.0228	D	1.457918
200	1700	15.475	0.68292	0.43995	2.1219	860.49	205.64	0.43928	0.023596	D	1.335605
200	1800	16.831	0.66484	0.46476	2.22	862.45	202.86	0.43331	0.024495	D	1.228010
200	1900	18.238	0.64762	0.48983	2.3192	866.65	200.05	0.42745	0.025501	D	1.133245
200	2000	19.689	0.6315	0.51424	2.4159	873.38	197.24	0.42172	0.026615	D	1.049786
200	2100	21.168	0.61672	0.53689	2.5056	882.88	194.47	0.41614	0.027833	D	0.976396
200	2200	22.661	0.60352	0.55671	2.5841	895.28	191.76	0.41076	0.029148	D	0.912066
200	2300	24.149	0.59208	0.57278	2.6475	910.62	189.15	0.4056	0.030543	D	0.855874
200	2400	25.614	0.5825	0.58459	2.6938	928.76	186.66	0.40069	0.032001	D	0.806941
200	2500	27.038	0.57481	0.59204	2.7225	949.49	184.31	0.39606	0.033501	D	0.764437
200	2600	28.408	0.56897	0.59541	2.7346	972.47	182.11	0.39172	0.035023	D	0.727567
200	2700	29.715	0.56487	0.59528	2.7324	997.32	180.08	0.38767	0.036547	D	0.695572
200	2800	30.952	0.56237	0.59238	2.7188	1023.7	178.2	0.3839	0.038058	D	0.667762
200	2900	32.119	0.5613	0.58745	2.6967	1051.1	176.48	0.3804	0.039545	D	0.643509
200	3000	33.215	0.5615	0.58115	2.6691	1079.4	174.9	0.37715	0.040999	D	0.622280
200	3100	34.243	0.5628	0.57401	2.6379	1108.2	173.46	0.37413	0.042415	D	0.603601

TABLE F.1 (continued)—CO_2 PROPERTIES AT VARIOUS TEMPERATURES, °F											
Temperature (°F)	Pressure (psia)	Density (lbm/ft³)	Compressibility Factor	Heat Capacity (Btu/lbm°F)	Heat Capacity Ratio (C_p/C_v)	Sonic Velocity (ft/s)	Enthalpy (Btu/lbm)	Entropy [Btu/(lbm°F)]	Viscosity (cp)	Phase	Gas Formation Volume Factor (res bbl/Mcf)
200	3200	35.206	0.56505	0.56643	2.6049	1137.2	172.14	0.37132	0.04379	D	0.587076
200	3300	36.11	0.56812	0.55869	2.5712	1166.3	170.92	0.36869	0.045124	D	0.572379
200	3400	36.959	0.57189	0.551	2.5377	1195.4	169.81	0.36624	0.046416	D	0.559231
200	3500	37.758	0.57626	0.54347	2.5049	1224.2	168.79	0.36393	0.047667	D	0.547404
200	3600	38.509	0.58115	0.53621	2.4732	1252.7	167.85	0.36177	0.04888	D	0.536714
200	3700	39.219	0.58649	0.52925	2.4427	1280.9	166.98	0.35973	0.050056	D	0.527007
200	3800	39.89	0.59221	0.52262	2.4136	1308.6	166.17	0.3578	0.051197	D	0.518143
200	3900	40.526	0.59825	0.51633	2.3859	1335.9	165.43	0.35597	0.052305	D	0.510006
200	4000	41.13	0.60458	0.51035	2.3596	1362.7	164.73	0.35423	0.053383	D	0.502517
200	4100	41.705	0.61116	0.5047	2.3346	1389.1	164.09	0.35258	0.054431	D	0.495597
200	4200	42.252	0.61796	0.49934	2.3108	1415.1	163.49	0.351	0.055453	D	0.489180
200	4300	42.775	0.62494	0.49426	2.2882	1440.5	162.93	0.34949	0.056449	D	0.483200
200	4400	43.274	0.63209	0.48944	2.2667	1465.6	162.4	0.34804	0.057422	D	0.477621
200	4500	43.753	0.63938	0.48487	2.2462	1490.2	161.91	0.34666	0.058372	D	0.472393
200	4600	44.212	0.64681	0.48054	2.2267	1514.4	161.46	0.34532	0.059301	D	0.467494
200	4700	44.653	0.65434	0.47642	2.2081	1538.2	161.03	0.34404	0.060211	D	0.462874
200	4800	45.077	0.66197	0.47251	2.1905	1561.6	160.62	0.3428	0.061102	D	0.458516
200	4900	45.486	0.6697	0.4688	2.1736	1584.6	160.24	0.34161	0.061976	D	0.454403
200	5000	45.88	0.6775	0.46527	2.1575	1607.3	159.89	0.34045	0.062833	D	0.450502
200	5100	46.26	0.68537	0.46191	2.1422	1629.5	159.55	0.33934	0.063675	D	0.446799
200	5200	46.627	0.6933	0.45871	2.1275	1651.4	159.24	0.33825	0.064503	D	0.443277
200	5300	46.983	0.70128	0.45566	2.1135	1673	158.94	0.33721	0.065316	D	0.439919
200	5400	47.327	0.70931	0.45276	2.1001	1694.2	158.66	0.33619	0.066116	D	0.436716
200	5500	47.661	0.71739	0.44999	2.0873	1715.1	158.4	0.3352	0.066904	D	0.433661
200	5600	47.985	0.7255	0.44735	2.0751	1735.6	158.15	0.33423	0.067681	D	0.430732
200	5700	48.3	0.73365	0.44483	2.0633	1755.8	157.92	0.3333	0.068446	D	0.427929
200	5800	48.605	0.74182	0.44242	2.0521	1775.8	157.7	0.33238	0.069201	D	0.425234
200	5900	48.903	0.75002	0.44012	2.0413	1795.4	157.49	0.33149	0.069946	D	0.422647
200	6000	49.192	0.75825	0.43791	2.0309	1814.8	157.3	0.33063	0.070681	D	0.420164
200	6100	49.474	0.76649	0.4358	2.0209	1833.9	157.11	0.32978	0.071407	D	0.417767
200	6200	49.749	0.77475	0.43378	2.0113	1852.7	156.94	0.32895	0.072125	D	0.415458
200	6300	50.017	0.78303	0.43184	2.0021	1871.3	156.77	0.32814	0.072834	D	0.413233
200	6400	50.279	0.79131	0.42998	1.9932	1889.6	156.62	0.32735	0.073536	D	0.411078
200	6500	50.535	0.79961	0.42819	1.9847	1907.6	156.48	0.32657	0.07423	D	0.408999
200	6600	50.785	0.80792	0.42647	1.9764	1925.4	156.34	0.32581	0.074917	D	0.406988
200	6700	51.029	0.81623	0.42482	1.9685	1943	156.21	0.32506	0.075597	D	0.405037
200	6800	51.268	0.82456	0.42323	1.9608	1960.4	156.09	0.32433	0.076271	D	0.403154
200	6900	51.502	0.83288	0.4217	1.9534	1977.5	155.98	0.32362	0.076939	D	0.401320
200	7000	51.731	0.84121	0.42022	1.9462	1994.5	155.87	0.32291	0.0776	D	0.399543
200	7100	51.955	0.84954	0.4188	1.9392	2011.2	155.78	0.32222	0.078256	D	0.397816
200	7200	52.175	0.85788	0.41742	1.9325	2027.7	155.68	0.32154	0.078906	D	0.396142
200	7300	52.391	0.86621	0.41609	1.9259	2044.1	155.6	0.32088	0.079551	D	0.394510
200	7400	52.602	0.87455	0.41481	1.9196	2060.2	155.52	0.32022	0.080191	D	0.392925
200	7500	52.81	0.88288	0.41357	1.9134	2076.2	155.44	0.31958	0.080826	D	0.391379
200	7600	53.014	0.89122	0.41237	1.9075	2091.9	155.38	0.31894	0.081456	D	0.389878
200	7700	53.214	0.89955	0.41121	1.9017	2107.6	155.31	0.31832	0.082082	D	0.388411

TABLE F.1 (continued)—CO_2 PROPERTIES AT VARIOUS TEMPERATURES, °F											
Temperature (°F)	Pressure (psia)	Density (lbm/ft^3)	Compressibility Factor	Heat Capacity (Btu/lbm°F)	Heat Capacity Ratio (C_p/C_v)	Sonic Velocity (ft/s)	Enthalpy (Btu/lbm)	Entropy [Btu/(lbm°F)]	Viscosity (cp)	Phase	Gas Formation Volume Factor (res bbl/Mcf)
200	7800	53.41	0.90787	0.41009	1.896	2123	155.26	0.3177	0.082704	D	0.386978
200	7900	53.604	0.9162	0.409	1.8906	2138.3	155.2	0.3171	0.083321	D	0.385585
200	8000	53.794	0.92452	0.40795	1.8852	2153.4	155.15	0.3165	0.083934	D	0.384223
200	8100	53.98	0.93284	0.40692	1.88	2168.4	155.11	0.31592	0.084544	D	0.382895
200	8200	54.164	0.94115	0.40593	1.875	2183.2	155.07	0.31534	0.08515	D	0.381595
200	8300	54.345	0.94946	0.40496	1.87	2197.8	155.04	0.31477	0.085752	D	0.380326
200	8400	54.523	0.95776	0.40403	1.8652	2212.3	155	0.3142	0.08635	D	0.379083
200	8500	54.698	0.96606	0.40312	1.8605	2226.7	154.98	0.31365	0.086945	D	0.377870
200	8600	54.871	0.97435	0.40223	1.856	2240.9	154.95	0.3131	0.087537	D	0.376681
200	8700	55.041	0.98264	0.40137	1.8515	2255	154.93	0.31256	0.088126	D	0.375519
200	8800	55.208	0.99092	0.40053	1.8471	2269	154.92	0.31203	0.088711	D	0.374380
200	8900	55.373	0.99919	0.39972	1.8428	2282.8	154.91	0.3115	0.089294	D	0.373263
200	9000	55.536	1.0075	0.39892	1.8387	2296.6	154.9	0.31098	0.089874	D	0.372186
200	9100	55.696	1.0157	0.39815	1.8346	2310.1	154.89	0.31046	0.090451	D	0.371092
200	9200	55.854	1.024	0.3974	1.8306	2323.6	154.89	0.30996	0.091025	D	0.370058
200	9300	56.01	1.0322	0.39666	1.8267	2337	154.89	0.30946	0.091597	D	0.369010
200	9400	56.164	1.0405	0.39595	1.8229	2350.2	154.89	0.30896	0.092166	D	0.368020
200	9500	56.316	1.0487	0.39525	1.8191	2363.3	154.9	0.30847	0.092732	D	0.367016
200	9600	56.466	1.0569	0.39456	1.8154	2376.3	154.9	0.30798	0.093297	D	0.366033
200	9700	56.614	1.0651	0.3939	1.8118	2389.2	154.92	0.3075	0.093859	D	0.365070
200	9800	56.76	1.0733	0.39325	1.8083	2402	154.93	0.30703	0.094418	D	0.364127
200	9900	56.904	1.0815	0.39261	1.8048	2414.7	154.95	0.30656	0.094976	D	0.363202
200	10000	57.047	1.0897	0.39199	1.8014	2427.3	154.96	0.3061	0.095531	D	0.362297
210	14.696	0.09021	0.99762	0.2195	1.2629	975.16	245.76	0.70089	0.018424	V	229.117037
210	100	0.62251	0.98373	0.22418	1.2797	967.92	244.28	0.61272	0.018466	D	33.202233
210	200	1.2662	0.96729	0.23006	1.3011	959.53	242.5	0.5795	0.01853	D	16.323680
210	300	1.9325	0.95066	0.23644	1.3245	951.27	240.68	0.55922	0.018613	D	10.695359
210	400	2.623	0.93386	0.24335	1.3501	943.14	238.82	0.5442	0.018716	D	7.879763
210	500	3.3395	0.91687	0.25087	1.3782	935.17	236.91	0.53202	0.018842	D	6.189123
210	600	4.0839	0.8997	0.25905	1.409	927.39	234.96	0.52162	0.018992	D	5.061018
210	700	4.8583	0.88235	0.26796	1.4428	919.85	232.95	0.51242	0.019171	D	4.254360
210	800	5.6648	0.86483	0.2777	1.48	912.59	230.9	0.50408	0.01938	D	3.648649
210	900	6.5058	0.84716	0.28833	1.5209	905.67	228.78	0.49638	0.019624	D	3.176979
210	1000	7.3837	0.82937	0.29995	1.5658	899.17	226.62	0.48915	0.019906	D	2.799237
210	1100	8.3012	0.81148	0.31264	1.6151	893.17	224.39	0.4823	0.020232	D	2.489869
210	1200	9.2606	0.79353	0.32649	1.6691	887.78	222.11	0.47574	0.020606	D	2.231894
210	1300	10.264	0.7756	0.34156	1.7281	883.13	219.77	0.46941	0.021033	D	2.013659
210	1400	11.314	0.75774	0.35789	1.7923	879.33	217.38	0.46327	0.021521	D	1.826769
210	1500	12.412	0.74004	0.37548	1.8615	876.57	214.94	0.45729	0.022074	D	1.665157
210	1600	13.559	0.72261	0.39428	1.9356	875.01	212.44	0.45143	0.022699	D	1.524317
210	1700	14.755	0.70557	0.41411	2.0139	874.86	209.91	0.44569	0.023403	D	1.400821
210	1800	15.997	0.68907	0.43471	2.0954	876.32	207.35	0.44007	0.02419	D	1.292059
210	1900	17.281	0.67328	0.45566	2.1783	879.62	204.77	0.43455	0.025065	D	1.196006
210	2000	18.603	0.65837	0.4764	2.2604	884.98	202.19	0.42916	0.026029	D	1.111044
210	2100	19.952	0.64453	0.49626	2.339	892.59	199.63	0.4239	0.027082	D	1.035894
210	2200	21.319	0.63194	0.5145	2.4111	902.57	197.11	0.41881	0.028218	D	0.969493

TABLE F.1 (continued)—CO_2 PROPERTIES AT VARIOUS TEMPERATURES, °F

Temperature (°F)	Pressure (psia)	Density (lbm/ft^3)	Compressibility Factor	Heat Capacity (Btu/lbm°F)	Heat Capacity Ratio (C_p/C_v)	Sonic Velocity (ft/s)	Enthalpy (Btu/lbm)	Entropy [Btu/(lbm°F)]	Viscosity (cp)	Phase	Gas Formation Volume Factor (res bbl/Mcf)
210	2300	22.691	0.62073	0.53044	2.474	915	194.66	0.41389	0.02943	D	0.910891
210	2400	24.053	0.61104	0.54356	2.5255	929.85	192.3	0.40917	0.030705	D	0.859310
210	2500	25.393	0.60291	0.55355	2.5644	947	190.03	0.40468	0.032031	D	0.813961
210	2600	26.699	0.59635	0.56037	2.5904	966.27	187.89	0.40042	0.033391	D	0.774140
210	2700	27.961	0.59134	0.56418	2.6042	987.42	185.88	0.39639	0.034771	D	0.739205
210	2800	29.172	0.58778	0.56533	2.6071	1010.2	183.99	0.39261	0.036158	D	0.708514
210	2900	30.327	0.58558	0.56427	2.601	1034.2	182.24	0.38907	0.037538	D	0.681522
210	3000	31.425	0.58462	0.56146	2.5877	1059.4	180.62	0.38575	0.038904	D	0.657724
210	3100	32.464	0.58477	0.55738	2.5691	1085.3	179.12	0.38264	0.040247	D	0.636671
210	3200	33.446	0.5859	0.5524	2.5468	1111.8	177.73	0.37974	0.041563	D	0.617966
210	3300	34.374	0.5879	0.54686	2.5222	1138.6	176.46	0.37701	0.042848	D	0.601286
210	3400	35.25	0.59067	0.54099	2.4963	1165.6	175.27	0.37446	0.0441	D	0.586351
210	3500	36.078	0.59409	0.535	2.4698	1192.7	174.18	0.37205	0.04532	D	0.572896
210	3600	36.86	0.5981	0.52899	2.4433	1219.7	173.17	0.36979	0.046506	D	0.560741
210	3700	37.6	0.6026	0.52308	2.4172	1246.6	172.24	0.36765	0.047661	D	0.549691
210	3800	38.302	0.60755	0.51731	2.3918	1273.2	171.37	0.36562	0.048784	D	0.539622
210	3900	38.968	0.61288	0.51174	2.3671	1299.5	170.57	0.36371	0.049877	D	0.530398
210	4000	39.602	0.61854	0.50637	2.3433	1325.5	169.82	0.36188	0.050941	D	0.521914
210	4100	40.205	0.62449	0.50123	2.3204	1351.2	169.12	0.36015	0.051979	D	0.514083
210	4200	40.78	0.6307	0.49631	2.2985	1376.5	168.47	0.35849	0.05299	D	0.506833
210	4300	41.33	0.63713	0.49161	2.2774	1401.4	167.86	0.35691	0.053978	D	0.500093
210	4400	41.856	0.64376	0.48712	2.2573	1425.9	167.29	0.35539	0.054943	D	0.493813
210	4500	42.359	0.65056	0.48283	2.238	1450.1	166.75	0.35394	0.055886	D	0.487940
210	4600	42.842	0.65752	0.47873	2.2196	1473.9	166.25	0.35254	0.056808	D	0.482439
210	4700	43.306	0.66461	0.47482	2.2019	1497.3	165.78	0.3512	0.057712	D	0.477266
210	4800	43.753	0.67183	0.47109	2.185	1520.4	165.34	0.3499	0.058597	D	0.472400
210	4900	44.183	0.67915	0.46752	2.1687	1543.1	164.93	0.34865	0.059465	D	0.467801
210	5000	44.597	0.68657	0.46411	2.1532	1565.5	164.54	0.34745	0.060317	D	0.463454
210	5100	44.997	0.69408	0.46085	2.1383	1587.6	164.17	0.34628	0.061153	D	0.459336
210	5200	45.383	0.70166	0.45774	2.124	1609.3	163.82	0.34515	0.061975	D	0.455423
210	5300	45.757	0.70931	0.45476	2.1102	1630.7	163.49	0.34406	0.062782	D	0.451702
210	5400	46.119	0.71703	0.45191	2.0971	1651.8	163.19	0.34299	0.063577	D	0.448162
210	5500	46.47	0.7248	0.44918	2.0844	1672.6	162.9	0.34196	0.064359	D	0.444782
210	5600	46.81	0.73261	0.44657	2.0723	1693.1	162.62	0.34096	0.06513	D	0.441546
210	5700	47.14	0.74048	0.44407	2.0607	1713.3	162.36	0.33999	0.065889	D	0.438460
210	5800	47.46	0.74838	0.44168	2.0495	1733.2	162.12	0.33904	0.066637	D	0.435497
210	5900	47.772	0.75631	0.43938	2.0387	1752.8	161.89	0.33811	0.067376	D	0.432652
210	6000	48.075	0.76428	0.43718	2.0283	1772.2	161.67	0.33721	0.068104	D	0.429925
210	6100	48.371	0.77227	0.43507	2.0184	1791.3	161.47	0.33633	0.068824	D	0.427298
210	6200	48.658	0.78029	0.43305	2.0088	1810.1	161.27	0.33547	0.069535	D	0.424772
210	6300	48.939	0.78833	0.4311	1.9995	1828.7	161.09	0.33463	0.070237	D	0.422337
210	6400	49.213	0.79639	0.42923	1.9906	1847	160.92	0.33381	0.070931	D	0.419988
210	6500	49.48	0.80447	0.42744	1.982	1865.1	160.75	0.33301	0.071618	D	0.417723
210	6600	49.741	0.81256	0.42571	1.9737	1883	160.6	0.33222	0.072297	D	0.415530
210	6700	49.996	0.82066	0.42404	1.9657	1900.6	160.46	0.33145	0.07297	D	0.413409
210	6800	50.245	0.82878	0.42244	1.958	1918.1	160.32	0.33069	0.073636	D	0.411360

TABLE F.1 (continued)—CO_2 PROPERTIES AT VARIOUS TEMPERATURES, °F

Temperature (°F)	Pressure (psia)	Density (lbm/ft^3)	Compressibility Factor	Heat Capacity (Btu/lbm°F)	Heat Capacity Ratio (C_p/C_v)	Sonic Velocity (ft/s)	Enthalpy (Btu/lbm)	Entropy [Btu/(lbm°F)]	Viscosity (cp)	Phase	Gas Formation Volume Factor (res bbl/Mcf)
210	6900	50.489	0.83691	0.4209	1.9505	1935.3	160.19	0.32996	0.074295	D	0.409375
210	7000	50.728	0.84504	0.41941	1.9433	1952.3	160.07	0.32923	0.074948	D	0.407447
210	7100	50.961	0.85318	0.41798	1.9362	1969.1	159.96	0.32852	0.075596	D	0.405577
210	7200	51.19	0.86132	0.41659	1.9295	1985.6	159.85	0.32782	0.076238	D	0.403760
210	7300	51.415	0.86947	0.41525	1.9229	2002.1	159.75	0.32713	0.076874	D	0.401997
210	7400	51.635	0.87763	0.41396	1.9165	2018.3	159.66	0.32646	0.077505	D	0.400287
210	7500	51.851	0.88578	0.41271	1.9103	2034.3	159.58	0.32579	0.078131	D	0.398617
210	7600	52.063	0.89394	0.4115	1.9043	2050.2	159.5	0.32514	0.078753	D	0.396996
210	7700	52.271	0.9021	0.41033	1.8985	2065.8	159.42	0.3245	0.07937	D	0.395417
210	7800	52.475	0.91026	0.4092	1.8928	2081.4	159.35	0.32387	0.079982	D	0.393878
210	7900	52.676	0.91842	0.4081	1.8873	2096.7	159.29	0.32325	0.08059	D	0.392379
210	8000	52.873	0.92657	0.40704	1.8819	2111.9	159.23	0.32263	0.081194	D	0.390913
210	8100	53.067	0.93473	0.40601	1.8767	2126.9	159.17	0.32203	0.081794	D	0.389487
210	8200	53.257	0.94288	0.40501	1.8716	2141.8	159.13	0.32144	0.08239	D	0.388091
210	8300	53.445	0.95103	0.40403	1.8666	2156.6	159.08	0.32085	0.082982	D	0.386730
210	8400	53.629	0.95918	0.40309	1.8618	2171.1	159.04	0.32028	0.083571	D	0.385400
210	8500	53.811	0.96733	0.40217	1.8571	2185.6	159	0.31971	0.084157	D	0.384102
210	8600	53.989	0.97547	0.40128	1.8525	2199.9	158.97	0.31915	0.084739	D	0.382831
210	8700	54.165	0.9836	0.40042	1.848	2214.1	158.94	0.31859	0.085317	D	0.381584
210	8800	54.339	0.99173	0.39957	1.8436	2228.1	158.92	0.31805	0.085893	D	0.380366
210	8900	54.51	0.99986	0.39875	1.8393	2242	158.9	0.31751	0.086466	D	0.379176
210	9000	54.678	1.008	0.39795	1.8351	2255.8	158.88	0.31697	0.087035	D	0.378015
210	9100	54.844	1.0161	0.39718	1.831	2269.4	158.87	0.31645	0.087602	D	0.376866
210	9200	55.007	1.0242	0.39642	1.827	2283	158.86	0.31593	0.088166	D	0.375741
210	9300	55.169	1.0323	0.39568	1.8231	2296.4	158.85	0.31542	0.088728	D	0.374640
210	9400	55.328	1.0404	0.39496	1.8192	2309.7	158.84	0.31491	0.089287	D	0.373563
210	9500	55.485	1.0485	0.39426	1.8155	2322.9	158.84	0.31441	0.089843	D	0.372509
210	9600	55.639	1.0566	0.39357	1.8118	2336	158.84	0.31391	0.090397	D	0.371476
210	9700	55.792	1.0647	0.39291	1.8082	2348.9	158.85	0.31342	0.090949	D	0.370465
210	9800	55.943	1.0728	0.39225	1.8046	2361.8	158.86	0.31294	0.091498	D	0.369474
210	9900	56.092	1.0808	0.39162	1.8012	2374.5	158.87	0.31246	0.092045	D	0.368469
210	10000	56.239	1.0889	0.39099	1.7978	2387.2	158.88	0.31199	0.09259	D	0.367519
220	14.696	0.088872	0.99774	0.22064	1.261	981.8	247.96	0.70415	0.018679	V	232.566350
220	100	0.61283	0.98457	0.22508	1.2771	974.94	246.52	0.61604	0.01872	D	33.726808
220	200	1.2453	0.969	0.23066	1.2973	967.02	244.8	0.58292	0.018783	D	16.596726
220	300	1.8988	0.9533	0.23667	1.3194	959.23	243.05	0.56273	0.018863	D	10.885214
220	400	2.5745	0.93746	0.24315	1.3434	951.6	241.25	0.5478	0.018963	D	8.028259
220	500	3.274	0.92147	0.25016	1.3696	944.15	239.42	0.53573	0.019084	D	6.313059
220	600	3.9987	0.90536	0.25774	1.3981	936.9	237.54	0.52545	0.019229	D	5.168907
220	700	4.7504	0.88911	0.26594	1.4292	929.9	235.62	0.51638	0.0194	D	4.350970
220	800	5.5308	0.87275	0.27484	1.4631	923.19	233.66	0.50818	0.019599	D	3.737047
220	900	6.3417	0.85629	0.28447	1.5001	916.82	231.65	0.50062	0.019829	D	3.259170
220	1000	7.185	0.83976	0.29492	1.5404	910.87	229.59	0.49356	0.020095	D	2.876629
220	1100	8.0626	0.82319	0.30622	1.5842	905.4	227.49	0.48688	0.020399	D	2.563516
220	1200	8.9763	0.80662	0.31844	1.6318	900.5	225.34	0.48052	0.020747	D	2.302589
220	1300	9.9277	0.79009	0.33161	1.6832	896.29	223.14	0.4744	0.021141	D	2.081910

TABLE F.1 (continued)—CO_2 PROPERTIES AT VARIOUS TEMPERATURES, °F

Temperature (°F)	Pressure (psia)	Density (lbm/ft³)	Compressibility Factor	Heat Capacity (Btu/lbm°F)	Heat Capacity Ratio (C_p/C_v)	Sonic Velocity (ft/s)	Enthalpy (Btu/lbm)	Entropy [Btu/(lbm°F)]	Viscosity (cp)	Phase	Gas Formation Volume Factor (res bbl/Mcf)
220	1400	10.918	0.77368	0.34576	1.7387	892.86	220.9	0.46848	0.021587	D	1.893050
220	1500	11.949	0.75744	0.36086	1.798	890.35	218.61	0.46274	0.02209	D	1.729759
220	1600	13.02	0.74147	0.37687	1.861	888.91	216.3	0.45714	0.022654	D	1.587458
220	1700	14.131	0.72587	0.39368	1.9273	888.68	213.94	0.45168	0.023284	D	1.462644
220	1800	15.281	0.71075	0.41109	1.9961	889.83	211.57	0.44633	0.023984	D	1.352611
220	1900	16.466	0.69624	0.42884	2.0662	892.53	209.18	0.4411	0.024757	D	1.255261
220	2000	17.682	0.68246	0.44658	2.1363	896.94	206.8	0.43599	0.025605	D	1.168896
220	2100	18.924	0.66956	0.46387	2.2046	903.22	204.42	0.43101	0.026528	D	1.092192
220	2200	20.183	0.65768	0.48021	2.2691	911.48	202.08	0.42617	0.027524	D	1.024049
220	2300	21.452	0.64692	0.49514	2.3279	921.79	199.78	0.42148	0.028587	D	0.963499
220	2400	22.719	0.6374	0.5082	2.3793	934.17	197.55	0.41696	0.02971	D	0.909765
220	2500	23.975	0.62918	0.51909	2.4218	948.57	195.39	0.41262	0.030884	D	0.862112
220	2600	25.21	0.62229	0.52764	2.4548	964.89	193.33	0.40848	0.032099	D	0.819876
220	2700	26.415	0.61673	0.53381	2.4782	982.97	191.37	0.40453	0.033343	D	0.782456
220	2800	27.584	0.61247	0.53772	2.4923	1002.6	189.51	0.40079	0.034605	D	0.749299
220	2900	28.711	0.60945	0.53958	2.4981	1023.7	187.76	0.39725	0.035875	D	0.719894
220	3000	29.792	0.60758	0.53967	2.4967	1045.9	186.13	0.39391	0.037144	D	0.693763
220	3100	30.825	0.60679	0.53831	2.4893	1069.1	184.6	0.39077	0.038403	D	0.670510
220	3200	31.81	0.60697	0.53581	2.4773	1093	183.18	0.3878	0.039647	D	0.649749
220	3300	32.747	0.60803	0.53245	2.4617	1117.5	181.85	0.38501	0.040871	D	0.631160
220	3400	33.637	0.60988	0.52848	2.4437	1142.4	180.62	0.38238	0.042072	D	0.614461
220	3500	34.483	0.61242	0.52411	2.424	1167.6	179.48	0.3799	0.043249	D	0.599391
220	3600	35.286	0.61558	0.5195	2.4033	1192.9	178.42	0.37756	0.044399	D	0.585748
220	3700	36.049	0.6193	0.51477	2.3821	1218.3	177.43	0.37534	0.045522	D	0.573361
220	3800	36.774	0.62349	0.51001	2.3608	1243.6	176.51	0.37324	0.046618	D	0.562049
220	3900	37.464	0.62811	0.50529	2.3398	1268.7	175.65	0.37124	0.047688	D	0.551696
220	4000	38.122	0.6331	0.50066	2.319	1293.7	174.85	0.36935	0.048733	D	0.542177
220	4100	38.749	0.63843	0.49614	2.2988	1318.4	174.11	0.36754	0.049753	D	0.533406
220	4200	39.348	0.64404	0.49177	2.2792	1342.9	173.41	0.36581	0.05075	D	0.525282
220	4300	39.92	0.64992	0.48753	2.2601	1367.1	172.75	0.36416	0.051724	D	0.517750
220	4400	40.469	0.65602	0.48346	2.2418	1391	172.14	0.36259	0.052676	D	0.510732
220	4500	40.995	0.66232	0.47953	2.224	1414.5	171.57	0.36107	0.053608	D	0.504178
220	4600	41.499	0.66881	0.47575	2.207	1437.8	171.03	0.35962	0.05452	D	0.498051
220	4700	41.984	0.67545	0.47213	2.1905	1460.8	170.52	0.35822	0.055414	D	0.492293
220	4800	42.451	0.68224	0.46864	2.1746	1483.4	170.04	0.35687	0.05629	D	0.486883
220	4900	42.9	0.68916	0.46529	2.1593	1505.7	169.59	0.35557	0.05715	D	0.481784
220	5000	43.334	0.69619	0.46207	2.1446	1527.8	169.17	0.35431	0.057993	D	0.476965
220	5100	43.752	0.70332	0.45898	2.1304	1549.5	168.77	0.3531	0.058822	D	0.472402
220	5200	44.156	0.71055	0.45601	2.1168	1571	168.39	0.35192	0.059636	D	0.468080
220	5300	44.547	0.71786	0.45316	2.1036	1592.1	168.03	0.35078	0.060437	D	0.463973
220	5400	44.926	0.72525	0.45042	2.0909	1613	167.7	0.34968	0.061224	D	0.460069
220	5500	45.292	0.7327	0.44778	2.0787	1633.6	167.38	0.34861	0.061999	D	0.456344
220	5600	45.648	0.74021	0.44525	2.0669	1653.9	167.08	0.34757	0.062763	D	0.452789
220	5700	45.992	0.74778	0.44282	2.0555	1674	166.8	0.34656	0.063515	D	0.449394
220	5800	46.327	0.7554	0.44048	2.0446	1693.8	166.53	0.34557	0.064257	D	0.446147
220	5900	46.653	0.76306	0.43823	2.034	1713.3	166.28	0.34462	0.064988	D	0.443032

TABLE F.1 (continued)—CO_2 PROPERTIES AT VARIOUS TEMPERATURES, °F											
Temperature (°F)	Pressure (psia)	Density (lbm/ft^3)	Compressibility Factor	Heat Capacity (Btu/lbm°F)	Heat Capacity Ratio (C_p/C_v)	Sonic Velocity (ft/s)	Enthalpy (Btu/lbm)	Entropy [Btu/(lbm°F)]	Viscosity (cp)	Phase	Gas Formation Volume Factor (res bbl/Mcf)
220	6000	46.97	0.77076	0.43607	2.0238	1732.6	166.04	0.34368	0.065709	D	0.440044
220	6100	47.278	0.7785	0.43399	2.014	1751.6	165.81	0.34277	0.066422	D	0.437177
220	6200	47.578	0.78627	0.43199	2.0045	1770.4	165.6	0.34188	0.067125	D	0.434419
220	6300	47.871	0.79407	0.43007	1.9953	1788.9	165.39	0.34101	0.06782	D	0.431764
220	6400	48.156	0.80189	0.42821	1.9865	1807.2	165.2	0.34016	0.068507	D	0.429204
220	6500	48.434	0.80974	0.42643	1.978	1825.3	165.02	0.33933	0.069186	D	0.426738
220	6600	48.706	0.81761	0.42471	1.9697	1843.2	164.85	0.33852	0.069858	D	0.424357
220	6700	48.972	0.8255	0.42305	1.9617	1860.8	164.69	0.33773	0.070523	D	0.422057
220	6800	49.231	0.8334	0.42146	1.954	1878.2	164.54	0.33695	0.071181	D	0.419830
220	6900	49.485	0.84132	0.41992	1.9465	1895.4	164.4	0.33619	0.071832	D	0.417677
220	7000	49.733	0.84925	0.41843	1.9393	1912.5	164.26	0.33544	0.072477	D	0.415591
220	7100	49.976	0.85719	0.417	1.9323	1929.3	164.13	0.33471	0.073116	D	0.413568
220	7200	50.214	0.86515	0.41561	1.9255	1945.9	164.01	0.33399	0.07375	D	0.411612
220	7300	50.448	0.87311	0.41427	1.9189	1962.4	163.9	0.33328	0.074378	D	0.409708
220	7400	50.676	0.88107	0.41298	1.9125	1978.6	163.8	0.33259	0.075001	D	0.407856
220	7500	50.9	0.88905	0.41172	1.9063	1994.7	163.7	0.3319	0.075618	D	0.406063
220	7600	51.12	0.89703	0.41051	1.9003	2010.6	163.61	0.33123	0.076231	D	0.404317
220	7700	51.336	0.90501	0.40934	1.8944	2026.3	163.52	0.33057	0.076839	D	0.402616
220	7800	51.548	0.913	0.4082	1.8888	2041.9	163.44	0.32993	0.077442	D	0.400963
220	7900	51.756	0.92099	0.4071	1.8832	2057.3	163.36	0.32929	0.078041	D	0.399353
220	8000	51.96	0.92898	0.40603	1.8778	2072.5	163.29	0.32866	0.078636	D	0.397782
220	8100	52.161	0.93697	0.405	1.8726	2087.6	163.23	0.32804	0.079227	D	0.396250
220	8200	52.358	0.94496	0.404	1.8675	2102.5	163.17	0.32743	0.079814	D	0.394756
220	8300	52.553	0.95295	0.40302	1.8625	2117.3	163.12	0.32683	0.080397	D	0.393297
220	8400	52.743	0.96094	0.40207	1.8577	2131.9	163.07	0.32624	0.080977	D	0.391873
220	8500	52.931	0.96893	0.40115	1.853	2146.4	163.02	0.32566	0.081552	D	0.390483
220	8600	53.116	0.97692	0.40026	1.8483	2160.8	162.98	0.32509	0.082125	D	0.389125
220	8700	53.298	0.9849	0.39939	1.8438	2175	162.94	0.32452	0.082694	D	0.387794
220	8800	53.477	0.99288	0.39855	1.8394	2189.1	162.91	0.32396	0.08326	D	0.386494
220	8900	53.654	1.0009	0.39772	1.8352	2203	162.88	0.32341	0.083823	D	0.385238
220	9000	53.828	1.0088	0.39692	1.831	2216.9	162.85	0.32286	0.084383	D	0.383965
220	9100	53.999	1.0168	0.39614	1.8269	2230.6	162.83	0.32233	0.084941	D	0.382757
220	9200	54.168	1.0248	0.39538	1.8228	2244.2	162.81	0.3218	0.085495	D	0.381575
220	9300	54.334	1.0327	0.39464	1.8189	2257.6	162.8	0.32127	0.086047	D	0.380382
220	9400	54.499	1.0407	0.39392	1.8151	2271	162.79	0.32076	0.086596	D	0.379251
220	9500	54.661	1.0487	0.39322	1.8113	2284.2	162.78	0.32024	0.087142	D	0.378143
220	9600	54.82	1.0566	0.39253	1.8076	2297.4	162.78	0.31974	0.087686	D	0.377023
220	9700	54.978	1.0646	0.39186	1.804	2310.4	162.77	0.31924	0.088228	D	0.375961
220	9800	55.134	1.0725	0.39121	1.8005	2323.3	162.77	0.31875	0.088767	D	0.374887
220	9900	55.287	1.0804	0.39057	1.797	2336.1	162.78	0.31826	0.089304	D	0.373833
220	10000	55.439	1.0884	0.38995	1.7936	2348.8	162.78	0.31778	0.089839	D	0.372835
230	14.696	0.087574	0.99785	0.22176	1.2592	988.4	250.17	0.70738	0.018933	V	236.014122
230	100	0.60346	0.98536	0.22599	1.2745	981.9	248.78	0.61934	0.018973	D	34.250491
230	200	1.2252	0.97062	0.23128	1.2938	974.41	247.11	0.58629	0.019035	D	16.869069
230	300	1.8664	0.95577	0.23695	1.3146	967.08	245.42	0.56619	0.019113	D	11.073987
230	400	2.5281	0.94082	0.24305	1.3371	959.91	243.69	0.55135	0.01921	D	8.175577

TABLE F.1 (continued)—CO_2 PROPERTIES AT VARIOUS TEMPERATURES, °F											
Temperature (°F)	Pressure (psia)	Density (lbm/ft^3)	Compressibility Factor	Heat Capacity (Btu/lbm°F)	Heat Capacity Ratio (C_p/C_v)	Sonic Velocity (ft/s)	Enthalpy (Btu/lbm)	Entropy [Btu/(lbm°F)]	Viscosity (cp)	Phase	Gas Formation Volume Factor (res bbl/Mcf)
230	500	3.2115	0.92577	0.2496	1.3616	952.94	241.92	0.53938	0.019327	D	6.435836
230	600	3.9179	0.91063	0.25665	1.3881	946.18	240.11	0.52921	0.019466	D	5.275487
230	700	4.6486	0.8954	0.26423	1.4168	939.68	238.27	0.52025	0.019629	D	4.446220
230	800	5.4051	0.88009	0.27239	1.4478	933.48	236.39	0.51217	0.019819	D	3.823922
230	900	6.1887	0.86474	0.28118	1.4815	927.61	234.48	0.50475	0.020038	D	3.339757
230	1000	7.0009	0.84935	0.29063	1.5179	922.15	232.52	0.49783	0.020289	D	2.952287
230	1100	7.8432	0.83396	0.30078	1.5572	917.16	230.52	0.49132	0.020575	D	2.635266
230	1200	8.7167	0.8186	0.31167	1.5994	912.71	228.49	0.48512	0.020899	D	2.371168
230	1300	9.6226	0.80333	0.32332	1.6448	908.89	226.41	0.47918	0.021265	D	2.147942
230	1400	10.562	0.78818	0.33572	1.6934	905.8	224.3	0.47346	0.021676	D	1.956903
230	1500	11.535	0.77323	0.34887	1.7449	903.54	222.16	0.46792	0.022137	D	1.791799
230	1600	12.542	0.75855	0.36272	1.7993	902.23	219.99	0.46254	0.022651	D	1.647920
230	1700	13.583	0.74421	0.37718	1.8562	901.99	217.8	0.4573	0.023221	D	1.521663
230	1800	14.656	0.7303	0.39213	1.9151	902.94	215.58	0.45219	0.023851	D	1.410265
230	1900	15.759	0.71693	0.40737	1.9752	905.23	213.36	0.4472	0.024543	D	1.311581
230	2000	16.888	0.70419	0.42268	2.0356	908.97	211.14	0.44234	0.025299	D	1.223860
230	2100	18.04	0.69219	0.43776	2.0951	914.3	208.93	0.43759	0.026119	D	1.145718
230	2200	19.209	0.68104	0.45227	2.1523	921.29	206.74	0.43297	0.027002	D	1.076024
230	2300	20.387	0.67083	0.46589	2.2058	930.03	204.58	0.42849	0.027944	D	1.013810
230	2400	21.569	0.66164	0.47827	2.2544	940.55	202.48	0.42416	0.028941	D	0.958258
230	2500	22.746	0.65354	0.48915	2.2969	952.84	200.43	0.41998	0.029988	D	0.908665
230	2600	23.911	0.64658	0.49832	2.3325	966.83	198.46	0.41596	0.031076	D	0.864412
230	2700	25.056	0.64076	0.5057	2.3608	982.45	196.56	0.41212	0.032197	D	0.824904
230	2800	26.175	0.63607	0.51126	2.3817	999.55	194.75	0.40845	0.033343	D	0.789621
230	2900	27.263	0.6325	0.51509	2.3956	1018	193.03	0.40495	0.034505	D	0.758114
230	3000	28.316	0.62998	0.51732	2.4029	1037.6	191.41	0.40163	0.035676	D	0.729924
230	3100	29.331	0.62847	0.51814	2.4045	1058.3	189.88	0.39848	0.036847	D	0.704685
230	3200	30.305	0.62788	0.51775	2.4012	1079.8	188.45	0.3955	0.038013	D	0.682022
230	3300	31.239	0.62815	0.51637	2.3938	1102.1	187.1	0.39268	0.039169	D	0.661639
230	3400	32.131	0.6292	0.51421	2.3832	1124.9	185.84	0.39	0.040311	D	0.643253
230	3500	32.984	0.63096	0.51146	2.3702	1148.1	184.66	0.38747	0.041435	D	0.626622
230	3600	33.798	0.63336	0.50827	2.3554	1171.6	183.56	0.38507	0.042541	D	0.611533
230	3700	34.575	0.63634	0.50478	2.3394	1195.4	182.53	0.38279	0.043625	D	0.597805
230	3800	35.316	0.63982	0.5011	2.3227	1219.2	181.57	0.38063	0.044688	D	0.585256
230	3900	36.023	0.64376	0.49732	2.3055	1243.1	180.67	0.37857	0.045728	D	0.573761
230	4000	36.699	0.64811	0.49349	2.2882	1266.8	179.83	0.37661	0.046747	D	0.563197
230	4100	37.345	0.65282	0.48968	2.2709	1290.5	179.04	0.37474	0.047744	D	0.553454
230	4200	37.963	0.65786	0.48592	2.2539	1314.1	178.3	0.37296	0.04872	D	0.544448
230	4300	38.555	0.66318	0.48223	2.2372	1337.4	177.6	0.37125	0.049676	D	0.536086
230	4400	39.123	0.66876	0.47862	2.2208	1360.5	176.95	0.36961	0.050612	D	0.528311
230	4500	39.667	0.67456	0.47511	2.2049	1383.4	176.34	0.36804	0.051529	D	0.521051
230	4600	40.191	0.68058	0.47171	2.1894	1406	175.76	0.36654	0.052427	D	0.514272
230	4700	40.694	0.68677	0.46842	2.1744	1428.4	175.22	0.36509	0.053308	D	0.507908
230	4800	41.178	0.69313	0.46523	2.1598	1450.5	174.71	0.36369	0.054172	D	0.501933
230	4900	41.645	0.69963	0.46215	2.1457	1472.4	174.23	0.36234	0.055021	D	0.496300
230	5000	42.096	0.70627	0.45918	2.132	1494	173.77	0.36104	0.055854	D	0.490990

TABLE F.1 (continued)—CO_2 PROPERTIES AT VARIOUS TEMPERATURES, °F											
Temperature (°F)	Pressure (psia)	Density (lbm/ft^3)	Compressibility Factor	Heat Capacity (Btu/lbm°F)	Heat Capacity Ratio (C_p/C_v)	Sonic Velocity (ft/s)	Enthalpy (Btu/lbm)	Entropy [Btu/(lbm°F)]	Viscosity (cp)	Phase	Gas Formation Volume Factor (res bbl/Mcf)
230	5100	42.531	0.71303	0.45631	2.1188	1515.3	173.34	0.35978	0.056673	D	0.485970
230	5200	42.951	0.7199	0.45353	2.1059	1536.4	172.94	0.35857	0.057477	D	0.481217
230	5300	43.358	0.72686	0.45085	2.0935	1557.2	172.56	0.35739	0.058269	D	0.476702
230	5400	43.751	0.73391	0.44827	2.0815	1577.7	172.19	0.35625	0.059048	D	0.472412
230	5500	44.133	0.74104	0.44577	2.0699	1598.1	171.85	0.35514	0.059814	D	0.468329
230	5600	44.503	0.74824	0.44336	2.0586	1618.1	171.52	0.35406	0.060569	D	0.464435
230	5700	44.862	0.75551	0.44104	2.0478	1637.9	171.22	0.35301	0.061313	D	0.460720
230	5800	45.21	0.76284	0.4388	2.0372	1657.5	170.93	0.352	0.062047	D	0.457170
230	5900	45.549	0.77022	0.43663	2.027	1676.9	170.65	0.35101	0.06277	D	0.453769
230	6000	45.878	0.77765	0.43454	2.0172	1696	170.39	0.35004	0.063484	D	0.450510
230	6100	46.199	0.78513	0.43253	2.0076	1714.9	170.14	0.3491	0.064189	D	0.447387
230	6200	46.511	0.79264	0.43058	1.9984	1733.5	169.91	0.34818	0.064884	D	0.444382
230	6300	46.815	0.80019	0.4287	1.9894	1751.9	169.69	0.34728	0.065572	D	0.441494
230	6400	47.112	0.80778	0.42689	1.9808	1770.2	169.48	0.34641	0.066251	D	0.438718
230	6500	47.401	0.81539	0.42514	1.9724	1788.2	169.28	0.34555	0.066922	D	0.436038
230	6600	47.683	0.82304	0.42345	1.9642	1806	169.09	0.34472	0.067586	D	0.433460
230	6700	47.959	0.8307	0.42182	1.9564	1823.5	168.92	0.3439	0.068243	D	0.430964
230	6800	48.229	0.83839	0.42025	1.9487	1840.9	168.75	0.3431	0.068893	D	0.428557
230	6900	48.492	0.8461	0.41873	1.9414	1858.1	168.59	0.34231	0.069536	D	0.426230
230	7000	48.75	0.85382	0.41726	1.9342	1875.1	168.44	0.34154	0.070173	D	0.423975
230	7100	49.002	0.86156	0.41583	1.9272	1891.9	168.3	0.34079	0.070804	D	0.421793
230	7200	49.249	0.86932	0.41446	1.9205	1908.5	168.17	0.34005	0.07143	D	0.419681
230	7300	49.491	0.87709	0.41313	1.914	1924.9	168.04	0.33932	0.072049	D	0.417631
230	7400	49.728	0.88487	0.41184	1.9076	1941.2	167.92	0.33861	0.072664	D	0.415642
230	7500	49.96	0.89265	0.4106	1.9014	1957.3	167.81	0.33791	0.073273	D	0.413706
230	7600	50.187	0.90045	0.40939	1.8954	1973.2	167.71	0.33722	0.073877	D	0.411830
230	7700	50.411	0.90826	0.40822	1.8896	1988.9	167.61	0.33654	0.074476	D	0.410007
230	7800	50.63	0.91607	0.40709	1.8839	2004.5	167.52	0.33588	0.075071	D	0.408231
230	7900	50.845	0.92388	0.40599	1.8784	2019.9	167.43	0.33523	0.075661	D	0.406500
230	8000	51.057	0.9317	0.40493	1.8731	2035.2	167.35	0.33458	0.076248	D	0.404816
230	8100	51.265	0.93953	0.40389	1.8678	2050.3	167.27	0.33395	0.07683	D	0.403179
230	8200	51.469	0.94736	0.40289	1.8627	2065.2	167.2	0.33333	0.077408	D	0.401581
230	8300	51.669	0.95519	0.40192	1.8578	2080	167.14	0.33271	0.077982	D	0.400022
230	8400	51.867	0.96302	0.40097	1.8529	2094.7	167.08	0.33211	0.078552	D	0.398500
230	8500	52.061	0.97085	0.40006	1.8482	2109.2	167.03	0.33151	0.079119	D	0.397013
230	8600	52.252	0.97868	0.39916	1.8436	2123.6	166.98	0.33092	0.079682	D	0.395562
230	8700	52.439	0.98651	0.3983	1.8391	2137.9	166.93	0.33034	0.080243	D	0.394143
230	8800	52.624	0.99435	0.39745	1.8347	2152	166.89	0.32977	0.080799	D	0.392761
230	8900	52.807	1.0022	0.39663	1.8305	2166	166.85	0.32921	0.081353	D	0.391414
230	9000	52.986	1.01	0.39583	1.8263	2179.8	166.82	0.32865	0.081904	D	0.390077
230	9100	53.163	1.0178	0.39505	1.8222	2193.6	166.79	0.32811	0.082452	D	0.388770
230	9200	53.337	1.0257	0.39429	1.8182	2207.2	166.76	0.32756	0.082996	D	0.387529
230	9300	53.509	1.0335	0.39355	1.8143	2220.7	166.74	0.32703	0.083539	D	0.386277
230	9400	53.678	1.0413	0.39283	1.8104	2234.1	166.72	0.3265	0.084078	D	0.385052
230	9500	53.845	1.0491	0.39213	1.8067	2247.4	166.71	0.32598	0.084615	D	0.383853
230	9600	54.009	1.0569	0.39145	1.803	2260.5	166.7	0.32546	0.08515	D	0.382679

TABLE F.1 (continued)—CO_2 PROPERTIES AT VARIOUS TEMPERATURES, °F											
Temperature (°F)	Pressure (psia)	Density (lbm/ft³)	Compressibility Factor	Heat Capacity (Btu/lbm°F)	Heat Capacity Ratio (C_p/C_v)	Sonic Velocity (ft/s)	Enthalpy (Btu/lbm)	Entropy [Btu/(lbm°F)]	Viscosity (cp)	Phase	Gas Formation Volume Factor (res bbl/Mcf)
230	9700	54.172	1.0647	0.39078	1.7994	2273.6	166.69	0.32496	0.085682	D	0.381529
230	9800	54.332	1.0725	0.39013	1.7959	2286.5	166.68	0.32445	0.086211	D	0.380402
230	9900	54.49	1.0803	0.38949	1.7924	2299.4	166.68	0.32395	0.086739	D	0.379298
230	10000	54.646	1.0881	0.38887	1.789	2312.1	166.68	0.32346	0.087264	D	0.378217
240	14.696	0.086313	0.99796	0.22287	1.2575	994.95	252.39	0.71058	0.019186	V	239.462648
240	100	0.59438	0.98611	0.2269	1.272	988.79	251.04	0.6226	0.019225	D	34.773560
240	200	1.2058	0.97214	0.23193	1.2903	981.72	249.43	0.58962	0.019285	D	17.140465
240	300	1.8353	0.95809	0.23729	1.3101	974.81	247.79	0.5696	0.019362	D	11.261827
240	400	2.4836	0.94398	0.24303	1.3313	968.08	246.12	0.55485	0.019456	D	8.321979
240	500	3.1519	0.92979	0.24917	1.3542	961.55	244.41	0.54297	0.019569	D	6.557505
240	600	3.8411	0.91555	0.25574	1.3788	955.25	242.68	0.5329	0.019703	D	5.380896
240	700	4.5524	0.90125	0.26278	1.4054	949.22	240.91	0.52404	0.019859	D	4.540159
240	800	5.2869	0.88691	0.2703	1.434	943.47	239.11	0.51608	0.020041	D	3.909429
240	900	6.0456	0.87256	0.27836	1.4648	938.07	237.27	0.50878	0.020249	D	3.418823
240	1000	6.8297	0.85821	0.28696	1.4979	933.06	235.41	0.50199	0.020487	D	3.026338
240	1100	7.6401	0.84388	0.29614	1.5333	928.5	233.51	0.49561	0.020756	D	2.705277
240	1200	8.478	0.82962	0.30593	1.5712	924.46	231.57	0.48956	0.02106	D	2.437933
240	1300	9.3439	0.81546	0.31632	1.6116	921	229.61	0.48378	0.021402	D	2.211990
240	1400	10.239	0.80145	0.32731	1.6545	918.22	227.62	0.47823	0.021784	D	2.018702
240	1500	11.162	0.78764	0.3389	1.6998	916.2	225.6	0.47287	0.022209	D	1.851656
240	1600	12.115	0.77409	0.35103	1.7474	915.03	223.56	0.46768	0.022681	D	1.706064
240	1700	13.096	0.76086	0.36364	1.7969	914.81	221.5	0.46263	0.023202	D	1.578264
240	1800	14.104	0.74803	0.37664	1.848	915.65	219.43	0.45772	0.023775	D	1.465448
240	1900	15.137	0.73568	0.38989	1.9001	917.65	217.35	0.45294	0.024401	D	1.365398
240	2000	16.194	0.72389	0.40322	1.9526	920.92	215.27	0.44828	0.025082	D	1.276340
240	2100	17.27	0.71273	0.41645	2.0046	925.54	213.2	0.44373	0.025818	D	1.196822
240	2200	18.361	0.7023	0.42932	2.0553	931.6	211.14	0.43931	0.026609	D	1.125703
240	2300	19.462	0.69267	0.44161	2.1035	939.16	209.12	0.43502	0.027453	D	1.061995
240	2400	20.569	0.68391	0.45307	2.1484	948.26	207.13	0.43086	0.028346	D	1.004874
240	2500	21.674	0.67607	0.46347	2.1891	958.91	205.19	0.42683	0.029285	D	0.953620
240	2600	22.773	0.66919	0.47266	2.2247	971.07	203.31	0.42295	0.030264	D	0.907611
240	2700	23.858	0.66331	0.48049	2.2549	984.69	201.49	0.41921	0.031278	D	0.866316
240	2800	24.926	0.65841	0.48691	2.2794	999.69	199.74	0.41563	0.032319	D	0.829206
240	2900	25.97	0.6545	0.49192	2.2981	1016	198.07	0.4122	0.033381	D	0.795858
240	3000	26.988	0.65155	0.49558	2.3113	1033.4	196.47	0.40892	0.034458	D	0.765862
240	3100	27.975	0.64951	0.49798	2.3194	1051.9	194.96	0.4058	0.035542	D	0.738836
240	3200	28.929	0.64833	0.49924	2.3228	1071.2	193.53	0.40282	0.036629	D	0.714447
240	3300	29.85	0.64797	0.4995	2.3222	1091.4	192.18	0.39999	0.037714	D	0.692412
240	3400	30.736	0.64837	0.49892	2.3181	1112.2	190.9	0.39729	0.038791	D	0.672462
240	3500	31.587	0.64946	0.49764	2.3112	1133.5	189.71	0.39473	0.039859	D	0.654347
240	3600	32.403	0.65119	0.4958	2.3021	1155.3	188.58	0.3923	0.040913	D	0.637865
240	3700	33.185	0.6535	0.49353	2.2912	1177.3	187.52	0.38998	0.041953	D	0.622827
240	3800	33.935	0.65634	0.49094	2.2789	1199.7	186.53	0.38777	0.042976	D	0.609073
240	3900	34.653	0.65965	0.48813	2.2658	1222.1	185.6	0.38566	0.043982	D	0.596448
240	4000	35.341	0.66339	0.48516	2.2521	1244.6	184.72	0.38365	0.04497	D	0.584834
240	4100	36.001	0.66752	0.4821	2.238	1267.2	183.9	0.38174	0.045939	D	0.574122

TABLE F.1 (continued)—CO_2 PROPERTIES AT VARIOUS TEMPERATURES, °F											
Temperature (°F)	Pressure (psia)	Density (lbm/ft^3)	Compressibility Factor	Heat Capacity (Btu/lbm°F)	Heat Capacity Ratio (C_p/C_v)	Sonic Velocity (ft/s)	Enthalpy (Btu/lbm)	Entropy [Btu/(lbm°F)]	Viscosity (cp)	Phase	Gas Formation Volume Factor (res bbl/Mcf)
240	4200	36.633	0.67199	0.479	2.2237	1289.7	183.12	0.3799	0.04689	D	0.564205
240	4300	37.24	0.67678	0.47589	2.2095	1312	182.4	0.37815	0.047823	D	0.555013
240	4400	37.823	0.68185	0.4728	2.1953	1334.3	181.71	0.37646	0.048739	D	0.546462
240	4500	38.383	0.68717	0.46976	2.1813	1356.4	181.07	0.37485	0.049637	D	0.538487
240	4600	38.922	0.69271	0.46677	2.1676	1378.3	180.46	0.37329	0.050518	D	0.531028
240	4700	39.441	0.69846	0.46384	2.1541	1400	179.88	0.3718	0.051384	D	0.524044
240	4800	39.941	0.70439	0.46099	2.141	1421.6	179.34	0.37036	0.052233	D	0.517483
240	4900	40.423	0.71049	0.45822	2.1282	1442.8	178.83	0.36897	0.053068	D	0.511312
240	5000	40.888	0.71674	0.45552	2.1157	1463.9	178.35	0.36762	0.053888	D	0.505493
240	5100	41.338	0.72312	0.45289	2.1036	1484.8	177.89	0.36633	0.054695	D	0.499993
240	5200	41.773	0.72963	0.45035	2.0917	1505.4	177.46	0.36507	0.055488	D	0.494793
240	5300	42.194	0.73624	0.44788	2.0802	1525.8	177.05	0.36386	0.056269	D	0.489855
240	5400	42.601	0.74296	0.44548	2.0691	1546	176.66	0.36268	0.057037	D	0.485172
240	5500	42.996	0.74977	0.44316	2.0582	1565.9	176.29	0.36154	0.057794	D	0.480717
240	5600	43.379	0.75666	0.44091	2.0476	1585.6	175.95	0.36043	0.058539	D	0.476471
240	5700	43.751	0.76362	0.43873	2.0374	1605.2	175.62	0.35935	0.059274	D	0.472418
240	5800	44.112	0.77066	0.43661	2.0274	1624.5	175.3	0.3583	0.059998	D	0.468553
240	5900	44.463	0.77776	0.43456	2.0177	1643.5	175.01	0.35728	0.060713	D	0.464855
240	6000	44.804	0.78491	0.43258	2.0083	1662.4	174.73	0.35628	0.061418	D	0.461310
240	6100	45.136	0.79212	0.43065	1.9992	1681.1	174.46	0.35531	0.062114	D	0.457915
240	6200	45.46	0.79938	0.42879	1.9903	1699.5	174.21	0.35437	0.062801	D	0.454659
240	6300	45.775	0.80668	0.42699	1.9817	1717.8	173.97	0.35344	0.06348	D	0.451528
240	6400	46.082	0.81402	0.42524	1.9733	1735.9	173.74	0.35254	0.064151	D	0.448517
240	6500	46.382	0.8214	0.42355	1.9652	1753.7	173.53	0.35166	0.064814	D	0.445620
240	6600	46.674	0.82881	0.42192	1.9573	1771.4	173.32	0.3508	0.06547	D	0.442828
240	6700	46.96	0.83625	0.42033	1.9496	1788.8	173.13	0.34996	0.066119	D	0.440134
240	6800	47.239	0.84372	0.4188	1.9422	1806.1	172.94	0.34914	0.066761	D	0.437535
240	6900	47.512	0.85121	0.41731	1.9349	1823.2	172.77	0.34833	0.067396	D	0.435022
240	7000	47.779	0.85873	0.41587	1.9279	1840.1	172.61	0.34754	0.068025	D	0.432596
240	7100	48.04	0.86626	0.41448	1.9211	1856.9	172.45	0.34676	0.068648	D	0.430243
240	7200	48.295	0.87382	0.41313	1.9145	1873.4	172.3	0.346	0.069265	D	0.427970
240	7300	48.545	0.88139	0.41182	1.908	1889.8	172.16	0.34526	0.069876	D	0.425764
240	7400	48.79	0.88898	0.41055	1.9017	1906	172.03	0.34453	0.070482	D	0.423627
240	7500	49.03	0.89658	0.40932	1.8957	1922.1	171.91	0.34381	0.071083	D	0.421552
240	7600	49.266	0.90419	0.40813	1.8897	1937.9	171.79	0.34311	0.071679	D	0.419537
240	7700	49.496	0.91182	0.40698	1.884	1953.7	171.68	0.34241	0.07227	D	0.417582
240	7800	49.723	0.91945	0.40586	1.8783	1969.2	171.58	0.34173	0.072856	D	0.415678
240	7900	49.945	0.92709	0.40477	1.8729	1984.6	171.48	0.34106	0.073438	D	0.413827
240	8000	50.164	0.93474	0.40371	1.8676	1999.9	171.39	0.3404	0.074015	D	0.412026
240	8100	50.378	0.9424	0.40269	1.8624	2015	171.31	0.33976	0.074589	D	0.410274
240	8200	50.589	0.95006	0.4017	1.8573	2029.9	171.23	0.33912	0.075158	D	0.408565
240	8300	50.796	0.95772	0.40073	1.8524	2044.7	171.15	0.33849	0.075724	D	0.406897
240	8400	50.999	0.96539	0.39979	1.8476	2059.4	171.08	0.33787	0.076285	D	0.405273
240	8500	51.2	0.97307	0.39888	1.8429	2073.9	171.02	0.33726	0.076843	D	0.403691
240	8600	51.396	0.98074	0.39799	1.8383	2088.3	170.96	0.33666	0.077398	D	0.402142
240	8700	51.59	0.98842	0.39713	1.8339	2102.6	170.91	0.33607	0.077949	D	0.400632

TABLE F.1 (continued)—CO_2 PROPERTIES AT VARIOUS TEMPERATURES, °F												
Temperature (°F)	Pressure (psia)	Density (lbm/ft^3)	Compressibility Factor	Heat Capacity (Btu/lbm°F)	Heat Capacity Ratio (C_p/C_v)	Sonic Velocity (ft/s)	Enthalpy (Btu/lbm)	Entropy [Btu/(lbm°F)]	Viscosity (cp)	Phase	Gas Formation Volume Factor (res bbl/Mcf)
240	8800	51.781	0.9961	0.39629	1.8295	2116.7	170.86	0.33549	0.078497	D	0.399157
240	8900	51.969	1.0038	0.39547	1.8252	2130.7	170.81	0.33491	0.079042	D	0.397723
240	9000	52.153	1.0115	0.39468	1.8211	2144.6	170.77	0.33434	0.079583	D	0.396321
240	9100	52.335	1.0191	0.3939	1.817	2158.4	170.73	0.33378	0.080122	D	0.394911
240	9200	52.515	1.0268	0.39315	1.813	2172	170.7	0.33323	0.080658	D	0.393570
240	9300	52.692	1.0345	0.39241	1.8091	2185.5	170.67	0.33269	0.081191	D	0.392258
240	9400	52.866	1.0422	0.3917	1.8053	2198.9	170.65	0.33215	0.081721	D	0.390973
240	9500	53.038	1.0499	0.391	1.8015	2212.2	170.62	0.33162	0.082249	D	0.389716
240	9600	53.207	1.0575	0.39031	1.7979	2225.4	170.6	0.33109	0.082774	D	0.388448
240	9700	53.374	1.0652	0.38965	1.7943	2238.5	170.59	0.33057	0.083297	D	0.387243
240	9800	53.539	1.0729	0.389	1.7908	2251.4	170.58	0.33006	0.083818	D	0.386062
240	9900	53.701	1.0805	0.38837	1.7873	2264.3	170.57	0.32955	0.084336	D	0.384869
240	10000	53.862	1.0882	0.38775	1.784	2277.1	170.56	0.32905	0.084852	D	0.383736
250	14.696	0.085088	0.99806	0.22397	1.2557	1001.5	254.63	0.71376	0.019437	V	242.909494
250	100	0.58559	0.98681	0.22782	1.2697	995.62	253.32	0.62583	0.019476	D	35.295596
250	200	1.1871	0.97357	0.23259	1.2871	988.93	251.75	0.59292	0.019535	D	17.411018
250	300	1.8053	0.96027	0.23768	1.3058	982.43	250.16	0.57297	0.019609	D	11.448777
250	400	2.441	0.94694	0.24309	1.3258	976.11	248.55	0.5583	0.019701	D	8.467388
250	500	3.095	0.93356	0.24886	1.3473	970	246.9	0.54651	0.01981	D	6.678197
250	600	3.7681	0.92015	0.25501	1.3703	964.13	245.23	0.53652	0.019939	D	5.485224
250	700	4.4612	0.90671	0.26155	1.395	958.52	243.53	0.52776	0.02009	D	4.632947
250	800	5.1753	0.89327	0.26852	1.4214	953.21	241.8	0.5199	0.020264	D	3.993740
250	900	5.9111	0.87983	0.27593	1.4497	948.23	240.04	0.51271	0.020462	D	3.496578
250	1000	6.6696	0.86642	0.28381	1.4799	943.63	238.26	0.50604	0.020688	D	3.098956
250	1100	7.4514	0.85306	0.29217	1.5121	939.47	236.45	0.49979	0.020943	D	2.773792
250	1200	8.2573	0.83979	0.30102	1.5463	935.79	234.61	0.49387	0.021229	D	2.503090
250	1300	9.0878	0.82663	0.31036	1.5826	932.68	232.74	0.48823	0.02155	D	2.274337
250	1400	9.9431	0.81364	0.3202	1.6208	930.18	230.85	0.48282	0.021906	D	2.078697
250	1500	10.824	0.80084	0.3305	1.661	928.38	228.94	0.47762	0.022301	D	1.909596
250	1600	11.729	0.7883	0.34124	1.703	927.36	227.02	0.47259	0.022737	D	1.762214
250	1700	12.658	0.77607	0.35236	1.7466	927.19	225.08	0.46771	0.023217	D	1.632822
250	1800	13.611	0.76421	0.36379	1.7914	927.97	223.13	0.46297	0.023741	D	1.518543
250	1900	14.585	0.75278	0.37543	1.8371	929.77	221.17	0.45837	0.024313	D	1.417103
250	2000	15.579	0.74184	0.38716	1.8831	932.69	219.22	0.45388	0.024932	D	1.326683
250	2100	16.59	0.73147	0.39883	1.9289	936.79	217.27	0.44952	0.025599	D	1.245846
250	2200	17.615	0.72173	0.41029	1.9739	942.14	215.34	0.44527	0.026315	D	1.173381
250	2300	18.649	0.71268	0.42135	2.0172	948.8	213.43	0.44114	0.027077	D	1.108291
250	2400	19.69	0.70437	0.43183	2.0583	956.81	211.55	0.43713	0.027884	D	1.049728
250	2500	20.731	0.69686	0.44157	2.0963	966.17	209.71	0.43325	0.028732	D	0.996994
250	2600	21.769	0.69018	0.45042	2.1306	976.88	207.92	0.4295	0.029619	D	0.949459
250	2700	22.798	0.68436	0.45826	2.1609	988.9	206.18	0.42587	0.030539	D	0.906584
250	2800	23.815	0.67942	0.46502	2.1868	1002.2	204.5	0.42238	0.031487	D	0.867895
250	2900	24.815	0.67533	0.47066	2.2081	1016.6	202.88	0.41903	0.032459	D	0.832923
250	3000	25.793	0.67211	0.47518	2.2249	1032.2	201.33	0.41581	0.033449	D	0.801320
250	3100	26.749	0.66971	0.47863	2.2373	1048.8	199.84	0.41273	0.034451	D	0.772702
250	3200	27.678	0.66811	0.48107	2.2456	1066.2	198.43	0.40978	0.035461	D	0.746767

TABLE F.1 (continued)—CO_2 PROPERTIES AT VARIOUS TEMPERATURES, °F											
Temperature (°F)	Pressure (psia)	Density (lbm/ft^3)	Compressibility Factor	Heat Capacity (Btu/lbm°F)	Heat Capacity Ratio (C_p/C_v)	Sonic Velocity (ft/s)	Enthalpy (Btu/lbm)	Entropy [Btu/(lbm°F)]	Viscosity (cp)	Phase	Gas Formation Volume Factor (res bbl/Mcf)
250	3300	28.578	0.66727	0.48259	2.2502	1084.5	197.09	0.40696	0.036474	D	0.723227
250	3400	29.45	0.66715	0.48328	2.2514	1103.5	195.82	0.40426	0.037487	D	0.701829
250	3500	30.292	0.66769	0.48325	2.2497	1123	194.61	0.40169	0.038495	D	0.682329
250	3600	31.103	0.66885	0.48262	2.2455	1143.1	193.47	0.39924	0.039496	D	0.664528
250	3700	31.884	0.67058	0.48148	2.2393	1163.5	192.4	0.3969	0.040487	D	0.648240
250	3800	32.636	0.67285	0.47993	2.2314	1184.3	191.38	0.39466	0.041467	D	0.633318
250	3900	33.359	0.67559	0.47806	2.2222	1205.4	190.43	0.39252	0.042433	D	0.619592
250	4000	34.053	0.67878	0.47594	2.212	1226.6	189.53	0.39048	0.043386	D	0.606954
250	4100	34.721	0.68236	0.47365	2.2011	1247.9	188.68	0.38852	0.044324	D	0.595274
250	4200	35.363	0.68631	0.47123	2.1897	1269.3	187.88	0.38665	0.045246	D	0.584464
250	4300	35.981	0.69059	0.46873	2.178	1290.7	187.12	0.38485	0.046153	D	0.574432
250	4400	36.575	0.69517	0.46619	2.1661	1312	186.41	0.38313	0.047045	D	0.565100
250	4500	37.147	0.70002	0.46363	2.1542	1333.2	185.73	0.38147	0.047922	D	0.556397
250	4600	37.699	0.70511	0.46108	2.1424	1354.4	185.1	0.37988	0.048783	D	0.548259
250	4700	38.23	0.71043	0.45855	2.1306	1375.4	184.5	0.37834	0.04963	D	0.540643
250	4800	38.743	0.71594	0.45605	2.119	1396.2	183.93	0.37686	0.050463	D	0.533485
250	4900	39.238	0.72164	0.4536	2.1075	1416.9	183.39	0.37544	0.051281	D	0.526758
250	5000	39.716	0.7275	0.4512	2.0963	1437.4	182.88	0.37406	0.052087	D	0.520415
250	5100	40.178	0.73351	0.44885	2.0854	1457.7	182.4	0.37273	0.052879	D	0.514426
250	5200	40.625	0.73966	0.44655	2.0746	1477.8	181.94	0.37144	0.053659	D	0.508763
250	5300	41.059	0.74593	0.44431	2.0641	1497.8	181.51	0.37019	0.054427	D	0.503395
250	5400	41.478	0.75232	0.44213	2.0538	1517.5	181.1	0.36898	0.055183	D	0.498306
250	5500	41.885	0.7588	0.44	2.0438	1537	180.71	0.3678	0.055928	D	0.493460
250	5600	42.28	0.76538	0.43793	2.0341	1556.4	180.34	0.36666	0.056663	D	0.488851
250	5700	42.663	0.77205	0.43591	2.0245	1575.5	179.99	0.36555	0.057387	D	0.484460
250	5800	43.036	0.77879	0.43395	2.0152	1594.5	179.66	0.36448	0.058101	D	0.480263
250	5900	43.398	0.78561	0.43204	2.0061	1613.2	179.34	0.36343	0.058806	D	0.476258
250	6000	43.75	0.79249	0.43018	1.9973	1631.8	179.04	0.3624	0.059501	D	0.472421
250	6100	44.093	0.79944	0.42838	1.9887	1650.2	178.76	0.36141	0.060188	D	0.468752
250	6200	44.427	0.80643	0.42662	1.9803	1668.4	178.48	0.36044	0.060866	D	0.465224
250	6300	44.753	0.81348	0.42491	1.9721	1686.4	178.23	0.35949	0.061536	D	0.461842
250	6400	45.07	0.82058	0.42326	1.9641	1704.3	177.98	0.35856	0.062198	D	0.458594
250	6500	45.379	0.82772	0.42165	1.9563	1721.9	177.75	0.35766	0.062853	D	0.455467
250	6600	45.681	0.83489	0.42008	1.9488	1739.4	177.53	0.35678	0.0635	D	0.452452
250	6700	45.976	0.84211	0.41856	1.9414	1756.7	177.32	0.35591	0.06414	D	0.449553
250	6800	46.264	0.84935	0.41709	1.9342	1773.9	177.12	0.35507	0.064773	D	0.446750
250	6900	46.546	0.85663	0.41565	1.9272	1790.8	176.94	0.35424	0.0654	D	0.444049
250	7000	46.821	0.86394	0.41426	1.9204	1807.6	176.76	0.35343	0.066021	D	0.441441
250	7100	47.091	0.87127	0.41291	1.9138	1824.2	176.59	0.35264	0.066635	D	0.438916
250	7200	47.354	0.87862	0.4116	1.9073	1840.7	176.43	0.35186	0.067244	D	0.436471
250	7300	47.612	0.88599	0.41033	1.901	1857	176.28	0.35109	0.067847	D	0.434103
250	7400	47.865	0.89339	0.40909	1.8949	1873.1	176.13	0.35034	0.068445	D	0.431814
250	7500	48.113	0.9008	0.40789	1.8889	1889	176	0.34961	0.069038	D	0.429590
250	7600	48.356	0.90822	0.40673	1.8831	1904.9	175.87	0.34889	0.069625	D	0.427429
250	7700	48.594	0.91566	0.40559	1.8775	1920.5	175.75	0.34818	0.070208	D	0.425334
250	7800	48.827	0.92312	0.4045	1.8719	1936	175.63	0.34748	0.070786	D	0.423302

TABLE F.1 (continued)—CO_2 PROPERTIES AT VARIOUS TEMPERATURES, °F											
Temperature (°F)	Pressure (psia)	Density (lbm/ft^3)	Compressibility Factor	Heat Capacity (Btu/lbm°F)	Heat Capacity Ratio (C_p/C_v)	Sonic Velocity (ft/s)	Enthalpy (Btu/lbm)	Entropy [Btu/(lbm°F)]	Viscosity (cp)	Phase	Gas Formation Volume Factor (res bbl/Mcf)
250	7900	49.057	0.93058	0.40343	1.8666	1951.4	175.52	0.3468	0.071359	D	0.421321
250	8000	49.282	0.93806	0.40239	1.8613	1966.6	175.42	0.34612	0.071928	D	0.419399
250	8100	49.503	0.94555	0.40138	1.8562	1981.6	175.33	0.34546	0.072493	D	0.417529
250	8200	49.72	0.95304	0.4004	1.8512	1996.6	175.24	0.34481	0.073054	D	0.415704
250	8300	49.933	0.96054	0.39945	1.8464	2011.4	175.15	0.34417	0.073611	D	0.413928
250	8400	50.143	0.96805	0.39852	1.8416	2026	175.08	0.34354	0.074164	D	0.412198
250	8500	50.349	0.97557	0.39762	1.837	2040.5	175	0.34291	0.074714	D	0.410513
250	8600	50.552	0.98308	0.39674	1.8324	2054.9	174.94	0.3423	0.07526	D	0.408863
250	8700	50.751	0.99061	0.39589	1.828	2069.1	174.87	0.3417	0.075802	D	0.407259
250	8800	50.947	0.99813	0.39506	1.8237	2083.3	174.81	0.3411	0.076342	D	0.405687
250	8900	51.14	1.0057	0.39425	1.8195	2097.3	174.76	0.34052	0.076878	D	0.404171
250	9000	51.331	1.0132	0.39346	1.8153	2111.1	174.71	0.33994	0.077411	D	0.402661
250	9100	51.518	1.0207	0.39269	1.8113	2124.9	174.67	0.33937	0.077941	D	0.401184
250	9200	51.702	1.0283	0.39195	1.8073	2138.5	174.63	0.3388	0.078468	D	0.399778
250	9300	51.884	1.0358	0.39122	1.8035	2152.1	174.59	0.33825	0.078992	D	0.398364
250	9400	52.063	1.0433	0.39051	1.7997	2165.5	174.56	0.3377	0.079514	D	0.396980
250	9500	52.24	1.0509	0.38981	1.796	2178.8	174.53	0.33716	0.080032	D	0.395662
250	9600	52.414	1.0584	0.38914	1.7924	2192	174.5	0.33662	0.080549	D	0.394335
250	9700	52.585	1.0659	0.38848	1.7888	2205	174.48	0.33609	0.081063	D	0.393035
250	9800	52.755	1.0735	0.38783	1.7853	2218	174.46	0.33557	0.081574	D	0.391799
250	9900	52.922	1.081	0.38721	1.7819	2230.9	174.45	0.33506	0.082083	D	0.390551
250	10000	53.086	1.0885	0.38659	1.7785	2243.7	174.43	0.33455	0.08259	D	0.389328
260	14.696	0.083897	0.99816	0.22505	1.2541	1007.9	256.87	0.7169	0.019688	V	246.357027
260	100	0.57707	0.98747	0.22873	1.2674	1002.4	255.6	0.62902	0.019726	D	35.816888
260	200	1.169	0.97491	0.23328	1.284	996.07	254.08	0.59618	0.019784	D	17.680660
260	300	1.7764	0.96233	0.2381	1.3017	989.94	252.54	0.5763	0.019856	D	11.635009
260	400	2.4	0.94972	0.24322	1.3206	984.01	250.98	0.5617	0.019945	D	8.611911
260	500	3.0405	0.93709	0.24865	1.3408	978.3	249.39	0.54999	0.020051	D	6.797907
260	600	3.6984	0.92445	0.25441	1.3624	972.83	247.78	0.54008	0.020176	D	5.588511
260	700	4.3746	0.91182	0.26052	1.3854	967.62	246.14	0.53142	0.020321	D	4.724708
260	800	5.0698	0.89919	0.26699	1.4099	962.7	244.48	0.52365	0.020487	D	4.076856
260	900	5.7845	0.8866	0.27384	1.436	958.11	242.79	0.51656	0.020677	D	3.573133
260	1000	6.5195	0.87405	0.28109	1.4637	953.9	241.08	0.50999	0.020892	D	3.170299
260	1100	7.2753	0.86158	0.28874	1.4931	950.1	239.35	0.50385	0.021134	D	2.840971
260	1200	8.0523	0.8492	0.29679	1.5242	946.76	237.6	0.49805	0.021405	D	2.566804
260	1300	8.851	0.83695	0.30526	1.5569	943.95	235.82	0.49253	0.021706	D	2.335179
260	1400	9.6714	0.82487	0.31412	1.5913	941.72	234.02	0.48726	0.022041	D	2.137083
260	1500	10.514	0.813	0.32337	1.6273	940.13	232.21	0.48219	0.022409	D	1.965908
260	1600	11.377	0.80137	0.33296	1.6647	939.26	230.39	0.4773	0.022815	D	1.816674
260	1700	12.262	0.79003	0.34286	1.7033	939.16	228.55	0.47257	0.023259	D	1.685616
260	1800	13.166	0.77903	0.35301	1.743	939.91	226.71	0.46799	0.023743	D	1.569805
260	1900	14.09	0.76843	0.36333	1.7834	941.59	224.86	0.46353	0.024268	D	1.466948
260	2000	15.03	0.75828	0.37373	1.8242	944.24	223.02	0.4592	0.024836	D	1.375193
260	2100	15.985	0.74863	0.38412	1.8648	947.95	221.18	0.45499	0.025446	D	1.293040
260	2200	16.952	0.73954	0.39435	1.9049	952.76	219.36	0.45089	0.026098	D	1.219279
260	2300	17.928	0.73105	0.40431	1.9439	958.72	217.56	0.44691	0.026792	D	1.152878

TABLE F.1 (continued)—CO_2 PROPERTIES AT VARIOUS TEMPERATURES, °F											
Temperature (°F)	Pressure (psia)	Density (lbm/ft^3)	Compressibility Factor	Heat Capacity (Btu/lbm°F)	Heat Capacity Ratio (C_p/C_v)	Sonic Velocity (ft/s)	Enthalpy (Btu/lbm)	Entropy [Btu/(lbm°F)]	Viscosity (cp)	Phase	Gas Formation Volume Factor (res bbl/Mcf)
260	2400	18.91	0.72321	0.41387	1.9812	965.87	215.78	0.44305	0.027526	D	1.092992
260	2500	19.895	0.71607	0.42289	2.0163	974.21	214.03	0.4393	0.028298	D	1.038914
260	2600	20.877	0.70965	0.43125	2.0487	983.76	212.33	0.43566	0.029105	D	0.989999
260	2700	21.855	0.70399	0.43885	2.0781	994.48	210.66	0.43215	0.029945	D	0.945729
260	2800	22.823	0.69908	0.44561	2.1041	1006.4	209.05	0.42875	0.030812	D	0.905592
260	2900	23.779	0.69495	0.4515	2.1265	1019.3	207.49	0.42548	0.031703	D	0.869200
260	3000	24.719	0.69158	0.45649	2.1452	1033.3	205.98	0.42233	0.032614	D	0.836152
260	3100	25.64	0.68897	0.46058	2.1603	1048.2	204.54	0.4193	0.033541	D	0.806125
260	3200	26.54	0.68707	0.4638	2.1719	1064.1	203.15	0.41639	0.034478	D	0.778780
260	3300	27.417	0.68588	0.46621	2.1802	1080.7	201.83	0.41359	0.035423	D	0.753873
260	3400	28.269	0.68536	0.46787	2.1854	1098	200.57	0.41092	0.036372	D	0.731145
260	3500	29.096	0.68546	0.46884	2.1878	1115.9	199.37	0.40835	0.03732	D	0.710359
260	3600	29.897	0.68616	0.4692	2.1877	1134.4	198.23	0.4059	0.038266	D	0.691332
260	3700	30.671	0.68742	0.46903	2.1855	1153.4	197.15	0.40355	0.039207	D	0.673883
260	3800	31.419	0.68919	0.46841	2.1815	1172.7	196.13	0.40129	0.040141	D	0.657839
260	3900	32.141	0.69143	0.46742	2.176	1192.4	195.16	0.39914	0.041066	D	0.643054
260	4000	32.838	0.69412	0.46611	2.1692	1212.3	194.24	0.39707	0.041981	D	0.629417
260	4100	33.509	0.69722	0.46457	2.1614	1232.4	193.37	0.39508	0.042884	D	0.616808
260	4200	34.157	0.70069	0.46283	2.1528	1252.6	192.55	0.39318	0.043775	D	0.605119
260	4300	34.781	0.70449	0.46094	2.1437	1272.9	191.77	0.39135	0.044653	D	0.594252
260	4400	35.383	0.70861	0.45896	2.1341	1293.3	191.03	0.3896	0.045518	D	0.584142
260	4500	35.964	0.71302	0.4569	2.1243	1313.6	190.34	0.38791	0.046371	D	0.574716
260	4600	36.524	0.71768	0.4548	2.1144	1333.9	189.68	0.38628	0.04721	D	0.565896
260	4700	37.065	0.72257	0.45268	2.1043	1354.1	189.05	0.38472	0.048036	D	0.557630
260	4800	37.588	0.72769	0.45056	2.0943	1374.2	188.46	0.38321	0.048849	D	0.549881
260	4900	38.093	0.733	0.44844	2.0843	1394.3	187.9	0.38175	0.04965	D	0.542590
260	5000	38.582	0.73848	0.44635	2.0744	1414.1	187.37	0.38034	0.050439	D	0.535714
260	5100	39.054	0.74413	0.44428	2.0647	1433.9	186.87	0.37898	0.051215	D	0.529228
260	5200	39.512	0.74993	0.44224	2.055	1453.4	186.39	0.37766	0.05198	D	0.523096
260	5300	39.956	0.75586	0.44024	2.0456	1472.9	185.93	0.37638	0.052734	D	0.517284
260	5400	40.386	0.76192	0.43828	2.0363	1492.1	185.5	0.37514	0.053477	D	0.511776
260	5500	40.804	0.76809	0.43636	2.0272	1511.2	185.09	0.37394	0.054209	D	0.506540
260	5600	41.209	0.77436	0.43448	2.0182	1530.1	184.7	0.37277	0.054931	D	0.501555
260	5700	41.603	0.78073	0.43265	2.0095	1548.9	184.33	0.37163	0.055644	D	0.496810
260	5800	41.985	0.78719	0.43085	2.0009	1567.5	183.98	0.37053	0.056347	D	0.492284
260	5900	42.358	0.79373	0.4291	1.9925	1585.9	183.65	0.36945	0.05704	D	0.487961
260	6000	42.72	0.80034	0.42738	1.9843	1604.1	183.33	0.3684	0.057725	D	0.483824
260	6100	43.072	0.80702	0.42571	1.9763	1622.2	183.03	0.36738	0.058402	D	0.479864
260	6200	43.416	0.81376	0.42408	1.9684	1640.1	182.74	0.36639	0.05907	D	0.476068
260	6300	43.75	0.82056	0.42249	1.9607	1657.8	182.46	0.36542	0.05973	D	0.472426
260	6400	44.077	0.82741	0.42094	1.9532	1675.4	182.2	0.36447	0.060383	D	0.468926
260	6500	44.395	0.83431	0.41942	1.9459	1692.8	181.96	0.36355	0.061028	D	0.465563
260	6600	44.706	0.84125	0.41795	1.9387	1710	181.72	0.36264	0.061666	D	0.462323
260	6700	45.01	0.84824	0.41651	1.9317	1727.1	181.5	0.36176	0.062297	D	0.459206
260	6800	45.306	0.85526	0.41511	1.9248	1744.1	181.28	0.36089	0.062922	D	0.456198
260	6900	45.596	0.86232	0.41375	1.9182	1760.8	181.08	0.36004	0.06354	D	0.453297

TABLE F.1 (continued)—CO_2 PROPERTIES AT VARIOUS TEMPERATURES, °F											
Temperature (°F)	Pressure (psia)	Density (lbm/ft^3)	Compressibility Factor	Heat Capacity (Btu/lbm°F)	Heat Capacity Ratio (C_p/C_v)	Sonic Velocity (ft/s)	Enthalpy (Btu/lbm)	Entropy [Btu/(lbm°F)]	Viscosity (cp)	Phase	Gas Formation Volume Factor (res bbl/Mcf)
260	7000	45.88	0.86942	0.41242	1.9116	1777.4	180.89	0.35921	0.064152	D	0.450501
260	7100	46.157	0.87654	0.41113	1.9053	1793.9	180.71	0.3584	0.064758	D	0.447793
260	7200	46.428	0.88369	0.40987	1.899	1810.2	180.53	0.3576	0.065358	D	0.445176
260	7300	46.694	0.89086	0.40865	1.893	1826.4	180.37	0.35682	0.065953	D	0.442640
260	7400	46.954	0.89806	0.40745	1.887	1842.4	180.21	0.35606	0.066542	D	0.440187
260	7500	47.209	0.90528	0.40629	1.8813	1858.2	180.07	0.35531	0.067127	D	0.437810
260	7600	47.459	0.91252	0.40517	1.8756	1873.9	179.93	0.35457	0.067706	D	0.435505
260	7700	47.704	0.91978	0.40407	1.8701	1889.5	179.79	0.35384	0.06828	D	0.433269
260	7800	47.945	0.92705	0.403	1.8647	1904.9	179.67	0.35313	0.06885	D	0.431095
260	7900	48.181	0.93434	0.40196	1.8595	1920.2	179.55	0.35243	0.069415	D	0.428985
260	8000	48.412	0.94164	0.40095	1.8543	1935.3	179.44	0.35174	0.069976	D	0.426932
260	8100	48.639	0.94896	0.39996	1.8493	1950.3	179.33	0.35107	0.070533	D	0.424939
260	8200	48.863	0.95629	0.399	1.8444	1965.2	179.24	0.3504	0.071085	D	0.422999
260	8300	49.082	0.96362	0.39807	1.8397	1979.9	179.14	0.34975	0.071634	D	0.421106
260	8400	49.297	0.97097	0.39716	1.835	1994.5	179.06	0.3491	0.072179	D	0.419267
260	8500	49.509	0.97832	0.39628	1.8304	2009	178.97	0.34847	0.07272	D	0.417471
260	8600	49.718	0.98568	0.39541	1.826	2023.3	178.9	0.34784	0.073257	D	0.415720
260	8700	49.923	0.99305	0.39457	1.8216	2037.5	178.83	0.34723	0.073792	D	0.414015
260	8800	50.124	1.0004	0.39376	1.8174	2051.6	178.76	0.34662	0.074323	D	0.412340
260	8900	50.323	1.0078	0.39296	1.8132	2065.6	178.7	0.34602	0.07485	D	0.410722
260	9000	50.518	1.0152	0.39218	1.8091	2079.5	178.64	0.34543	0.075375	D	0.409141
260	9100	50.711	1.0226	0.39143	1.8051	2093.2	178.59	0.34485	0.075896	D	0.407595
260	9200	50.9	1.03	0.39069	1.8012	2106.8	178.54	0.34428	0.076415	D	0.406082
260	9300	51.087	1.0374	0.38997	1.7974	2120.3	178.5	0.34371	0.076931	D	0.404601
260	9400	51.27	1.0447	0.38927	1.7937	2133.7	178.45	0.34315	0.077444	D	0.403114
260	9500	51.452	1.0521	0.38858	1.79	2147	178.42	0.3426	0.077954	D	0.401696
260	9600	51.63	1.0595	0.38792	1.7864	2160.2	178.39	0.34206	0.078462	D	0.400307
260	9700	51.806	1.0669	0.38726	1.7829	2173.2	178.36	0.34152	0.078967	D	0.398948
260	9800	51.98	1.0743	0.38663	1.7794	2186.2	178.33	0.34099	0.07947	D	0.397616
260	9900	52.151	1.0817	0.38601	1.7761	2199.1	178.31	0.34047	0.07997	D	0.396310
260	10000	52.32	1.0891	0.3854	1.7728	2211.8	178.29	0.33995	0.080468	D	0.395031
270	14.696	0.08274	0.99825	0.22613	1.2524	1014.3	259.13	0.72001	0.019937	V	249.802743
270	100	0.5688	0.9881	0.22964	1.2652	1009.1	257.89	0.63218	0.019975	D	36.337741
270	200	1.1515	0.97618	0.23398	1.281	1003.1	256.42	0.5994	0.020031	D	17.949689
270	300	1.7486	0.96426	0.23856	1.2978	997.36	254.92	0.57959	0.020102	D	11.820339
270	400	2.3606	0.95233	0.24341	1.3158	991.8	253.41	0.56506	0.020189	D	8.755572
270	500	2.9882	0.94041	0.24853	1.3348	986.46	251.88	0.55342	0.020291	D	6.916785
270	600	3.6319	0.92849	0.25394	1.3551	981.35	250.32	0.54359	0.020412	D	5.690927
270	700	4.2922	0.91659	0.25966	1.3766	976.51	248.74	0.53501	0.020552	D	4.815419
270	800	4.9697	0.90473	0.26569	1.3994	971.97	247.14	0.52732	0.020712	D	4.158972
270	900	5.6649	0.89291	0.27205	1.4235	967.74	245.52	0.52032	0.020894	D	3.648566
270	1000	6.3783	0.88116	0.27874	1.4491	963.88	243.88	0.51385	0.021099	D	3.240498
270	1100	7.1102	0.86949	0.28577	1.476	960.42	242.22	0.50781	0.021329	D	2.906892
270	1200	7.8611	0.85793	0.29315	1.5044	957.4	240.54	0.50212	0.021586	D	2.629224
270	1300	8.6311	0.84652	0.30086	1.5341	954.87	238.85	0.49672	0.02187	D	2.394699
270	1400	9.4202	0.83526	0.3089	1.5653	952.89	237.14	0.49156	0.022185	D	2.194071

TABLE F.1 (continued)—CO_2 PROPERTIES AT VARIOUS TEMPERATURES, °F

Temperature (°F)	Pressure (psia)	Density (lbm/ft^3)	Compressibility Factor	Heat Capacity (Btu/lbm°F)	Heat Capacity Ratio (C_p/C_v)	Sonic Velocity (ft/s)	Enthalpy (Btu/lbm)	Entropy [Btu/(lbm°F)]	Viscosity (cp)	Phase	Gas Formation Volume Factor (res bbl/Mcf)
270	1500	10.228	0.82422	0.31725	1.5977	951.5	235.42	0.48661	0.022531	D	2.020733
270	1600	11.055	0.81341	0.32589	1.6312	950.77	233.68	0.48185	0.02291	D	1.869591
270	1700	11.9	0.80288	0.33477	1.6658	950.76	231.94	0.47725	0.023324	D	1.736836
270	1800	12.763	0.79267	0.34386	1.7013	951.51	230.19	0.47279	0.023773	D	1.619485
270	1900	13.641	0.78282	0.35309	1.7373	953.09	228.44	0.46847	0.024259	D	1.515184
270	2000	14.534	0.77339	0.36239	1.7736	955.56	226.7	0.46428	0.024782	D	1.422085
270	2100	15.44	0.7644	0.37169	1.8099	958.96	224.96	0.4602	0.025343	D	1.338623
270	2200	16.357	0.75591	0.38088	1.8458	963.35	223.23	0.45624	0.025942	D	1.263585
270	2300	17.282	0.74796	0.38988	1.881	968.76	221.53	0.45239	0.026578	D	1.195935
270	2400	18.213	0.74059	0.39858	1.9149	975.23	219.84	0.44865	0.027251	D	1.134811
270	2500	19.147	0.73383	0.40689	1.9471	982.76	218.18	0.44502	0.027958	D	1.079475
270	2600	20.08	0.72771	0.4147	1.9774	991.37	216.55	0.4415	0.028697	D	1.029300
270	2700	21.01	0.72225	0.42194	2.0054	1001	214.97	0.43808	0.029466	D	0.983741
270	2800	21.934	0.71747	0.42853	2.0307	1011.7	213.42	0.43478	0.030263	D	0.942329
270	2900	22.848	0.71337	0.43442	2.0532	1023.5	211.92	0.43159	0.031083	D	0.904636
270	3000	23.749	0.70996	0.43959	2.0727	1036.1	210.46	0.42851	0.031924	D	0.870301
270	3100	24.636	0.70722	0.44403	2.0893	1049.7	209.06	0.42554	0.032781	D	0.838977
270	3200	25.505	0.70514	0.44774	2.1029	1064.1	207.71	0.42267	0.033652	D	0.810368
270	3300	26.356	0.7037	0.45075	2.1136	1079.3	206.42	0.41992	0.034532	D	0.784207
270	3400	27.186	0.70288	0.45309	2.1217	1095.1	205.17	0.41727	0.03542	D	0.760255
270	3500	27.995	0.70266	0.4548	2.1272	1111.6	203.99	0.41473	0.036311	D	0.738302
270	3600	28.781	0.70299	0.45594	2.1304	1128.7	202.86	0.41228	0.037203	D	0.718131
270	3700	29.545	0.70385	0.45657	2.1315	1146.2	201.78	0.40993	0.038094	D	0.699577
270	3800	30.285	0.70521	0.45674	2.1308	1164.2	200.75	0.40768	0.038981	D	0.682483
270	3900	31.001	0.70703	0.45652	2.1284	1182.5	199.77	0.40551	0.039863	D	0.666700
270	4000	31.695	0.70929	0.45595	2.1247	1201.2	198.85	0.40343	0.040738	D	0.652110
270	4100	32.366	0.71195	0.45509	2.1198	1220.1	197.97	0.40143	0.041604	D	0.638591
270	4200	33.015	0.71499	0.45399	2.1139	1239.2	197.13	0.39951	0.042462	D	0.626048
270	4300	33.642	0.71836	0.45271	2.1073	1258.4	196.34	0.39766	0.043309	D	0.614371
270	4400	34.248	0.72206	0.45127	2.1	1277.8	195.58	0.39588	0.044146	D	0.603501
270	4500	34.834	0.72605	0.44971	2.0923	1297.2	194.87	0.39417	0.044972	D	0.593350
270	4600	35.401	0.73031	0.44807	2.0843	1316.6	194.19	0.39251	0.045787	D	0.583857
270	4700	35.948	0.73481	0.44636	2.076	1336	193.55	0.39092	0.04659	D	0.574955
270	4800	36.479	0.73954	0.44462	2.0676	1355.4	192.94	0.38938	0.047383	D	0.566601
270	4900	36.992	0.74447	0.44285	2.0591	1374.7	192.36	0.3879	0.048164	D	0.558738
270	5000	37.488	0.7496	0.44107	2.0505	1393.9	191.81	0.38646	0.048934	D	0.551336
270	5100	37.97	0.7549	0.43929	2.042	1413	191.28	0.38507	0.049693	D	0.544348
270	5200	38.436	0.76036	0.43753	2.0335	1432	190.79	0.38373	0.050442	D	0.537741
270	5300	38.889	0.76596	0.43578	2.0251	1450.9	190.31	0.38242	0.05118	D	0.531480
270	5400	39.328	0.7717	0.43404	2.0168	1469.7	189.86	0.38116	0.051908	D	0.525547
270	5500	39.754	0.77756	0.43234	2.0086	1488.3	189.44	0.37993	0.052627	D	0.519910
270	5600	40.168	0.78354	0.43066	2.0005	1506.7	189.03	0.37874	0.053336	D	0.514553
270	5700	40.571	0.78962	0.42901	1.9926	1525	188.64	0.37758	0.054035	D	0.509449
270	5800	40.962	0.79579	0.42739	1.9848	1543.2	188.27	0.37645	0.054726	D	0.504577
270	5900	41.343	0.80205	0.4258	1.9771	1561.2	187.92	0.37535	0.055408	D	0.499927
270	6000	41.714	0.8084	0.42424	1.9695	1579.1	187.59	0.37428	0.056081	D	0.495487

TABLE F.1 (continued)—CO_2 PROPERTIES AT VARIOUS TEMPERATURES, °F

Temperature (°F)	Pressure (psia)	Density (lbm/ft^3)	Compressibility Factor	Heat Capacity (Btu/lbm°F)	Heat Capacity Ratio (C_p/C_v)	Sonic Velocity (ft/s)	Enthalpy (Btu/lbm)	Entropy [Btu/(lbm°F)]	Viscosity (cp)	Phase	Gas Formation Volume Factor (res bbl/Mcf)
270	6100	42.075	0.81481	0.42271	1.9621	1596.8	187.27	0.37324	0.056746	D	0.491228
270	6200	42.427	0.8213	0.42121	1.9549	1614.4	186.97	0.37222	0.057404	D	0.487155
270	6300	42.771	0.82785	0.41975	1.9477	1631.8	186.68	0.37123	0.058054	D	0.483246
270	6400	43.105	0.83446	0.41831	1.9408	1649.1	186.4	0.37026	0.058696	D	0.479493
270	6500	43.432	0.84112	0.41691	1.9339	1666.2	186.14	0.36932	0.059332	D	0.475885
270	6600	43.751	0.84784	0.41554	1.9272	1683.2	185.89	0.36839	0.05996	D	0.472419
270	6700	44.062	0.8546	0.4142	1.9206	1700	185.65	0.36749	0.060582	D	0.469078
270	6800	44.367	0.86141	0.41288	1.9141	1716.7	185.42	0.3666	0.061197	D	0.465863
270	6900	44.664	0.86825	0.4116	1.9078	1733.2	185.21	0.36574	0.061806	D	0.462757
270	7000	44.955	0.87513	0.41035	1.9016	1749.6	185	0.36489	0.062409	D	0.459760
270	7100	45.24	0.88205	0.40913	1.8956	1765.9	184.81	0.36406	0.063007	D	0.456869
270	7200	45.519	0.88899	0.40794	1.8897	1782	184.62	0.36325	0.063598	D	0.454069
270	7300	45.792	0.89597	0.40677	1.8838	1798	184.45	0.36245	0.064184	D	0.451365
270	7400	46.059	0.90298	0.40564	1.8782	1813.8	184.28	0.36167	0.064765	D	0.448749
270	7500	46.321	0.91001	0.40453	1.8726	1829.5	184.12	0.3609	0.065341	D	0.446213
270	7600	46.577	0.91706	0.40344	1.8672	1845.1	183.97	0.36015	0.065912	D	0.443753
270	7700	46.829	0.92413	0.40239	1.8619	1860.5	183.83	0.35941	0.066478	D	0.441367
270	7800	47.076	0.93123	0.40136	1.8567	1875.8	183.69	0.35868	0.067039	D	0.439055
270	7900	47.318	0.93834	0.40036	1.8516	1891	183.56	0.35797	0.067596	D	0.436808
270	8000	47.556	0.94547	0.39938	1.8466	1906	183.44	0.35727	0.068149	D	0.434625
270	8100	47.789	0.95261	0.39843	1.8417	1920.9	183.33	0.35658	0.068697	D	0.432501
270	8200	48.018	0.95977	0.3975	1.837	1935.7	183.22	0.3559	0.069242	D	0.430438
270	8300	48.243	0.96694	0.39659	1.8323	1950.4	183.12	0.35523	0.069782	D	0.428429
270	8400	48.464	0.97412	0.3957	1.8278	1964.9	183.02	0.35457	0.070319	D	0.426472
270	8500	48.682	0.98131	0.39484	1.8233	1979.3	182.93	0.35393	0.070852	D	0.424565
270	8600	48.896	0.98852	0.394	1.8189	1993.6	182.84	0.35329	0.071381	D	0.422711
270	8700	49.106	0.99573	0.39318	1.8147	2007.7	182.76	0.35266	0.071907	D	0.420900
270	8800	49.313	1.0029	0.39238	1.8105	2021.8	182.69	0.35205	0.07243	D	0.419114
270	8900	49.517	1.0102	0.3916	1.8064	2035.7	182.62	0.35144	0.072949	D	0.417421
270	9000	49.717	1.0174	0.39084	1.8024	2049.5	182.56	0.35084	0.073466	D	0.415725
270	9100	49.914	1.0246	0.3901	1.7985	2063.2	182.5	0.35024	0.073979	D	0.414066
270	9200	50.108	1.0319	0.38938	1.7947	2076.8	182.44	0.34966	0.074489	D	0.412484
270	9300	50.3	1.0391	0.38867	1.7909	2090.2	182.39	0.34909	0.074997	D	0.410896
270	9400	50.488	1.0464	0.38798	1.7872	2103.6	182.34	0.34852	0.075501	D	0.409380
270	9500	50.674	1.0536	0.38731	1.7836	2116.9	182.3	0.34796	0.076003	D	0.407858
270	9600	50.857	1.0609	0.38665	1.7801	2130	182.26	0.3474	0.076503	D	0.406406
270	9700	51.038	1.0682	0.38601	1.7766	2143.1	182.22	0.34686	0.077	D	0.404984
270	9800	51.216	1.0754	0.38538	1.7732	2156	182.19	0.34632	0.077494	D	0.403553
270	9900	51.391	1.0827	0.38477	1.7699	2168.9	182.17	0.34578	0.077986	D	0.402189
270	10000	51.564	1.09	0.38417	1.7666	2181.6	182.14	0.34526	0.078476	D	0.400852
280	14.696	0.081614	0.99834	0.22719	1.2509	1020.7	261.4	0.72309	0.020185	V	253.249076
280	100	0.56077	0.98869	0.23055	1.2631	1015.7	260.19	0.63531	0.020222	D	36.857738
280	200	1.1345	0.97738	0.23469	1.2782	1010.1	258.76	0.60259	0.020278	D	18.218054
280	300	1.7217	0.96608	0.23905	1.2942	1004.7	257.31	0.58284	0.020347	D	12.004951
280	400	2.3227	0.95479	0.24365	1.3112	999.47	255.85	0.56838	0.020431	D	8.898492
280	500	2.9381	0.94353	0.24849	1.3292	994.47	254.36	0.5568	0.020531	D	7.034840

TABLE F.1 (continued)—CO_2 PROPERTIES AT VARIOUS TEMPERATURES, °F

Temperature (°F)	Pressure (psia)	Density (lbm/ft^3)	Compressibility Factor	Heat Capacity (Btu/lbm°F)	Heat Capacity Ratio (C_p/C_v)	Sonic Velocity (ft/s)	Enthalpy (Btu/lbm)	Entropy [Btu/(lbm°F)]	Viscosity (cp)	Phase	Gas Formation Volume Factor (res bbl/Mcf)
280	600	3.5682	0.93228	0.25359	1.3482	989.72	252.86	0.54705	0.020648	D	5.792468
280	700	4.2136	0.92107	0.25895	1.3683	985.23	251.33	0.53854	0.020783	D	4.905272
280	800	4.8746	0.90991	0.26458	1.3896	981.03	249.79	0.53093	0.020937	D	4.240109
280	900	5.5516	0.89881	0.2705	1.4121	977.14	248.23	0.52401	0.021111	D	3.723008
280	1000	6.245	0.88779	0.27671	1.4357	973.61	246.66	0.51763	0.021308	D	3.309625
280	1100	6.9551	0.87687	0.2832	1.4605	970.46	245.07	0.51168	0.021527	D	2.971742
280	1200	7.6821	0.86606	0.28998	1.4865	967.73	243.46	0.50609	0.021771	D	2.690514
280	1300	8.426	0.8554	0.29705	1.5137	965.47	241.84	0.50078	0.022041	D	2.452982
280	1400	9.1868	0.8449	0.30439	1.5421	963.71	240.21	0.49573	0.022338	D	2.249810
280	1500	9.9644	0.83461	0.31198	1.5715	962.52	238.56	0.49089	0.022664	D	2.074249
280	1600	10.758	0.82455	0.31981	1.6018	961.93	236.91	0.48624	0.02302	D	1.921169
280	1700	11.568	0.81475	0.32784	1.633	962	235.25	0.48175	0.023406	D	1.786669
280	1800	12.393	0.80526	0.33603	1.6648	962.77	233.59	0.47742	0.023825	D	1.667755
280	1900	13.232	0.7961	0.34435	1.6972	964.3	231.93	0.47322	0.024277	D	1.562006
280	2000	14.084	0.78732	0.35272	1.7298	966.63	230.27	0.46914	0.024763	D	1.467540
280	2100	14.947	0.77895	0.3611	1.7624	969.8	228.62	0.46519	0.025282	D	1.382798
280	2200	15.82	0.77103	0.3694	1.7948	973.85	226.98	0.46135	0.025836	D	1.306523
280	2300	16.7	0.76359	0.37756	1.8265	978.83	225.36	0.45761	0.026423	D	1.237659
280	2400	17.585	0.75666	0.3855	1.8574	984.75	223.76	0.45398	0.027042	D	1.175325
280	2500	18.474	0.75029	0.39313	1.887	991.62	222.18	0.45046	0.027693	D	1.118813
280	2600	19.363	0.74448	0.4004	1.9151	999.46	220.63	0.44704	0.028374	D	1.067452
280	2700	20.25	0.73926	0.40721	1.9414	1008.3	219.11	0.44372	0.029083	D	1.020709
280	2800	21.131	0.73464	0.41352	1.9656	1018	217.63	0.44051	0.029817	D	0.978104
280	2900	22.006	0.73063	0.41929	1.9876	1028.7	216.18	0.4374	0.030575	D	0.939222
280	3000	22.871	0.72724	0.42447	2.0073	1040.2	214.78	0.43439	0.031353	D	0.903702
280	3100	23.724	0.72446	0.42904	2.0245	1052.6	213.42	0.43148	0.032148	D	0.871207
280	3200	24.564	0.72228	0.43302	2.0392	1065.8	212.11	0.42867	0.032958	D	0.841442
280	3300	25.387	0.72068	0.43639	2.0515	1079.7	210.85	0.42596	0.033779	D	0.814136
280	3400	26.194	0.71966	0.43919	2.0615	1094.3	209.64	0.42334	0.034609	D	0.789073
280	3500	26.982	0.71919	0.44143	2.0692	1109.5	208.47	0.42083	0.035446	D	0.766027
280	3600	27.751	0.71923	0.44316	2.0749	1125.3	207.35	0.4184	0.036286	D	0.744790
280	3700	28.5	0.71978	0.4444	2.0785	1141.6	206.28	0.41606	0.037128	D	0.725215
280	3800	29.229	0.72081	0.44521	2.0804	1158.3	205.26	0.41382	0.037969	D	0.707141
280	3900	29.937	0.72228	0.44562	2.0808	1175.4	204.29	0.41165	0.038807	D	0.690414
280	4000	30.624	0.72417	0.44568	2.0797	1192.9	203.36	0.40957	0.039642	D	0.674915
280	4100	31.291	0.72646	0.44544	2.0773	1210.6	202.47	0.40756	0.040471	D	0.660536
280	4200	31.938	0.72911	0.44493	2.0739	1228.6	201.62	0.40563	0.041293	D	0.647161
280	4300	32.564	0.7321	0.4442	2.0696	1246.8	200.82	0.40376	0.042109	D	0.634703
280	4400	33.171	0.73542	0.44328	2.0645	1265.2	200.06	0.40197	0.042916	D	0.623091
280	4500	33.76	0.73903	0.44221	2.0588	1283.6	199.33	0.40024	0.043714	D	0.612235
280	4600	34.329	0.74291	0.44101	2.0527	1302.2	198.64	0.39857	0.044503	D	0.602070
280	4700	34.882	0.74705	0.43972	2.0461	1320.7	197.98	0.39695	0.045282	D	0.592544
280	4800	35.417	0.75142	0.43835	2.0393	1339.3	197.35	0.39539	0.046052	D	0.583593
280	4900	35.935	0.756	0.43693	2.0322	1357.9	196.76	0.39389	0.046812	D	0.575167
280	5000	36.438	0.76078	0.43547	2.025	1376.4	196.19	0.39243	0.047563	D	0.567228
280	5100	36.926	0.76575	0.43399	2.0177	1394.9	195.65	0.39102	0.048304	D	0.559739

TABLE F.1 (continued)—CO_2 PROPERTIES AT VARIOUS TEMPERATURES, °F											
Temperature (°F)	Pressure (psia)	Density (lbm/ft³)	Compressibility Factor	Heat Capacity (Btu/lbm°F)	Heat Capacity Ratio (C_p/C_v)	Sonic Velocity (ft/s)	Enthalpy (Btu/lbm)	Entropy [Btu/(lbm°F)]	Viscosity (cp)	Phase	Gas Formation Volume Factor (res bbl/Mcf)
280	5200	37.399	0.77088	0.43249	2.0104	1413.3	195.14	0.38965	0.049035	D	0.552652
280	5300	37.859	0.77617	0.43099	2.0031	1431.6	194.65	0.38832	0.049756	D	0.545946
280	5400	38.305	0.7816	0.42949	1.9958	1449.9	194.18	0.38704	0.050469	D	0.539584
280	5500	38.739	0.78716	0.428	1.9885	1468	193.74	0.38579	0.051172	D	0.533542
280	5600	39.16	0.79285	0.42653	1.9813	1485.9	193.32	0.38457	0.051866	D	0.527803
280	5700	39.57	0.79864	0.42506	1.9742	1503.8	192.91	0.38339	0.052552	D	0.522330
280	5800	39.969	0.80454	0.42362	1.9671	1521.5	192.53	0.38224	0.05323	D	0.517116
280	5900	40.358	0.81053	0.4222	1.9602	1539.1	192.16	0.38112	0.053899	D	0.512136
280	6000	40.736	0.81662	0.4208	1.9533	1556.6	191.81	0.38003	0.05456	D	0.507385
280	6100	41.105	0.82278	0.41942	1.9466	1574	191.48	0.37897	0.055214	D	0.502831
280	6200	41.464	0.82902	0.41806	1.9399	1591.2	191.16	0.37793	0.05586	D	0.498473
280	6300	41.815	0.83532	0.41673	1.9334	1608.2	190.86	0.37692	0.056499	D	0.494289
280	6400	42.157	0.8417	0.41542	1.9269	1625.2	190.57	0.37594	0.05713	D	0.490282
280	6500	42.491	0.84813	0.41413	1.9206	1642	190.29	0.37497	0.057755	D	0.486427
280	6600	42.817	0.85462	0.41287	1.9143	1658.7	190.03	0.37403	0.058374	D	0.482723
280	6700	43.136	0.86116	0.41163	1.9082	1675.2	189.78	0.37311	0.058986	D	0.479157
280	6800	43.447	0.86774	0.41042	1.9022	1691.6	189.54	0.37221	0.059591	D	0.475718
280	6900	43.752	0.87437	0.40923	1.8963	1707.9	189.31	0.37133	0.060191	D	0.472405
280	7000	44.05	0.88105	0.40806	1.8905	1724.1	189.1	0.37046	0.060785	D	0.469214
280	7100	44.341	0.88776	0.40692	1.8848	1740.1	188.89	0.36962	0.061373	D	0.466129
280	7200	44.627	0.89451	0.4058	1.8792	1756	188.69	0.36879	0.061956	D	0.463150
280	7300	44.906	0.90129	0.40471	1.8737	1771.8	188.51	0.36797	0.062533	D	0.460267
280	7400	45.18	0.9081	0.40364	1.8683	1787.4	188.33	0.36718	0.063105	D	0.457478
280	7500	45.448	0.91494	0.40259	1.863	1803	188.16	0.36639	0.063672	D	0.454778
280	7600	45.711	0.9218	0.40156	1.8578	1818.3	188	0.36563	0.064235	D	0.452159
280	7700	45.969	0.9287	0.40056	1.8528	1833.6	187.84	0.36487	0.064792	D	0.449628
280	7800	46.222	0.93561	0.39958	1.8478	1848.8	187.7	0.36413	0.065345	D	0.447166
280	7900	46.47	0.94255	0.39862	1.8429	1863.8	187.56	0.36341	0.065894	D	0.444781
280	8000	46.713	0.9495	0.39769	1.8381	1878.7	187.43	0.36269	0.066438	D	0.442459
280	8100	46.952	0.95647	0.39677	1.8334	1893.5	187.3	0.36199	0.066979	D	0.440205
280	8200	47.187	0.96346	0.39588	1.8288	1908.1	187.19	0.3613	0.067515	D	0.438014
280	8300	47.418	0.97047	0.395	1.8243	1922.7	187.07	0.36062	0.068047	D	0.435886
280	8400	47.644	0.97749	0.39415	1.8199	1937.1	186.97	0.35995	0.068576	D	0.433812
280	8500	47.867	0.98452	0.39332	1.8156	1951.4	186.87	0.35929	0.069101	D	0.431792
280	8600	48.086	0.99157	0.3925	1.8113	1965.6	186.78	0.35864	0.069622	D	0.429827
280	8700	48.302	0.99862	0.39171	1.8072	1979.7	186.69	0.35801	0.07014	D	0.427907
280	8800	48.514	1.0057	0.39093	1.8031	1993.6	186.61	0.35738	0.070655	D	0.426044
280	8900	48.722	1.0128	0.39018	1.7991	2007.5	186.53	0.35676	0.071166	D	0.424231
280	9000	48.927	1.0198	0.38944	1.7952	2021.2	186.46	0.35615	0.071674	D	0.422417
280	9100	49.129	1.0269	0.38871	1.7914	2034.8	186.39	0.35555	0.07218	D	0.420683
280	9200	49.328	1.034	0.38801	1.7877	2048.4	186.33	0.35495	0.072682	D	0.418988
280	9300	49.524	1.0411	0.38732	1.784	2061.8	186.27	0.35437	0.073182	D	0.417328
280	9400	49.717	1.0483	0.38664	1.7804	2075.1	186.21	0.35379	0.073678	D	0.415744
280	9500	49.907	1.0554	0.38599	1.7768	2088.3	186.16	0.35322	0.074172	D	0.414154
280	9600	50.095	1.0625	0.38534	1.7734	2101.4	186.12	0.35266	0.074664	D	0.412597
280	9700	50.279	1.0696	0.38471	1.77	2114.4	186.08	0.3521	0.075152	D	0.411072

TABLE F.1 (continued)—CO_2 PROPERTIES AT VARIOUS TEMPERATURES, °F

Temperature (°F)	Pressure (psia)	Density (lbm/ft³)	Compressibility Factor	Heat Capacity (Btu/lbm°F)	Heat Capacity Ratio (C_p/C_v)	Sonic Velocity (ft/s)	Enthalpy (Btu/lbm)	Entropy [Btu/(lbm°F)]	Viscosity (cp)	Phase	Gas Formation Volume Factor (res bbl/Mcf)
280	9800	50.462	1.0767	0.3841	1.7666	2127.3	186.04	0.35155	0.075639	D	0.409579
280	9900	50.641	1.0839	0.3835	1.7633	2140.2	186.01	0.35101	0.076123	D	0.408153
280	10000	50.818	1.091	0.38291	1.7601	2152.9	185.98	0.35048	0.076605	D	0.406718
290	14.696	0.080519	0.99842	0.22824	1.2493	1027	263.67	0.72615	0.020433	V	256.693456
290	100	0.55298	0.98925	0.23146	1.261	1022.3	262.5	0.63842	0.020469	D	37.377197
290	200	1.1181	0.97851	0.23541	1.2754	1017	261.11	0.60575	0.020523	D	18.485702
290	300	1.6957	0.9678	0.23957	1.2907	1011.9	259.71	0.58605	0.020591	D	12.188915
290	400	2.2862	0.95712	0.24394	1.3068	1007	258.28	0.57165	0.020673	D	9.040804
290	500	2.8899	0.94646	0.24852	1.3238	1002.4	256.85	0.56014	0.02077	D	7.152089
290	600	3.5072	0.93585	0.25333	1.3418	997.94	255.39	0.55045	0.020883	D	5.893261
290	700	4.1384	0.92528	0.25837	1.3607	993.78	253.92	0.54201	0.021014	D	4.994314
290	800	4.784	0.91477	0.26365	1.3806	989.9	252.43	0.53448	0.021162	D	4.320386
290	900	5.4441	0.90434	0.26918	1.4015	986.33	250.93	0.52764	0.02133	D	3.796557
290	1000	6.119	0.89399	0.27495	1.4235	983.09	249.42	0.52134	0.021518	D	3.377795
290	1100	6.8089	0.88375	0.28097	1.4464	980.23	247.89	0.51547	0.021728	D	3.035550
290	1200	7.5139	0.87364	0.28723	1.4704	977.78	246.35	0.50996	0.02196	D	2.750755
290	1300	8.234	0.86367	0.29374	1.4954	975.76	244.79	0.50475	0.022217	D	2.510182
290	1400	8.9691	0.85387	0.30047	1.5213	974.23	243.23	0.49979	0.022498	D	2.304435
290	1500	9.7191	0.84426	0.30741	1.5481	973.21	241.66	0.49505	0.022806	D	2.126599
290	1600	10.484	0.83489	0.31455	1.5757	972.76	240.08	0.4905	0.023141	D	1.971560
290	1700	11.262	0.82576	0.32185	1.6039	972.92	238.5	0.48612	0.023504	D	1.835294
290	1800	12.053	0.81692	0.32929	1.6328	973.73	236.92	0.48188	0.023897	D	1.714777
290	1900	12.857	0.80839	0.33682	1.662	975.22	235.33	0.47779	0.024319	D	1.607563
290	2000	13.672	0.80021	0.34441	1.6915	977.45	233.76	0.47382	0.024772	D	1.511731
290	2100	14.497	0.7924	0.352	1.721	980.44	232.19	0.46998	0.025255	D	1.425692
290	2200	15.331	0.78501	0.35954	1.7502	984.23	230.63	0.46624	0.025769	D	1.348196
290	2300	16.171	0.77805	0.36697	1.7791	988.85	229.08	0.46261	0.026314	D	1.278146
290	2400	17.016	0.77156	0.37423	1.8072	994.33	227.56	0.45908	0.026887	D	1.214672
290	2500	17.864	0.76556	0.38125	1.8344	1000.7	226.05	0.45566	0.02749	D	1.157017
290	2600	18.713	0.76006	0.38798	1.8604	1007.9	224.57	0.45233	0.02812	D	1.104524
290	2700	19.56	0.7551	0.39437	1.885	1016	223.12	0.4491	0.028776	D	1.056675
290	2800	20.404	0.75068	0.40035	1.908	1024.9	221.69	0.44597	0.029456	D	1.012972
290	2900	21.243	0.7468	0.4059	1.9291	1034.7	220.31	0.44294	0.030159	D	0.972987
290	3000	22.073	0.74348	0.41008	1.9484	1045.3	218.96	0.43999	0.030881	D	0.936373
290	3100	22.894	0.74071	0.41556	1.9657	1056.7	217.65	0.43715	0.03162	D	0.902791
290	3200	23.704	0.73849	0.41964	1.9809	1068.9	216.38	0.43439	0.032375	D	0.871957
290	3300	24.5	0.73681	0.42322	1.9941	1081.7	215.15	0.43173	0.033142	D	0.843611
290	3400	25.283	0.73565	0.4263	2.0053	1095.2	213.96	0.42915	0.03392	D	0.817510
290	3500	26.049	0.735	0.4289	2.0145	1109.2	212.82	0.42667	0.034705	D	0.793451
290	3600	26.799	0.73484	0.43104	2.0219	1123.9	211.72	0.42427	0.035496	D	0.771243
290	3700	27.532	0.73514	0.43274	2.0275	1139	210.67	0.42195	0.036291	D	0.750704
290	3800	28.247	0.7359	0.43404	2.0315	1154.6	209.66	0.41972	0.037087	D	0.731705
290	3900	28.944	0.73708	0.43496	2.0339	1170.6	208.69	0.41756	0.037883	D	0.714086
290	4000	29.623	0.73867	0.43554	2.035	1187	207.76	0.41548	0.038678	D	0.697736
290	4100	30.283	0.74063	0.43582	2.0348	1203.6	206.88	0.41348	0.039469	D	0.682524
290	4200	30.924	0.74296	0.43582	2.0335	1220.6	206.03	0.41154	0.040257	D	0.668370

TABLE F.1 (continued)—CO_2 PROPERTIES AT VARIOUS TEMPERATURES, °F											
Temperature (°F)	Pressure (psia)	Density (lbm/ft³)	Compressibility Factor	Heat Capacity (Btu/lbm°F)	Heat Capacity Ratio (C_p/C_v)	Sonic Velocity (ft/s)	Enthalpy (Btu/lbm)	Entropy [Btu/(lbm°F)]	Viscosity (cp)	Phase	Gas Formation Volume Factor (res bbl/Mcf)
290	4300	31.547	0.74562	0.43559	2.0313	1237.8	205.22	0.40967	0.041039	D	0.655164
290	4400	32.153	0.7486	0.43514	2.0282	1255.1	204.45	0.40787	0.041815	D	0.642832
290	4500	32.741	0.75187	0.43452	2.0244	1272.7	203.71	0.40612	0.042585	D	0.631293
290	4600	33.311	0.75541	0.43375	2.02	1290.3	203.01	0.40444	0.043347	D	0.620477
290	4700	33.865	0.75921	0.43285	2.0152	1308.1	202.34	0.40281	0.044101	D	0.610330
290	4800	34.403	0.76324	0.43186	2.0099	1325.9	201.7	0.40124	0.044848	D	0.600787
290	4900	34.924	0.7675	0.43078	2.0043	1343.7	201.1	0.39971	0.045586	D	0.591811
290	5000	35.431	0.77196	0.42964	1.9984	1361.5	200.52	0.39824	0.046315	D	0.583345
290	5100	35.924	0.77661	0.42845	1.9924	1379.4	199.96	0.39681	0.047036	D	0.575352
290	5200	36.402	0.78144	0.42722	1.9862	1397.1	199.44	0.39542	0.047749	D	0.567797
290	5300	36.867	0.78642	0.42597	1.9799	1414.9	198.93	0.39408	0.048453	D	0.560634
290	5400	37.318	0.79156	0.42471	1.9736	1432.5	198.45	0.39277	0.049149	D	0.553848
290	5500	37.758	0.79683	0.42344	1.9673	1450.1	198	0.3915	0.049836	D	0.547399
290	5600	38.186	0.80224	0.42216	1.9609	1467.6	197.56	0.39027	0.050516	D	0.541274
290	5700	38.602	0.80776	0.42089	1.9546	1485	197.14	0.38907	0.051187	D	0.535437
290	5800	39.007	0.81339	0.41962	1.9483	1502.3	196.74	0.3879	0.05185	D	0.529873
290	5900	39.402	0.81912	0.41837	1.9421	1519.4	196.37	0.38677	0.052506	D	0.524561
290	6000	39.787	0.82495	0.41712	1.9359	1536.5	196	0.38566	0.053155	D	0.519490
290	6100	40.162	0.83086	0.41589	1.9298	1553.5	195.66	0.38458	0.053796	D	0.514634
290	6200	40.528	0.83686	0.41468	1.9238	1570.3	195.33	0.38353	0.05443	D	0.509990
290	6300	40.885	0.84293	0.41348	1.9178	1587	195.01	0.3825	0.055057	D	0.505535
290	6400	41.233	0.84907	0.41229	1.9119	1603.6	194.71	0.38149	0.055678	D	0.501261
290	6500	41.574	0.85528	0.41113	1.9061	1620.1	194.42	0.38051	0.056292	D	0.497159
290	6600	41.906	0.86154	0.40998	1.9004	1636.5	194.15	0.37956	0.0569	D	0.493210
290	6700	42.232	0.86786	0.40885	1.8947	1652.7	193.88	0.37862	0.057501	D	0.489413
290	6800	42.549	0.87424	0.40774	1.8892	1668.8	193.63	0.3777	0.058097	D	0.485761
290	6900	42.86	0.88066	0.40665	1.8837	1684.8	193.39	0.3768	0.058687	D	0.482236
290	7000	43.164	0.88713	0.40558	1.8783	1700.7	193.17	0.37592	0.059271	D	0.478839
290	7100	43.462	0.89364	0.40452	1.873	1716.5	192.95	0.37506	0.05985	D	0.475560
290	7200	43.753	0.90019	0.40348	1.8677	1732.2	192.74	0.37422	0.060423	D	0.472392
290	7300	44.039	0.90678	0.40247	1.8626	1747.7	192.54	0.37339	0.060992	D	0.469332
290	7400	44.319	0.9134	0.40147	1.8575	1763.1	192.35	0.37258	0.061555	D	0.466369
290	7500	44.593	0.92005	0.40049	1.8525	1778.4	192.17	0.37179	0.062113	D	0.463501
290	7600	44.861	0.92673	0.39953	1.8476	1793.6	192	0.37101	0.062667	D	0.460723
290	7700	45.125	0.93344	0.39858	1.8428	1808.7	191.84	0.37024	0.063216	D	0.458033
290	7800	45.383	0.94018	0.39766	1.8381	1823.7	191.68	0.36949	0.063761	D	0.455425
290	7900	45.637	0.94694	0.39675	1.8335	1838.5	191.53	0.36875	0.064301	D	0.452893
290	8000	45.886	0.95372	0.39587	1.8289	1853.2	191.39	0.36802	0.064837	D	0.450434
290	8100	46.131	0.96053	0.395	1.8244	1867.9	191.26	0.36731	0.065369	D	0.448050
290	8200	46.371	0.96735	0.39415	1.82	1882.4	191.14	0.3666	0.065897	D	0.445729
290	8300	46.607	0.97419	0.39331	1.8157	1896.8	191.02	0.36591	0.066421	D	0.443472
290	8400	46.838	0.98105	0.3925	1.8114	1911.1	190.9	0.36523	0.066942	D	0.441278
290	8500	47.066	0.98792	0.3917	1.8073	1925.3	190.79	0.36456	0.067459	D	0.439141
290	8600	47.29	0.99481	0.39092	1.8032	1939.4	190.69	0.3639	0.067972	D	0.437061
290	8700	47.51	1.0017	0.39015	1.7992	1953.3	190.6	0.36326	0.068482	D	0.435030
290	8800	47.727	1.0086	0.38941	1.7952	1967.2	190.51	0.36262	0.068989	D	0.433049

TABLE F.1 (continued)—CO_2 PROPERTIES AT VARIOUS TEMPERATURES, °F

Temperature (°F)	Pressure (psia)	Density (lbm/ft³)	Compressibility Factor	Heat Capacity (Btu/lbm°F)	Heat Capacity Ratio (C_p/C_v)	Sonic Velocity (ft/s)	Enthalpy (Btu/lbm)	Entropy [Btu/(lbm°F)]	Viscosity (cp)	Phase	Gas Formation Volume Factor (res bbl/Mcf)
290	8900	47.94	1.0156	0.38868	1.7914	1980.9	190.42	0.36199	0.069493	D	0.431155
290	9000	48.15	1.0225	0.38796	1.7876	1994.6	190.34	0.36137	0.069993	D	0.429261
290	9100	48.357	1.0294	0.38726	1.7839	2008.2	190.27	0.36076	0.070491	D	0.427409
290	9200	48.56	1.0364	0.38658	1.7802	2021.6	190.2	0.36015	0.070985	D	0.425638
290	9300	48.76	1.0433	0.38591	1.7766	2034.9	190.13	0.35956	0.071477	D	0.423864
290	9400	48.958	1.0503	0.38525	1.7731	2048.2	190.07	0.35897	0.071966	D	0.422169
290	9500	49.152	1.0573	0.38461	1.7697	2061.3	190.02	0.35839	0.072452	D	0.420509
290	9600	49.344	1.0643	0.38399	1.7663	2074.4	189.97	0.35782	0.072935	D	0.418884
290	9700	49.532	1.0713	0.38338	1.7629	2087.3	189.92	0.35726	0.073416	D	0.417292
290	9800	49.719	1.0783	0.38278	1.7597	2100.2	189.87	0.3567	0.073895	D	0.415733
290	9900	49.902	1.0852	0.38219	1.7565	2113	189.83	0.35616	0.074371	D	0.414167
290	10000	50.083	1.0922	0.38162	1.7533	2125.6	189.8	0.35561	0.074845	D	0.412670
300	14.696	0.079453	0.9985	0.22928	1.2478	1033.3	265.96	0.72918	0.020679	V	260.138384
300	100	0.54541	0.98978	0.23236	1.259	1028.9	264.82	0.64149	0.020715	D	37.896071
300	200	1.1022	0.97958	0.23615	1.2728	1023.9	263.47	0.60887	0.020768	D	18.752770
300	300	1.6706	0.96943	0.24011	1.2874	1019.1	262.1	0.58923	0.020834	D	12.372308
300	400	2.2509	0.95931	0.24426	1.3027	1014.5	260.72	0.57488	0.020914	D	9.182364
300	500	2.8435	0.94923	0.24861	1.3188	1010.1	259.33	0.56343	0.021008	D	7.268704
300	600	3.4487	0.9392	0.25315	1.3358	1006	257.92	0.5538	0.021118	D	5.993249
300	700	4.0666	0.92924	0.2579	1.3536	1002.2	256.5	0.54543	0.021244	D	5.082593
300	800	4.6976	0.91934	0.26287	1.3722	998.59	255.07	0.53797	0.021387	D	4.399889
300	900	5.3418	0.90952	0.26804	1.3918	995.31	253.62	0.5312	0.021549	D	3.869236
300	1000	5.9994	0.8998	0.27343	1.4122	992.36	252.16	0.52497	0.02173	D	3.445097
300	1100	6.6706	0.8902	0.27903	1.4335	989.77	250.69	0.51918	0.02193	D	3.098492
300	1200	7.3554	0.88072	0.28483	1.4557	987.57	249.21	0.51375	0.022152	D	2.810038
300	1300	8.0536	0.87138	0.29084	1.4787	985.78	247.71	0.50862	0.022397	D	2.566373
300	1400	8.7653	0.86222	0.29705	1.5025	984.45	246.22	0.50375	0.022664	D	2.358010
300	1500	9.4901	0.85325	0.30343	1.5271	983.61	244.71	0.4991	0.022956	D	2.177913
300	1600	10.228	0.8445	0.30997	1.5523	983.29	243.2	0.49464	0.023273	D	2.020855
300	1700	10.978	0.83599	0.31665	1.5781	983.54	241.69	0.49035	0.023615	D	1.882815
300	1800	11.739	0.82774	0.32344	1.6043	984.39	240.18	0.48621	0.023985	D	1.760666
300	1900	12.511	0.81979	0.33031	1.6309	985.87	238.67	0.48221	0.024381	D	1.651979
300	2000	13.294	0.81217	0.33722	1.6577	988.02	237.16	0.47834	0.024805	D	1.554793
300	2100	14.085	0.80489	0.34414	1.6845	990.88	235.67	0.47459	0.025257	D	1.467482
300	2200	14.883	0.79798	0.35101	1.7111	994.46	234.18	0.47095	0.025736	D	1.388752
300	2300	15.687	0.79148	0.3578	1.7374	998.8	232.71	0.46741	0.026243	D	1.317552
300	2400	16.496	0.78539	0.36446	1.7631	1003.9	231.25	0.46398	0.026777	D	1.252938
300	2500	17.308	0.77975	0.37093	1.7881	1009.8	229.81	0.46064	0.027338	D	1.194183
300	2600	18.12	0.77457	0.37717	1.8122	1016.5	228.39	0.4574	0.027924	D	1.140625
300	2700	18.932	0.76987	0.38313	1.8351	1024	227	0.45425	0.028534	D	1.091715
300	2800	19.741	0.76566	0.38878	1.8567	1032.3	225.64	0.4512	0.029166	D	1.046968
300	2900	20.546	0.76195	0.39407	1.8769	1041.3	224.31	0.44823	0.029819	D	1.005968
300	3000	21.345	0.75873	0.39898	1.8956	1051.1	223.01	0.44536	0.030492	D	0.968326
300	3100	22.135	0.75602	0.40348	1.9125	1061.7	221.74	0.44257	0.031182	D	0.933742
300	3200	22.917	0.75381	0.40757	1.9278	1072.9	220.51	0.43987	0.031887	D	0.901919
300	3300	23.687	0.75208	0.41123	1.9414	1084.8	219.32	0.43726	0.032605	D	0.872581

TABLE F.1 (continued)—CO_2 PROPERTIES AT VARIOUS TEMPERATURES, °F											
Temperature (°F)	Pressure (psia)	Density (lbm/ft^3)	Compressibility Factor	Heat Capacity (Btu/lbm°F)	Heat Capacity Ratio (C_p/C_v)	Sonic Velocity (ft/s)	Enthalpy (Btu/lbm)	Entropy [Btu/(lbm°F)]	Viscosity (cp)	Phase	Gas Formation Volume Factor (res bbl/Mcf)
300	3400	24.445	0.75084	0.41447	1.9533	1097.3	218.17	0.43472	0.033334	D	0.845520
300	3500	25.19	0.75008	0.41729	1.9635	1110.4	217.05	0.43227	0.034071	D	0.820531
300	3600	25.92	0.74976	0.4197	1.972	1124	215.98	0.4299	0.034816	D	0.797398
300	3700	26.636	0.74989	0.42173	1.979	1138.1	214.94	0.42761	0.035567	D	0.775981
300	3800	27.336	0.75044	0.42339	1.9845	1152.7	213.94	0.4254	0.03632	D	0.756115
300	3900	28.019	0.75139	0.42471	1.9886	1167.7	212.99	0.42326	0.037076	D	0.737660
300	4000	28.687	0.75272	0.4257	1.9914	1183	212.07	0.42119	0.037831	D	0.720492
300	4100	29.338	0.75442	0.4264	1.993	1198.7	211.19	0.41919	0.038586	D	0.704506
300	4200	29.972	0.75647	0.42683	1.9935	1214.7	210.34	0.41726	0.039339	D	0.689601
300	4300	30.59	0.75884	0.42702	1.9931	1230.9	209.53	0.41539	0.040088	D	0.675674
300	4400	31.191	0.76152	0.427	1.9917	1247.4	208.76	0.41358	0.040833	D	0.662650
300	4500	31.776	0.76449	0.42678	1.9897	1264	208.02	0.41183	0.041573	D	0.650451
300	4600	32.345	0.76773	0.4264	1.9869	1280.8	207.31	0.41014	0.042308	D	0.639008
300	4700	32.898	0.77122	0.42587	1.9836	1297.7	206.64	0.4085	0.043037	D	0.628255
300	4800	33.437	0.77496	0.42523	1.9798	1314.7	205.99	0.40692	0.043759	D	0.618150
300	4900	33.96	0.77891	0.42448	1.9755	1331.8	205.37	0.40538	0.044474	D	0.608621
300	5000	34.469	0.78307	0.42364	1.971	1348.9	204.78	0.40389	0.045182	D	0.599634
300	5100	34.964	0.78743	0.42274	1.9661	1366.1	204.22	0.40245	0.045883	D	0.591149
300	5200	35.445	0.79196	0.42178	1.9611	1383.2	203.68	0.40105	0.046576	D	0.583117
300	5300	35.914	0.79666	0.42078	1.9559	1400.4	203.17	0.39969	0.047262	D	0.575510
300	5400	36.37	0.80152	0.41975	1.9506	1417.4	202.68	0.39837	0.04794	D	0.568298
300	5500	36.813	0.80652	0.41869	1.9451	1434.5	202.21	0.39708	0.048611	D	0.561446
300	5600	37.246	0.81166	0.41762	1.9397	1451.5	201.76	0.39583	0.049275	D	0.554934
300	5700	37.667	0.81692	0.41654	1.9341	1468.4	201.33	0.39462	0.049931	D	0.548732
300	5800	38.077	0.82229	0.41545	1.9286	1485.2	200.92	0.39344	0.05058	D	0.542816
300	5900	38.477	0.82777	0.41436	1.9231	1501.9	200.53	0.39228	0.051222	D	0.537172
300	6000	38.867	0.83335	0.41327	1.9176	1518.6	200.15	0.39116	0.051857	D	0.531780
300	6100	39.248	0.83903	0.41219	1.9121	1535.1	199.8	0.39007	0.052485	D	0.526627
300	6200	39.619	0.84479	0.41111	1.9067	1551.6	199.45	0.389	0.053107	D	0.521690
300	6300	39.982	0.85063	0.41005	1.9013	1567.9	199.13	0.38795	0.053722	D	0.516958
300	6400	40.336	0.85654	0.40899	1.896	1584.2	198.82	0.38694	0.054331	D	0.512417
300	6500	40.682	0.86253	0.40795	1.8907	1600.3	198.52	0.38594	0.054934	D	0.508062
300	6600	41.02	0.86858	0.40692	1.8855	1616.4	198.23	0.38497	0.055531	D	0.503873
300	6700	41.35	0.87469	0.4059	1.8803	1632.3	197.96	0.38402	0.056122	D	0.499844
300	6800	41.674	0.88085	0.40489	1.8752	1648.1	197.7	0.38309	0.056707	D	0.495962
300	6900	41.99	0.88707	0.4039	1.8702	1663.8	197.45	0.38217	0.057287	D	0.492226
300	7000	42.3	0.89334	0.40292	1.8652	1679.4	197.21	0.38128	0.057861	D	0.488623
300	7100	42.603	0.89965	0.40195	1.8603	1694.9	196.98	0.38041	0.05843	D	0.485144
300	7200	42.9	0.90601	0.401	1.8554	1710.3	196.76	0.37955	0.058994	D	0.481788
300	7300	43.191	0.91241	0.40006	1.8506	1725.6	196.55	0.37871	0.059553	D	0.478545
300	7400	43.476	0.91884	0.39914	1.8459	1740.8	196.36	0.37789	0.060107	D	0.475405
300	7500	43.755	0.92531	0.39824	1.8412	1755.9	196.17	0.37708	0.060657	D	0.472369
300	7600	44.029	0.93181	0.39734	1.8367	1770.8	195.99	0.37629	0.061201	D	0.469428
300	7700	44.298	0.93835	0.39647	1.8321	1785.7	195.81	0.37551	0.061742	D	0.466584
300	7800	44.562	0.94491	0.3956	1.8277	1800.5	195.65	0.37474	0.062278	D	0.463822
300	7900	44.821	0.9515	0.39476	1.8233	1815.1	195.49	0.37399	0.06281	D	0.461145

TABLE F.1 (continued)—CO_2 PROPERTIES AT VARIOUS TEMPERATURES, °F

Temperature (°F)	Pressure (psia)	Density (lbm/ft^3)	Compressibility Factor	Heat Capacity (Btu/lbm°F)	Heat Capacity Ratio (C_P/C_V)	Sonic Velocity (ft/s)	Enthalpy (Btu/lbm)	Entropy [Btu/(lbm°F)]	Viscosity (cp)	Phase	Gas Formation Volume Factor (res bbl/Mcf)
300	8000	45.075	0.95811	0.39392	1.819	1829.7	195.34	0.37325	0.063338	D	0.458544
300	8100	45.324	0.96475	0.39311	1.8147	1844.1	195.2	0.37253	0.063861	D	0.456021
300	8200	45.569	0.9714	0.39231	1.8105	1858.5	195.07	0.37181	0.064381	D	0.453565
300	8300	45.81	0.97808	0.39152	1.8064	1872.7	194.94	0.37111	0.064898	D	0.451182
300	8400	46.047	0.98478	0.39075	1.8024	1886.9	194.82	0.37042	0.06541	D	0.448865
300	8500	46.279	0.9915	0.38999	1.7984	1900.9	194.7	0.36974	0.065919	D	0.446611
300	8600	46.508	0.99823	0.38925	1.7945	1914.8	194.59	0.36907	0.066425	D	0.444414
300	8700	46.733	1.005	0.38852	1.7906	1928.7	194.49	0.36841	0.066927	D	0.442285
300	8800	46.954	1.0117	0.3878	1.7869	1942.4	194.39	0.36777	0.067426	D	0.440174
300	8900	47.172	1.0185	0.3871	1.7831	1956.1	194.3	0.36713	0.067922	D	0.438154
300	9000	47.386	1.0253	0.38642	1.7795	1969.6	194.22	0.3665	0.068414	D	0.436178
300	9100	47.597	1.0321	0.38575	1.7759	1983.1	194.13	0.36588	0.068904	D	0.434246
300	9200	47.804	1.0389	0.38509	1.7724	1996.4	194.06	0.36527	0.069391	D	0.432356
300	9300	48.009	1.0457	0.38444	1.7689	2009.7	193.99	0.36466	0.069875	D	0.430506
300	9400	48.21	1.0526	0.38381	1.7655	2022.8	193.92	0.36407	0.070356	D	0.428737
300	9500	48.409	1.0594	0.38319	1.7621	2035.9	193.86	0.36348	0.070834	D	0.426965
300	9600	48.604	1.0662	0.38259	1.7588	2048.9	193.8	0.3629	0.07131	D	0.425229
300	9700	48.797	1.0731	0.382	1.7556	2061.8	193.75	0.36233	0.071784	D	0.423569
300	9800	48.987	1.08	0.38142	1.7524	2074.6	193.7	0.36177	0.072255	D	0.421942
300	9900	49.174	1.0868	0.38085	1.7493	2087.2	193.65	0.36121	0.072723	D	0.420310
300	10000	49.359	1.0937	0.38029	1.7462	2099.9	193.61	0.36066	0.07319	D	0.418749
310	14.696	0.078415	0.99857	0.2303	1.2463	1039.5	268.26	0.73219	0.020924	V	263.581222
310	100	0.53805	0.99028	0.23327	1.2571	1035.4	267.15	0.64453	0.020959	D	38.414316
310	200	1.0867	0.9806	0.23689	1.2703	1030.7	265.83	0.61197	0.021011	D	19.019408
310	300	1.6463	0.97096	0.24067	1.2842	1026.1	264.51	0.59237	0.021076	D	12.554956
310	400	2.2169	0.96137	0.24463	1.2988	1021.9	263.17	0.57808	0.021154	D	9.323214
310	500	2.7989	0.95184	0.24875	1.3141	1017.8	261.82	0.56668	0.021246	D	7.384635
310	600	3.3924	0.94237	0.25306	1.3301	1014	260.45	0.55711	0.021352	D	6.092637
310	700	3.9977	0.93296	0.25754	1.3469	1010.4	259.08	0.5488	0.021474	D	5.170114
310	800	4.615	0.92363	0.26221	1.3645	1007.1	257.69	0.5414	0.021613	D	4.478609
310	900	5.2443	0.91439	0.26707	1.3828	1004.1	256.29	0.5347	0.021769	D	3.941160
310	1000	5.8858	0.90526	0.27211	1.4019	1001.4	254.89	0.52854	0.021942	D	3.511627
310	1100	6.5396	0.89623	0.27734	1.4217	999.09	253.47	0.52282	0.022135	D	3.160544
310	1200	7.2056	0.88734	0.28274	1.4423	997.12	252.04	0.51746	0.022347	D	2.868428
310	1300	7.8837	0.8786	0.28832	1.4636	995.55	250.61	0.51241	0.022581	D	2.621700
310	1400	8.5738	0.87003	0.29406	1.4856	994.41	249.17	0.50761	0.022835	D	2.410690
310	1500	9.2757	0.86164	0.29994	1.5082	993.73	247.73	0.50304	0.023113	D	2.228280
310	1600	9.9888	0.85346	0.30597	1.5313	993.55	246.28	0.49866	0.023413	D	2.069180
310	1700	10.713	0.84551	0.3121	1.5549	993.89	244.83	0.49446	0.023737	D	1.929323
310	1800	11.447	0.83782	0.31833	1.579	994.78	243.39	0.4904	0.024086	D	1.805566
310	1900	12.191	0.8304	0.32463	1.6032	996.27	241.94	0.48649	0.024459	D	1.695387
310	2000	12.944	0.82329	0.33096	1.6277	998.37	240.5	0.48271	0.024857	D	1.596827
310	2100	13.704	0.81649	0.33729	1.6521	1001.1	239.07	0.47904	0.025281	D	1.508227
310	2200	14.471	0.81005	0.34359	1.6765	1004.5	237.65	0.47549	0.025731	D	1.428316
310	2300	15.243	0.80397	0.34983	1.7005	1008.6	236.24	0.47204	0.026205	D	1.355961
310	2400	16.019	0.79827	0.35595	1.7241	1013.4	234.85	0.46869	0.026705	D	1.290249

TABLE F.1 (continued)—CO_2 PROPERTIES AT VARIOUS TEMPERATURES, °F												
Temperature (°F)	Pressure (psia)	Density (lbm/ft^3)	Compressibility Factor	Heat Capacity (Btu/lbm°F)	Heat Capacity Ratio (C_p/C_v)	Sonic Velocity (ft/s)	Enthalpy (Btu/lbm)	Entropy [Btu/(lbm°F)]	Viscosity (cp)	Phase	Gas Formation Volume Factor (res bbl/Mcf)
310	2500	16.798	0.79298	0.36193	1.7472	1019	233.47	0.46543	0.027228	D	1.230431
310	2600	17.578	0.78811	0.36771	1.7694	1025.2	232.12	0.46227	0.027775	D	1.175841
310	2700	18.357	0.78367	0.37328	1.7908	1032.2	230.78	0.4592	0.028344	D	1.125912
310	2800	19.135	0.77968	0.37858	1.8111	1039.9	229.48	0.45622	0.028934	D	1.080173
310	2900	19.909	0.77614	0.3836	1.8302	1048.4	228.19	0.45332	0.029545	D	1.038191
310	3000	20.677	0.77305	0.38831	1.848	1057.5	226.94	0.45051	0.030173	D	0.999589
310	3100	21.439	0.77042	0.39268	1.8645	1067.3	225.72	0.44778	0.030818	D	0.964053
310	3200	22.193	0.76825	0.3967	1.8796	1077.7	224.53	0.44513	0.031478	D	0.931296
310	3300	22.938	0.76653	0.40036	1.8932	1088.8	223.38	0.44256	0.032151	D	0.901053
310	3400	23.673	0.76526	0.40366	1.9054	1100.5	222.25	0.44007	0.032836	D	0.873102
310	3500	24.396	0.76441	0.4066	1.9161	1112.7	221.17	0.43766	0.03353	D	0.847215
310	3600	25.107	0.764	0.40919	1.9254	1125.4	220.12	0.43532	0.034233	D	0.823239
310	3700	25.804	0.76399	0.41144	1.9333	1138.6	219.1	0.43306	0.034941	D	0.800979
310	3800	26.488	0.76438	0.41336	1.9399	1152.3	218.13	0.43087	0.035654	D	0.780299
310	3900	27.158	0.76515	0.41496	1.9453	1166.4	217.18	0.42875	0.036371	D	0.761057
310	4000	27.813	0.76628	0.41628	1.9494	1180.8	216.28	0.42669	0.037089	D	0.743126
310	4100	28.453	0.76776	0.41731	1.9524	1195.6	215.4	0.42471	0.037808	D	0.726401
310	4200	29.079	0.76958	0.41809	1.9544	1210.6	214.57	0.42278	0.038527	D	0.710787
310	4300	29.689	0.7717	0.41863	1.9554	1226	213.76	0.42092	0.039244	D	0.696170
310	4400	30.284	0.77413	0.41896	1.9556	1241.6	212.99	0.41911	0.039958	D	0.682490
310	4500	30.865	0.77684	0.4191	1.955	1257.3	212.25	0.41736	0.040669	D	0.669660
310	4600	31.43	0.77981	0.41906	1.9537	1273.3	211.54	0.41567	0.041376	D	0.657606
310	4700	31.981	0.78303	0.41887	1.9518	1289.4	210.86	0.41403	0.042079	D	0.646272
310	4800	32.518	0.78649	0.41855	1.9493	1305.7	210.21	0.41243	0.042776	D	0.635605
310	4900	33.041	0.79017	0.41811	1.9464	1322	209.59	0.41089	0.043468	D	0.625546
310	5000	33.55	0.79406	0.41756	1.9431	1338.5	208.99	0.40939	0.044154	D	0.616053
310	5100	34.046	0.79814	0.41694	1.9394	1354.9	208.42	0.40794	0.044834	D	0.607077
310	5200	34.529	0.80241	0.41624	1.9354	1371.4	207.87	0.40653	0.045507	D	0.598588
310	5300	35	0.80684	0.41548	1.9313	1388	207.35	0.40516	0.046174	D	0.590536
310	5400	35.458	0.81144	0.41467	1.9269	1404.5	206.85	0.40382	0.046835	D	0.582905
310	5500	35.905	0.81618	0.41383	1.9224	1421	206.37	0.40253	0.047489	D	0.575650
310	5600	36.341	0.82106	0.41296	1.9177	1437.4	205.91	0.40127	0.048136	D	0.568751
310	5700	36.765	0.82607	0.41206	1.913	1453.8	205.47	0.40004	0.048777	D	0.562182
310	5800	37.179	0.8312	0.41115	1.9083	1470.2	205.05	0.39884	0.049411	D	0.555920
310	5900	37.583	0.83644	0.41022	1.9034	1486.5	204.65	0.39768	0.050038	D	0.549943
310	6000	37.978	0.84178	0.40929	1.8986	1502.7	204.27	0.39654	0.05066	D	0.544230
310	6100	38.363	0.84723	0.40836	1.8938	1518.8	203.9	0.39543	0.051275	D	0.538774
310	6200	38.739	0.85276	0.40742	1.8889	1534.9	203.55	0.39435	0.051884	D	0.533544
310	6300	39.106	0.85838	0.40649	1.8841	1550.9	203.21	0.39329	0.052487	D	0.528535
310	6400	39.465	0.86408	0.40556	1.8793	1566.7	202.89	0.39226	0.053083	D	0.523732
310	6500	39.815	0.86985	0.40463	1.8746	1582.5	202.58	0.39125	0.053675	D	0.519118
310	6600	40.158	0.87569	0.40371	1.8698	1598.2	202.28	0.39027	0.05426	D	0.514685
310	6700	40.494	0.88159	0.4028	1.8651	1613.8	202	0.3893	0.05484	D	0.510419
310	6800	40.822	0.88755	0.4019	1.8605	1629.3	201.73	0.38836	0.055415	D	0.506313
310	6900	41.143	0.89357	0.401	1.8559	1644.7	201.47	0.38744	0.055984	D	0.502359
310	7000	41.458	0.89965	0.40012	1.8513	1660.1	201.22	0.38653	0.056548	D	0.498552

TABLE F.1 (continued)—CO_2 PROPERTIES AT VARIOUS TEMPERATURES, °F											
Temperature (°F)	Pressure (psia)	Density (lbm/ft³)	Compressibility Factor	Heat Capacity (Btu/lbm°F)	Heat Capacity Ratio (C_p/C_v)	Sonic Velocity (ft/s)	Enthalpy (Btu/lbm)	Entropy [Btu/(lbm°F)]	Viscosity (cp)	Phase	Gas Formation Volume Factor (res bbl/Mcf)
310	7100	41.766	0.90577	0.39925	1.8468	1675.3	200.99	0.38565	0.057107	D	0.494874
310	7200	42.067	0.91194	0.39838	1.8423	1690.4	200.76	0.38478	0.057661	D	0.491325
310	7300	42.363	0.91815	0.39753	1.8379	1705.4	200.54	0.38393	0.058211	D	0.487894
310	7400	42.653	0.9244	0.39669	1.8335	1720.3	200.33	0.38309	0.058756	D	0.484578
310	7500	42.937	0.93069	0.39585	1.8292	1735.2	200.14	0.38227	0.059296	D	0.481370
310	7600	43.216	0.93701	0.39503	1.825	1749.9	199.95	0.38147	0.059832	D	0.478262
310	7700	43.49	0.94337	0.39423	1.8207	1764.5	199.77	0.38068	0.060363	D	0.475255
310	7800	43.758	0.94976	0.39343	1.8166	1779.1	199.59	0.3799	0.060891	D	0.472340
310	7900	44.022	0.95618	0.39264	1.8125	1793.5	199.43	0.37914	0.061414	D	0.469513
310	8000	44.28	0.96263	0.39187	1.8084	1807.9	199.27	0.37839	0.061934	D	0.466772
310	8100	44.534	0.9691	0.39111	1.8044	1822.1	199.12	0.37765	0.062449	D	0.464108
310	8200	44.784	0.9756	0.39036	1.8005	1836.3	198.98	0.37693	0.062961	D	0.461523
310	8300	45.029	0.98212	0.38962	1.7966	1850.4	198.85	0.37622	0.063469	D	0.459009
310	8400	45.27	0.98866	0.3889	1.7928	1864.3	198.72	0.37552	0.063974	D	0.456565
310	8500	45.507	0.99522	0.38819	1.789	1878.2	198.59	0.37483	0.064475	D	0.454188
310	8600	45.74	1.0018	0.38749	1.7853	1892	198.48	0.37415	0.064972	D	0.451874
310	8700	45.969	1.0084	0.3868	1.7816	1905.7	198.37	0.37348	0.065467	D	0.449623
310	8800	46.195	1.015	0.38612	1.778	1919.3	198.26	0.37283	0.065958	D	0.447423
310	8900	46.416	1.0216	0.38546	1.7744	1932.8	198.17	0.37218	0.066446	D	0.445273
310	9000	46.635	1.0283	0.38481	1.7709	1946.2	198.07	0.37154	0.066931	D	0.443213
310	9100	46.85	1.0349	0.38417	1.7675	1959.6	197.98	0.37091	0.067413	D	0.441156
310	9200	47.061	1.0416	0.38354	1.7641	1972.8	197.9	0.37029	0.067892	D	0.439186
310	9300	47.27	1.0483	0.38292	1.7607	1985.9	197.82	0.36968	0.068369	D	0.437258
310	9400	47.475	1.055	0.38232	1.7574	1999	197.75	0.36908	0.068842	D	0.435371
310	9500	47.677	1.0617	0.38172	1.7542	2012	197.68	0.36848	0.069313	D	0.433524
310	9600	47.877	1.0684	0.38114	1.751	2024.9	197.62	0.3679	0.069782	D	0.431716
310	9700	48.073	1.0751	0.38057	1.7479	2037.6	197.56	0.36732	0.070248	D	0.429944
310	9800	48.267	1.0818	0.38001	1.7448	2050.3	197.5	0.36675	0.070711	D	0.428209
310	9900	48.458	1.0886	0.37947	1.7418	2063	197.45	0.36618	0.071173	D	0.426548
310	10000	48.646	1.0953	0.37893	1.7388	2075.5	197.4	0.36563	0.071631	D	0.424882
320	14.696	0.077404	0.99864	0.23132	1.2449	1045.7	270.57	0.73517	0.021168	V	267.024539
320	100	0.53089	0.99075	0.23416	1.2553	1041.8	269.49	0.64755	0.021202	D	38.931886
320	200	1.0717	0.98156	0.23763	1.2679	1037.4	268.21	0.61503	0.021254	D	19.285381
320	300	1.6227	0.97242	0.24125	1.2812	1033.1	266.92	0.59548	0.021317	D	12.737201
320	400	2.184	0.96333	0.24502	1.2951	1029.1	265.62	0.58124	0.021393	D	9.463602
320	500	2.7558	0.95431	0.24894	1.3096	1025.3	264.31	0.5699	0.021483	D	7.499993
320	600	3.3383	0.94535	0.25303	1.3248	1021.8	262.98	0.56038	0.021586	D	6.191313
320	700	3.9317	0.93647	0.25727	1.3407	1018.5	261.65	0.55212	0.021704	D	5.256990
320	800	4.5359	0.92768	0.26168	1.3572	1015.5	260.31	0.54478	0.021838	D	4.556691
320	900	5.1512	0.91897	0.26625	1.3744	1012.7	258.96	0.53814	0.021989	D	4.012363
320	1000	5.7776	0.91038	0.27098	1.3923	1010.3	257.6	0.53204	0.022156	D	3.577372
320	1100	6.4152	0.9019	0.27587	1.4108	1008.2	256.24	0.52639	0.022341	D	3.221863
320	1200	7.0637	0.89356	0.28091	1.43	1006.5	254.86	0.5211	0.022545	D	2.926064
320	1300	7.7232	0.88536	0.2861	1.4497	1005.1	253.48	0.51612	0.022768	D	2.676196
320	1400	8.3934	0.87733	0.29143	1.4701	1004.1	252.1	0.51139	0.023011	D	2.462500
320	1500	9.0741	0.86948	0.29689	1.491	1003.6	250.71	0.50689	0.023275	D	2.277769

TABLE F.1 (continued)—CO_2 PROPERTIES AT VARIOUS TEMPERATURES, °F											
Temperature (°F)	Pressure (psia)	Density (lbm/ft³)	Compressibility Factor	Heat Capacity (Btu/lbm°F)	Heat Capacity Ratio (C_p/C_v)	Sonic Velocity (ft/s)	Enthalpy (Btu/lbm)	Entropy [Btu/(lbm°F)]	Viscosity (cp)	Phase	Gas Formation Volume Factor (res bbl/Mcf)
320	1600	9.7649	0.86183	0.30246	1.5123	1003.5	249.32	0.50259	0.023561	D	2.116620
320	1700	10.465	0.85441	0.30812	1.5341	1004	247.93	0.49846	0.023868	D	1.974962
320	1800	11.175	0.84722	0.31386	1.5561	1004.9	246.55	0.49448	0.024198	D	1.849546
320	1900	11.893	0.8403	0.31966	1.5784	1006.4	245.16	0.49065	0.024551	D	1.737889
320	2000	12.619	0.83366	0.32548	1.6009	1008.5	243.79	0.48694	0.024927	D	1.637949
320	2100	13.351	0.82731	0.33131	1.6233	1011.1	242.41	0.48336	0.025326	D	1.548069
320	2200	14.09	0.82129	0.33711	1.6456	1014.4	241.05	0.47988	0.025749	D	1.466950
320	2300	14.833	0.81561	0.34285	1.6677	1018.3	239.7	0.47651	0.026194	D	1.393465
320	2400	15.579	0.81028	0.3485	1.6895	1022.9	238.37	0.47323	0.026663	D	1.326677
320	2500	16.328	0.80532	0.35403	1.7107	1028.1	237.05	0.47005	0.027154	D	1.265814
320	2600	17.079	0.80075	0.35941	1.7313	1034	235.75	0.46696	0.027666	D	1.210222
320	2700	17.828	0.79657	0.3646	1.7512	1040.6	234.47	0.46396	0.028199	D	1.159315
320	2800	18.577	0.7928	0.36958	1.7702	1047.8	233.22	0.46104	0.028752	D	1.112620
320	2900	19.322	0.78944	0.37432	1.7883	1055.7	231.98	0.45821	0.029324	D	1.069701
320	3000	20.063	0.7865	0.37881	1.8053	1064.2	230.78	0.45546	0.029913	D	1.030194
320	3100	20.798	0.78398	0.38301	1.8211	1073.4	229.6	0.45278	0.030518	D	0.993767
320	3200	21.527	0.78187	0.38692	1.8358	1083.2	228.45	0.45019	0.031137	D	0.960121
320	3300	22.248	0.78018	0.39053	1.8493	1093.6	227.33	0.44766	0.03177	D	0.929014
320	3400	22.96	0.77891	0.39383	1.8615	1104.5	226.24	0.44522	0.032414	D	0.900222
320	3500	23.662	0.77803	0.39683	1.8724	1115.9	225.19	0.44285	0.033069	D	0.873514
320	3600	24.353	0.77755	0.39951	1.8822	1127.9	224.16	0.44054	0.033732	D	0.848725
320	3700	25.032	0.77745	0.4019	1.8907	1140.3	223.17	0.43831	0.034401	D	0.825681
320	3800	25.7	0.77772	0.40399	1.898	1153.1	222.21	0.43614	0.035077	D	0.804231
320	3900	26.355	0.77835	0.4058	1.9042	1166.3	221.29	0.43405	0.035757	D	0.784245
320	4000	26.997	0.77932	0.40735	1.9094	1179.9	220.39	0.43201	0.03644	D	0.765592
320	4100	27.626	0.78063	0.40864	1.9135	1193.9	219.53	0.43004	0.037125	D	0.748174
320	4200	28.241	0.78225	0.40969	1.9166	1208.1	218.7	0.42812	0.03781	D	0.731876
320	4300	28.842	0.78417	0.41052	1.9188	1222.6	217.91	0.42627	0.038496	D	0.716610
320	4400	29.43	0.78638	0.41115	1.9203	1237.4	217.14	0.42447	0.03918	D	0.702298
320	4500	30.004	0.78886	0.41158	1.9209	1252.4	216.4	0.42272	0.039862	D	0.688857
320	4600	30.565	0.7916	0.41184	1.9209	1267.6	215.69	0.42103	0.040542	D	0.676222
320	4700	31.112	0.79458	0.41194	1.9202	1283	215.01	0.41939	0.041219	D	0.664326
320	4800	31.646	0.7978	0.4119	1.9189	1298.5	214.36	0.41779	0.041891	D	0.653122
320	4900	32.167	0.80124	0.41174	1.9172	1314.1	213.73	0.41625	0.04256	D	0.642551
320	5000	32.675	0.80488	0.41147	1.915	1329.9	213.13	0.41474	0.043223	D	0.632561
320	5100	33.17	0.80871	0.4111	1.9124	1345.7	212.56	0.41328	0.043882	D	0.623109
320	5200	33.654	0.81273	0.41065	1.9095	1361.5	212.01	0.41186	0.044535	D	0.614164
320	5300	34.125	0.81692	0.41012	1.9063	1377.4	211.48	0.41048	0.045183	D	0.605682
320	5400	34.585	0.82127	0.40953	1.9028	1393.4	210.97	0.40914	0.045825	D	0.597632
320	5500	35.033	0.82577	0.4089	1.8992	1409.3	210.48	0.40784	0.046461	D	0.589981
320	5600	35.471	0.83041	0.40822	1.8954	1425.2	210.02	0.40657	0.047092	D	0.582701
320	5700	35.898	0.83518	0.40751	1.8915	1441.2	209.57	0.40533	0.047717	D	0.575767
320	5800	36.315	0.84007	0.40677	1.8874	1457	209.14	0.40412	0.048336	D	0.569153
320	5900	36.722	0.84509	0.406	1.8833	1472.9	208.73	0.40294	0.048949	D	0.562850
320	6000	37.119	0.85021	0.40522	1.8791	1488.7	208.34	0.4018	0.049556	D	0.556822
320	6100	37.508	0.85543	0.40443	1.8749	1504.4	207.96	0.40068	0.050157	D	0.551056

TABLE F.1 (continued)—CO_2 PROPERTIES AT VARIOUS TEMPERATURES, °F											
Temperature (°F)	Pressure (psia)	Density (lbm/ft³)	Compressibility Factor	Heat Capacity (Btu/lbm°F)	Heat Capacity Ratio (C_p/C_v)	Sonic Velocity (ft/s)	Enthalpy (Btu/lbm)	Entropy [Btu/(lbm°F)]	Viscosity (cp)	Phase	Gas Formation Volume Factor (res bbl/Mcf)
320	6200	37.887	0.86074	0.40363	1.8707	1520	207.6	0.39959	0.050753	D	0.545534
320	6300	38.258	0.86615	0.40283	1.8664	1535.6	207.26	0.39852	0.051343	D	0.540249
320	6400	38.621	0.87163	0.40202	1.8621	1551.1	206.93	0.39748	0.051928	D	0.535172
320	6500	38.975	0.8772	0.40121	1.8579	1566.6	206.61	0.39646	0.052507	D	0.530306
320	6600	39.322	0.88284	0.40041	1.8536	1581.9	206.3	0.39546	0.053081	D	0.525629
320	6700	39.662	0.88854	0.3996	1.8494	1597.2	206.01	0.39448	0.05365	D	0.521127
320	6800	39.994	0.89431	0.3988	1.8452	1612.4	205.73	0.39353	0.054213	D	0.516798
320	6900	40.319	0.90014	0.39801	1.841	1627.5	205.47	0.39259	0.054772	D	0.512628
320	7000	40.638	0.90602	0.39722	1.8369	1642.5	205.21	0.39168	0.055326	D	0.508606
320	7100	40.95	0.91196	0.39643	1.8327	1657.4	204.96	0.39078	0.055875	D	0.504730
320	7200	41.256	0.91794	0.39565	1.8287	1672.3	204.73	0.3899	0.056419	D	0.500983
320	7300	41.556	0.92397	0.39488	1.8246	1687	204.5	0.38904	0.056959	D	0.497366
320	7400	41.85	0.93004	0.39412	1.8206	1701.7	204.29	0.3882	0.057494	D	0.493868
320	7500	42.139	0.93616	0.39337	1.8166	1716.2	204.08	0.38737	0.058025	D	0.490490
320	7600	42.422	0.94231	0.39262	1.8127	1730.7	203.89	0.38655	0.058552	D	0.487216
320	7700	42.7	0.9485	0.39188	1.8088	1745.1	203.7	0.38575	0.059075	D	0.484047
320	7800	42.972	0.95472	0.39115	1.8049	1759.4	203.52	0.38497	0.059593	D	0.480975
320	7900	43.24	0.96098	0.39043	1.8011	1773.7	203.35	0.38419	0.060108	D	0.478001
320	8000	43.503	0.96726	0.38972	1.7973	1787.8	203.18	0.38344	0.060619	D	0.475110
320	8100	43.761	0.97357	0.38902	1.7936	1801.8	203.02	0.38269	0.061126	D	0.472306
320	8200	44.015	0.97991	0.38832	1.7899	1815.8	202.87	0.38196	0.06163	D	0.469584
320	8300	44.264	0.98627	0.38764	1.7862	1829.7	202.73	0.38124	0.06213	D	0.466938
320	8400	44.509	0.99266	0.38696	1.7826	1843.5	202.6	0.38053	0.062626	D	0.464368
320	8500	44.75	0.99907	0.3863	1.7791	1857.2	202.47	0.37983	0.063119	D	0.461869
320	8600	44.987	1.0055	0.38565	1.7755	1870.8	202.34	0.37914	0.063609	D	0.459436
320	8700	45.22	1.0119	0.385	1.7721	1884.3	202.23	0.37847	0.064096	D	0.457046
320	8800	45.45	1.0184	0.38437	1.7686	1897.8	202.12	0.3778	0.064579	D	0.454755
320	8900	45.676	1.0249	0.38374	1.7653	1911.1	202.01	0.37714	0.06506	D	0.452515
320	9000	45.898	1.0314	0.38313	1.7619	1924.4	201.91	0.3765	0.065537	D	0.450325
320	9100	46.116	1.0379	0.38252	1.7586	1937.6	201.82	0.37586	0.066012	D	0.448183
320	9200	46.332	1.0444	0.38193	1.7554	1950.7	201.73	0.37523	0.066483	D	0.446088
320	9300	46.544	1.051	0.38134	1.7522	1963.7	201.64	0.37461	0.066952	D	0.444080
320	9400	46.753	1.0575	0.38077	1.7491	1976.6	201.57	0.374	0.067419	D	0.442073
320	9500	46.959	1.0641	0.38021	1.7459	1989.5	201.49	0.3734	0.067882	D	0.440149
320	9600	47.162	1.0707	0.37965	1.7429	2002.3	201.42	0.37281	0.068343	D	0.438266
320	9700	47.362	1.0773	0.37911	1.7399	2015	201.36	0.37222	0.068802	D	0.436422
320	9800	47.559	1.0838	0.37857	1.7369	2027.6	201.3	0.37164	0.069258	D	0.434575
320	9900	47.753	1.0904	0.37805	1.734	2040.1	201.24	0.37107	0.069712	D	0.432805
320	10000	47.945	1.0971	0.37753	1.7311	2052.5	201.19	0.37051	0.070164	D	0.431109
330	14.696	0.076419	0.9987	0.23232	1.2435	1051.9	272.89	0.73812	0.021411	V	270.465629
330	100	0.52393	0.99121	0.23506	1.2535	1048.2	271.83	0.65054	0.021445	D	39.449532
330	200	1.0572	0.98247	0.23839	1.2656	1044	270.59	0.61806	0.021495	D	19.550843
330	300	1.5999	0.97379	0.24185	1.2783	1040.1	269.33	0.59856	0.021557	D	12.918742
330	400	2.1522	0.96518	0.24544	1.2916	1036.3	268.07	0.58437	0.021631	D	9.603389
330	500	2.7143	0.95664	0.24918	1.3054	1032.8	266.8	0.57307	0.021718	D	7.614733
330	600	3.2863	0.94817	0.25306	1.3198	1029.5	265.52	0.56361	0.021819	D	6.289428

TABLE F.1 (continued)—CO_2 PROPERTIES AT VARIOUS TEMPERATURES, °F											
Temperature (°F)	Pressure (psia)	Density (lbm/ft^3)	Compressibility Factor	Heat Capacity (Btu/lbm°F)	Heat Capacity Ratio (C_p/C_v)	Sonic Velocity (ft/s)	Enthalpy (Btu/lbm)	Entropy [Btu/(lbm°F)]	Viscosity (cp)	Phase	Gas Formation Volume Factor (res bbl/Mcf)
330	700	3.8682	0.93978	0.25708	1.3348	1026.5	264.22	0.5554	0.021934	D	5.343236
330	800	4.4602	0.93149	0.26125	1.3504	1023.7	262.93	0.54811	0.022064	D	4.634089
330	900	5.0623	0.92329	0.26556	1.3666	1021.2	261.62	0.54153	0.022209	D	4.082929
330	1000	5.6744	0.9152	0.27001	1.3834	1019	260.31	0.53549	0.02237	D	3.642438
330	1100	6.2967	0.90723	0.27459	1.4007	1017.1	258.99	0.5299	0.022548	D	3.282471
330	1200	6.929	0.89939	0.27931	1.4186	1015.6	257.66	0.52467	0.022744	D	2.982929
330	1300	7.5711	0.8917	0.28416	1.437	1014.4	256.33	0.51975	0.022958	D	2.729930
330	1400	8.223	0.88417	0.28913	1.4559	1013.6	255	0.51509	0.023191	D	2.513529
330	1500	8.8841	0.87683	0.2942	1.4753	1013.2	253.67	0.51066	0.023443	D	2.326485
330	1600	9.5544	0.86967	0.29937	1.4951	1013.3	252.33	0.50642	0.023715	D	2.163270
330	1700	10.233	0.86273	0.30462	1.5152	1013.8	251	0.50236	0.024007	D	2.019771
330	1800	10.92	0.85601	0.30993	1.5355	1014.8	249.67	0.49846	0.02432	D	1.892703
330	1900	11.615	0.84955	0.31529	1.5561	1016.3	248.34	0.4947	0.024655	D	1.779555
330	2000	12.316	0.84334	0.32067	1.5767	1018.4	247.02	0.49106	0.02501	D	1.678220
330	2100	13.023	0.83742	0.32605	1.5974	1021	245.7	0.48754	0.025388	D	1.587085
330	2200	13.735	0.8318	0.3314	1.618	1024.2	244.4	0.48414	0.025786	D	1.504778
330	2300	14.452	0.82648	0.33671	1.6383	1027.9	243.1	0.48084	0.026206	D	1.430147
330	2400	15.172	0.8215	0.34194	1.6584	1032.3	241.82	0.47763	0.026647	D	1.362299
330	2500	15.894	0.81686	0.34708	1.6781	1037.2	240.56	0.47452	0.027109	D	1.300420
330	2600	16.617	0.81257	0.35208	1.6972	1042.8	239.31	0.4715	0.027591	D	1.243837
330	2700	17.34	0.80865	0.35693	1.7157	1049	238.08	0.46856	0.028092	D	1.191991
330	2800	18.061	0.8051	0.36161	1.7336	1055.8	236.87	0.4657	0.028612	D	1.144374
330	2900	18.78	0.80193	0.36608	1.7506	1063.3	235.68	0.46293	0.029149	D	1.100562
330	3000	19.496	0.79914	0.37034	1.7667	1071.3	234.52	0.46023	0.029703	D	1.060176
330	3100	20.206	0.79674	0.37437	1.7819	1079.9	233.38	0.45761	0.030272	D	1.022895
330	3200	20.911	0.79472	0.37814	1.796	1089.1	232.27	0.45506	0.030855	D	0.988417
330	3300	21.609	0.79308	0.38166	1.8092	1098.8	231.19	0.45258	0.031451	D	0.956487
330	3400	22.299	0.79183	0.38492	1.8212	1109.1	230.13	0.45018	0.032058	D	0.926892
330	3500	22.981	0.79094	0.38791	1.8322	1119.9	229.11	0.44784	0.032676	D	0.899397
330	3600	23.653	0.79043	0.39063	1.8421	1131.1	228.11	0.44558	0.033303	D	0.873850
330	3700	24.314	0.79027	0.3931	1.851	1142.8	227.15	0.44338	0.033937	D	0.850061
330	3800	24.966	0.79046	0.3953	1.8588	1154.9	226.21	0.44124	0.034577	D	0.827890
330	3900	25.606	0.79098	0.39725	1.8656	1167.4	225.3	0.43916	0.035223	D	0.807192
330	4000	26.234	0.79183	0.39896	1.8715	1180.3	224.43	0.43715	0.035872	D	0.787858
330	4100	26.85	0.793	0.40045	1.8764	1193.4	223.58	0.43519	0.036524	D	0.769778
330	4200	27.455	0.79446	0.40171	1.8804	1206.9	222.76	0.43329	0.037179	D	0.752833
330	4300	28.047	0.79621	0.40276	1.8836	1220.7	221.97	0.43145	0.037834	D	0.736945
330	4400	28.626	0.79824	0.40362	1.886	1234.7	221.21	0.42966	0.038489	D	0.722033
330	4500	29.193	0.80052	0.4043	1.8877	1249	220.48	0.42792	0.039143	D	0.708004
330	4600	29.747	0.80306	0.40481	1.8887	1263.5	219.78	0.42623	0.039796	D	0.694810
330	4700	30.289	0.80584	0.40516	1.8891	1278.1	219.1	0.42459	0.040447	D	0.682381
330	4800	30.819	0.80884	0.40537	1.8889	1292.9	218.45	0.423	0.041095	D	0.670653
330	4900	31.336	0.81206	0.40546	1.8882	1307.9	217.82	0.42145	0.04174	D	0.659581
330	5000	31.842	0.81548	0.40542	1.887	1322.9	217.22	0.41995	0.042381	D	0.649112
330	5100	32.335	0.81909	0.40529	1.8854	1338.1	216.64	0.41849	0.043018	D	0.639201
330	5200	32.817	0.82288	0.40506	1.8835	1353.4	216.08	0.41706	0.043651	D	0.629810

TABLE F.1 (continued)—CO_2 PROPERTIES AT VARIOUS TEMPERATURES, °F											
Temperature (°F)	Pressure (psia)	Density (lbm/ft^3)	Compressibility Factor	Heat Capacity (Btu/lbm°F)	Heat Capacity Ratio (C_p/C_v)	Sonic Velocity (ft/s)	Enthalpy (Btu/lbm)	Entropy [Btu/(lbm°F)]	Viscosity (cp)	Phase	Gas Formation Volume Factor (res bbl/Mcf)
330	5300	33.288	0.82684	0.40475	1.8812	1368.7	215.55	0.41568	0.04428	D	0.620900
330	5400	33.748	0.83096	0.40438	1.8786	1384	215.04	0.41433	0.044904	D	0.612439
330	5500	34.197	0.83524	0.40394	1.8758	1399.4	214.55	0.41302	0.045522	D	0.604400
330	5600	34.636	0.83966	0.40344	1.8728	1414.8	214.08	0.41174	0.046136	D	0.596749
330	5700	35.064	0.84421	0.40291	1.8696	1430.2	213.62	0.41049	0.046744	D	0.589457
330	5800	35.483	0.84888	0.40234	1.8663	1445.6	213.19	0.40928	0.047348	D	0.582498
330	5900	35.892	0.85368	0.40173	1.8629	1461	212.77	0.40809	0.047946	D	0.575863
330	6000	36.292	0.85859	0.4011	1.8593	1476.3	212.37	0.40694	0.048539	D	0.569522
330	6100	36.682	0.8636	0.40045	1.8557	1491.6	211.99	0.40581	0.049126	D	0.563455
330	6200	37.064	0.8687	0.39979	1.852	1506.9	211.62	0.40471	0.049709	D	0.557640
330	6300	37.438	0.8739	0.39911	1.8483	1522.1	211.27	0.40363	0.050286	D	0.552074
330	6400	37.804	0.87919	0.39842	1.8446	1537.2	210.93	0.40258	0.050858	D	0.546738
330	6500	38.162	0.88455	0.39773	1.8408	1552.3	210.6	0.40155	0.051425	D	0.541608
330	6600	38.512	0.89	0.39703	1.837	1567.3	210.29	0.40054	0.051987	D	0.536688
330	6700	38.855	0.89551	0.39633	1.8333	1582.3	209.99	0.39956	0.052545	D	0.531951
330	6800	39.19	0.90109	0.39563	1.8295	1597.1	209.71	0.39859	0.053097	D	0.527394
330	6900	39.519	0.90673	0.39493	1.8257	1611.9	209.43	0.39765	0.053645	D	0.523004
330	7000	39.842	0.91243	0.39423	1.822	1626.6	209.17	0.39672	0.054188	D	0.518773
330	7100	40.157	0.91818	0.39354	1.8182	1641.2	208.91	0.39582	0.054727	D	0.514690
330	7200	40.467	0.92399	0.39284	1.8145	1655.8	208.67	0.39493	0.055261	D	0.510753
330	7300	40.771	0.92985	0.39216	1.8108	1670.3	208.44	0.39406	0.055791	D	0.506951
330	7400	41.069	0.93575	0.39147	1.8071	1684.7	208.22	0.3932	0.056317	D	0.503274
330	7500	41.361	0.94169	0.39079	1.8035	1699	208	0.39236	0.056839	D	0.499716
330	7600	41.648	0.94768	0.39012	1.7999	1713.2	207.8	0.39154	0.057356	D	0.496277
330	7700	41.929	0.9537	0.38945	1.7963	1727.4	207.6	0.39073	0.05787	D	0.492944
330	7800	42.206	0.95976	0.38879	1.7927	1741.4	207.42	0.38994	0.05838	D	0.489716
330	7900	42.477	0.96585	0.38813	1.7892	1755.4	207.24	0.38916	0.058886	D	0.486585
330	8000	42.744	0.97198	0.38748	1.7857	1769.3	207.07	0.38839	0.059388	D	0.483552
330	8100	43.006	0.97813	0.38684	1.7822	1783.2	206.9	0.38763	0.059887	D	0.480604
330	8200	43.263	0.98432	0.3862	1.7788	1796.9	206.75	0.38689	0.060382	D	0.477748
330	8300	43.516	0.99053	0.38557	1.7754	1810.6	206.6	0.38616	0.060874	D	0.474969
330	8400	43.765	0.99676	0.38495	1.772	1824.2	206.46	0.38545	0.061362	D	0.472267
330	8500	44.01	1.003	0.38434	1.7687	1837.7	206.32	0.38474	0.061848	D	0.469633
330	8600	44.25	1.0093	0.38373	1.7654	1851.1	206.19	0.38404	0.062329	D	0.467087
330	8700	44.487	1.0156	0.38313	1.7621	1864.5	206.07	0.38336	0.062808	D	0.464600
330	8800	44.72	1.0219	0.38254	1.7589	1877.7	205.95	0.38269	0.063284	D	0.462170
330	8900	44.949	1.0283	0.38196	1.7557	1890.9	205.84	0.38202	0.063757	D	0.459839
330	9000	45.175	1.0346	0.38139	1.7525	1904.1	205.73	0.38137	0.064227	D	0.457516
330	9100	45.397	1.041	0.38082	1.7494	1917.1	205.63	0.38072	0.064694	D	0.455287
330	9200	45.616	1.0474	0.38026	1.7464	1930	205.54	0.38009	0.065158	D	0.453107
330	9300	45.832	1.0538	0.37971	1.7433	1942.9	205.45	0.37946	0.06562	D	0.450974
330	9400	46.044	1.0602	0.37917	1.7403	1955.7	205.37	0.37885	0.066079	D	0.448886
330	9500	46.253	1.0666	0.37864	1.7374	1968.5	205.29	0.37824	0.066535	D	0.446842
330	9600	46.46	1.0731	0.37811	1.7344	1981.1	205.21	0.37764	0.066989	D	0.444882
330	9700	46.663	1.0795	0.3776	1.7315	1993.7	205.14	0.37704	0.067441	D	0.442922
330	9800	46.863	1.086	0.37709	1.7287	2006.2	205.07	0.37646	0.06789	D	0.441042

TABLE F.1 (continued)—CO_2 PROPERTIES AT VARIOUS TEMPERATURES, °F											
Temperature (°F)	Pressure (psia)	Density (lbm/ft^3)	Compressibility Factor	Heat Capacity (Btu/lbm°F)	Heat Capacity Ratio (C_p/C_v)	Sonic Velocity (ft/s)	Enthalpy (Btu/lbm)	Entropy [Btu/(lbm°F)]	Viscosity (cp)	Phase	Gas Formation Volume Factor (res bbl/Mcf)
330	9900	47.061	1.0925	0.37659	1.7259	2018.6	205.01	0.37588	0.068337	D	0.439200
330	10000	47.256	1.099	0.3761	1.7231	2030.9	204.96	0.37531	0.068781	D	0.437395
340	14.696	0.075458	0.99877	0.23332	1.2422	1058	275.21	0.74105	0.021652	V	273.909872
340	100	0.51715	0.99164	0.23595	1.2517	1054.5	274.19	0.6535	0.021686	D	39.966432
340	200	1.043	0.98333	0.23914	1.2634	1050.6	272.97	0.62107	0.021735	D	19.815755
340	300	1.5778	0.9751	0.24245	1.2755	1046.9	271.75	0.60161	0.021796	D	13.099938
340	400	2.1215	0.96694	0.24589	1.2882	1043.4	270.53	0.58746	0.021869	D	9.742735
340	500	2.6742	0.95885	0.24946	1.3014	1040.1	269.29	0.57621	0.021954	D	7.728977
340	600	3.2361	0.95084	0.25315	1.3151	1037.1	268.05	0.56679	0.022052	D	6.387009
340	700	3.8071	0.94291	0.25697	1.3293	1034.3	266.79	0.55863	0.022163	D	5.428921
340	800	4.3875	0.93508	0.26091	1.344	1031.8	265.54	0.5514	0.022289	D	4.710859
340	900	4.977	0.92735	0.26498	1.3593	1029.5	264.27	0.54487	0.022429	D	4.152814
340	1000	5.5758	0.91974	0.26918	1.3751	1027.5	263	0.53888	0.022585	D	3.706862
340	1100	6.1838	0.91224	0.27349	1.3914	1025.9	261.73	0.53335	0.022757	D	3.342395
340	1200	6.8009	0.90488	0.27792	1.4081	1024.5	260.45	0.52818	0.022945	D	3.039143
340	1300	7.4268	0.89766	0.28246	1.4253	1023.5	259.17	0.52332	0.02315	D	2.782979
340	1400	8.0615	0.8906	0.2871	1.443	1022.9	257.88	0.51872	0.023373	D	2.563870
340	1500	8.7046	0.88372	0.29183	1.461	1022.7	256.6	0.51435	0.023614	D	2.374459
340	1600	9.3558	0.87702	0.29665	1.4793	1022.8	255.31	0.51017	0.023874	D	2.209179
340	1700	10.015	0.87052	0.30153	1.498	1023.4	254.03	0.50618	0.024153	D	2.063817
340	1800	10.681	0.86425	0.30646	1.5168	1024.5	252.75	0.50234	0.024451	D	1.935121
340	1900	11.354	0.8582	0.31143	1.5358	1026	251.47	0.49864	0.024769	D	1.820439
340	2000	12.032	0.85241	0.31642	1.5549	1028.1	250.2	0.49507	0.025106	D	1.717750
340	2100	12.717	0.84688	0.3214	1.574	1030.6	248.94	0.49162	0.025464	D	1.625339
340	2200	13.405	0.84162	0.32637	1.5931	1033.7	247.68	0.48828	0.025841	D	1.541824
340	2300	14.098	0.83666	0.33129	1.6119	1037.3	246.44	0.48504	0.026238	D	1.466096
340	2400	14.793	0.83201	0.33616	1.6305	1041.5	245.21	0.4819	0.026655	D	1.397200
340	2500	15.49	0.82767	0.34093	1.6487	1046.3	244	0.47885	0.02709	D	1.334316
340	2600	16.188	0.82365	0.3456	1.6665	1051.6	242.8	0.47588	0.027545	D	1.276764
340	2700	16.886	0.81998	0.35013	1.6838	1057.5	241.61	0.47301	0.028017	D	1.223998
340	2800	17.583	0.81664	0.35452	1.7005	1063.9	240.45	0.47021	0.028507	D	1.175477
340	2900	18.278	0.81366	0.35874	1.7165	1071	239.31	0.46749	0.029013	D	1.130801
340	3000	18.97	0.81103	0.36278	1.7318	1078.6	238.19	0.46484	0.029535	D	1.089575
340	3100	19.657	0.80875	0.36662	1.7462	1086.7	237.09	0.46227	0.030072	D	1.051463
340	3200	20.339	0.80683	0.37025	1.7598	1095.4	236.01	0.45977	0.030622	D	1.016186
340	3300	21.016	0.80527	0.37366	1.7726	1104.6	234.97	0.45734	0.031185	D	0.983488
340	3400	21.685	0.80405	0.37684	1.7844	1114.3	233.94	0.45497	0.031759	D	0.953115
340	3500	22.347	0.80318	0.3798	1.7952	1124.4	232.95	0.45267	0.032343	D	0.924882
340	3600	23.001	0.80266	0.38252	1.8052	1135.1	231.98	0.45044	0.032936	D	0.898608
340	3700	23.645	0.80247	0.38501	1.8142	1146.1	231.04	0.44827	0.033537	D	0.874115
340	3800	24.28	0.8026	0.38727	1.8223	1157.5	230.12	0.44616	0.034145	D	0.851250
340	3900	24.905	0.80305	0.38931	1.8295	1169.4	229.23	0.44411	0.034758	D	0.829888
340	4000	25.52	0.8038	0.39114	1.8358	1181.6	228.38	0.44212	0.035377	D	0.809896
340	4100	26.124	0.80486	0.39275	1.8413	1194.1	227.54	0.44018	0.035998	D	0.791185
340	4200	26.717	0.80619	0.39417	1.846	1206.9	226.74	0.4383	0.036623	D	0.773623
340	4300	27.298	0.8078	0.39539	1.8499	1219.9	225.96	0.43647	0.037249	D	0.757141

TABLE F.1 (continued)—CO_2 PROPERTIES AT VARIOUS TEMPERATURES, °F												
Temperature (°F)	Pressure (psia)	Density (lbm/ft^3)	Compressibility Factor	Heat Capacity (Btu/lbm°F)	Heat Capacity Ratio (C_p/C_v)	Sonic Velocity (ft/s)	Enthalpy (Btu/lbm)	Entropy [Btu/(lbm°F)]	Viscosity (cp)	Phase	Gas Formation Volume Factor (res bbl/Mcf)
340	4400	27.868	0.80968	0.39644	1.8531	1233.3	225.21	0.43469	0.037876	D	0.741655
340	4500	28.427	0.8118	0.39731	1.8556	1246.9	224.49	0.43297	0.038503	D	0.727073
340	4600	28.974	0.81417	0.39802	1.8575	1260.7	223.79	0.43129	0.03913	D	0.713343
340	4700	29.51	0.81677	0.39859	1.8588	1274.6	223.12	0.42965	0.039756	D	0.700395
340	4800	30.035	0.81958	0.39901	1.8595	1288.8	222.47	0.42806	0.04038	D	0.688163
340	4900	30.548	0.82261	0.39931	1.8597	1303.1	221.84	0.42652	0.041001	D	0.676611
340	5000	31.049	0.82583	0.39949	1.8594	1317.5	221.24	0.42501	0.041621	D	0.665675
340	5100	31.54	0.82924	0.39956	1.8587	1332.1	220.66	0.42355	0.042237	D	0.655317
340	5200	32.02	0.83282	0.39954	1.8576	1346.7	220.11	0.42212	0.042849	D	0.645489
340	5300	32.489	0.83658	0.39943	1.8562	1361.5	219.57	0.42074	0.043458	D	0.636170
340	5400	32.948	0.8405	0.39924	1.8544	1376.3	219.06	0.41939	0.044064	D	0.627314
340	5500	33.397	0.84457	0.39899	1.8524	1391.1	218.56	0.41807	0.044664	D	0.618891
340	5600	33.835	0.84878	0.39867	1.8501	1406	218.09	0.41678	0.045261	D	0.610870
340	5700	34.264	0.85312	0.39831	1.8477	1420.9	217.63	0.41553	0.045853	D	0.603221
340	5800	34.683	0.85759	0.39789	1.845	1435.8	217.19	0.41431	0.046441	D	0.595927
340	5900	35.093	0.86218	0.39744	1.8422	1450.7	216.77	0.41312	0.047024	D	0.588962
340	6000	35.494	0.86689	0.39696	1.8393	1465.6	216.36	0.41196	0.047602	D	0.582310
340	6100	35.887	0.8717	0.39645	1.8363	1480.5	215.97	0.41082	0.048176	D	0.575942
340	6200	36.271	0.87661	0.39591	1.8332	1495.3	215.6	0.40971	0.048745	D	0.569844
340	6300	36.647	0.88162	0.39536	1.83	1510.1	215.24	0.40863	0.049309	D	0.564004
340	6400	37.014	0.88671	0.39479	1.8268	1524.9	214.89	0.40757	0.049868	D	0.558397
340	6500	37.375	0.89188	0.39421	1.8235	1539.6	214.56	0.40653	0.050423	D	0.553012
340	6600	37.727	0.89714	0.39362	1.8202	1554.3	214.24	0.40552	0.050973	D	0.547845
340	6700	38.073	0.90247	0.39302	1.8168	1568.9	213.94	0.40452	0.051519	D	0.542874
340	6800	38.411	0.90786	0.39242	1.8135	1583.4	213.65	0.40355	0.052061	D	0.538086
340	6900	38.743	0.91333	0.39181	1.8101	1597.9	213.36	0.4026	0.052598	D	0.533482
340	7000	39.068	0.91885	0.3912	1.8068	1612.3	213.09	0.40166	0.05313	D	0.529039
340	7100	39.387	0.92443	0.39059	1.8034	1626.6	212.83	0.40075	0.053659	D	0.524756
340	7200	39.7	0.93007	0.38998	1.8	1640.9	212.59	0.39985	0.054183	D	0.520624
340	7300	40.007	0.93575	0.38937	1.7967	1655.1	212.35	0.39897	0.054703	D	0.516628
340	7400	40.308	0.94149	0.38876	1.7933	1669.2	212.12	0.39811	0.055219	D	0.512773
340	7500	40.603	0.94727	0.38816	1.79	1683.3	211.9	0.39726	0.055731	D	0.509042
340	7600	40.893	0.95309	0.38755	1.7867	1697.2	211.69	0.39643	0.05624	D	0.505431
340	7700	41.178	0.95895	0.38695	1.7834	1711.1	211.49	0.39562	0.056744	D	0.501934
340	7800	41.458	0.96485	0.38636	1.7802	1725	211.29	0.39481	0.057245	D	0.498547
340	7900	41.733	0.97079	0.38577	1.7769	1738.7	211.11	0.39402	0.057742	D	0.495267
340	8000	42.002	0.97676	0.38518	1.7737	1752.4	210.93	0.39325	0.058236	D	0.492084
340	8100	42.268	0.98276	0.3846	1.7705	1766	210.76	0.39249	0.058726	D	0.488994
340	8200	42.529	0.98879	0.38402	1.7673	1779.5	210.6	0.39174	0.059213	D	0.485995
340	8300	42.785	0.99485	0.38344	1.7641	1793	210.44	0.391	0.059697	D	0.483082
340	8400	43.037	1.0009	0.38288	1.761	1806.4	210.29	0.39028	0.060177	D	0.480234
340	8500	43.285	1.0071	0.38231	1.7579	1819.7	210.15	0.38956	0.060654	D	0.477524
340	8600	43.529	1.0132	0.38176	1.7548	1832.9	210.02	0.38886	0.061128	D	0.474830
340	8700	43.769	1.0194	0.38121	1.7518	1846.1	209.89	0.38817	0.061599	D	0.472244
340	8800	44.005	1.0255	0.38066	1.7488	1859.2	209.77	0.38749	0.062067	D	0.469672
340	8900	44.238	1.0317	0.38012	1.7458	1872.2	209.65	0.38682	0.062533	D	0.467202

TABLE F.1 (continued)—CO_2 PROPERTIES AT VARIOUS TEMPERATURES, °F											
Temperature (°F)	Pressure (psia)	Density (lbm/ft³)	Compressibility Factor	Heat Capacity (Btu/lbm°F)	Heat Capacity Ratio (C_p/C_v)	Sonic Velocity (ft/s)	Enthalpy (Btu/lbm)	Entropy [Btu/(lbm°F)]	Viscosity (cp)	Phase	Gas Formation Volume Factor (res bbl/Mcf)
340	9000	44.467	1.038	0.37959	1.7428	1885.1	209.54	0.38616	0.062995	D	0.464832
340	9100	44.692	1.0442	0.37906	1.7399	1898	209.43	0.38551	0.063455	D	0.462470
340	9200	44.914	1.0504	0.37854	1.737	1910.8	209.33	0.38486	0.063912	D	0.460159
340	9300	45.133	1.0567	0.37803	1.7341	1923.6	209.24	0.38423	0.064366	D	0.457942
340	9400	45.349	1.063	0.37752	1.7313	1936.2	209.15	0.38361	0.064818	D	0.455771
340	9500	45.561	1.0693	0.37702	1.7285	1948.8	209.06	0.38299	0.065267	D	0.453646
340	9600	45.771	1.0756	0.37653	1.7257	1961.3	208.98	0.38238	0.065714	D	0.451566
340	9700	45.977	1.0819	0.37604	1.7229	1973.7	208.91	0.38178	0.066158	D	0.449528
340	9800	46.18	1.0883	0.37556	1.7202	1986.1	208.84	0.38119	0.0666	D	0.447573
340	9900	46.381	1.0946	0.37509	1.7175	1998.4	208.77	0.38061	0.06704	D	0.445617
340	10000	46.579	1.101	0.37463	1.7149	2010.6	208.71	0.38003	0.067478	D	0.443740
350	14.696	0.074522	0.99883	0.2343	1.2408	1064.1	277.55	0.74396	0.021893	V	277.351819
350	100	0.51056	0.99204	0.23683	1.25	1060.8	276.55	0.65644	0.021926	D	40.482541
350	200	1.0293	0.98416	0.2399	1.2612	1057.2	275.37	0.62404	0.021975	D	20.080490
350	300	1.5563	0.97634	0.24308	1.2729	1053.7	274.18	0.60463	0.022034	D	13.280622
350	400	2.0917	0.9686	0.24637	1.285	1050.4	272.99	0.59052	0.022105	D	9.881504
350	500	2.6354	0.96094	0.24977	1.2975	1047.4	271.79	0.57931	0.022188	D	7.842686
350	600	3.1876	0.95336	0.25328	1.3106	1044.6	270.58	0.56994	0.022283	D	6.484019
350	700	3.7484	0.94587	0.25691	1.3241	1042	269.36	0.56183	0.022392	D	5.514066
350	800	4.3176	0.93848	0.26065	1.338	1039.7	268.14	0.55464	0.022514	D	4.787112
350	900	4.8953	0.93119	0.26451	1.3525	1037.7	266.92	0.54816	0.02265	D	4.222157
350	1000	5.4814	0.92402	0.26847	1.3673	1035.9	265.69	0.54222	0.0228	D	3.770682
350	1100	6.076	0.91696	0.27254	1.3827	1034.4	264.46	0.53674	0.022966	D	3.401702
350	1200	6.6787	0.91004	0.2767	1.3984	1033.3	263.22	0.53162	0.023147	D	3.094695
350	1300	7.2896	0.90327	0.28097	1.4145	1032.4	261.98	0.52682	0.023345	D	2.835390
350	1400	7.9083	0.89665	0.28532	1.431	1032	260.74	0.52227	0.023559	D	2.613566
350	1500	8.5346	0.89019	0.28974	1.4478	1031.9	259.5	0.51796	0.02379	D	2.421754
350	1600	9.1682	0.88392	0.29424	1.4649	1032.1	258.27	0.51385	0.024038	D	2.254403
350	1700	9.8086	0.87784	0.29879	1.4823	1032.8	257.03	0.50991	0.024305	D	2.107196
350	1800	10.456	0.87197	0.30339	1.4998	1034	255.8	0.50613	0.024589	D	1.976822
350	1900	11.108	0.86632	0.30801	1.5174	1035.5	254.57	0.50249	0.024891	D	1.860644
350	2000	11.767	0.86091	0.31266	1.5351	1037.6	253.35	0.49898	0.025213	D	1.756574
350	2100	12.429	0.85574	0.31729	1.5529	1040.1	252.13	0.49559	0.025552	D	1.662881
350	2200	13.096	0.85084	0.32191	1.5705	1043.1	250.93	0.4923	0.02591	D	1.578206
350	2300	13.767	0.8462	0.3265	1.588	1046.6	249.73	0.48913	0.026287	D	1.501356
350	2400	14.439	0.84185	0.33102	1.6053	1050.7	248.55	0.48604	0.026681	D	1.431403
350	2500	15.114	0.8378	0.33548	1.6222	1055.2	247.38	0.48305	0.027094	D	1.367537
350	2600	15.789	0.83405	0.33984	1.6388	1060.3	246.22	0.48014	0.027523	D	1.309053
350	2700	16.464	0.83061	0.34409	1.655	1065.9	245.08	0.47732	0.02797	D	1.255371
350	2800	17.138	0.82749	0.34821	1.6706	1072.1	243.96	0.47457	0.028433	D	1.205989
350	2900	17.811	0.82469	0.35219	1.6857	1078.8	242.86	0.4719	0.028912	D	1.160463
350	3000	18.48	0.82222	0.35602	1.7001	1086	241.78	0.46931	0.029405	D	1.118421
350	3100	19.146	0.82007	0.35967	1.7139	1093.7	240.72	0.46678	0.029912	D	1.079513
350	3200	19.808	0.81826	0.36315	1.7269	1101.9	239.68	0.46432	0.030432	D	1.043470
350	3300	20.464	0.81678	0.36644	1.7391	1110.6	238.67	0.46193	0.030965	D	1.010019
350	3400	21.114	0.81562	0.36953	1.7506	1119.8	237.67	0.45961	0.031508	D	0.978921

TABLE F.1 (continued)—CO_2 PROPERTIES AT VARIOUS TEMPERATURES, °F											
Temperature (°F)	Pressure (psia)	Density (lbm/ft^3)	Compressibility Factor	Heat Capacity (Btu/lbm°F)	Heat Capacity Ratio (C_p/C_v)	Sonic Velocity (ft/s)	Enthalpy (Btu/lbm)	Entropy [Btu/(lbm°F)]	Viscosity (cp)	Phase	Gas Formation Volume Factor (res bbl/Mcf)
350	3500	21.757	0.81478	0.37242	1.7612	1129.5	236.71	0.45735	0.032062	D	0.949972
350	3600	22.393	0.81427	0.37511	1.7711	1139.5	235.76	0.45515	0.032624	D	0.923006
350	3700	23.021	0.81406	0.37759	1.7801	1150	234.85	0.45301	0.033195	D	0.897828
350	3800	23.64	0.81416	0.37988	1.7883	1160.9	233.96	0.45093	0.033772	D	0.874309
350	3900	24.25	0.81456	0.38197	1.7958	1172.1	233.09	0.4489	0.034356	D	0.852309
350	4000	24.851	0.81525	0.38387	1.8024	1183.7	232.25	0.44693	0.034945	D	0.831705
350	4100	25.442	0.81621	0.38557	1.8083	1195.5	231.44	0.44502	0.035537	D	0.812375
350	4200	26.023	0.81745	0.3871	1.8135	1207.7	230.65	0.44315	0.036134	D	0.794238
350	4300	26.594	0.81895	0.38845	1.8179	1220.2	229.88	0.44134	0.036732	D	0.777191
350	4400	27.155	0.8207	0.38963	1.8217	1232.9	229.14	0.43958	0.037333	D	0.761150
350	4500	27.705	0.82268	0.39066	1.8249	1245.8	228.43	0.43786	0.037934	D	0.746031
350	4600	28.244	0.82491	0.39153	1.8274	1259	227.74	0.43619	0.038536	D	0.731791
350	4700	28.773	0.82735	0.39227	1.8294	1272.4	227.07	0.43457	0.039137	D	0.718340
350	4800	29.291	0.83	0.39287	1.8309	1285.9	226.43	0.43298	0.039738	D	0.705627
350	4900	29.799	0.83285	0.39334	1.8318	1299.6	225.81	0.43144	0.040337	D	0.693600
350	5000	30.296	0.8359	0.3937	1.8323	1313.4	225.21	0.42994	0.040935	D	0.682218
350	5100	30.783	0.83913	0.39396	1.8324	1327.4	224.63	0.42848	0.04153	D	0.671425
350	5200	31.26	0.84254	0.39412	1.8321	1341.5	224.08	0.42706	0.042122	D	0.661189
350	5300	31.727	0.84611	0.39419	1.8314	1355.6	223.54	0.42567	0.042712	D	0.651463
350	5400	32.184	0.84984	0.39417	1.8304	1369.9	223.02	0.42432	0.043299	D	0.642217
350	5500	32.631	0.85372	0.39409	1.8291	1384.2	222.53	0.423	0.043882	D	0.633419
350	5600	33.068	0.85774	0.39394	1.8275	1398.6	222.05	0.42171	0.044461	D	0.625038
350	5700	33.496	0.86189	0.39373	1.8258	1413	221.59	0.42045	0.045037	D	0.617043
350	5800	33.916	0.86617	0.39347	1.8238	1427.4	221.15	0.41923	0.045609	D	0.609416
350	5900	34.326	0.87057	0.39316	1.8216	1441.9	220.72	0.41803	0.046177	D	0.602130
350	6000	34.728	0.87509	0.39282	1.8193	1456.3	220.31	0.41686	0.04674	D	0.595169
350	6100	35.121	0.87971	0.39244	1.8168	1470.8	219.92	0.41572	0.0473	D	0.588502
350	6200	35.506	0.88444	0.39203	1.8142	1485.2	219.54	0.41461	0.047855	D	0.582124
350	6300	35.883	0.88926	0.3916	1.8116	1499.6	219.17	0.41352	0.048406	D	0.576006
350	6400	36.252	0.89417	0.39114	1.8088	1514	218.82	0.41245	0.048953	D	0.570136
350	6500	36.614	0.89916	0.39067	1.806	1528.4	218.49	0.41141	0.049496	D	0.564498
350	6600	36.969	0.90424	0.39019	1.8031	1542.7	218.16	0.41039	0.050034	D	0.559086
350	6700	37.316	0.90939	0.38969	1.8002	1556.9	217.85	0.40939	0.050568	D	0.553878
350	6800	37.657	0.91461	0.38918	1.7973	1571.1	217.55	0.40841	0.051098	D	0.548865
350	6900	37.991	0.9199	0.38866	1.7943	1585.3	217.27	0.40745	0.051624	D	0.544039
350	7000	38.319	0.92526	0.38814	1.7913	1599.4	216.99	0.40651	0.052146	D	0.539392
350	7100	38.64	0.93067	0.38761	1.7883	1613.4	216.73	0.40559	0.052664	D	0.534904
350	7200	38.955	0.93614	0.38708	1.7853	1627.4	216.47	0.40468	0.053178	D	0.530575
350	7300	39.265	0.94166	0.38655	1.7823	1641.3	216.23	0.4038	0.053688	D	0.526393
350	7400	39.568	0.94724	0.38602	1.7793	1655.2	215.99	0.40293	0.054195	D	0.522356
350	7500	39.867	0.95286	0.38548	1.7763	1669	215.77	0.40207	0.054697	D	0.518449
350	7600	40.159	0.95852	0.38495	1.7733	1682.7	215.55	0.40123	0.055196	D	0.514667
350	7700	40.447	0.96423	0.38442	1.7703	1696.3	215.34	0.40041	0.055692	D	0.511009
350	7800	40.729	0.96998	0.38388	1.7673	1709.9	215.14	0.3996	0.056184	D	0.507466
350	7900	41.007	0.97576	0.38336	1.7643	1723.4	214.95	0.3988	0.056672	D	0.504028
350	8000	41.28	0.98158	0.38283	1.7614	1736.9	214.77	0.39802	0.057157	D	0.500696

TABLE F.1 (continued)—CO_2 PROPERTIES AT VARIOUS TEMPERATURES, °F											
Temperature (°F)	Pressure (psia)	Density (lbm/ft³)	Compressibility Factor	Heat Capacity (Btu/lbm°F)	Heat Capacity Ratio (C_p/C_v)	Sonic Velocity (ft/s)	Enthalpy (Btu/lbm)	Entropy [Btu/(lbm°F)]	Viscosity (cp)	Phase	Gas Formation Volume Factor (res bbl/Mcf)
350	8100	41.548	0.98744	0.3823	1.7584	1750.3	214.6	0.39725	0.057639	D	0.497467
350	8200	41.812	0.99332	0.38178	1.7555	1763.6	214.43	0.3965	0.058118	D	0.494327
350	8300	42.071	0.99924	0.38126	1.7526	1776.8	214.27	0.39575	0.058593	D	0.491281
350	8400	42.326	1.0052	0.38075	1.7497	1790	214.11	0.39502	0.059065	D	0.488328
350	8500	42.577	1.0112	0.38024	1.7468	1803.1	213.97	0.3943	0.059534	D	0.485464
350	8600	42.824	1.0172	0.37973	1.744	1816.1	213.83	0.39359	0.060001	D	0.482666
350	8700	43.067	1.0232	0.37923	1.7411	1829.1	213.69	0.3929	0.060464	D	0.479932
350	8800	43.306	1.0292	0.37873	1.7383	1842	213.56	0.39221	0.060924	D	0.477261
350	8900	43.542	1.0353	0.37823	1.7355	1854.9	213.44	0.39153	0.061382	D	0.474695
350	9000	43.774	1.0414	0.37774	1.7328	1867.6	213.33	0.39086	0.061837	D	0.472187
350	9100	44.002	1.0475	0.37726	1.73	1880.3	213.22	0.39021	0.062289	D	0.469733
350	9200	44.227	1.0536	0.37678	1.7273	1893	213.11	0.38956	0.062739	D	0.467333
350	9300	44.449	1.0597	0.3763	1.7246	1905.5	213.01	0.38892	0.063186	D	0.464985
350	9400	44.667	1.0659	0.37583	1.7219	1918	212.92	0.38829	0.063631	D	0.462730
350	9500	44.883	1.0721	0.37537	1.7193	1930.5	212.83	0.38767	0.064073	D	0.460522
350	9600	45.095	1.0782	0.37491	1.7166	1942.8	212.74	0.38705	0.064513	D	0.458318
350	9700	45.304	1.0844	0.37445	1.714	1955.1	212.66	0.38645	0.06495	D	0.456201
350	9800	45.51	1.0907	0.37401	1.7115	1967.3	212.59	0.38585	0.065385	D	0.454169
350	9900	45.714	1.0969	0.37356	1.7089	1979.5	212.51	0.38526	0.065818	D	0.452137
350	10000	45.915	1.1031	0.37313	1.7064	1991.6	212.45	0.38468	0.066249	D	0.450146
360	14.696	0.073609	0.99888	0.23527	1.2395	1070.2	279.9	0.74684	0.022132	V	280.791367
360	100	0.50413	0.99243	0.23771	1.2484	1067.1	278.92	0.65935	0.022165	D	40.998641
360	200	1.0159	0.98494	0.24066	1.2592	1063.6	277.77	0.62699	0.022213	D	20.344609
360	300	1.5355	0.97752	0.24371	1.2703	1060.4	276.62	0.60761	0.022271	D	13.460896
360	400	2.0628	0.97018	0.24686	1.2819	1057.4	275.45	0.59354	0.022341	D	10.019866
360	500	2.5979	0.96292	0.25011	1.2939	1054.6	274.29	0.58238	0.022421	D	7.955908
360	600	3.1409	0.95575	0.25346	1.3063	1052	273.11	0.57305	0.022514	D	6.580557
360	700	3.6917	0.94867	0.25692	1.3192	1049.6	271.93	0.56498	0.02262	D	5.598694
360	800	4.2504	0.94169	0.26047	1.3324	1047.5	270.75	0.55784	0.022738	D	4.862813
360	900	4.8168	0.93482	0.26413	1.3461	1045.7	269.56	0.5514	0.02287	D	4.290966
360	1000	5.391	0.92806	0.26788	1.3601	1044.1	268.37	0.54552	0.023016	D	3.833943
360	1100	5.9728	0.92142	0.27172	1.3745	1042.8	267.18	0.54008	0.023176	D	3.460466
360	1200	6.5621	0.91492	0.27565	1.3893	1041.9	265.98	0.53501	0.023351	D	3.149716
360	1300	7.1588	0.90855	0.27966	1.4045	1041.2	264.79	0.53026	0.023541	D	2.887188
360	1400	7.7625	0.90234	0.28374	1.4199	1040.9	263.59	0.52577	0.023747	D	2.662636
360	1500	8.3731	0.89629	0.2879	1.4357	1040.9	262.39	0.52151	0.023968	D	2.468464
360	1600	8.9903	0.89041	0.29211	1.4517	1041.3	261.2	0.51744	0.024207	D	2.299003
360	1700	9.6136	0.88472	0.29637	1.4678	1042.1	260	0.51356	0.024461	D	2.149941
360	1800	10.243	0.87923	0.30066	1.4842	1043.2	258.82	0.50983	0.024733	D	2.017900
360	1900	10.877	0.87395	0.30498	1.5006	1044.8	257.63	0.50625	0.025022	D	1.900214
360	2000	11.516	0.86889	0.30932	1.5171	1046.9	256.45	0.50279	0.025328	D	1.794752
360	2100	12.16	0.86407	0.31364	1.5336	1049.4	255.29	0.49946	0.025651	D	1.699805
360	2200	12.806	0.85949	0.31795	1.55	1052.3	254.12	0.49623	0.025992	D	1.613941
360	2300	13.456	0.85516	0.32223	1.5663	1055.8	252.97	0.49311	0.02635	D	1.535993
360	2400	14.108	0.8511	0.32646	1.5824	1059.7	251.83	0.49008	0.026724	D	1.465004
360	2500	14.762	0.84732	0.33062	1.5982	1064.1	250.71	0.48714	0.027116	D	1.400158

TABLE F.1 (continued)—CO_2 PROPERTIES AT VARIOUS TEMPERATURES, °F											
Temperature (°F)	Pressure (psia)	Density (lbm/ft^3)	Compressibility Factor	Heat Capacity (Btu/lbm°F)	Heat Capacity Ratio (C_p/C_v)	Sonic Velocity (ft/s)	Enthalpy (Btu/lbm)	Entropy [Btu/(lbm°F)]	Viscosity (cp)	Phase	Gas Formation Volume Factor (res bbl/Mcf)
360	2600	15.416	0.84382	0.33471	1.6137	1069	249.59	0.48428	0.027523	D	1.340745
360	2700	16.07	0.8406	0.3387	1.6288	1074.4	248.5	0.48151	0.027947	D	1.286161
360	2800	16.723	0.83769	0.34258	1.6435	1080.2	247.42	0.47881	0.028385	D	1.235933
360	2900	17.375	0.83507	0.34633	1.6577	1086.6	246.35	0.47619	0.028839	D	1.189582
360	3000	18.024	0.83276	0.34996	1.6713	1093.5	245.31	0.47364	0.029306	D	1.146748
360	3100	18.67	0.83075	0.35343	1.6843	1100.9	244.28	0.47116	0.029787	D	1.107078
360	3200	19.311	0.82905	0.35675	1.6968	1108.7	243.28	0.46874	0.03028	D	1.070287
360	3300	19.948	0.82765	0.35991	1.7085	1117	242.3	0.46639	0.030785	D	1.036102
360	3400	20.58	0.82656	0.3629	1.7196	1125.7	241.34	0.4641	0.0313	D	1.004304
360	3500	21.206	0.82577	0.36571	1.73	1134.9	240.4	0.46188	0.031826	D	0.974677
360	3600	21.825	0.82528	0.36835	1.7396	1144.4	239.48	0.45971	0.03236	D	0.947040
360	3700	22.436	0.82508	0.37081	1.7486	1154.4	238.59	0.4576	0.032902	D	0.921221
360	3800	23.04	0.82517	0.37309	1.7568	1164.7	237.72	0.45555	0.033452	D	0.897076
360	3900	23.636	0.82553	0.3752	1.7643	1175.4	236.88	0.45355	0.034008	D	0.874456
360	4000	24.223	0.82617	0.37713	1.7712	1186.4	236.05	0.4516	0.034569	D	0.853255
360	4100	24.802	0.82707	0.37889	1.7773	1197.8	235.26	0.44971	0.035135	D	0.833351
360	4200	25.371	0.82823	0.38049	1.7828	1209.4	234.48	0.44787	0.035704	D	0.814650
360	4300	25.931	0.82964	0.38193	1.7877	1221.3	233.73	0.44607	0.036277	D	0.797060
360	4400	26.482	0.83128	0.38322	1.7919	1233.4	233.01	0.44432	0.036852	D	0.780484
360	4500	27.023	0.83316	0.38436	1.7956	1245.8	232.3	0.44262	0.037429	D	0.764866
360	4600	27.554	0.83525	0.38537	1.7987	1258.3	231.62	0.44096	0.038007	D	0.750116
360	4700	28.075	0.83756	0.38624	1.8012	1271.1	230.96	0.43934	0.038585	D	0.736186
360	4800	28.587	0.84008	0.38698	1.8033	1284.1	230.33	0.43777	0.039163	D	0.723018
360	4900	29.089	0.84278	0.3876	1.8049	1297.2	229.71	0.43624	0.03974	D	0.710539
360	5000	29.581	0.84568	0.38812	1.806	1310.5	229.12	0.43474	0.040316	D	0.698724
360	5100	30.063	0.84875	0.38853	1.8068	1323.9	228.54	0.43328	0.040891	D	0.687510
360	5200	30.536	0.85199	0.38884	1.8071	1337.4	227.99	0.43186	0.041464	D	0.676863
360	5300	30.999	0.8554	0.38907	1.8071	1351.1	227.46	0.43048	0.042035	D	0.666750
360	5400	31.453	0.85896	0.38921	1.8067	1364.8	226.94	0.42912	0.042603	D	0.657126
360	5500	31.898	0.86266	0.38927	1.8061	1378.6	226.44	0.4278	0.043169	D	0.647958
360	5600	32.334	0.86651	0.38927	1.8052	1392.5	225.97	0.42652	0.043731	D	0.639227
360	5700	32.761	0.87049	0.38921	1.804	1406.4	225.5	0.42526	0.04429	D	0.630897
360	5800	33.179	0.8746	0.38909	1.8026	1420.4	225.06	0.42403	0.044847	D	0.622947
360	5900	33.589	0.87882	0.38892	1.801	1434.4	224.63	0.42283	0.045399	D	0.615343
360	6000	33.99	0.88316	0.38871	1.7993	1448.4	224.22	0.42166	0.045948	D	0.608076
360	6100	34.384	0.88761	0.38846	1.7973	1462.4	223.82	0.42052	0.046493	D	0.601121
360	6200	34.769	0.89216	0.38817	1.7953	1476.5	223.44	0.4194	0.047035	D	0.594457
360	6300	35.147	0.89681	0.38785	1.7931	1490.5	223.07	0.4183	0.047573	D	0.588071
360	6400	35.517	0.90155	0.38751	1.7908	1504.5	222.72	0.41723	0.048107	D	0.581942
360	6500	35.88	0.90637	0.38714	1.7885	1518.5	222.38	0.41618	0.048637	D	0.576052
360	6600	36.236	0.91128	0.38675	1.786	1532.4	222.05	0.41515	0.049164	D	0.570397
360	6700	36.585	0.91626	0.38635	1.7836	1546.3	221.73	0.41415	0.049686	D	0.564955
360	6800	36.927	0.92132	0.38593	1.781	1560.2	221.43	0.41316	0.050205	D	0.559720
360	6900	37.263	0.92644	0.38551	1.7784	1574.1	221.14	0.4122	0.05072	D	0.554674
360	7000	37.592	0.93163	0.38507	1.7758	1587.9	220.86	0.41125	0.051231	D	0.549813
360	7100	37.915	0.93689	0.38462	1.7731	1601.6	220.59	0.41033	0.051738	D	0.545130

TABLE F.1 (continued)—CO_2 PROPERTIES AT VARIOUS TEMPERATURES, °F

Temperature (°F)	Pressure (psia)	Density (lbm/ft³)	Compressibility Factor	Heat Capacity (Btu/lbm°F)	Heat Capacity Ratio (C_p/C_v)	Sonic Velocity (ft/s)	Enthalpy (Btu/lbm)	Entropy [Btu/(lbm°F)]	Viscosity (cp)	Phase	Gas Formation Volume Factor (res bbl/Mcf)
360	7200	38.233	0.9422	0.38417	1.7705	1615.3	220.33	0.40942	0.052242	D	0.540605
360	7300	38.544	0.94756	0.38371	1.7678	1628.9	220.08	0.40852	0.052743	D	0.536233
360	7400	38.85	0.95298	0.38325	1.7651	1642.5	219.84	0.40765	0.053239	D	0.532012
360	7500	39.151	0.95845	0.38279	1.7624	1656	219.61	0.40679	0.053732	D	0.527932
360	7600	39.446	0.96396	0.38232	1.7597	1669.5	219.39	0.40594	0.054222	D	0.523980
360	7700	39.736	0.96952	0.38185	1.757	1682.9	219.17	0.40511	0.054708	D	0.520158
360	7800	40.021	0.97512	0.38138	1.7543	1696.2	218.97	0.4043	0.055191	D	0.516456
360	7900	40.301	0.98075	0.38091	1.7515	1709.5	218.77	0.40349	0.055671	D	0.512862
360	8000	40.576	0.98643	0.38045	1.7488	1722.7	218.59	0.40271	0.056147	D	0.509385
360	8100	40.847	0.99214	0.37998	1.7462	1735.9	218.41	0.40193	0.056621	D	0.506008
360	8200	41.113	0.99788	0.37951	1.7435	1749	218.23	0.40117	0.057091	D	0.502729
360	8300	41.375	1.0037	0.37905	1.7408	1762	218.07	0.40042	0.057558	D	0.499569
360	8400	41.633	1.0095	0.37859	1.7381	1775	217.91	0.39968	0.058022	D	0.496474
360	8500	41.886	1.0153	0.37812	1.7355	1787.9	217.76	0.39896	0.058484	D	0.493452
360	8600	42.136	1.0212	0.37767	1.7329	1800.7	217.61	0.39824	0.058942	D	0.490548
360	8700	42.381	1.027	0.37721	1.7302	1813.5	217.47	0.39754	0.059398	D	0.487664
360	8800	42.623	1.033	0.37676	1.7276	1826.2	217.34	0.39684	0.05985	D	0.484939
360	8900	42.861	1.0389	0.37631	1.725	1838.8	217.21	0.39616	0.060301	D	0.482229
360	9000	43.096	1.0448	0.37586	1.7225	1851.4	217.09	0.39549	0.060748	D	0.479579
360	9100	43.327	1.0508	0.37542	1.7199	1863.9	216.98	0.39483	0.061193	D	0.477033
360	9200	43.555	1.0568	0.37498	1.7174	1876.4	216.87	0.39417	0.061635	D	0.474542
360	9300	43.779	1.0628	0.37454	1.7148	1888.8	216.76	0.39353	0.062075	D	0.472105
360	9400	44	1.0689	0.37411	1.7123	1901.1	216.67	0.39289	0.062513	D	0.469763
360	9500	44.218	1.0749	0.37368	1.7098	1913.4	216.57	0.39226	0.062948	D	0.467427
360	9600	44.433	1.081	0.37325	1.7074	1925.6	216.48	0.39165	0.063381	D	0.465183
360	9700	44.645	1.087	0.37283	1.7049	1937.7	216.4	0.39103	0.063811	D	0.462943
360	9800	44.854	1.0931	0.37242	1.7025	1949.8	216.32	0.39043	0.06424	D	0.460790
360	9900	45.06	1.0992	0.372	1.7001	1961.9	216.24	0.38984	0.064666	D	0.458681
360	10000	45.263	1.1054	0.3716	1.6977	1973.8	216.17	0.38925	0.06509	D	0.456656

Glossary

API: American Petroleum Institute.
ASME: American Society of Mechanical Engineers.
ASTM: American Society for Testing and Materials.
Azeotrope: constant boiling-point mixture.
Base permeability: effective oil permeability at connate water saturation. (See also *Relative permeability.*)
Black oil: an oil whose physical properties (for practical purposes) are a function of pressure only.
Black-oil model: a reservoir simulator used for predicting the performance of reservoirs that contain black oils. Such simulators typically account for three components: separator oil, separator gas, and water.
BLM: U.S. Bureau of Land Management.
Carbon sequestration: the capture and secure storage of carbon that otherwise would be emitted to or remain in the atmosphere.
Chase water: water injected after solvent injection is discontinued.
CO_2 source: any CO_2 supply location from which CO_2 can be transported to CO_2 floods. Sources may be natural (e.g., subsurface CO_2 reservoirs) or industrial (e.g., power plants).
CO_2 utilization ratio: the amount of CO_2 injected (gross CO_2 utilization ratio) or purchased (net CO_2 utilization ratio) per incremental barrel of oil recovered. The term may be applied on an instantaneous or ultimate basis.
CO_2 utilization factor: the volume of CO_2 injected, Mscf/STB incremental oil recovered.
Component: a chemical compound in the CO_2/oil/water system, or a group of similar chemical components that have been grouped together as a pseudocomponent to simplify engineering calculations.
Condensing/vaporizing mechanism: a series of multiple contacts between a gas phase and oil, in which the solvent (gas) condenses into the oil and vaporizes components of the oil. This mechanism is common in many enriched gas/oil systems and in CO_2/oil systems in the three-hydrocarbon-phase region. (See also *Multiple-contact miscibility.*)
Conformance: a measure of where injected fluids (water or CO_2) are entering the pay zone of an injection well.
Connate water saturation: the initial water saturation found in the oil reservoir upon discovery.
Continuous CO_2: a large slug of injected CO_2 uninterrupted by water cycles. A water chase may be used.
Conventional facilities: surface facilities associated with primary and/or secondary (waterflood) recovery.
Conventional WAG injection: WAG injection in which the half-cycle slug sizes of CO_2 and water remain constant. (See also *Tapered WAG.*)
Cycle: in a CO_2 flood, the period of CO_2 injection plus the following period of water injection.
Cycling: the situation that occurs when CO_2 moves through high-permeability zones and is produced quickly without recovering much oil. Cycling causes higher than expected CO_2 production rates at the producers, prevents efficient flooding of the less permeable zones, and increases gas handling costs.
Dead-end pores: within a porous medium, those pores that connect to the remainder of the porous medium through only one opening. Fluids in dead-end pores can become stagnant.
Diffusion coefficient: a measure of how quickly molecules of a chemical component will move on their own throughout a phase to equalize the concentration of the component within the phase. This motion takes place regardless of whether the phase is flowing or not.
Dispersion coefficient: a measure of how quickly a component within a phase will move by convection in a moving phase to equalize the concentration of the component within the phase. The dispersion coefficient equals the dispersivity multiplied by the superficial velocity.
Dispersivity: a parameter used to calculate dispersion. There are longitudinal and transverse components of dispersivity (in the direction of flow and perpendicular to the direction of flow).
Drill stem test: a short-term pressure buildup test conducted before completing the well.
Dykstra-Parsons coefficient: a statistical measure of reservoir heterogeneity defined as the standard deviation from the mean permeability divided by the mean permeability. This coefficient approaches unity for extremely heterogeneous rock and zero for extremely homogeneous rock.
Effective wellbore radius: the actual radius plus the effect of skin, whether skin is a positive quantity (reflecting wellbore damage) or a negative quantity (representing fracturing).
Elastomers: flexible compounds that resemble rubber.
Enhanced oil recovery (EOR): the use of recovery methods that economically recover oil that is incremental to oil produced by continued primary or conventional improved recovery methods.
EOR tax credit: a U.S. federal tax credit extended to owners of fields, as specified by Sec. 43 of the U.S. Internal Revenue Code (IRS). Currently, Sec. 43 allows an EOR credit percent of 15% of the qualified EOR costs incurred in a tax year, including qualified tertiary injectant expenses that are paid or incurred in connection with a qualified EOR project.
ESP: electrical submersible pump.
Equation of state (EOS): a mathematical model used to predict phase densities and the partitioning of chemical components among various phases.
Fick's law of diffusion: a mathematical relationship to model diffusion of chemical species on a molecular level. Fick's law of diffusion calls for the rate of diffusion per unit area to equal the product of a diffusion coefficient and the spatial derivative of the concentration of the chemical species.
First-contact miscibility: a condition that occurs when two phases mix immediately to form one phase with no interface separating the two phases (e.g., water and ethanol).

Flood pattern: a grouping of an injection (or production) well and the surrounding production (or injection) wells. Examples of symmetrical flood patterns are line-drive, five-spot, and nine-spot.
Fully compositional model: a reservoir simulator that uses an EOS to predict fluid properties and partitioning of chemical components among the various phases, the development of miscibility between CO_2 and oil, the final oil saturation left after CO_2 flooding, and the overall reservoir performance (saturations and pressures distributed throughout the reservoir, as well as rates and pressures at wells).
Gas/liquid ratio (GLR): the amount of gas that is produced per barrel of fluid, scf/bbl.
Gel polymer: a colloid in which polymers form a semisolid material that can be used to block thief zones in a CO_2 flood.
Gravity tonguing: a phenomenon that occurs when a lighter phase (e.g., CO_2-rich phase) overrides a heavier phase (e.g. oil or water).
HDPE: high-density polyethylene.
High-velocity (k/ϕ) layers: reservoir layers with a high ratio of permeability to porosity. These layers are most prone to gas cycling because k/ϕ is a measure of how quickly fluids will move through the layer.
Hydrocarbon pore volume (HCPV): in the reservoir, the pore volume initially filled with oil. Calculated as $(1.0-S_{wi})(\phi)(h)(A)$,HCPV often is used when normalizing one reservoir with another. (See also *Movable pore volume.*)
Immiscible: said of two phases, that they are incapable of mixing. The two phases are said to be immiscible with one another. (See also *Miscible.*)
Immiscibility: the state in which two phases in contact with one another with a distinct interface. Even when two phases are immiscible, there still can be an exchange of some of the chemical components.
Incremental oil rate and reserves: oil rate and reserves produced by CO_2 flooding, minus the oil rates and reserves that would have been produced from continuation of operations that existed before the CO_2 flood.
Injection/withdrawal ratio (IWR): the injection rate in RB/D, divided by the total fluid production rate in RB/D.
Interstitial velocity: the actual velocity through the pores, equal to the volumetric flow rate divided by the product of the porosity and the cross-sectional area open to flow.
K-value: the partitioning coefficient of a chemical component at a state of chemical equilibrium. Mathematically, the K-value of a particular chemical component is the mole fraction of that component in the vapor phase, divided by its mole fraction in the oil phase.
LACT: lease automated custody transfer.
Lorenz coefficient: a statistical measure of reservoir heterogeneity. The coefficient is determined by first plotting the fraction of total flow capacity of all the intervals (in order of decreasing permeability/porosity ratio) vs. the corresponding fraction of the total PV for all the intervals. The Lorenz coefficient then is obtained by integrating the area under the resulting curve [the curve begins at coordinates of (0.0, 0.0) and ends at the coordinates of (1.0, 1.0)] and subtracting 0.5. The Lorenz coefficient equals zero if the formation is homogeneous, and approaches unity as the formation becomes more heterogeneous.
Majority purchase contract: a CO_2-purchase contract that covers the volume of CO_2 to be delivered to a project, not including the CO_2 that will be supplied in-kind by oilfield owners who have their own CO_2 supplies.
Minimum miscibility pressure (MMP): the minimum pressure at which CO_2 can become miscible or near-miscible with oil. When determined in the laboratory through a slim-tube experiment, it is referred to as the thermodynamic miscibility. Carbon dioxide flooding below the thermodynamic MMP is considered immiscible flooding. Dispersion in the reservoir can have an adverse enough affect on the development of miscibility that pressure would have to be increased above the thermodynamic MMP in order to develop miscibility or near-miscibility. Also referred to as thermodynamic MMP when there is no reservoir mixing.
Miscibility mechanism: the physical process by which a solvent such as CO_2 develops miscibility with oil.
Miscibility: the state in which two phases are miscible.
Miscible: said of two phases, that they are capable of mixing to the extent that there is no interface separating them, i.e., of becoming a single phase. The two phases are said to be miscible with one another.
Miscible flooding: the situation in which CO_2 mixes with oil to form a miscible zone that moves through the reservoir and is produced. For miscible flooding to occur, the reservoir pressure must remain above the thermodynamic MMP.
Mobility control: a process used to improve areal and vertical sweep efficiency. Fluid entry into the fast (high-permeability) layers is slowed to prevent CO_2 from cycling through these layers and to ensure more efficient volumetric sweep of the entire reservoir. (See also *Cycling.*)
Mobility: the speed at which a phase can move through a porous medium when a pressure gradient is applied. Phase mobility is defined as the relative permeability of the phase divided by its viscosity.
Modified black-oil model: a black-oil reservoir simulator that has been modified to handle CO_2 as an additional (fourth) component. Compositional effects such as the development of miscibility between the CO_2 and oil are not simulated; instead, the thermodynamic MMP and final oil saturation are specified by the user.
Movable oil saturation: the difference between the oil saturation at the beginning of a flood and the residual oil saturation after the flood.
Movable pore volume: the pore volume through which flow can occur, calculated as $(1.0-S_{wirr}-S_{om})(\phi)(h)(A)$. (See also *Hydrocarbon pore volume.*)
Multiple-contact miscibility: a state of miscibility that occurs when two phases (e.g., a gas phase containing CO_2 and an oil phase) initially are not miscible but become so after numerous equilibrium contacts between components of the two phases. In the case of CO_2 and oil, the oil "vaporizes" into the CO_2 phase and the CO_2 "condenses" into the oil phase, until a single phase results. At that point, the gas containing the CO_2 has developed miscibility with the oil. (See also *Condensing/vaporizing mechanism.*)
NACE: National Association of Corrosion Engineers.
Net overburden pressure: the pressure difference resulting from the weight of the overburden formations above the reservoir of interest and the reservoir's pressure.
Net utilization factor: the cumulative CO_2 purchase volume in scf, divided by cumulative tertiary oil recovered in bbl.
NGL: natural gas liquid.
Non-spec oil: oil that fails to meet specifications for sale (also referred to as "slop oil" or "bad oil").
NPC: National Petroleum Council, oil and natural gas advisory committee to the U.S. Secretary of Energy.
OSHA: Occupational Safety & Health Administration, U.S. Dept. of Labor.
Peclet number: the ratio of convective to dispersive transport.
Pilot: a small-scale CO_2 flood test area with a small number of field wells, designed to assess the technical and operational aspects of a potential field-scale flood and to determine whether to start up a field-scale flood.
Prototype area: a small area within an oil field that serves as a model for a larger area.
Pseudocomponent: a group of similar chemical components that have been lumped together to simplify engineering calculations.
P-test: a qualitative test in which a sample of oil is titrated with varying amounts of solvent (CO_2) until asphaltene flocculation (or precipitation) occurs.
Purchase requirement: the total amount of CO_2 that must be purchased for a CO_2 flood.The purchase requirement is equal to CO_2 injection minus the CO_2 production that will be recycled.
Recycled CO_2: produced CO_2 that is recompressed and reinjected. Recycled CO_2 may contain impurities, depending on the gas processing procedure used.
Relative permeability: the effective permeability of a phase divided by the base permeability. Relative permeability is a function of the saturation of the phase in the porous medium (rock). (See also *Base permeability.*)

Relative permeability hysteresis: a change in the relative permeability of a particular phase, triggered by a change in the direction of saturation (increasing saturation to decreasing saturation or vice versa).
Repeat formation tester (RFT): a wireline tool that is positioned against the sandface to first collect a small sample of fluid from the formation and then to conduct a short-term pressure buildup test.
Reservoir simulator: a computer program that solves fluid-flow equations and predicts how quickly fluid will flow through the reservoir as a function of time. It also predicts rates and the bottomhole pressures at the wells.
Retention: the amount of CO_2 remaining in the reservoir at any given time, which equals the amount of CO_2 injected less the amount of CO_2 produced. This normally is expressed as a percentage: CO_2 retention=[1.0−(cumulative CO_2 production)]/(cumulative CO_2 injection).
RTU: remote telemetry unit.
Scaleup: a simple mathematical technique that uses dimensionless variables to scale the performance of one reservoir to another reservoir that may have different spacing, pore volume, or permeability. The two reservoirs must have similar geology and operations.
Shape factor: a pattern dependent coefficient that corrects the Darcy's radial flow equation to the pattern geometry.
Skin factor: a mathematical concept to describe how much a well can produce or inject because of its level of stimulation. A stimulated well has a negative skin. A well in which formation damage restricts production has a positive skin. A zero skin implies the well is neither stimulated nor affected by formation damage.
Slim-tube test: a test in which a long narrow tube filled with very homogeneous sand is used to measure the displacement of oil at various temperatures and compositions by CO_2 or enriched gas. The oil recovery results often are used in conjunction with a high-pressure sight glass to determine when miscibility occurs.
Slug volume: the cumulative volume of CO_2 injected. The slug volume usually is expressed as a percentage of the hydrocarbon pore volume (% HCPV).
Spreading coefficient: a quantitative measure of the ability of a phase to be spread or displaced as a thin film rather than forming a residual blob trapped in the pore space. If the spreading coefficient has a small positive value, then spreading as a thin film is likely.
Superficial velocity: the velocity calculated as if a rock's porosity were 100%. The superficial velocity equals the volumetric flow rate divided by the cross-sectional area.
Tapered WAG: a WAG ratio that changes over time. Tapered WAG injection schemes start with a very low WAG ratio (possibly a large first cycle of CO_2) and then increase the WAG ratio gradually.
Tertiary Oil Recovery Information System (TORIS): a database developed by the U.S. DOE for characterizing U.S. oil resources. The goals of TORIS were to estimate potential domestic oil reserves, to project U.S. oil production potential, and to target research and development efforts on enhanced exploration, drilling, completion, and production technologies for exploiting the existing domestic resource.
Transition zone: the vertical zone between the depth at which 100% oil is produced during primary production and the depth at which no oil exists. This sometimes is called the capillary transition zone.
Upscaling: a process to create a set of pseudorelative permeability curves that can be used in a coarse grid numerical simulation to yield predictions equivalent to a fine grid numerical simulation, and that includes detailed reservoir heterogeneities and uses laboratory-measured relative permeability curves.
Viscous fingering: a laboratory phenomenon that takes place in homogeneous media, in which "fingers" of the less viscous fluid (CO_2) move into the more viscous fluid (oil) ahead of flood front.
WAG (water alternating gas): water injection alternating with gas injection in a tertiary recovery project.
WAG cycle: an injection cycle that includes one period of CO_2 (or water) injection followed by a period of water (or CO_2) injection. A WAG cycle may be measured in terms of either HCPV or time. (See also *WAG half cycle.*)
WAG half cycle: in a WAG cycle, the period in which either CO_2 or water is injected. It may be measured in terms of either HCPV or time. (See also *WAG cycle.*)
WAG ratio: ratio of injected water to CO_2 (solvent). WAG ratios may be expressed in terms of reservoir injection (i.e., barrels of water injected at reservoir conditions to barrels of CO_2 injected, measured at reservoir conditions) or in terms of duration, (i.e., the time over which injection takes place).
Water blocking: a phenomenon that occurs when water in a reservoir shields residual oil from injected CO_2.
Wet gas: liquid-saturated gas at a specific temperature and pressure.

Nomenclature

A = area, L^2, ft^2
h = pay height, L, ft
k = permeability, L^2, md
S_{om} = miscible residual oil saturation, fraction
S_{wi} = interstitial water saturation, fraction
S_{wirr} = irreducible water saturation, fraction
ϕ = porosity, fraction

SI Metric Conversion Factors

bbl × 1.589 873	E − 01 = m^3
ft × 3.048*	E − 01 = m
ft^2 × 9.290 304*	E − 02 = m^2
ft^3 × 2.831 685	E − 02 = m^3

*Conversion factor is exact.

Author Index

A

Aasum, Y., 75
Ali, N., 87
Ammot wettability test, 44
Arya, A., 20-21

B

Baker, L.E., 85
Ball, B., 151
Bangia, V.K., 139, 153
Bennett, C.O., 78
Bradley, R.T., 151, 153
Brokmeyer, R., 34
Brownlee, M.H., 148

C

Campbell, J.M. Jr., 66
Chang, Y., 18, 22, 28, 85
Chopra, A.K., 21-22, 24, 29, 70
  conditional simulation and, 76-77
  geostatistics and, 76-77
  model of, 30-31
  operations and, 140
  permeability/porosity plots and, 74
  relative permeability and, 81-82, 85
  viscosity and, 81
Christman, P.G., 140
Cox, M.T., 149
Craig, F.F. Jr., 1, 44-45
Crane, S.D., 21, 86-87

D

Dagan, G., 75
Deitz, D.N., 35
Dria, D.E., 32, 85
Dumore, J.M., 35
Dykstra-Parsons coefficient, 20, 44-45, 78

E

Ehrlich, R., 32
El-Saleh, M., 47

F

Feinberg, D., 58
Fick's law, 18
Flanders, W.F., 153
Fleming, E.A., 148, 153
Folger, L.K., 50
Fong, W.S., 85
Fullbright, G.D., 34

G

Gantt charts, 55
Gardner, J.W., 86
Gorell, S.B., 140
Grisak, G.E., 20-21
Gupta, S.P., 20

H

Hadlow, R.E., 47
Haldorsen, H.H., 77
Hallenbeck, L.D., 16
Hansen, P.W., 149, 153
Harpole, K.J., 16, 29, 70, 81
Harrell, B., 151
Hendrickson, S.G., 153
Henry, R.L., 25-26, 31
Hewett, T.A., 77, 81
Hill, A.D., 45
Hill, S., 35
Hill, W.J., 87
Hohn, M.E., 77
Holm, L.W., 15-16, 40
Hsu, C-F., 31, 50, 71, 84-87

J

Jarrell, Perry M., 117, 121
Jerauld, G.R., 84, 86
Joffe, J., 84
Johns, R.T., 15-16
Johnson, J.P., 40
Johnston, O.C., 19
Josendal, V.A., 15-16, 40

K

Kane, A.V., 145
Kelkar, B.G., 20
Khan, S.A., 85
Kittridge, M.G., 20-21, 45
Klins, M.A., 1, 13
Koinis, R.L., 123

L

Lake, L.W., 13, 76-77
Lantz, R.B., 85
Lorenz-Bray-Clark viscosity, 85
Lorenz coefficient, 44-45

M

Magyari, D., 53
Malik, M., 76
Masoner, L.O., 145, 153
Melzer, L.S., 150
Metcalfe, R.S., 14-16, 25-26, 31, 40
Moody, L.F., 100
Morgenthaler, L.N., 148
Moshell, M.K., 153
Mungan, N., 40
Murray, L.R., 87

N

Negahban, S., 19, 86

O

Orbey-Sandler viscosity, 85
Orr, F.M. Jr., 15-16
Owens, W.W., 24

P

Pariani, G.J., 87
Patel, P.D., 140
Peclet number, 19-20
Pederson, K.S., 84
Peng, D.Y., 84
Perez, G., 21
Perkins, T.K., 19
Pickens, J.R., 20-21
Pittaway, K.R., 145
Pollack, N.R., 18
Pollin, J.S., 40
Prieditis, J., 29, 70, 81
Prophet model, 80

R

Ramirez, W.F., 21, 86
Redlich-Kwong equations of state, 84
Reynolds number, 98
Richardson, J.G., 77
Robinson, D.B., 84
Roper, M.K. Jr., 27, 86
Rose, S.C., 1
Runyon, E.E., 145
Ryan-Holmes process, 110

S

Schneider, F.N., 22, 24
Sebastian, H.M., 40
Shelton, J.L., 22
Shiralkar, G.S., 40-41
Shuler, P.J., 148
Shyeh-Yung, J.J., 28-29, 70, 81
Soave, G., 84
Spivak, A., 85
Srivastava, V.A., 149
Stalkup, F.I. Jr., 1, 13, 21, 86-87
Stein, M.H., 40-41
Stephenson, D.J., 33
Stevens, D.G., 87
Stiles, L., 63
Stretford process, 59
Sugg, L.A., 148

T

Taber, J.J., 58
Tanner, C.S., 145
Thompson, R.G., 151
Tiffin, D.L., 22
Todd-Longstaff mixing parameter, 80
Turek, E.A., 69

W

Wackowski, R.K., 145, 153
Wang, Y., 15-16
Warden, G.W., 44-45
Wegener, D.C., 29, 70, 81
Wehner, S.C., 78, 89
Wendel, D.J., 70
Wolcott, D.S., 76-77

Y

Yellig, W.F., 14-16, 19, 22, 40
Young, L.C., 84

Z

Zick, A.A., 13
Zudkevitch, D., 84

Subject Index

A

acrylonitrile monomers, 95
adsorption, 112
affidavit of publication, 131
agreements, 125, 127, 129-130
air compressors. *See* compressors
air permits, 131-132
allocation methods, 141
alloys, 97
alternating CO_2 flooding, 3-4
aluminum, 97
Ammot wettability test, 44
Amoco, 119-120
API gravity, 4, 40-41
areal sweep efficiency, 3
asphaltenes, 129, 147, 149
atmospheric hazards, 156-157
auxiliary systems, 101
azeotropes, 110

B

backpressure valves, 115
batteries, 152
  flare system and, 103-104
  piping and, 103
  separators and, 102-103
  tanks and, 103
best-practice review, 127
black-oil simulation, 80-84
blowdown, 108
bottomhole pressure, 88-89, 139
  history matching and, 79
  performance prediction and, 79-87
  relative permeability and, 25-28
  scaleup and, 48
boundary conditions, 72
breakthrough, 143-144
  material selection and, 148-150
  reinjection and, 150-151
bubblepoint pressure, 15-16
Buna-N, 95
butadiene, 95

C

capillary forces, 140
carbon dioxide
  atmospheric hazard control and, 156-157
  breakthrough and, 143-144, 148-151
  dangers of, 156-158
  density of, 13, 46-47
  elastomers and, 95
  erratic volumes and, 151
  flare systems and, 103-104
  fluid properties of, 13
  handling considerations for, 153
  history matching and, 71-79
  impurities and, 70
  injection systems and, 108-109
  lubricating oils and, 151
  oil density and, 81
  price of, 130
  purchase requirements and, 51
  rich-phase flow and, 70
  securing source of, 6-9, 56-58
  switching water cycles and, 120
  transmission costs and, 57-58
  venting of, 156-157
  viscosity and, 81
carbon dioxide flooding, 36-38
  business factors and, 5
  Chopra model and, 30-31
  CO_2 sources and, 6-9
  design aspects of, 3-4, 69-92
  facility optimization and, 153
  feasibility of, 124
  field data and, 95-96
  fluid mobility and, 24-28
  implementation and, 123-137
  increased production and, 6
  injection designs and, 3-4 (*see also* injection)
  management of, 4, 138-155
  manufactured, 8-9
  miscibility and, 13-18, 28-32 (*see also* miscibility)
  NPC study and, 10
  objectives for, 1-2
  performance prediction and, 79-87
  projected potential and, 9-11
  projects of, 5-6
  recovery predictions for, 47-51
  relative permeability and, 22-32
  safety and, 120, 128, 132-137, 140-141, 153, 156-158

scaleup and, 47-50
screening processes and, 39-68
sequestration and, 8
small-scale reservoir mixing and, 18-22
surface facilities design and, 93-114
surveillance methods and, 140-141
sweep aspects of, 32-35, 40-41, 46-47, 143-144
swelling and, 16-17
technological factors and, 4
types of, 123
U.S. DOE study and, 10-11
well design and, 41, 115-122
Carbon Dioxide Flooding: Basic Mechanisms and Project Design (Klins), 1
carbonic acid, 153
carbon steel, 97, 108, 153
cement, 149
chase fluid, 3
chemicals
injection systems, 101
breakthrough and, 148-149
chromium ions, 34
Claus process, 59
compressors, 96
batteries and, 103
breakthroughs and, 143-144
centrifugal units, 104
coolers and, 105
design of, 104-105
dump valves and, 101
equations for, 105
horsepower determination and, 105
location of, 104
lubricating oils and, 151
number of, 104
pressure swings and, 151
scrubbers and, 105
variable-speed, 104-105
computers, 71, 77, 100
condensation, 13
contracts, 129-132
control skids, 108-109
convective dispersion, 18-21
conversions, 117-118, 147
coolers, 105
core samples, 70
correlations, 15
history matching and, 72-73
Stone's, 26
corrosion, 63, 106, 153
gas-gathering systems and, 97
material selection and, 118-121
vessel, 100-101
well design and, 115-120
cost. *See* economic issues
couplings, 116-117, 121-122
crossflow, 46-47, 136

D

data
breakthrough and, 149-150
monitoring and, 132-137
oil/gas composition, 128
preinjection, 128-129
See also design
deadbands, 139-140
decline curve analysis, 48
dehydration, 110, 151
density
algorithms, 109
atmospheric hazards and, 156-157
CO_2, 13, 46-47
EOS and, 14
gas, 98
oil, 81
prediction and, 69-70
Denver Unit, 116-117, 119, 123
depletion, 41
design
batteries and, 102-104
compressors, 104-105
drip vessels, 98-99
flood optimization and, 87-89
Gantt chart and, 55
gas processing and, 97-99, 109-112
implementation and, 123-137
infill drilling and, 88
initial steps for, 93
iteration and, 56
liquid gathering and, 101-102
metering and, 98
multi-unit fields and, 56
pattern modification and, 88
pressure, 98
for redundancy, 101
reinjection and, 55-56, 58-60
slug size and, 87-88
stakeholder involvement and, 57
surface facilities and, 93-114
wells and, 115-122
Design Engineering Aspects of Waterflooding, The (Rose), 1
dewpoint, 16
Diethanolamine, 59
diffusion, 18
Dimethylethylamine, 59
dispersion
black-oil simulation and, 79-87
convective, 18-21, 86
gas-release models and, 106
longitudinal, 19-21
model of, 18-19
porosity and, 18-21
transverse component of, 19
downhole equipment, 116, 119
drip vessels, 98-99
dump valves, 101
Dykstra-Parsons coefficient, 20, 44-45, 78

E

economic issues, 5
chemical usage, 63
CO_2 sources and, 9-11, 56-58, 130
design planning and, 54-57
heterogeneity and, 117-118
hydrogen sulfide removal and, 59-60
implementation and, 123-137
incremental recovery and, 64-66
injection and, 46, 60-63, 115
labor, 63
net present value and, 66
NPC study and, 10
operating costs and, 51-66, 123-124, 145-155
packer/pump conversion and, 117-118
partners and, 129
performance prediction and, 79-87
power costs, 63
production facilities and, 60-63
projected CO_2 potential, 9-11
recovery predictions and, 47-51

reinjection plant and, 58-60
reservoir screening and, 53
risk and, 123
safety monitoring and, 132-137
scaleup and, 47-50
scoping and, 39, 51-66
simulation and, 69 (*see also* simulation)
staging and, 124-128
tapered applications and, 3-4
timing and, 123-124
transmission costs and, 57-58
upfront costs and, 57
U.S. DOE study and, 10-11
well design and, 52-53, 63, 118, 122
workovers, 63
elastomers, 95
electric feed-through (EFT) connections, 147, 149
electric submersible pumping, 146-147, 149
emergency shutdown (ESD) systems, 103-106
engineering, 1
data requirements and, 69-71
design optimization and, 87-89
fluid properties and, 69-70
geostatistics and, 75-77
grid sensitivity study and, 73
history matching and, 71-79
implementation and, 123-137
incrementing recovery and, 39-47, 64-66
model study area and, 71-72
performance prediction and, 79-87
phase equilibrium and, 69-70
reservoir description and, 73-75
screening processes and, 39-66 (*see also* screening processes)
seismic applications and, 77-78
WAG injection tests and, 70-71
well design and, 115-122
enhanced oil recovery (EOR)
defined, 1
increased production and, 6
NPC study and, 10
pattern balancing and, 140-142
process choice for, 1-2
production methods for, 145-147
projected CO_2 potential and, 9-11
projects on, 5-6
surveillance and, 140-141
tax incentives for, 132
U.S. DOE study and, 10-11
enriched-gas flooding, 1
environmental, health, and safety (EHS) plan, 156
EPDM, 95
epichlorohydrine, 95
equations
adiabatic work, 105
average reservoir pressure, 41
cash flow, 65
Chopra, 30
compressor design, 105
cost of delivery, 58
critical gas velocity, 35
discharge temperature, 105
effective wellbore radius, 78
flow rate, 97
formation flow conservation, 79
gravity displacement rate, 35
heterogeneity, 78
ID flow, 19
incremental oil rate, 49
injection well cost, 63
line diameter, 97-98
longitudinal dispersion, 19
multiphase flow, 99
orifice meters, 98
Peclet number, 19
pipeline efficiency, 97
producing well cost, 63
required duty, 105
retention time, 100
solvent relative permeability, 30
spreading coefficient, 34
thermodynamic MMP impurities, 16
time scaling, 48-49
trunkline cost, 58
viscosity, 81
viscosity/gravity ratio, 48
water relative permeability, 70
equations of state (EOS), 14, 16
complex phase behavior and, 85
component lumping and, 84-85
density algorithms and, 109
drip design and, 98-99
fine-tuning of, 84
fully compositional reservoirs and, 84-87
miscibility and, 84
Redlich-Kwong, 84
equilibrium
miscibility effects and, 18
phase, 15-16, 69-70
ethane, 110
extrusion, 119
Exxon, 116

F

facility management, 150-153
fence diagrams, 74
fiberglass, 97, 122, 149-150
Fick's law, 18
flare systems, 103-104
flowing wells, 117-119
flue gas, 16
fluid properties, 3
black-oil simulation and, 80-81
CO_2, 13
dispersion and, 18-21
history matching and, 72-73
prediction and, 69
relative permeability and, 22-32
flumping, 118
Fluoroline, 118
foam injection, 150
fracturing, 41
friction, 98
frostbite, 157

G

Gantt charts, 55
Gardner, J.W., 86
gas, 109
busters, 116
continuous CO_2 injection and, 3
dehydration and, 110
density, 98
drip vessels and, 98-99
flue, 16
gathering systems and, 97-99
history matching and, 71-79
hydrogen sulfide removal and, 110-112
locking, 147
metering and, 98
natural gas anchor, 116

NGL recovery and, 110
poor-boy separators and, 116
processing implementation, 130
production methods for, 145-147
REDOX system and, 111-112
regenerative adsorption systems and, 112
scavenger systems and, 112
solvent recirculation and, 111-112
sour reinjection and, 110-112
sulfur recovery and, 112
sweep aspects and, 32-34
gas-oil ratio (GOR)
black-oil simulation and, 82-84
breakthroughs and, 143-144
gel polymers, 34, 150
geostatistics, 75-77
government, 132
graph studies, 142
gravity
continuous CO_2 injection and, 3
crossflow and, 46-47
dipping reservoirs and, 34, 46
miscible flooding and, 4
oil recovery and, 34-35, 39-47, 64-66
pattern floods and, 88
scaleup and, 48
stable displacement rate and, 35
sweep aspects and, 33-35
swelling and, 16-17
tongues, 34
vertical permeability and, 34
viscosity and, 48, 140
grid studies, 73
black-oil simulation and, 82-84
geostatistics and, 75-77
kriging and, 76
pseudoflow and, 77
pseudorelative permeability and, 86
GSLIB program, 77

H

heterogeneity, 44
geostatistics and, 76
improved recovery and, 117-118
interwell continuity and, 45
history matching
boundary conditions and, 72
cross sections and, 72
fluid properties for, 72-73
geostatistic applications and, 75-77
grid sensitivity study and, 73
layering and, 74-75
model study area and, 71-72
operational effects and, 78-79
permeability/porosity plots and, 74-75
refinement of, 78-79
relative permeability and, 72-73
reservoir description data and, 72-75
seismic applications and, 77-78
time period for, 72
variables used in, 78
horizontal wells, 88
hydrocarbons, 1, 4, 157
Chopra model and, 31
contamination and, 139
NGL recovery and, 110
pore volumes (HCPVs), 15
reinjection, 55-56, 58-60, 150-151
thermal relief and, 158
thermodynamic MMP model and, 15-16
three-phase, 17, 25-26, 70
hydrogen sulfide, 4, 51, 66, 96
atmospheric hazards and, 157
CO_2 reinjection and, 108
plant costs and, 59-60
sour gas reinjection and, 110-112
hysteresis, 23-24

I

ice plugs, 157
implementation
agreements and, 125, 127, 129-130
contracts and, 129-130
flood monitoring and, 132-137
government incentives and, 132
permits and, 130-132
planning and, 123-128
preinjection data gathering and, 128-129
safety and, 128
schedule of, 125-128
impurities, 16, 70, 89, 139
incentives, 132
infill drilling, 88
infrared monitors, 157
injection
alternating CO_2, 3-4
blowdown capability and, 108
bottomhole pressure and, 25-28, 88-89
chemical systems and, 101
CO_2 distribution systems and, 3-4, 108-109
CO_2 sources and, 6-9
corrosion and, 106 (*see also* corrosion)
cumulative, 50
design planning and, 54-57
economic issues and, 46, 60-63, 115
facilities, 127
foam, 150
gravity and, 46-47
hydrogen sulfide and, 51
hysteresis and, 23
incremental oil recovery and, 39-47, 64-66
interwell continuity and, 41
investment cost of, 61-63
layering and, 44-45
management of, 144-145
monitoring and, 136
optimizing of, 138-139
packer assembly and, 120-122
performance prediction and, 79-87
permits for, 130-131
pressure, 26, 106-107, 122, 139-140
profiles, 45, 129
pumps, 107
rate, 3, 106-107, 139-140
recovery and, 141-142
reinjection and, 51, 55-56, 58-60, 110-112, 150-151
relative permeability and, 22-32
repressurization and, 106-107
scaleup and, 47-50
SIE, 150
slim-tube tests and, 14-15
solvent, 23-24
sour reinjection and, 110-112
stream impurities and, 89
surveillance methods and, 140-141
sweep aspects and, 32-34
temperature and, 45-46
tubing for, 120-122
WAG projects and, 24-28, 70-71
water, 21-22, 106-108

well design and, 45-46, 119-122
withdrawal ratio and, 45
interwell continuity, 45
iteration, 56

K

Kalrez, 95
knockout, 152
kriging, 76

L

labor costs, 63
layering, 44-45, 74-75, 78-79
lift equipment, 115-117
liners, 116-117
liquid gathering, 101-102
LO-CAT process, 59
Lorenz-Bray-Clark viscosity, 85
Lorenz coefficient, 44-45
lubricating oils, 151

M

material selection
alloys, 97
aluminum, 97
asphaltenes, 129, 147, 149
breakthrough and, 148-150
Buna-N, 95
carbon steel, 97, 108, 115
cement, 149
corrosion and, 97, 118-121
Duoline, 118
EPDM, 95
epichlorohydrine, 95
fiberglass, 97, 149-150
field tests and, 93-95
flowing wells and, 118-119
Fluoroline, 118
foam, 150
gas-gathering and, 97
injection wells and, 121
Kalrez, 95
paraffins, 63, 119, 147-150
polymers, 34, 148, 150
stainless steel, 97, 120
Teflon, 120
tubing and, 120-122
valves and, 94-95
Viton, 95
workover and, 148-150
yellow band tubing, 118
mathematical models, 15-16
boundary conditions, 72
Chopra, 30-31
CO_2-rich phase, 70
convective dispersion, 18-20, 86
history matching, 71-79
kriging, 76
Prophet, 80
recovery predictions, 47-51
scaleup, 47-50
mechanisms
CO_2/oil miscibility and, 13-18
condensing, 13
fluid mobility and, 24-28
hysteresis and, 23
miscibility and, 28-32
oil-wet, 21-22
relative permeability and, 22-32
small-scale reservoir mixing, 18-22
sweep aspects and, 32-35
vaporizing, 13
water-wet, 21-22
membrane systems, 110
metering, 98
control skids and, 108-109
injection and, 108-109
monitoring and, 136
separation and, 151-153
methane, 13, 16
Methyl Diethanolamine, 59
mineral oil, 42
miscibility, 2
black-oil simulation and, 79-87
CO_2 with oil, 13-18
correlation use and, 15
displacement and, 1
equations of state and, 84
gravity and, 4
impurity effects and, 16
mechanisms for, 13
NPC study and, 10
Peclet number and, 19-20
potential projections and, 9-11
pressure effects and, 13-14
projects on, 5-6
relative permeability and, 22-32
residual oil phase and, 17-18
solvents and, 23-24, 28-32
swelling and, 16-17
temperature and, 13
thermodynamic MMP and, 14-17
[*see also* thermodynamic minimum miscibility pressure (MMP)]
three-phase hydrocarbon region and, 17
water effects on, 18
Miscible Displacement (Stalkup), 1
monitoring, 132-137
monoethanolamine, 59
multiphase flow, 40, 85-86, 102
multiple-contact experiment, 70

N

National Petroleum Council (NPC), 10
natural flow, 147
natural gas anchor, 116
neutron logs, 144
NGL recovery, 110
nitrile, 95
nitrogen, 4, 16
noise levels, 157
nonisothermal simulation, 87

O

oil
API gravity crudes, 4
black-oil simulation and, 80-84
Chopra model and, 30-31
density, 81
fluid mobility and, 24-28
gravity and, 34-35
history matching and, 71-79
hysteresis and, 23
incremental recovery and, 39-47, 64-66

miscibility and, 13-18, 28-32
Peclet number and, 19-20
price, 1, 123
recovery predictions for, 47-51
relative permeability and, 22-32, 70
reservoir mixing and, 18-22
residual phase and, 17-18, 41-42, 80
saturation and, 4, 80
scaleup and, 47-50
solvents and, 28-32
sweep aspects and, 32-35
swelling and, 16-17
three-phase hydrocarbon region and, 17
viscosity and, 81
water blocking and, 21-22
wet mechanisms and, 21-22
wettability tests and, 42-44
operations, 154-155
facility management and, 150-153
reservoir management and, 138-145
philosophy of, 133-134
well management and, 145-150
See also economic issues
Orbey-Sandler viscosity, 85
orifice meters, 98
original oil in place (OOIP), 26

P

packer assembly, 120-122
paraffins, 63, 119, 147-150
pattern balancing, 88, 139-142
Peclet number, 19-20
Permian Basin, 5-6
permits, 130-132
phase relative permeability, 70
phenolic polymers, 34
phosphonate inhibitors, 148
pipelines, 94, 100
ice plugs and, 157
injection and, 107-108
liquid gathering and, 101-102
planning
startup and, 123-128
timing and, 123-124
plugbacks, 149
plungers, 119
politics, 5
polymers, 34, 148, 150
porosity
convective dispersion and, 18-21
relative permeability and, 22-32
viscous fingering and, 22
wet mechanisms and, 21-22
power costs, 63
prediction
composition and, 85
core samples and, 70
density and, 69-70
design optimization and, 87-89
flood performance and, 79-87
fluid properties and, 69
history matching and, 71-79
phase equilibrium and, 69-70
relative permeability and, 70
slim-tube tests and, 69
viscosity and, 69-70
WAG injection tests and, 70-71
See also simulation
pressure
average, 40
bottomhole, 25-28, 48, 79-89, 139
bubblepoint, 15-16
compressor inlet/outlet, 151
design, 98
EOS and, 14
fiberglass and, 97
flowing wells and, 118
formation parting, 128-129
history matching and, 79
Holm-Josendal, 15
implementation and, 124, 128-129
injection and, 26, 106-109, 122, 139-140 (*see also* injection)
miscibility effects and, 18
monitoring and, 136
multiphase flow errors and, 40
oil recovery and, 13-14
performance prediction and, 79-87
production flowlines and, 100
relative permeability and, 22-32
relief systems, 98
safety and, 120, 157-158
scaleup and, 48
shape factor and, 40
slim-tube tests and, 15
sponge core, 42
stuffing boxes and, 115
transient tests and, 74, 78
WAG projects and, 24-28
pressure/volume/temperature (PVT) data, 16
production, 3
auxiliary systems and, 101
battery design and, 102-104
emergency shutdown and, 105-106
facilities and, 127-128
field compression and, 104-105
flare system and, 103-104
flowlines and, 99-100
fluid phases and, 101-102
gas-gathering systems and, 97-99
hydrogen sulfide and, 51
injection and, 46
IWR and, 45
manifold, 100
method choice and, 145-147
multiphase flow and, 99-100, 102
optimization of, 145-147
permeability plots and, 74-75
piping and, 103
refinement of, 79
satellite batteries and, 99
selective equipment for, 150
separation and, 96, 100-101, 152
tanks and, 103
vapor recovery and, 103
vessels, 100-101
wettability and, 42-44
See also economic issues
profile nipple, 119, 121
Prophet model, 80
pseudoflow, 77
pseudorelative permeability, 73, 77
puddle-packing, 122
pulsed neutron logs, 144
pumps, 107
electric submersible, 146-147, 149
packer conversion and, 117-118
production methods and, 145-147
rod, 115-117, 145-146

R

radioactivity, 45
Rangely wells, 147
recycling. *See* reinjection
Redlich-Kwong equations of state, 84
REDOX systems, 111-112
regenerative adsorption systems, 112
regulations, 130-132
reinjection
  CO_2 purchase requirements and, 51
  management of, 150-151
  plant design and, 58-60
  project costs and, 55-56
  sour gas, 110-112
relative permeability
  averaging of, 75
  bottomhole pressure and, 25-28
  Chopra model and, 30-31
  correlations and, 26
  dispersion and, 18-21
  distribution refinement of, 79
  fluid mobility and, 24-28
  history matching and, 72-73
  hysteresis, 23-24
  layering and, 44-45
  multiphase, 85-86
  oil and, 22-23, 70
  phase, 70
  porosity plots and, 74-75
  pseudo, 73, 77
  solubility and, 27-28
  solvents and, 23-24, 28-32
  sweep aspects and, 32-35
  tests for, 42-44
  three-phase, 82
  vertical, 77
  WAG projects and, 24-28
  water and, 23, 70
repeat formation tester (RFT), 40
Reservoir Engineering Aspects of Waterflooding, The (Craig), 1
reservoirs
  black-oil simulation and, 79-87
  convective dispersion and, 18-21
  cross section applications and, 72
  dipping, 46
  flood performance prediction and, 47-51
  fracturing and, 26-27
  fully compositional, 84-87
  geostatistics and, 75-77
  gravity and, 34 (*see also* gravity)
  heterogeneity and, 44-45
  history matching and, 71-79
  implementation and, 123-137
  incremental oil recovery and, 39-47, 64-66
  layering and, 44-45, 74-75
  management of, 138-155
  operating costs and, 51-66
  pressure and, 124 (*see also* pressure)
  relative permeability and, 22-32
  scaleup and, 47-50
  solvents and, 28-32
  stratification degree and, 48
  water blocking and, 21-22
Reynolds number, 98
right-of-way (ROW), 97-98
rock properties, 3
  fence diagrams and, 74
  fracturing and, 26-27
  history matching and, 71-79
  porosity, 18-32
  pseudo, 77
  wettability and, 42-44
rod pumps, 115-117, 145-146
Ryan-Holmes process, 110

S

SACROC, 140, 145
safety, 120
  atmospheric hazard control and, 156-157
  CO_2 dangers and, 156-158
  EHS plan and, 156
  flood monitoring and, 132-137
  frostbite and, 157
  handling considerations and, 153
  high pressure and, 157-158
  hydrates and, 157
  ice plugs and, 157
  implementation and, 128
  monitors for, 157
  noise levels and, 157
  surface water protection and, 158
  surveillance methods and, 140-141
  thermal relief and, 158
  wellheads and, 158
  workovers and, 158
San Andres field, 17-18, 20, 28-29, 124
satellites, 96, 99-100
saturation, 4, 129
  Chopra model and, 30-31
  CO_2-rich phase, 70
  incremental oil recovery and, 39-47
  Peclet number and, 19-20
  relative permeability and, 22-32
  residual oil and, 17-18, 41-42, 80
  solubility and, 27-28
  solvents and, 28-32
  tests for, 41-44
  waterflood endpoint and, 42-43
  wettability and, 42-44
scale formation, 63, 148
scaleup
  assumptions of, 47-48
  connate water and, 49
  incremental rates and, 48
  simplification of, 49-50
  time and, 48-49
scatter plots, 142
scavenger systems, 112
scoping, 39
  cost justification and, 64-66
  operating costs and, 51-66
  recovery predictions and, 47-51
  See also simulation
screening processes, 67-68
  flood performance prediction and, 47-51
  incremental oil recovery and, 39-47, 64-66
  operating costs and, 51-66
  reservoirs and, 53-54
  skin effects and, 41
scrubbers, 105
seals, 120
seismic imaging, 77-78
selective injection equipment (SIE), 150
Selectox process, 59
separation
  batteries and, 102-103
  compression and, 96
  downhole, 116
  facility management and, 151-153
  flare systems and, 103-104
  metering and, 151-153

multiphase, 96, 102-103
poor-boy, 116
production, 152
quality, 100
satellite number and, 96
vessel design and, 100-101
sequestration, 8
shape factors, 40
Shell, 119
simulation
black-oil, 79-87
data requirements for, 69-71
design optimization and, 87-89
equations of state and, 84
fully compositional, 84-87
geostatistics and, 75-77
history matching and, 71-79
kriging and, 76
miscibility effects and, 18
nonisothermal, 87
performance prediction and, 79-87
recovery predictions and, 47-51
relative permeability and, 22-32
seismic imaging and, 77-78
vertical permeability and, 77
single-well tracer tests, 42
skin effects, 41
slim-tube tests, 14
correlations and, 15
dispersion and, 20
mathematical models and, 15-16
prediction and, 69
slugs, 3-4
half-size optimation and, 138-139
ultimate size for, 140
viscosity and, 140
WAG optimization and, 87-88
software, 77, 100
solubility, 27-28
solvents
black-oil simulation and, 81-82
Chopra model and, 30-31
miscibility and, 28-32
recirculation systems and, 111-112
reinjection and, 59
relative permeability and, 23-24, 28-32
sour gas reinjection, 110-112
sponge core, 42
SRU process, 59
staging, 124-128
stainless steel, 97, 120
startup, 123-128
steam flooding, 1-2
steel, 97, 108, 120-121, 153
Stretford process, 59
stuffing boxes, 115
sulfur, 112
surface facilities, 113-114
elastomers and, 95
field tests and, 93-95
gas processing and, 109-112
injection and, 106-109
material selection and, 93-96
piping and, 94
production and, 96-106
valve applications and, 94
surveillance, 140-141
sweep aspects
CO_2 density and, 46-47
efficiency and, 3, 143-144
gravity and, 34-35, 46-47
improvement of, 32-34
vertical permeability and, 34
viscosity and, 40-41

T

tapered alternating CO_2 flooding, 3-4
tax incentives, 132
Teflon, 120
temperature
discharge, 105
Holm-Josendal, 15
ice plugs and, 157
injection and, 45-46
oil recovery and, 13-14
three-phase hydrocarbon region and, 17
Tertiary Oil Recovery Information System (TORIS), 10
tests
field tests, 93-94
monitoring and, 132-137
relative permeability and, 42-44
repeat formation, 40
separators and, 152
slim-tube, 14-16, 20, 69
step-rate, 136
WAG injection, 70-71
waterflood endpoint, 42-43
thermal conductivity monitors, 157
thermal displacement, 1
thermal relief, 158
thermodynamic minimum miscibility pressure (MMP)
black-oil simulation and, 80-84, 86
bottomhole injection pressure and, 88-89
Chopra model and, 31
convective dispersion and, 18-21
correlations used and, 15
implementation and, 124, 128-129
impurities and, 16, 70
incremental oil recovery and, 39-47
mathematical determination of, 15-16
multiphase relative permeability and, 85-86
oil displacement by CO_2 and, 16-17
prediction of, 15-16, 69
pressure effects and, 13-14, 26
relative permeability and, 23-24
reservoir screening and, 53
scaleup and, 48
slim-tube measurement of, 14-15
three-phase hydrocarbons and, 25-26
viscosity and, 40-41
time
Gantt chart and, 55
grid studies and, 84
history matching and, 71-79
implementation and, 123-124
lapse logging and, 144
scaleup and, 48-49
WAG and, 138
wettability and, 41
Todd-Longstaff mixing parameter, 80
tracer tests, 20
transmission costs, 57-58
Triethanolamine, 59
tubing, 118-119
couplings and, 121-122
injection wells and, 120-122
material selection and, 120-122
reaction monitors, 157
safety and, 158

U

United States
- carbon sequestration and, 8
- CO_2 sources and, 6-9
- Department of Energy (DOE), 8, 10-11, 80
- Department of Transportation, 132
- EOR projects in, 1
- Federal tax incentives and, 132
- NPC study and, 10
- projects in, 5-6
- State tax incentives and, 132

unsteady-state measurements, 70

V

valves, 100-101, 108, 120
- backpressure, 115
- facilities design and, 94-95
- fail-close, 106
- well design and, 115

vaporization, 13
- hydrogen sulfide and, 51
- residual oil and, 17-18
- three-phase hydrocarbon region and, 17

variable-speed drives (VSDs), 104-105
vertical sweep efficiency, 3
vessels, 100-101
vinyl polymers, 34
viscosity
- CO_2, 13, 81
- equation for, 81
- fingering and, 22, 80
- gravity and, 140
- layering and, 44-45
- Lorenz-Bray-Clark, 85
- oil and, 81
- Orbey-Sandler, 85
- prediction and, 69-70
- relative permeability and, 22-32
- scaleup and, 48
- sweep effects and, 40-41

Viton, 95

W

Wasson field, 85-88, 118, 120, 123
water
- black-oil simulation and, 79-87
- connate, 49
- filtering of, 153
- fractional flow plots and, 44
- handling considerations for, 153
- history matching and, 71-79
- hysteresis and, 23
- injection and, 106-108
- knockout, 152
- miscibility effects and, 18
- protection of near-surface, 158
- quality, 136
- relative permeability and, 23-28, 70
- reservoir blocking and, 21-22
- slim-tube tests and, 14
- solvents and, 28-32
- sweep aspects and, 32-34
- switching CO_2 cycles and, 120
- use permits and, 131
- wet mechanisms and, 21-22

water alternating gas (WAG) ratio, 40
- alternating CO_2 flooding and, 3-4
- black-oil simulation and, 82-84
- bottomhole pressure and, 89
- depletion stages and, 41
- economic issues and, 46
- injection and, 70-71, 106-108
- material selection and, 119-122
- miscibility effects and, 18
- monitoring of, 132-137
- optimization of, 87-88, 138-139
- relative permeability and, 24-28
- reservoir management and, 138-145
- scaleup and, 48
- slug size and, 87-88, 140
- solubility and, 27-28
- sweep aspects and, 32-34
- three-phase hydrocarbons and, 25-26
- time and, 138
- water blocking and, 22

waterflood endpoint test, 42-43
wellbore, 3
- conformance and, 45-46
- damage to, 41
- gravity and, 46-47
- interwell continuity and, 45
- radius variation and, 49
- recompletions and, 46
- scaleup and, 47-50

well design
- carbon steel and, 115
- CO_2/water cycles and, 120
- corrosion and, 115, 118-119
- couplings and, 121-122
- downhole equipment and, 116, 119
- flowing wells and, 117-119
- flumping and, 118
- hydrates and, 119
- injection and, 119-122
- lifting equipment and, 115-117, 119
- liners and, 116-117
- packer assembly and, 120-122
- paraffins and, 119
- plungers and, 119
- producing well and, 115-119
- pumps and, 116-117
- rods and, 116-117
- safety and, 158
- tubing and, 118-122
- wellbore modifications and, 122
- wellhead and, 115, 118-120

well patterns, 41
well servicing, 63
Wertz Tensleep project, 26-27, 88
wettability, 41-44
workovers, 63
- handling considerations and, 153
- safety and, 158
- selection of, 147-150

www.ingramcontent.com/pod-product-compliance
Ingram Content Group UK Ltd.
Pitfield, Milton Keynes, MK11 3LW, UK
UKHW051138260726
13967UKWH00010B/3124

9 781555 630966